Advances in microbial food safety

Related titles:

Viruses in food and water
(ISBN 978-0-85709-430-8)

Protective cultures, antimicrobial metabolites and bacteriophages for food and beverage biopreservation
(ISBN 978-1-84569-669-6)

Foodborne pathogens
(ISBN 978-1-84569-362-6)

Details of these books and a complete list of titles from Woodhead Publishing can be obtained by:

- visiting our web site at www.woodheadpublishing.com
- contacting Customer Services (e-mail: sales@woodheadpublishing.com; fax: +44 (0) 1223 832819; tel.: +44 (0) 1223 499140 ext. 130; address: Woodhead Publishing Limited, 80, High Street, Sawston, Cambridge CB22 3HJ, UK)
- in North America, contacting our US office (e-mail: usmarketing@ woodheadpublishing.com; tel.: (215) 928 9112; address: Woodhead Publishing, 1518 Walnut Street, Suite 1100, Philadelphia, PA 19102-3406, USA)

If you would like e-versions of our content, please visit our online platform: www. woodheadpublishingonline.com. Please recommend it to your librarian so that everyone in your institution can benefit from the wealth of content on the site.

We are always happy to receive suggestions for new books from potential editors. To enquire about contributing to our Food Science, Technology and Nutrition series, please send your name, contact address and details of the topic/s you are interested in to nell. holden@woodheadpublishing.com. We look forward to hearing from you.

The team responsible for publishing this book:

Commissioning Editor: Nell Holden
Publications Coordinator: Steven Matthews
Project Editor: Diana Paulding
Editorial and Production Manager: Mary Campbell
Production Editor: Mandy Kingsmill
Project Manager: Annette Wiseman, RCL
Copyeditor: Jonathan Webley
Proofreader: Jo Egre
Cover Designer: Terry Callanan

Woodhead Publishing Series in Food Science, Technology and Nutrition:
Number 259

Advances in microbial food safety

Volume 1

Edited by
John Sofos

Oxford Cambridge Philadelphia New Delhi

Published by Woodhead Publishing Limited,
80 High Street, Sawston, Cambridge CB22 3HJ, UK
www.woodheadpublishing.com
www.woodheadpublishingonline.com

Woodhead Publishing, 1518 Walnut Street, Suite 1100, Philadelphia, PA 19102-3406, USA

Woodhead Publishing India Private Limited, 303, Vardaan House, 7/28 Ansari Road, Daryaganj, New Delhi – 110002, India
www.woodheadpublishingindia.com

First published 2013, Woodhead Publishing Limited

British Library Cataloguing in Publication Data
A catalogue record for this book is available from the British Library.

Library of Congress Control Number: 2013937983

ISBN 978-0-85709-438-4 (print)
ISBN 978-0-85709-874-0 (online)
ISSN 2042-8049 Woodhead Publishing Series in Food Science, Technology and Nutrition (print)
ISSN 2042-8057 Woodhead Publishing Series in Food Science, Technology and Nutrition (online)

The publisher's policy is to use permanent paper from mills that operate a sustainable forestry policy, and which has been manufactured from pulp which is processed using acid-free and elemental chlorine-free practices. Furthermore, the publisher ensures that the text paper and cover board used have met acceptable environmental accreditation standards.

Typeset by RefineCatch Limited, Bungay, Suffolk
Printed by Lightning Source

Contents

Contributor contact details

(* = main contact)

Editor

Prof. John Sofos
Department of Animal Sciences
Center for Meat Safety and Quality
and Food Safety Cluster
Colorado State University
Fort Collins
CO 80523-1171
USA

Email: john.sofos@colostate.edu

Chapter 1

Dr Bruce Tompkin
1319 West 54th Street
LaGrange
IL 60525
USA

Email: r.tompkin@comcast.net

Chapter 2

Dr Tine Hald
Technical University of Denmark
National Food Institute
Mørkhøj Bygade 28
Building H
2860 Søborg
Denmark

Email: tiha@food.dtu.dk

Chapter 3

Dr Jessica C. Chen
Texas Tech University
Lubbock
TX 79404
USA

Email: JLC232@gmail.com

Kendra K. Nightingale*
Department of Animal and Food
Sciences (Room 201)
Texas Tech University
Lubbock
TX 79409
USA

Email: kendra.nightingale@ttu.edu

Chapter 4

Dr Frédéric Carlin* and Dr Christophe Nguyen-The
UMR408 Sécurité et qualité des produits d'origine végétale
INRA
Site Agroparc
CS40509
84914 Avignon cedex 9
France

Email: frederic.carlin@avignon.inra.fr

Chapter 5

Dr Mark Strom*, Dr R. N. Paranjpye, Dr W. B. Nilsson, Dr J. W. Turner and G. K. Yanagida
Northwest Fisheries Science Center
National Marine Fisheries Service
National Oceanic and Atmospheric Administration
2725 Montlake Blvd. E.
Seattle
WA 98112
USA

Email: mark.strom@noaa.gov

Chapter 6

Dr Ynes Ortega
Center for Food Safety
University of Georgia
Griffin
GA 30223
USA

Email: ortega@uga.edu

Chapter 7

Dr John Threlfall
HPA
61 Colindale Avenue
London
NW9 5EQ
UK

Email: e.j.threlfall@btinternet.com

Chapter 8

Dr Benno ter Kuile*
Office for Risk Assessment and Research
Netherlands Food and Consumer Product Safety Authority
Catharijnesingel 59
3511 GG Utrecht
The Netherlands

and

Department of Molecular Biology & Microbial Food Safety
Swammerdam Institute of Life Sciences
University of Amsterdam
Science Park 904
1098 XH Amsterdam
The Netherlands

Email: benno.ter.kuile@vwa.nl

Stanley Brul
Department of Molecular Biology & Microbial Food Safety
Swammerdam Institute of Life Sciences
University of Amsterdam
Science Park 904
1098 XH Amsterdam
The Netherlands

Chapter 9

Dr Soo Hwan Suh and Lee-Ann Jaykus
Department of Food, Bioprocessing and Nutrition Sciences
Campus Box 7624/313 Schaub Hall
400 Dan Allen Drive
North Carolina State University
Raleigh
NC 27695-7624
USA

Email: ssuh@ncsu.edu; leeann_jaykus@ncsu.edu

Dr Byron Brehm-Stecher
Department of Food Science & Human Nutrition
2312 Food Sciences Building
Iowa State University
Ames
IA 50011
USA

Email: byron@iastate.edu

Chapter 10

Dr Luca Cocolin* and Kalliopi Rantsiou
DISAFA-University of Turin
Via Leonardo da Vinci 44
10095 Grugliasco
Turin
Italy

Email: lucasimone.cocolin@unito.it; kalliopi.rantsiou@unitto.it

Chapter 11

L. N. Kahyaoglu and Dr Joseph Irudayaraj*
Agricultural and Biological Engineering
Bindley Bioscience Center
Birck Nanotechnology Center
225 S. University Street
Purdue University
West Lafayette
IN 47907
USA

Email: josephi@purdue.edu

Chapter 12

Dr Kendra K. Nightingale*
International Center for Food Industry Excellence
Department of Animal and Food Sciences
Texas Tech University
301 Animal and Food Sciences Building
Lubbock
TX 79409
USA

Email: kendra.nightingale@ttu.edu

Jacob R. Elder
Department of Veterinary Microbiology and Pathology
Washington State University
Bustad 402
Pullman
WA 99164
USA

Email: jelder@vetmed.wsu.edu

Chapter 13

Dr Rob Lake*
ESR – Institute of Environmental Science and Research
PO Box 29-181
Christchurch 8540
New Zealand

Email: rob.lake@esr.cri.nz

Prof. A. H. Havelaar
Centre for Infectious Disease Control
National Institute for Public Health and the Environment (RIVM)
PO Box 1
Bilthoven
The Netherlands

Email: arie.havelaar@rivm.nl

and

Division Veterinary Public Health
Institute for Risk Assessment Sciences
Faculty of Veterinary Medicine
Utrecht University
PO Box 80 175
3508 TD Utrecht
The Netherlands

Email: a.h.havelaar@uu.nl

Tanja Kuchenmüller
Information, Evidence, Research and Innovation
WHO Regional Office for Europe
Scherfigsvej 8
DK-2100 Copenhagen Ø
Denmark

Email: tku@euro.who.int

Chapter 14

Todd R. Callaway*, Robin C. Anderson, Tom S. Edrington, Kenneth J. Genovese, Roger B. Harvey, Toni L. Poole and David J. Nisbet
Food and Feed Safety Research Unit
Agricultural Research Service/USDA
2881 F&B Road
College Station
TX 77845
USA

Email: todd.callaway@ars.usda.gov

Chapter 15

Dr Jean-Louis Cordier
Nestlé Quality Assurance Center
Nestec Ltd
Avenue Nestlé 55
CH-1800 Vevey
Switzerland

Email: jean-louis.cordier@nestle.com

Chapter 16

Dr Jochen Klumpp and Prof. Martin J. Loessner*
Institute of Food, Nutrition and Health, ETH Zurich
Schmelzbergstrasse 7
8092 Zurich
Switzerland

Email: martin.loessner@ethz.ch

Chapter 17

Frank Devlieghere*, An Vermeulen, Peter Ragaert, Andreja Rajkovic, Simbarashe Samapundo and Francisco Lopez-Galvez
Department of Food Safety and Food Quality
Ghent University
Coupure Links 653
Ghent 9000
Belgium

Email: Frank.Devlieghere@UGent.be

Chapter 18

Prof. Kostas Koutsoumanis*
Department of Food Science & Technology
Laboratory of Food Microbiology & Hygiene
Agricultural School, Aristotle University of Thessaloniki
256, Thessaloniki, 541 24
Greece

Email: kkoutsou@agro.auth.gr

Prof. Panagiotis N. Skandamis
Department of Food Science and Technology
Agricultural University of Athens
Iera Odos 75, 118 55
Athens
Greece

Email: pskan@aua.gr

Chapter 19

Dr Kalmia E. Kniel
Microbial Food Safety
College of Agriculture and Natural Resources
University of Delaware
044 Townsend Hall
531 South College Avenue
Newark
DE 19716
USA

Email: kniel@udel.edu

Chapter 20

Dr Chris Griffith
Mayne House
1A Martel Close
Broadmayne
Dorset
DT2 8PL
UK

Email: cgriffith@cardiffmet.ac.uk

Chapter 21

Dr William Sperber
Retired
5814 Oakview Circle
Minnetonka
MN 55345
USA

Email: brsperber@comcast.net

Chapter 22

Dr Gary Barker
Institute of Food Research
Norwich Research Park
Colney
Norwich
NR4 7UA
UK

Email: gary.barker@ifr.ac.uk

Chapter 23

Dr Byron Brehm-Stecher
Department of Food Science & Human Nutrition
2312 Food Sciences Building
Iowa State University
Ames
IA 50011
USA

Email: byron@iastate.edu

Chapter 24

Dr Pina Fratamico* and Dr Nereus W. Gunther IV
USDA, Agricultural Research Service
Eastern Regional Research Center
600 East Mermaid Lane
Wyndmoor
PA 19038
USA

Email: pina.fratamico@ars.usda.gov

Chapter 25

Dr Carmen Pin*, Dr Aline Metris and Dr Jozsef Baranyi
Institute of Food Research
Norwich Research Park
Colney
Norwich
NR4 7UA
UK

Email: carmen.pin@bbsrc.ac.uk; aline.metris@bbsrc.ac.uk; jozsef.baranyi@bbsrc.ac.uk

Editorial advisors

Woodhead Publishing Series in Food Science, Technology and Nutrition

1 **Chilled foods: a comprehensive guide** *Edited by C. Dennis and M. Stringer*
2 **Yoghurt: science and technology** *A. Y. Tamime and R. K. Robinson*
3 **Food processing technology: principles and practice** *P. J. Fellows*
4 **Bender's dictionary of nutrition and food technology Sixth edition** *D. A. Bender*
5 **Determination of veterinary residues in food** *Edited by N. T. Crosby*
6 **Food contaminants: sources and surveillance** *Edited by C. Creaser and R. Purchase*
7 **Nitrates and nitrites in food and water** *Edited by M. J. Hill*
8 **Pesticide chemistry and bioscience: the food-environment challenge** *Edited by G. T. Brooks and T. Roberts*
9 **Pesticides: developments, impacts and controls** *Edited by G. A. Best and A. D. Ruthven*
10 **Dietary fibre: chemical and biological aspects** *Edited by D. A. T. Southgate, K. W. Waldron, I. T. Johnson and G. R. Fenwick*
11 **Vitamins and minerals in health and nutrition** *M. Tolonen*
12 **Technology of biscuits, crackers and cookies Second edition** *D. Manley*
13 **Instrumentation and sensors for the food industry** *Edited by E. Kress-Rogers*
14 **Food and cancer prevention: chemical and biological aspects** *Edited by K. W. Waldron, I. T. Johnson and G. R. Fenwick*
15 **Food colloids: proteins, lipids and polysaccharides** *Edited by E. Dickinson and B. Bergenstahl*
16 **Food emulsions and foams** *Edited by E. Dickinson*
17 **Maillard reactions in chemistry, food and health** *Edited by T. P. Labuza, V. Monnier, J. Baynes and J. O'Brien*
18 **The Maillard reaction in foods and medicine** *Edited by J. O'Brien, H. E. Nursten, M. J. Crabbe and J. M. Ames*
19 **Encapsulation and controlled release** *Edited by D. R. Karsa and R. A. Stephenson*
20 **Flavours and fragrances** *Edited by A. D. Swift*
21 **Feta and related cheeses** *Edited by A. Y. Tamime and R. K. Robinson*

22 **Biochemistry of milk products** *Edited by A. T. Andrews and J. R. Varley*
23 **Physical properties of foods and food processing systems** *M. J. Lewis*
24 **Food irradiation: a reference guide** *V. M. Wilkinson and G. Gould*
25 **Kent's technology of cereals: an introduction for students of food science and agriculture Fourth edition** *N. L. Kent and A. D. Evers*
26 **Biosensors for food analysis** *Edited by A. O. Scott*
27 **Separation processes in the food and biotechnology industries: principles and applications** *Edited by A. S. Grandison and M. J. Lewis*
28 **Handbook of indices of food quality and authenticity** *R. S. Singhal, P. K. Kulkarni and D. V. Rege*
29 **Principles and practices for the safe processing of foods** *D. A. Shapton and N. F. Shapton*
30 **Biscuit, cookie and cracker manufacturing manuals Volume 1: ingredients** *D. Manley*
31 **Biscuit, cookie and cracker manufacturing manuals Volume 2: biscuit doughs** *D. Manley*
32 **Biscuit, cookie and cracker manufacturing manuals Volume 3: biscuit dough piece forming** *D. Manley*
33 **Biscuit, cookie and cracker manufacturing manuals Volume 4: baking and cooling of biscuits** *D. Manley*
34 **Biscuit, cookie and cracker manufacturing manuals Volume 5: secondary processing in biscuit manufacturing** *D. Manley*
35 **Biscuit, cookie and cracker manufacturing manuals Volume 6: biscuit packaging and storage** *D. Manley*
36 **Practical dehydration Second edition** *M. Greensmith*
37 **Lawrie's meat science Sixth edition** *R. A. Lawrie*
38 **Yoghurt: science and technology Second edition** *A. Y. Tamime and R. K. Robinson*
39 **New ingredients in food processing: biochemistry and agriculture** *G. Linden and D. Lorient*
40 **Benders' dictionary of nutrition and food technology Seventh edition** *D. A. Bender and A. E. Bender*
41 **Technology of biscuits, crackers and cookies Third edition** *D. Manley*
42 **Food processing technology: principles and practice Second edition** *P. J. Fellows*
43 **Managing frozen foods** *Edited by C. J. Kennedy*
44 **Handbook of hydrocolloids** *Edited by G. O. Phillips and P. A. Williams*
45 **Food labelling** *Edited by J. R. Blanchfield*
46 **Cereal biotechnology** *Edited by P. C. Morris and J. H. Bryce*
47 **Food intolerance and the food industry** *Edited by T. Dean*
48 **The stability and shelf-life of food** *Edited by D. Kilcast and P. Subramaniam*
49 **Functional foods: concept to product** *Edited by G. R. Gibson and C. M. Williams*
50 **Chilled foods: a comprehensive guide Second edition** *Edited by M. Stringer and C. Dennis*
51 **HACCP in the meat industry** *Edited by M. Brown*
52 **Biscuit, cracker and cookie recipes for the food industry** *D. Manley*
53 **Cereals processing technology** *Edited by G. Owens*
54 **Baking problems solved** *S. P. Cauvain and L. S. Young*
55 **Thermal technologies in food processing** *Edited by P. Richardson*
56 **Frying: improving quality** *Edited by J. B. Rossell*
57 **Food chemical safety Volume 1: contaminants** *Edited by D. Watson*

58 **Making the most of HACCP: learning from others' experience** *Edited by T. Mayes and S. Mortimore*
59 **Food process modelling** *Edited by L. M. M. Tijskens, M. L. A. T. M. Hertog and B. M. Nicolaï*
60 **EU food law: a practical guide** *Edited by K. Goodburn*
61 **Extrusion cooking: technologies and applications** *Edited by R. Guy*
62 **Auditing in the food industry: from safety and quality to environmental and other audits** *Edited by M. Dillon and C. Griffith*
63 **Handbook of herbs and spices Volume 1** *Edited by K. V. Peter*
64 **Food product development: maximising success** *M. Earle, R. Earle and A. Anderson*
65 **Instrumentation and sensors for the food industry Second edition** *Edited by E. Kress-Rogers and C. J. B. Brimelow*
66 **Food chemical safety Volume 2: additives** *Edited by D. Watson*
67 **Fruit and vegetable biotechnology** *Edited by V. Valpuesta*
68 **Foodborne pathogens: hazards, risk analysis and control** *Edited by C. de W. Blackburn and P. J. McClure*
69 **Meat refrigeration** *S. J. James and C. James*
70 **Lockhart and Wiseman's crop husbandry Eighth edition** *H. J. S. Finch, A. M. Samuel and G. P. F. Lane*
71 **Safety and quality issues in fish processing** *Edited by H. A. Bremner*
72 **Minimal processing technologies in the food industries** *Edited by T. Ohlsson and N. Bengtsson*
73 **Fruit and vegetable processing: improving quality** *Edited by W. Jongen*
74 **The nutrition handbook for food processors** *Edited by C. J. K. Henry and C. Chapman*
75 **Colour in food: improving quality** *Edited by D. MacDougall*
76 **Meat processing: improving quality** *Edited by J. P. Kerry, J. F. Kerry and D. A. Ledward*
77 **Microbiological risk assessment in food processing** *Edited by M. Brown and M. Stringer*
78 **Performance functional foods** *Edited by D. Watson*
79 **Functional dairy products Volume 1** *Edited by T. Mattila-Sandholm and M. Saarela*
80 **Taints and off-flavours in foods** *Edited by B. Baigrie*
81 **Yeasts in food** *Edited by T. Boekhout and V. Robert*
82 **Phytochemical functional foods** *Edited by I. T. Johnson and G. Williamson*
83 **Novel food packaging techniques** *Edited by R. Ahvenainen*
84 **Detecting pathogens in food** *Edited by T. A. McMeekin*
85 **Natural antimicrobials for the minimal processing of foods** *Edited by S. Roller*
86 **Texture in food Volume 1: semi-solid foods** *Edited by B. M. McKenna*
87 **Dairy processing: improving quality** *Edited by G. Smit*
88 **Hygiene in food processing: principles and practice** *Edited by H. L. M. Lelieveld, M. A. Mostert, B. White and J. Holah*
89 **Rapid and on-line instrumentation for food quality assurance** *Edited by I. Tothill*
90 **Sausage manufacture: principles and practice** *E. Essien*
91 **Environmentally-friendly food processing** *Edited by B. Mattsson and U. Sonesson*
92 **Bread making: improving quality** *Edited by S. P. Cauvain*
93 **Food preservation techniques** *Edited by P. Zeuthen and L. Bøgh-Sørensen*
94 **Food authenticity and traceability** *Edited by M. Lees*

95 **Analytical methods for food additives** *R. Wood, L. Foster, A. Damant and P. Key*
96 **Handbook of herbs and spices Volume 2** *Edited by K. V. Peter*
97 **Texture in food Volume 2: solid foods** *Edited by D. Kilcast*
98 **Proteins in food processing** *Edited by R. Yada*
99 **Detecting foreign bodies in food** *Edited by M. Edwards*
100 **Understanding and measuring the shelf-life of food** *Edited by R. Steele*
101 **Poultry meat processing and quality** *Edited by G. Mead*
102 **Functional foods, ageing and degenerative disease** *Edited by C. Remacle and B. Reusens*
103 **Mycotoxins in food: detection and control** *Edited by N. Magan and M. Olsen*
104 **Improving the thermal processing of foods** *Edited by P. Richardson*
105 **Pesticide, veterinary and other residues in food** *Edited by D. Watson*
106 **Starch in food: structure, functions and applications** *Edited by A.-C. Eliasson*
107 **Functional foods, cardiovascular disease and diabetes** *Edited by A. Arnoldi*
108 **Brewing: science and practice** *D. E. Briggs, P. A. Brookes, R. Stevens and C. A. Boulton*
109 **Using cereal science and technology for the benefit of consumers: proceedings of the 12th International ICC Cereal and Bread Congress, 24–26th May, 2004, Harrogate, UK** *Edited by S. P. Cauvain, L. S. Young and S. Salmon*
110 **Improving the safety of fresh meat** *Edited by J. Sofos*
111 **Understanding pathogen behaviour: virulence, stress response and resistance** *Edited by M. Griffiths*
112 **The microwave processing of foods** *Edited by H. Schubert and M. Regier*
113 **Food safety control in the poultry industry** *Edited by G. Mead*
114 **Improving the safety of fresh fruit and vegetables** *Edited by W. Jongen*
115 **Food, diet and obesity** *Edited by D. Mela*
116 **Handbook of hygiene control in the food industry** *Edited by H. L. M. Lelieveld, M. A. Mostert and J. Holah*
117 **Detecting allergens in food** *Edited by S. Koppelman and S. Hefle*
118 **Improving the fat content of foods** *Edited by C. Williams and J. Buttriss*
119 **Improving traceability in food processing and distribution** *Edited by I. Smith and A. Furness*
120 **Flavour in food** *Edited by A. Voilley and P. Etievant*
121 **The Chorleywood bread process** *S. P. Cauvain and L. S. Young*
122 **Food spoilage microorganisms** *Edited by C. de W. Blackburn*
123 **Emerging foodborne pathogens** *Edited by Y. Motarjemi and M. Adams*
124 **Benders' dictionary of nutrition and food technology Eighth edition** *D. A. Bender*
125 **Optimising sweet taste in foods** *Edited by W. J. Spillane*
126 **Brewing: new technologies** *Edited by C. Bamforth*
127 **Handbook of herbs and spices Volume 3** *Edited by K. V. Peter*
128 **Lawrie's meat science Seventh edition** *R. A. Lawrie in collaboration with D. A. Ledward*
129 **Modifying lipids for use in food** *Edited by F. Gunstone*
130 **Meat products handbook: practical science and technology** *G. Feiner*
131 **Food consumption and disease risk: consumer–pathogen interactions** *Edited by M. Potter*
132 **Acrylamide and other hazardous compounds in heat-treated foods** *Edited by K. Skog and J. Alexander*

133 **Managing allergens in food** *Edited by C. Mills, H. Wichers and K. Hoffman-Sommergruber*
134 **Microbiological analysis of red meat, poultry and eggs** *Edited by G. Mead*
135 **Maximising the value of marine by-products** *Edited by F. Shahidi*
136 **Chemical migration and food contact materials** *Edited by K. Barnes, R. Sinclair and D. Watson*
137 **Understanding consumers of food products** *Edited by L. Frewer and H. van Trijp*
138 **Reducing salt in foods: practical strategies** *Edited by D. Kilcast and F. Angus*
139 **Modelling microorganisms in food** *Edited by S. Brul, S. Van Gerwen and M. Zwietering*
140 **Tamime and Robinson's Yoghurt: science and technology Third edition** *A. Y. Tamime and R. K. Robinson*
141 **Handbook of waste management and co-product recovery in food processing Volume 1** *Edited by K. W. Waldron*
142 **Improving the flavour of cheese** *Edited by B. Weimer*
143 **Novel food ingredients for weight control** *Edited by C. J. K. Henry*
144 **Consumer-led food product development** *Edited by H. MacFie*
145 **Functional dairy products Volume 2** *Edited by M. Saarela*
146 **Modifying flavour in food** *Edited by A. J. Taylor and J. Hort*
147 **Cheese problems solved** *Edited by P. L. H. McSweeney*
148 **Handbook of organic food safety and quality** *Edited by J. Cooper, C. Leifert and U. Niggli*
149 **Understanding and controlling the microstructure of complex foods** *Edited by D. J. McClements*
150 **Novel enzyme technology for food applications** *Edited by R. Rastall*
151 **Food preservation by pulsed electric fields: from research to application** *Edited by H. L. M. Lelieveld and S. W. H. de Haan*
152 **Technology of functional cereal products** *Edited by B. R. Hamaker*
153 **Case studies in food product development** *Edited by M. Earle and R. Earle*
154 **Delivery and controlled release of bioactives in foods and nutraceuticals** *Edited by N. Garti*
155 **Fruit and vegetable flavour: recent advances and future prospects** *Edited by B. Brückner and S. G. Wyllie*
156 **Food fortification and supplementation: technological, safety and regulatory aspects** *Edited by P. Berry Ottaway*
157 **Improving the health-promoting properties of fruit and vegetable products** *Edited by F. A. Tomás-Barberán and M. I. Gil*
158 **Improving seafood products for the consumer** *Edited by T. Børresen*
159 **In-pack processed foods: improving quality** *Edited by P. Richardson*
160 **Handbook of water and energy management in food processing** *Edited by J. Klemeš, R. Smith and J.-K. Kim*
161 **Environmentally compatible food packaging** *Edited by E. Chiellini*
162 **Improving farmed fish quality and safety** *Edited by Ø. Lie*
163 **Carbohydrate-active enzymes** *Edited by K.-H. Park*
164 **Chilled foods: a comprehensive guide Third edition** *Edited by M. Brown*
165 **Food for the ageing population** *Edited by M. M. Raats, C. P. G. M. de Groot and W. A. Van Staveren*

166 **Improving the sensory and nutritional quality of fresh meat** *Edited by J. P. Kerry and D. A. Ledward*
167 **Shellfish safety and quality** *Edited by S. E. Shumway and G. E. Rodrick*
168 **Functional and speciality beverage technology** *Edited by P. Paquin*
169 **Functional foods: principles and technology** *M. Guo*
170 **Endocrine-disrupting chemicals in food** *Edited by I. Shaw*
171 **Meals in science and practice: interdisciplinary research and business applications** *Edited by H. L. Meiselman*
172 **Food constituents and oral health: current status and future prospects** *Edited by M. Wilson*
173 **Handbook of hydrocolloids Second edition** *Edited by G. O. Phillips and P. A. Williams*
174 **Food processing technology: principles and practice Third edition** *P. J. Fellows*
175 **Science and technology of enrobed and filled chocolate, confectionery and bakery products** *Edited by G. Talbot*
176 **Foodborne pathogens: hazards, risk analysis and control Second edition** *Edited by C. de W. Blackburn and P. J. McClure*
177 **Designing functional foods: measuring and controlling food structure breakdown and absorption** *Edited by D. J. McClements and E. A. Decker*
178 **New technologies in aquaculture: improving production efficiency, quality and environmental management** *Edited by G. Burnell and G. Allan*
179 **More baking problems solved** *S. P. Cauvain and L. S. Young*
180 **Soft drink and fruit juice problems solved** *P. Ashurst and R. Hargitt*
181 **Biofilms in the food and beverage industries** *Edited by P. M. Fratamico, B. A. Annous and N. W. Gunther*
182 **Dairy-derived ingredients: food and neutraceutical uses** *Edited by M. Corredig*
183 **Handbook of waste management and co-product recovery in food processing Volume 2** *Edited by K. W. Waldron*
184 **Innovations in food labelling** *Edited by J. Albert*
185 **Delivering performance in food supply chains** *Edited by C. Mena and G. Stevens*
186 **Chemical deterioration and physical instability of food and beverages** *Edited by L. H. Skibsted, J. Risbo and M. L. Andersen*
187 **Managing wine quality Volume 1: viticulture and wine quality** *Edited by A. G. Reynolds*
188 **Improving the safety and quality of milk Volume 1: milk production and processing** *Edited by M. Griffiths*
189 **Improving the safety and quality of milk Volume 2: improving quality in milk products** *Edited by M. Griffiths*
190 **Cereal grains: assessing and managing quality** *Edited by C. Wrigley and I. Batey*
191 **Sensory analysis for food and beverage quality control: a practical guide** *Edited by D. Kilcast*
192 **Managing wine quality Volume 2: oenology and wine quality** *Edited by A. G. Reynolds*
193 **Winemaking problems solved** *Edited by C. E. Butzke*
194 **Environmental assessment and management in the food industry** *Edited by U. Sonesson, J. Berlin and F. Ziegler*
195 **Consumer-driven innovation in food and personal care products** *Edited by S. R. Jaeger and H. MacFie*

196 **Tracing pathogens in the food chain** *Edited by S. Brul, P. M. Fratamico and T. A. McMeekin*
197 **Case studies in novel food processing technologies: innovations in processing, packaging, and predictive modelling** *Edited by C. J. Doona, K. Kustin and F. E. Feeherry*
198 **Freeze-drying of pharmaceutical and food products** *T.-C. Hua, B.-L. Liu and H. Zhang*
199 **Oxidation in foods and beverages and antioxidant applications Volume 1: understanding mechanisms of oxidation and antioxidant activity** *Edited by E. A. Decker, R. J. Elias and D. J. McClements*
200 **Oxidation in foods and beverages and antioxidant applications Volume 2: management in different industry sectors** *Edited by E. A. Decker, R. J. Elias and D. J. McClements*
201 **Protective cultures, antimicrobial metabolites and bacteriophages for food and beverage biopreservation** *Edited by C. Lacroix*
202 **Separation, extraction and concentration processes in the food, beverage and nutraceutical industries** *Edited by S. S. H. Rizvi*
203 **Determining mycotoxins and mycotoxigenic fungi in food and feed** *Edited by S. De Saeger*
204 **Developing children's food products** *Edited by D. Kilcast and F. Angus*
205 **Functional foods: concept to product Second edition** *Edited by M. Saarela*
206 **Postharvest biology and technology of tropical and subtropical fruits Volume 1: Fundamental issues** *Edited by E. M. Yahia*
207 **Postharvest biology and technology of tropical and subtropical fruits Volume 2: Açai to citrus** *Edited by E. M. Yahia*
208 **Postharvest biology and technology of tropical and subtropical fruits Volume 3: Cocona to mango** *Edited by E. M. Yahia*
209 **Postharvest biology and technology of tropical and subtropical fruits Volume 4: Mangosteen to white sapote** *Edited by E. M. Yahia*
210 **Food and beverage stability and shelf life** *Edited by D. Kilcast and P. Subramaniam*
211 **Processed Meats: improving safety, nutrition and quality** *Edited by J. P. Kerry and J. F. Kerry*
212 **Food chain integrity: a holistic approach to food traceability, safety, quality and authenticity** *Edited by J. Hoorfar, K. Jordan, F. Butler and R. Prugger*
213 **Improving the safety and quality of eggs and egg products Volume 1** *Edited by Y. Nys, M. Bain and F. Van Immerseel*
214 **Improving the safety and quality of eggs and egg products Volume 2** *Edited by F. Van Immerseel, Y. Nys and M. Bain*
215 **Animal feed contamination: effects on livestock and food safety** *Edited by J. Fink-Gremmels*
216 **Hygienic design of food factories** *Edited by J. Holah and H. L. M. Lelieveld*
217 **Manley's technology of biscuits, crackers and cookies Fourth edition** *Edited by D. Manley*
218 **Nanotechnology in the food, beverage and nutraceutical industries** *Edited by Q. Huang*
219 **Rice quality: A guide to rice properties and analysis** *K. R. Bhattacharya*
220 **Advances in meat, poultry and seafood packaging** *Edited by J. P. Kerry*
221 **Reducing saturated fats in foods** *Edited by G. Talbot*

222 **Handbook of food proteins** *Edited by G. O. Phillips and P. A. Williams*
223 **Lifetime nutritional influences on cognition, behaviour and psychiatric illness** *Edited by D. Benton*
224 **Food machinery for the production of cereal foods, snack foods and confectionery** *L.-M. Cheng*
225 **Alcoholic beverages: sensory evaluation and consumer research** *Edited by J. Piggott*
226 **Extrusion problems solved: food, pet food and feed** *M. N. Riaz and G. J. Rokey*
227 **Handbook of herbs and spices Second edition Volume 1** *Edited by K. V. Peter*
228 **Handbook of herbs and spices Second edition Volume 2** *Edited by K. V. Peter*
229 **Breadmaking: improving quality Second edition** *Edited by S. P. Cauvain*
230 **Emerging food packaging technologies: principles and practice** *Edited by K. L. Yam and D. S. Lee*
231 **Infectious disease in aquaculture: prevention and control** *Edited by B. Austin*
232 **Diet, immunity and inflammation** *Edited by P. C. Calder and P. Yaqoob*
233 **Natural food additives, ingredients and flavourings** *Edited by D. Baines and R. Seal*
234 **Microbial decontamination in the food industry: novel methods and applications** *Edited by A. Demirci and M.O. Ngadi*
235 **Chemical contaminants and residues in foods** *Edited by D. Schrenk*
236 **Robotics and automation in the food industry: current and future technologies** *Edited by D. G. Caldwell*
237 **Fibre-rich and wholegrain foods: improving quality** *Edited by J. A. Delcour and K. Poutanen*
238 **Computer vision technology in the food and beverage industries** *Edited by D.-W. Sun*
239 **Encapsulation technologies and delivery systems for food ingredients and nutraceuticals** *Edited by N. Garti and D. J. McClements*
240 **Case studies in food safety and authenticity** *Edited by J. Hoorfar*
241 **Heat treatment for insect control: developments and applications** *D. Hammond*
242 **Advances in aquaculture hatchery technology** *Edited by G. Allan and G. Burnell*
243 **Open innovation in the food and beverage industry** *Edited by M. Garcia Martinez*
244 **Trends in packaging of food, beverages and other fast-moving consumer goods (FMCG)** *Edited by N. Farmer*
245 **New analytical approaches for verifying the origin of food** *Edited by P. Brereton*
246 **Microbial production of food ingredients, enzymes and nutraceuticals** *Edited by B. McNeil, D. Archer, I. Giavasis and L. Harvey*
247 **Persistent organic pollutants and toxic metals in foods** *Edited by M. Rose and A. Fernandes*
248 **Cereal grains for the food and beverage industries** *E. Arendt and E. Zannini*
249 **Viruses in food and water: risks, surveillance and control** *Edited by N. Cook*
250 **Improving the safety and quality of nuts** *Edited by L. J. Harris*
251 **Metabolomics in food and nutrition** *Edited by B. C. Weimer and C. Slupsky*
252 **Food enrichment with omega-3 fatty acids** *Edited by C. Jacobsen, N. S. Nielsen, A. F. Horn and A.-D. M. Sørensen*
253 **Instrumental assessment of food sensory quality: a practical guide** *Edited by D. Kilcast*

254 **Food microstructures: microscopy, measurement and modelling** *Edited by V. J. Morris and K. Groves*
255 **Handbook of food powders: processes and properties** *Edited by B. R. Bhandari, N. Bansal, M. Zhang and P. Schuck*
256 **Functional ingredients from algae for foods and nutraceuticals** *Edited by H. Domínguez*
257 **Satiation, satiety and the control of food intake: theory and practice** *Edited by J. E. Blundell and F. Bellisle*
258 **Hygiene in food processing: principles and practice Second edition** *Edited by H. L. M. Lelieveld, J. Holah and D. Napper*
259 **Advances in microbial food safety: Volume 1** *Edited by J. Sofos*
260 **Global safety of fresh produce: A handbook of best practice, innovative commercial solutions and case studies** *Edited by J. Hoorfar*
261 **Human milk biochemistry and infant formula manufacturing technology** *Edited by M. Guo*

Part I

Expert interview with Dr R. Bruce Tompkin

1

Interview with a food safety expert: Dr R. Bruce Tompkin

R. B. Tompkin, Retired, USA and J. Sofos, Colorado State University, USA

DOI: 10.1533/9780857098740.1.3

Abstract: This chapter is a series of questions and answers between John Sofos and Bruce Tompkin, giving Bruce's perspective as a food microbiologist on food safety. Progress depends on epidemiologic studies by public health agencies and research by universities and government laboratories. Industry responded to microbial hazards and food-pathogen combinations with good manufacturing practices (GMPs) and hazard analysis and critical control point (HACCP) systems. Examples demonstrate improved consumer protection. Success depends on control measures, regulatory approval and consumer acceptance. Existing weaknesses and the underlying reasons are described. The challenge for the food safety community will be to find acceptable solutions and achieve the higher level of protection that consumers expect.

Key words: food safety, food safety management system, foodborne illness, microbial hazards in food, hazard analysis and critical control point (HACCP), good manufacturing practices (GMPs).

Dr Bruce Tompkin received a PhD in microbiology from Ohio State University in 1963 and started with Swift & Company in 1964. He became chief microbiologist in 1966 and retained that position until 1993 when he became vice president for product safety for ConAgra Refrigerated Prepared Foods (formerly Armour Swift-Eckrich, Inc). His efforts were directed toward improving the microbiological quality and safety of food. He approached these goals through research, publications, presentations and serving on a wide variety of committees at national and international level. Areas of research include control of pathogens in food processing environments and in a wide variety of foods, the use of additives and new processes to improve food safety and the role of sodium nitrite in controlling *Clostridium botulinum*. A major contribution was sharing best practices for pathogen control with others in industry, government and academia. He has contributed more than 185 publications, numerous presentations and 30 book chapters. Bruce was a member of the National Advisory Committee on Microbiological Criteria for Foods for five terms and the International Commission on Microbiological Criteria for Foods for 20 years, serving as an advisor for an additional seven years. He helped define the principles of hazard analysis and critical control point (HACCP), the concept of a food safety objective and the role of microbiological testing in food safety management systems. He retired from ConAgra Refrigerated Prepared Foods in 2002 and continues to promote food safety through participation on committees and other means.

1.1 Food safety: past and current

1.1.1 Could you briefly describe some of the significant scientific advances that have had a positive or negative impact on microbial food safety?

Thermal processing and the use of heat in combination with pH control were extremely important advances, which since the late 1920s have enabled the production of safe shelf-stable canned foods. The scientific basis for thermal processing was so well done that the research has served as a hallmark for subsequent generations of food scientists and microbiologists.

Historically, salt was commonly used to preserve a wide variety of foods. The use of salt and other solutes was placed on a better footing in the early 1950s with the introduction of water activity (a_w) as the scientific basis for preventing pathogens from multiplying in food. It could be said that through the use of heat, pH, a_w and combinations thereof, many foods have been converted from perishable to shelf stable. Subsequent research revealed the mechanisms and the information needed to establish critical limits for the control of foodborne pathogens.

Sodium nitrite and other more modern antimicrobial agents have played an important role in the preservation and safety of food. In addition there are a wide variety of bactericidal agents that are now applied as dips, rinses or ingredients to improve the safety of food. In addition to temperature control, including low-temperature storage, newer technologies have been developed to improve the safety of food. Irradiation and ultrahigh pressure processing are two examples that are used to inactivate pathogens in selected foods.

It is important to understand that there is a considerable amount of science underlying each of the above technologies. The underlying science affecting each factor is essential because it forms the basis for the validation of critical control points in hazard analysis and critical control point plans as applied in modern food safety management systems. The accumulated science is now available through regulations, guidance documents and published scientific literature. For many foods it is a regulatory requirement to provide the scientific basis for the critical limits that are used at selected steps in the food chain.

During the 1960s, 1970s and 1980s a considerable amount of research was conducted on cell injury, a condition in which microbial cells are injured and cannot multiply in foods unless conditions allow their resuscitation. The cells can recover when appropriate cultural conditions (e.g., temperature, nutrients, absence of certain inhibitory ingredients) are applied. While cell injury does occur, I do not recall an incident where consumer illness was confirmed as originating from the presence of injured cells, although that would be difficult to prove. Also, the possibility that viable but unrecoverable cells can exist in a food and subsequently cause disease has yet to be demonstrated. Both concepts were extensively studied and led to a much greater understanding of the factors affecting microbial survival and death.

In addition to the above technologies, there have been numerous other significant advances in food safety that resulted from many years of investigation and research. The resulting progress in food safety is now taken for granted and current scientists do not know the researchers who contributed to solving the problems. In fact, the successes are no longer considered newsworthy in today's environment with frequent reports of outbreaks and food recalls. The reports leave a very negative impression about the status of our food supply and existing systems for food control. Yet, a list of the past successes is impressive and can be used as templates for addressing current and future problems. It is of interest that they are often associated with specific food-pathogen combinations. I think the following partial list is impressive and a tribute to all who were involved:

- Use of epidemiologic studies and other research to elucidate the relationship between specific contaminated foods and the conditions that led to foodborne disease.
- Pasteurization of milk for direct consumption and for further processing into dairy products.
- Pasteurization of liquid eggs, particularly albumen, that are used in liquid or dried form as ingredients in a variety of foods that previously had been associated with salmonellosis.
- Identifying the conditions that contribute to the presence of *Salmonella* Enteritidis in shell eggs and how risk can be managed during production, processing and distribution.
- In the 1960s and 1970s, identifying the conditions needed for control of salmonellae in the environment in which low-moisture foods are produced.
- Identification and documentation of many currently important pathogens as foodborne.
- Progress in microbial analysis of foods and in detection of pathogens, including molecular and rapid methods.
- Developing technologies to destroy enteric pathogens in nuts, spices and other dry foods and ingredients.
- Defining the multiple hurdle technology concept for microbial inactivation and control in foods.
- In the 1980s and 1990s identifying the conditions needed for control of *Listeria monocytogenes* in the environment in which refrigerated ready-to-eat (RTE) foods are chilled and prepared for final packaging.

- Defining the conditions that lead to an increased risk of hazards associated with seafood and developing monitoring systems as control measures, such as:
  - monitoring the water in which mollusks are harvested because of the risks of enteric pathogens and paralytic shellfish poisoning.
- Pasteurization of seafood (e.g., crab meat, surimi) to control psychrotrophic strains of *Clostridium botulinum*.
- Defining the conditions that lead to an increased risk of mycotoxins in a variety of commodities (e.g., peanuts, corn, grains) and defining the procedures that minimize their presence in food (e.g., electronic sorting to eliminate peanuts more likely to contain aflatoxin).
- Developing commercial starter cultures for cheese and fermented meat that control *Staphylococcus aureus* and hasten the destruction of enteric pathogens and *L. monocytogenes*.
- Validating the conditions necessary to destroy enteric pathogens in cheeses and fermented meats.
- Mathematical models for predicting microbial behavior under different conditions and parameters.
- Establishing the conditions necessary to control *Trichinella spiralis*:
  - in hogs throughout Europe about the turn of the 20th century and later in the USA after the 1960s
  - during the production of dry cured and fermented pork products, which involved two different approaches to managing food safety in Europe and the USA.
- Establishing the parameters required for preventing germination and outgrowth of *C. botulinum* in:
  - processed cheese products
  - canned shelf-stable cured meats given less than a 'botulinal process'
  - onions, herbs, etc., in oil
  - smoked fish.
- Identifying honey as a significant source of infant botulism, recommending that honey is not fed to infants less than one year of age and developing an antitoxin for infants that neutralizes the toxin and hastens patient recovery.
- Establishing measures that control the presence of enteric pathogens during the production of sprouts, fresh citrus juice and apple cider.
- In addition to the above, implementing procedures for fast food outlets and their suppliers to successfully reduce the risk of *Escherichia coli* O157:H7 in ground beef.
- Developing the information required to manage the risk of bovine spongiform encephalopathy (BSE) in cattle, which apparently has led to its containment.
- Establishing the conditions necessary for chilling and holding foods at the food-service level where the majority of *C. perfringens* outbreaks have occurred.

- Developing bacterial DNA fingerprinting procedures and programs that have contributed to identification and containment of foodborne outbreaks.

Fortunately, there have been fewer *significant* scientific advances that have had a negative impact on microbiological food safety because consideration will probably have been given to the impact, pros and cons of a new process or procedure. If a concept is open for others to examine, someone will likely ask the 'what if' question(s). If the concept is not open for outside input, there is a greater possibility for error. For example, a newly patented process by a major competitor was developed in around 1960 with no input from its own microbiologists or outside scientists. The new process virtually guaranteed an enterotoxigenic product made from Italian-style salami. Commercial starter cultures were rarely used at that time and the fermentation conditions described in the patent favored the growth of *S. aureus*. That fiasco led to a loss of several million dollars to the company, the dry sausage operation was closed for months causing customers to seek other suppliers and resulted in major changes throughout the dry sausage industry, including the use of commercial cultures. It is worth noting that patent development is a closed process between the developers and the patent office.

Another type of example worth noting is that during the 1950s a major change was made in the procedures used to produce ham in the USA. New technologies were implemented that dramatically reduced the time and labor needed to produce ham. They involved injecting a brine solution into the ham rather than coating hams with salt and allowing the salt to migrate into the center of the hams over weeks in refrigerated rooms during dry curing. The newer hams had lower, more uniform levels of salt throughout the ham but also were more prone to microbial spoilage. Consumers found the new hams to be more tender and moist, easier to prepare and more preferable because they were less salty. Unfortunately, consumers did not understand the increased perishability of the new hams and the importance of refrigeration after cooking. During the transitional period, meat companies encountered increased complaints of foodborne illness due to enterotoxins produced by *Staphylococcus aureus*, a pathogen that is relatively salt tolerant and multiplies rapidly at warm temperatures, especially on ham which lacks other competitive microflora after cooking.

Several new concepts of processing, packaging and distribution have been perceived to have a negative impact and, thus, extensively studied over the past few decades. Examples include:

- Converting frozen cooked uncured vacuum-packaged poultry rolls and breast products into refrigerated products:
  - Risk: The absence of sodium nitrite in these products will increase the risk of outgrowth and toxin production by *C. botulinum*, if temperature abused.
  - Current status: The conversion occurred in the mid-1980s and provided the low fat and low salt products desired by consumers since that time. There have been no incidents of botulism.

- Packaging fresh fish and seafood in vacuum, a modified atmosphere or a controlled atmosphere:
  - The Food and Drug Administration (FDA) prohibits the packaging of fresh fish and seafood in vacuum, a modified atmosphere or a controlled atmosphere for sale at retail outlets.
  - Risk: The outgrowth and toxin production by psychrotrophic *C. botulinum* during refrigerated storage.
  - Current status: The FDA continues to prohibit retail sale but does permit such packaging for imported products provided the manufacturer, shipper or importer submits documentation that the established control system will reduce the potential for production of *C. botulinum* toxin as specified in Import Alert 16-125. There have been no incidents of botulism attributed to the imported products.
- Vacuum packaging of cooked entrees that are distributed and stored refrigerated until reheated for serving (i.e., sous vide foods). Consumers perceive these products to be fresher than canned or frozen products and very easy to prepare. The primary pathogen of concern is *C. botulinum*. This concept was studied extensively by scientists in North America and Europe during the 1990s and led to numerous research papers and guidance documents.
  - Risk: The outgrowth and toxin production by psychrotrophic *C. botulinum* during refrigerated storage or mesophilic *C. botulinum*, if temperature abused.
  - Current status: Numerous precooked refrigerated entrees and side dishes are now available at retail outlets. In one incident, two cases of botulism occurred in 1994 when a carton containing a plastic bag of precooked refrigerated clam chowder was mistakenly placed in a cupboard among the shelf-stable items for three weeks. The patients noted the soup had an unusual odor and flavor but still ate it.

1.1.2 Could you describe the role and impact of the following groups on food safety: industry, trade associations, regulatory and public health agencies, academia, international organizations, food safety advocates and consumers?

Industry

The food industry has primary responsibility for food safety but shares this responsibility with others. Companies cannot be successful if they cannot consistently deliver food that is safe. To be successful, business managers must be able to predict and plan how well their products will perform throughout the year in the marketplace. It has been my experience that managers will support efforts to ensure incidents of food contamination, illness, recalls and lawsuits do not occur. They know they cannot meet their goals if such unexpected events occur. Consumer protection is not only morally correct but it is essential to being successful in the food industry.

Trade associations

There are many trade associations throughout the food industry with each focusing on certain food groups such as nuts, meat, poultry, eggs, produce and fruits, dairy and sprouts. Trade associations serve a variety of functions that are important to industry. They represent their membership before government as new regulations and policies are developed. One of their roles is to ensure that new regulations and policies are based on good science, that they are achievable by their member companies and that they are cost effective in delivering the desired improvements in food safety.

It is at trade association meetings that companies share their experiences and develop best practices that can be applied throughout their segment of the food industry. Since the early 1990s, there has been a greater willingness among competitor companies to collaborate and share their food safety experiences. The reason is that all companies are impacted financially, if one of them produces a product that is contaminated or implicated in foodborne disease.

Some trade associations have a long history of supporting research at universities and elsewhere to develop the information needed to improve the control of foodborne pathogens. The resulting information is often incorporated into guidance documents and made available to others. An example has been the need for improved control of enteric pathogens (e.g., *E. coli* O157, salmonellae, *Campylobacter jejuni*) in raw agricultural commodities.

Government: regulatory agencies

The primary role of government is to verify that industry is fulfilling its responsibilities in the production and handling of foods and in preventing the release of potentially hazardous foods to the market. Unfortunately, there is considerable variability in the knowledge and experience among business managers throughout the food industry. Some are ignorant of the factors that must be controlled to ensure their food products are not contaminated and could possibly lead to foodborne illness. It is for this reason that governments play an important role in establishing and enforcing regulations and policies designed to ensure foods are safe. In addition, governments can provide guidance through documents that incorporate the best available science and industrial best practices. This information is available to companies that lack the technical staff that exists in larger companies.

Government agencies also maintain research programs that generate information needed to support the government's role as a regulator for food control. Government research programs have historically provided information about foodborne pathogens and their metabolites and how they can be controlled.

The National Advisory Committee on Microbiological Criteria for Foods (NACMCF) was founded by the federal agencies responsible for food control. Members are food microbiologists with expertise on topics of interest to the agencies and represent a cross section of industry, government, academia and consumer groups. The agencies decide upon the issues they wish the committee to consider and typically provide specific questions about these issues. The committee has proven to be very effective in generating a wide variety of documents such as guidance to small businesses, the principles of HACCP and the current status of knowledge on specific pathogens of concern.

NACMCF also provides a forum for sharing knowledge and current best practices among the committee members. It is a great opportunity to both share and gain insight on topics that would otherwise not be available to the members during their normal jobs.

Government: public health agencies

Federal, state and, to a lesser degree, local agencies maintain personnel who focus primarily on foodborne disease. Throughout my 40 years as a microbiologist in the food industry, I have found the epidemiological investigations performed by public health agencies to be of immense value. They define the relationship between the foods produced by industry, their risk to consumers and identify improvements that are critical for consumer protection. Epidemiologic studies (e.g., trace back and case-control studies) have been used to identify the food(s) and pathogen(s) responsible for consumer illness. This information is used by industry as the scientific basis for the hazards that are significant and must be controlled in their food operations. Ideally, public health agencies generate and publish information that describes the conditions (e.g., temperature abuse, cross contamination) that have led to consumer illness but unfortunately this is not always possible. In some incidents the information needed to describe why an illness occurred is not reliable or not known, or the agency lacked the necessary resources to investigate and report the conditions leading to the illness. Occasionally, information is not available due to a pending lawsuit or the agency believes its role is limited to bringing the outbreak to a conclusion so that additional cases do not occur. Each incident that is successfully investigated and reported is an opportunity for the food industry, consumers and others to learn how to prevent similar incidents in the future.

Another role for governments is to document trends in foodborne disease and use this information to assess the effectiveness of regulations and policies and identify where changes may be necessary. The data are most helpful when related to specific food groups because industry and regulatory agencies can use the data to measure their ability and progress in controlling microbial hazards in specific foods.

In addition to the traditional passive surveillance of disease, newer forms of active surveillance (e.g., FoodNet, CaliciNet, SalmNet) have been developed at the national and international levels. Such data can be used by governments to establish public health goals such as Healthy People 2000, 2010 and 2020. Although industry does not participate in establishing the goals, it is ultimately industry that must implement the changes that will lead to compliance with new regulations and policies that are intended to reduce foodborne disease. It is unfortunate that to date, public health agencies and governments have claimed credit for the improvements in consumer protection. It is as though they led the way and forced changes upon an unwilling industry when in fact industry desires to participate in improved consumer protection. I guess that is a sensitive issue for me. In addition, changes in consumer behavior such as food selection, storage, preparation and handling also deserve recognition.

Academia

The most basic function of universities is to teach new scientists and other professionals, who are subsequently employed by the food industry, governments and academia. All three rely on universities to provide a continuous stream of well-educated professionals who understand and can address current food safety concerns and are prepared to address future concerns. Universities maintain and make available information about past and current knowledge and investigate new ideas and systems to improve food safety in the future.

Universities conduct research to generate most of the available information about infectious and toxigenic agents. This information is critical for understanding the factors that contribute to pathogenicity and foodborne disease. Universities also contribute immensely by generating the scientific information needed to develop and implement effective, reliable control measures. Academia has a long history of research into new food processing technologies for controlling foodborne pathogens and their metabolites.

A very effective combination occurs when university, government and industry work together to generate information that can enhance food safety. It can be a very rewarding experience for all who participate in such cooperative programs.

International organizations

We live on a world where the estimated population will increase from 6.8 to 9.2 billion between 2010 and 2050. Clearly, food security and international trade will become increasingly important issues facing future generations. Two organizations supported through the United Nations provide the majority of guidance to developing countries on food safety: the World Health Organization (WHO) and the Food and Agriculture Organization (FAO). They bring together experts to generate guidance documents on food pathogens and toxins and best practices for their control. They also conduct risk assessments that lead to an understanding of the factors that have the greatest impact on preventing foodborne illness. The FAO, in particular, plays an important role by providing educational programs and commodity experts to countries that do not have the resources or knowledge base to improve their industry's food safety programs. Improvements in food safety education and government control procedures can lead to greater marketability and acceptance of foods from such countries.

The *Codex Alimentarius* Commission (CAC) brings together official representatives from countries throughout the world to develop guidance on what is considered acceptable for foods in international trade. This work is typically done by the Codex Committee of Food Hygiene (CCFH). The guidance is often used by member countries or regions as they consider new regulations. The CCFH also develops guidance on how to manage hazards in food and the use of risk analysis as a framework for managing food safety at the country level. An underlying goal of the CAC is to ensure consistency with the requirements of the World Trade Organization (WTO) for foods in international commerce. Briefly, this means that restrictions placed on foods in international commerce must be supported by existing science in addition to other considerations specified in the WTO agreement.

Food safety advocates

Food safety advocate organizations each have a particular focus (e.g., enterotoxigenic *E. coli*, regulatory policies, allergens). They contribute to food safety by participating in the debates on issues of importance to their members. They were particularly active during the 1990s as the pathogen reduction and HACCP regulations for meat and poultry products were being developed and subsequently as policies evolved. Industry seeks guidance from some of these organizations as they modify their control measures to reduce consumer risk, particularly for allergen control.

Consumers

Industry has the primary responsibility for producing safe foods. Consumer advocates have repeatedly stated that this responsibility should not be shifted to consumers, who must allow for the inadequate control of foodborne hazards upstream along the food chain. However, consumers and others in food service play a critical role by properly handling and preparing food. Food safety is a shared responsibility of industry and consumers with oversight provided by government. Consumers and others involved in food handling and preparation, however, must be provided with the information they need to understand and manage risk. This can be achieved in many ways, such as providing labels and instructions that accompany food products. I frequently reviewed the instructions on the products we produced as new products were developed and introduced into the market. Each year, I reviewed the instruction leaflet that accompanied our Butterball-branded turkeys. In some years modifications to the cooking procedure and/or time-temperatures for cooking were re-validated. I saw many changes over 35 years in how turkeys should be handled, cooked and chilled. These activities are commonplace in industry.

Government agencies and academia at the local school and college levels also make very important contributions by providing information to food handlers in the home. Also, there has been an increased amount of guidance provided at retail, food service and institutional levels. Each land-grant university has extension agents who provide guidance throughout the food chain with an emphasis on the foods produced in their state. Over the past two decades numerous studies have been conducted to measure the effectiveness of educational programs that lead to changes in the behavior of food handlers. These efforts are, to a degree, in response to emerging concerns over food safety. Since there is a continuous stream of new food handlers, many of whom have no knowledge about how foods are produced and their inherent risks, there is a continuing need to improve educational programs and materials.

1.1.3 What has been the impact of the food safety concerns of recent years on food producers, processors, food service providers, regulators, academia and public health agencies?

Consumers are better informed today of recalls of foods implicated in outbreaks. In addition to traditional media (e.g., television, radio and newspapers) there has been an increase in the number of websites that inform consumers of current recalls. Continual reports in the media of outbreaks at the multi-state and national levels have contributed to changes in the choices consumers make regarding the foods they purchase at retail outlets and in restaurants. Some consumers respond in the short term to publicity about a recall by avoiding that particular food when shopping. In the longer term some consumers have purchased more organic and/or locally grown foods believing such foods are safer. As this trend continues it will be interesting to learn whether states will have the resources to detect and report smaller clusters of cases of foodborne illness. It is possible that such outbreaks will be of short duration, difficult to detect and under-reported.

In at least two relatively recent major outbreaks, the impact was national in scope. The time between detection of an outbreak and accurate identification of the offending food can lead agencies to warn consumers to avoid certain foods (e.g., spinach or a certain type of tomato). In one outbreak involving spinach, the lag time was almost one week, during which it was recommended that consumers avoid spinach. This had a tremendous impact on the spinach industry and led to further strengthening of its control measures at the grower and processing levels.

Agencies have responded with various options (e.g., increased inspection, new regulations or policies, supporting research by academia and providing guidance documents). Agencies have selected the option(s) to bring about the changes needed to manage each food-pathogen combination and with regard to the agency's resources and ability to oversee the responsible segment of the food industry. In some cases an outbreak can have as much or more of an impact on an agency than on the affected industry segment.

While the above is important, the greatest impact is to affected consumers, their families and friends. Depending on the hazard and the consumer's health status, the impact can be loss of life or lifelong in nature. These factors understandably influence the degree of precaution when an agency must inform the public about an ongoing outbreak and the food source is uncertain.

1.2 Food safety management systems

1.2.1 Is HACCP an effective system for managing food safety?

HACCP is based on a logical thought process that was applied long before it reached its current level of maturity. Today, however, the concept is fully developed as a set of basic principles as defined by the NACMCF and the CCFH. It is one of only two systems available to industry to manage food safety along the food chain. The other system is known by three different names (i.e., good manufacturing practices – GMPs, sanitation standard operating practices – SSOPs and good hygiene practices – GHPs). All three address similar factors that must be controlled, primarily in food manufacturing facilities, and are sometimes referred to as prerequisite programs. HACCP was introduced in the 1960s as a formal system of control during the production of food for astronauts. The concept was further developed through the efforts of the International Commission on Microbiological Specifications for Foods (ICMSF) in the early 1980s and later by the NACMCF and CCFH in the 1990s. In the 1990s in the United States the Food Safety and Inspection Service (FSIS), as well as corresponding agencies from other countries, recognized the value of HACCP as a food control system and began to make its use mandatory for meat and poultry products and subsequently for other foods by the FDA.

Unfortunately, the perception that HACCP would prevent foodborne illness developed among some regulators and public media. Outbreaks, however, continued to occur due to at least two factors: (1) the presence of pathogens in raw agricultural commodities that were consumed uncooked (e.g., leafy greens, melons, sprouts, apple cider, fresh citrus juice) and (2) enteric pathogens in or on raw foods, which contaminated ready-to-eat (RTE) foods in the kitchen or were inadequately cooked (e.g., raw meat and poultry, shell eggs). Improvements have continued to be made to reduce consumer exposure through these vehicles.

1.2.2 What are the relative roles of HACCP, GMPs and targeted control measures in food safety management?

During the 1960s the Centers for Disease Control and Prevention (CDC) embarked on a major campaign to eradicate salmonellae from RTE foods and food ingredients. Epidemiologic investigations revealed that a wide variety of foods contained salmonellae. Initially, the results were reported through a weekly report that was available to the public. One common finding was the presence of salmonellae in dry foods and food ingredients. This is noteworthy because similar revelations occurred to the surprise of many between 1995 and 2012.

In response to the CDC investigations in the 1960s, questions were raised about the tolerance level for salmonellae in foods and whether a positive lot could be resampled to verify the lot was contaminated. A report from the National Research Council of the National Academy of Sciences in the United States by a committee of food microbiologists chaired by Professor E. M. Foster of the Food Research Institute, the University of Wisconsin, established the first microbiological sampling plans based on risk to consumers. This was a mega-leap above the sampling plan used at that time by the food industry, which was based on military standard 106D for sampling materials purchased by the Department of Defense. The concept was subsequently expanded by the ICMSF in several of its books and publications and by the CAC. The FDA decided that retesting a positive lot was unacceptable and such a lot must be considered adulterated.

Both in the 1960s and more recently since 1995, the presence of salmonellae in dry processed foods has been found to be due to the establishment and multiplication of salmonellae in the processing environment, which leads to product contamination. The same can be said for *Cronobacter sakazakii* in powdered infant formulae and *Listeria monocytogenes* in or on RTE foods that have been cooked and recontaminated.

This brings us to the second system of food safety management: GMPs, etc. Contamination of RTE foods in food processing facilities by pathogens, such as salmonellae, *L. monocytogenes* or, in the case of powdered infant formulae, *Cronobacter sakazakii*, must be controlled through GMPs/SSOPs/GHPs, not HACCP. This difference must be understood when conducting the hazard analysis for a food and how it is processed.

Modern food safety management systems consider the pathogens of concern as determined by the hazard analysis and how, when and where the pathogens enter into the food chain. This should lead to a targeted approach for control of the hazard(s). In a targeted approach certain factors are more important to control. HACCP systems identify critical control points in the process, which must be controlled, whereas to prevent post-process contamination of RTE foods certain GMPs are of greater importance for controlling a targeted pathogen. This concept was adopted, for example, by the CCFH in its guidance for preventing contamination of RTE foods with *L. monocytogenes* and by the FDA and FSIS in a variety of recent guidance documents.

It is uncertain whether all regulatory inspectors and independent auditors understand and accept the targeted approach as they conduct their evaluations of food processes and facilities. This may be a future source of conflict.

1.2.3 Can you describe the role of microbiological sampling and testing in food safety management?

There are many reasons for microbiological testing of food ingredients, foods in-process, the processing environment and finished products. This has been discussed by the ICMSF in its recent book (ICMSF, 2011), which recommends sampling and testing when the results can provide useful information. Microbiological sampling and testing is most useful when applied at selected steps and the information is used to detect trends and assess controls.

Testing also is conducted to ensure compliance with customer specifications and regulatory requirements (e.g., FSIS performance standards for salmonellae in or on raw meat and poultry products) or to demonstrate that a food is not adulterated within the limits of a sampling plan). In specific cases a sampling plan can be used as a control measure with the understanding that a negative result does not guarantee the absence of the pathogen of concern. In the case of *E. coli* O157:H7 the severity of the disease, the possibility of cross contamination in the kitchen and a history of undercooking are used to justify pre-testing of beef trimmings intended for ground beef. Sampling is used as a final control measure, which supplements and evaluates the prior measures that are applied before, during and after slaughter, to reduce consumer risk.

When recontamination of RTE foods is of concern, sampling is essential to assess control of the environment because visual inspection is unreliable. Furthermore, the results can be used to provide a record of the level of control and trends.

1.2.4 Would you care to comment on the impact of globalization on food safety?

According to data from the FAO the quantity of food shipped internationally each year has continued to increase and is now worth well over 1 trillion US dollars. With that increase in international trade there is greater concern about food safety. Reports of international outbreaks involving certain lots of food or food from a region increase this concern. A very long-term attempt at improving the safety of foods in international commerce has been ongoing since 1963 through the efforts of the CAC. A major goal of the CAC is to provide an international basis for judging equivalency and to ensure compliance with the WTO agreement. Aside from CAC activities, regulatory agencies in importing countries apply a variety of methods to evaluate foods from other countries (e.g., microbiological testing at port of entry and audits of facilities in the exporting countries). In some instances education is provided to the exporting companies and control authorities by experts from the FAO, or from the control authority of the importing country. Importing companies also play a role by providing guidance and/or enforcing purchase specifications through product testing and on-site audits.

In many countries there is no infrastructure to detect, document and track foodborne disease. Thus, importing countries and companies have difficulty estimating the level of risk associated with certain foods. This is made more complicated by emerging pathogens that are associated with certain foods from the exporting countries. It is wise to understand the food-pathogen combinations that may exist in each exporting country. A microbiological sampling plan is one means of evaluating ingredients originating from countries where the level of control is uncertain or unreliable (ICMSF, 2002).

1.2.5 What has been the impact of risk assessment and risk categorization on food safety?

The concepts are very useful because they provide a scientific approach to generating information needed by risk managers, who then can make more informed decisions. While other factors (e.g., social and political) also must be considered, having scientific input is essential when food safety issues are involved. Risk assessments and risk categorizations, however, can require extensive resources and time to develop. Furthermore, by the time sufficient information becomes available to conduct a risk assessment the outputs are generally predictable and seldom yield surprising results.

One goal of the risk assessment process is to use the best available data to provide quantitative estimates that can be used by risk managers as guidance when justifying and establishing policy. These concepts have met with mixed success. When the FSIS, for example, finalized its HACCP rule for pathogen reduction in 1996, the agency estimated that 43–72% of the total number of cases of salmonellosis in the USA was attributable to meat and poultry products and the majority of cases resulted from cross contamination in kitchens in homes and in food service outlets. The expectation was that the performance standards for salmonellae on raw meat and poultry would contribute to a reduction in salmonellosis in the USA. Unfortunately, despite industry's efforts to modernize and improve its facilities and processes to comply with the standards there has been no decrease in human salmonellosis through 2010/2011 as measured by two separate systems reported by the CDC (i.e., FoodNet and the Summary of Notifiable Disease).

Another unfortunate reality is that after four years of effort to develop a risk categorization for foods associated with listeriosis in the USA, FSIS policies do not reflect the results of that effort. This continues despite an FAO/WHO risk assessment for *L. monocytogenes* in RTE foods, which concluded that foods in which *L. monocytogenes* can multiply are the highest risk and foods in which growth cannot occur are the lowest risk. Thus, the FSIS, for example, continues to take action against imported dry fermented meats in which *L. monocytogenes* has been detected. To my knowledge there has not been a reported instance of commercially prepared dry fermented meat being found to be a source of listeriosis. An outbreak in Philadelphia that was attributed to dry fermented meat proved to be false.

1.3 Future efforts to further control food safety concerns

1.3.1 What do you think are the most important aspects in educating and training future food safety professionals?

First, it is worth noting the increase in membership of the International Association for Food Protection and the large number of young professionals attending its annual meetings. Similarly, there has been a very large increase in the number of younger members of the American Meat Science Association, many of whom are involved in improving the safety of meat and poultry products. Clearly, there is now much greater interest in food safety and this has led to an increase in more professional opportunities.

When I graduated with a PhD in 1964 it was my plan to work in the food industry for three years then find a job in government for three years and, finally, work in academia. I felt that the experience in industry and government would help me understand how each segment thinks and operates and my students would benefit from the experience. It seemed to be a good plan but I found industry to be so interesting that I didn't leave. It is possible the same may have occurred if I had started in government or academia. Why? From each different perspective it is possible to find a niche that can be fascinating and professionally rewarding. There is a great need to enhance the safety and quality of foods and contribute to ensuring an adequate supply of food for the ever-growing world population. As previously mentioned industry, government and academia play different roles but there are interesting challenges to be met through each of their perspectives.

I have been teaching a graduate-level course on HACCP and food safety management and find the quantity of information available today about the hazards to food safety is immense. It is important to learn where and how to access the specific information needed to address the hazards associated with a food operation. While HACCP plans tend to become locked-in when finalized and signed by plant management the plans must accommodate new information as emerging issues develop.

It is better perhaps to emphasize the scientific basis of food safety management and spend minimal time on regulations until the information is needed. Quite frankly it is difficult to stay current with all the regulations and resources emanating from the federal government. The science is applicable to most settings but regulations may change and agencies may reorganize. Exposure to food operations along the food chain affords an opportunity to learn the conditions of processing, why certain hazards are of concern and how they are controlled.

1.3.2 What do you consider to be the most important food safety challenges of the future?

Raw agricultural commodities will continue to be a significant source of foodborne disease. Many foods are eaten raw or mildly processed. The absence of a reliable kill step for enteric pathogens is the greatest hurdle to overcome. Current controls involve multiple steps that reduce but cannot eliminate enteric pathogens on raw foods derived from plants and animals. Even the weather, seasonality and the environment in which they are grown influence the level of risk associated with many of these foods. It is realistic to use the term 'best practices' to describe the combination of control measures that are applied because HACCP is inappropriate.

Clearly, there continues to be a need for improved technologies that reduce the risk associated with raw agricultural commodities. Irradiation is available for raw meat and poultry but the technology has been rejected by the majority of consumers. In-shell pasteurization of eggs is a promising technology but at added cost to consumers. Pasteurized shell eggs and liquid albumen in retail cartons are options that have become available for individuals whose health status places them at greater risk.

The role that food handlers play in contamination and foodborne disease in the home and especially in food service outlets, where large numbers of people can be exposed, is a continuing concern for the transmission of enteric pathogens, including viruses. The CDC has estimated that 31 pathogens are the cause of 9.4 million cases of foodborne illness each year in the USA. Of these it was estimated that 5.5 million (59%) foodborne illnesses are caused by viruses, 3.6 million by bacteria (39%) and 0.2 million by parasites (2%). In a study of data from six states it was estimated that about 50% of all foodborne outbreaks in the USA are due to noroviruses with salads, sandwiches and fresh produce serving as the primary vehicles. Advances have been made in detecting viruses, and methods for inactivating viruses on hands and food contact surfaces have improved. The education of food handlers, however, will be an important ongoing need to manage the risk of norovirus in RTE foods of the type described in the recent study.

A major concern that has been receiving considerable attention is the transmission of pathogens from animals to humans (i.e., zoonotic disease). Livestock, poultry and wild animals that are in close contact with humans have historically been a source of certain human pathogens (e.g., *Salmonella* and *Campylobacter*) but also represent a source of emerging human pathogens such as avian influenza. While avian influenza has not been demonstrated to be foodborne, the virus has caused severe illness and death particularly among farmers in Southeast Asia and had an enormous impact on the poultry industry in the affected region. In some cases pathogens may not cause disease in the animal population but can be deadly to susceptible humans (e.g., enterohemorrhagic *E. coli* in ground beef and *Salmonella* Enteritidis in shell eggs).

Perhaps, more attention will be devoted in the future to the control of *Toxoplasma gondii*. The most recent report from the CDC of foodborne illness due to major pathogens estimates that *T. gondii* is responsible for 8% of hospitalizations due to foodborne illnesses and 24% of deaths. This means that toxoplasmosis is the second leading cause of death after nontyphoidal *Salmonella* spp (28%) and ahead of *L. monocytogenes* (19%) and norovirus (11%). An analysis of sera among persons 12–49 years of age in the year 1999–2000 indicated that 15.8% were infected with *T. gondii*, a prevalence similar to the years 1988–1994. Infection can range from asymptomatic or a mild flu-like disease to congenital and acquired disease, resulting in blindness, neurological illness (e.g., schizophrenia) or death. The United States Department of Agriculture has estimated that about one half of the *T. gondii* infections occurring each year are due to eating raw or undercooked infected meat. Livestock, humans and wildlife become infected from a variety of sources. Cats and other felids are the only hosts in which the parasite can complete its entire life cycle. Thus, vaccination of cats has been proposed as one option for control.

The challenge of providing an adequate supply of safe foods to an increasing population on a world faced with climate change and decreasing supplies of freshwater will require long-term research with repeated breakthroughs in technology. Even the lack of toilets is a major concern, which can have international consequences considering the movement of food ingredients and foods around the world. A consortium of three international organizations, including the WHO, issued an assessment of the global water supply and sanitation for the year 2000. It was estimated that 40% of the world's population lacked access to improved sanitation. That means that 40% do not have access to: (1) a facility with a connection to a public sewer or septic system, (2) a pour-flush latrine, (3) a simple pit latrine or (4) a ventilated improved pit latrine. Instead, they must use service or bucket latrines where excreta are manually removed, public latrines or an open latrine. It is difficult to comprehend that about 40% of the world's population of six billion do not have access to what is considered to be 'improved sanitation.'

Food safety concerns will likely continue for several reasons that have more to do with human nature than the many other aspects that have been discussed. One is ignorance among upper management in some food companies of the hazards (e.g., sources, prevalence, severity and risk to consumers) associated with the foods for which they are responsible. A second reason is that although upper management may be aware of the hazards, they are complacent and do not commit the attention or resources required to ensure the hazards are controlled. Management commitment is the foundation for food safety management and the absence of a commitment to food safety will quickly permeate a business and increase consumer risk. It can occur, for example, with new management or when a company is purchased by a firm with little or no food industry experience. In some instances the technical staff is reduced and memberships in trade associations are discontinued to help pay for the loans used for the purchase. These and other changes result in a loss of access to scientific knowledge and the support needed to understand and properly manage food safety.

A third reason is a lack of oversight of food operations by individuals (e.g., inspectors, auditors and corporate technologists) who have the knowledge and experience to correctly identify food safety concerns and inform management of changes that are necessary for control. The fourth reason is much less likely; however, the desire to knowingly produce products under unacceptable conditions, ignore warning signals of ongoing contamination and maximize profit does exist and will continue in the absence of critical oversight. There does not seem to be a concerted effort to address these four weaknesses in management throughout the food chain so it is likely they will continue to impact food safety well into the future.

1.3.3 How can we stay ahead of emerging foodborne pathogens and newly recognized food–pathogen combinations?

I have seen a pattern evolve as new hazards are recognized and have found that it consists of five steps. With some modification the steps also apply to new food–pathogen combinations. Let me briefly describe them.

Detection – In this first step a patient is observed by a local doctor or other healthcare professional. The symptoms and circumstances leading to illness appear to be unusual and warrant further examination. Another trigger is the occurrence of an unusually large number of patients in a community or region. The unusual characteristics lead the healthcare provider to alert a supervisor or the local or state public health agency.

Investigation – The discovery of what appears to be a new emerging microbiological hazard or food–pathogen combination leads to an investigation by public health agencies to determine how widespread the hazard is in their domain and the severity of the disease.

Knowledge acquisition – Academia, governments and public health agencies investigate the emerging foodborne pathogen or metabolite. Initially, a definition is established based on the symptoms that identify individuals who can be included as a case. If possible, methods are developed to detect the pathogen or metabolite. This enables the accumulation of additional information such as where the hazard originates, how it enters the food supply, the conditions that lead to consumer exposure, where and how interventions can be implemented to prevent exposure, the infectious dose, the mode of action that leads to specific symptoms and illness and effective treatments for the disease. Industry can play a role, depending on the severity of the disease and methods of detection. For bacterial pathogens, surveys may be conducted of food processing establishments and products to determine the prevalence of the hazard. The discovery that listeriosis could be foodborne, for example, led to such surveys for *L. monocytogenes* during the 1980s and 1990s as well as efforts to validate interventions at certain steps in processing and to identify targeted GMPs.

Risk assessment – As the body of knowledge increases each group undertakes some form of risk assessment (RA). In my case, I mentally considered and discussed with my colleagues the available knowledge, the likely occurrence of the hazard in our foods and how and where we could intervene to minimize or prevent consumer exposure through HACCP and 'GMPs'. Government agencies likewise performed RAs, which over time became more formal and shifted from qualitative to quantitative RAs following the principles of RA as set forth by the federal agencies and CAC.

Regulatory response – Governments use information as knowledge is acquired to establish regulations, policies and guidance that industry should follow to ensure consumer protection. Microbiological criteria may be established in the form of performance objectives at specific steps in the food process or chain, including end-product testing. As mentioned elsewhere the initial response may not be consistent with information that eventually becomes available from a RA or risk categorization. At the international level, guidance is developed through the CAC for internal use by member countries and for foods in international trade.

1.3.4 Are changes in food regulations and inspection needed to improve food safety?

Changes in the current system of governing by multiple agencies with differing approaches and resources for regulation and inspection could be very beneficial in the USA. The first discussion I recall about placing all aspects of food safety under one agency was at the first National Conference on Food Protection in Denver in 1971. Can the challenges of the future be met with the existing system? One food safety agency would be a much better legacy to leave for future generations of consumers and food safety professionals but as is often said: 'it would take an act of Congress' and that is not likely.

1.4 References

ICMSF, 2002. *Microorganisms in Foods 7. Microbiological Testing in Food Safety Management*. Springer.

ICMSF, 2011. *Microorganisms in Foods 8: Use of Data for Assessing Process Control and Product Acceptance*. Springer.

Part II

Pathogen updates

2

Pathogen update: *Salmonella*

T. Hald, DTU, Denmark

DOI: 10.1533/9780857098740.2.25

Abstract: This chapter gives an overview of the recent research on the epidemiology of human salmonellosis. It begins by describing the occurrence of human salmonellosis including a discussion on the burden of illness studies, before describing the most important risk factors for human disease. Subtyping of *Salmonella* is increasingly used for tracking sources of infection and an update on classification and characterisation is given. Finally, the chapter reviews source attribution studies based on the application of microbial subtyping data and foodborne disease outbreak data and concludes with a discussion of the observed differences.

Key words: disease burden, epidemiology, *Salmonella*, source attribution, foodborne disease outbreaks.

2.1 Introduction

Salmonella is an important cause of foodborne disease in humans throughout the world and is a significant cause of morbidity, mortality and economic loss (Roberts and Sockett, 1994; Mead *et al.*, 1999; Adak *et al.*, 2002; Voetsch *et al.*, 2004; Schroeder *et al.*, 2005; Scallan *et al.*, 2011). Illness can range from mild to severe gastroenteritis and in some people invasive disease that can be fatal. Long-term sequelae such as reactive arthritis and irritable bowel syndrome have also been described as outcomes of salmonellosis (Doorduyn *et al.*, 2008; Haagsma *et al.*, 2010).

The majority of infections are transmitted from healthy carrier animals to humans via contaminated food. The main reservoir of zoonotic *Salmonella* is the gastrointestinal tract of warm-blooded animals including food-producing animals. Although other sources are recognised, its transmission to humans in most parts of the world occurs mainly through the ingestion of contaminated food (Baker *et al.*, 2007; O'Reilly *et al.*, 2007). Implicated foods are normally beef, pork, poultry, dairy products, eggs and fresh produce, and reports confirm the transmission of strains from the animal reservoir through the food chain and to the human population (Newell *et al.*, 2010).

Since the early 1990s, surveillance programs and control methods for *Salmonella* in the farm-to-fork continuum have been implemented in an increasing number of countries worldwide (EFSA, 2012a; Guo *et al.*, 2011). The growing awareness of a global food market has led to international initiatives towards a global control of *Salmonella*. Still, the incidence of human salmonellosis in many industrialised countries has remained high (EFSA, 2010). The effort to reduce human salmonellosis is challenged by the widespread occurrence of *Salmonella* in a variety of food-producing animals and the ability of *Salmonella* to adapt to and survive changing environmental conditions.

2.2 Incidence and burden of human salmonellosis

Non-typhoid *Salmonella enterica* is considered one of the leading causes of gastroenteritis and bacteremia in the world (Scallan *et al.*, 2011; Hendriksen *et al.*, 2011). Typically, the onset of symptoms is 12 to 72 hours after exposure, and the duration is 4–5 days, often followed by a period of fatigue. Symptoms are mainly from the gastrointestinal tract, often accompanied by fever, headache and muscle or joint pain. The infection is usually self-limiting and clinically indistinguishable from other common bacterial gastrointestinal infections (Mølbak *et al.*, 2006). Sequelae are observed in 1–2% of patients, and the mortality rate is usually low (<1%) for many serovars (Scallan *et al.*, 2011; EFSA, 2012a). In contrast, infections with the host-adapted serovars (e.g. *S.* Dublin and *S.* Choleraesuis) often lead to septicaemia, requiring antibiotic treatment and with case fatality rates of 20–30% (Mølbak *et al.*, 2006).

Human salmonellosis is the second ranking foodborne disease in the EU and most European countries, only exceeded by campylobacteriosis. The EU notification rate in 2010 was 21.5 cases per 100 000 population, ranging from 1.9 in Portugal to 91.1 confirmed cases per 100 000 population in Slovakia. The case fatality rate was 0.13% among the confirmed cases for which this information was reported (EFSA, 2012a). As in previous years, *S.* Enteritidis and *S.* Typhimurium were the most frequently reported serovars constituting 67.4% of all reported cases where the information on serovars was provided. During the past five years, a decreasing trend of human salmonellosis has been observed (Fig. 2.1), which is mainly explained by a decrease in the number of *S.* Enteritidis infections (EFSA, 2012a).

Statistics for the incidence of human salmonellosis (and other foodborne infections) are notoriously difficult to compare between countries and sometimes even within a country, as they depend on the definition of a case, the diagnostic method used and how the information is collected and analysed. In addition, the subjective reactions of the patients and general practitioners will influence whether a case will be diagnosed and reported. There have been attempts to calibrate *Salmonella* surveillance data at national surveillance institutes, and some research

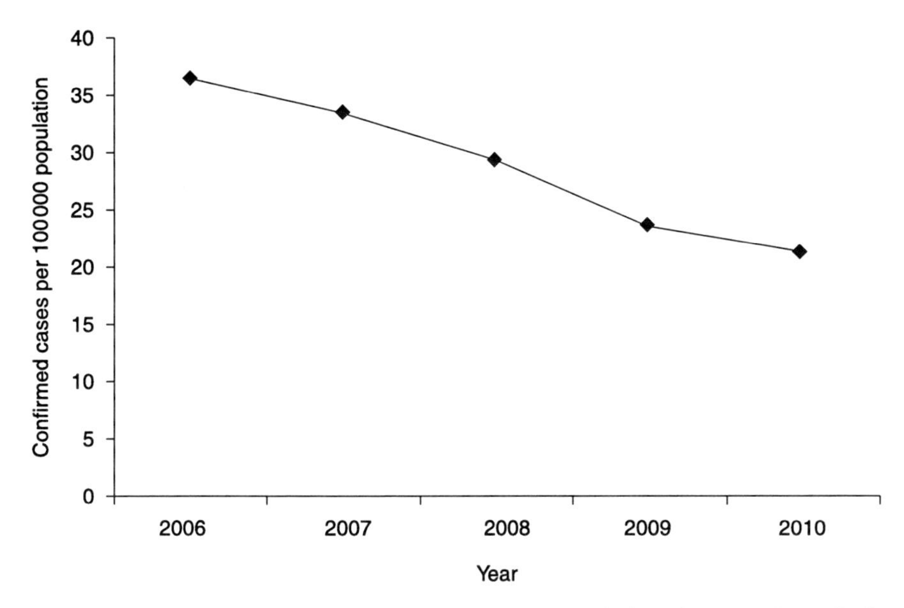

Fig. 2.1 Trend in reported confirmed cases per 100 000 population of human salmonellosis in the EU, 2006–2010 (EFSA, 2012a).

groups have attempted to equate disease in the population to what appears in official statistics (Wheeler *et al.*, 1999; Mead *et al.*, 1999; Gallay *et al.*, 2000; de Wit *et al.*, 2001; Voetsch *et al.*, 2004; van Pelt *et al.*, 2003; O'Brien *at al.*, 2010; Scallan *et al.*, 2011). The studies suggest that for every reported case of salmonellosis, between 3.8 and 38 people in the population fell ill (Mølbak *et al.*, 2006).

Under-reporting factors for human salmonellosis in the European Union (EU) Member states (MSs) have recently been estimated employing information on the risk from Swedish travellers returning to Sweden after a trip to a specific state in the EU in 2009 (Havelaar *et al.*, 2012). The risk of salmonellosis in returning Swedish travellers in the EU was 8.44 per 100 000 travels (95% confidence interval (CI): 8.18–8.71), ranging between 0.13 for travel to Finland to 94.3 for travel to Bulgaria. Based on these risk estimates, the true incidence of salmonellosis in 2009 was estimated as 6.2 (95% CI: 1.0–19) million cases in the EU and the under-reporting factor at the EU level was 57.5 (95% CI: 9.0–172), ranging from 0.4 in Finland to more than 2000 in Portugal. Based on these estimates, the European Food Safety Authority (EFSA) estimated the disease burden of salmonellosis and its sequelae as 0.23 (0.05–0.6) million disability-adjusted life years (DALYs) per year and total annual costs were estimated at 2 (0.3–4) billion euros (EFSA, 2011a).

A study by Majowicz *et al.* (2010) had the ambitious aim of estimating the global burden of human salmonellosis. By synthesising existing data from multiple studies and surveillance reports (including prospective population-based studies, 'multiplier studies', disease notifications and data on returning travellers), incidence estimates were calculated for each of the 21 global burden of disease (GBD) regions, which were then summed to provide a global number of cases. It was estimated that 61.8–131.6 million (ranging from 40/100 000 cases in the Pacific regions and Central Asia to 3 980/100 000 in East and Southeast Asia) cases of gastroenteritis due to *Salmonella* occur globally each year, with 39 000–303 000 deaths. The study obviously involves significant uncertainty. For instance, for 11 of the 21 GBD regions the risk estimates applied were based on data for returning Swedish travellers (Ekdahl *et al.*, 2005) in a comparable manner to the study at the EU level described above. Both studies, therefore, assume that the incidence rate in the local population is comparable with that of Swedish travellers who acquired their infection in that region. This assumption is debatable, as travellers (mostly tourists) may consume and behave very differently from the local population and may also be more susceptible to infection due to a general low exposure level in their home country. Still, the resulting estimates at least for the EU MSs are believed to provide a better reflection of the true human salmonellosis incidence than the actual reported number of cases (EFSA, 2011a).

2.3 Epidemiology and disease transmission in humans

Salmonella can affect mammals, birds, reptiles and insects (Mølbak *et al.*, 2006). There are numerous transmission pathways through which humans can be exposed to *Salmonella* including a wide range of domestic and wild animals and a variety of foodstuffs covering food of both animal and plant origin. Infected animals will have *Salmonella* in their faeces and the usual route of infection is through faecal-oral transmission. The epidemiology of *Salmonella* is, therefore, primarily due to direct or indirect faecal contamination of live animals, food or humans.

Despite the many efforts to prevent and control foodborne salmonellosis during the last 20 years, this pathogen continues to be one of the leading causes of human gastroenteritis. The many factors for this include the adaptive ability of the pathogen itself, the changing characteristics of the population, the increasing globalisation of the food trade and changes in industrial structure and in consumer behaviour.

The occurrence of antimicrobial resistance among zoonotic *Salmonella* is an increasing problem. Antimicrobial-resistant *Salmonella* involved in human disease are mostly spread through foods, predominantly poultry, eggs, pork and beef (Hald *et al.*, 2007). Meat is recognised as a source of human exposure to fluoroquinolone-resistant *Salmonella*, and high levels of extended-spectrum beta-lactamase (ESBL)-producing *Salmonella* have also been reported, particularly in poultry and meat in some EU MSs (EFSA, 2012b). Such resistant strains may be associated with a significant level of human infection, depending on the pathogenicity of the strains involved and the opportunity for them to contaminate the food chain (Butaye *et al.*, 2006; EFSA, 2011b; de Jong *et al.*, 2012; Rodriguez *et al.*, 2012). The control of antimicrobial-resistant bacteria in food is complicated by the fact that resistance mechanisms can be located on mobile genetic elements such as plasmids and thereby transferred between different bacterial species, for instance between generally apathogenic *Escherichia coli* and *Salmonella* spp.

The use of antimicrobials in food animals is a major contributing factor for the selection and dissemination of resistance for *Salmonella* (Emborg *et al.*, 2007; van den Bogaard and Stobberingh, 1999), but also the increasing use of antimicrobials, particularly fluoroquinolones, in humans has recently been associated with an increased incidence of infections caused by drug-resistant *Salmonella* (Koningstein *et al.*, 2010). Compared with patients infected with susceptible *Salmonella* strains, patients with multi-drug-resistant infections may be more likely to have a protracted course of disease, which in addition is more severe, often requires hospitalisation and may lead to excess mortality (Helms *et al.*, 2002; Varma *et al.*, 2005). This observed increase in severity can be a consequence of incorrect choice of antibiotic treatment for patients with extraintestinal infection resulting in reduced efficacy of early empirical treatment (Mølbak *et al.*, 2006).

Children and elderly people are considered to be more at risk from an infection with *Salmonella* than the average adult. It is generally accepted that immunocompromised people suffering from underlying diseases, e.g. cancer, AIDS or chronic bowel disorders, are more prone to become infected than people in good health (Helms *et al.*, 2006). People receiving antacids have also been reported as having an increased risk of infection due to the increased pH level in the ventricle (Neal *et al.*, 1994). The number of both elderly and chronically diseased people is growing, which may partially explain the continuing high level of human salmonellosis.

The rapidly growing international trade in live animals (including breeding animals), animal feed stuffs, raw materials and processed foods has increased the length and complexity of the food chain and has facilitated the introduction of new *Salmonella* types in importing countries (Aarestrup *et al.*, 2007; Hendriksen *et al.*, 2008). Concurrently, there has been an increase in consolidation in the food industry, including primary production and mass distribution. This trend toward greater geographic distribution of products from large centralised food processors carries a risk for more widespread outbreaks affecting more people (Crump *et al.*, 2002). The dissemination of *S.* Enteritidis in the table-egg industry (Thorns, 2000; EFSA, 2010) and the occurrence of multi-state and international foodborne outbreaks (Isaacs *et al.*, 2005; Werber *et al.*, 2005; Pezzoli *et al.*, 2008) are examples of this.

Traditionally, foods implicated in foodborne outbreaks have been poultry products including eggs, red meat and unpasteurised milk. In recent years, new types of food previously thought to be safe have emerged as sources of outbreaks. These include in particular fresh produce, which may be contaminated with animal or human faeces during growth, harvest and distribution. In particular, alfalfa sprouts have been implicated in large multi-state and international outbreaks (Mahon *et al.*, 1997; van Beneden *et al.*, 1999; Sivapalasingam *et al.*, 2004; Emberland *et al.*, 2007), and sprouts are recognised as a special problem because of the potential for pathogen growth during the sprouting process (EFSA, 2011c).

International travel has grown rapidly during recent decades. In countries with a low prevalence of *Salmonella* in their domestic livestock and food this fact influences the national statistics for humans markedly. In Sweden and Norway, for instance, it is estimated that approximately 70–80% of all human *Salmonella* infections are acquired abroad (Kapperud and Hasseltvedt, 1999; Ekdahl *et al.*, 2005; EFSA, 2010). Overall, in 2010 around 63% of human *Salmonella* cases in the EU were reported to be acquired domestically and 11% abroad. The proportion of cases with an unknown location of origin represented around 26% of confirmed cases, but in some MSs this proportion is 100%. Data on domestic versus travel-related cases are, therefore, often incomplete, but should as far as possible be ascertained, since a high proportion of travel-associated cases is likely to reduce the expected effect of national intervention strategies.

2.4 Classification and subtypes

The genus *Salmonella* belongs to the family Enterobacteriaceae. *Salmonella* is a gram-negative, facultative anaerobic motile rod. The bacteria are catalase positive and cytochrome oxidase negative. They produce gas from glucose and are able to reduce nitrate. *Salmonella* is closely related to *Escherichia coli* and is believed to share a common ancestor. During evolution *E. coli* has become lactose positive and is closely associated with mammals, while *Salmonella* has become lactose negative and is closely associated with reptiles and birds. The acquisition of pathogenicity islands, with SPI-1 and SP-2 being the most prominent, has conferred virulence to the bacteria.

Currently the genus *Salmonella* is divided into two species: *S. enterica* and *S. bongori*. The species *S. enterica* consist of six subspecies: *S. enterica enterica, S. enterica salamae, S. enterica arizonae, S. enterica diarizonae, S. enterica houtenae* and *S. enterica indica*, whereas no subspecies has been assigned to *S. bongori* (Su and Chiu, 2007). Within the species and subspecies, more than 2500 different serovars have been identified (Guibordenche *et al.*, 2010). The vast majority of zoonotic serovars associated with human illness belong to the subspecies *S. enterica enterica*.

All *Salmonella* serovars are considered potentially pathogenic for humans, but the degree of host adaptation varies, which affects the pathogenicity. Some serovars, such as *S.* Typhi, *S.* Paratyphi and *S.* Sendai, are highly adapted to humans (Mølbak *et al.*, 2006). They cause severe systemic illness in humans characterised by fever and abdominal symptoms, such as enteric fever (typhoid fever and paratyphoid fever) (Miller *et al.*, 1995). These serovars are usually not pathogenic to animals and are not considered to have a zoonotic potential.

Non-typhoid, ubiquitous serovars, such as *S.* Typhimurium and *S.* Infantis, affect a wide range of animals and humans. Although these serovars in principle are not host adapted, strong associations between certain serovars or subtypes within a serovar and a given animal reservoir may occur, e.g. *S.* Enteritidis in laying hens. In contrast, there exists a group of serovars that are highly adapted to an animal host, e.g. *S.* Choleraesuis in pigs, *S.* Dublin in cattle, *S.* Abortus-ovis in sheep and *S.* Gallinarum in poultry. These serovars only occasionally infect humans, where they may produce no, mild or serious disease (Acha and Szyfres, 2001; Mølbak *et al.*, 2006). The non-host-adapted serovars are those with principal zoonotic significance and their ability to infect animals and eventually infect humans via food seems to vary (Hald *et al.*, 2006; Pires and Hald, 2010).

A subtyping system based on lysis of *Salmonella* from a panel of *Salmonella* bacteriophages (phage-typing) is available. Phage typing is routinely used in some countries for the more common serovars, *S.* Enteritidis and *S.* Typhimurium (EFSA, 2012a). Phage typing subdivides serovars into phage types (PTs) in *S.* Enteritidis or definitive types (DTs) in *S.* Typhimurium. Antimicrobial susceptibility testing may also be used to characterise *Salmonella* isolates and studies have indicated that there seems to be a strong association between some phage types and antimicrobial resistance patterns, in particular for *S.* Typhimurium strains (Emborg *et al.*, 2007).

Molecular methods based on characterisation of bacterial DNA (e.g., pulsed field gel electrophoresis or PFGE) have a considerably higher discriminatory power than the above-mentioned phenotypic methods. The most recently developed methods typically target specific areas or genes of the genome and include the multiple-locus variable number tandem repeat analysis (or MLVA typing), where the numbers of repeat elements in specific loci are measured, and multi-locus sequence typing (MLST), where DNA sequences of specific genes are determined (Lindstedt *et al.*, 2004; Dingle *et al.*, 2001). For the sequence-based methods in particular, a whole new research area has become available as the latest technology makes it possible to perform large-scale sequencing at an affordable price.

Subtyping of *Salmonella* is used in epidemiological investigations. The high differentiation of strains obtained from genotyping is particularly useful in the investigation of outbreaks, as it helps to define groups of cases that have been infected with the same strain from the same source (Mølbak *et al.*, 2006; Torpdahl *et al.*, 2007). Subtyping is also increasingly being used to trace the sources of sporadic *Salmonella* cases, i.e. for source attribution purposes (Pires *et al.*, 2009) as discussed in the next section.

2.5 Tracing the sources of human salmonellosis – source attribution

Source attribution is defined as the partitioning of the human disease burden of one or more foodborne infections to specific sources, where the term source includes animal reservoirs and vehicles, e.g. foods. Source attribution methods attempt to attribute the burden of disease at the population level, and do not describe causation of disease at the individual level. Methods for source attribution of foodborne diseases include microbiological approaches, epidemiological approaches, intervention studies and expert elicitations. For a thorough review of source attribution methods see Pires *et al.* (2009). In the following, recent source attribution studies for human salmonellosis are presented and discussed.

2.5.1 Source attribution using microbial subtyping

The microbial subtyping approach characterises pathogen isolates using phenotypic and/or genotypic subtyping methods. The distribution of subtypes in potential sources (e.g. animals or food) is compared with the subtype distribution in humans, which is possible because of the identification of strong associations between some of the dominant subtypes and a specific reservoir or source, providing a heterogeneous distribution of subtypes among the sources. Subtypes exclusively or almost exclusively isolated from one source are regarded as indicators for the human health impact of that particular source, assuming that all human infections with these subtypes originate only from that source. Human infections caused by subtypes found in different reservoirs are then distributed relative to the prevalence of the indicator types. This approach requires a collection of temporally and spatially related isolates from various sources and humans, and is consequently facilitated by an integrated foodborne disease surveillance programme focused on the collection of isolates from the major food animal reservoirs of foodborne pathogens and from humans (Pires *et al.*, 2009).

The principle of comparing the distribution of *Salmonella* subtypes found in animal and food sources with those found in humans to make inferences about the most important sources of human disease has been applied by several research groups (Van Pelt *et al.*, 1999; Sarwari *et al.*, 2001). A Bayesian model developed to attribute human salmonellosis in Denmark (Hald *et al.*, 2004) has regularly being improved to include data on antimicrobial susceptibility (Hald *et al.*, 2007) as well as data from multiple time periods (Pires and Hald, 2010). The model attributes domestically acquired laboratory-confirmed human *Salmonella* infections caused by different *Salmonella* subtypes (e.g. serotypes, phage types and antimicrobial-resistant profiles) as a function of the prevalence of these subtypes in animal and food sources and the amount of each food source consumed. This approach has proved to be a valuable tool in focusing food safety interventions on the appropriate animal reservoir in Denmark (Fig. 2.2) (Wegener *et al.*, 2003; Korsgaard *et al.*, 2009), and the model has recently been adapted to attribute human salmonellosis in other EU countries (Pires *et al.*, 2008; Wahlström *et al.*, 2011; Valkenburgh *et al.*, 2007), as well as in the United States (Guo *et al.*, 2011), New Zealand (Mullner *et al.*, 2009) and Japan (Toyofuku *et al.*, 2011).

The microbial subtyping approach has also been adapted to accommodate data from EU countries in a model that utilised data from the European Centre for Disease Prevention and Control (ECDC) and the European Food Safety Authority (EFSA) (Pires *et al.*, 2011a). The model has been applied to data from 24 MSs and attributed human sporadic salmonellosis to four animal reservoirs: pigs, broilers, layers and turkeys. Overall estimates for the EU showed that the laying hen

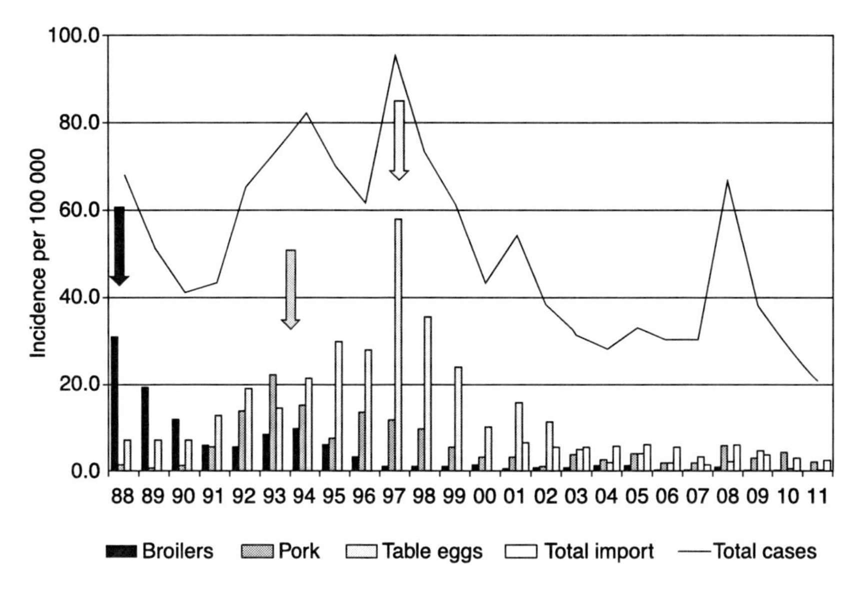

Fig. 2.2 Effects of *Salmonella* control programmes in Denmark as estimated by application of the microbiological subtyping approach on an annual basis. The arrows indicate the initiation of a new control programme in broiler chickens, in pigs and pork, and laying hens, respectively. Remaining cases were attributable to beef, imported food products, infections acquired while travelling abroad and unknown sources.

reservoir (eggs) was estimated to be the most important source of salmonellosis (43.8% of cases), followed by pigs (26.9%). Turkeys (4.0%) and broilers (3.4%) were estimated to be less important sources of *Salmonella*. Around 9.2% of all salmonellosis cases were reported as being travel related, 3.6% were reported as being part of an outbreak with an unknown source, and 9% of the cases could not be attributed to any source included in the model (Table 2.1).

Results varied substantially according to EU region (Table 2.1), revealing differences in the epidemiology of *Salmonella* among regions, in the relative contribution of sources for disease, and potentially in the efficiency of surveillance systems, data availability and representativeness. Layers were the most important source in eastern, northern and southern Europe, contributing with between 30% and 59.4% of human salmonellosis, whereas pigs were the major source of salmonellosis in southern Europe (43.6%). Turkeys and broilers contributed with varying but lower proportions of reported cases. In northern EU, a large proportion of the reported *Salmonella* infections was reported as acquired abroad.

Adapting the Danish approach, Hald *et al.* (2004) and Guo *et al.* (2011) estimated the relative proportions of domestically acquired sporadic *Salmonella* infections resulting from contamination of six food sources processed in the United States from 1998 through 2003. Results suggested that broiler chickens were the most important food source of domestic sporadic cases of salmonellosis

Table 2.1 Proportion (%) of *Salmonella* cases attributed to food sources in the EU regions,[a] 2007–2009, median and 95% credibility interval (%)

	EU	Eastern EU	Northern EU	Western EU	Southern EU
Broilers	3.4 (3.1–3.7)	7.0 (6.4–7.6)	1.2 (1.0–1.4)	2.1 (1.8–2.5)	3.1 (2.6–3.6)
Pigs	26.9 (26.3–27.6)	22.7 (21.5–23.9)	10.6 (10.0–11.1)	34.1 (33.5–34.7)	43.6 (42.5–44.8)
Turkeys	4.0 (3.8–4.3)	2.2 (2.0–2.5)	7.4 (6.9–8.0)	4.1 (3.8–4.3)	7.6 (6.8–8.4)
Layers	43.8 (43.2–44.4)	59.4 (58.1–60.6)	30.0 (29.4–30.6)	41.8 (41.3–42.3)	28.4 (27.5–29.3)
Outbreak[b]	3.6 (n.a.)[c]	5.4 (n.a.)	4.0 (n.a.)	2.2 (n.a.)	4.2 (n.a.)
Travel	9.2 (n.a.)	0.8 (n.a.)	34.5 (n.a.)	4.8 (n.a.)	0.7 (n.a.)
Unknown	9.0 (8.7–9.3)	2.5 (1.9–3.1)	12.4 (11.8–13.0)	10.9 (10.5–11.4)	12.5 (11.4–13.5)

Notes:
[a]EU regions as defined by the United Nations (Pires *et al.*, 2011a). Eastern Europe: Czech Republic, Hungary, Poland and Slovakia. Northern Europe: Denmark, Estonia, Finland, Ireland, Latvia, Lithuania, Sweden and the United Kingdom. Southern Europe: Cyprus, Greece, Italy, Portugal, Slovenia, Spain. Western Europe: Austria, Belgium, France, Germany, Luxembourg and the Netherlands.
[b]Includes outbreaks with unknown source. Outbreaks for which the source was identified were assigned to the relevant animal sources.
[c]n.a. – not applicable: the proportions of outbreak-related and travel-related cases were derived directly from the reported data (i.e. they were not estimated and consequently no credibility intervals were calculated).

Source: Adapted from Pires *et al.* (2011).

(48%) for all study years. Additional estimated sources of foodborne illness in the US were ground beef (28%), turkey (17%), egg products (6%), intact beef (1%) and pork (<1%). Both the EU and US analyses utilised food source data collected from points of food processing (farm and slaughter), but the US model assumed that all estimated sporadic illnesses were associated with the modelled food sources and did not attribute illnesses to travel or unknown sources.

The Danish approach was also adapted to national surveillance data from Japan collected between 1998 and 2007 to estimate the number of human *Salmonella* illnesses attributable to each of the major animal-food sources (Toyofuku *et al.*, 2011). In this analysis, eggs were estimated to be the most important source of disease, being responsible for over 50% of the cases in most years. Broilers and swine were the second most important sources, depending on the year, while cattle/beef was seldom associated with human salmonellosis. The New Zealand adaptation of the Hald model included some modifications to the original approach, allowing for instance the model to be more adaptable to countries with less intensive surveillance systems (Mullner *et al.*, 2009). The model attributed the majority of the *Salmonella* illnesses to pork (60%), followed by poultry (21%) and beef and veal (12%). Eggs (3%) and lamb (1%) were estimated as minor sources of infection.

The phenotypic typing methods currently applied to *Salmonella* isolates included in source attribution studies have limitations in their power to identify the origin of a given isolate, particularly for commonly occurring subtypes. Molecular methods based on characterisation of bacterial DNA (e.g. MLVA and

MLST) have a considerably higher discriminatory power than the phenotypic methods and are increasingly being applied in outbreak investigation for pinpointing a particular source (Torpdahl *et al.*, 2007). Still, the value of the DNA-based methods for source attribution of human salmonellosis needs to be assessed, and will undoubtedly challenge the optimal strategy: 'one typing method that fits all needs'. Very discriminatory methods are not necessarily the best solution for source attribution, where we are not looking for a single source for a particular outbreak, but rather want to relate groups of *Salmonella* strains with particular reservoirs/sources and then attribute sporadic human cases to these sources. The process must allow for some genetic diversity between strains from human and food sources even if they are epidemiologically related. It is, therefore, expected that serotyping, phage typing and susceptibility testing will remain useful tools for source attribution for some time and will strengthen global *Salmonella* surveillance in general.

2.5.2 Source attribution using outbreak data

Another way of trying to assess the proportion of human infections that is likely to be foodborne, and the foods implicated in causing human disease, is to use data from outbreak investigations. One advantage is that these data are observed at the public health endpoint and are often available in countries with little or no surveillance of sporadic cases (Pires *et al.*, 2009). A simple descriptive analysis or summary of outbreak data is useful for attributing illnesses to foods, but often the implicated food is a 'complex' food containing several food items, where any of the items could be the actual source of the infection.

An alternative method for conducting an analysis of data from outbreak investigations was developed in the United States. In this method, food items are categorised into a hierarchical scheme according to their ingredients (Fig. 2.3) (Painter *et al.*, 2009). Foods that contain ingredients from only a single food category are considered 'simple foods', while foods that contain ingredients from multiple food categories are considered 'complex foods'. For example, steak is a simple food whereas meat loaf is a complex food. Each implicated food is assigned to one or more mutually exclusive food categories, according to its ingredients. For outbreaks with a simple implicated food item, all illnesses are attributed to that single category. For outbreaks with a complex implicated food item, illnesses are partitioned to each category in the complex food according to the proportion of illnesses attributed to each of those categories in outbreaks caused by simple foods. As a result, illnesses in an outbreak due to a complex food item are attributed to a category in the implicated complex food, only if that category has been implicated in at least one outbreak due to a simple food. The number of illnesses attributed to each category are then summed and used to determine the percentage of disease attributed to each category (Painter *et al.*, 2006, 2009).

This method has been adapted to attribute human salmonellosis in Europe (Pires *et al.*, 2010). Based on foodborne outbreak data reported by the EFSA for the years 2005 and 2006, the authors estimated that the most important food

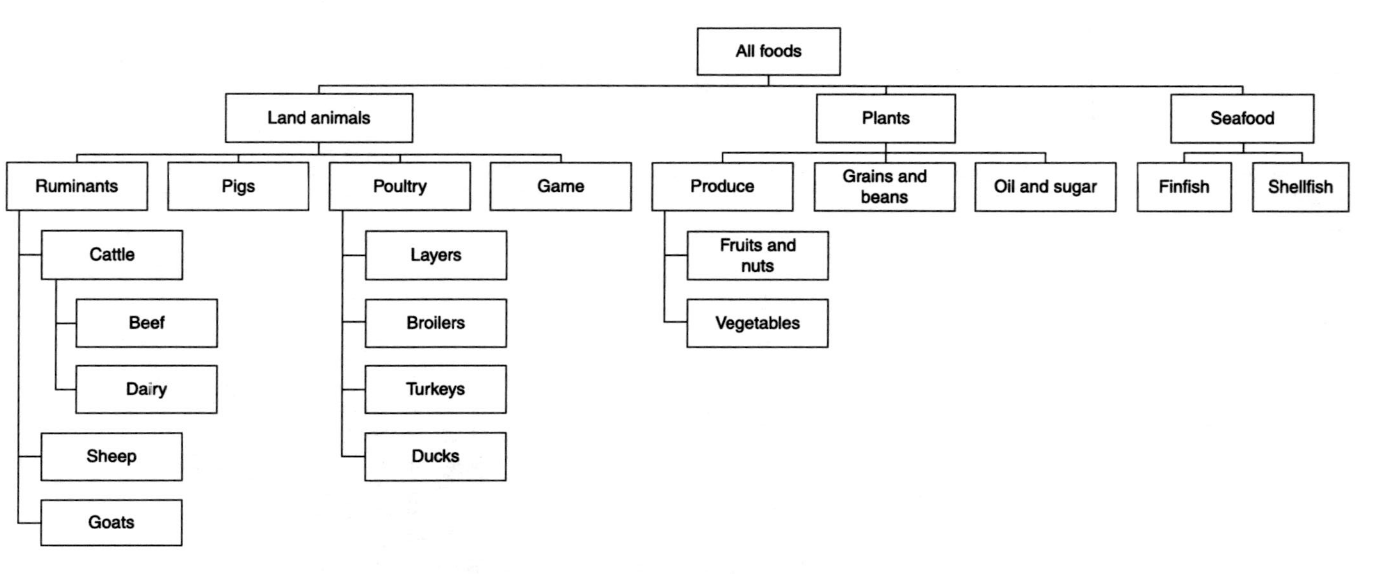

Fig. 2.3 Hierarchical scheme for categorising food items into commodities (Pires *et al.*, 2011).

sources were eggs (32%) and meat and poultry and meat (15%), but also that a large proportion of cases could not be linked to any source. The EU model was recently updated with foodborne outbreak data reported in the period between 2007 and 2009 (Pires *et al.*, 2011a). In this study, eggs were still the most important source of disease followed by pork, chicken, the general category 'meat and poultry', and dairy products (Table 2.2). In both studies, the proportion of *Salmonella* outbreaks attributed to an unknown source varied substantially between regions. When comparing source attribution results between EU regions excluding outbreaks with an unknown source (i.e. only accounting for outbreaks with known sources), the proportion of disease attributed to food sources varied. Source attribution estimates for eggs were higher in eastern Europe (84.3%) and southern Europe (73.8%). Pork followed in importance in western Europe (16.6%), whereas vegetables were estimated to be a major contributor for salmonellosis in northern Europe (18.5%). Chicken and dairy products were of similar but lesser importance in the regions (Pires *et al.*, 2011a).

A similar model based on outbreak data, which is able to consider complex foods, was also applied to attribute human foodborne illnesses to specific sources in Latin America and the Caribbean (Pires *et al.*, 2012). Data from 20 countries covering the period from 1993 and 2010 were collected. In general, eggs, meat products, vegetables, chicken, grains and beans, and pork were the most important sources of salmonellosis in the whole period. When excluding outbreaks with an

Table 2.2 Proportion (%) of *Salmonella* outbreaks attributed to food sources in the EU, 2007–2009, by year, median and 95% credibility interval (%)

	2007		2008		2009	
Eggs	56.1	(56.0–56.3)	61.6	(61.4–61.7)	34.5	(34.3–34.6)
Dairy	2.2	(2.1–2.3)	2.9	(2.7–3.0)	1.2	(1.1–1.3)
Goat milk	0	(0–0)	0	(0–0)	0	(0–0)
Meat	1.1	(1.1–1.12)	3.3	(3.3–3.4)	1.3	(1.3–1.4)
Poultry	0	(0–0)	0.2	(0.2–0.2)	10.9	(10.9–10.9)
Chicken	3.5	(3.5–3.5)	2.3	(2.2–2.3)	2.6	(2.5–2.6)
Ducks	0	(0–0)	0	(0–0)	0	(0–0)
Turkey	0.5	(0.5–0.5)	0.04	(0.03–0.06)	0.3	(0.3–0.3)
Beef	0.5	(0.5–0.6)	0.6	(0.5–0.7)	0.6	(0.6–0.6)
Pork	5.4	(5.4–5.4)	6.1	(6.1–6.1)	1.8	(1.8–1.8)
Lamb	0.2	(0.2–0.2)	0		0	
Mutton	0		0		0	
Game	0		0		0	
Fruits and nuts	0.2	(0.2–0.2)	0.2	(0.2–0.2)	0.01	(0.005–0.03)
Vegetables	2.4	(2.3–2.4)	1.0	(0.8–1.1)	1.5	(1.4–1.6)
Grains and beans	0.8	(0.7–0.9)	0.7	(0.5–0.8)	0.4	(0.3–0.4)
Oils and sugar	0.9	(0.8–1.0)	0.2	(0.1–0.3)	0.8	(0.7–0.8)
Seafood	0.8	(0.8–0.8)	2.4	(2.4–2.5)	0.9	(0.8–0.9)
Water	0.5	(0.5–0.5)	0		0	
Unknown	25.0		18.5		43.5	

Source: Pires *et al.* (2011).

unknown source, results showed a substantial increase from the 1990s to the 2000s in the proportion of disease attributed to the sources eggs (16.8% to 43.3%) and pork (3.7% to 9.1%), and minor increases in the relative contribution of vegetables (10.2% to 11.6%) in the same period. In contrast, the proportion of disease attributed to meat products (29.2% to 8.9%) and chicken (12% to 5.6%) decreased in the same period. The same method for analysis of outbreak data was applied to achieve source attribution estimates for Japan (Pires *et al.*, 2011b). Data covered the period from 2000 to 2009 and attributed disease to food sources and water. *Salmonella* source attribution estimates suggested that eggs were the most important food source during the whole study period, and that the proportion of disease attributed to this source increased in the second half of the decade. Among outbreaks with a known source, vegetables followed eggs in importance causing 13.2% of illnesses from 2000 to 2004 and 16.8% from 2005 to 2009, and with grains and beans causing 11.8% and 12.4%, respectively, in the same time periods. All remaining food sources were of minor importance for salmonellosis in Japan. Over 80% of reported outbreaks could not be attributed to any source.

A statistical analysis of data collected from 1996 to 2005 through the Canadian foodborne outbreak surveillance system has also been performed (Ravel *et al.*, 2009). Only data from outbreak investigations identifying both the agent and the food vehicle were used. The results indicated that produce was the most frequent cause of *Salmonella* outbreaks (28.9%), followed by poultry (14.5%), other meats (14.5%), dairy products (9.2%) and seafood (6.6%). Eggs caused 5.3% of all *Salmonella* outbreaks, and in 13.2% the implicated food was multi-ingredient, where the responsible ingredient was not identified or estimated through modelling.

The limitations of using of outbreak data for attribution include that the quality of evidence varies between data sources and classification schemes for the data are not consistently used. Also, large outbreaks, outbreaks associated with point sources, outbreaks that have short incubation periods and outbreaks that cause serious illness, are more likely to be investigated. Likewise, certain food vehicles are more likely to be associated with reported outbreaks than others, which can lead to an overestimation of the proportion of human illnesses attributed to a specific food. An important factor to consider is that illnesses included in data from outbreak investigations may not be representative of all foodborne illnesses. The fraction of the burden of foodborne disease that is associated with outbreaks varies between pathogens but is typically smaller than that corresponding to sporadic disease. Consequently, the extrapolation of source attribution estimates obtained through an analysis of data from outbreaks to the overall burden of disease should be made with care. Pires *et al.* (2010), however, concluded that the approach seemed useful for attributing human salmonellosis, but not campylobacteriosis (Pires *et al.*, 2010). The latter because there are relatively few reported outbreaks of campylobacteriosis and the relative importance of the implicated sources seems to differ between outbreaks (e.g. water) and sporadic cases (e.g. poultry and meat).

2.6 Discussion on sources of human salmonellosis

Source attribution approaches are increasingly being applied to surveillance data in an effort to inform control and intervention. In this chapter, we have focused on studies based on microbial subtyping and outbreak data. Each approach has different data requirements and method uncertainties, and they attribute illnesses to different points in the farm-to-consumption continuum. Methodological differences and differences in data availability and quality contribute to the observed variability in source attribution across the studies. However, the variation in the relative importance of different sources between countries and regions undoubtedly also reflects differences in the epidemiology of salmonellosis including differences in animal and food prevalence, consumption patterns and preferences, and animal and food production systems.

In the EU, the overall incidence of human salmonellosis decreased from 2006 to 2010, which is mainly explained by a decrease in the number of *S.* Enteritidis infections, presumably as a result of improved surveillance and control of *S.* Enteritidis in laying hens in EU MSs (Fig. 2.4) (EFSA, 2012a; Korsgaard *et al.* 2009). In contrast, the incidence of *S.* Typhimurium infections has changed little indicating the need for improved monitoring and control of *Salmonella* in the major sources of these infections, particular pigs and pig meat, which recently has been pointed out as the major source in many EU countries (Hald *et al.*, 2012).

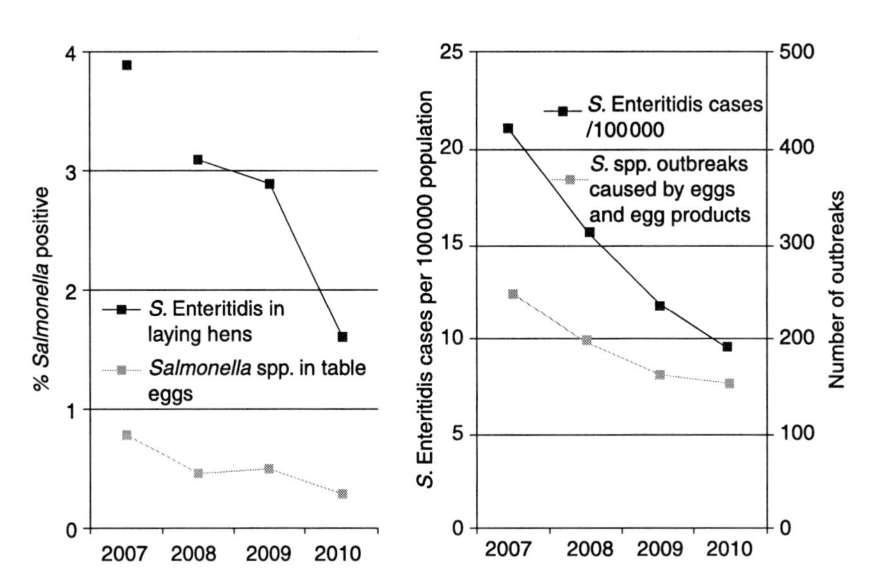

Fig. 2.4 *Salmonella* in human cases, eggs and laying hens and the number of *Salmonella* outbreaks caused by eggs within the EU, 2007–2010 (EFSA, 2012a). Data for laying hens and table eggs are shown only for samples ≥ 25. For laying hens only data from sampling during the production period were included.

Despite the decreasing trend of *S.* Enteritidis cases, eggs from laying hens are still considered one of the most important sources of *S.* Enteritidis infections in many EU MSs. This is supported by source attribution analysis based on subtyping data and outbreak data (Pires *et al.*, 2010, 2011a). A certain proportion of human *S.* Enteritidis infections are also attribued to broilers, particularly in countries with a high *S.* Enteritidis prevalence in broiler flocks.

In the US model described by Guo *et al.* (2011), chickens were found to be the most important source, whereas egg products were of less importance. This is in contrast to the EU study and may be explained by differences in consumption patterns (the average US citizen eats 2–3 times more chicken meat than the average EU citizen) and/or prevalence in the two regions. Also, in the US model, representative data on shell egg contamination were not available and egg products, which presumably are associated with a lower foodborne risk, were included instead (Guo *et al.*, 2011). In New Zealand, eggs were also found to contribute only a few per cent of the human cases, but here the reason is probably due to the fact that *S.* Enteritidis has never been established in the national shell-egg production (Mullner *et al.*, 2009).

Human *S.* Typhimurium infections represent 10–20% of all cases in the EU and the majority of these cases are likely to be associated with pig meat consumption (Pires *et al.*, 2011a; Hald *et al.*, 2012). Certainly broilers and beef also contribute to these infections, but the contribution is assessed to be low due to low prevalence and/or lower impact through the food production chain. The latter is because some of the dominant *S.* Typhimurium phage types in broilers only occur in low frequencies in humans. In New Zealand, pork was found to be the major contributor, probably reflecting the fact that *S.* Typhimurium is the single most important serovar found in domestically acquired salmonellosis cases (Lim *et al.*, 2011; Hendriksen *et al.*, 2011). In the US, the contribution from pork was very low, which cannot be readily explained by the data available.

Travel appears to be an important 'source' of sporadic salmonellosis particularly in northern Europe. However, travel data are lacking from many countries, so the role may be underestimated in other parts of the EU. Attribution studies from the US and Japan do not consider travel.

Only a few countries systematically collect data on *Salmonella* in imported food. Experience from Denmark indicates that the relative importance of this source increases when efforts to control domestic sources are successful. This will, however, depend on the amount and origin of the imported food.

Fresh produce is described as an increasing source of foodborne outbreaks in the US and Canada, including salmonellosis. In Europe, multi-state outbreaks caused by fresh produce have also been observed (e.g. Pezzoli *et al.* 2008; EFSA, 2011d), but based on an attribution study using outbreak data, fresh produce does not seem to be a major source in the EU in general. Fresh produce has not been included in the attribution studies based on subtyping due to lack of appropriate surveillance data. Still, from a risk management point of view, it can be argued that fresh produce is included indirectly in attribution estimates since food-producing animals constitute a reservoir for the contamination of fresh products.

Reducing *Salmonella* occurrence in food-producing animals will also, therefore, to some extent reduce the contamination level of fresh produce produced in the same country or region (Hald *et al.*, 2004).

In conclusion, salmonellosis continues to represent a considerable foodborne disease burden in the majority of countries. There is considerable under-reporting, and the true number of cases of illness is likely to be much higher than what is reported.

The successful control of *Salmonella* (and other zoonotic pathogens) requires knowing about the most important sources or reservoirs as well as the principal routes of transmission, and these are very likely to vary between countries and regions. The identification of sources should whenever possible be based on several source attribution approaches, as the combined data will increase our confidence in the results. In addition to evaluating the trends and dynamics of sources of human infections, the results will support risk managers in allocating resources to achieve the highest possible benefit. The fact that we are living in a 'global village' furthermore means that food-safety surveillance and efforts to reduce transmission of *Salmonella* by food and other routes should be implemented on a global scale.

Finally, it is stressed that the successful surveillance and control of *Salmonella* (and other zoonoses) requires collaboration between all experts in the food-production chain, i.e. between microbiologists and epidemiologists, across veterinary and public health borders and between food-safety authorities, scientists and the food industry.

2.7 References

AARESTRUP, F.M., HENDRIKSEN, R.S., LOCKETT, J., GAY, K., TEATES, K. *et al.* (2007). International spread of multidrug-resistant *Salmonella* Schwarzengrund in food products. *Emerg. Infect. Dis.*, 13, 726–731.

ACHA, P.N. and SZYFRES, B. (2001). *Salmonella*. In: *Zoonoses and Communicable Diseases Common to Man and Animals*. Acha, P.N. and Szyfres, B. (Eds). PAHO Scientific and Technical Publications, No. 580.

ADAK, G.K., LONG, S.M. and O'BRIEN, S.J. (2002). Trends in indigenous foodborne disease and deaths, England and Wales: 1992 to 2000. *Gut*, 51, 832–841.

BAKER, M.G., THORNLEY, C.N., LOPEZ, L.D., GARRETT, N.K. and NICOL, C.M. (2007). A recurring salmonellosis epidemic in New Zealand linked to contact with sheep. *Epidemiol. Infect.*, 135, 76–83.

BUTAYE, P., MICHAEL, G.B., SCHWARZ, S., BARRETT, T.J., BRISABOIS, A., *et al.* (2006). The clonal spread of multidrug-resistant non-typhi *Salmonella* serotypes. *Microbes Infect.*, 8(7), 1891–1897. Available from www.ncbi.nlm.nih.gov/pubmed/16714135.

CRUMP, J.A., GRIFFIN, P.M. and ANGULO, F.J. (2002). Bacterial contamination of animal feed and its relationship to human foodborne illness. *Clin. Infect. Dis.*, 35, 859–865.

DE JONG, A., STEPHAN, B. and SILLEY, P. (2012). Fluoroquinolone resistance of *Escherichia coli* and *Salmonella* from healthy livestock and poultry in the EU. *J. Appl. Microbiol.*, 112(2), 239–245. Available from www.ncbi.nlm.nih.gov/pubmed/22066763.

DE WIT, M.A., KOOPMANS, M.P., KORTBEEK, L.M. *et al.* (2001). Sensor, a population-based cohort study on gastroenteritis in the Netherlands: Incidence and etiology. *Am. J. Epidem.*, 154(7), 666–674.

DINGLE, K.E., COLLES, F.M., WAREING, D.R., *et al.* (2001). Multilocus sequence typing system for *Campylobacter jejuni*. *J. Clin. Microbiol.*, 39, 14–23.

DOORDUYN, Y., VAN PELT, W., SIEZEN, C.L.E., VAN DER HORST, F., VAN DUYNHOVEN, Y.T H.P., *et al.* (2008). Novel insight in the association between salmonellosis or campylobacteriosis and chronic illness, and the role of host genetics in susceptibility to these diseases. *Epidemiol. Infect.*, 136, 1225–1234.

EFSA (2010). EFSA panel on biological hazards (BIOHAZ); Scientific opinion on a quantitative estimate of the public health impact of setting a new target for the reduction of *Salmonella* in laying hens. *EFSA J.*, 8(4), 1546 (86 pp.).

EFSA (2011a). EFSA panel on biological hazards (BIOHAZ); Scientific opinion on a quantitative estimation of the public health impact of setting a new target for the reduction of *Salmonella* in broilers. *EFSA J.*, 9(7), 2106, (94 pp.).

EFSA (2011b). EFSA panel on biological hazards (BIOHAZ); Scientific opinion on the public health risks of bacterial strains producing extended-spectrum β-lactamases and/or AmpC β-lactamases in food and food-producing animals. *EFSA J.*, 9(8), (95 pp.).

EFSA (2011c). EFSA panel on biological hazards (BIOHAZ); Scientific opinion on the risk posed by Shiga toxin-producing *Escherichia coli* (STEC) and other pathogenic bacteria in seeds and sprouted seeds. *EFSA J.*, 9(11), 2424 (101 pp.).

EFSA (2011d). Shiga toxin-producing *E. coli* (STEC) O104:H4 2011 outbreaks in Europe: Taking stock. *EFSA Journal* 9(10), 2390 (22 pp.).

EFSA (2012a). The European Union summary report on trends and sources of zoonoses, zoonotic agents and food-borne outbreaks in 2010. *EFSA J.*, 10(3), 2597 (442 pp.).

EFSA (2012b). The European Union summary report on antimicrobial resistance in zoonotic and indicator bacteria from humans, animals and food in 2010. *EFSA J.*, 10(3), (233 pp.).

EKDAHL, K., DE JONG, B., WOLLIN, R. and ANDERSSON, Y. (2005). Travel-associated non-typhoidal salmonellosis: geographical and seasonal differences and serotype distribution. *Clin. Microbiol. Infect.*, 11, 138–144.

EMBERLAND, K.E., ETHELBERG, S., KUUSI, M., VOLD, L., JENSVOLL, L., *et al.* (2007). Outbreak of *Salmonella* Weltevreden infections in Norway, Denmark and Finland associated with alfalfa sprouts, July–October. *Eurosurveillance*, 12(48). Available at: www.eurosurveillance.org/ViewArticle.aspx?ArticleId=3321.

EMBORG, H.D., VIGRE, H., JENSEN, V.F., *et al.* (2007). Tetracycline consumption and occurrence of tetracycline resistance in *Salmonella* Typhimurium phage types from Danish pigs. *Microb. Drug Resist.*, 13, 289–294.

GALLAY, A., VAILLANT, V., BOUVET, P., *et al.* (2000). How many foodborne outbreaks of *Salmonella* infection occurred in France in 1995? Application of the capture-recapture method to three surveillance systems. *Am. J. Epidemiol.*, 152, 171–177.

GUIBOURDENCHE, M., ROGGENTIN, P., MIKOLEIT, *et al.* (2010). Supplement 2003–2007 (No. 47) to the White-Kauffmann-Le Minor scheme. *Research in Microbiol.*, 161, 26–29.

GUO, C., HOEKSTRA, R.B., SCHROEDER, C.M., PIRES, S.M., ONG, K.L., *et al.* (2011). Application of Bayesian techniques to model the burden of human salmonellosis attributable to US food commodities at the point of processing: Adaptation of a Danish model. *Foodborne Pathogens and Disease*, 8(4), 509–516.

HAAGSMA, J.A., SIERSEMA, P.D., DE WIT, N.J. and HAVELAAR, A.H. (2010). Disease burden of post-infectious irritable bowel syndrome in the Netherlands. *Epidemiol. Infect.*, 138, 1650–1656.

HALD, T., VOSE, D., WEGENER, H.C. and KOUPEEV, T. (2004). A Bayesian approach to quantify the contribution of animal-food sources to human salmonellosis. *Risk Anal.*, 24, 255–269.

HALD, T., WINGSTRAND, A., BRØNDSTED, T. and LO FO WONG, D.M.A. (2006). Human health impact of *Salmonella* contamination in imported soybean products: A semiquantitative risk assessment. *Foodborne Pathogens and Disease*, 3(4), 422–431.

HALD, T., LO FO WONG, D. and AARESTRUP, F.M. (2007). The attribution of human infections with antimicrobial resistant *Salmonella* bacteria in Denmark to sources of animal origin. *Foodborne Pathogens and Disease*, 4, 313–326.

HALD, T., PIRES, S.M. and DE KNEGT, L. (2012). Development of a *Salmonella* source-attribution model for evaluating targets in the turkey meat production. Supporting Publications 2012:EN-259. (35 pp.). Available online: www.efsa.europa.eu/en/supporting/doc/259e.pdf.

HAVELAAR, A.H., IVARSSON, S., LÖFDAHL, M. and NAUTA, M.J. (2012). Estimating the true incidence of campylobacteriosis and salmonellosis in the European Union, 2009. *Epidem. Infect.*, Published online 13 April 2012.

HELMS, M., SIMONSEN, J. and MØLBAK, K. (2006). Foodborne bacterial infection and hospitalization: A registry-based study. *Clin. Infect. Dis.*, 42(4), 498–506.

HELMS, M., VASTRUP, P., GERNER-SMIDT, P. and MØLBAK, K. (2002). Short and long term mortality associated with foodborne bacterial gastrointestinal infections: Registry based study. *BMJ*, 15, 326(7385): 357.

HENDRIKSEN, R.S., BANGTRAKULNONTH, A., PULSRIKARN, C., PORNREONGWONG, S., HASMAN, H., *et al.* (2008). Antimicrobial resistance and molecular epidemiology of *Salmonella* Rissen from animals, food products and patients in Thailand and Denmark. *Foodborne Pathogens and Disease*, 5, 605–619.

HENDRIKSEN, R.S., VIEIRA, A.R., KARLSMOSE, S., LO FO WONG, D.M.A., JENSEN, A.B., *et al.* (2011). Global monitoring of *Salmonella* serovar distribution from the World Health Organization global foodborne infections network country data bank: Results of quality assured laboratories from 2001 to 2007. *Foodborne Pathogens and Disease*, 8(8), 887–900.

ISAACS, S., ARAMINI, J., CIEBIN, B., *et al.* (2005). An international outbreak of salmonellosis associated with raw almonds contaminated with a rare phage type of *Salmonella* Enteritidis. *J. Food Prot.*, 68(1), 191–198(8).

KAPPERUD, G. and HASSELTVEDT, V. (1999). Status og udviklingstendenser for fødevarebårne zoonoser i Norge. *Zoonose-Nyt.*, 5, 9–14.

KONINGSTEIN, M., SIMONSEN, J., HELMS, M. and MØLBAK, K. (2010). The interaction between prior antimicrobial drug exposure and resistance in human *Salmonella* infections. *J. Antimicrob. Chemother.*, 17 May, doi: 10.1093/jac/dkq176.

KORSGAARD, H., MADSEN, M., FELD, N.C., MYGIND, J. and HALD, T. (2009). The effects, costs and benefits of *Salmonella* control in the Danish table-egg sector. *Epidemiol. Infect.*, 137, 828–836.

LIM, E., LOPEZ, L., CRESSEY, P. and PIRIE, R. (2011). Foodborne disease in New Zealand. Annual report. Institute of Environmental Science & Research Limited, Christchurch Science Centre. Available at: www.foodsafety.govt.nz/elibrary/industry/FBI-report-2011.pdf.

LINDSTEDT, B.-A., VARDUND, T., AAS, L. and KAPPERUD, G. (2004). Multiple-locus variable-number tandem-repeats analysis of *Salmonella enterica* subsp. *enterica* serovar Typhimurium using PCR multiplexing and multicolour capillary electrophoresis. *J. Microbiol. Meth.*, 59, 163–172.

MAHON, B.E., PONKA, A., HALL, W.N., *et al.* (1997). An international outbreak of *Salmonella* infections caused by alfalfa sprouts grown from contaminated seeds. *J. Infect. Dis.*, 175, 876–882.

MAJOWICZ, S.E., MUSTO, J., SCALLAN, E., *et al.* (2010). The global burden of nontyphoidal *Salmonella* gastroenteritis. *Clin. Infect. Dis.*, 50(6), 882–889.

MEAD, P.S., SLUTSKER, L., DIETZ, V., *et al.* (1999). Food-related illness and death in the United States. *Emerg. Infect. Dis.*, 5, 607–625.

MILLER, S.I., HOHMANN, E.L. and PEGUES, D.A. (1995). *Salmonella* (including *Salmonella* Typhi). In: *Principles and Practice of Infectious Diseases*. Mandell, G.L., Bennett, J.E. and Dolin, R. (Eds). Churchill Livingstone, New York, pp. 2013–2033.

MØLBAK, K., OLSEN, J.E. and WEGENER, H.C. (2006). *Salmonella* infections. In: *Foodborne Infections and Intoxications*. Third edition. Rieman, H.P and D.O. Cliver. (Eds). School of Veterinary Medicine, University of California, Davis. Academic Press, Elsevier.

MULLNER, P., JONES, G., NOBLE, A., SPENCER, S.E., HATHAWAY, S., *et al.* (2009). Source attribution of food-borne zoonoses in New Zealand: A modified Hald model. *Risk Anal.*, 29(7), 970–984.

NEAL, K.R., BRIJI, S.O., SLACK, R.C., *et al.* (1994). Recent treatment with H2 antagonists and antibiotics and gastric surgery as risk factors for *Salmonella* infection. *BMJ*, 308, 176.

NEWELL, D.G., KOOPMANS, M., VERHOEF, L., DUIZER, E., AIDARA-KANE, A., *et al.* (2010). Food-borne diseases — The challenges of 20 years ago still persist while new ones continue to emerge. *Int. J. Food Microbiol.*, 139, S3–S15.

O'BRIEN, S.J., RAIT, G., HUNTER, P.R., GRAY, J.J., BOLTON, F.J., *et al.* (2010). Methods for determining disease burden and calibrating national surveillance data in the United Kingdom: The second study of infectious intestinal disease in the community (IID2 study). *BMC Med. Res. Methodol.*, 10, 39.

O'REILLY, C.E., BOWEN, A.B., PEREZ, N.E., *et al.* (2007). A waterborne outbreak of gastroenteritis with multiple etiologies among resort island visitors and residents: Ohio, 2004. *Clin. Infect. Dis.*, 44, 506–512.

PAINTER, J. (2006). Estimating attribution of illnesses to food vehicle from reports of foodborne outbreak investigations. Society for Risk Analysis, 2006 Annual Meeting, Baltimore, MD, 3–6 December.

PAINTER J.A., AYERS T., WOODRUFF R., BLANTON E., PEREZ N. *et al.* (2009). Recipes for foodborne outbreaks: A scheme for categorizing and grouping implicated foods. *Foodborne Pathog. Dis.*, 6(10), 1259–1264. doi: 10.1089/fpd.2009.0350.

PEZZOLI, L., ELSON, R., LITTLE, C., *et al.* (2008). Packed with *Salmonella* – Investigation of an international outbreak of *Salmonella* Senftenberg infection linked to contamination of prepacked basil in 2007. *Foodborne Pathog. Dis.*, 5(5), 661–668. doi: 10.1089/fpd.2008.0103.

PIRES, S.M. and HALD, T. (2010). Assessing the differences in public health impact of *Salmonella* subtypes using a Bayesian microbial subtyping approach for source attribution. *Foodborne Pathogens and Disease*, 7(2), 143–151.

PIRES, S.M., NICHOLS, G., WHALSTRÖM, H., KAESBOHRER, A., DAVID, J. *et al.* (2008). *Salmonella* source attribution in different European countries. *Proceedings of FoodMicro 2008*, Aberdeen, Scotland.

PIRES, S.M., EVERS, E.G., VAN PELT, W., AYERS, T., SCALLAN, E., *et al.* (2009). Attributing the human disease burden of foodborne infections to specific sources. *Foodborne Pathogens and Disease*, 6, 417–424.

PIRES, S.M., VIGRE, H., MAKELA, P. and HALD, T. (2010). Using outbreak data for source attribution of human salmonellosis and campylobacteriosis in Europe. *Foodborne Pathogens and Disease*, 7(11): 1351–1361.

PIRES, S., DE KNEGT, L. and HALD, T. (2011a). Estimation of the relative contribution of different food and animal sources to human *Salmonella* infections in the European Union. Question No. EFSA-Q-2010-00685. Published as an external scientific report on 28 July 2011: www.efsa.europa.eu/en/supporting/pub/184e.htm.

PIRES, S.M., TOYOFUKU, H., KASUGA, F. and HALD, T. (2011b). Attributing foodborne disease in Japan using outbreak data. International Association for Food Protection Annual Meeting, Milwaukee, Wisconsin, 31 July 31–3 August.

PIRES, S.M., VIEIRA, A., PEREZ, E., LO FO WONG, D. and HALD, T. (2012). Attributing human foodborne illness to food sources and water in Latin America and the Caribbean using data from outbreak investigations. *Int. J. Food Microbiol.*, 152(3), 129–138.

RAVEL, A., GREIG, J., TINGA, C., TODD, E., CAMPBELL, G., *et al.* (2009). Exploring historical Canadian foodborne outbreak data sets for human illness attribution. *J. Food Prot.*, 72(9), 1963–1976.

ROBERTS, J.A. and SOCKETT, P.N. (1994). The socio-economic impact of human *Salmonella* Enteritidis infection. *Int. J. Food Microbiol.*, 21, 117–129.

RODRIGUEZ, I., JAHN, S., SCHROETER, A., MALORNY, B., HELMUTH, R., *et al.* (2012). Extended-spectrum beta-lactamases in German isolates belonging to the emerging monophasic *Salmonella enterica* subsp. *enterica* serovar Typhimurium 4,[5],12:i:- European clone. *J. Antimicrob. Chemother.*, 67(2), 505–508. Available from www.ncbi.nlm.nih.gov/pubmed/22058374.

SARWARI, A.R., MAGDER, L.S., LEVINE, P., MCNAMARA, A.M., KNOWER, S., *et al.*. (2001). Serotype distribution of *Salmonella* isolates from food animals after slaughter differs from that of isolates found in humans. *J. Infect. Dis.*, 183, 1295–1299.

SCALLAN, E., HOEKSTRA, R.M., ANGULO, F.J., TAUXE, R.V., WIDDOWSON, A., *et al.* (2011). Foodborne illness acquired in the United States – Major pathogens. *Emerg. Infect. Dis.*, 17(1), 7–15.

SCHROEDER, C.M., NAUGLE, A.L., SCHLOSSER, W.D., HOGUE, A.T., ANGULO, F.J., *et al.* (2005). Estimate of illness from *Salmonella* Enteritidis in eggs, United States, 2000. *Emerg. Infect. Dis.*, 11, 113–115.

SIVAPALASINGAM, S., FRIEDMAN, C.R., COHEN, L. and TAUXE, R.V. (2004). Fresh produce: A growing cause of outbreaks of foodborne illness in the United States, 1973 through 1997. *J. Food Prot.*, 67, 10, 2342–2353.

SU, L.-H. and CHIU C.-H. (2007). *Salmonella*: Clinical importance and evolution of nomenclature. *Chang Gung Med. J.*, 30 (3), 210–219.

THORNS, C.J. (2000). Bacterial food-borne zoonoses. *Rev. Sci. Tech. Off. Int. Epiz.*, 19, 226–239.

TORPDAHL, M., SØRENSEN, G., LINDSTEDT, B.-A. and NIELSEN, E.M. (2007). Tandem repeat analysis for surveillance of human *Salmonella* Typhimurium infections. *Emerg. Infect. Dis.*, 13, 388–395.

TOYOFUKU, H., PIRES, S.M. and HALD, T. (2011). *Salmonella* source attribution in Japan by a microbiological subtyping approach. *EcoHealth*, 7, Suppl. 1, S22–S23. Available at: www.ecohealth.net/pdf/journal_pdf/Vol_7/Vol7_S1/ECH_7_S1_Abstracts.pdf.

VALKENBURGH, S., OOSTEROM, R., VAN STENVERS, O., AALTEN, M., BRAKS, M., *et al.* (2007). Zoonoses and zoonotic agents in humans, food, animals and feed in the Netherlands 2003–2006. RIVM report: 330152001 ISBN-13, 978-90-6960-184-7.

VAN BENEDEN, C.A., KEENE, W.E., STRANG, R.A., *et al.* (1999). Multinational outbreak of *Salmonella enterica* serotype Newport infections due to contaminated alfalfa sprouts. *JAMA*, 281, 158–162.

VAN DEN BOGAARD, A.E. and STOBBERINGH, E.E. (1999). Antibiotic usage in animals: Impact on bacterial resistance and public health. *Drugs*, 58, 589–607.

VAN PELT, W., VAN DE GIESSEN, A.W., LEEUWEN, W.J., WANNET, W., HENKEN, A.M., *et al.* (1999). Oorsprong, omvang en kosten van humane salmonellose. Deel 1. Oorsprong van humane salmonellose met betrekking tot varken, rund, kip, ei en overige bronnen. *Infectieziekten Bull.*, 240–243.

VAN PELT, W., DE WIT, M.A.S., WANNET, W.J.B., LIGTVOET, E.J.J., WIDDOWSON, M.A., *et al.* (2003). Laboratory surveillance of bacterial gastroenteric pathogens in the Netherlands, 1999–2001. *Epidemiol. infect.*, 130, 431–441.

VARMA, J.K., MOLBAK, K., BARRETT, T.J., BEEBE, J.L., JONES, T.F., *et al.* (2005). Antimicrobial-resistant non-typhoidal *Salmonella* is associated with excess bloodstream infections and hospitalizations. *J. Infect. Dis.*, 191(4), 554–561.

VOETSCH, A.C., VAN GILDER, T.J., ANGULO, F.J., *et al.* (2004). Emerging infections program FoodNet working group. FoodNet estimate of the burden of illness caused by nontyphoidal *Salmonella* infections in the United States. *Clin. Infect. Dis.*, 38, Suppl 3, S127–134.

WAHLSTRÖM, H., ANDERSSON, Y., PLYM-FORSHELL, L. and PIRES, S.M. (2011). Source attribution of human *Salmonella* cases in Sweden. *Epidem. Infect.*, 139(8), 1246–1253.

WEGENER, H.C., HALD, T., LO FO WONG, .D., MADSEN, M., KORSGAARD, H., *et al.* (2003). *Salmonella* control programs in Denmark. *Emerg. Infect. Dis.*, 9, 774–780.

WERBER, D., DREESMAN, J., FEIL, F., *et al.* (2005). International outbreak of *Salmonella* Oranienburg due to German chocolate. *BMC Infect. Dis.*, 5, 7 doi: 10.1186/1471-2334-5-7.

WHEELER, J.G., SETHI, D., COWDEN, J.M., WALL, P.G., RODRIGUES, L.C., *et al.* (1999) – on behalf of the Infectious Intestinal Disease Study Executive. Study of infectious intestinal disease in England: Rates in the community, presenting to general practice and reported to national surveillance. *BMJ*, 318, 1046–1050.

3

Pathogen update: *Listeria monocytogenes*

J. Chen and K. Nightingale, Texas Tech, USA

DOI: 10.1533/9780857098740.2.47

Abstract: Recent advances regarding the human foodborne pathogen *Listeria monocytogenes* are discussed with specific emphasis on epidemiology, ecology and virulence. Recent routine surveillance and outbreak investigation studies indicate a shift in *L. monocytogenes* serotypes linked to listeriosis in some countries. There is growing evidence to show that *L. monocytogenes* isolates can be divided beyond the serotype level into at least two epidemiologically and genetically distinct subpopulations differentiated by biologically meaningful markers, including strains responsible for the majority of listeriosis cases and strains carrying virulence attenuating mutations in the key virulence gene *inlA*, which codes for internalin A.

Key words: *Listeria monocytogenes*, epidemiology, virulence, ecology, animal models.

3.1 The genus *Listeria*, *L. monocytogenes* and listeriosis

The *Listeria* genus contains (i) species almost exclusively restricted to a saprophytic lifestyle (i.e., *L. innocua*, *L. seeligeri*, *L. welshimeri* and *L. grayi*), and (ii) species classified as opportunistic pathogens (i.e., *L. ivanovii* and *L. monocytogenes*). Recent research described two proposed novel species, *L. marthii* and *L. rocourtiae*, both of which have been determined to be non-pathogens (Graves *et al.*, 2010; Leclercq *et al.*, 2010). Although *L. ivanovii*, *L. grayi*, *L. seeligeri* and *L. innocua* have been associated with human disease on a few rare occasions, *L. monocytogenes* is the only human pathogen of public health significance in the *Listeria* genus (Guillet *et al.*, 2010; Perrin *et al.*, 2003; Rocourt *et al.*, 1986; Rapose *et al.*, 2008).

Exposure to *L. monocytogenes* may manifest as mild gastroenteritis in healthy individuals, which is not well understood, or as invasive listeriosis in high-risk host populations. Those that are particularly susceptible to invasive listeriosis include the elderly, pregnant women, neonates and individuals undergoing immunosuppressive treatments. Invasive listeriosis may lead to septicemia, meningitis, encephalitis, as well as spontaneous abortions or still births in pregnant women (Drevets and

Bronze, 2008). Listeriosis has an exceptionally long incubation period, frequently greater than 30 days, which is a hindrance to traditional case-control studies and microbial source-tracking investigations (Swaminathan and Gerner-Smidt, 2007). However, molecular subtyping methods have proved a valuable tool for epidemiological investigation of listeriosis outbreaks, where application of molecular subtyping has provided insight into ecology and transmission of *L. monocytogenes* along with food attribution of listeriosis. The most recent large multistate outbreak of listeriosis in the United States prompted a rapid source-tracking investigation that resulted in a nationwide recall of cantaloupes from a grower in the Rocky Ford region of Colorado. This investigation was aided by analyses of pulsed field gel electrophoresis (PFGE) patterns from human patients, food and environmental samples included in the PulseNet database. Ultimately, four distinct PFGE patterns were associated with the outbreak and used in formulating a case definition; the identification of shared PFGE patterns between human clinical cases and cantaloupes from one farm in south-east Colorado facilitated identification of the food vehicle responsible for the outbreak (CDC, 2011a, 2011b).

3.2 Listeriosis: epidemiology, virulence factors and evolution

3.2.1 Epidemiology of listeriosis

A recent estimate of the burden of foodborne disease in the US found there are approximately 1600 domestically acquired foodborne illnesses due to invasive *L. monocytogenes* each year. Of these cases, approximately 94% were predicted to result in hospitalization and 15% were estimated to be fatal. Furthermore, listeriosis cases were suggested to account for nearly 20% of total deaths attributed to known pathogens in the US each year (Scallan *et al.*, 2011). While the incidence of human listeriosis in the US declined from 5 cases/million people to 3 cases/million people, based on a comparison of the finalized 2006–2008 FoodNet data and the baseline 1996–1998 data, the majority of this reduction was observed prior to 2002 (CDC, 2009). In five European countries, Belgium, Denmark, England, Wales and Finland, increases in the incidence of listeriosis have been observed over recent years. Collectively, these countries had a median listeriosis incidence of 4.7 cases/million people in 2000, which increased to 6.3 cases/million people by 2006. The increase in the incidence of listeriosis in Europe has been especially noticeable in bacteremic cases among the elderly (Goulet *et al.*, 2008). An increase in listeriosis was observed in Germany between 2001 and 2005, most drastically in those over the age of 60 (Koch and Stark, 2006). An increase in listeriosis cases among adults over 70 years of age in France was also observed in 2006 (Goulet *et al.*, 2008). A similar increase of listeriosis cases was also noticed in England and Wales between 2001 and 2004, also in individuals 60 years of age or older (Gillespie *et al.*, 2006). It is not yet known what has led to this increase in incidence; however, improvements in awareness and detection of listeriosis are not considered to be major contributing factors.

New data suggests that the increased incidence of listeriosis in England and Wales is occurring among cancer patients and those receiving treatment to suppress stomach acid (Gillespie *et al.*, 2009). A recent case-control study in England highlighted differences in food exposure between elderly cases and controls. Specifically, cases were more likely than controls to consume certain cooked meats, cold cooked fish (e.g. smoked salmon), shellfish, certain cheese types, milk and mixed salads. On the other hand, cases were less likely than controls to consume some types of seafood, other dairy products, sandwiches and raw vegetables (Gillespie *et al.*, 2010).

Worldwide, the majority of listeriosis outbreaks have been attributed to *L. monocytogenes* belonging to serotype 4b (Swaminathan and Gerner-Smidt, 2007); however, serotypes 1/2a and 1/2b have also been linked to epidemics, including the most recent outbreak caused by cantaloupes (Bille *et al.*, 2006; CDC, 2011b; Cokes *et al.*, 2011; Fretz *et al.*, 2010; Gilmour *et al.*, 2010; Swaminathan and Gerner-Smidt, 2007). In several countries, cases of listeriosis linked to a 1/2a serotype strain are on the rise. In Finland between 1991 and 2001, the number of clinical cases caused by a 1/2 a strain was greater than the number of clinical cases caused by 4b strains (Lukinmaa *et al.*, 2003) and a similar trend was observed in Sweden between 2000 and 2007 (Parihar *et al.*, 2008).

The recent United States cantaloupe outbreak was the largest observed in this country with 146 illnesses, 30 deaths and one miscarriage reported at the time this chapter was prepared (December, 2011). Of 144 individuals for whom medical information was available, 142 or 99% were hospitalized. This outbreak was unique for two reasons: first of all, this was the first documented outbreak of listeriosis associated with cantaloupes. Secondly, four PFGE patterns, which were classified into two serotypes (i.e., 1/2a and 1/2b), were associated with the outbreak (CDC, 2011b, 2011c).

3.2.2 *Listeria* virulence factors

L. monocytogenes and *L. ivanovii* are both facultative intracellular pathogens, and the ability of *L. monocytogenes* to induce its uptake by non-phagocytic cells is primarily mediated by a family of surface proteins known as the internalins. Many internalin and internalin-like proteins have been described through genome sequencing studies, and up to 28 have been identified in a single *L. monocytogenes* strain (den Bakker *et al.*, 2010a); however, only two internalins (i.e., internalin A and internalin B) have been extensively studied. Internalin A (InlA), encoded by *inlA*, plays a key role in crossing the intestinal barrier during the initial stages of infection. *L. monocytogenes* utilizes InlA to induce its own uptake into non-professional phagocytes through interaction of the leucine-rich repeat region functional domain with host cell receptor E-cadherin (Lecuit *et al.*, 1997). The InlA/E-cadherin interaction is also known to play a role in the crossing of the placenta in materno-fetal cases of listeriosis (Lecuit *et al.*, 2004). Internalin B (InlB), encoded by *inlB*, on the other hand is more important in the later stages of infection through its interaction with the hepatic receptor Met and gC1q-R (Shen *et al.*, 2000; Braun *et al.*, 2000).

Recently, the roles of a few other internalins, notably InlC, InlJ and InlK have been elucidated. InlC, a secreted internalin, has been documented to be involved in the spread between epithelial cells by disrupting the apical junction (Rajabian *et al.*, 2009). InlJ acts as an adhesion and Δ*inlJ* mutants demonstrated attenuated virulence in both intraperitoneally infected mice and orally infected transgenic mice (Sabet *et al.*, 2005, 2008). InlK has been found to interact with the major vault protein (MVP) found in the cytoplasm of mammalian cells. MVP is thought to play a role in a number of cell processes including immunity and signal cascades. Through its interaction with MVP, *L. monocytogenes* avoids recognition by autophagic machinery, aiding its intracellular survival (Dortet *et al.*, 2012). Lastly, *Listeria* adhesion protein B, an LPXTG surface protein related to the internalins, has been found to be important in adhesion and entry into both Caco-2 and Hep-2 cells (Reis *et al.*, 2010).

Once inside the cell, pathogenic *Listeria* species use a suite of genes clustered in a pathogenicity island designated *Listeria* pathogenicity island one (LIPI-1) to survive and replicate intracellularly. These genes are necessary for the intracellular lifecycle of this pathogen, simultaneously allowing for cell-to-cell spread and evasion of humoral immune responses (Vázquez-Boland *et al.*, 2001). In *L. monocytogenes*, this locus consists of six genes: *prfA*, *plcA*, *plcB*, *hly*, *mpl* and *actA* (Vázquez-Boland *et al.*, 2001). Hemolysin (*hly)* encodes *Listeria* lysin-O (LLO), a pore-forming toxin, which in conjunction with two phospholipases encoded by *plcA* and *plcB*, is responsible for disruption of the vacuole (Camilli *et al.*, 1993; Gedde *et al.*, 2000). After *L. monocytogenes* escapes into the cytosol, where it can replicate freely, it recruits host cell actin using ActA encoded by *actA*, in order to become motile and spread into adjacent cells (Tilney and Portnoy, 1989).

A number of newly discovered virulence factors are involved in the interplay of *L. monocytogenes* and the host immune system and assist this pathogen in establishing deep infection. OatA, an o-acetyltransferase, has recently been found to be involved in the modification of *L. monocytogenes* peptidoglycan, which may lead to a dampening of the innate immune response, thus enhancing *L. monocytogenes* survival in mice (Aubry *et al.*, 2011). A *L. monocytogenes* isogenic mutant lacking *Listeria* phosphatase A, or *lipA*, elicited less cytokine production, which promotes inflammation, and demonstrated lower levels of *L. monocytogenes* in mice (Kastner *et al.*, 2011). *Listeria* nuclear targeted protein A or LntA has been shown to associate with BAHD1, a nuclear repressor of type III interferons; however, *lntA* knockout mutants elicit the production of interferon-γ supporting a role for LntA in redirecting the host's immune response in the mouse model. A *L. monocytogenes* Δ*lntA* strain was found at lower levels in the spleen compared to its parent strain, and *lntA* was highly expressed in the wild-type parent strain when isolated from spleens (Lebreton *et al.*, 2011).

3.2.3 Evolution of *Listeria* to contain pathogens and non-pathogens

As full genome sequencing becomes more cost effective, more genome sequences will be available for evolutionary and phylogenetic analysis of *Listeria*.

Comparative genomic analyses between *L. welshimeri*, *L. innocua* and *L. monocytogenes* revealed that a number of genes in *L. monocytogenes* are absent from *L. welshimeri* and *L. innocua*. Some of the genes were likely lost over the course of evolution; however, there is evidence that some genes present in *L. monocytogenes*, but not the strictly saprophytic species, may have been introduced by horizontal gene transfer (Hain *et al.*, 2006). Interestingly, a full genome sequence of *L. ivanovii* subsp. *ivanovii* was recently released and results indicate a recent bottleneck in this species, and revealed a large number of pseudo-genes that could possibly explain the adaptation of this opportunistic pathogen to infecting ruminants (Buchrieser *et al.*, 2011).

More recent work demonstrated that the pan-genome of *Listeria* is conserved, showing little acquisition of new genetic material, although horizontal gene transfer events have been documented (Dunn *et al.*, 2009; Orsi *et al.*, 2008b). However, the evolution of the *Listeria* genus from a pathogenic common ancestor has been punctuated by periods of gene loss, through the transition from a pathogenic to a saprophytic lifestyle (den Bakker *et al.*, 2010a; Doumith *et al.*, 2004; Vázquez-Boland *et al.*, 2001). Another study by den Bakker *et al.* (2010b), which used stochastic character mapping in addition to sequence data from the gene *hly* to study the phylogenic relationships between *Listeria* species, suggested that the most recent common ancestor of the *Listeria* genus possessed the LIPI-1 pathogenicity island and the cluster was lost five times over the course of the evolution of species within the genus *Listeria*. These losses occurred once each in *L. welshimeri* and *L. marthii*, twice in *L. innocua* and again very recently in some strains of *L. seeligeri* (den Bakker *et al.*, 2010b). Both *L. ivanovii* and *L. seeligeri* usually contain the LIPI-1 pathogenicity island; however, in *L. seeligeri* it is non-functional due to an insertion between *plcA* and *prfA*. This virulence island is typically absent in *L. innocua, L. welshimeri* and *L. grayi* (Vázquez-Boland *et al.*, 2001). Some atypical hemolytic *L. innocua* isolates were reported to contain at least a portion of this LIPI-1 island, specifically the genes *hly* and *plcA*, which are expressed, making these strains hemolytic. Hemolytic *L. innocua* has been demonstrated to be avirulent in the mouse model and it was initially suggested that this attenuated virulence phenotype could be explained by an apparent lack of key invasion genes such as the internalins (Johnson *et al.*, 2004). However, Volokhov and coworkers in 2007 identified *inlA* in a hemolytic *L. innocua* strain, and den Bakker *et al.* confirmed the presence of *inlA* and showed this hemolytic *L. innocua* strain efficiently invades Caco-2 cells (Volokhov *et al.*, 2007; den Bakker *et al.*, 2010a). It is suggested that hemolytic *L. innocua* strains represent an evolutionary intermediate between *L. monocytogenes* and non-hemolytic *L. innocua*.

3.3 *In vitro* and *in vivo* models to assess virulence

3.3.1 *In vitro* models to assess virulence

There are a number of *in vitro* models to assess virulence and typically these assays are directed towards assessing the functionality of one specific virulence factor at a time. Therefore, no single *in vitro* assay can predict the ability to cause disease *in vivo*; however, these assays can suggest the functionality of certain virulence factors and may facilitate identification of markers within specific virulence genes associated with defined virulence phenotypes (e.g., Van Stelten and Nightingale, 2008). A number of cell-culture-based methods have also been developed to assess the ability of *Listeria monocytogenes* to invade various cell types. Caco-2 and HepG2 cells are human colonic and hepatic cell lines, respectively, and invasion into these cell types is mediated chiefly by internalins, specifically InlA and InlB (Kim *et al.*, 2005; Roberts *et al.*, 2009; Van Langendonck *et al.*, 1998). The cell-to-cell spread can be assessed by the size of plaques formed in mouse L2 cells, and this ability is mediated by ActA (Smith *et al.*, 1995; Sun *et al.*, 1990). The ability to grow inside mouse macrophages can also be assessed *in vitro* using a mouse macrophage cell line, J774 (Portnoy *et al.*, 1988; Roberts *et al.*, 2005).

3.3.2 *In vivo* models to assess virulence

The mouse has been used extensively to probe the virulence of *L. monocytogenes*. Early experiments found that it was difficult to establish a systemic infection in the mouse via oral inoculation. When mice were challenged orally, even with doses $>5 \times 10^{10}$ CFU, lethality was not observed (Lecuit *et al.*, 1999, 2001). However, intravenous (iv) inoculation of the mouse with *L. monocytogenes* routinely resulted in mouse mortality where the LD_{50} for these experiments was observed around 1×10^5 CFU (Lecuit *et al.*, 2001). Given that *L. monocytogenes* is a foodborne pathogen, the inability to induce natural infections with orally administered doses poses a problem with the mouse model. E-cadherin, an extracellular receptor of enterocytes, interacts with InlA, which facilitates crossing of the intestinal barrier by *L. monocytogenes*. Mouse E-cadherin differs from human E-cadherin, by a single amino acid at position 16, where the substitution of a proline in the human isoform for a glutamic acid in rats and mice is sufficient to abolish efficient invasion (Lecuit *et al.*, 1999). The creation of a transgenic mouse line, which expresses the human isoform of E-cadherin, alleviated some of the problems with the mouse model. In this transgenic mouse model, the human isoform of E-cadherin is expressed in enterocytes, and this expression is limited to the small intestines (Lecuit *et al.*, 2001). Disson and coworkers used a knock-in approach in mice to substitute the glutamic acid to a proline at position 16. This allowed for efficient translocation of *L. monocytogenes* across not only the small intestines, but the cecum and the colon as well (Disson *et al.*, 2008).

The creation of a transgenic mouse model and subsequent use of that model for virulence studies can be costly, as a result, research groups have successfully developed 'murinized' *L. monocytogenes* InlA, which more effectively binds with mouse E-cadherin so that oral infections can be studied in this model. Regions of InlA that prevent interaction with mouse E-cadherin were identified and modified (Wollert *et al.*, 2007). This model has since been improved to reflect current knowledge of codon usage by *L. monocytogenes*. Monk and colleagues performed intragastric infections of mice with both the wild-type standard laboratory control strain EGD-e and the murinized EDG-e InlA^{m*} strain. Through both bioluminescent imaging as well as bacterial counts from livers and spleens, they were able to demonstrate that the murinized strain was capable of infecting deep tissues at significantly higher levels than wild-type EGD-e following oral inoculation (Monk *et al.*, 2010).

Substantial amounts of work have been done to study virulence of *L. monocytogenes* in the guinea pig model (Garner *et al.*, 2006; Lecuit *et al.*, 2001; Nightingale *et al.*, 2008; Oliver *et al.*, 2010; Van Stelten *et al.*, 2011; Williams *et al.*, 2007). Guinea pigs, like humans, carry a proline at position 16 in E-cadherin, in contrast to rats and mice (Lecuit *et al.*, 1999), which make them a suitable rodent model for studying InlA mediated virulence. Previous work done with iv infections of EGD-e in guinea pigs showed an LD_{50} of $\sim 5 \times 10^7$ CFU (Lecuit *et al.*, 2001). The guinea pig model has also been used to study infections during pregnancy. With this model an LD_{50} (defined as the dose at which 50% fetal mortality was observed) of 2×10^7 CFU was observed, which is lower than observed for a fully virulent strain in the non-pregnant model, supporting the idea that pregnancy increases the risk of contracting listeriosis (Williams *et al.*, 2007). Van Stelten and others infected juvenile male guinea pigs and used log logistic regression with combined organ data to determine strain-specific infectious doses. Two strains were used in this infection study and an ID_{50} of 2.57×10^7 CFU was observed for an epidemic clone strain previously associated with an outbreak and an ID_{50} of 5.50×10^8 CFU was observed for a strain producing truncated InlA. Information gained using the guinea pig model, however, is limited when studying infections of deeper tissues by *L. monocytogenes* due to a lack of species specificity between InlB and the isoform of Met found in the guinea pig (Disson *et al.*, 2008).

Recent work has suggested that the gerbil may be an ideal model for studying listeriosis, as gerbils have been found to express both the human form of E-cadherin and Met (Blanot *et al.*, 1997; Disson *et al.*, 2008). Orally inoculating gerbils with doses $>2 \times 10^5$ CFU of a bioluminescent EGD-e led to feto-placental listeriosis in this host (Disson *et al.*, 2008). Another study noted the development of rhombencephalitis and circling syndrome in gerbils following inoculation of the middle ear (Blanot *et al.*, 1997). Another animal model, which is used to simulate human infections of listeriosis, is the primate model. Specifically, the rhesus monkey is similar in reproductive characteristics to humans and has been used to study *L. monocytogenes* stillbirths. Infections in primates have been studied and oral inoculations resulted in stillbirths, suggesting that InlA interacts with the primate isoform of E-cadherin (Smith *et al.*, 2003).

3.4 Ecology, transmission and genetic diversity of *L. monocytogenes*

3.4.1 Ecology and transmission of *L. monocytogenes*

Listeria spp. including *L. monocytogenes* primarily lead a saprophytic life; however, pathogens like *L. monocytogenes* are able to switch between being an environmental microorganism and an opportunistic pathogen. As early as 1971 it was noted that *Listeria* could be isolated from soil in vegetation from both farm and non-farm environments (Welshimer and Donker-Voet, 1971). Weis and Seeliger described the widespread isolation of *Listeria* from the natural environment in 1975 in Germany. Sources of *Listeria* included soil and vegetation from fields and feces from both deer and birds (Weis and Seeliger, 1975).

L. monocytogenes has been shown to infect more than 40 species of animals and birds (Lecuit, 2007). Listeriosis has been noted to be an important veterinary concern in large farm ruminants such as cows, sheep and goats (Low and Donachie, 1997). A study aimed at elucidating the transmission of *L. monocytogenes* on ruminant farms showed that approximately 25% of asymptomatic cattle shed *L. monocytogenes* in their feces, regardless of whether or not those cattle were maintained on a farm with a history of listeriosis (Nightingale *et al.*, 2004). Poorly fermented silage has been shown to allow the growth of *L. monocytogenes* and may be involved in the transmission of this pathogen from the farm environment to animals (Ruxton and Gibson, 1995). Listeriosis can also occur in swine, and healthy pigs can also carry and shed *L. monocytogenes* in their feces (Hellström *et al.*, 2010; Long and Dukes, 1972). Listeriosis cases occur in many species of fowl (Akanbi *et al.*, 2008; Cooper *et al.*, 1989). Horses and non-domesticated ruminants such as deer occasionally develop listeriosis (Evans *et al.*, 2004; Gudmundsdottir *et al.*, 2004; Rütten *et al.*, 2006; Schwaiger *et al.*, 2005; Tham *et al.*, 1999).

The common presence of *Listeria* and *L. monocytogenes* in nature and production agriculture systems presents a significant challenge to the ready-to-eat (RTE) food industry as transmission of *Listeria* from these environments into the human food supply can occur. Numerous studies have described the molecular ecology and transmission of *Listeria* in food-associated environments, including the food-processing plant and retail environments. In particular, *L. monocytogenes* has been isolated from RTE smoked fish plants, RTE meat plants, and delicatessens (Autio *et al.*, 1999; Eklund *et al.*, 1995; Gibbons *et al.*, 2006; Hoelzer *et al.*, 2011; Hoffman *et al.*, 2003; Hudson and Mott, 1993; Lappi *et al.*, 2004; Lundén *et al.*, 2003; Norton *et al.*, 2001; Rørvik *et al.*, 1995; Samelis and Metaxopoulos, 1999; Sauders *et al.*, 2004, 2009; Thimothe *et al.*, 2002, 2004; Williams *et al.*, 2011). It has also been noted that *L. monocytogenes* may persist for an extended period in a given food-processing plant. Comparative genomics of *L. monocytogenes* food and human clinical isolates associated with a sporadic case of listeriosis in 1988 and a listeriosis outbreak in 2000, which were linked to the same food-processing plant, showed that the same *L. monocytogenes* strain caused both the sporadic case and the outbreak, suggesting the persistence of this strain for 12 years in the plant environment. Minimal changes between the 1998 and the 2000 isolates were observed: one single nucleotide polymorphism (SNP) in the tRNA-Thr-4 prophage region and substantial rearrangement in a phage residing in *comK* (Orsi *et al.*, 2008a). Interestingly, it has been noted that *comK* prophages may serve as mutational hotspots and contribute to rapid adaptation to novel environments, and may thus be involved in the persistence of *L. monocytogenes* in a given environmental niche (Verghese *et al.*, 2011).

It is estimated that over 99% of all listeriosis cases are attributable to exposure through contaminated foods (Mead *et al.*, 1999). *L. monocytogenes* is easily inactivated by cooking above 60 °C and pasteurization (Golden *et al.*, 1998; Van der Veen *et al.*, 2009). Because *L. monocytogenes* is easily inactivated by lethality treatments incorporated into most processes, cross-contamination by *L. monocytogenes* in food-associated environments (i.e., processing plant, retail or possibly even the home kitchen environment) following the lethality step in the process represents the primary mode of finished RTE product contamination (Shank *et al.*, 1996). *L. monocytogenes* is capable of growing at 0–45 °C, which encompasses the processing plant environment, the retail environment and shelf storage (Farber *et al.*, 1988). Growth rates of *L. monocytogenes* at refrigeration temperature have been found to vary substantially. In one study, the growth of 25 genetically diverse *L. monocytogenes* isolates in a broth culture at 4 °C ranged from 0.28 log CFU/ml/day to 0.43 log CFU/ml/day (Lianou *et al.*, 2006). Growth of *L. monocytogenes* has been shown to vary based on both extrinsic and intrinsic factors of food production such as temperature, pH, presence or absence of antimicrobials, water activity, etc. (Koutsoumanis *et al.*, 2004; Lianou *et al.*, 2007). Many of these stresses are used to control pathogens in foods and in the food-processing environment; therefore, RTE foods that have an extended shelf-life have the greatest likelihood of harboring *L. monocytogenes*.

Several studies have investigated the potential of various antimicrobials to inhibit and reduce *L. monocytogenes* both *in vitro* and on RTE food products. Work by Samelis and others demonstrated that sodium lactate (1.8%) alone inhibited *L. monocytogenes* growth in hot dogs for up to 50 days and that in combination with sodium acetate, sodium diacetate or glucono-d-lactone, growth was inhibited throughout 120 days of storage (Samelis *et al.*, 2002). The reformulation of a sliced bologna product with sodium lactate and sodium diacetate slowed growth rates to 0.009 log CFU/cm^2/day and 0.084 log CFU/cm^2/day at 4 and 10 °C, respectively, when inoculated with 3–4 log CFU/cm^2 of a cocktail of *L. monocytogenes* strains. The uninoculated product showed growth rates of 0.609 log CFU/cm^2/day and 1.34 log CFU/cm^2/day for these two temperatures (Barmpalia *et al.*, 2005). The inhibition of the growth of *L. monocytogenes* by formulating RTE meat products with these antimicrobials demonstrates a significant potential to reduce the number of listeriosis cases by preventing the growth of the pathogen to levels capable of causing a systemic infection.

In a 2003 *L. monocytogenes* risk assessment, the predicted median cases of listeriosis for 23 RTE food categories were calculated on a per annum and per serving basis and deli meats were attributed to nearly 90% of cases on a per annum basis (FDA/FSIS/CDC, 2003). A risk assessment involving deli meats suggested that reformulation of deli meats with growth inhibitors will decrease cases caused by these foods by 2.5–7.8-fold depending on the specific food product (Pradhan *et al.*, 2009). Recent trends in the incidence of listeriosis and the prevalence of *L. monocytogenes* in deli meats have not paralleled each other as closely as one might expect assuming deli meats account for the overwhelming majority of cases, suggesting that other foods or strain-specific virulence traits might play a more significant role in the attribution of listeriosis. Specifically, the prevalence of *L. monocytogenes* in RTE deli meats declined by nearly sevenfold, while the incidence of listeriosis decreased by less than twofold based on finalized data from 2008 when compared to baseline data from 1996–1998 (CDC, 2009; USDA/FSIS, 2011). However, recent outbreaks have suggested that we may not have identified all food products that are a significant vehicle for listeriosis. Small outbreaks associated with sprouts, celery and a very recent large multistate outbreak associated with cantaloupes indicates a critical need to examine the ability of this pathogen to grow and survive in fresh produce (FDA, 2010; CDC, 2011b). In addition, a case-control study by CDC investigators showed that melons and hummus prepared at a commercial establishment, but not including foods currently considered to represent the greatest risk of acquiring listeriosis (e.g., deli meats), were significantly associated with sporadic listeriosis cases across nine states (Varma *et al.*, 2007).

3.4.2 Genetic diversity of *L. monocytogenes*

L. monocytogenes isolates can be grouped into two major genetic lineages (i.e., lineages I and II), which have distinct epidemiological and genetic characteristics. Lineage I isolates are over-represented among human listeriosis cases despite their underrepresentation in RTE foods, while some lineage II strains are over-represented among RTE food isolates but are rarely or never associated with human listeriosis (Gray *et al.*, 2004). Several molecular epidemiology studies have suggested that lineage I strains may have an increased ability to cause disease compared to lineage II strains; furthermore, data from *in vitro* virulence phenotype assays also support this (Chen *et al.*, 2006, 2011; Gray *et al.*, 2004; Nightingale *et al.*, 2006; Wiedmann *et al.*, 1997). Lineage II isolates have been found to commonly serve as recipients of recombination events, while this is rarely the case with lineage I isolates. This is also supported by evidence demonstrating that lineage II isolates are more genetically divergent than lineage I isolates (Orsi *et al.*, 2008b). A third lineage is described, containing two distinct subgroups: IIIA and IIIC (Liu *et al.*, 2006; Roberts *et al.*, 2006). Recent evidence suggests that what was formerly classified as lineage IIIB (Roberts *et al.*, 2006) constitutes its own lineage termed lineage IV (Orsi *et al.*, 2008b; Ward *et al.*, 2008). Lineage III and IV isolates are rare in general and although these isolates have been associated with human disease they are more frequently associated with ruminant listeriosis (Jeffers *et al.*, 2001; Roberts *et al.*, 2006). A recent sequence-based study demonstrated significant diversity among lineage III/IV isolates, which appeared to be the result of frequent recombination events involving strains in these lineages (Tsai *et al.*, 2011).

Epidemic clone strains

There is an emerging body of evidence supporting that strain-specific virulence differences exist with *L. monocytogenes*. Early epidemiological data from the

1980s in the United Kingdom showed that only three of 13 serotypes (i.e., 1/2a, 1/2b and 4b) caused nearly 90% of listeriosis cases (McLauchlin, 1990). The majority of listeriosis outbreaks worldwide have been associated with *L. monocytogenes* strains belonging to serotype 4b (Table 3.1). Through a large-scale molecular epidemiology study of >1000 *L. monocytogenes* isolates, two highly clonal serotype 4b strains termed epidemic clones I and Ia, belonging to EcoRI ribotypes 1038B and 1042B, respectively, were found to be over-represented among isolates from human listeriosis cases in the US (Gray *et al.*, 2004). Epidemic clones I and Ia have also been implicated in listeriosis outbreaks reported on multiple continents (Table 3.1). A comparative genomics study by Nelson *et al.* (2004), which compared lineage I 4b strains to lineage II 1/2a strains from human listeriosis cases and foods, found 51 genes that were found exclusively in 4b strains and 83 genes found exclusively in 1/2a strains, suggesting some level of genetic diversity between these serotypes. Interestingly, the 1/2a strains contained some unique sugar-transport pathways and a pathway for rhamnose biosynthesis, which may allow for the survival of 1/2a strains in particular environmental niches where synthesis, transport and the use of these specific sugars are important. The authors also noted that all the major virulence determinants were conserved between the strains studied, suggesting that the virulence of a particular strain cannot be predicted solely by probing the presence or absence of these genes (Nelson *et al.*, 2004).

Virulence-attenuated strains carrying a mutation leading to a premature stop codon in inlA

Collectively, at least 18 unique single nucleotide polymorphisms in the virulence gene *inlA* that lead to a premature stop codon (PMSC) have been described worldwide to date (Felício *et al.*, 2007; Handa-Miya *et al.*, 2007; Jacquet *et al.*, 2004; Jonquières *et al.*, 1998; Nightingale *et al.*, 2005; Olier *et al.*, 2002, 2003; Orsi *et al.*, 2007; Ragon *et al.*, 2008; Rousseaux *et al.*, 2004, Van Stelten and Nightingale, 2008; Ward *et al.*, 2010). These PMSC mutations in *inlA* occur upstream of the LPXTG membrane anchoring motif and result in production of an InlA that is secreted rather than anchored to the bacterial cell wall or loss of InlA production; thus eliminating the InlA-mediated interaction between *L. monocytogenes* and host cells that express the human isoform of E-cadherin (Jacquet *et al.*, 2004; Nightingale *et al.*, 2005; Orsi *et al.*, 2008b). Characterization of isogenic strains with and without a PMSC in *inlA* demonstrated that these mutations are causally associated with reduced invasion of intestinal and hepatic cells *in vitro* and attenuated virulence as demonstrated using an intragastric guinea pig challenge model (Nightingale *et al.*, 2008). Development and application of a SNP genotyping assay to detect the presence or absence of these 18 mutations leading to a PMSC in *inlA* demonstrated that 45% of >500 RTE food isolates carry these mutations, while only 5% of >500 human clinical cases were caused by a *L. monocytogenes* isolate carrying a PMSC in *inlA* (Van Stelten *et al.*, 2010).

Table 3.1 Summary of listeriosis outbreaks worldwide[a,b]

Year	Location	No. of cases	No. of peri-natal cases	No. of deaths	Food vehicle	Serotype	Source
1981	Nova Scotia, Canada	41	34	18	Coleslaw	4b	Schlech *et al.*, 1983
1983	Massachusetts, USA	49	7	14	Pasteurized milk	4b	Fleming *et al.*, 1985
1985	California, USA	142	93	48	Mexican-style cheese	4b	Linnan *et al.*, 1988
1983–1987	Switzerland	122	65	34	Vacherin Mont d'Or cheese	4b	Büla *et al.*, 1995
1987–1989	United Kingdom	366	NA[c]	NA	Paté	4bx	McLauchlin *et al.*, 1991
1989–1990	Denmark	26	3	7	Blue mold cheese	4b	Jensen *et al.*, 1994
1992	France	279	0	85	Pork tongue in jelly	4b	Jacquet *et al.*, 1995
1993	France	38	31	10	Rillettes	4b	Goulet *et al.*, 1998
1998–1999	Multiple states, USA	108	NA	14	Hot dogs	4b	Mead *et al.*, 2006
1999	Finland	25	0	6	Butter	3a	Lyytikäinen *et al.*, 2000
1999–2000	France	10	3	3	Rillettes	4b	de Valk *et al.*, 2001
1999–2000	France	32	9	10	Pork tongue in aspic	4b	de Valk *et al.*, 2001
2000	Multiple states, USA	30	8	7	Delicatessen turkey ready-to-eat meats	1/2a	Olsen *et al.*, 2005
2000	North Carolina, USA	13	11	5	Home-made Mexican-style cheese	4b	MacDonald *et al.*, 2005
2002	Multiple states USA	54	3	8	Delicatessen Turkey	4b	Cottueb *et al.*, 2006
2002	Quebec, Canada	17	2	0	Cheese made from raw milk	NA	Gaulin *et al.*, 2003
2003	Texas, USA	12	NA	NA	Mexican-style cheese	4b	Swaminathan and Gerner-Smidt, 2007
2005	Switzerland	10	NA	3	Tomme soft cheese	1/2a	Bille *et al.*, 2006
2006–2007	Germany	189	NA	NA	Acid curd cheese from pasteurized milk	4b	Koch *et al.*, 2010
2006–2008	Germany	16	NA	5	Ready-to-eat sausages	4b	Winter *et al.*, 2009
2008	NY, USA	5	NA	3	Tuna salad	1/2a	Cokes *et al.*, 2011
2008	Canada	56	NA	21	Ready-to-eat meat products	1/2a	Ontario Ministry of Health and Long Term Care, 2009; Gilmour *et al.*, 2010
2009	Austria, Germany, Czech Republic	34	NA	7	Acid curd cheese	1/2a (2 clones)	Fretz *et al.*, 2010
2011	USA	146	1	30	Cantaloupes	1/2a and 1/2b	CDC, 2011c

[a] This table was adapted from and expanded upon from a similar table published in Swaminathan and Gerner-Smidt, 2007.
[b] Additional outbreaks associated with produce, specifically sprouts and celery, have been observed in the US. To date there are no published articles on these outbreaks, although they have been widely discussed on blogs and food safety listservs.
[c] NA: not available.

3.5 Regulations and risk assessments

Due to the severity of the disease listeriosis, and the common presence of *L. monocytogenes* in nature, this pathogen is subject to strict regulatory policies and controls worldwide. The Codex Alimentarius Commission (a joint effort of the Food and Agriculture Organization of the United Nations and the World Health Organization) strives to set international standards for food safety. A recent update to the Codex that went into effect in 2009 recommends a risk-based sampling program where a 100 CFU/g limit of *L. monocytogenes* in RTE foods that do not promote the growth of the pathogen is imposed. In foods that do allow the growth of *L. monocytogenes*, a zero tolerance policy is recommended. This update to the Codex also stresses the importance of environmental monitoring as well as process verification to ensure functionality of hazard analysis and critical control point (HACCP) procedures (Codex Alimentarius Commission, 2009). These guidelines mirror the policies currently used in much of Europe (EC, 2005) and Canada (Health Canada, 2011). The US has adhered to a strict policy of zero tolerance for detection of *L. monocytogenes* in all RTE foods since 1985 (Shank *et al.*, 1996). In 2008, the US Food and Drug Administration published a document discussing the possibility of allowing a 100 CFU/g limit of *L. monocytogenes* in RTE foods that do not support the growth of *L. monocytogenes* in the form of a draft Compliance Policy Guide. However, no regulatory changes resulted from this draft Compliance Policy Guide (FDA, 2008).

The 2003 *L. monocytogenes* risk assessment was based on data from mouse challenge studies (through the iv or intraperitoneal infection route) and while this risk assessment allows some heterogeneity in virulence between strains it does not mechanistically relate virulence to strain-specific genetic characteristics (FDA/FSIS/CDC, 2003). Recently considerable work has been performed in the guinea pig model to define the range of infectious doses with strains with defined virulence characteristics. Previously, the LD_{50} for fetuses in the pregnant guinea pig had been established at approximately 2×10^7 CFU using a single strain shown to cause abortions in rhesus monkeys (Williams *et al.*, 2007). In a second study performed by this group, the virulence potential of a fish processing plant strain, a human clinical strain and a monkey clinical strain were examined by orally inoculating pregnant guinea pigs with approximately 1×10^8 CFU of each strain. Of fetuses from dams inoculated with the monkey clinical strain, 22% were positive for *L. monocytogenes*, while this number was 56% for the processing plant strain, and 0% for the human clinical strain. This suggests that there may be significant strain-to-strain variation in terms of ability to cause infection in the pregnant guinea pig model (Jensen *et al.*, 2008). A study by Van Stelten and others examined the dose response relationship when juvenile male guinea pigs were administered varying levels of a virulence-attenuated strain with an *inlA* PMSC mutation or an outbreak-associated epidemic clone strain. There was an approximate 1.2 $\log_{10}$ CFU increase in the median infectious dose for the *inlA* PMSC strain compared to the epidemic clone strain (Van Stelten *et al.*, 2011). A study by Chen *et al.* reported that *inlA* PMSC strains are found at higher levels in RTE foods compared to strains without a PMSC. Chen *et al.* then went on to model the *r* value (the probability of a single cell causing disease) of strains with and without an *inlA* PMSC and found that strains with a PMSC had an *r* value approximately 2 orders of magnitude lower than strains without a PMSC (Chen *et al.*, 2011).

Currently we do not entirely understand the impact of prior exposure to *L. monocytogenes* through consumption of contaminated RTE foods on host immune status. This effect may be significant as the 2003 *L. monocytogenes* risk assessment indicates that the average individual in the US consumes foods contaminated with $<1 \times 10^3$ CFU of *L. monocytogenes* 19 times a year, 1×10^3 to 1×10^6 CFU 2.4 times a year, 1×10^6 to 1×10^9 CFU once every other year, and $>1 \times 10^9$ CFU approximately every three years (FDA/FSIS/CDC, 2003). Furthermore, a study by Nightingale and others (2008) found that *inlA* PMSC strains have the potential to confer protection against a subsequent challenge by fully virulent strains in a guinea pig model.

3.6 Conclusions

Recent studies have extended existing knowledge regarding the epidemiology, ecology and virulence of *L. monocytogenes*. Increases in the incidence of listeriosis in several European countries have been noted as well as an increase in 1/2a serotype-associated cases in multiple countries. A recent large multistate listeriosis outbreak, epidemiologically associated clusters of listeriosis and a case-control analyses of sporadic listeriosis cases in the US have highlighted the need to look beyond RTE meats as the primary source of *L. monocytogenes* infections, since fresh produce may represent an important food vehicle. Technological advances have fueled whole genome sequencing of diverse *Listeria* strains, which has led to a better understanding of the evolution of virulence within the genus and has facilitated the discovery of novel virulence factors. The discovery and development of improved animal models and murinized *L. monocytogenes* strains will facilitate generation of empirical data to enhance current and develop future risk assessments. Lastly, it is now understood that at least two subpopulations of *L. monocytogenes* exist, with varying pathogenic potential, which highlights the importance of understanding strain-to-strain differences in virulence.

3.7 References

AKANBI, O. B., BREITHAUPT, A., POLSTER, U., ALTER, T., QUANDT, A. *et al.* 2008. Systemic listeriosis in caged canaries (*Serinus canarius*). *Avian Pathol*, 37, 329–32.

AUBRY, C., GOULARD, C., NAHORI, M. A., CAYET, N., DECALF, J. *et al.* 2011. OatA, a peptidoglycan O-acetyltransferase involved in *Listeria monocytogenes* immune escape, is critical for virulence. *J Infect Dis*, 204, 731–40.

AUTIO, T., HIELM, S., MIETTINEN, M., SJÖBERG, A. M., AARNISALO, K. *et al.* 1999. Sources of *Listeria monocytogenes* contamination in a cold-smoked rainbow trout processing plant detected by pulsed-field gel electrophoresis typing. *Appl Environ Microbiol*, 65, 150–5.

BARMPALIA, I. M., KOUTSOUMANIS, K. P., GEONARAS, I., BELK, K. E., SCANGA, J. A. *et al.* 2005. Effect of antimicrobials as ingredients of pork bologna for *Listeria monocytogenes* control during storage at 4 or 10 °C. *Food Microbiology*, 22, 205–11.

BILLE, J., BLANC, D. S., SCHMID, H., BOUBAKER, K., BAUMGARTNER, A. *et al.* 2006. Outbreak of human listeriosis associated with tomme cheese in northwest Switzerland, 2005. *Euro Surveill*, 11, 91–3.

BLANOT, S., JOLY, M. M., VILDE, F., JAUBERT, F., CLEMENT, O. *et al.* 1997. A gerbil model for rhombencephalitis due to *Listeria monocytogenes*. *Microb Pathog*, 23, 39–48.

BRAUN, L., GHEBREHIWET, B., and COSSART, P. 2000. gC1q-R/p32, a C1q-binding protein, is a receptor for the InlB invasion protein of *Listeria monocytogenes*. *EMBO J*, 19, 1458–66.

BUCHRIESER, C., RUSNIOK, C., GARRIDO, P., HAIN, T., SCORTTI, M. *et al.* 2011. Complete genome sequence of the animal pathogen *Listeria ivanovii*, which provides insights into host specificities and evolution of the genus *Listeria*. *J Bacteriol*, 193, 6787–8.

BÜLA, C. J., BILLE, J. and GLAUSER, M. P. 1995. An epidemic of food-borne listeriosis in western Switzerland: Description of 57 cases involving adults. *Clin Infect Dis*, 20, 66–72.

CAMILLI, A., TILNEY, L. G. and PORTNOY, D. A. 1993. Dual roles of *plcA* in *Listeria monocytogenes* pathogenesis. *Mol Microbiol*, 8, 143–57.

CDC 2009. *Preliminary FoodNet Data on the Incidence of Infection with Pathogens Transmitted Commonly Through Food – 10 States, 2008* [Online]. Available: www.cdc.gov/mmwr/preview/mmwrhtml/mm5813a2.htm?s_cid=mm5813a2_e [Accessed 14 December 2012].

CDC 2011a. *Deadly Listeria Outbreak Halted in Record Time* [Online]. Available: www.cdc.gov/24-7/SavingLives/listeria [Accessed 23 November 2011].

CDC 2011b. Multistate outbreak of listeriosis associated with Jensen Farms Cantaloupe – United States, August–September 2011. *MMWR*, 60, 1357–8.

CDC 2011c. *Multistate Outbreak of Listeriosis Linked to Whole Cantaloupes from Jensen Farms, Colorado* [Online]. Available: www.cdc.gov/listeria/outbreaks/cantaloupes-jensen-farms/120811/index.html [Accessed 14 December 2011].

CHEN, Y., ROSS, W. H., GRAY, M. J., WIEDMANN, M., WHITING, R. C. *et al.* 2006. Attributing risk to *Listeria monocytogenes* subgroups: Dose response in relation to genetic lineages. *J Food Prot*, 69, 335–44.

CHEN, Y., ROSS, W. H., WHITING, R. C., VAN STELTEN, A., NIGHTINGALE, K. K. *et al.* 2011. Variation in *Listeria monocytogenes* dose responses in relation to subtypes encoding a full-length or truncated internalin A. *Appl Environ Microbiol*, 77, 1171–80.

CODEX ALIMENTARIUS COMMISSION 2009. Proposed Draft Microbiological Criteria for *Listeria monocytogenes* in ready-to-eat foods.

COKES, C., FRANCE, A. M., REDDY, V., HANSON, H., LEE, L. *et al.* 2011. Serving high-risk foods in a high-risk setting: Survey of hospital food service practices after an outbreak of listeriosis in a hospital. *Infect Control Hosp Epidemiol*, 32, 380–6.

COOPER, G. L. 1989. An encephalitic form of listeriosis in broiler chickens. *Avian Dis*, 33, 182–5.

DE VALK, H., VAILLANT, V., JACQUET, C., ROCOURT, J., LE QUERREC, F. *et al.* 2001. Two consecutive nationwide outbreaks of listeriosis in France, October 1999–February 2000. *Am J Epidemiol*, 154, 944–50.

DEN BAKKER, H. C., BUNDRANT, B. N., FORTES, E. D., ORSI, R. H. and WIEDMANN, M. 2010a. A population genetics-based and phylogenetic approach to understanding the evolution of virulence in the genus *Listeria*. *Appl Environ Microbiol*, 76, 6085–100.

DEN BAKKER, H. C., CUMMINGS, C. A., FERREIRA, V., VATTA, P., ORSI, R. H. *et al.* 2010b. Comparative genomics of the bacterial genus *Listeria*: Genome evolution is characterized by limited gene acquisition and limited gene loss. *BMC Genomics*, 11, 688.

DISSON, O., GRAYO, S., HUILLET, E., NIKITAS, G., LANGA-VIVES, F. *et al.* 2008. Conjugated action of two species-specific invasion proteins for fetoplacental listeriosis. *Nature*, 455, 1114–18.

DORTET, L., MOSTOWY, S. and COSSART, P. 2012. *Listeria* and autophagy escape: Involvement of InlK, an internalin-like protein. *Autophagy*, 8 (1), 132–4.

DOUMITH, M., CAZALET, C., SIMOES, N., FRANGEUL, L., JACQUET, C. *et al.* 2004. New aspects regarding evolution and virulence of *Listeria monocytogenes* revealed by comparative genomics and DNA arrays. *Infect Immun*, 72, 1072–83.

DREVETS, D. A. and BRONZE, M. S. 2008. *Listeria monocytogenes*: Epidemiology, human disease, and mechanisms of brain invasion. *FEMS Immunol Med Microbiol*, 53, 151–65.

DUNN, K. A., BIELAWSKI, J. P., WARD, T. J., URQUHART, C. and GU, H. 2009. Reconciling ecological and genomic divergence among lineages of *Listeria* under an 'extended mosaic genome concept'. *Mol Biol Evol*, 26, 2605–15.

EC 2005. Commission Regulation (EC) No 2073/2005. *Official Journal of the European Union.*

EKLUND, M. W., POYSKY, F. T., PARANJPYE, R. N., LASHBROOK, L. C., PETERSON, M. E. *et al.* 1995. Incidence and sources of *Listeria monocytogenes* in cold-smoked fishery products and processing plants. *J Food Prot*, 58, 502–8.

EVANS, K., SMITH, M., MCDONOUGH, P. and WIEDMANN, M. 2004. Eye infections due to *Listeria monocytogenes* in three cows and one horse. *J Vet Diagn Invest*, 16, 464–9.

FARBER, J. M., SANDERS, G. W., SPEIRS, J. I., D'AOUST, J. Y., EMMONS, D. B. *et al.* 1988. Thermal resistance of *Listeria monocytogenes* in inoculated and naturally contaminated raw milk. *Int J Food Microbiol*, 7, 277–86.

FDA 2008. *Compliance Policy Guide Sec. 555.320 Listeria monocytogenes* [Online]. Available: www.fda.gov/ICECI/ComplianceManuals/CompliancePolicyGuidanceManual/ucm136694.htm [Accessed 12 December 2011].

FDA 2010. *DSHS Orders Sangar Produce to Close, Recall Products* [Online]. Available: www.fda.gov/Safety/Recalls/ArchiveRecalls/2010/ucm230709.htm.

FDA/FSIS/CDC 2003. *Quantitative Assessment of Relative Risk to Public Health from Foodborne Listeria monocytogenes Among Selected Categories of Ready-to-Eat Food* [Online]. Available: www.fda.gov/downloads/Food/ScienceResearch/ResearchAreas/RiskAssessmentSafetyAssessment/UCM197330.pdf.

FELÍCIO, M. T., HOGG, T., GIBBS, P., TEIXEIRA, P. and WIEDMANN, M. 2007. Recurrent and sporadic *Listeria monocytogenes* contamination in alheiras represents considerable diversity, including virulence-attenuated isolates. *Appl Environ Microbiol*, 73, 3887–95.

FLEMING, D. W., COCHI, S. L., MACDONALD, K. L., BRONDUM, J., HAYES, P. S. *et al.* 1985. Pasteurized milk as a vehicle of infection in an outbreak of listeriosis. *N Engl J Med*, 312, 404–7.

FRETZ, R., PICHLER, J., SAGEL, U., MUCH, P., RUPPITSCH, W. *et al.* 2010. Update: Multinational listeriosis outbreak due to 'Quargel', a sour milk curd cheese, caused by two different *L. monocytogenes* serotype 1/2a strains, 2009–2010. *Euro Surveill*, 15, 31–2.

GARNER, M. R., NJAA, B. L., WIEDMANN, M. and BOOR, K. J. 2006. Sigma B contributes *to Listeria monocytogenes* gastrointestinal infection but not to systemic spread in the guinea pig infection model. *Infect Immun*, 74, 876–86.

GAULIN, C., RAMSAY, D., RINGUETTE, L. and ISMAÏL, J. 2003. First documented outbreak of *Listeria monocytogenes* in Quebec, 2002. *Can Commun Dis Rep*, 29, 181–6.

GEDDE, M. M., HIGGINS, D. E., TILNEY, L. G. and PORTNOY, D. A. 2000. Role of listeriolysin O in cell-to-cell spread of *Listeria monocytogenes*. *Infect Immun*, 68, 999–1003.

GIBBONS, I. S., ADESIYUN, A., SEEPERSADSINGH, N. and RAHAMAN, S. 2006. Investigation for possible source(s) of contamination of ready-to-eat meat products with *Listeria spp.* and other pathogens in a meat processing plant in Trinidad. *Food Microbiol*, 23, 359–66.

GILLESPIE, I. A., MCLAUCHLIN, J., GRANT, K. A., LITTLE, C. L., MITHANI, V. *et al.* 2006. Changing pattern of human listeriosis, England and Wales, 2001–2004. *Emerg Infect Dis*, 12, 1361–6.

GILLESPIE, I. A., MCLAUCHLIN, J., LITTLE, C. L., PENMAN, C., MOOK, P. *et al.* 2009. Disease presentation in relation to infection foci for non-pregnancy-associated human listeriosis in England and Wales, 2001 to 2007. *J Clin Microbiol*, 47, 3301–7.

GILLESPIE, I. A., MOOK, P., LITTLE, C. L., GRANT, K. and ADAK, G. K. 2010. *Listeria monocytogenes* infection in the over-60s in England between 2005 and 2008: A retrospective case-control study utilizing market research panel data. *Foodborne Pathog Dis*, 7, 1373–9.

GILMOUR, M. W., GRAHAM, M., VAN DOMSELAAR, G., TYLER, S., KENT, H. *et al.* 2010. High-throughput genome sequencing of two *Listeria monocytogenes* clinical isolates during a large foodborne outbreak. *BMC Genomics*, 11, 120.

GOLDEN, D. A., BEUCHAT, L. R. and BRACKETT, R. E. 1988. Inactivation and injury of *Listeria monocytogenes* as affected by heating and freezing. *Food Microb*, 5, 17–23.

GOTTLIEB, S. L., NEWBERN, E. C., GRIFFIN, P. M., GRAVES, L. M., HOEKSTRA, R. M., *et al.* 2006. Multi State Outbreak of Listeriosis Linked to Turkey Deli Meat and Subsequent Changes in US Regulatory Policy. *Clin Infect Dis*, 42, 29–36.

GOULET, V., HEDBERG, C., LE MONNIER, A. and DE VALK, H. 2008. Increasing incidence of listeriosis in France and other European countries. *Emerg Infect Dis*, 14, 734–40.

GOULET, V., ROCOURT, J., REBIERE, I., JACQUET, C., MOYSE, C. *et al.* 1998. Listeriosis outbreak associated with the consumption of rillettes in France in 1993. *J Infect Dis*, 177, 155–60.

GRAVES, L. M., HELSEL, L. O., STEIGERWALT, A. G., MOREY, R. E., DANESHVAR, M. I. *et al.* 2010. *Listeria marthii* sp. nov., isolated from the natural environment, Finger Lakes National Forest. *Int J Syst Evol Microbiol*, 60, 1280–8.

GRAY, M. J., ZADOKS, R. N., FORTES, E. D., DOGAN, B., CAI, S. *et al.* 2004. *Listeria monocytogenes* isolates from foods and humans form distinct but overlapping populations. *Appl Environ Microbiol*, 70, 5833–41.

GUDMUNDSDOTTIR, K. B., SVANSSON, V., AALBAEK, B., GUNNARSSON, E. and SIGURDARSON, S. 2004. *Listeria monocytogenes* in horses in Iceland. *Vet Rec*, 155, 456–9.

GUILLET, C., JOIN-LAMBERT, O., LE MONNIER, A., LECLERCQ, A., MECHAÏ, F. *et al.* 2010. Human listeriosis caused by *Listeria ivanovii*. *Emerg Infect Dis*, 16, 136–8.

HAIN, T., STEINWEG, C., KUENNE, C. T., BILLION, A., GHAI, R. *et al.* 2006. Whole-genome sequence of *Listeria welshimeri* reveals common steps in genome reduction with *Listeria innocua* as compared to *Listeria monocytogenes*. *J Bacteriol*, 188, 7405–15.

HANDA-MIYA, S., KIMURA, B., TAKAHASHI, H., SATO, M., ISHIKAWA, T. *et al.* 2007. Nonsense-mutated *inlA* and *prfA* not widely distributed in *Listeria monocytogenes* isolates from ready-to-eat seafood products in Japan. *Int J Food Microbiol*, 117, 312–8.

HEALTH CANADA 2011. Policy on *Listeria* monocytogenes in ready-to-eat foods [Online] Available: www.hc-sc.gc.ca/fn-an/legislation/pol/policy_listeria_monocytogenes_2011-eng.php.

HELLSTRÖM, S., LAUKKANEN, R., SIEKKINEN, K. M., RANTA, J., MAIJALA, R. *et al.* 2010. *Listeria monocytogenes* contamination in pork can originate from farms. *J Food Prot*, 73, 641–8.

HOELZER, K., SAUDERS, B. D., SANCHEZ, M. D., OLSEN, P. T., PICKETT, M. M. *et al.* 2011. Prevalence, distribution, and diversity of *Listeria monocytogenes* in retail environments, focusing on small establishments and establishments with a history of failed inspections. *J Food Prot*, 74, 1083–95.

HOFFMAN, A. D., GALL, K. L., NORTON, D. M. and WIEDMANN, M. 2003. *Listeria monocytogenes* contamination patterns for the smoked fish processing environment and for raw fish. *J Food Prot*, 66, 52–60.

HUDSON, J. A. and MOTT, S. J. 1993. Presence of *Listeria monocytogenes*, motile aeromonads and *Yersinia enterocolitica* in environmental samples taken from a supermarket delicatessen. *Int J Food Microbiol*, 18, 333–7.

JACQUET, C., CATIMEL, B., BROSCH, R., BUCHRIESER, C., DEHAUMONT, P. *et al.* 1995. Investigations related to the epidemic strain involved in the French listeriosis outbreak in 1992. *Appl Environ Microbiol*, 61, 2242–6.

JACQUET, C., DOUMITH, M., GORDON, J. I., MARTIN, P. M., COSSART, P. *et al.* 2004. A molecular marker for evaluating the pathogenic potential of foodborne *Listeria monocytogenes*. *J Infect Dis*, 189, 2094–100.

JEFFERS, G. T., BRUCE, J. L., MCDONOUGH, P. L., SCARLETT, J., BOOR, K. J. *et al.* 2001. Comparative genetic characterization of *Listeria monocytogenes* isolates from human and animal listeriosis cases. *Microbiology*, 147, 1095–104.

JENSEN, A., FREDERIKSEN, W. and GERNER-SMIDT, P. 1994. Risk factors for listeriosis in Denmark, 1989–1990. *Scand J Infect Dis*, 26, 171–8.

JENSEN, A., WILLIAMS, D., IRVIN, E.A., GRAM, L. and SMITH, M.A. 2008. A processing plant persistent strain of *Listeria monocytogenes* crosses the fetoplacental barrier in a pregnant guinea pig model. *J Food Prot*, 74, 1028–34.

JOHNSON, J., JINNEMAN, K., STELMA, G., SMITH, B. G., LYE, D. *et al.* 2004. Natural atypical *Listeria* innocua strains with *Listeria monocytogenes* pathogenicity island 1 genes. *Appl Environ Microbiol*, 70, 4256–66.

JONQUIÈRES, R., BIERNE, H., MENGAUD, J. and COSSART, P. 1998. The *inlA* gene of *Listeria monocytogenes* LO28 harbors a nonsense mutation resulting in release of internalin. *Infect Immun*, 66, 3420–2.

KASTNER, R., DUSSURGET, O., ARCHAMBAUD, C., KERNBAUER, E., SOULAT, D. *et al.* 2011. LipA, a tyrosine and lipid phosphatase involved in the virulence of *Listeria monocytogenes*. *Infect Immun*, 79, 2489–98.

KIM, H., MARQUIS, H. and BOOR, K. J. 2005. SigmaB contributes to *Listeria monocytogenes* invasion by controlling expression of *inlA* and *inlB*. *Microbiology*, 151, 3215–22.

KOCH, J., DWORAK, R., PRAGER, R., BECKER, B., BROCKMANN, S. *et al.* 2010. Large listeriosis outbreak linked to cheese made from pasteurized milk, Germany, 2006–2007. *Foodborne Pathog Dis*, 7, 1581–4.

KOCH, J. and STARK, K. 2006. Significant increase of listeriosis in Germany – epidemiological patterns 2001–2005. *Euro Surveill*, 11, 85–8.

KOUTSOUMANIS, K. P., ASHTON, L. V., GEORNARAS, I., BELK, K. E., SCANGA, J. A. *et al.* 2004. Effect of single or sequential hot water and lactic acid decontamination treatments on the survival and growth of *Listeria monocytogenes* and spoilage microflora during aerobic storage of fresh beef at 4, 10, and 25 degrees C. *J Food Prot*, 67, 2703–11.

LAPPI, V. R., THIMOTHE, J., NIGHTINGALE, K. K., GALL, K., SCOTT, V. N. *et al.* 2004. Longitudinal studies on *Listeria* in smoked fish plants: Impact of intervention strategies on contamination patterns. *J Food Prot*, 67, 2500–14.

LEBRETON, A., LAKISIC, G., JOB, V., FRITSCH, L., THAM, T. N. *et al.* 2011. A bacterial protein targets the BAHD1 chromatin complex to stimulate type III interferon response. *Science*, 331, 1319–21.

LECLERCQ, A., CLERMONT, D., BIZET, C., GRIMONT, P. A., LE FLÈCHE-MATÉOS, A. *et al.* 2010. *Listeria rocourtiae* sp. nov. *Int J Syst Evol Microbiol*, 60, 2210–4.

LECUIT, M. 2007. Human listeriosis and animal models. *Microbes Infect*, 9, 1216–25.

LECUIT, M., DRAMSI, S., GOTTARDI, C., FEDOR-CHAIKEN, M., GUMBINER, B. *et al.* 1999. A single amino acid in E-cadherin responsible for host specificity towards the human pathogen *Listeria monocytogenes*. *EMBO J*, 18, 3956–63.

LECUIT, M., NELSON, D. M., SMITH, S. D., KHUN, H., HUERRE, M. *et al.* 2004. Targeting and crossing of the human maternofetal barrier by *Listeria monocytogenes*: Role of internalin interaction with trophoblast E-cadherin. *Proc Natl Acad Sci USA*, 101, 6152–7.

LECUIT, M., OHAYON, H., BRAUN, L., MENGAUD, J. and COSSART, P. 1997. Internalin of *Listeria monocytogenes* with an intact leucine-rich repeat region is sufficient to promote internalization. *Infect Immun*, 65, 5309–19.

LECUIT, M., VANDORMAEL-POURNIN, S., LEFORT, J., HUERRE, M., GOUNON, P. *et al.* 2001. A transgenic model for listeriosis: Role of internalin in crossing the intestinal barrier. *Science*, 292, 1722–5.

LIANOU, A., GEORNARAS, I., KENDALL, P. A., SCANGA, J. A. and SOFOS, J. N. 2007. Behavior of *Listeria monocytogenes* at 7 degrees C in commercial turkey breast, with or without antimicrobials, after simulated contamination for manufacturing, retail and consumer settings. *Food Microbiol*, 24, 433–43.

LIANOU, A., STOPFORTH, J. D., YOON, Y., WIEDMANN, M. and SOFOS, J. N. 2006. Growth and stress resistance variation in culture broth among *Listeria monocytogenes* strains of various serotypes and origins. *J Food Prot*, 69, 2640–7.

LINNAN, M. J., MASCOLA, L., LOU, X. D., GOULET, V., MAY, S. *et al.* 1988. Epidemic listeriosis associated with Mexican-style cheese. *N Engl J Med*, 319, 823–8.

LIU, D., LAWRENCE, M. L., WIEDMANN, M., GORSKI, L., MANDRELL, R. E. *et al.* 2006. *Listeria monocytogenes* subgroups IIIA, IIIB, and IIIC delineate genetically distinct populations with varied pathogenic potential. *J Clin Microbiol*, 44, 4229–33.

LONG, J. and DUKES, T. 1972. Listeriosis in newborn swine. *Can Vet J*, 13, 49–52.

LOW, J. C. and DONACHIE, W. 1997. A review of *Listeria monocytogenes* and listeriosis. *Vet J*, 153, 9–29.

LUKINMAA, S., MIETTINEN, M., NAKARI, U. M., KORKEALA, H. and SIITONEN, A. 2003. *Listeria monocytogenes* isolates from invasive infections: Variation of sero- and genotypes during an 11-year period in Finland. *J Clin Microbiol*, 41, 1694–700.

LUNDÉN, J. M., AUTIO, T. J., SJÖBERG, A. M. and KORKEALA, H. J. 2003. Persistent and nonpersistent *Listeria monocytogenes* contamination in meat and poultry processing plants. *J Food Prot*, 66, 2062–9.

LYYTIKÄINEN, O., AUTIO, T., MAIJALA, R., RUUTU, P., HONKANEN-BUZALSKI, T. *et al.* 2000. An outbreak of *Listeria monocytogenes* serotype 3a infections from butter in Finland. *J Infect Dis*, 181, 1838–41.

MACDONALD, P. D., WHITWAM, R. E., BOGGS, J. D., MACCORMACK, J. N., ANDERSON, K. L. *et al.* 2005. Outbreak of listeriosis among Mexican immigrants as a result of consumption of illicitly produced Mexican-style cheese. *Clin Infect Dis*, 40, 677–82.

MCLAUCHLIN, J. 1990. Distribution of serovars of *Listeria monocytogenes* isolated from different categories of patients with listeriosis. *Eur J Clin Microbiol Infect Dis*, 9, 210–3.

MCLAUCHLIN, J., HALL, S. M., VELANI, S. K. and GILBERT, R. J. 1991. Human listeriosis and paté: A possible association. *BMJ*, 303, 773–5.

MEAD, P. S., DUNNE, E. F., GRAVES, L., WIEDMANN, M., PATRICK, M. *et al.* 2006. Nationwide outbreak of listeriosis due to contaminated meat. *Epidemiol Infect*, 134, 744–51.

MEAD, P. S., SLUTSKER, L., DIETZ, V., MCCAIG, L. F., BRESEE, J. S. *et al.* 1999. Food-related illness and death in the United States. *Emerg Infect Dis*, 5, 607–25.

MONK, I. R., CASEY, P. G., HILL, C. and GAHAN, C. G. 2010. Directed evolution and targeted mutagenesis to murinize *Listeria monocytogenes internalin A* for enhanced infectivity in the murine oral infection model. *BMC Microbiol*, 10, 318.

NELSON, K. E., FOUTS, D. E., MONGODIN, E. F., RAVEL, J., DEBOY, R. T. *et al.* 2004. Whole genome comparisons of serotype 4b and 1/2a strains of the food-borne pathogen *Listeria monocytogenes* reveal new insights into the core genome components of this species. *Nucleic Acids Res*, 32, 2386–95.

NIGHTINGALE, K. K., IVY, R. A., HO, A. J., FORTES, E. D., NJAA, B. L. *et al.* 2008. *inlA* premature stop codons are common among *Listeria monocytogenes* isolates from foods and yield virulence-attenuated strains that confer protection against fully virulent strains. *Appl Environ Microbiol*, 74, 6570–83.

NIGHTINGALE, K. K., LYLES, K., AYODELE, M., JALAN, P., NIELSEN, R. *et al.* 2006. Novel method to identify source-associated phylogenetic clustering shows that *Listeria monocytogenes* includes niche-adapted clonal groups with distinct ecological preferences. *J Clin Microbiol*, 44, 3742–51.

NIGHTINGALE, K. K., SCHUKKEN, Y. H., NIGHTINGALE, C. R., FORTES, E. D., HO, A. J. *et al.* 2004. Ecology and transmission of *Listeria monocytogenes* infecting ruminants and in the farm environment. *Appl Environ Microbiol*, 70, 4458–67.

NIGHTINGALE, K. K., WINDHAM, K., MARTIN, K. E., YEUNG, M. and WIEDMANN, M. 2005. Select *Listeria monocytogenes* subtypes commonly found in foods carry distinct nonsense mutations in *inlA*, leading to expression of truncated and secreted internalin A, and are associated with a reduced invasion phenotype for human intestinal epithelial cells. *Appl Environ Microbiol*, 71, 8764–72.

NORTON, D. M., MCCAMEY, M. A., GALL, K. L., SCARLETT, J. M., BOOR, K. J. *et al.* 2001. Molecular studies on the ecology of *Listeria monocytogenes* in the smoked fish processing industry. *Appl Environ Microbiol*, 67, 198–205.

OLIER, M., PIERRE, F., LEMAÎTRE, J. P., DIVIES, C., ROUSSET, A. *et al.* 2002. Assessment of the pathogenic potential of two *Listeria monocytogenes* human faecal carriage isolates. *Microbiology*, 148, 1855–62.

OLIER, M., PIERRE, F., ROUSSEAUX, S., LEMAÎTRE, J. P., ROUSSET, A. *et al.* 2003. Expression of truncated internalin A is involved in impaired internalization of some *Listeria monocytogenes* isolates carried asymptomatically by humans. *Infect Immun*, 71, 1217–24.

OLIVER, H. F., ORSI, R. H., WIEDMANN, M. and BOOR, K. J. 2010. *Listeria monocytogenes* σB has a small core regulon and a conserved role in virulence but makes differential contributions to stress tolerance across a diverse collection of strains. *Appl Environ Microbiol*, 76, 4216–32.

OLSEN, S. J., PATRICK, M., HUNTER, S. B., REDDY, V., KORNSTEIN, L. *et al.* 2005. Multistate outbreak of *Listeria monocytogenes* infection linked to delicatessen turkey meat. *Clin Infect Dis*, 40, 962–7.

ONTARIO MINISTRY OF HEALTH AND LONG TERM CARE 2009. *Chief Medical Officer of Health's Report on the Management of the 2008 Listeriosis Outbreak in Ontario* [Online]. Available: www.health.gov.on.ca/en/public/publications/disease/docs/listeriosis_outbreak_rep.pdf.

ORSI, R. H., BOROWSKY, M. L., LAUER, P., YOUNG, S. K., NUSBAUM, C. *et al.* 2008a. Short-term genome evolution of *Listeria monocytogenes* in a non-controlled environment. *BMC Genomics*, 9, 539.

ORSI, R. H., RIPOLL, D. R., YEUNG, M., NIGHTINGALE, K. K. and WIEDMANN, M. 2007. Recombination and positive selection contribute to evolution of *Listeria monocytogenes inlA*. *Microbiology*, 153, 2666–78.

ORSI, R. H., SUN, Q. and WIEDMANN, M. 2008b. Genome-wide analyses reveal lineage specific contributions of positive selection and recombination to the evolution of *Listeria monocytogenes*. *BMC Evol Biol*, 8, 233.

PARIHAR, V. S., LOPEZ-VALLADARES, G., DANIELSSON-THAM, M. L., PEIRIS, I., HELMERSSON, S. *et al.* 2008. Characterization of human invasive isolates of *Listeria monocytogenes* in Sweden 1986–2007. *Foodborne Pathog Dis*, 5, 755–61.

PERRIN, M., BEMER, M. and DELAMARE, C. 2003. Fatal case of *Listeria innocua* bacteremia. *J Clin Microbiol*, 41, 5308–9.

PRADHAN, A. K., IVANEK, R., GRÖHN, Y. T., GEORNARAS, I., SOFOS, J. N. *et al.* 2009. Quantitative risk assessment for *Listeria monocytogenes* in selected categories of deli meats: Impact of lactate and diacetate on listeriosis cases and deaths. *J Food Prot*, 72, 978–89.

PORTNOY, D. A., JACKS, P. S. and HINRICHS, D. J. 1988. Role of hemolysin for the intracellular growth of *Listeria monocytogenes*. *J Exp Med*, 167, 1459–71.

RAGON, M., WIRTH, T., HOLLANDT, F., LAVENIR, R., LECUIT, M. *et al.* 2008. A new perspective on *Listeria monocytogenes* evolution. *PLoS Pathog*, 4, e1000146.

RAJABIAN, T., GAVICHERLA, B., HEISIG, M., MÜLLER-ALTROCK, S., GOEBEL, W. *et al.* 2009. The bacterial virulence factor InlC perturbs apical cell junctions and promotes cell-to-cell spread of *Listeria*. *Nat Cell Biol*, 11, 1212–8.

RAPOSE, A., LICK, S. D. and ISMAIL, N. 2008. *Listeria grayi* bacteremia in a heart transplant recipient. *Transpl Infect Dis*, 10, 434–6.

REIS, O., SOUSA, S., CAMEJO, A., VILLIERS, V., GOUIN, E. *et al.* 2010. LapB, a novel *Listeria monocytogenes* LPXTG surface adhesin, required for entry into eukaryotic cells and virulence. *J Infect Dis*, 202, 551–62.

ROBERTS, A., CHAN, Y. and WIEDMANN, M. 2005. Definition of genetically distinct attenuation mechanisms in naturally virulence-attenuated *Listeria monocytogenes* by comparative cell culture and molecular characterization. *Appl Environ Microbiol*, 71, 3900–10.

ROBERTS, A., NIGHTINGALE, K., JEFFERS, G., FORTES, E., KONGO, J. M. *et al.* 2006. Genetic and phenotypic characterization of *Listeria monocytogenes* lineage III. *Microbiology*, 152, 685–93.

ROBERTS, A. J., WILLIAMS, S. K., WIEDMANN, M. and NIGHTINGALE, K. K. 2009. Some *Listeria monocytogenes* outbreak strains demonstrate significantly reduced invasion, *inlA* transcript levels, and swarming motility *in vitro*. *Appl Environ Microbiol*, 75, 5647–58.

ROCOURT, J., HOF, H., SCHRETTENBRUNNER, A., MALINVERNI, R. and BILLE, J. 1986. Acute purulent *Listeria seelingeri* meningitis in an immunocompetent adult. *Schweiz Med Wochenschr*, 116, 248–251.

RØRVIK, L. M., CAUGANT, D. A. and YNDESTAD, M. 1995. Contamination pattern of *Listeria monocytogenes* and other *Listeria* spp. in a salmon slaughterhouse and smoked salmon processing plant. *Int J Food Microbiol*, 25, 19–27.

ROUSSEAUX, S., OLIER, M., LEMAÎTRE, J. P., PIVETEAU, P. and GUZZO, J. 2004. Use of PCR-restriction fragment length polymorphism of *inlA* for rapid screening of *Listeria monocytogenes* strains deficient in the ability to invade Caco-2 cells. *Appl Environ Microbiol*, 70, 2180–5.

RÜTTEN, M., LEHNER, A., POSPISCHIL, A. and SYDLER, T. 2006. Cerebral listeriosis in an adult Freiberger gelding. *J Comp Pathol*, 134, 249–53.

RUXTON, G. D. and GIBSON, G. J. 1995. A mathematical model of the aerobic deterioration of big-bale silage and its implications for the growth of *Listeria monocytogenes*. *Grass Forage Sci*, 50, 331–44.

SABET, C., LECUIT, M., CABANES, D., COSSART, P. and BIERNE, H. 2005. LPXTG protein InlJ, a newly identified internalin involved in *Listeria monocytogenes* virulence. *Infect Immun*, 73, 6912–22.

SABET, C., TOLEDO-ARANA, A., PERSONNIC, N., LECUIT, M., DUBRAC, S. *et al.* 2008. The *Listeria monocytogenes* virulence factor InlJ is specifically expressed *in vivo* and behaves as an adhesin. *Infect Immun*, 76, 1368–78.

SAMELIS, J., BEDIE, G. K., SOFOS, J. N., BELK, K. E., SCANGA, J. A. *et al.* 2002. Control of *Listeria monocytogenes* with combined antimicrobials after postprocess contamination and extended storage of frankfurters at 4 degrees C in vacuum packages. *J Food Prot*, 65, 299–307.

SAMELIS, J. and METAXOPOULOS, J. 1999. Incidence and principal sources of *Listeria* spp. and *Listeria monocytogenes* contamination in processed meats and a meat processing plant. *Food Microb*, 16, 465–77.

SAUDERS, B. D., MANGIONE, K., VINCENT, C., SCHERMERHORN, J., FARCHIONE, C. M. *et al.* 2004. Distribution of *Listeria monocytogenes* molecular subtypes among human and food isolates from New York State shows persistence of human disease-associated *Listeria monocytogenes* strains in retail environments. *J Food Prot*, 67, 1417–28.

SAUDERS, B. D., SANCHEZ, M. D., RICE, D. H., CORBY, J., STICH, S. *et al.* 2009. Prevalence and molecular diversity of *Listeria monocytogenes* in retail establishments. *J Food Prot*, 72, 2337–49.

SCALLAN, E., HOEKSTRA, R. M., ANGULO, F. J., TAUXE, R. V., WIDDOWSON, M. A. *et al.* 2011. Foodborne illness acquired in the United States – major pathogens. *Emerg Infect Dis*, 17, 7–15.

SCHLECH, W. F., LAVIGNE, P. M., BORTOLUSSI, R. A., ALLEN, A. C., HALDANE, E. V. *et al.* 1983. Epidemic listeriosis – evidence for transmission by food. *N Engl J Med*, 308, 203–6.

SCHWAIGER, K., STIERSTORFER, B., SCHMAHL, W., LEHMANN, S., GALLIEN, P. *et al.* 2005. The incidence of bacterial CNS infections in roe deer (*Capreolus capreolus*), red deer (*Cervus elaphus*) and chamois (*Rupicapra rupicapra*) in Bavaria. *Berl Munch Tierarztl Wochenschr*, 118, 45–51.

SHANK, F. R., ELLIOT, E. L., WACHSMUTH, I. K. and LOSIKOFF, M. E. 1996. US position on *Listeria monocytogenes* in foods. *Food Control*, 7, 229–234.

SHEN, Y., NAUJOKAS, M., PARK, M. and IRETON, K. 2000. InlB-dependent internalization of *Listeria* is mediated by the Met receptor tyrosine kinase. *Cell*, 103, 501–10.

SMITH, G. A., MARQUIS, H., JONES, S., JOHNSTON, N. C., PORTNOY, D. A. *et al.* 1995. The two distinct phospholipases C of *Listeria monocytogenes* have overlapping roles in escape from a vacuole and cell-to-cell spread. *Infect Immun*, 63, 4231–7.

SMITH, M. A., TAKEUCHI, K., BRACKETT, R. E., MCCLURE, H. M., RAYBOURNE, R. B. *et al.* 2003. Nonhuman primate model for *Listeria monocytogenes*-induced stillbirths. *Infect Immun*, 71, 1574–9.

SUN, A. N., CAMILLI, A. and PORTNOY, D. A. 1990. Isolation of *Listeria monocytogenes* small-plaque mutants defective for intracellular growth and cell-to-cell spread. *Infect Immun*, 58, 3770–8.

SWAMINATHAN, B. and GERNER-SMIDT, P. 2007. The epidemiology of human listeriosis. *Microbes Infect*, 9, 1236–43.

THAM, W., BANNERMAN, E., BILLE, J., DANIELSSON-THAM, M. L., ELD, K. *et al.* 1999. *Listeria monocytogenes* subtypes associated with mortality among fallow deer (*Dama dama*). *J Zoo Wildl Med*, 30, 545–9.

THIMOTHE, J., NIGHTINGALE, K. K., GALL, K., SCOTT, V. N. and WIEDMANN, M. 2004. Tracking of *Listeria monocytogenes* in smoked fish processing plants. *J Food Prot*, 67, 328–41.

THIMOTHE, J., WALKER, J., SUVANICH, V., GALL, K. L., MOODY, M. W. *et al.* 2002. Detection of *Listeria* in crawfish processing plants and in raw, whole crawfish and processed crawfish (*Procambarus* spp.). *J Food Prot*, 65, 1735–9.

TILNEY, L. G. and PORTNOY, D. A. 1989. Actin filaments and the growth, movement, and spread of the intracellular bacterial parasite, *Listeria monocytogenes*. *J Cell Biol*, 109, 1597–608.

TSAI, Y. H., MARON, S. B., MCGANN, P., NIGHTINGALE, K. K., WIEDMANN, M. *et al.* 2011. Recombination and positive selection contributed to the evolution of *Listeria monocytogenes* lineages III and IV, two distinct and well supported uncommon *L. monocytogenes* lineages. *Infect Genet Evol*, 11(8), 1881–90.

USDA/FSIS 2011. *The FSIS Microbiological Testing Program for Ready-to-Eat (RTE) Meat and Poultry Products, 1990–2011* [Online]. Available: www.fsis.usda.gov/Science/Micro_Testing_RTE/index.asp.

VAN DER VEEN, S., WAGENDORP, A., ABEE, T. and WELLS-BENNIK, M. H. 2009. Diversity assessment of heat resistance of *Listeria monocytogenes* strains in a continuous-flow heating system. *J Food Prot*, 72, 999–1004.

VAN LANGENDONCK, N., BOTTREAU, E., BAILLY, S., TABOURET, M., MARLY, J. *et al.* 1998. Tissue culture assays using Caco-2 cell line differentiate virulent from non-virulent *Listeria monocytogenes* strains. *J Appl Microbiol*, 85, 337–46.

VAN STELTEN, A. and NIGHTINGALE, K. K. 2008. Development and implementation of a multiplex single-nucleotide polymorphism genotyping assay for detection of virulence-attenuating mutations in the *Listeria monocytogenes* virulence-associated gene *inlA*. *Appl Environ Microbiol*, 74, 7365–75.

VAN STELTEN, A., SIMPSON, J. M., CHEN, Y., SCOTT, V. N., WHITING, R. C. *et al.* 2011. Significant shift in median guinea pig infectious dose shown by an outbreak-associated *Listeria monocytogenes* epidemic clone strain and a strain carrying a premature stop codon mutation in *inlA*. *Appl Environ Microbiol*, 77, 2479–87.

VAN STELTEN, A., SIMPSON, J. M., WARD, T. J. and NIGHTINGALE, K. K. 2010. Revelation by single-nucleotide polymorphism genotyping that mutations leading to a premature stop codon in *inlA* are common among *Listeria monocytogenes* isolates from ready-to-eat foods but not human listeriosis cases. *Appl Environ Microbiol*, 76, 2783–90.

VARMA, J. K., SAMUEL, M. C., MARCUS, R., HOEKSTRA, R. M., MEDUS, C. *et al.* 2007. *Listeria monocytogenes* infection from foods prepared in a commercial establishment: A case-control study of potential sources of sporadic illness in the United States. *Clin Infect Dis*, 44, 521–8.

VÁZQUEZ-BOLAND, J. A., DOMÍNGUEZ-BERNAL, G., GONZÁLEZ-ZORN, B., KREFT, J. and GOEBEL, W. 2001. Pathogenicity islands and virulence evolution in *Listeria*. *Microbes Infect*, 3, 571–84.

VERGHESE, B., LOK, M., WEN, J., ALESSANDRIA, V., CHEN, Y. *et al.* 2011. *comK* prophage junction fragments as markers for *Listeria monocytogenes* genotypes unique to individual meat and poultry processing plants and a model for rapid niche-specific adaptation, biofilm formation, and persistence. *Appl Environ Microbiol*, 77, 3279–92.

VOLOKHOV, D. V., DUPERRIER, S., NEVEROV, A. A., GEORGE, J., BUCHRIESER, C. *et al.* 2007. The presence of the internalin gene in natural atypically hemolytic *Listeria innocua* strains suggests descent from *L. monocytogenes*. *Appl Environ Microbiol*, 73, 1928–39.

WARD, T. J., EVANS, P., WIEDMANN, M., USGAARD, T., ROOF, S. E. *et al.* 2010. Molecular and phenotypic characterization of *Listeria monocytogenes* from US Department of Agriculture Food Safety and Inspection Service surveillance of ready-to-eat foods and processing facilities. *J Food Prot*, 73, 861–9.

WARD, T. J., DUCEY, T. F., USGAARD, T., DUNN, K. A. and BIELAWSKI, J. P. 2008. Multilocus genotyping assays for single nucleotide polymorphism-based subtyping of *Listeria monocytogenes* isolates. *Appl Environ Microbiol*, 74, 7629–42.

WEIS, J. and SEELIGER, H. P. 1975. Incidence of *Listeria monocytogenes* in nature. *Appl Microbiol*, 30, 29–32.

WELSHIMER, H. J. and DONKER-VOET, J. 1971. *Listeria monocytogenes* in nature. *Appl Microbiol*, 21, 516–9.

WIEDMANN, M., BRUCE, J. L., KEATING, C., JOHNSON, A. E., MCDONOUGH, P. L. *et al.* 1997. Ribotypes and virulence gene polymorphisms suggest three distinct *Listeria monocytogenes* lineages with differences in pathogenic potential. *Infect Immun*, 65, 2707–16.

WILLIAMS, D., IRVIN, E. A., CHMIELEWSKI, R. A. and FRANK, J. F. 2007. Dose-response of *Listeria monocytogenes* after oral exposure in pregnant guinea pigs. *J Food Prot*, 70, 1122–8.

WILLIAMS, S. K., ROOF, S., BOYLE, E. A., BURSON, D., THIPPAREDDI, H. *et al.* 2011. Molecular ecology of *Listeria monocytogenes* and other *Listeria* species in small and very small ready-to-eat meat processing plants. *J Food Prot*, 74, 63–77.

WINTER, C. H., BROCKMANN, S. O., SONNENTAG, S. R., SCHAUPP, T., PRAGER, R. *et al.* 2009. Prolonged hospital and community-based listeriosis outbreak caused by ready-to-eat scalded sausages. *J Hosp Infect*, 73, 121–8.

WOLLERT, T., PASCHE, B., ROCHON, M., DEPPENMEIER, S., VAN DEN HEUVEL, J. *et al.* 2007. Extending the host range of *Listeria monocytogenes* by rational protein design. *Cell*, 129, 891–902.

4

Pathogen update: *Bacillus* species

F. Carlin, INRA, France and C. Nguyen-The, INRA, France

DOI: 10.1533/9780857098740.2.70

Abstract: The aerobic spore-forming bacteria of the *Bacillus* genus have been identified in many environmental niches, including food production. Their ubiquity and the resistance of their endospores to the various operations used in food processing contribute to their presence in foods, while their ability to adapt to a wide range of temperatures, pH and nutrient sources promotes their multiplication. Several *Bacillus* species, mainly *Bacillus cereus*, have been implicated in foodborne gastroenteritis. New virulence factors have been identified and several phylogenetic groups of *B. cereus sensu lato* have been defined. The control of *Bacillus* spp. requires the inactivation of spores or the inhibition of cell growth using a variety of techniques.

Key words: *Bacillus cereus*, spore, virulence, food, taxonomy.

4.1 Introduction

Aerobic spore-forming bacteria can be found in a wide range of environmental niches such as food production. These include bacteria belonging to the *Bacillus* genus and the genera defined over the last two decades, such as *Geobacillus*, *Alicyclobacillus* and *Paenibacillus* (Logan and De Vos, 2009). This ubiquity, combined with the resistance of their endospores to the physical and chemical treatments implemented in food-processing operations, means that they are a common source of contamination in food commodities. The ability of such bacteria to adapt to differences in temperature, pH and nutrient sources promotes their multiplication in foods and their ability to cause food spoilage. A few of these species are known to be pathogenic to animals. In particular, several *Bacillus* species, particularly *B. cereus* and other species such as *B. licheniformis*, *B. subtilis* and *B. pumilus*, have been implicated in foodborne gastroenteritis (Kramer and Gilbert, 1989; Granum and Baird-Parker, 2000; EFSA, 2005). This chapter focuses on recent scientific developments that have increased our understanding of the biology of the foodborne pathogenic *B. cereus* and other *Bacillus* species.

4.2 *Bacillus cereus* in food: characterization and taxonomy

4.2.1 Ecological niches, routes of food contamination, prevalence in foods, tools for detection and characterization

Soil is regarded as a major habitat for spore-forming bacteria; however, the spores can be dispersed from soil to colonize very diverse environmental niches. *B. cereus* and *B. subtilis* spores are common in soil, as well as in the gastro-intestinal tracts of invertebrates and vertebrates including mammals, and bacterial spores may contaminate the food production chain from all of these sources (Jensen *et al.*, 2003; Earl *et al.*, 2008; Swiecicka, 2008). Using a random amplification of polymorphic DNA (RAPD) derived method for typing *B. cereus*, the strains present in the soil used for courgette cultivation were also detected in a processed courgette purée after prolonged storage, demonstrating a transfer from soil to courgette and the persistence of the strains during processing operations (Guinebretiere *et al.*, 2003; Guinebretiere and Nguyen-The, 2003). In a dairy farm, spore concentrations in milk were correlated to spore concentrations in cow faeces, which were also correlated to the concentration of *B. cereus* in the cows' feed (Magnusson *et al.*, 2007).

Some dispersion routes are the result of specific properties of the individual bacterial species. *B. thuringiensis* contains genes for an insecticidal toxin, but is otherwise genetically undistinguishable from *B. cereus*: both species contain the enterotoxin genes. The presence of *B. cereus* in some horticultural crops could therefore be a consequence of the spreading of *B. thuringiensis*, which is widely applied because of the insecticidal properties of its toxic crystal protein (Frederiksen *et al.*, 2006). Similarly, spores of *B. cereus*, *B. licheniformis* and *B. subtilis* strains are administered because of their probiotics properties in human, veterinary and aquaculture applications, and the bacteria may then spread after ingestion (Cutting, 2011). The prevalence of *B. cereus* in foods is well documented because it is known as a foodborne pathogen and the efficacy of selective media and confirmation methods, which persist despite the development of chromogenic media (Peng *et al.*, 2001; Reissbrodt *et al.*, 2004; AFNOR, 2005, 2006; Fricker *et al.*, 2008; Williams *et al.*, 2009).

B. cereus has been detected in almost every type of food. The concentration before storage is usually lower than 100 *B. cereus* cfu/g, but herbs and spices may sometimes be contaminated with concentrations higher than 10^3 cfu/g (EFSA, 2005). Surveys of the prevalence of *B. cereus* in foods increasingly include indications of the isolates' ability to grow at low temperatures, which is pertinent if the contaminated foodstuff is to form part of a refrigerated food. These surveys also contain information about the virulence profiles of isolated strains, supported by the detection of the main toxins and/or of the presence of their genes (Table 4.1).

The detailed characterization of the *B. cereus* strains isolated from foods has been enhanced in recent years thanks to the development of several biochemical and molecular tools. The main advances in the field are:

- Toxin genes have been sequenced and primers have been designed for the main enterotoxin genes *nhe*, *hbl*, and *cytK*. These complement the immunological

Table 4.1 Selected surveys associating *B. cereus* prevalence and/or counts in foods with a phenotypic characterization of *B. cereus* in relation to its growth behaviour and its ability to form toxins

Food	Number of samples	*B. cereus* prevalence and/or concentrations	Behaviour of isolates at low temperature	Toxin and/or toxin genes	Source
Cooked dishes, fresh vegetables, fresh meat	575	374 positive samples[a]	2.6% growing at $\leq 7\,°C$ and 87.9% growing at 10 °C, $n = 380$	0% harbouring the *ces* gene; 52.5% harbouring *hblA, hb D, hblC, nheA nheB, nheC* and *cytK;* $n = 80$	Samapundo *et al.*, 2011
Oil(s) and fat(s), fish, meat, milk, vegetable(s) and their products, flavourings, ready-to-eat foods, pastry	33 787	0.24% > 10^5 *B. cereus* cfu/g	4.4% of psychrotrophic strains, $n = 796$[b]	97%, 66% and 50% harbouring the genes for NHE, HBL and CytK; 8.2% positive for cereulide-like production	Wijnands *et al.*, 2006

Notes:
[a] At least one presumptive *B. cereus* in 25 g.
[b] Based on the detection 16S ribosomal DNA signature.

detection tests for toxins haemolysin BL (Hbl) and non-haemolytic enterotoxin (Nhe) (Stenfors Arnesen *et al.*, 2008).

- The sequence of the *ces* gene, implicated in the synthesis of the cereulide that triggers the *B. cereus* emetic syndrome, has been identified (Ehling-Schulz *et al.*, 2005b). A liquid chromatography–mass spectrometry (LC-MS) assay has been designed for the detection and quantification of the cereulide. Both methods for characterizing emetic *B. cereus* are more specific than the biological tests that were previously used (Jääskelainen *et al.*, 2003). Some studies have also used a prototype enzyme-linked immunosorbent assay (ELISA) to detect and quantify the cereulide (Oh *et al.*, 2012).
- The sequence of the *panC* gene allows a *B. cereus* strain to affiliate rapidly to one of the phylogenetic groups determined by Guinebretiere *et al.* (2008, 2010), which correspond to temperature growth domains and virulence potentials. This has been used, for instance, to characterize *B. cereus* surimi isolates (Coton *et al.*, 2011).

Psychrotrophic strains are quite common while emetic strains are generally rare in environmental and food samples, representing less than 4% of the isolates in each of two panels in which several thousand strains were tested (Svensson *et al.*, 2006; Hoton *et al.*, 2009). The presence of *B. subtilis*, *B. licheniformis* and *B. pumilus* strains is much less widely documented. An analysis of the prevalence, concentrations or percentage proportion of these strains compared to those of *B. cereus* in a collection of strains isolated from foods suggests that they are also common in foods at concentrations similar to that of *B. cereus* (Sutherland and Murdoch, 1994; te Giffel *et al.*, 1996; Cosentino *et al.*, 1997; Ostensvik *et al.*, 2004). They are also particularly abundant, along with *B. cereus*, in indigenous fermented foods made using spontaneous fermentation (Ouoba *et al.*, 2008).

4.2.2 Taxonomy

Taxonomy of Bacillus cereus

Bacillus cereus is part of the genus *Bacillus*, in the Bacillaceae family (Logan and De Vos, 2009). It is a Gram-positive, facultatively anaerobic bacterium, which usually occurs in the form of motile rods of size 1.0–1.2 by 3.0–5.0 μm, and forms ellipsoidal, subterminal, spores which do not swell the sporangia (Logan and De Vos, 2009). Characteristics that distinguish *B. cereus* from other members of the *Bacillus* species can be found in Logan and De Vos (2009). *B. cereus* is phenotypically similar to *B. anthracis*, *B. mycoides*, *B. thuringiensis*, *B. weihenstephanensis* and *B. pseudomycoides*. These species are genetically very close and form the *Bacillus cereus* group, also known as *Bacillus cereus sensu lato* (Guinebretiere *et al.*, 2008). *B. anthracis* differs from *B. cereus* in that the former is non-haemolytic; it also differs from the virulent strains in terms of the presence of the virulence genes carried by the two plasmids pXO1 and pXO2 (Logan and De Vos, 2009). Virulent strains of *B. anthracis* cause anthrax in warm-blooded animals. *B. thuringiensis* differs from other members of the *B. cereus* group as it contains a parasporal crystal consisting of insecticidal proteins, while *B. mycoides* and *B. pseudomycoides* are characterized by the formation of rhizoid colonies (Logan and De Vos, 2009). *B. weihenstephanensis* can be distinguished by its ability to grow at 7 °C, its inability to grow at 43 °C and the presence of certain DNA signatures (Lechner *et al.*, 1998). However, some *B. cereus* strains other than *B. weihenstephanensis* are able to grow at 7 °C (Stenfors and Granum, 2001; Guinebretiere *et al.*, 2008).

Sequencing of a strain of *B. cereus*, NVH 391-98, isolated from a severe outbreak (Lund *et al.* 2000), revealed that the level of identity in the protein-coding regions between this strain and the type strain of *B. cereus* was surprisingly low for two strains of the same species (Lapidus *et al.*, 2008). It has therefore been proposed that strain NVH 391–98 represents a new species named *B. cytotoxicus* (Lapidus *et al.*, 2008; Guinebretiere *et al.*, 2012), belonging to the *B. cereus* group, and phenotypically distinguishable by a particularly high-temperature growth domain (Guinebretiere *et al.*, 2008). Procedures to enumerate *B. cereus* in foods or during investigations of outbreaks may not allow a clear distinction to be drawn between *B. cereus*, *B. thuringiensis* and *B. weihenstephanensis*, or the newly proposed *B. cytotoxicus*. For instance, according to the EN-ISO standards 7932 and 21871 (AFNOR, 2005, 2006) colonies on the counting media are identified as *B. cereus* on the basis of the absence of D-mannitol fermentation and the production of lecithinase, and are confirmed by the haemolysis on sheep blood agar.

Phylogenetic analysis has shown that the species currently categorized as the *B. cereus* group are not true genomic species (Helgason *et al.*, 2004; Hill *et al.*, 2004; Priest *et al.*, 2004; Guinebretiere *et al.*, 2008). Guinebretière *et al.* (2008) defined seven phylogenetic groups (I to VII), which provide a robust description of the genetic structure of the whole *B. cereus* group. These have been used by Tourasse *et al.* (2010, 2011) to form the structure of the HyperCat, a wide database of *B. cereus* strains (http://mlstoslo.uio.no/cgi-bin/mlstdb/mlstdbnet4.pl?dbase=hyperdb&page=hyperindex&file=bcereusgrp_isolates.xml). A procedure to assign a *B. cereus* strain to one of the seven phylogenetic groups using the sequence of the *panC* gene is described in Guinebretière *et al.* (2010) and is available at https://www.tools.symprevius.org/Bcereus/. The correspondence between the seven phylogenetic groups and the current species is given in Table 4.2.

With respect to food safety, these seven groups have different growth temperature domains; groups III and IV contain the majority of the strains involved in foodborne poisoning incidents (Table 4.2). In particular, it is worth noting that *B. thuringiensis* strains are in the same phylogenetic clusters as *B. cereus* strains implicated in foodborne outbreaks. All strains that have been isolated from emetic syndrome events to date belong to phylogenetic group III (Guinebretiere *et al.*, 2008, 2010). *B. anthracis* appears as a lineage included in phylogenetic group III. Only a small number of strains from group VII are present in the database (Table 4.2). Since it was published, additional strains have been isolated from foodborne outbreaks, resulting in a large proportion of the known

Table 4.2 Phylogenetic groups defined by Guinebretière *et al.* (2008) within *Bacillus cereus sensu lato*

Phylogenetic group	Current species	Number of foodborne poisoning isolates/ total strain number	Growth temperature domain[c]	Resistance to moist heat[f]
I	*B. pseudomycoides*	0/28[d]–0/45[e]	M	?
II	*B. cereus* II *B. thuringiensis* II	7/33–9/123	P	++
III	*B. cereus* III *B. thuringiensis* III *B. anthracis*	29/96–62/520	M	+++
IV	*B. cereus* IV *B. thuringiensis* IV	22/101–40/816	M	++
V	*B. cereus* V *B. thuringiensis* V	2/17–2/207	MP	++
VI	*B. weihenstephanensis* *B. mycoides* *B. thuringiensis* VI	0/143–1/464	P	+
VII	*B. cereus* VII[a] or *B. cytotoxicus*[b]	1/2–2/4	MT	+++

[a] No strains from phylogenetic group VII producing the parasporal crystal of *B. thuringiensis* have been described so far.
[b] *B. cytotoxicus* has been proposed as a new species corresponding to phylogenetic group VII (Lapidus *et al.*, 2008, Guinebretière *et al.*, 2012).
[c] M: mesophilic, P: psychrotolerant, MP: moderately psychrotolerant, MT: moderately thermotolerant.
[d] Data from Guinebretière *et al.* (2008).
[e] Data from the HyperCat database (Tourasse *et al.*, 2010) consulted on 5 January 2012, available at http://mlstoslo.uio.no/cgi-bin/mlstdb/mlstdbnet4.pl?dbase=hyperdb&page=hyperindex&file=bcereus grp_isolates.xml.
[f] Ranking from Afchain *et al.* (2008) and Carlin *et al.* (2006).

strains from group VII being associated with foodborne diseases (Guinebretiere *et al.*, 2012).

Taxonomy of other Bacillus *spp. implicated in outbreaks of foodborne poisoning*

The episodes of foodborne illness linked to *Bacillus* species other than *B. cereus* that were described at the end of the 1980s were attributed to *B. subtilis*, *B. licheniformis* and to a lesser extent *B. pumilus* (Kramer and Gilbert, 1989). More recent foodborne outbreaks have also been attributed to *B. licheniformis* (Salkinoja-Salonen *et al.*, 1999), *B. subtilis* (From *et al.*, 2005; Apetroaie-Constantin *et al.*, 2009) and *B. pumilus* (From *et al.*, 2007a). In addition to these three species, *B. mojavensis* has also been associated with one instance of a foodborne outbreak (Apetroaie-Constantin *et al.*, 2009), but this species was defined on the basis of *B. subtilis*-like strains isolated from soil (Roberts *et al.*, 1994). All these episodes of foodborne illness have therefore been associated with species that are phylogenetically close to *B. subtilis* (Logan and De Vos, 2009). No other species or subspecies defined from, or close to, *B. subtilis* have been reported as a cause of foodborne illness; these include *B. amyloliquefaciens* (Priest *et al.*, 1987), *B. atrophaeus* (Nakamura, 1989), *B. vallismortis* (Roberts *et al.*, 1996), *B. subtilis* subsp. *spizizenii* (Nakamura *et al.*, 1999) and *B. subtilis* subsp. *subtilis*. However, all these species differ by few or no phenotypic characters, and have very high similarities in their 16S rRNA sequences; some of the *B. subtilis* isolates associated with past foodborne illness events may therefore have been misidentified. The sequence of the gyrase A (Chun and Bae, 2000) or gyrase B genes (Wang *et al.*, 2007) may also be required, in addition to standard phenotypic features and 16S rRNA sequences, in order to distinguish between these species. For instance, gyrase A sequences were used by From *et al.* (2005) and Apetroaie-Constantin *et al.* (2009) to identify isolates associated with foodborne illness, but earlier studies did not use this technique.

4.3 Poisoning caused by *B. cereus* and other *Bacillus* spp.

4.3.1 *Bacillus cereus*

Diarrhoeic and emetic syndromes

B. cereus causes two types of foodborne diseases: the emetic (vomiting) type is caused by a heat-stable toxin called cereulide, while the diarrhoeic type is caused by several heat-labile proteins with enterotoxic activities (EFSA, 2005). Emetic syndrome usually occurs 0.5 to 5 h after ingestion of the contaminated food, and is characterised by vomiting followed in some cases by diarrhoea after 8 to 16 h. Diarrhoeic illness usually occurs 8 to 16 h after ingestion of the contaminated food and is characterised by watery diarrhoea and abdominal pain. Both types of diseases are usually self-limiting, although some fatal or severe cases have been reported (EFSA, 2005). The short incubation time of the emetic poisoning is compatible with the ingestion of a toxin produced in the food (intoxication), whereas the longer period before symptoms of diarrhoeic disease develop is consistent with the production of enterotoxins during the development of the bacteria in the intestine (toxico-infection) (Granum and Baird-Parker, 2000; Stenfors Arnesen *et al.*, 2008).

Recent reports of significant outbreaks

B. cereus outbreaks have been linked to a wide variety of foods, mostly heat-treated foods such as stews, fried rice and pasta dishes. However, some rare occurrences of outbreaks caused by fresh produce (sprouted seeds and orange juice) have been reported, demonstrating the ubiquity of this bacterium (EFSA, 2005). Detailed descriptions of *B. cereus* outbreaks can be found in a number of studies, including Kramer and Gilbert (1989), Granum and Baird-Parker (2000) and in EFSA (2005). Typical *B. cereus* emetic intoxication occurs after ingestion of a cooked starchy food (rice or pasta) left at room temperature for some hours before consumption; this allows the spores that have survived cooking to multiply, leading to production of the emetic toxin (cereulide) in the food. As cereulide is heat stable, intoxication has been found to occur even if the food receives a second heat treatment before consumption, e.g. fried rice that was boiled in advance and left at ambient temperature before frying in restaurants (Kramer and Gilbert, 1989; EFSA, 2005; Stenfors Arnesen *et al.*, 2008). Diarrhoeic diseases are caused by a wider variety of factors, but are usually linked to a cooked food left unrefrigerated before consumption (EFSA, 2005). Globally, *Bacillus* spp. outbreaks, most of which are caused by *B. cereus*, increased by 42% in 2007 and by 18% in 2008 in the EU (EFSA, 2009, 2010). It is not known whether this is the result of improved investigation and reporting or from a true increase in the number of outbreaks.

The increase in the number of fatal or very severe emetic outbreaks since 2000 is worthy of note (Dierick *et al.*, 2005; Posfay-Barbe *et al.*, 2008; Ichikawa *et al.*, 2010; Shiota *et al.*, 2010; Naranjo *et al.*, 2011). The cause of death or illness was attributed to liver failure (Dierick *et al.*, 2005; Posfay-Barbe *et al.*, 2008) as previously described (Mahler *et al.*, 1997), but also to acute encephalopathy (Ichikawa *et al.*, 2010) and presumably acute cardiac insufficiency (Naranjo *et al.*, 2011). The most significant diarrhoeic *B. cereus* outbreak caused the death of three people out of a group of six who had bloody diarrhoea (Lund *et al.*, 2000). This was the first fatal *B. cereus* diarrhoeic disease, and was caused by a very unusual strain (NVH 391–98), which did not seem to produce the known enterotoxins. A potent cytotoxin, named CytK, was purified from the culture supernatant of this strain, leading to the identification of a new virulence factor of *B. cereus* (Lund *et al.*, 2000). NVH 391-98 was the first strain ever characterized for phylogenetic group VII (Table 4.2).

Virulence factors

The virulence factors involved in the gastro-intestinal infections caused by *B. cereus* have recently been presented in detail in Stenfors Arnesen *et al.* (2008). It is currently accepted that the diarrhoeic disease is caused by three toxins, Hbl, Nhe and CytK, all three of which are active on intestinal epithelial cell cultures. The enterotoxic activity of purified Hbl was verified in the ileal loop in rabbits (Beecher *et al.*, 1995). Hbl and Nhe both consist of three proteins (B, L_1, L_2 for Hbl and A, B, C for Nhe) whose genes are arranged in operons, whereas CytK consists of a single protein. Two forms of CytK have been described (CytK1 and CytK2). CytK1 is more toxic to cell cultures (Fagerlund *et al.*, 2004) and so far only CytK1 has been found in phylogenetic group VII (Guinebretiere *et al.*, 2010). Group VII has caused fatal cases of diarrhoeic disease (Lund *et al.*, 2000). Nhe was first described in a strain involved in a severe diarrhoeic outbreak that did not produce Hbl or CytK (Lund and Granum, 1996; Ehling-Schulz *et al.*, 2005a; Stenfors Arnesen *et al.*, 2008; Guinebretiere *et al.*, 2010). Nhe is presumably present in all *B. cereus* strains, unlike Hbl and CytK (Guinebretière *et al.*, 2002, 2010; Moravek *et al.*, 2006), and the amount of Nhe produced is positively correlated with the global cytotoxic activity of the culture supernatants of the strains (Moravek *et al.*, 2006). Nhe can therefore be considered to be the major virulence factor of *B. cereus* for diarrhoeic disease. Hbl and CytK certainly play an important role in virulence for some strains: for Hbl a deletion mutant abolished most cytotoxic activity (Lindback *et al.*, 1999), and CytK (form CytK1) was co-purified with the cytotoxic activity of strain NVH 391–98 (Lund *et al.*, 2000).

CytK is a ß-barrel toxin that assembles to form pores in the membrane of the target cells (Lund *et al.*, 2000). All three components of Hbl are necessary for cell lysis and can act after the binding of any of the components to the target cell surface (Beecher and Wong, 1997). In contrast, Nhe acts only after the NheB component has bound to the cell surface, with an adequate ratio of NheB/NheC components (Lindback *et al.*, 2004). In the case of Hbl, the secretion of a fourth component (HblB', coded by a gene located downstream of the operon) was observed (Clair *et al.*, 2010), but its role has not been identified.

The existence of other putative enterotoxins, enterotoxin T (BceT) and enterotoxin FM (entFM), has previously been proposed. Although their role is unconfirmed biologically, they have often been used as virulence markers for strain characterization. In 2003 it was shown that BceT was the result of a cloning artefact (Hansen *et al.*, 2003) and should no longer be considered as a virulence factor. EntFM has recently been identified as an endopeptidase (Tran *et al.*, 2010), which did not show direct toxic activity on epithelial cells, although it may contribute to virulence through its role in adhesion to epithelial cells. *B. cereus* also produces a toxin very similar to CytK, haemolysin II (HlyII), which has cytotoxic properties (Andreeva *et al.*, 2006). In addition, HlyII was found more frequently among strains involved in foodborne diseases than in strains that are unrelated to human diseases (Cadot *et al.*, 2010). However, HlyII induced apoptosis in macrophages (Tran *et al.*, 2011) and may also be involved in the non-food-related clinical infections caused by *B. cereus* (Bottone, 2010). Several metalloproteases (InhA1, InhA2 and InhA3) are produced by *B. cereus*, and seem to play important roles in interactions between the bacterium and the host immune system (Guillemet *et al.*, 2010). They may be more significant in invasive clinical infections than in foodborne toxico-infections, which are restricted to the intestinal environment.

As mentioned above, the emetic disease is caused by a heat-stable cyclic peptide called cereulide (Agata *et al.*, 1994, 1995), which is resistant to temperatures as high as 121 °C for 15 min (Shinagawa *et al.*, 1995; Rajkovic *et al.*, 2008). No other virulence factor contributing to the emetic syndrome of *B. cereus* has been found. In particular, no other toxic peptides, such as the lipopeptides produced by *B. subtilis* (see the relevant section of this chapter), have been identified. In addition to causing emesis in a primate model, cereulide has also been demonstrated to be toxic in various mammalian cell lines (Andersson *et al.*, 2007). Purified cereulide injected into mice was toxic (Shinagawa *et al.*, 1996) and caused degeneration of hepatocytes in mice, although this degeneration was reversible (Yokoyama *et al.*, 1999). Cereulide is produced non-ribosomically by a peptide synthase, the genetic determinants of which have been identified (Yokoyama *et al.*, 1999; Ehling-Schulz *et al.*, 2004, 2005b) and are located on a 200 kd plasmid (Hoton *et al.*, 2005; Ehling-Schulz *et al.*, 2006).

The virulence of *B. cereus* with respect to foodborne diseases varies significantly between different strains. Some strains have been used as probiotics (Lestradet, 1995; EFSA, 2007; Williams *et al.*, 2009) whereas others have caused fatal or severe foodborne outbreaks. With respect to the emetic disease, the gene for cereulide peptide synthase (*ces*) is restricted to a small number of clusters of *B. cereus* (Ehling-Schulz *et al.*, 2005a). Most cereulide-producing strains are distributed in two sub-groups of phylogenetic group III (Guinebretiere *et al.*, 2010); however, five *B. cereus* strains producing cereulide and belonging to phylogenetic group VI have recently been identified (Hoton *et al.*, 2009; Thorsen *et al.*, 2009a), but these were not connected to emetic diseases, which have to date all been caused by cereulide-producing strains from phylogenetic group III (Guinebretiere *et al.*, 2008, 2010) (Table 4.2). In contrast to the *ces* gene, the genes of the diarrhoeic toxins are widespread among *B. cereus* strains (Ehling-Schulz *et al.*, 2005a; Moravek *et al.*, 2006; Guinebretiere *et al.*, 2010). Cereulide-producing strains may also possess enterotoxin genes. Due to the sequence diversity of *hbl* and *nhe* operons (Stenfors Arnesen *et al.*, 2008), polymerase chain reaction (PCR) amplification may underestimate their prevalence and must be completed by Southern hybridization for PCR negative strains (Guinebretière *et al.*, 2002). However, the production of diarrhoeic toxins and global enterotoxic activity differs significantly among *B. cereus* strains (Guinebretière *et al.*, 2002), and it may be assumed that not all strains have the same potential to cause diarrhoea. This 'diarrhoeic potential' seems to differ among phylogenetic groups (Guinebretiere *et al.*, 2010), with group VI having a low potential while groups IV and III have a high potential: this is consistent with the prevalence of foodborne disease strains in the groups (Table 4.2).

Dose response assessment

The number of *B. cereus* in foods implicated in diarrhoeal diseases is usually over 10^4–10^5 cfu/g, with some rare outbreaks where the food was assumed to contain only 10^3 cfu /g (EFSA, 2005; Stenfors Arnesen *et al.*, 2008). Among the factors limiting the infection of the intestine with ingested *B. cereus* cells are the low pH conditions of the stomach and the inhibitory effects of bile salts (Clavel *et al.*, 2004, 2007; Wijnands *et al.*, 2006). Spores are more resistant to stomach conditions, meaning that the dose response for spores may be different than that for vegetative cells. In addition, some food matrices have been shown to have a protective effect (Clavel *et al.*, 2004, 2007), which may also influence the dose response.

For cereulide, 8–10 µg/kg of body weight is sufficient to cause emesis, according to the results of experiments on animal models (Shinagawa *et al.*, 1995) and analysis of foods linked to emetic outbreaks (Jääskelainen *et al.*, 2003). Cereulide is produced in the stationary phase of growth, when the culture reaches at least 10^5 cells per ml or g of substrate (Haggblom *et al.*, 2002). However, as cereulide is resistant to most treatments applied to foods, it may persist even if *B. cereus* cells and spores have been killed. The multiplication of *B. cereus* in foods must reach sufficient numbers, or must cause the production of cereulide in order to cause either type of disease (EFSA, 2005), which is consistent with the description of *B. cereus* foodborne outbreaks.

Factors modulating toxin production

The transcription of the three diarrhoeal toxin genes or operons, *hbl*, *nhe* and *cytK*, is dependent on, and positively regulated by, PlcR (Gohar *et al.*, 2002, 2008; Brillard and Lereclus, 2004), a pleiotropic regulator of *B. cereus* (Agaisse *et al.*, 1999). PlcR production is itself triggered by a quorum-sensing signal (Slamti and Lereclus, 2002), which may explain why enterotoxin production occurs at the end of the exponential growth phase. The low oxido-reduction potential (between −100 mV and −200 mV) values found in some parts of the human intestine causes a marked increase in the production of Nhe and Hbl (Zigha *et al.*, 2006). The production of both Hbl and Nhe was also found to be dependent on the red-ox sensor-regulators FnR and ResDE (Duport *et al.*, 2006; Zigha *et al.*, 2007; Esbelin *et al.*, 2008, 2009), and is modified by the nature of the carbon sources (Ouhib-Jacobs *et al.*, 2009); specifically, it is repressed by the presence of high glucose concentrations, which is consistent with the regulation of both operons by the catabolite control regulator CcpA (van der Voort *et al.*, 2008). The prevailing conditions in the human intestine can therefore influence the production of enterotoxins.

It has been found that cereulide production in foods ceases in the absence of oxygen (Jääskelainen *et al.*, 2004). The optimal temperature for cereulide production is around 23–25 °C (Haggblom *et al.*, 2002; Apetroaie-Constantin *et al.*, 2008), lower than the optimal temperature for growth of *B. cereus*. Cereulide production was not observed at 4–5 °C and 8 °C, or the amounts were so low that they were below the quantification limit, even for the psychrotrophic cereulide-producing strains from phylogenetic group VI (Thorsen *et al.*, 2009a; Delbrassinne *et al.*, 2011). Keeping foods refrigerated should therefore prevent any risk of emetic poisoning.

4.3.2 Other *Bacillus* species

In 1989, Kramer and Gilbert reviewed several foodborne outbreaks with symptoms of gastroenteritis where the implicated food contained a large population of *Bacillus subtilis*, *Bacillus licheniformis* or *Bacillus pumilus* and no other known foodborne pathogens (Kramer and Gilbert, 1989). A wide range of foods was implicated, mostly meat dishes but also pastries, bread, pizzas and vegetable dishes. The numbers of *Bacillus* ranged from 10^5 to 10^9 cfu/g of food, and the symptoms included vomiting, diarrhoea, nausea and abdominal pain. The incubation period was usually short, from a few minutes to 14 h. Since then, several outbreaks have confirmed the implication of *Bacillus* spp. other than *B. cereus* in foodborne illness.

Recent outbreaks

Eleven episodes of foodborne illness in the 1990s associated with *B. licheniformis* (Salkinoja-Salonen *et al.*, 1999) were characterized by nausea, abdominal pain, vomiting and diarrhoea occurring from 5 to 12 h after meal consumption. When determined, the population of *B. licheniformis* in the food ranged from 10^5 to 10^8 cfu/g. One outbreak involved 111 people with six hospitalized patients. Two deaths were reported, one caused by contaminated infant food. In 2005 *B. subtilis* was isolated from fried, marinated chicken associated with an outbreak involving several people, with nausea, vomiting, stomach pain and diarrhoea occurring 6 h after ingestion of the meal (From *et al.*, 2005). In 2007, boiled rice that was cooled to room temperature and then reheated was the most likely cause of an outbreak involving three people. Acute symptoms started during the meal, and were followed by stomach pain and diarrhoea, which lasted for two weeks. The rice contained 10^5 cfu per g of *B. pumilus* (From *et al.*, 2007a); however, although *B. pumilus* was dominant, it was not the only *Bacillus* species found in the rice. These more recent reports show that *Bacillus* species other than *B. cereus* may cause severe illnesses, with symptoms lasting for several days. The illnesses occasionally required hospitalization, and proved fatal in two cases. The delay before the onset of symptoms is short and similar to that of emetic *B. cereus* illness. It is compatible with intoxication caused by toxins preformed in the foods.

Mechanisms of foodborne poisoning and possible toxins

Some isolates of *Bacillus* species from a variety of origins, including foods associated with illness, were toxic to cells classically used to detect the activity of cereulide, the emetic toxin of *B. cereus*: boar sperm (Salkinoja-Salonen *et al.*, 1999; From *et al.*, 2005; Apetroaie-Constantin *et al.*, 2009) and Hep2 cells (Taylor *et al.*, 2005). These toxins were as heat stable as the emetic toxin of *B. cereus* but had a different physiological impact on sperm cells. These toxic isolates were a minority, making up only 1% to 6% of the tested isolates (From *et al.*, 2005; Apetroaie-Constantin *et al.*, 2009), which could be consistent with the rare occurrence of foodborne illness associated with *Bacillus* species other than *B. cereus*.

The fractionation of culture extracts from toxigenic *Bacillus* isolates using high-performance liquid chromatography, followed by the identification of compounds present in the toxic fractions using mass spectrometry, revealed several types of lipopeptides: amylosin produced by an isolate of *B. amyloliquefaciens* (Mikkola *et al.*, 2007); surfactins produced by isolates of *B. subtilis* and *B. mojavensis* (From *et al.*, 2007b); pumilacidins produced by an isolate of *B. pumilus* (From *et al.*, 2007a) and lichenysin produced by an isolate of *B. licheniformis* (Nieminen *et al.*, 2007). In the case of a toxigenic strain of *B. mojavensis*, the deletion of the locus involved in the biosynthesis of surfactin (From *et al.*, 2007b) suppressed both surfactin production and toxicity to sperm cells and vero cells.

It is very probable that the lipopeptides were the cause of the *in vitro* toxicity of the *Bacillus* isolates. When these toxigenic isolates are associated with foodborne illness (From *et al.*, 2007a, 2007b) it may also be assumed that the lipopeptides were the cause of the human disease. However, the illness could not be reproduced in animal models, nor could the *in vivo* role of lipopeptides, as was carried out for cereulide (Agata *et al.*, 1995). Pumilacidin killed Caco 2 cells, a model more relevant to gastroenteritis than the sperm cells used for its detection (From *et al.*, 2007a). Apetroaie-Constantin *et al.* (2009) also found that *B. subtilis* isolates toxic to sperm cells killed Caco 2 cells. Lipopeptides are currently the factors most likely to be responsible for gastroenteritis caused by *Bacillus* other than *B. cereus*. They are mostly found among *Bacillus* species close to *Bacillus subtilis* (Raaijmakers *et al.*, 2010), i.e. with the *Bacillus* species involved in foodborne illness. However, these foodborne illnesses are rare, and few isolates have been characterized. The production of lipopeptides and toxicity among *Bacillus* species associated with foodborne illness should be investigated in a wider set of isolates.

A further difficulty results from the wide range of different lipopeptides produced by *Bacillus* species. They consist of a peptidic ring with a fatty acid tail (Ongena and Jacques, 2008). The amino acid composition of the ring determines the type of lipopeptide (e.g. surfactin), with each type comprising several forms that differ in their fatty acid tail. The mode of action on a cell membrane differs among the various lipopeptides; their action is also different on different types of membrane (Ongena and Jacques, 2008). For instance, surfactins are haemolytic, but this is not the case for all lipopeptides from *Bacillus*. The *B. pumilus* isolate implicated in a foodborne outbreak (From *et al.*, 2007a) produced seven forms of pumilacidins with different toxic activities on sperm and Caco 2 cells.

The production of heat-stable toxins has also been observed in *Bacillus* species other than *B. subtilis*, *B. mojavensis*, *B. amyloliquefaciens*, *B. licheniformis* and *B. pumilus* (Taylor *et al.*, 2005) but the toxic factors have not been identified and the isolates belong to species that have so far not been associated with foodborne diseases. The production of heat-labile toxins that are active against cell cultures, and the production of proteins that react with immunological tests designed to detect *B. cereus* enterotoxins have been described for various *Bacillus* species others than *B. cereus* (Hoult and Tuxford, 1991; Beattie and Williams, 1999; Rowan *et al.*, 2001, 2003; Phelps and McKillip, 2002), but again, these strains have so far not been associated with foodborne diseases.

Toxin production

The structure, biological and chemical properties, and biosynthesis of some *Bacillus* lipopeptides have been extensively studied due to their very potent surfactant properties and their role in protecting plants as antifungal agents or plant defence elicitors (Ongena and Jacques, 2008; Raaijmakers *et al.*, 2010). They are produced non-ribosomically, and identification of the synthesis operon permitted the development of PCR methods to detect producer isolates for surfactin (Tapi *et al.*, 2010). Their production kinetics has also been investigated in laboratory media. For a *B. pumilus* outbreak associated with a foodborne illness, the production of pumilacidin started after 48 h, reached a maximum after 96 h and occurred from 12 °C to 32 °C, with maximal production at 15 °C (From *et al.*, 2007a). Temperature ranges for the production of amylosin by *B. mojavensis* and *B. subtilis* varied for different isolates (Apetroaie-Constantin *et al.*, 2009).

Dose response assessment

The toxins produced by *Bacillus* spp. other than *B. cereus*, which is responsible for foodborne illness, were heat stable. Therefore, the toxins could be present in a food in which the bacterial cells have been killed by a heat treatment. However, in all of these outbreaks the number of *Bacillus* in the food was high, equal to or above 10^5 cfu/g. The relation between the dose of the toxic lipopeptides and the risk of human illness is not known.

4.4 Control of *Bacillus* species in foods and food processing

4.4.1 Inactivation by heat

Heat is still the most common method for bacterial spore inactivation during food processing. The determination of heat-resistance parameters has been the subject of a great deal of research over the years. The heat resistance of *B. cereus* is highly variable, as illustrated in the work of Dufrenne *et al.* (1994) in which the decimal reduction time at 90 °C ($D_{90°C}$) of 31 strains ranged from 4.6 min to >200 min. Overall the highest heat sensitivity among *B. cereus* strains is $D_{90°C}$ values close to 1 min, while $D_{100°C}$ values close to 20 min have also been reported (EFSA, 2005). The sterilizing value, F_0, is the equivalent amount of time, in minutes at 121 °C, during which the product undergoes sterilization. In practice, a treatment of $F_0 = 3$ used to control *Clostridium botulinum* will also control *B. cereus*, but heat treatments at a lower sterilizing value will allow *B. cereus* spores to persist. This variability in heat resistance is linked to the phylogeny of the group: the *B. cereus* emetic strains are for instance more resistant on average than the other *B. cereus* strains (Carlin *et al.*, 2006). More generally there is a strong negative correlation between the heat resistance parameters and the minimal growth temperature of the strains within the *B. cereus* group, confirming at a species level previous observations among *Bacillus* spp. (Warth, 1978; Afchain *et al.*, 2008).

Some variability in the *z* value (the increase in temperature leading to a 1 log reduction in the decimal reduction time, *D*) has also been reported. The median of the *z* values reported in a series of papers was 8.3 °C, with extreme values ranging between 6.5 °C and 14 °C (Afchain *et al.*, 2008). It remains unclear whether this variability is due to the microbiological methods used for the determination of the parameter or whether it has genetic support. The sporulation environment causes some variation in the heat resistance of spores: for instance, a tenfold increase in the decimal reduction time of several *B. cereus* strains was caused by an increase in temperature of 25 °C (Leguerinel *et al.*, 2007). Some data suggest that the preparation of spores in conditions that allow optimal growth should result in the production of the most resistant spores, as has been shown for *B. weihenstephanensis* and *B. licheniformis* (Baril *et al.*, 2012a).

4.4.2 Inactivation by other physical processes

Because of the high resistance of their spores, *B. cereus* and *B. subtilis* strains are commonly used as biological indicators to test the inactivation efficiency of decontamination technologies. A high hydrostatic pressure treatment induces spore germination, and the mechanisms that operate at moderately high pressure (50 MPa to 300 MPa) may differ from those at very high pressure (400 MPa to 800 MPa). A significant log reduction is obtained only in combination with heat treatments at high temperature (60 °C to 80 °C) (Margosch *et al.*, 2004; Black *et al.*, 2007). Pathogenic spore-forming bacteria do not present any particular resistance to ionizing radiation or UV radiation: decimal reduction doses of 1.6 kGy have been reported for *B. cereus* spores, meaning that an irradiation treatment of 10 kGy, which is considered to be safe for food applications, can theoretically result in an approximately 5 log reduction of the bacterium (Farkas, 2006). A 4 log reduction was obtained with UV doses (fluences) of approximately 60 mJ/cm^2 (Coohill and Sagripanti, 2008). The use of a pulsed light treatment that delivered white light with a high percentage of UV led to a 5 log reduction in spores of *B. cereus*, *B. licheniformis*, *B. subtilis* and *B. pumilus* with fluences lower than 2 J/cm^2 (Levy *et al.*, 2012) delivered in one or two light pulses. The method of delivery (short treatments with a high peak power versus longer treatments with lower peak power) does not seem to affect the efficiency of the UV treatment (Rice and Ewell, 2001; Levy *et al.*, 2012). It should be noted the difference between the resistance of spores and that of vegetative cells that is observed in moist heat treatments is not present in this type of treatment. The *D* values of spores reported after irradiation treatment are ten times higher than those of vegetative cells. In contrast, spores require a temperature approximately 45 °C higher than vegetative cells for inactivation to occur (Warth, 1978; Coleman *et al.*, 2007). This means, assuming *z* values of 10 °C, that moist heat treatments of spores should be at least 1000-fold longer to achieve the same inactivation as that of vegetative cells, regardless of the temperature. Several log reductions of *B. cereus* spores have been obtained using pulsed electric fields (Marquez *et al.*, 1997), with excimer-UV irradiation applied to spores suspended in water or spread on packaging material (Warriner *et al.*, 2002; Wang *et al.*, 2010), and with plasma technologies. In the last of these, the differences between spores and vegetative cells are less extensive than those observed with wet heat treatments (Moisan *et al.*, 2001; Muranyi *et al.*, 2007). The relevance of these treatments under conditions that could be used in food processing has yet to be established.

4.4.3 Chemical elimination

Many *Bacillus species*, notably *B. cereus*, possess an exosporium, the outermost structure in the bacteria spore architecture. The exosporium can be regarded as a shell surrounding the spore coat. It is made of a basal layer supporting a filamentous hairy nap, and is composed of protein, lipids and carbohydrates (Kailas *et al.*, 2011; Lequette *et al.*, 2011). The exosporium, together with other physical and chemical properties of the spore, such as hydrophobicity, contributes to the strong adhering properties of the spore and its ability to form biofilms. The surfaces of the equipment used in food-processing installations must be cleaned, but spores are generally highly resistant to chemical biocides (Maillard, 2011). When *B. cereus* spores were exposed to a 500 ppm sodium hypochlorite or peroxyacetic acid solution for 10 min at 20 °C, only a 1 log reduction was obtained (Hilgren *et al.*, 2009), while the same concentrations would result in a several log reduction of vegetative cells with a much shorter exposure time. Higher log reductions of spores are obtained only with higher concentrations of biocides. The elimination of spores is therefore often limited to their removal from surfaces using detergents in heated cleaning solutions (Sundberg *et al.*, 2011). A common procedure involves the use of NaOH solutions at high temperature. Treatments with 0.5% NaOH at 60 °C or 2% NaOH at 80 °C caused damage to the structure of the exosporium, and prevented further adhesion during a second redeposition downstream in the processing line. Additionally, the treatment with 2% NaOH at 80 °C also affected the germination capacity and delayed the recovery of the treated spores (Faille *et al.*, 2010).

4.4.4 Control of growth

The control of the growth of *B. cereus* in foods has stimulated extensive research on a number of possible methods, including the application of modified atmospheres, and the use of antimicrobial compounds such as essential oils, bacteriocins including nisin, and organic acids, often combined with refrigeration temperatures. Most foods with pH and a_w values within their growth limits support the growth of *B. cereus*. *B. cereus* does not seem to be particularly pH resistant, with only a few strains able to grow at pH 4.3 (Carlin *et al.*, 2013). The minimal pH of growth varies according to the genetic group: psychrotrophic strains of group VI are the most pH sensitive, with some strains in this group unable to grow at a pH of 4.8, in contrast to all the strains in the other groups. The a_w growth limits are also subject to variation according to the genetic group. The few strains that were able to grow in a brain-heart infusion (BHI) with 10% NaCl as humectant (giving an a_w of 0.929) belonged to groups VII, II and IV, while none of the group I strains was able to grow in BHI with 6% NaCl (a_w = 0.960). In general terms, the relatively wide range of lower limits of vegetative cell growth reported for *B. cereus* by Kramer and Gilbert (1989) (upper limits at pH 4.35–4.90 and minimum a_w at 0.912–0.950) generally reflects the genetic diversity within the *B. cereus* group. These limits for the growth of *B. subtilis*, *B. licheniformis* and *B. pumilus* are much less widely documented and it is almost impossible to recommend a pH or a_w that could guarantee the safety of foods with regards to these species.

Many *B. cereus* strains are able to grow at low temperatures. The minimal growth temperature of these psychrotrophic strains is generally accepted to be 4 °C. However, this ability is restricted to some phylogenetic groups, mainly groups II and VI (EFSA, 2005; Guinebretiere *et al.*, 2008). The optimal growth temperature is between 30 °C and 37 °C, and the maximal growth temperature never exceeds 43 °C. These bacteria are naturally a concern in refrigerated foods, and growth of *B. cereus* in foods, such as milk, stored at temperatures below 10 °C has been reported (Larsen and Jorgensen, 1999). The maximal growth temperature of mesophilic strains can be higher than 50 °C (EFSA, 2005), while some *B. cereus* strains, belonging to group VII, are moderately thermophilic. *B. pumilus* and *B. subtilis* strains are also able to grow at temperatures lower than 10 °C, while *B. licheniformis* is decidedly mesophilic, with no *B. licheniformis* strain able to grow at temperatures lower than 10 °C (Logan and De Vos, 2009). Recent data suggest that the spore yield is also markedly reduced as culture pH, a_w, and incubation temperature approach the limits for growth (Baril *et al.*, 2012b).

B. cereus is a facultative anaerobe, like *B. licheniformis*, while *B. subtilis* and *B. pumilus* are strict aerobes. However, low O_2 concentration and anaerobiosis, as found, for instance, in vacuum-packed foods and foods stored under modified atmospheres, reduce *B. cereus* growth rates and maximal population density, lower toxin production by emetic strains, and extend the lag phase; this inhibitory effect is reinforced when incubation temperatures are close to the minimal growth limit (Jääskelainen *et al.*, 2004; Samapundo *et al.*, 2011; de Sarrau *et al.*, 2012). However, incubation in an anaerobic environment could also favour resistance to stress such as heat or acid (Mols *et al.*, 2009). CO_2 concentrations have to be fairly high (50% CO_2 in a modified atmosphere), or combined with low oxygen concentrations (practically at the level of anaerobiosis), in order to control *B. cereus* in a range of foods or culture media (Bennik *et al.*, 1995; Thorsen *et al.*, 2009b).

4.5 Conclusion

Extensive work has been carried out in recent years to establish the genetic structure of *B. cereus sensu lato* and its diversity, and has clearly shown that the health risks are not identical for all strains. Many molecular and biochemical tools have been developed, which have proved useful in evaluating the risk presented by *B. cereus* strains. In contrast, this risk is still unclear for other *Bacillus* spp., and in the absence of a significant number of foodborne illness events that would reveal the actual virulence potential of these bacteria, there is little chance that our understanding of these species will progress as it did for *B. cereus*. Foodborne pathogenic *Bacillus* spp. need to grow in foods in order to cause disease. Some can multiply at refrigeration temperatures, but they do not have specific resistance to other environmental conditions. The elimination of *Bacillus* spp. spores from foods requires severe heat treatment. A combination of treatments to inactivate spores, including non-thermal technologies, may provide interesting alternatives; the prevention of spore germination and growth has therefore been employed in order to lower the *Bacillus* spp. risk and ensure food safety.

4.6 References

AFCHAIN A L, CARLIN F, NGUYEN-THE C and ALBERT I (2008). Improving quantitative exposure assessment by considering genetic diversity of *B. cereus* in cooked, pasteurised and chilled foods. *Int. J. Food Microbiol.*, 128, 165–173. doi 10.1016/j.ijfoodmicro.2008.07.028.

AFNOR (2005). NF EN ISO 7932 – Méthode horizontale pour le dénombrement de *Bacillus cereus* présomptifs – Technique par comptage des colonies à 30 °C. La Plaine Saint-Denis: AFNOR.

AFNOR (2006). NF EN ISO 21871 – Microbiologie des aliments. Méthode horizontale pour le dénombrement de *Bacillus cereus* présumés en petit nombre – Technique du nombre le plus probable et méthode de recherche. La Plaine Saint-Denis: AFNOR.

AGAISSE H, GOMINET M, OKSTAD O A, KOLSTO A B and LERECLUS D (1999). PlcR is a pleiotropic regulator of extracellular virulence factor gene expression in *Bacillus thuringiensis*. *Mol. Microbiol.*, 32, 1043–1053. doi 10.1046/j.1365-2958.1999.01419.x.

AGATA N, MORI M, OHTA M, SUWAN S, OHTANI I *et al.* (1994). A novel dodecadepsipeptide, cereulide, isolated from *Bacillus cereus* causes vacuole formation in HEp-2 cells. *FEMS Microbiol. Lett.*, 121, 31–34. doi 10.1111/j.1574-6968.1994.tb07071.x.

AGATA N, OHTA M, MORI M and ISOBE M (1995). A novel dodecadepsipeptide, cereulide, is an emetic toxin of *Bacillus cereus*. *FEMS Microbiol. Lett.*, 121, 31–34. doi 10.1111/j.1574-6968.1995.tb07550.x.

ANDERSSON M A, HAKULINEN P, HONKALAMPI-HAMALAINEN U, HOORNSTRA D, LHUGUENOT J C *et al.* (2007). Toxicological profile of cereulide, the *Bacillus cereus* emetic toxin, in functional assays with human, animal and bacterial cells. *Toxicon*, 49, 351–367. doi 10.1016/j.toxicon.2006.10.006.

ANDREEVA Z I, NESTERENKO V F, YURKOV I S, BUDARINA Z I, SINEVA E V and SOLONIN A S (2006). Purification and cytotoxic properties of *Bacillus cereus* hemolysin II. *Prot. Exp. Purif.*, 47, 186–193. doi 10.1016/j.pep.2005.10.030.

APETROAIE-CONSTANTIN C, MIKKOLA R, ANDERSSON M A, TEPLOVA V, SUOMINEN I *et al.* (2009). *Bacillus subtilis* and *B. mojavensis* strains connected to food poisoning produce the heat stable toxin amylosin. *J. Appl. Microbiol.*, 106, 1976–1985. doi 10.1111/j.1365-2672.2009.04167.x.

APETROAIE-CONSTANTIN C, SHAHEEN R, ANDRUP L, SMIDT L, RITA H and SALLINOJA-SALONEN M (2008). Environment driven cereulide production by emetic strains of *Bacillus cereus*. *Int. J. Food Microbiol.*, 127, 60–67. doi 10.1016/j.ijfoodmicro.2008.06.006.

BARIL E, COROLLER L, COUVERT O, LEGUÉRINEL I, POSTOLLEC F *et al.* (2012a). Modeling heat resistance of *Bacillus weihenstephanensis* and *Bacillus licheniformis* spores as function of sporulation temperature and pH. *Food Microbiol.*, 30, 29–36. doi 10.1016/j.fm.2011.09.017.

BARIL E, COROLLER L, EL JABRI M, LEGUERINEL I, POSTOLLEC F *et al.* (2012b). Sporulation boundaries and spore formation kinetics of *Bacillus* spp. as a function of temperature, pH and *aw*. *Food Microbiol.*, 32, 79–86. doi 10.1016/j.fm.2012.04.011.

BEATTIE S H and WILLIAMS A G (1999). Detection of toxigenic strains of *Bacillus cereus* and other *Bacillus* spp. with an improved cytotoxicity assay. *Lett. Appl. Microbiol.*, 28, 221–225. doi 10.1046/j.1365-2672.1999.00498.x.

BEECHER D J and WONG A C L (1997). Tripartite hemolysin BL from *Bacillus cereus* – Hemolytic analysis of component interactions and a model for its characteristic paradoxical zone phenomenon. *J. Biol. Chem.*, 272, 233–239.

BEECHER D J, SCHOENI J L and WONG A C (1995). Enterotoxic activity of hemolysin BL from *Bacillus cereus*. *Infect. Immun.*, 63, 4423–4428.

BENNIK M H J, SMID E J, ROMBOUTS F M and GORRIS L G M (1995). Growth of psychrotrophic foodborne pathogens in a solid surface model system under the influence of carbon dioxide and oxygen. *Food Microbiol.*, 12, 509–519. doi 10.1006/fmic.1995.0066.

BLACK E P, SETLOW P, HOCKING A D, STEWART C M, KELLY A L *et al.* (2007). Response of spores to high-pressure processing. *Compr. Rev. Food Sci. Food Saf.*, 6, 103–119. doi 10.1111/j.1541-4337.2007.00021.x.

BOTTONE E J (2010). *Bacillus cereus*, a volatile human pathogen. *Clin. Microbiol. Rev.*, 23, 382–398. doi 10.1128/cmr.00073-09.

BRILLARD J and LERECLUS D (2004). Comparison of cytotoxin *cytK* promoters from *Bacillus cereus* strains ATCC 14579 and from a *Bacillus cereus* food-poisoning strain. *Microbiology-SGM*, 150, 2699–2705. doi 10.1099/mic.0.27069-0.

CADOT C, TRAN S L, VIGNAUD M L, DE BUYSER M L, KOLSTO A B *et al.* (2010). InhA1, NprA, and HlyII as candidates for markers to differentiate pathogenic from nonpathogenic *Bacillus cereus* strains. *J. Clin. Microbiol.*, 48, 1358–1365. doi 10.1128/jcm.02123-09.

CARLIN F, ALBAGNAC C, RIDA A, COUVERT O, GUINEBRETIERE M H *et al.* (2013). Variation of cardinal growth parameters according to phylogenetic affiliation in the *Bacillus cereus* group and consequences for risk assessment. *Food Microbiol.*, 33, 69–76. doi 10.1016/j.fm.2012.08.014.

CARLIN F, FRICKER M, PIELAAT A, HEISTERKAMP S, SHAHEEN R *et al.* (2006). Emetic toxin-producing strains of *Bacillus cereus* show distinct characteristics within the *Bacillus cereus* group. *Int. J. Food Microbiol.*, 109, 132–138. doi 10.1016/j.ijfoodmicro.2006.01.022.

CHUN J and BAE K S (2000). Phylogenetic analysis of *Bacillus subtilis* and related taxa based on partial *gyrA* gene sequences. *Anton. Leeuw.*, 78, 123–127. doi 10.1023/A:1026555830014.

CLAIR G, ROUSSI S, ARMENGAUD J and DUPORT C (2010). Expanding the known repertoire of virulence factors produced by *Bacillus cereus* through early secretome profiling in three redox conditions. *Mol. Cell. Proteomics*, 9, 1486–1498. doi 10.1074/mcp.M000027-MCP201.

CLAVEL T, CARLIN F, DARGAIGNARATZ C, LAIRON D, NGUYEN-THE C *et al.* (2007). Effects of porcine bile on survival of *Bacillus cereus* vegetative cells and haemolysin BL enterotoxin production in reconstituted human small intestine media. *J. Appl. Microbiol.*, 103, 1568–1575. doi 10.1111/j.1365-2672.2007.03410.x.

CLAVEL T, CARLIN F, LAIRON D, NGUYEN-THE C and SCHMITT P (2004). Survival of *Bacillus cereus* spores and vegetative cells in acid media simulating human stomach. *J. Appl. Microbiol.*, 97, 214–219. doi 10.1111/j.1365-2672.2004.02292.x.

COLEMAN W H, CHEN D, LI Y-Q, COWAN A E and SETLOW P (2007). How moist heat kills spores of *Bacillus subtilis*. *J. Bacteriol.*, 189, 8458–8466. doi 10.1128/JB.01242-07.

COOHILL T P and SAGRIPANTI J L (2008). Overview of the inactivation by 254 nm ultraviolet radiation of bacteria with particular relevance to biodefense. *Photochem. Photobiol.*, 84, 1084–1090. doi 10.1111/j.1751-1097.2008.00387.x.

COSENTINO S, MULARGIA A F, PISANO B, TUVERI P and PALMAS F (1997). Incidence and biochemical characteristics of *Bacillus* flora in Sardinian dairy products. *Int. J. Food Microbiol.*, 38, 235–238. doi 10.1016/S0168-1605(97)00107-4.

COTON M, DENIS C, CADOT P and COTON E (2011). Biodiversity and characterization of aerobic spore-forming bacteria in surimi seafood products. *Food Microbiol.*, 28, 252–260. doi 10.1016/j.fm.2010.03.017.

CUTTING S M (2011). *Bacillus* probiotics. *Food Microbiol.*, 28, 214–220. doi 10.1016/j.fm.2010.03.007.

DE SARRAU B, CLAVEL T, CLERTÉ C, CARLIN F, GINIES C *et al.* (2012). Influence of anaerobiosis and low temperature on *Bacillus cereus* growth, metabolism, and membrane properties. *Appl. Environ. Microbiol.*, 78, 1715–1723. doi 10.1128/aem.06410-11.

DELBRASSINNE L, ANDJELKOVIC M, RAJKOVIC A, BOTTLEDOORN N, MAHILLON J *et al.* (2011). Follow-up of the *Bacillus cereus* emetic toxin production in penne pasta under household conditions using liquid chromatography coupled with mass spectrometry. *Food Microbiol.*, 28, 1105–1109. doi 10.1016/j.fm.2011.02.014.

DIERICK K, VAN COILLIE E, SWIECICKA I, MEYFROIDT G, DEVLIEGER H *et al.* (2005). Fatal family outbreak of *Bacillus cereus*-associated food poisoning. *J. Clin. Microbiol.*, 43, 4277–4279. doi 10.1128/JCM.43.8.4277-4279.2005.

DUFRENNE J, SOENTORO P, TATINI S, DAY T and NOTERMANS S (1994). Characteristics of *Bacillus cereus* related to safe food production. *Int. J. Food Microbiol.*, 23, 99–109. doi 10.1016/0168-1605(94)90225-9.

DUPORT C, ZIGHA A, ROSENFELD E and SCHMITT P (2006). Control of enterotoxin gene expression in *Bacillus cereus* F4430/73 involves the redox-sensitive ResDE signal transduction system. *J. Bacteriol.*, 188, 6640–6651. doi 10.1128/jb.00702-06.

EARL A M, LOSICK R and KOLTER R (2008). Ecology and genomics of *Bacillus subtilis*. *Tr. Microbiol.*, 16, 269–275. doi 10.1016/j.tim.2008.03.004.

EFSA (2005). Opinion of the scientific panel on biological hazards on *Bacillus cereus* and other *Bacillus* spp in foodstuffs. *EFSA J.*, 175, 1–48.

EFSA (2007). Introduction of a Qualified Presumption of Safety (QPS) approach for assessment of selected microorganisms referred to EFSA. Opinion of the Scientific Committee. *EFSA J.*, 587, Appendix B – Assessment of the *Bacillus* species.

EFSA (2009). The Community summary report on food-borne outbreaks in the European Union in 2007. *EFSA J.*, 271, 1–128.

EFSA (2010). The Community summary report on trends and sources of zoonoses, zoonotic agents and food-borne outbreaks in the European Union in 2008. *EFSA J.*, 8, 1496.

EHLING-SCHULZ M, FRICKER M, GRALLERT H, RIECK P, WAGNER M *et al.* (2006). Cereulide synthetase gene cluster from emetic *Bacillus cereus*: Structure and location on a mega virulence plasmid related to *Bacillus anthracis* toxin plasmid pXOI. *BMC Microbiol.*, 6, 20. doi 10.1186/1471-2180-6-20.

EHLING-SCHULZ M, FRICKER M and SCHERER S (2004). Identification of emetic toxin producing *Bacillus cereus* strains by a novel molecular assay. *FEMS Microbiol. Lett.*, 232, 189–195. doi 10.1016/S0378-1097(04)00066-7.

EHLING-SCHULZ M, SVENSSON B, GUINEBRETIERE M-H, LINDBACK T, ANDERSSON M *et al.* (2005a). Emetic toxin formation of *Bacillus cereus* is restricted to a single evolutionary lineage of closely related strains. *Microbiol.-SGM*, 151, 183–197. doi 10.1099/mic.0.27607-0.

EHLING-SCHULZ M, VUKOV N, SCHULZ A, SHAHEEN R, ANDERSSON M *et al.* (2005b). Identification and partial characterization of the nonribosomal peptide synthetase gene responsible for cereulide production in emetic *Bacillus cereus*. *Appl. Environ. Microbiol.*, 71, 105–113. doi 10.1128/AEM.71.1.105-113.2005.

ESBELIN J, ARMENGAUD J, ZIGHA A and DUPORT C (2009). ResDE-dependent regulation of enterotoxin gene expression in *Bacillus cereus*: Evidence for multiple modes of binding for ResD and interaction with Fnr. *J. Bacteriol.*, 191, 4419–4426. doi 10.1128/JB.00321-09.

ESBELIN J, JOUANNEAU Y, ARMENGAUD J and DUPORT C (2008). ApoFnr binds as a monomer to promoters regulating the expression of enterotoxin genes of *Bacillus cereus*. *J. Bacteriol.*, 190, 4242–4251. doi 10.1128/jb.00336-08.

FAGERLUND A, WEEN A, LUND T, HARDY S P and GRANUM P E (2004). Genetic and functional analysis of the cytK family of genes in *Bacillus cereus*. *Microbiol.-SGM*, 150, 2689–2697. doi 10.1099/mic.0.26975-0.

FAILLE C, SYLLA Y, LE GENTIL C, BÉNÉZECH T, SLOMIANNY C *et al.* (2010). Viability and surface properties of spores subjected to a cleaning-in-place procedure: Consequences on their ability to contaminate surfaces of equipment. *Food Microbiol.*, 27, 769–776. doi 10.1016/j.fm.2010.04.001.

FARKAS J (2006). Irradiation for better foods. *Trends Food Sci. Technol.*, 17, 148–152. doi 10.1016/j.tifs.2005.12.003.

FREDERIKSEN K, ROSENQUIST H, JORGENSEN K and WILCKS A (2006). Occurrence of natural *Bacillus thuringiensis* contaminants and residues of *Bacillus thuringiensis*-based insecticides on fresh fruits and vegetables. *Appl. Environ. Microbiol.*, 72, 3435–3440. doi 10.1128/AEM.72.5.3435-3440.2006.

FRICKER M, REISSBRODT R and EHLING-SCHULZ M (2008). Evaluation of standard and new chromogenic selective plating media for isolation and identification of *Bacillus cereus*. *Int. J. Food Microbiol.*, 121, 27–34. doi 10.1016/j.ijfoodmicro.2007.10.012.

FROM C, HORMAZABAL V and GRANUM P E (2007a). Food poisoning associated with pumilacidin-producing *Bacillus pumilus* in rice. *Int. J. Food Microbiol.*, 115, 319–324. doi 10.1016/j.ijfoodmicro.2006.11.005.

FROM C, HORMAZABAL V, HARDY S P and GRANUM P E (2007b). Cytotoxicity in *Bacillus mojavensis* is abolished following loss of surfactin synthesis: Implications for assessment of toxicity and food poisoning potential. *Int. J. Food Microbiol.*, 117, 43–49. doi 10.1016/j.ijfoodmicro.2007.01.013.

FROM C, PUKALL R, SCHUMANN P, HORMAZABAL V and GRANUM P E (2005). Toxin-producing ability among *Bacillus* spp. outside the *Bacillus cereus* group. *Appl. Environ. Microbiol.*, 71, 1178–1183. doi 10.1128/AEM.71.3.1178-1183.2005.

GOHAR M, FAEGRI K, PERCHAT S, RAVNUM S, OKSTAD O A *et al.* (2008). The PlcR virulence regulon of *Bacillus cereus*. *PLoS ONE*, 3, doi 10.1371/journal.pone.0002793.

GOHAR M, OKSTAD O A, GILOIS N, SANCHIS V, KOLSTO A B *et al.* (2002). Two-dimensional electrophoresis analysis of the extracellular proteome of *Bacillus cereus* reveals the importance of the PlcR regulon. *Proteomics*, 2, 784–791. doi 10.1002/1615-9861(200206)2:6<784::aid-prot784>3.0.co;2-r.

GRANUM P E and BAIRD-PARKER T C (2000), *Bacillus* species, in Lund B M, Baird-Parker T C and Gould G W, *The Microbiological Safety and Quality of Food*, Gaithersburg: Aspen Publishers, 1029–1056.

GUILLEMET E, CADOT C, TRAN S L, GUINEBRETIERE M H, LERECLUS D *et al.* (2010). The InhA metalloproteases of *Bacillus cereus* contribute concomitantly to virulence. *J. Bacteriol.*, 192, 286–294. doi 10.1128/jb.00264-09.

GUINEBRETIERE M H and NGUYEN-THE C (2003). Sources of *Bacillus cereus* contamination in a pasteurized zucchini purée processing line, differentiated by two PCR-based methods. *FEMS Microbiol. Ecol.*, 43, 207–215. doi 10.1111/j.1574-6941.2003.tb01060.x.

GUINEBRETIERE M H, AUGER S, GALLERON N, CONTZEN M, DE SARRAU B *et al.* (2012). *Bacillus cytotoxicus* sp. nov. is a new thermotolerant species of the *Bacillus cereus* group occasionally associated with food poisoning. *Int. J. Syst. Evol. Microbiol.* 63, 31–40 doi 10.1099/ijs.0.030627-0.

GUINEBRETIÈRE M H, BROUSSOLLE V and NGUYEN-THE C (2002). Enterotoxigenic profiles of food-poisoning and food-borne *Bacillus cereus* strains. *J. Clin. Microbiol.*, 40, 3053–3056. doi 10.1128/JCM.40.8.3053-3056.2002.

GUINEBRETIERE M H, GIRARDIN H, DARGAIGNARATZ C, CARLIN F and NGUYEN-THE C (2003). Contamination flows of *Bacillus cereus* and spore-forming aerobic bacteria in a cooked, pasteurized and chilled zucchini purée processing line. *Int. J. Food Microbiol.*, 82, 223–232. doi 10.1016/S0168-1605(02)00307-0.

GUINEBRETIERE M H, THOMPSON F L, SOROKIN A, NORMAND P, DAWYNDT P *et al.* (2008). Ecological diversification in the *Bacillus cereus* group. *Environ. Microbiol.*, 10, 851–865. doi 10.1111/j.1462-2920.2007.01495.x.

GUINEBRETIÈRE M H, VELGE P, COUVERT O, CARLIN F, DEBUYSER M L *et al.* (2010). Ability of *Bacillus cereus* group strains to cause food poisoning varies according to phylogenetic affiliation (Groups I to VII) rather than species affiliation. *J. Clin. Microbiol.*, 48, 3388–3391. doi 10.1128/jcm.00921-10.

HAGGBLOM M M, APETROAIE C, ANDERSSON M A and SALKINOJA-SALONEN M S (2002). Quantitative analysis of cereulide, the emetic toxin of *Bacillus cereus*, produced under various conditions. *Appl. Environ. Microbiol.*, 68, 2479–2483. doi 10.1128/AEM.68.5.2479-2483.2002.

HANSEN B M, HOIBY P E, JENSEN G B and HENDRIKSEN N B (2003). The *Bacillus cereus bceT* enterotoxin sequence reappraised. *FEMS Microbiol. Lett.*, 223, 21–24. doi 10.1016/s0378-1097(03)00249-0.

HELGASON E, TOURASSE N J, MEISAL R, CAUGANT D A and KOLSTO A B (2004). Multilocus sequence typing scheme for bacteria of the *Bacillus cereus* group. *Appl. Environ. Microbiol.*, 70, 191–201. doi 10.1128/AEM.70.1.191-201.2004.

HILGREN J, SWANSON K M J, DIEZ-GONZALEZ F and CORDS B (2009). Susceptibilities of *Bacillus subtilis*, *Bacillus cereus*, and avirulent *Bacillus anthracis* spores to liquid biocides. *J. Food Prot.*, 72, 360–364.

HILL K K, TICKNOR L O, OKINAKA R T, ASAY M, BLAIR H *et al.* (2004). Fluorescent amplified fragment length polymorphism analysis of *Bacillus anthracis*, *Bacillus cereus*, and *Bacillus thuringiensis* isolates. *Appl. Environ. Microbiol.*, 70, 1068–1080. doi 10.1128/AEM.70.2.1068-1080.2004.

HOTON F M, ANDRUP L, SWIECICKA I and MAHILLON J (2005). The cereulide genetic determinants of emetic *Bacillus cereus* are plasmid-borne. *Microbiol.-SGM*, 151, 2121–2124. doi 10.1099/mic.0.28069-0.

HOTON F M, FORNELOS N, N'GUESSAN E, HU X M, SWIECICKA I *et al.* (2009). Family portrait of *Bacillus cereus* and *Bacillus weihenstephanensis* cereulide-producing strains. *Environ. Microbiol. Rep.*, 1, 177–183. doi 10.1111/j.1758-2229.2009.00028.x.

HOULT B and TUXFORD A F (1991). Toxin production by *Bacillus pumilus*. *J. Clin. Pathol.*, 44, 455–458. doi 10.1136/jcp.44.6.455.

ICHIKAWA K, GAKUMAZAWA M, INABA A, SHIGA K, TAKESHITA S *et al.* (2010). Acute encephalopathy of *Bacillus cereus* mimicking Reye syndrome. *Brain. Dev.*, 32, 688–690. doi 10.1016/j.braindev.2009.09.004.

JÄÄSKELAINEN E L, HAGGBLOM M M, ANDERSSON M A and SALKINOJA-SALONEN M S (2004). Atmospheric oxygen and other conditions affecting the production of cereulide by *Bacillus cereus* in food. *Int. J. Food Microbiol.*, 96, 75–83. doi 10.1016/j.ijfoodmicro.2004.03.011.

JÄÄSKELAINEN E L, TEPLOVA V, ANDERSSON M A, ANDERSSON L C, TAMMELA P *et al.* (2003). *In vitro* assay for human toxicity of cereulide, the emetic mitochondrial toxin produced by food poisoning *Bacillus cereus*. *Toxicol. in Vitro*, 17, 737–744. doi 10.1016/S0887-2333(03)00096-1.

JENSEN G B, HANSEN B M, EILENBERG J and MAHILLON J (2003). The hidden lifestyles of *Bacillus cereus* and relatives. *Environ. Microbiol.*, 5, 631–640. doi 10.1046/j.1462-2920.2003.00461.x.

KAILAS L, TERRY C, ABBOTT N, TAYLOR R, MULLIN N *et al.* (2011). Surface architecture of endospores of the *Bacillus cereus/anthracis/thuringiensis* family at the subnanometer scale. *P. Nat. Acad. Sci. USA*, 108, 16014–16019. doi 10.1073/pnas.1109419108.

KRAMER J M and GILBERT R J (1989), *Bacillus cereus* and other *Bacillus* species, in Doyle M P, *Foodborne Bacterial Pathogens*, New York: Marcel Dekker, 21–70.

LAPIDUS A, GOLTSMAN E, AUGER S, GALLERON N, SEGURENS B *et al.* (2008). Extending the *Bacillus cereus* group genomics to putative food-borne pathogens of different toxicity. *Chem.-Biol. Interact.*, 171, 236–249. doi 10.1016/j.cbi.2007.03.003.

LARSEN H D and JORGENSEN K (1999). Growth of *Bacillus cereus* in pasteurized milk products. *Int. J. Food Microbiol.*, 46, 173–176. doi 10.1016/S0168-1605(98)00188-3.

LECHNER S, MAYR R, FRANCIS K P, PRUSS B M, KAPLAN T *et al.* (1998). *Bacillus weihenstephanensis* sp. nov. is a new psychrotolerant species of the *Bacillus cereus* group. *Int. J. Syst. Microbiol.*, 48, 1373–1382.

LEGUERINEL I, COUVERT O and MAFART P (2007). Modelling the influence of the sporulation temperature upon the bacterial spore heat resistance, application to heating process calculation. *Int. J. Food Microbiol.*, 114, 100–104. doi 10.1016/j.ijfoodmicro.2006.10.035.

LEQUETTE Y, GARENAUX E, TAUVERON G, DUMEZ S, PERCHAT S *et al.* (2011). Role played by exosporium glycoproteins in the surface properties of *Bacillus cereus* spores and in their adhesion to stainless steel. *Appl. Environ. Microbiol.*, 77, 4905–4911. doi 10.1128/AEM.02872-10.

LESTRADET H (1995). Probiotiques III – Le *Bacillus* CIP 5832* chez l'homme et chez l'animal. *Med. Chir. Dig.*, 24, 37–39.

LEVY C, AUBERT X, LACOUR B and CARLIN F (2012). Relevant factors affecting microbial surface decontamination by pulsed light. *Int. J. Food Microbiol.*, 152, 168–174. doi 10.1016/j.ijfoodmicro.2011.08.022.

LINDBACK T, FAGERLUND A, RODLAND M S and GRANUM P E (2004). Characterization of the *Bacillus cereus* Nhe enterotoxin. *Microbiol.-SGM*, 150, 3959–3967. doi 10.1099/mic.0.27359-0.

LINDBACK T, OKSTAD O A, RISHOVD A L and KOLSTO A B (1999). Insertional inactivation of hblC encoding the L-2 component of *Bacillus cereus* ATCC 14579 haemolysin BL strongly reduces enterotoxigenic activity, but not the haemolytic activity against human erythrocytes. *Microbiol.-SGM*, 145, 3139–3146.

LOGAN N A and DE VOS P (2009). Genus I. *Bacillus* Cohn 1872, 174[AL], in De Vos P, Garrity G M, Jones D, Krieg N R, Ludwig W *et al.*, *Bergey's Manual of Systematic Bacteriology. Second Edition. Volume Three. The Firmicutes*, Dordrecht: Springer, 21–128.

LUND T and GRANUM P E (1996). Characterisation of a non-haemolytic enterotoxin complex from *Bacillus cereus* isolated after a foodborne outbreak. *FEMS Microbiol. Lett.*, 141, 151–156. doi 10.1016/0378-1097(96)00208-X.

LUND T, DE BUYSER M L and GRANUM P E (2000). A new cytotoxin from *Bacillus cereus* that may cause necrotic enteritis. *Mol. Microbiol.*, 38, 254–261. doi 10.1046/j.1365-2958.2000.02147.x.

MAGNUSSON M, CHRISTIANSSON A and SVENSSON B (2007). *Bacillus cereus* spores during housing of dairy cows: Factors affecting contamination of raw milk. *J. Dairy Sci.*, 90, 2745–2754. doi 10.3168/jds.2006-754.

MAHLER H, PASI A, KRAMER J M, SCHULTE P, SCOGING A C *et al.* (1997). Fulminant liver failure in association with the emetic toxin of *Bacillus cereus*. *New Eng. J. Med.*, 336, 1142–1148. doi 10.1056/NEJM199704173361604.

MAILLARD J-Y (2011). Innate resistance to sporicides and potential failure to decontaminate. *J. Hosp. Infect.*, 77, 204–209. doi 10.1016/j.jhin.2010.06.028.

MARGOSCH D, GANZLE M G, EHRMANN M A and VOGEL R F (2004). Pressure inactivation of *Bacillus* endospores. *Appl. Environ. Microbiol.*, 70, 7321–7328. doi 10.1128/aem.70.12.7321-7328.2004.

MARQUEZ V O, MITTAL G S and GRIFFITHS M W (1997). Destruction and inhibition of bacterial spores by high voltage pulsed electric field. *J. Food Sci.*, 62, 399–401. doi 10.1111/j.1365-2621.1997.tb04010.x.

MIKKOLA R, ANDERSSON M A, TEPLOVA V, GRIGORIEV P, KUEHN T *et al.* (2007). Amylosin from *Bacillus amyloliquefaeiens*, a K^+ and Na^+ channel-forming toxic peptide containing a polyene structure. *Toxicon*, 49, 1158–1171. doi 10.1016/j.toxicon.2007.02.010.

MOISAN M, BARBEAU J, MOREAU S, PELLETIER J, TABRIZIAN M *et al.* (2001). Low-temperature sterilization using gas plasmas: A review of the experiments and an analysis of the inactivation mechanisms. *Int. J. Pharmaceut.*, 226, 1–21. doi 10.1016/s0378-5173(01)00752-9.

MOLS M, PIER I, ZWIETERING M H and ABEE T (2009). The impact of oxygen availability on stress survival and radical formation of *Bacillus cereus*. *Int. J. Food Microbiol.*, 135, 303–311. doi 10.1016/j.ijfoodmicro.2009.09.002.

MORAVEK M, DIETRICH R, BUERK C, BROUSSOLLE V, GUINEBRETIERE M-H *et al.* (2006). Determination of the toxic potential of *Bacillus cereus* isolates by quantitative enterotoxin analyses. *FEMS Microbiol. Lett.*, 257, 293–298. doi 10.1111/j.1574-6968.2006.00185.x.

MURANYI P, WUNDERLICH J and HEISE M (2007). Sterilization efficiency of a cascaded dielectric barrier discharge. *J. Appl. Microbiol.*, 103, 1535–1544. doi 10.1111/j.1365-2672.2007.03385.x.

NAKAMURA L K (1989). Taxonomic relationship of black-pigmented *Bacillus subtilis* strains and a proposal for *Bacillus atrophaeus* sp. nov. *Int. J. Syst. Microbiol.*, 39, 295–300.

NAKAMURA L K, ROBERTS M S and COHAN F M (1999). Relationship of *Bacillus subtilis* clades associated with strains 168 and W23: A proposal for *Bacillus subtilis* subsp. *subtilis* subsp. nov. and *Bacillus subtilis* subsp. *spizizenii* subsp. nov. *Int. J. Syst. Microbiol.*, 49, 1211–1215.

NARANJO M, DENAYER S, BOTTELDOORN N, DELBRASSINNE L, VEYS J *et al.* (2011). Sudden death of a young adult associated with *Bacillus cereus* food poisoning. *J. Clin. Microbiol.*, 49, 4379–4381. doi 10.1128/jcm.05129-11.

NIEMINEN T, RINTALUOMA N, ANDERSSON M, TAIMISTO A M, ALI-VEHMAS T *et al.* (2007). Toxinogenic *Bacillus pumilus* and *Bacillus licheniformis* from mastitic milk. *Vet. Microbiol.*, 124, 329–339. doi 10.1016/j.vetmic.2007.05.015.

OH M-H, HAM J-S and COX J M (2012). Diversity and toxigenicity among members of the *Bacillus cereus* group. *Int. J. Food Microbiol.*, 152, 1–8. doi 10.1016/j.ijfoodmicro.2011.09.018.

ONGENA M and JACQUES P (2008). *Bacillus* lipopeptides: Versatile weapons for plant disease biocontrol. *Tr. Microbiol.*, 16, 115–125. doi 10.1016/j.tim.2007.12.009.

OSTENSVIK O, FROM C, HEIDENREICH B, O'SULLIVAN K and GRANUM P E (2004). Cytotoxic *Bacillus* spp. belonging to the *B. cereus* and *B. subtilis* groups in Norwegian surface waters. *J. Appl. Microbiol.*, 96, 987–993. doi 10.1111/j.1365-2672.2004.02226.x.

OUHIB-JACOBS O, LINDLEY N D, SCHMITT P and CLAVEL T (2009). Fructose and glucose mediates enterotoxin production and anaerobic metabolism of *Bacillus cereus* ATCC14579(T). *J. Appl. Microbiol.*, 107, 821–829. doi 10.1111/j.1365-2672.2009.04254.x.

OUOBA L I I, THORSEN L and VARNAM A H (2008). Enterotoxins and emetic toxins production by *Bacillus cereus* and other species of *Bacillus* isolated from soumbala and bikalga, African alkaline fermented food condiments. *Int. J. Food Microbiol.*, 124, 224–230. doi 10.1016/j.ijfoodmicro.2008.03.026.

PENG H, FORD V, FRAMPTON E W, RESTAINO L, SHELEF L A *et al.* (2001). Isolation and enumeration of *Bacillus cereus* from foods on a novel chromogenic plating medium. *Food Microbiol.*, 18, 231–238. doi 10.1006/fmic.2000.0369.

PHELPS R J and MCKILLIP J L (2002). Enterotoxin production in natural isolates of Bacillaceae outside the *Bacillus cereus* group. *Appl. Environ. Microbiol.*, 68, 3147–3151. doi 10.1128/AEM.68.6.3147-3151.2002.

POSFAY-BARBE K M, SCHRENZEL J, FREY J, STUDER R, KROFF C *et al.* (2008). Food poisoning as a cause of acute liver failure. *Ped. Infect. Dis. J.*, 27, 846–847. doi 10.1097/INF.0b013e318170f2ae.

PRIEST F G, BARKER M, BAILLIE L W J, HOLMES E C and MAIDEN M C J (2004). Population structure and evolution of the *Bacillus cereus* group. *J. Bacteriol.*, 186, 7959–7970. doi 10.1128/JB.186.23.7959-7970.2004.

PRIEST F G, GOODFELLOW M, SHUTE L A and BERKELEY R C W (1987). *Bacillus amyloliquefaciens* sp. nov, nom. rev. *Int. J. Syst. Microbiol.*, 37, 69–71.

RAAIJMAKERS J M, DE BRUIJN I, NYBROE O and ONGENA M (2010). Natural functions of lipopeptides from *Bacillus* and *Pseudomonas*: More than surfactants and antibiotics. *FEMS Microbiol. Rev.*, 34, 1037–1062. doi 10.1111/j.1574-6976.2010.00221.x.

RAJKOVIC A, UYTTENDAELE M, VERMEULEN A, ANDJELKOVIC M, FITZ-JAMES *et al.* (2008). Heat resistance of *Bacillus cereus* emetic toxin, cereulide. *Lett. Appl. Microbiol.*, 46, 536–541. doi 10.1111/j.1472-765X.2008.02350.x.

REISSBRODT R, RASSBACH A, BURGHARDT B, RIENACKER I, MIETKE H *et al.* (2004). Assessment of a new selective chromogenic *Bacillus cereus* group plating medium and use of enterobacterial autoinducer of growth for cultural identification of *Bacillus* species. *J. Clin. Microbiol.*, 42, 3795–3798. doi 10.1128/JCM.42.8.3795-3798.2004.

RICE J K and EWELL M (2001). Examination of peak power dependence in the UV inactivation of bacterial spores. *Appl. Environ. Microbiol.*, 67, 5830–5832. doi 10.1128/AEM.67.12.5830-5832.2001.

ROBERTS M S, NAKAMURA L K and COHAN F M (1994). *Bacillus mojavensis* sp. nov., distinguishable from *Bacillus subtilis* by sexual isolation, divergence in DNA-sequence, and differences in fatty-acid composition. *Int. J. Syst. Microbiol.*, 44, 256–264.

ROBERTS M S, NAKAMURA L K and COHAN F M (1996). *Bacillus vallismortis* sp. nov., a close relative of *Bacillus subtilis*, isolated from soil in Death Valley, California. *Int. J. Syst. Microbiol.*, 46, 470–475.

ROWAN N J, CALDOW G, GEMMELL C G and HUNTER I S (2003). Production of diarrheal enterotoxins and other potential virulence factors by veterinary isolates of *Bacillus* species associated with nongastrointestinal infections. *Appl. Environ. Microbiol.*, 69, 2372–2376. doi 10.1128/AEM.69.4.2372-2376.2003.

ROWAN N J, DEANS K, ANDERSON J G, GEMMELL C G, HUNTER I S *et al.* (2001). Putative virulence factor expression by clinical and food Isolates of *Bacillus* spp. after growth in reconstituted infant milk formulae. *Appl. Environ. Microbiol.*, 67, 3873–3881. doi 10.1128/AEM.67.9.3873-3881.2001.

SALKINOJA-SALONEN M S, VUORIO R, ANDERSSON M A, KAMPFER P, ANDERSSON M C *et al.* (1999). Toxigenic strains of *Bacillus licheniformis* related to food poisoning. *Appl. Environ. Microbiol.*, 65, 4637–4645.

SAMAPUNDO S, EVERAERT H, WANDUTU J N, RAJKOVIC A, UYTTENDAELE M *et al.* (2011). The influence of headspace and dissolved oxygen level on growth and haemolytic BL enterotoxin production of a psychrotolerant *Bacillus weihenstephanensis* isolate on potato based ready-to-eat food products. *Food Microbiol.*, 28, 298–304. doi 10.1016/j.fm.2010.04.013.

SHINAGAWA K, KONUMA H, SEKITA H and SUGII S (1995). Emesis of rhesus-monkeys induced by intragastric administration with the HEp-2 vacuolation factor (cereulide) produced by *Bacillus cereus*. *FEMS Microbiol. Lett.*, 130, 87–90. doi 10.1111/j.1574-6968.1995.tb07703.x.

SHINAGAWA K, UENO Y, HU D L, UEDA S and SUGII S (1996). Mouse lethal activity of a HEp-2 vacuolation factor, cereulide, produced by *Bacillus cereus* isolated from vomiting-type food poisoning. *J. Vet. Med. Sci.*, 58, 1027–1029.

SHIOTA M, SAITOU K, MIZUMOTO H, MATSUSAKA M, AGATA N *et al.* (2010). Rapid detoxification of cereulide in *Bacillus cereus* food poisoning. *Pediatrics*, 125, E951–E955. doi 10.1542/peds.2009-2319.

SLAMTI L and LERECLUS D (2002). A cell-cell signaling peptide activates the PlcR virulence regulon in bacteria of the *Bacillus cereus* group. *Embo J.*, 21, 4550–4559. doi 10.1093/emboj/cdf450.

STENFORS L P and GRANUM P E (2001). Psychrotolerant species from the *Bacillus cereus* group are not necessarily *Bacillus weihenstephanensis*. *FEMS Microbiol. Lett.*, 197, 223–228. doi 10.1111/j.1574-6968.2001.tb10607.x.

STENFORS ARNESEN L P, FAGERLUND A and GRANUM P E (2008). From soil to gut: *Bacillus cereus* and its food poisoning toxins. *FEMS Microbiol. Rev.*, 32, 579–606. doi 10.1111/j.1574-6976.2008.00112.x.

SUNDBERG M, CHRISTIANSSON A, LINDAHL C, WAHLUND L and BIRGERSSON C (2011). Cleaning effectiveness of chlorine-free detergents for use on dairy farms. *J. Dairy Res.*, 78, 105–110. doi 10.1017/s0022029910000762.

SUTHERLAND A D and MURDOCH R (1994). Seasonal occurrence of psychrotrophic *Bacillus* species in raw milk, and studies on the interactions with mesophilic *Bacillus* sp. *Int. J. Food Microbiol.*, 21, 279–292. doi 10.1016/0168-1605(94)90058-2.

SVENSSON B, MONTHAN A, SHAHEEN R, ANDERSSON M A, SALKINOJA-SALONEN M *et al.* (2006). Occurrence of emetic toxin producing *Bacillus cereus* in the dairy production chain. *Int. Dairy J.*, 16, 740–749. doi 10.1016/j.idairyj.2005.07.002.

SWIECICKA I (2008). Natural occurrence of *Bacillus thuringiensis* and *Bacillus cereus* in eukaryotic organisms: A case for symbiosis. *Biocont. Sci. Technol.*, 18, 221–239. doi 10.1080/09583150801942334.

TAPI A, CHOLLET-IMBERT M, SCHERENS B and JACQUES P (2010). New approach for the detection of non-ribosomal peptide synthetase genes in *Bacillus* strains by polymerase chain reaction. *Appl. Microbiol. Biotechnol.*, 85, 1521–1531. doi 10.1007/s00253-009-2176-4.

TAYLOR J M W, SUTHERLAND A D, AIDOO K E and LOGAN N A (2005). Heat-stable toxin production by strains of *Bacillus cereus*, *Bacillus firmus*, *Bacillus megaterium*, *Bacillus simplex* and *Bacillus licheniformis*. *FEMS Microbiol. Lett.*, 242, 313–317. doi 10.1016/j.femsle.2004.11.022.

TE GIFFEL M C, BEUMER R R, LEIJENDEKKERS S and ROMBOUTS F M (1996). Incidence of *Bacillus cereus* and *Bacillus subtilis* in foods in the Netherlands. *Food Microbiol.*, 13, 53–58.

THORSEN L, BUDDE B B, HENRICHSEN L, MARTINUSSEN T and JAKOBSEN M (2009a). Cereulide formation by *Bacillus weihenstephanensis* and mesophilic emetic *Bacillus cereus* at temperature abuse depends on pre-incubation conditions. *Int. J. Food Microbiol.*, 134, 133–139. doi 10.1016/j.ijfoodmicro.2009.03.023.

THORSEN L, BUDDE B B, KOCH A G and KLINGBERG T D (2009b). Effect of modified atmosphere and temperature abuse on the growth from spores and cereulide production of *Bacillus weihenstephanensis* in a cooked chilled meat sausage. *Int. J. Food Microbiol.*, 130, 172–178. doi 10.1016/j.ijfoodmicro.2009.01.009.

TOURASSE N J, HELGASON E, KLEVAN A, SYLVESTRE P, MOYA M *et al.* (2011). Extended and global phylogenetic view of the *Bacillus cereus* group population by combination of MLST, AFLP, and MLEE genotyping data. *Food Microbiol.*, 28, 236–244. doi 10.1016/j.fm.2010.06.014.

TOURASSE N J, ØKSTAD O A and KOLSTØ A-B (2010). HyperCAT: an extension of the SuperCAT database for global multi-scheme and multi-datatype phylogenetic analysis of the *Bacillus cereus* group population. *Database*, 2010, July. doi 10.1093/database/baq017.

TRAN S L, GUILLEMET E, GOHAR M, LERECLUS D and RAMARAO N (2010). CwpFM (EntFM) is a *Bacillus cereus* potential cell wall peptidase implicated in adhesion, biofilm formation, and virulence. *J. Bacteriol.*, 192, 2638–2642. doi 10.1128/jb.01315-09.

TRAN S L, GUILLEMET E, NGO-CAMUS M, CLYBOUW C, PUHAR A *et al.* (2011). Haemolysin II is a *Bacillus cereus* virulence factor that induces apoptosis of macrophages. *Cell. Microbiol.*, 13, 92–108. doi 10.1111/j.1462-5822.2010.01522.x.

VAN DER VOORT M, KUIPERS O P, BUIST G, DE VOS W M and ABEE T (2008). Assessment of CcpA-mediated catabolite control of gene expression in *Bacillus cereus* ATCC 14579. *BMC Microbiol.*, 8, 62. doi 10.1186/1471-2180-8-62.

WANG D, OPPENLÄNDER T, EL-DIN M G and BOLTON J R (2010). Comparison of the disinfection effects of vacuum-UV (VUV) and UV light on *Bacillus subtilis* spores in aqueous suspensions at 172, 222 and 254 nm. *Photochem. Photobiol.*, 86, 176–181. doi 10.1111/j.1751-1097.2009.00640.x.

WANG L-T, LEE F-L, TAI C-J, YOKOTA A and KUO H-P (2007). Reclassification of *Bacillus axarquiensis* Ruiz-Garcia *et al.* 2005 and *Bacillus malacitensis* Ruiz-Garcia *et al.* 2005 as later heterotypic synonyms of *Bacillus mojavensis* Roberts *et al.* 1994. *Int. J. Syst. Evol. Microbiol.*, 57, 1663–1667.

WARRINER K, KOLSTAD J, RUMSBY P and WAITES W M (2002). Carton sterilization by u.v.-C excimer laser light: Recovery of *Bacillus subtilis* spores on vegetable extracts and food simulation matrices. *J. Appl. Microbiol.*, 92, 1051–1057. doi 10.1046/j.1365-2672.2002.01641.x.

WARTH A D (1978). Relationship between the heat resistance of spores and the optimum and maximum growth temperatures of *Bacillus* species. *J. Bacteriol.*, 134, 699–705.

WIJNANDS L M, DUFRENNE J B, ZWIETERING M H and VAN LEUSDEN F M (2006). Spores from mesophilic *Bacillus cereus* strains germinate better and grow faster in simulated gastro-intestinal conditions than spores from psychrotrophic strains. *Int. J. Food Microbiol.*, 112, 120–128. doi 10.1016/j.ijfoodmicro.2006.06.015.

WILLIAMS L D, BURDOCK G A, JIMENEZ G and CASTILLO M (2009). Literature review on the safety of Toyocerin (R), a non-toxigenic and non-pathogenic *Bacillus cereus* var. *toyoi* preparation. *Regul. Toxicol. Pharm.*, 55, 236–246. doi 10.1016/j.yrtph.2009.07.009.

YOKOYAMA K, ITO M, AGATA N, ISOBE M, SHIBAYAMA K *et al.* (1999). Pathological effect of synthetic cereulide, an emetic toxin of *Bacillus cereus*, is reversible in mice. *FEMS Immunol. Med. Microbiol.*, 24, 115–120. doi 10.1016/S0928-8244(99)00017-6.

ZIGHA A, ROSENFELD E, SCHMITT P and DUPORT C (2006). Anaerobic cells of *Bacillus cereus* F4430/73 respond to low oxidoreduction potential by metabolic readjustments and activation of enterotoxin expression. *Arch. Microbiol.*, 185, 222–233. doi 10.1007/s00203-006-0090-z.

ZIGHA A, ROSENFELD E, SCHMITT P and DUPORT C (2007). The redox regulator fnr is required for fermentative growth and enterotoxin synthesis in *Bacillus cereus* F4430/73. *J. Bacteriol.*, 189, 2813–2824. doi 10.1128/jb.01701-06.

5

Pathogen update: *Vibrio* species

M. Strom, R. N. Paranjpye, W. B. Nilsson, J. W. Turner and G. K. Yanagida, National Oceanic and Atmospheric Administration, USA

DOI: 10.1533/9780857098740.2.97

Abstract: Members of the genus *Vibrio* are common inhabitants of the marine environment, associated with vertebrate and invertebrate animals. Some species cause foodborne illness from seafood such as undercooked or raw fish and shellfish. This review focuses on the current state of knowledge for the human pathogenic *Vibrio* species. *V. parahaemolyticus* is the leading cause of bacterial seafood-borne illness worldwide, while *V. vulnificus* is the leading cause of seafood-related deaths. While much research has ascertained potential mechanisms of virulence and contributed to a variety of mitigation strategies, there is a need to improve methods to assess risk and to develop improved early warning systems for public health and resource managers.

Key words: *Vibrio*, seafood, shellfish, oysters, virulence, prediction.

5.1 Introduction

Foodborne pathogens include many naturally occurring marine bacteria that have the capability of causing opportunistic disease in humans. The most common and best studied are members of the genus *Vibrio*. Vibrios are natural members of the marine and estuarine bacterial communities (Feldhusen, 2000). A broad range of environmental parameters, including temperature and salinity as well as biotic factors such as the presence of phytoplankton and zooplankton, influence the density of marine *Vibrio* spp. in the water column. These bacteria are frequently isolated from various types of fish, shellfish and aquatic flora, and are also found in sediment. Not all vibrios cause illness and within a given species not all strains are equally pathogenic. As the water temperature increases (i.e. during late spring to summer) so does the population of these bacteria. *Vibrio* illnesses tend to follow this trend and peak during the summer months (CDC, 2006).

By far the best-known pathogenic *Vibrio* spp. is *Vibrio cholerae*, which causes serious disease epidemics and is spread via the oral/fecal route through contaminated drinking water. Cholera epidemics are now primarily confined to the developing world where sanitation is problematic. Even though *V. cholerae* is the leading cause of *Vibrio*-associated illnesses (Faruque *et al.*, 1998), it is not generally a major threat to human health through seafood consumption, which is the focus of this chapter. Non-cholera *Vibrio*-related disease is usually associated with the consumption of raw or undercooked shellfish, or exposure of open wounds to water or seafood. Among the many members of the *Vibrio* genus responsible for seafood-associated illnesses, two species, *V. parahaemolyticus* and *V. vulnificus*, account for the majority of cases.

V. parahaemolyticus is the leading cause of seafood-associated bacterial illness in the US (Newton *et al.*, 2012). Infections caused by this bacterium were sporadic until 1996, after which an increasing number of outbreaks were shown to be caused by strains belonging to a single pandemic clonal complex. This complex initially emerged in India and quickly spread to Asia and the North and South American continents (Ansaruzzaman *et al.*, 2005; Chowdhury *et al.*, 2000; DePaola *et al.*, 2000; Fuenzalida *et al.*, 2006, 2007; Gonzalez-Escalona *et al.*, 2005; Martinez-Urtaza *et al.*, 2005; Matsumoto *et al.*, 2000; Okuda *et al.*, 1997; Quilici *et al.*, 2005). Although *V. parahaemolyticus* infections are generally mild, outbreaks cause considerable loss of consumer confidence in commercial shellfish, and therefore may have a significant economic effect as well as a public health impact.

Infections caused by *V. vulnificus* also represent a significant percentage of *Vibrio*-related illnesses (Bonner *et al.*, 1983; Hlady, 1997). Though not as common as those caused by *V. parahaemolyticus, V. vulnificus* infections are usually more serious and often lead to death. Even though there are only about 50 cases annually of seafood-related *V. vulnificus* infections, the severity of the disease with a ~50% mortality rate makes it the leading cause of seafood-associated deaths in the US (Morris, 1988; Morris and Black, 1985).

5.2 Sources of infection and types of pathology

A significant percentage of infections due to *V. parahaemolyticus* and *V. vulnificus* are acquired by consumption of raw or undercooked shellfish, such as oysters. *V. parahaemolyticus*, when ingested, causes acute gastroenteritis characterized by diarrhea that is occasionally bloody, nausea, vomiting, abdominal cramping, low-grade fever and headache, usually 4–90 h after the person has consumed the contaminated seafood (Morris, 2003). Oral rehydration, especially in people who are not immunocompromised, is usually sufficient to treat this infection (Su and Liu, 2007). Septicemia with *V. parahaemolyticu*s rarely occurs, except in immunocompromised patients with underlying conditions such as cirrhosis, cancer, diabetes or alcoholism, who are at highest risk (Broberg *et al.*, 2011). Infections can also occur if wounds come in contact with water harboring the bacteria. This form of infection is usually limited to cellulitis; however, necrosis of tissue can develop, which requires surgical debridement (Hlady and Klontz, 1996).

People infected with *V. vulnificus* may develop primary septicemia: the patient exhibits fever and shock, has a history of recently consuming raw shellfish, and there is no apparent primary site of infection. These patients can develop secondary cutaneous lesions that may become necrotic (necrotizing fasciitis) and thus require the surgical removal of contaminated tissue. Individuals most susceptible to these infections usually suffer from chronic diseases that affect the liver (e.g. cirrhosis), have elevated serum iron levels (e.g. due to hemochromatosis or thalassemia) or are immune compromised due to other chronic diseases. Systemic infection caused by *V. vulnificus* can be lethal with a greater than 50% chance of fatality (Feldhusen, 2000; Hlady and Klontz, 1996). Wound infections can also occur from handling fish or shellfish or by infection of a preexisting wound. Symptoms include inflammation of the wound, which can become septicemic and can also develop into necrotizing fasciitis accompanied by chills and fever. The fatality rate of these infections is around 25% (Strom and Paranjpye, 2000; Jones and Oliver, 2009). Gastroenteritis due to *V. vulnificus* is less common, is self-limiting and is often unreported (Strom and Paranjpye, 2000; Paranjpye and Strom, 2005). Symptoms include fever, diarrhea, abdominal cramps and vomiting but patients do not suffer from systemic shock nor will they have localized cellulitis.

Susceptible individuals may also be prone to *Vibrio* infections after natural disasters as was experienced after Hurricane Katrina in the US in 2005 when 18 cases of wound-associated *Vibrio* infections were reported. Of these 18 cases, 14 were infected with *V. vulnificus*, three of which resulted in death. Three individuals were infected with *V. parahaemolyticus*, two of whom died. Out of the 18 patients, 13 of them had some form of underlying condition, which might have increased their risk to a more severe infection (CDC, 2005).

5.3 Virulence and strain variability

5.3.1 *V. vulnificus* virulence factors

Multiple virulence-associated factors have been attributed to *V. vulnificus* host colonization, immune evasion, rapid in-host growth, cytotoxicity and necrosis. Colonization of host cells has been shown to be mediated by the pilin, PilA, and the prepilin peptidase, PilD (Paranjpye *et al.*, 1998, 2007; Paranjpye and Strom, 2005), and the flagellar proteins FlgC and FlgE (Lee *et al.*, 2004a). Host immune evasion is facilitated by a protective capsular polysaccharide (CPS) (Simonson and Siebeling, 1993), the presence of which has been strongly correlated with virulence in a mouse model (Wright *et al.*, 1990; Yoshida *et al.*, 1985). Host resistance is also facilitated by the production of factors related to acid tolerance (encoded by the *cad*BA operon) (Rhee *et al.*, 2005). Iron-scavenging siderophores such as vulnibactin aid rapid in-host growth (Wright *et al.*, 1981; Helms *et al.*, 1984; Zakaria-Meehan *et al.*, 1988). Cytotoxicity and cellular damage have been attributed to an extracellular hemolysin (VvhA) (Lee *et al.*, 2004b) and the RtxA1 toxin (Kim *et al.*, 2008; Lee *et al.*, 2007), while an endotoxic lipopolysaccharide (LPS) is thought to be responsible for the endotoxic shock characteristic of the disease caused by this bacterium (McPherson *et al.*, 1991). Necrosis of host tissue is the result of an extracellular protease (VvpA) (Lee *et al.*, 2004b) and CPS (Powell *et al.*, 1997). Genomic investigations suggest that GGDEF proteins and a Flp pilin could also be associated with pathogenesis (Gulig *et al.*, 2010). Regardless of the cumulative research contributing to our current understanding of *V. vulnificus*, the pathogenesis of this bacterium remains far from resolved (Gulig *et al.*, 2005; Jones and Oliver, 2009; Strom and Paranjpye, 2000).

Techniques for the classification of *V. vulnificus* include biochemical characterization (biotypes) and genetic characterization (genotypes). *V. vulnificus* is typically classified according to biotype (1, 2 or 3), where biotype 1 is commonly virulent to humans and associated with the consumption of raw oysters (Blake *et al.*, 1979; Coleman *et al.*, 1996), biotype 2 is predominantly a pathogen of eels (Tison *et al.*, 1982) and biotype 3 is primarily associated with wound infections in humans working in close association with tilapia farms in Israel (Bisharat *et al.*, 1999). Based on variations in the 16S rRNA sub-unit (Aznar *et al.*, 1994), biotype 1 can be further divided into genotypes where genotype A strains are predominantly environmental while genotype B strains are predominantly clinical (Nilsson *et al.*, 2003). Similarly sequence variations in the virulence correlated gene (*vcg)* also divide biotype 1 strains into predominantly environmental (*vcg*E) and clinical (*vcg*C) genotypes (Warner and Oliver, 2007). Although biotype 2 is notable as an eel pathogen, this biotype is also an opportunistic pathogen of humans (Amaro and Biosca, 1996) and recent multilocus sequence typing (MLST) analyses revealed that biotype 2 is polyphyletic (Sanjuan *et al.*, 2011).

At the time of publication, a total of three *V. vulnificus* closed genomes (YJO16, CMCP6 and MO624/O) have been deposited in GenBank. As with *V. cholerae* and *V. parahaemolyticus, V. vulnificus* contains a larger chromosome I and a smaller chromosome II comprising ~3.4 Mb and ~1.9 Mb, respectively, and genome comparisons have revealed that gene content and position are more conserved on chromosome I (Chen *et al.*, 2003). Regardless, chromosome I is also a source of genetic variability between strains owing largely to duplication and transposition events on a large super-integron (Chen *et al.*, 2003; Gulig *et al.*, 2010). Lateral gene transfer, as evidenced by the presence of a ~48 kb conjugative plasmid in strain YJO16, represents another mechanism of introducing genetic variability (Chen *et al.*, 2003). Further, at least 14 genomic islands, regions of aberrant guanosine:cytosine content containing integrase genes, appear to be non-uniformly distributed in *V. vulnificus*, suggesting that many strains harbor an abundance of strain-specific DNA (Quirke *et al.*, 2006). Genomic diversity and variability may provide a means to survive and adapt to the pressures of a harsh and fluctuating marine environment.

5.3.2 *V. parahaemolyticus* virulence factors

As for all pathogenic vibrios, only a small sub-population of *V. parahaemolyticus* strains is pathogenic to humans. Pili and the polysaccharide capsule expressed on its surface have been associated with initial adherence of the bacterium to epithelial cells (Hsieh *et al.*, 2003; Nakasone and Iwanaga, 1990), and recent studies suggest that type IV pili contribute to virulence in a zebrafish model (Paranjpye *et al.*, unpublished). Recently MAM7, a multivalent adhesion molecule, was shown to be necessary for initial attachment of several Gram-negative pathogens, including *V. parahaemolyticus*, to epithelial cells (Krachler *et al.*, 2011).

Pathogenicity has primarily been associated with strains carrying the thermostable direct hemolysin (*tdh*) (Nishibuchi *et al.*, 1985; Shirai *et al.*, 1990) and/or the *tdh*-related hemolysin (*trh*) genes (Nishibuchi *et al.*, 1989; Honda *et al.*, 1988). The thermostable direct hemolysin (TDH) is a pore-forming toxin (Honda *et al.*, 1992) known to exhibit cytotoxic and enterotoxic activity in *in vitro* systems (Ansaruzzaman *et al.*, 2005; Raimondi *et al.*, 2000; Chowdhury *et al.*, 2000; DePaola *et al.*, 2000; Fuenzalida *et al.*, 2006, 2007; Gonzalez-Escalona *et al.*, 2005; Martinez-Urtaza *et al.*, 2005; Matsumoto *et al.*, 2000; Okuda *et al.*, 1997; Quilici *et al.*, 2005). TRH is biochemically, immunologically and physiologically similar to TDH and is also associated with gastroenteritis in humans (Nishibuchi *et al.*, 1989; Honda *et al.*, 1988). Subsequent studies have shown that effector proteins secreted through type 3 secretion systems (T3SSs) play a significant role in both enterotoxicity and cytotoxicity in epithelial cells and will be described below (Park *et al.*, 2004; Broberg *et al.*, 2011).

Genome sequencing and analysis of a clinical strain of serotype O3:K6, RIMD2210633, revealed the presence of two pathogenicity islands on each of the two chromosomes, that include two type 3 secretion systems, T3SS1 and T3SS2 (Makino *et al.*, 2003). These secretion systems resemble the flagellar apparatus, have a basal body that spans the inner and outer membranes and a needle complex that delivers proteins (effectors) directly into the cytoplasm of eukaryotic cells. T3SS1 on chromosome 1 has been identified in all pathogenic and non-pathogenic *V. parahaemolyticus* strains and causes cytotoxicity in cultured human epithelial cells, but does not contribute to enterotoxicity in a rabbit ileal loop model (Park *et al.*, 2004; Lynch *et al.*, 2005). Four potential effectors and their *in vitro* activities have been identified in this secretion system: VopQ (VP1680) induces autophagy and blocks phagocytosis of the bacterium by macrophages (Burdette *et al.*, 2009; Ono *et al.*, 2006); VopS (VP1686) causes disruption of the actin cytoskeleton (Ono *et al.*, 2006; Yarbrough *et al.*, 2009); the mechanism of action of VopR (VP1683) has yet to be determined (Ono *et al.*, 2006); and finally VPA0450 disrupts the association of actin-binding proteins with the membrane to induce blebbing and host cell lysis (Ono *et al.*, 2006; Broberg *et al.*, 2011).

The presence of T3SS2 has predominantly been associated with clinical strains and two clusters T3SS2α and T3SS2β with some genetic rearrangement have been detected in strains suggesting that these islands may have been acquired more recently and at different times (Makino *et al.*, 2003; Okada *et al.*, 2010). T3SS2 is part of the pathogenicity island Vp-PA1 (or VPA7) on chromosome 2 and has been associated with enterotoxicity in a rabbit ileal loop model as well as tight junction integrity in cultured cell monolayers. Although the presence of *tdh* is also associated with T3SS2, it does not appear to be necessary for the pathogenic effects observed in these *in vitro* systems (Caburlotto *et al.*, 2010; Lynch *et al.*, 2005). To date, four effectors have been identified in T3SS2: VopC (VPA1321), homologous to cytotoxic necrotizing factor 1 (CNF1) in some pathogenic *Escherichia coli* strains, may be involved with the disruption of the actin cytoskeleton and apoptosis although its specific action even *in vitro* has yet to be determined (Kodama *et al.*, 2007). VopA/P (VPA1346) inhibits MAPK signaling, which prevents cytokine signaling (Kodama *et al.*, 2007; Lawrence *et al.*, 2008; Trosky *et al.*, 2004, 2007). VopL (VPA1370), like VopC, could also be involved in actin polymerization (Kodama *et al.*, 2007). VopT (VPA1327) is homologous to the adenosine diphosphate ribosyltransferase of *Pseudomonas aeruginosa* and is partially responsible for the cytotoxicity observed in cultured cells (Kodama *et al.*, 2007). While the roles of the toxins, secretion systems and their effectors *in vitro* are continuing to be investigated, the concerted effects of these proteins in their natural environment as well as the accidental human host remains to be determined. The use of appropriate animal models will be crucial in examining the mechanism of pathogenesis *in vivo*.

A sudden increase since the mid-1990s of *V. parahaemolyticus*-related infections led to the emergence and identification of a new serotype, O3:K6, which subsequently spread to North and South America, Asia and Europe. It has been responsible for the majority of illnesses worldwide (Nair *et al.*, 2007; Chowdhury *et al.*, 2000; Matsumoto *et al.*, 2000). This serotype has evolved and diverged to include at least 11 different pathogenic serovariants (Gonzalez-Escalona *et al.*, 2011). Strains belonging to this clonal complex carry *tdh*, but not *trh*, and can be distinguished using the group-specific polymerase chain reaction (GS-PCR) based on the presence of mismatched nucleotides at seven base positions of the *toxRS* gene sequence and an ORF8 sequence from the f237 phage (Matsumoto *et al.*, 2000; Nasu *et al.*, 2000). However, strains of diverse serogroups continue to be identified in clinical cases from the US making it difficult to be used as a predictor of virulence (Jones *et al.*, 2012).

Several molecular and genetic techniques including ribotyping, pulse field gel electrophoresis (PFGE), multilocus sequence typing (MLST), multilocus variable number tandem repeat analysis (MLVA) and arbitrarily primed PCR (AP-PCR) have been useful as typing methods for epidemiological purposes (Chowdhury *et al.*, 2004; Gonzalez-Escalona and Martinez-Urtaza, 2008; Kimura *et al.*, 2008; Wong *et al.*, 2007; Wong and Lin, 2001; Okura *et al.*, 2003; Suffredini *et al.*, 2011). These methods have also been useful in analyzing the emergence and spread of the pandemic clone O3:K6 (Okuda *et al.*, 1997). However, typing systems that use genetic and virulence information are needed to differentiate the pathogenic and non-pathogenic strains of these bacteria.

5.4 Current risk management

The majority of non-cholera *Vibrio* infections worldwide are caused by *V. parahaemolyticus*, followed by *V. vulnificus* and other vibrios. Reports of the incidence of sporadic infections due to both *V. parahaemolyticus* and *V. vulnificus* have been increasing since the 1990s in temperate regions around the world partly attributed to increases in seawater temperature (Baker-Austin *et al.*, 2010; Harth, 2009; Julie *et al.*, 2010; Vezzulli *et al.*, 2009). There is also growing evidence that changes in populations of vibrios may also be a result of climate change (Marques *et al.*, 2010; Baker-Austin *et al.*, 2012).

In the US raw oysters are considered to be the major source of *Vibrio*-related infections and current risk management systems used by the US Food and Drug Administration (US FDA) are targeted at reducing the probability of illnesses caused by the two pathogenic species of concern, *V. parahaemolyticus* and *V. vulnificus*, after consumption of raw oysters. Risk assessment is calculated using data from several federal and state organizations such as the Centers for Disease Control and Prevention (CDC), state shellfish protection agencies, shellfish growers and the Interstate Shellfish Sanitation Conference (ISSC). The predicted risk of illness takes into consideration abiotic parameters such as air and water temperature, the time from harvest until the oysters are placed under refrigeration, the time it takes the oysters to cool and the time between refrigeration and consumption as well as regional and seasonal variation. Differences due to harvesting methods and the number of oysters consumed per serving are also factored into the model (FAO/WHO, 2011). These models are also based on the assumption that total numbers of *V. parahaemolyticus* and *V. vulnificus* can be estimated based on water temperature, and that the pathogenic and total vibrios grow at the same rate. Presently for *V. parahaemolyticus* the risk is also estimated based on the concentration of strains carrying *tdh*, which is considered to be the marker for virulence. However, an increasing number of illnesses in the Pacific Northwest of the US and regions in the Southern Hemisphere have been reported without the detection of *tdh* or *trh*, underscoring the need for additional markers for detection of pathogenic *V. parahaemolyticus* strains (Washington State Department of Health, personal communication).

Post-harvest treatment of oysters is increasingly promoted as a means to obtain a safer product, which has the characteristics of a raw oyster on a half-shell. In the US, post-harvest treatments are validated by the ISSC, which requires that after treatment the levels of *V. vulnificus* and *V. parahaemolyticus* are non-detectable (<30 MPN/g), a minimum 3.52 log reduction is achieved, the characteristics of a raw oyster are maintained and consumer acceptance is received. Post-harvest treated oysters presently comprise 10% of domestic raw oyster sales. Approved methods include high hydrostatic pressure (HHP) processing, individual quick freezing and low-temperature or cool pasteurization.

The non-thermal treatment of raw foods, such as HHP treatment, has been used effectively in maintaining the characteristics of a raw product including appearance, flavor, texture and nutritional qualities while inactivating several bacteria including *Vibrio* spp. (Styles *et al.*, 1991; Berlin *et al.*, 1999). HHP treatment of oysters was initiated in 1999 in Houma, LA, US, and has proven to be effective in achieving at least a 5 log reduction in the concentration of vibrios using pressures of 35 000–40 000 psi for 3–5 mins (Motivatit Seafood Inc, Houma,

LA, US). In this process, cleaned oysters are banded and treated under ultra-high pressure in stainless steel containers for the optimized time and temperature. The process can be used both for half-shell and shucked oysters and produces a product that also maintains the qualities of the raw oyster. However, the equipment is expensive and not easily affordable for the majority of smaller oyster growers.

In another method, individual oysters are cryogenically or flash frozen and stored frozen until ready to eat. This process is very convenient but freezing oysters, especially those harvested in the summer months, tends to compromise both texture and appearance. However, the product's appeal to consumers as a non-processed product is enticing, and the process is currently being used by several companies within the US, Australia, Canada and New Zealand. A third method, cool pasteurization, involves a mild heat treatment of 126°F (52°C) in water for 24 mins, followed by 15 mins at 40°F (4°C), after which the oysters are packed for half-shell or shucked and shipped on ice. The treatment has been shown to be effective in reducing *V. vulnificus* to non-detectable levels. The process has also been validated by the ISSC, it has been patented and it is currently being used only by a single company (AmeriPure Processing Co, Franklin, LA, US).

The use of low-dose gamma irradiation has also been shown to be effective in inactivating both *V. vulnificus* and *V. parahaemolyticus* (Andrews *et al.*, 2012) and has been recently approved by the FDA to treat raw oysters (FDA, 2011). However, the process is awaiting validation by the ISSC and studies on consumer acceptance and is currently not being used commercially.

Illnesses attributed to *V. vulnificus* and *V. parahaemolyticus* from consumption of raw oysters continue to pose challenges both to public health and the shellfish industry in the US as well as other countries. Recent studies suggest that in addition to variation in environmental conditions due to climate change, strain variation may contribute significantly to the number of illnesses in different geographic locations. Consequently different risk models may need to be developed or modified based on regional environmental differences as well as variations in the strains responsible for illnesses, especially for *V. parahaemolyticus*.

5.5 Other human foodborne pathogenic vibrios

5.5.1 *V. cholerae*

While less likely, other members of the *Vibrio* genus can cause foodborne illness. Although *V. cholera* is typically transmitted through contamination of the water supply, it is autochthonous in the marine environment and like all vibrios can be concentrated by filter-feeding shellfish and other seafood. Between 2003 and 2007, eight cases of severe diarrhea were linked to strains of *V. cholerae* belonging to the serogroup O75, which produce cholera toxin (CT). All eight individuals had eaten raw oysters that were traced to the US Gulf Coast (Tobin D'Angelo *et al.*, 2008). More recently, a second outbreak attributed to consumption of oysters harboring strains of the same serogroup was identified in Florida in the spring of 2011 (Onifade *et al.*, 2011). Seafood-associated, non-O1, non-O139 strains of

V. cholerae (i.e. strains that do not produce CT) have also been reported to be responsible for an illnesses (Altekruse et al., 2000).

5.5.2 *V. mimicus*

V. mimicus is a species that is closely related to *V. cholerae* and clinical isolates are often found to carry genes related to the cholera toxin gene in *V. cholerae. V. mimicus* was identified as the cause of a sizeable 2004 foodborne outbreak in Thailand (Chitov *et al.*, 2008). Over 300 people attending a dinner at which a number of seafood dishes were served reported symptoms that included diarrhea, abdominal pain and vomiting between 6 and 36 h after the meal. Rectal swabs were recovered from 24 patients. *V. mimicus* was recovered from all 24 specimens. No *Salmonella* spp. were recovered. A more limited outbreak was recently reported in Washington State (CDC, 2010). *V. mimicus* was isolated from the cultures of stool specimens of two individuals who had consumed crayfish at a party. Although the crayfish was cooked, it was served in the original unwashed container and infections were probably due to cross-contamination. Isolated strains were found to be positive for genes encoding the cholera toxin.

5.5.3 *V. hollisae*

V. hollisae, which has been reclassified as *Grimontia hollisae* (Thompson, 2003), has been blamed for a number of gastroenteritis cases associated with seafood. This bacterium cannot be cultured on the typical medium used for isolation of vibrios, thiosulfate-citrate-bile salts-sucrose (TCBS) agar (Massad and Oliver, 1987), and therefore is likely to be underreported. The majority of reports of *G. hollisae*-related illnesses have been in patients with an underlying disease. However, *G. hollisae* was isolated from an otherwise healthy male in Pennsylvania suffering from severe gastroenteritis and acute renal failure, who had recently consumed shellfish (Hinestrosa *et al.*, 2007). More recently in France, *G. hollisae* was isolated from an elderly patient with a history of liver disease who had eaten raw oysters. The patient developed severe gastroenteritis and subsequent bacteremia (Edouard *et al.*, 2008). Lowry *et al.* (1986) reported isolation of *G. hollisae* from the blood of a 65-year-old man with a history of alcoholism. The only seafood he had consumed was catfish at lunch and dinner. Symptoms began by day's end. The fish was either undercooked or infection was due to cross-contamination. Gras-Rouzet *et al.* (1996) reported a case of septicemia due to *G. hollisae* that was linked to consumption of raw cockles by an elderly man in Brittany, France. This patient, who recovered fully, is unique in that he apparently had no history of liver disease or any other underlying condition.

5.5.4 *V. fluvialis*

V. fluvialis has also been identified as a cause of seafood-associated gastroenteritis (Igbinosa and Okoh, 2010). This species was found to be responsible for *ca.* 10%

Published by Woodhead Publishing Limited, 2013

of cases of gastroenteritis in a 10-year survey of the US Gulf region (Altekruse *et al.*, 2000). A 2006 report describes cholera-like gastroenteritis in a 72-year-old male in Mississippi who had consumed shellfish within the previous week (Allton *et al.*, 2006). *V. fluvialis* was the sole isolate from stool on TCBS agar.

5.5.5 Other *Vibrio* spp.

V. alginolyticus, V. furnissii and *Photobacterium (Vibrio) damsela* have infrequently been reported to cause gastroenteritis in association with seafood consumption (Altekruse *et al.*, 2000), typically in immunocompromised individuals. Several of these vibrios have also been responsible for life-threatening cases of septicemia due to the consumption of raw or undercooked seafood. For example, after consuming raw fish in 2008, a cirrhotic Korean patient died from septic shock due to *V. alginolyticus* (Lee *et al.*, 2008). Recently, *V. furnissii* was isolated from the blood of a diabetic patient presenting with skin lesions (Derber *et al.*, 2011). The source of the infection was speculated to be shrimps, apparently undercooked, consumed about a week prior to hospital admission. Finally, there is a single report of a fatal infection due to septicemia caused by *P.* (*Vibrio*) *damsela* (Shin *et al.*, 1996). The victim had a history of alcoholic liver disease and diabetes and had consumed raw eel the day before becoming symptomatic.

5.6 Current and future trends

5.6.1 Prediction of incidence and risk

The US FDA has published a *V. parahaemolyticus* risk assessment for consumption of raw oysters that predicts densities of the bacterium based on water temperature (FAO/WHO, 2011). However, this risk assessment tool has been less effective for areas outside of the Gulf of Mexico, particularly in shellfish harvest areas in Washington State (Washington State Department of Health, personal communications), where temperature alone has not been a good predictor of *V. parahaemolyticus* prevalence in oysters. There have been several efforts to develop more refined tools that not only incorporate temperature, but also other environmental parameters such as salinity, chlorophyll concentrations and other factors. In one example, remotely sensed sea surface temperature, turbidity and chlorophyll concentrations were obtained for a section of the Gulf of Mexico by National Oceanic and Atmospheric Administration (NOAA) and National Aeronautics and Space Administration (NASA) satellites (Phillips *et al.*, 2007). After validation of the remotely sensed data with similar measurements gathered *in situ*, a new model was developed that integrates the FDA temperature model with the remotely sensed data. It graphically displays the predicted *V. parahaemolyticus* density in oysters against spatial variations in water temperature. Efforts to incorporate remotely sensed environmental data into *Vibrio* risk models elsewhere in the world are still in the nascent stage, and in some circumstances, such as the US Pacific Northwest, obtaining year-round satellite data is encumbered by cloud cover. It is envisioned that more accurate risk models or early warning systems will require a better understanding of the combinations of environmental parameters that are needed to predict increased prevalence of *V. parahaemolyticus* as well as what factors determine whether there are sufficient numbers of virulent strains present in the shellfish. When these factors are better understood, it is foreseeable that future early warning systems will be developed, which will integrate satellite sensors with *in situ* oceanographic buoys and novel biosensors.

5.6.2 Climate change and the emergence of new virulent strains

Future climate change resulting in elevated ocean temperatures or other environmental variability may play an important role in the prevalence of vibrios. For example, in 2004 there was a significant outbreak of *V. parahaemolyticus* gastroenteritis caused by consumption of oysters harvested in Prince William Sound, Alaska, with over 400 confirmed cases (McLaughlin *et al.*, 2005). Increased water temperature was considered to be the primary cause of the emergence of *V. parahaemolyticus* in an area where it had not been a problem in the past, as the temperature remained above 15°C for a two-month period. However, there was also evidence that a more virulent strain was involved in this outbreak, as phylogenetic characterization showed the majority of clinically isolated strains to be distinct from the *V. parahaemolyticus* pandemic complex (Gonzalez-Escalona and Martinez-Urtaza, 2008). Strains of *V. parahaemolyticus* phylogenetically related to the pandemic complex are prevalent in environmental samples from the same geographical area, but have yet to be isolated from clinical disease (Turner *et al.*, 2013). This suggests that local environmental factors may play a strong role in the epidemiology of this pathogen. Such factors may include changes in the densities of phytoplankton and zooplankton that *Vibrio* spp. are known to attach to, or changes in water chemistry brought about by increased carbon dioxide and subsequent acidification of marine environments. Indeed, since ocean acidification significantly alters the ocean's carbonate chemistry, the resulting impacts on plankton exoskeleton composition or bivalve shell deposition could significantly change *Vibrio* spp. habitat, possibly leading to the expansion of seasonal and geographic distributions or the emergence of new or hypervirulent strains.

5.7 References

ALLTON, D.R., FORGIONE, M.A. and GROS, S.P., 2006. Cholera-like presentation in *Vibrio fluvialis* enteritis. *Southern Medical Journal*, 99(7), 765–767.

ALTEKRUSE, S.F. *et al.*, 2000. *Vibrio* gastroenteritis in the US Gulf of Mexico region: The role of raw oysters. *Epidemiology and Infection*, 124(3), 489–495.

AMARO, C. and BIOSCA, E.G., 1996. *Vibrio vulnificus* biotype 2, pathogenic for eels, is also an opportunistic pathogen for humans. *Applied and Environmental Microbiology*, 62(4), 1454–1457.

ANDREWS, L., JAHNCKE, M. and MALLIKARJUNAN, K., 2012. Low dose gamma irradiation to reduce pathogenic vibrios in live oysters (*Crassostrea virginica*). *Journal of Aquatic Food Product Technology*, 12(3), 71–82.

ANSARUZZAMAN, M. *et al.*, 2005. Pandemic serovars (O3: K6 and O4: K68) of *Vibrio parahaemolyticus* associated with diarrhea in Mozambique: Spread of the pandemic into the African continent. *Journal of Clinical Microbiology*, 43(6), 2559–2562.

AZNAR, R. *et al.*, 1994. Sequence determination of rRNA genes of pathogenic *Vibrio* species and whole-cell identification of *Vibrio vulnificus* with rRNA-targeted oligonucleotide probes. *International Journal of Systematic Bacteriology*, 44(2), 330–337.

BAKER-AUSTIN, C. *et al.*, 2010. Environmental occurrence and clinical impact of *Vibrio vulnificus* and *Vibrio parahaemolyticus*: A European perspective. *Environmental Microbiology Reports*, 2(1), 7–18.

BAKER-AUSTIN, C. *et al.*, 2012. Emerging *Vibrio* risk at high latitudes in response to ocean warming. *Nature Climate Change*, 2(8), 1–5.

BERLIN, D.L. *et al.*, 1999. Response of pathogenic *Vibrio* species to high hydrostatic pressure. *Applied and Environmental Microbiology*, 65(6), 2776–2780.

BISHARAT, N. *et al.*, 1999. Clinical, epidemiological, and microbiological features of *Vibrio vulnificus* biogroup 3 causing outbreaks of wound infection and bacteraemia in Israel. Israel *Vibrio* Study Group. *The Lancet*, 354(9188), 1421–1424.

BLAKE, P.A. *et al.*, 1979. Disease caused by a marine *Vibrio*. Clinical characteristics and epidemiology. *New England Journal of Medicine*, 300(1), 1–5.

BONNER, J.R. *et al.*, 1983. Spectrum of *Vibrio* infections in a Gulf-Coast community. *Annals of Internal Medicine*, 99(4), 464–469.

BROBERG, C.A., CALDER, T.J. and ORTH, K., 2011. *Vibrio parahaemolyticus* cell biology and pathogenicity determinants. *Microbes and Infection*, 13(12–13), 992–1001.

BURDETTE, D.L., YARBROUGH, M.L. and ORTH, K., 2009. Not without cause: *Vibrio parahaemolyticus* induces acute autophagy and cell death. *Autophagy*, 5(1), 100–102.

CABURLOTTO, G. *et al.*, 2010. Effect on human cells of environmental *Vibrio parahaemolyticus* strains carrying type III secretion system 2. *Infection and Immunity*, 78(7), 3280–3287.

CDC, 2005. *Vibrio* illnesses after Hurricane Katrina – multiple states, August–September 2005. Centers for Disease Control and Prevention. *Morbidity and Mortality Weekly Report*, 54(37), 928–931.

CDC, 2006. *Vibrio parahaemolyticus* infections associated with consumption of raw shellfish – three states, 2006. Centers for Disease Control and Prevention. *Morbidity and Mortality Weekly Report*, 55(31), 854–856.

CDC, 2010. Notes from the field: *Vibrio mimicus* infection from consuming crayfish – Spokane, Washington, June 2010. Centers for Disease Control and Prevention. *Morbidity and Mortality Weekly Report*, 59(42), 1374.

CHEN, C.-Y.C. *et al.*, 2003. Comparative genome analysis of *Vibrio vulnificus*, a marine pathogen. *Genes and Development*, 13(12), 2577–2587.

CHITOV, T. *et al.*, 2008. An incidence of large foodborne outbreak associated with *Vibrio mimicus*. *European Journal of Clinical Microbiology and Infectious Diseases*, 28(4), 421–424.

CHOWDHURY, N.R.N. *et al.*, 2000. Molecular evidence of clonal *Vibrio parahaemolyticus* pandemic strains. *Emerging Infectious Diseases*, 6(6), 631–636.

CHOWDHURY, N.R.N. *et al.*, 2004. Assessment of evolution of pandemic *Vibrio parahaemolyticus* by multilocus sequence typing. *Journal of Clinical Microbiology*, 42(3), 1280–1282.

COLEMAN, S.S. *et al.*, 1996. Detection of *Vibrio vulnificus* biotypes 1 and 2 in eels and oysters by PCR amplification. *Applied and Environmental Microbiology*, 62(4), 1378–1382.

DEPAOLA, A. *et al.*, 2000. Environmental investigations of *Vibrio parahaemolyticus* in oysters after outbreaks in Washington, Texas, and New York (1997 and 1998). *Applied and Environmental Microbiology*, 66(11), 4649–4654.

DERBER, C. *et al.*, 2011. *Vibrio furnissii*: An unusual cause of bacteremia and skin lesions after ingestion of seafood. *Journal of Clinical Microbiology*, 49(6), 2348–2349.

EDOUARD, S. *et al.*, 2008. *Grimontia hollisae*, a potential agent of gastroenteritis and bacteraemia in the Mediterranean area. *European Journal of Clinical Microbiology and Infectious Diseases*, 28(6), 705–707.

FARUQUE, S.M., ALBERT, M.J. and MEKALANOS, J.J., 1998. Epidemiology, genetics, and ecology of toxigenic *Vibrio cholerae*. *Microbiology and Molecular Biology Reviews*, 62(4), 1301–1314.

FELDHUSEN, F., 2000. The role of seafood in bacterial foodborne diseases. *Microbes and Infection*, 2(13), 1651–1660.

FAO/WHO, 2011. Risk assessment of *Vibrio parahaemolyticus* in seafood: Interpretative summary and technical report. Food and Agriculture Organization of the United Nations/ World Health Organization Microbiological Risk Assessment Series No 16, Rome, 193.

FDA, 2011. *Ionizing radiation for the treatment of food*. US Food and Drug Administration, Code of Federal Regulations, 21 CFR 179.26. US Government Printing Office, Washington, DC.

FUENZALIDA, L. *et al.*, 2006. *Vibrio parahaemolyticus* in shellfish and clinical samples during two large epidemics of diarrhoea in southern Chile. *Environmental Microbiology*, 8(4), 675–683.

FUENZALIDA, L. *et al.*, 2007. *Vibrio parahaemolyticus* strains isolated during investigation of the summer 2006 seafood related diarrhea outbreaks in two regions of Chile. *International Journal of Food Microbiology*, 117(3), 270–275.

GONZALEZ-ESCALONA, N. and MARTINEZ-URTAZA, J., 2008. Determination of molecular phylogenetics of *Vibrio parahaemolyticus* strains by multilocus sequence typing. *Journal of Bacteriology*, 190(8), 2831–2840.

GONZALEZ-ESCALONA, N. *et al.*, 2005. *Vibrio parahaemolyticus* diarrhea, Chile, 1998 and 2004. *Emerging Infectious Diseases*, 11(1), 129–131.

GONZALEZ-ESCALONA, N. *et al.*, 2011. Genome sequence of the clinical O4:K12 serotype *Vibrio parahaemolyticus* strain 10329. *Journal of Bacteriology*, 193(13), 3405–3406.

GRAS-ROUZET, S. *et al.*, 1996. First European case of gastroenteritis and bacteremia due to *Vibrio hollisae*. *European Journal of Clinical Microbiology and Infectious Diseases*, 15(11), 864–866.

GULIG, P.A.P., BOURDAGE, K.L.K. and STARKS, A.M.A., 2005. Molecular pathogenesis of *Vibrio vulnificus*. *Journal of Microbiology*, 43(5), 118–131.

GULIG, P.A. *et al.*, 2010. SOLiD sequencing of four *Vibrio vulnificus* genomes enables comparative genomic analysis and identification of candidate clade-specific virulence genes. *BMC Genomics*, 11(1), 512–512.

HARTH, E., 2009. Epidemiology of *Vibrio parahaemolyticus* outbreaks, southern Chile. *Emerging Infectious Diseases*, 15(2), 163–168.

HELMS, S.D., OLIVER, J.D. and TRAVIS, J.C., 1984. Role of heme compounds and haptoglobin in *Vibrio vulnificus* pathogenicity. *Infection and Immunity*, 45(2), 345–349.

HINESTROSA, F., MADEIRA, R.G. and BOURBEAU, P.P., 2007. Severe gastroenteritis and hypovolemic shock caused by *Grimontia (Vibrio) hollisae* Infection. *Journal of Clinical Microbiology*, 45(10), 3462–3463.

HLADY, W.G., 1997. *Vibrio* infections associated with raw oyster consumption in Florida, 1981–1994. *Journal of Food Protection*, 60(4), 353–357.

HLADY, W.G. and KLONTZ, K.C., 1996. The epidemiology of *Vibrio* infections in Florida, 1981–1993. *Journal of Infectious Diseases*, 173(5), 1176–1183.

HONDA, T., NI, Y.X. and MIWATANI, T., 1988. Purification and characterization of a hemolysin produced by a clinical isolate of Kanagawa phenomenon-negative *Vibrio parahaemolyticus* and related to the thermostable direct hemolysin. *Infection and Immunity*, 56(4), 961–965.

HONDA, T. *et al.*, 1992. The thermostable direct hemolysin of *Vibrio parahaemolyticus* is a pore-forming toxin. *Canadian Journal of Microbiology*, 38(11), 1175–1180.

HSIEH, Y.C. *et al.*, 2003. Study of capsular polysaccharide from *Vibrio parahaemolyticus*. *Infection and Immunity*, 71(6), 3329–3336.

IGBINOSA, E.O. and OKOH, A.I., 2010. *Vibrio fluvialis*: An unusual enteric pathogen of increasing public health concern. *International Journal of Environmental Research and Public Health*, 7(10), 3628–3643.

JONES, J.L. *et al.*, 2012. Biochemical, serological, and virulence characterization of clinical and oyster *Vibrio parahaemolyticus* isolates. *Journal of Clinical Microbiology*, 50(7), 2343–2352.

JONES, M.K. and OLIVER, J.D., 2009. *Vibrio vulnificus*: Disease and pathogenesis. *Infection and Immunity*, 77(5), 1723–1733.

JULIE, D. *et al.*, 2010. Ecology of pathogenic and non-pathogenic *Vibrio parahaemolyticus* on the French Atlantic coast. Effects of temperature, salinity, turbidity and chlorophyll a. *Environmental Microbiology*, 12(4), 929–937.

KIM, Y.R. *et al.*, 2008. *Vibrio vulnificus* RTX toxin kills host cells only after contact of the bacteria with host cells. *Cellular Microbiology*, 10(4), 848–862.

KIMURA, B. *et al.*, 2008. Multiple-locus variable-number of tandem-repeats analysis distinguishes *Vibrio parahaemolyticus* pandemic O3:K6 strains. *Journal of Microbiological Methods*, 72(3), 313–320.

KODAMA, T. *et al.*, 2007. Identification and characterization of VopT, a novel ADP-ribosyltransferase effector protein secreted via the *Vibrio parahaemolyticus* type III secretion system 2. *Cellular Microbiology*, 9(11), 2598–2609.

KRACHLER, A.M., HAM, H. and ORTH, K., 2011. Outer membrane adhesion factor multivalent adhesion molecule 7 initiates host cell binding during infection by Gram-negative pathogens. *Proceedings of the National Academy of Sciences*, 108(28), 11614–11619.

LAWRENCE, M.C. *et al.*, 2008. The roles of MAPKs in disease. *Cell Research*, 18(4), 436–442.

LEE, D.-Y. *et al.*, 2008. Septic shock due to *Vibrio alginolyticus* in a cirrhotic patient: The first case in Korea. *Yonsei Medical Journal*, 49(2), 329–332.

LEE, J.-H. *et al.*, 2004a. Role of flagellum and motility in pathogenesis of *Vibrio vulnificus*. *Infection and Immunity*, 72(8), 4905–4910.

LEE, J.H. *et al.*, 2007. Identification and characterization of the *Vibrio vulnificus* rtxA essential for cytotoxicity *in vitro* and virulence in mice. *Journal of Microbiology*, 45(2), 146–152.

LEE, S.E. *et al.*, 2004b. Production of *Vibrio vulnificus* hemolysin *in vivo* and its pathogenic significance. *Biochemical and Biophysical Research Communications*, 324(1), 86–91.

LOWRY, P.W., MCFARLAND, L.M. and THREEFOOT, H.K., 1986. *Vibrio hollisae* septicemia after consumption of catfish. *Journal of Infectious Diseases*, 154(4), 730–731.

LYNCH, T. *et al.*, 2005. *Vibrio parahaemolyticus* disruption of epithelial cell tight junctions occurs independently of toxin production. *Infection and Immunity*, 73(3), 1275–1283.

MAKINO, K. *et al.*, 2003. Genome sequence of *Vibrio parahaemolyticus*: A pathogenic mechanism distinct from that of *V. cholerae*. *The Lancet*, 361(9359), 743–749.

MARQUES, A. *et al.*, 2010. Climate change and seafood safety: Human health implications. *Food Research International*, 43(7), 1766–1779.

MARTINEZ-URTAZA, J.J. *et al.*, 2005. Pandemic *Vibrio parahaemolyticus* O3:K6, Europe. *Emerging Infectious Diseases*, 11(8), 1319–1320.

MASSAD, G.G. and OLIVER, J.D.J., 1987. New selective and differential medium for *Vibrio cholerae* and *Vibrio vulnificus*. *Applied and Environmental Microbiology*, 53(9), 2262–2264.

MATSUMOTO, C. *et al.*, 2000. Pandemic spread of an O3: K6 clone of *Vibrio parahaemolyticus* and emergence of related strains evidenced by arbitrarily primed PCR and toxRS sequence analyses. *Journal of Clinical Microbiology*, 38(2), 578–585.

MCLAUGHLIN, J.B. *et al.*, 2005. Outbreak of *Vibrio parahaemolyticus* gastroenteritis associated with Alaskan oysters. *The New England Journal of Medicine*, 353(14), 1463–1470.

MCPHERSON, V.L. *et al.*, 1991. Physiological effects of the lipopolysaccharide on *Vibrio vulnificus* on mice and rats. *Microbios*, 67(272–273), 141–149.

MORRIS, J.G.J., 1988. *Vibrio vulnificus* – A new monster of the deep? *Annals of Internal Medicine*, 109(4), 261–263.

MORRIS, J.G., 2003. Cholera and other types of vibriosis: A story of human pandemics and oysters on the half shell. *Clinical Infectious Diseases*, 37(2), 272–280.

MORRIS, J.G.J. and BLACK, R.E., 1985. Cholera and other vibrioses in the United States. *New England Journal of Medicine*, 312(6), 343–350.

NAIR, G.B. *et al.*, 2007. Global dissemination of *Vibrio parahaemolyticus* serotype O3:K6 and its serovariants. *Clinical Microbiology Reviews*, 20(1), 39–48.

NAKASONE, N. and IWANAGA, M., 1990. Pili of a *Vibrio parahaemolyticus* strain as a possible colonization factor. *Infection and Immunity*, 58(1), 61–69.

NASU, H. *et al.*, 2000. A filamentous phage associated with recent pandemic *Vibrio parahaemolyticus* O3: K6 strains. *Journal of Clinical Microbiology*, 38(6), 2156–2161.

NEWTON, A. *et al.*, 2012. Increasing rates of vibriosis in the United States, 1996–2010: Review of surveillance data from 2 systems. *Clinical Infectious Diseases*, 54(5), S391–S395.

NILSSON, W.B. *et al.*, 2003. Sequence polymorphism of the 16S rRNA gene of *Vibrio vulnificus* is a possible indicator of strain virulence. *Journal of Clinical Microbiology*, 41(1), 442–446.

NISHIBUCHI, M. *et al.*, 1985. Detection of the thermostable direct hemolysin gene and related DNA sequences in *Vibrio parahaemolyticus* and other vibrio species by the DNA colony hybridization test. *Infection and Immunity*, 49(3), 481–486.

NISHIBUCHI, M. *et al.*, 1989. Cloning and nucleotide sequence of the gene (*trh*) encoding the hemolysin related to the thermostable direct hemolysin of *Vibrio parahaemolyticus*. *Infection and Immunity*, 57(9), 2691–2697.

OKADA, N. *et al.*, 2010. Presence of genes for type III secretion system 2 in *Vibrio mimicus* strains. *BMC Microbiology*, 10, 302.

OKUDA, J. *et al.*, 1997. Emergence of a unique O3: K6 clone of *Vibrio parahaemolyticus* in Calcutta, India, and isolation of strains from the same clonal group from Southeast Asian travelers arriving in Japan. *Journal of Clinical Microbiology*, 35(12), 3150–3155.

OKURA, M. *et al.*, 2003. Genotypic analyses of *Vibrio parahaemolyticus* and development of a pandemic group-specific multiplex PCR assay. *Journal of Clinical Microbiology*, 41(10), 4676–4682.

ONIFADE, T.J.M. *et al.*, 2011. Toxin producing *Vibrio cholerae* O75 outbreak, United States, March to April 2011. *Eurosurveillance*, 16(20), 19870.

ONO, T. *et al.*, 2006. Identification of proteins secreted via *Vibrio parahaemolyticus* type III secretion system 1. *Infection and Immunity*, 74(2), 1032–1042.

PARANJPYE, R.N. and STROM, M.S., 2005. A *Vibrio vulnificus* Type IV pilin contributes to biofilm formation, adherence to epithelial cells, and virulence. *Infection and Immunity*, 73(3), 1411–1422.

PARANJPYE, R.N. *et al.*, 1998. The type IV leader peptidase/N-methyltransferase of *Vibrio vulnificus* controls factors required for adherence to HEp-2 cells and virulence in iron-overloaded mice. *Infection and Immunity*, 66(12), 5659–5668.

PARANJPYE, R.N. *et al.*, 2007. Role of type IV pilins in persistence of *Vibrio vulnificus* in *Crassostrea virginica* oysters. *Applied and Environmental Microbiology*, 73(15), 5041–5044.

PARK, K.S. *et al.*, 2004. Cytotoxicity and enterotoxicity of the thermostable direct hemolysin-deletion mutants of *Vibrio parahaemolyticus*. *Microbiology and Immunology*, 48(4), 313–318.

PHILLIPS, A.M.B. *et al.*, 2007. An evaluation of the use of remotely sensed parameters for prediction of incidence and risk associated with *Vibrio parahaemolyticus* in Gulf Coast oysters (*Crassostrea virginica*). *Journal of Food Protection*, 70(4), 879–884.

POWELL, J.L.J. *et al.*, 1997. Release of tumor necrosis factor alpha in response to *Vibrio vulnificus* capsular polysaccharide in *in vivo* and *in vitro* models. *Infection and Immunity*, 65(9), 3713–3718.

QUILICI, M.-L.M. *et al.*, 2005. Pandemic *Vibrio parahaemolyticus* O3:K6 spread, France. *Emerging Infectious Diseases*, 11(7), 1148–1149.

QUIRKE, A.M. *et al.*, 2006. Genomic island identification in *Vibrio vulnificus* reveals significant genome plasticity in this human pathogen. *Bioinformatics*, 22(8), 905–910.

RAIMONDI, F. *et al.*, 2000. Enterotoxicity and cytotoxicity of *Vibrio parahaemolyticus* thermostable direct hemolysin in *in vitro* systems. *Infection and Immunity*, 68(6), 3180–3185.

RHEE, J.E., KIM, K.-S. and CHOI, S.H., 2005. CadC activates pH-dependent expression of the *Vibrio vulnificus* cadBA operon at a distance through direct binding to an upstream region. *Journal of Bacteriology*, 187(22), 7870–7875.

SANJUAN, E., GONZALEZ-CANDELAS, F. and AMARO, C., 2011. Polyphyletic origin of *Vibrio vulnificus* biotype 2 as revealed by sequence-based analysis. *Applied and Environmental Microbiology*, 77(2), 688–695.

SHIN, J.H.J. *et al.*, 1996. Primary *Vibrio damsela* septicemia. *Clinical Infectious Diseases*, 22(5), 856–857.

SHIRAI, H. *et al.*, 1990. Molecular epidemiologic evidence for association of thermostable direct hemolysin (Tdh) and Tdh-related hemolysin of *Vibrio parahaemolyticus* with gastroenteritis. *Infection and Immunity*, 58(11), 3568–3573.

SIMONSON, J.G. and SIEBELING, R.J., 1993. Immunogenicity of *Vibrio vulnificus* capsular polysaccharides and polysaccharide-protein conjugates. *Infection and Immunity*, 61(5), 2053–2058.

STROM, M.S. and PARANJPYE, R.N., 2000. Epidemiology and pathogenesis of *Vibrio vulnificus*. *Microbes and Infection*, 2(2), 177–188.

STYLES, M.F., HOOVER, D.G. and FARKAS, D.F., 1991. Response of *Listeria monocytogenes* and *Vibrio parahaemolyticus* to high hydrostatic pressure. *Journal of Food Science*, 56(5), 1404–1407.

SU, Y.-C. and LIU, C., 2007. *Vibrio parahaemolyticus*: A concern of seafood safety. *Food Microbiology*, 24(6), 549–558.

SUFFREDINI, E. *et al.*, 2011. Pulsed-field gel electrophoresis and PCR characterization of environmental *Vibrio parahaemolyticus* strains of different origins. *Applied and Environmental Microbiology*, 77(17), 6301–6304.

THOMPSON, F.L., 2003. Reclassification of *Vibrio hollisae* as *Grimontia hollisae* gen. nov., comb. nov. *International Journal of Systematic and Evolutionary Microbiology*, 53(5), 1615–1617.

TISON, D.L. *et al.*, 1982. *Vibrio vulnificus* biogroup 2: New biogroup pathogenic for eels. *Applied and Environmental Microbiology*, 44(3), 640–646.

TOBIN D'ANGELO, M. *et al.*, 2008. Severe diarrhea caused by cholera toxin-producing *Vibrio cholerae* serogroup O75 infections acquired in the southeastern United States. *Clinical Infectious Diseases*, 47(8), 1035–1040.

TROSKY, J.E. *et al.*, 2004. Inhibition of MAPK signaling pathways by VopA from *Vibrio parahaemolyticus*. *Journal of Biological Chemistry*, 279(50), 51953–51957.

TROSKY, J.E. *et al.*, 2007. VopA inhibits ATP binding by acetylating the catalytic loop of MAPK kinases. *Journal of Biological Chemistry*, 282(47), 34299–34305.

TURNER, J.W. *et al.*, 2013. Population structure of clinical and environmental *Vibrio parahaemolyticus* from the Pacific Northwest Coast of the United States. Accepted for PLoS ONE.

VEZZULLI, L. *et al.*, 2009. Benthic ecology of *Vibrio* spp. and pathogenic *Vibrio* species in a coastal Mediterranean environment (La Spezia Gulf, Italy). *Microbial Ecology*, 58(4), 808–818.

WARNER, E. and OLIVER, J.D., 2007. Population structures of two genotypes of *Vibrio vulnificus* in oysters (*Crassostrea virginica*) and seawater. *Applied and Environmental Microbiology*, 74(1), 80–85.

WONG, H.C. and LIN, C.H., 2001. Evaluation of typing of *Vibrio parahaemolyticus* by three PCR methods using specific primers. *Journal of Clinical Microbiology*, 39(12), 4233–4240.

WONG, H.-C. *et al.*, 2007. A pulsed-field gel electrophoresis typing scheme for *Vibrio parahaemolyticus* isolates from fifteen countries. *International Journal of Food Microbiology*, 114(3), 280–287.

WRIGHT, A.C., SIMPSON, L.M. and OLIVER, J.D., 1981. Role of iron in the pathogenesis of *Vibrio vulnificus* infections. *Infections and Immunity*, 34(2), 503–507.

WRIGHT, A.C., SIMPSON, L.M. and OLIVER, J.D., 1990. Phenotypic evaluation of acapsular transposon mutants of *Vibrio vulnificus*. *Infections and Immunity*, 58(6), 1769–1773.

YARBROUGH, M.L. *et al.*, 2009. AMPylation of Rho GTPases by *Vibrio* VopS disrupts effector binding and downstream signaling. *Science*, 323(5911), 269–272.

YOSHIDA, S., OGAWA, M. and MIZUGUCHI, Y., 1985. Relation of capsular materials and colony opacity to virulence of *Vibrio vulnificus*. *Infections and Immunity*, 47(2), 446–451.

ZAKARIA-MEEHAN, Z. *et al.*, 1988. Ability of *Vibrio vulnificus* to obtain iron from hemoglobin-haptoglobin complexes. *Infections and Immunity*, 56(1), 275–277.

6

Emerging parasites in food

Y. Ortega, University of Georgia, USA

DOI: 10.1533/9780857098740.2.114

Abstract: Foodborne parasites have affected humans since antiquity. The complexities of their life cycles play a significant role in their prevalence in certain locations worldwide. Good agricultural practices and food processing have limited or eradicated some parasites from select countries. Still, the emergence or re-emergence of foodborne parasites is occurring worldwide. This chapter will address protozoa and helminths of public health relevance and that are identified in the food supply. Available pre- and post-harvest measures to control parasites in foods are discussed.

Key words: parasite, Protozoa, Cestoda, nematode, Trematoda, helminth.

6.1 Introduction

Accounts of food- and waterborne parasitic infections affecting humans have been reported throughout human history. The presence of parasites varies because of many factors. Among them are appropriate environmental conditions, sanitation and economic conditions, and the presence of intermediary hosts or reservoirs. The life cycles of some of these parasites are so complex that controlling the contamination of foods or water can only be achieved by interrupting these cycles. Human behavior, including social and alimentary habits and sanitary practices, has an important role in the eradication and control of parasitic diseases in large regions of the world, and changing behavior is theoretically achievable. Several industrialized nations had eliminated some of these infections, but in recent years there are reports of parasites that are emerging or re-emerging. This rise in the number of parasitic infections could be associated with the increased travel to tropical regions, increased availability and consumption of fresh fruits and vegetables year round, and the emergence of other types of eating preferences or cuisines that include fresh or uncooked products. As a result of these changes, meats, fruits and vegetables are now constantly imported to industrialized nations from countries where these parasites are still endemic. Although not a rule, there have been instances where trace-back investigations have determined that some products might not have been produced following good agricultural practices.

Parasites comprise various and quite distinct groups of organisms. Some are unicellular (the Protozoa), and others are multicellular (the helminths). The helminths of most relevance to public health belong to the phylum Platyhelminthes. These include the Cestoda or tapeworms and the Trematoda or flukes. Also relevant is the phylum Nematoda, known as roundworms. This chapter will focus on parasites that are being more frequently reported in industrialized nations, particularly the US. The diversity of these foodborne parasites demonstrates their complex involvement with human health and the need for further studies to better understand their biology, including its presence in food products, and their ability to overcome some production and processing practices. Additional knowledge would lead to better strategies for their detection, control and prevention.

6.2 Protozoa

Protozoa are unicellular parasites. Overall, they tend to have two defined developmental stages: the cyst stage that is environmentally resistant and the reproductive stage that infect the host cells. Some are capable of reproducing sexually (*Cyclospora, Cryptosporidium, Sarcocystis* and *Toxoplasma*) whereas others only multiply asexually (*Giardia* and *Trypanosoma*).

6.2.1 *Cyclospora cayetanensis*

Earlier reports of organisms that could have been *Cyclospora* date from the early 1970s. However, it was not until 1992 that *C. cayetanensis* was fully identified and formally described as a new species. *Cyclospora cayetanensis* seems to be highly host specific and only infects humans. People acquire the infection by ingestion of contaminated water or foods containing sporulated oocysts (Fig. 6.1(a)). The oocysts, thanks to their cyst wall, survive the acidity of the gastric environment and excyst in the small intestine. There, the sporozoites are released and infect intestinal epithelial cells where they start multiplying asexually (type I and II meronts) in an exponential fashion, followed by differentiation into sexual stages or gamonts.[1] Fertilization of the macrogametocyte by the microgametocyte results in the formation of immature or unsporulated oocysts, which are excreted to the environment. Once excreted, it takes between 7–15 days in the environment for these oocysts to sporulate and become infectious.[2]

Patients with cyclosporiasis usually present with acute diarrhea. Anorexia, flatulence, abdominal pain and nausea are also present. Symptoms may persist for 19 days to 1.5 months although there have been reports of prolonged infections in HIV+ patients for up to 199 days. Severe and prolonged diarrhea is usually noted in immunocompromised individuals, children and the elderly. A few reports have indicated that biliary disease after cyclosporiasis can occur. Acalculous

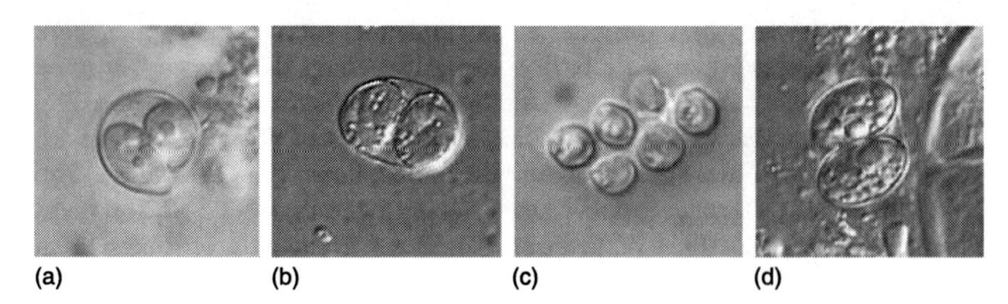

Fig. 6.1 Coccidian oocysts: (a) *Cyclospora cayetanensis* (8–10 μm); (b) *Toxoplasma gondii* (10–12 μm); (c) *Cryptosporidium parvum* (4–6 μm). Oocysts containing sporocysts: (d) *Sarcocystis* spp. (15–20 μm). Sources: (a) and (c) Y. Ortega, Center for Food Safety – University of Georgia; (b) and (d) www.dpd.cdc.gov/dpdx/HTML/Image_Library.htm.

cholecystitis, Guillain–Barré syndrome and Reiter syndrome have also been anecdotally reported following *Cyclospora* infection.[3]

Regardless of the setting, *Cyclospora* infections are very seasonal, and usually occur in specific months of the year. The seasonal pattern tends to vary according to the endemic area where people go to or produce is imported from. Interestingly, it is unknown how or where the oocysts of *Cyclospora* survive until the next infectious season. Ambient temperature, relative humidity and light intensity seem to play a key role in sporulation and in the marked seasonality so characteristic of this parasite. In Haiti and Peru, cases of cyclosporiasis occur during the months of December to March.[2,4–6] In Nepal, the high season falls during the months of June and July,[7–9] and in Guatemala and Mexico from May to August[7] (and also based on unpublished data supplied by Orozco-Mosqueda).

In endemic areas, young children usually have diarrhea when infected with *Cyclospora*. After repeated exposures to the parasite, the cyclosporiasis-associated diarrhea is of shorter duration and less severe.[2,5,8,9] In non-endemic areas, individuals of all ages are susceptible to *Cyclospora* infection. The first cases of *Cyclospora* in the US were found in travelers returning from *Cyclospora*-endemic locations. Since 1995, cases and outbreaks of cyclosporiasis became more frequently reported in the US and were mostly associated with the consumption of raw imported berries.[10,11] From the year 2000 and forward, most outbreak reports have implicated lettuce, basil and sugar peas.[3]

6.2.2 *Toxoplasma gondii*

In the US, *Toxoplasma* is considered the second cause of death (24%) and the fourth cause of hospitalizations (8%) associated with foodborne illnesses. Toxoplasmosis can be acquired by ingestion of meats containing tissue cysts or by ingestion of fruits, vegetables or water contaminated with feline feces containing *Toxoplasma* oocysts.[12]

Toxoplasmosis has a complex life cycle, requiring definitive (cats), and intermediate hosts. When a cat ingests infected tissues from an intermediate host,

Toxoplasma multiplies sexually and asexually in the intestinal epithelium of the cat. Following gamont fertilization, the zygotes form and differentiate into the immature oocysts that are excreted in cat feces. These oocysts (Fig. 6.1(b)) are infectious to susceptible intermediate hosts, which include most warm-blooded vertebrates. The intermediate hosts will acquire the infection after ingestion of foods or drinking water that is contaminated with cat feces containing *Toxoplasma* oocysts. In the intermediate hosts (including humans), oocysts excyst and sporozoites are released, crossing the intestinal wall and migrating to various tissues where they will encyst and multiply and differentiate into bradyzoites. If an intermediate host or cat ingests infected meats with *Toxoplasma* cysts, the bradyzoites are released upon digestion and transform into tachyzoites, which multiply rapidly and infect various types of tissue. In the case of humans, expectant mothers are at higher risk because *Toxoplasma* can cross the placental tissues and infect the fetus. Infection during the first months of gestation is particularly severe, resulting in malformation, hydrocephaly, encephalopathy, retardation and even death. It has been estimated that the economic impact of congenital toxoplasmosis is as high as $7.7 billion a year.[13]

The acute phase of human infections is characterized by fever, lymph node enlargement, asthenia and headaches. Infection can also be asymptomatic. In immunocompetent individuals, 10–20% of infected people will have symptoms that may include chorioretinitis, lymphadenitis, myocarditis or polymyositis.[14]

Human infections with *Toxoplasma* have been associated with ingestion of undercooked meats from pigs, goats or sheep. Infections are not restricted to specific areas of the world and *Toxoplasma* has been described worldwide. Outbreaks have been associated with contaminated water and foods. Raw goat milk, raw meat and drinking creek water have been attributed as sources of infection.[15] There was an outbreak of 110 cases of acute toxoplasmosis in Canada and municipal drinking water was implicated.[16] In Brazil, 426 people[17] and in India 248 individuals[18] acquired the infection by drinking municipal drinking water. Major *Toxoplasma* foodborne outbreaks have not been reported, probably because most infected individuals are asymptomatic. *Toxoplasma* oocysts have also been reported in vegetables, suggesting another potential source of infection.

6.2.3 *Giardia*, *Cryptosporidium* and *Sarcocystis*

Infections with *Giardia* and *Cryptosporidium* occur most frequently through ingestion of contaminated drinking or recreational water. *Giardia* is a flagellate and the vegetative stage, the trophozoite, divides by binary fission inside the small intestine of the host. Once in the gastrointestinal tract, the trophozoites attach to the microvilli by their ventral disk. As they are displaced to distal sections of the enteric tract and enter the colon, the trophozoites mature and develop to form cysts. This stage is environmentally resistant and is excreted by the infected host. When cysts are ingested by another susceptible host, two trophozoites are released from the cyst and start asexual multiplication in the proximal small bowel.

At the DNA level, it is now known that there are several types of morphologically identical *Giardia* species, suggesting that *Giardia duodenalis* (synonyms *G. lamblia* and *G. intestinalis*) is a species complex. *Giardia duodenalis* (assemblage A) and *G. enterica* (assemblage B) can infect humans, other primates, dogs, cats, livestock and other wild mammals. Other species that can infect animals but are not infectious to humans are: *G. canis, G. bovis, G. cati, G. simondi, G. microti, G. psittaci, G. ardeae, G. muris* and *G. agilis*. Two other species that infect pinnipeds and marsupials have yet to be named.[19]

Cryptosporidium is another coccidian parasite with a life cycle similar to that of *Cyclospora*. A significant difference is that *Cryptosporidium* resides in a vacuole inside the cell although it has an extracytoplasmic localization. The parasitic vacuole communicates with the infected cell via a feeding organelle. Another difference is that when oocysts are excreted they are already infectious and have four sporozoites. Unlike *Cyclospora* or *Toxoplasma*, the oocysts of *Cryptosporidium* lack sporocysts; thus the sporozoites are only protected by the oocyst wall (Fig. 6.1(c)). *Cryptosporidium* was initially considered to be a single species: *C. parvum*. With the development of molecular tools, now it is known that *C. parvum* is indeed a species complex. Several defined species, genotypes and subtypes are now clearly recognized. Seven genotypes have been reported in humans: *C. parvum, C. hominis, C. meleagridis, C. felis, C. canis, C. ubiquitum* and *C. muris*. Humans are more frequently infected by the anthroponotic species, *C. hominis*. *C. parvum* is a zoonotic species and is most frequently found in young calves. Within *C. hominis*, subtype IbA10G2 is the most frequently associated with waterborne outbreaks. *C. parvum*, subtype IIaA15G2R1, is also frequent and has been primarily identified in calves and humans in Portugal, Slovenia and the Netherlands. In Ireland, the IIaA15G2R1 subtype of *C. parvum* is the one most frequently found in humans and calves.[20] *Cryptosporidium* with human and bovine genotypes have also been identified in oysters from six of seven commercial harvesting sites in Chesapeake Bay. These oocysts were infectious to mice, confirming its viability and suggesting a potential source of infection. Nevertheless, human infections have yet to be associated with consumption of shellfish.[21]

Cryptosporidium and *Giardia* infections are associated with acute persistent diarrhea, abdominal cramps, nausea and anorexia. *Giardia* infections are also characterized with foul flatulence and the recurrence of infection is common. In endemic regions, children can present asymptomatic cryptosporidiosis and giardiasis, however, stunting and cognitive impairment have been reported in children that had repeated infections.[22,23]

Although the mechanism of transmission is mostly waterborne (from drinking water and recreational treated or natural water such as pools, play parks and artificial lakes), *Cryptosporidium* and *Giardia* have been isolated from a variety of foods including ice cubes, fresh produce (mostly herbs and vegetables grown close to the ground), and shellfish such as mussels, oysters, cockles and clams.[21,24,25] Cysts and oocysts have also been reported in raw meats including chicken breasts, minced beef and pork chops from retail outlets. Reports indicated that *Cryptosporidium* contamination of tripe, meat and liver could occur during processing at a slaughterhouse. Outbreaks have also been associated with milk and apple cider, more likely contaminated with cattle feces.[26]

Foodborne outbreaks have also been attributed to food handlers who prepared ready-to-eat foods while ill. In 1998, on a Washington university campus, 88 students complained of gastrointestinal illness. An ill food handler, who processed fruits and vegetables that were served raw, was the source of the outbreak. Four days before the onset of symptoms, the food handler cared for his 18-month-old niece, who was ill with diarrhea and vomiting. Molecular and epidemiological studies later demonstrated that this food handler was the source of the *Cryptosporidium* outbreak.[27]

Sarcocystis spp. have been implicated in foodborne illnesses in humans, particularly travelers returning from endemic areas. *Sarcocystis* infection can present two forms of illness: muscular and intestinal sarcocystosis. Many species of *Sarcocystis* can cause muscular sarcocystosis. When oocysts (Fig. 6.1(d)) or sporocysts of *Sarcocystis* spp. are ingested with water or foods, the parasite will survive gastric digestion, and migrate and multiply asexually within the vascular endothelium, followed by dissemination into the skeletal, cardiac and smooth muscle.[28] In muscular sarcocystosis, sarcocysts form, which can result in eosinophilic myositis. There is no proven treatment for muscular sarcocystosis and symptoms resolve in a few weeks to months.[29]

Most cases have been reported in persons living in tropical or subtropical countries in Asia and South-East Asia. In 2011, muscular sarcocystosis was reported in 32 travelers returning from the Tioman Island off the east coast of Peninsular Malaysia. Patients reported fever and muscular pain, often severe and prolonged. All had peripheral eosinophilia and several had elevated serum creatinine.[29] Humans can serve as definitive hosts of *Sarcocystis hominis* and *S. suihominis* when ingesting meats containing sarcocysts. The parasites reproduce sexually in the intestine of humans and cause gastrointestinal illness. The few cases of intestinal sarcocystosis have mostly been reported in Europe.[28]

Diagnosis of intestinal sarcocystosis can be made by finding the oocysts or sporocysts in the feces of infected individuals. Diagnosis of muscular sarcocystosis is by clinical presentation, travel history and muscle biopsies to identify the sarcocysts.[28]

6.2.4 *Trypanosoma cruzi*

The parasite *Trypanosoma cruzi* causes Chagas' disease, and its transmission is mainly through a triatomine vector. Humans become infected by introduction of vector stages of the parasite. Interestingly, *T. cruzi* is not directly introduced through triatomine bites. Humans become infected when they scratch the bite area, and physically contaminate the small wound with the excreta from the infected triatomine that just bit them. Other modes of transmission are the congenital route and organ transplantations. However, there are several reports indicating that *T. cruzi* infections are epidemiologically associated with the consumption of foods contaminated with either the guts or feces of infected triatomines. Once orally ingested by people, the trypomastigotes (the motile form found in the vector) invade the cells of a variety of tissues and develop into amastigotes (the non-motile form), which multiply by binary fission. This process continues leading to colonization in various organs.[30]

Chagas' disease has three defined phases of clinical manifestation. In the initial acute phase people can be either asymptomatic or present mild symptoms, such as fever malaise and the swelling of one eye (Romaña's sign) or at the site of the bite (in cases of vector involvement). This is followed by the indeterminate phase, where the parasite survives for years in muscles or other organs of the host. Twenty or more years after the initial infection, the individual may present with the chronic phase. This is characterized by irreversible manifestations, such as constipation, digestive problems and cardiomyopathy. Other manifestations can be enlargement of the liver, spleen and lymph nodes. The enlargement of the colon and esophagus and abnormal cardiac function may result in sudden death.[32]

The first evidence of oral transmission was reported in 1965. Transmission can occur when foods are contaminated with triatomine bugs or their feces, from uncooked meats from infested animals, and from the urine or anal secretions of infected marsupials.[30] It has been estimated that oral transmission is the current most relevant transmission route in Brazil. Some notable documented outbreaks include Teutônia with 17 cases of acute Chagas, Paraíba with 26 cases associated with ingestion of sugar cane juice, Mazagão with 15 cases and Santa Catarina with 24 cases. In 2006, 94 cases with 6 fatal cases were associated with consumption of acai or sugar cane juice. Overall, 845 cases of oral trypanosomiasis have been reported in South America, specifically Argentina, Brazil, Colombia, Ecuador and Venezuela.[30]

The southern United States have enzootic *T. cruzi* cycles involving at least 11 triatomine species and hosts such as raccoons, opossums and domestic dogs, which can carry the parasite. Seven autochthonous vector-borne human infections have been reported in Texas, California, Tennessee and Louisiana. However, most cases in the US are imported, where the affected people are immigrants from endemic areas in Latin America. In 1992 it was estimated that about 370 000 people living in the US were infected with *T. cruzi* but there are no reports of foodborne transmission in the US.[33]

6.3 Cestoda

Tapeworms are widely recognized as a public health burden particularly in developing countries. Infections are meat-borne, fish-borne and others crustacean-borne. Human taeniasis is caused by *Taenia saginata* (cattle tapeworm) and *Taenia solium* (pig tapeworm). Infections occur when people consume infected meats that have not been properly cooked and which contain tissue with cysticerci, the intermediate stage of the parasite. Cysticercosis, however, is not the result of the consumption of contaminated meat products, but from the ingestion of *Taenia solium* eggs. In this case, the intermediate stage develops in the person. In the US, cysticercosis is more frequently identified in travelers and immigrants from endemic areas. However, domestic cases have also being reported. These are more likely caused by taeniasic patients who have had direct contact with people or worked as food handlers.

6.3.1 *Taenia* spp.

Cysticercosis is caused by the larval stage of *Taenia solium* that is found in infected tissues. Transmission occurs when the eggs of tapeworms are excreted in the feces of an infected individual (a carrier) into the environment and pigs or humans ingest the food or water contaminated with the *Taenia* eggs. The oncosphere, which is the embryo inside the egg, is released and penetrates the intestinal wall and migrates via the bloodstream to various tissues where it will encyst to form the cysticercus.[34]

Another way of acquiring the infection is when an individual consumes raw or undercooked pork meat that contains a cysticercus. The cysticercus membrane breaks open and the scolex will attach to the small bowel. The individual will acquire taeniasis, which is the disease caused by the adult stage of *Taenia solium* (Fig. 6.2(a)). Individuals harboring the adult stage of *Taenia solium* can be asymptomatic, but the number of eggs shed in the environment is high. *Taenia saginata* is transmitted in a similar way, except that people ingest undercooked meats containing a *Taenia saginata* cysticercus.

The symptoms of *T. solium* cysticercosis can vary according to the location of the cysticerci. They infect muscle tissue but can also infect the central nervous system (CNS) where they will cause severe neurological disease. The most common presentations include epilepsy and headaches. Although uncommon, confusion, brain swelling and hydrocephalus can also occur.[35]

In the US, most of the cases have been reported in immigrants from Latin America. This is not a reportable disease in the US and many domestic cases go

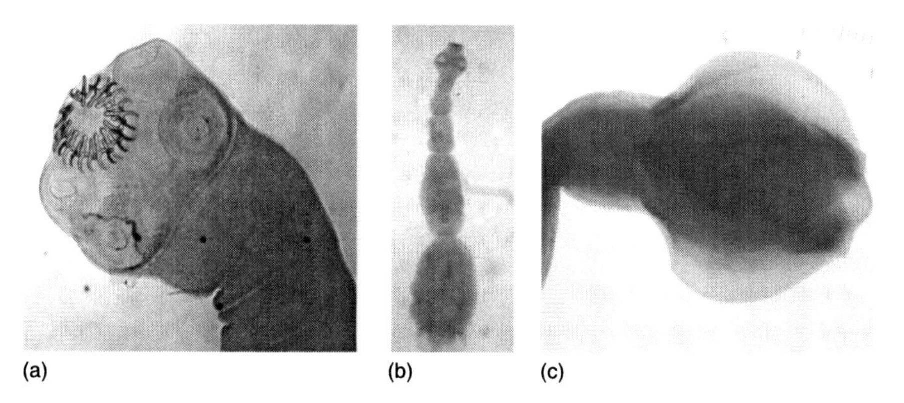

Fig. 6.2 Cestoda. (a) Scolex of *Taenia solium*. (b) Adult *Echinococcus granulosus*. (c) Scolex of *Diphyllobothrium pacificum*. Source: H. Garcia and J. Jimenez, Instituto Nacional de Ciencias Neurológicas, Peru.

unreported. Because of the large migration from endemic locations, many cases are travel related. In a study from 1990 to 2002, 221 individuals with fatal cysticercosis were reported in the US. Of those 84.6% were Latinos and it was most commonly identified in those 25–34 years old. For the country of origin, 62% were born in Mexico and 15% in the US. Of those US born, 36.4% were white and 51.5% were Latino.[36] An outbreak of cysticercosis in an Orthodox Jewish community in New York implicated domestic employees from Mexico who were infected with *Taenia*.[37] It is estimated that the mortality rate is between 6% to 10%. Cysticercosis is found worldwide and is common in developing countries where pigs roam free and eat human feces. Cysticercosis is prevalent in Latin America, Asia and parts of Africa. It is estimated that there are about 50 million individuals with cysticercosis worldwide.[34]

6.3.2 *Echinococcus granulosus* and hydatid disease

The larval stage of this cestode is responsible for illness (hydatidosis) in humans and animals. Humans are accidental hosts. *Echinococcus granulosus* (2–7 mm long) causes hepatic echinococcosis and the definitive hosts are dogs and other canids. *E. multilocularis* (1–4 mm long) causes alveolar echinococcosis and the definitive hosts are dogs, cats, coyotes and wolves. *E. vogeli* (5.6 mm long) causes polycystic echinococcosis and it is found in bush dogs. *E. oligarthrus* (up to 2.9 mm long) is less frequently identified in humans. Human cases have been reported in individuals who are in frequent contact with rodents (pacas) and wild felines.

The definitive hosts of *E. granulosus* are canids; the adult worm (Fig. 6.2(b)) resides in their intestines. The proglottids that are gravid (full of eggs) release their eggs, which are excreted in the feces. When eggs are ingested via contaminated pasture, water or foods by susceptible hosts (animals and humans), they hatch and the oncosphere penetrates the intestinal wall and migrates via the circulatory system. The oncosphere can localize in various tissues and organs, preferentially the liver and lungs. The oncosphere will then form a cyst that can reach an average size of 12 cm in diameter and is full of hydatid sand, which is composed of protoscoleces and the cystic fluid. When a canid such as a dog eats raw infected organs containing these hydatid cysts, the scoleces will take residence and mature in the intestines to form the adult worm, which excretes eggs. Ungulates such as sheep, goats and cattle serve as intermediary hosts of *Echinococcus* spp.[38]

Hydatid cysts may affect the functionality of an infected organ. However, infected individuals are usually asymptomatic. When the cyst ruptures and the hydatid sand is released, the infected individual will present with fever, urticaria, pruritus, eosinophilia and anaphylaxis. If the cyst is localized in the biliary tree, jaundice, cholestasis and cholecystitis can occur. If the cyst is localized in the lungs, then pain, fever, dyspnea, cough and hemoptysis can occur. About 20–40% of infected individuals have cysts in more than one location.[38]

Hydatidosis has a worldwide distribution, but the greatest prevalence is in temperate areas, including Eurasia, Australia, America (principally South America) and north and east Africa. There are 10 genotypes of *Echinococcus granulosus* and the one most frequently associated with humans is G1 from sheep. This genotype is commonly reported in sheep-farming communities in Morocco, Tunisia, Kenya, Kazakhstan, China and Argentina. In the US, reports from the 1960s described the sheep strain in sporadic autochthonous echinococcosis in Arizona, California, New Mexico and Utah, mostly coinciding with areas where sheep are slaughtered. These areas are inhabited primarily by Basque Americans, Mormons, and Navajo and Zuni Indians from New Mexico and Arizona. Cervid hydatidosis (G8) is also common in the Americas and is present in the wildlife of Canada, Alaska and Minnesota in cervids, coyotes, wolves and dogs. The most endemic areas in Europe are the Mediterranean regions with an annual incidence of 4–8 cases per 100 000 inhabitants. In Salamanca, Spain, the number of surgical cases was 10.8/100 000 inhabitants between 1980 and 2000.[39]

6.3.3 Other cestodes: *Diphyllobothrium*

The head or scolex of *Diphyllobothrium latum* has two sucking grooves or bothria. Each proglottid of the body or strobila contains eggs. Eggs are more fully developed in the proglottids further from the scolex. The unembryonated eggs along with the proglottids are excreted in the feces. When the eggs reach freshwater they embryonate and the coracidium forms, which is released when the egg operculum opens. The coracidia are ingested by copepods or water fleas. Then the procercoid (a larval stage) forms and when a fish ingests the copepod, the procercoid will mature to form a plerocercoid cyst (or sparganum). This process can continue along the food chain until a definitive host (humans, dogs, cats, foxes, pigs, etc.) ingests fish meat containing the larvae. A *Diphyllobothrium* larva will grow to form an adult worm in the intestine reaching up to 20 cm in length. Most infections are asymptomatic but diarrhea, abdominal distension and general malaise can be observed.[40] Outbreaks have been associated with a high consumption of marinated, raw or undercooked fish (salmon and sushi).

The life cycle of *Diphyllobothrium pacificum* (Fig. 6.2(c)) is similar to *D. latum* but the intermediary hosts are saltwater fish.

Most infections with *Diphyllobothrium* are asymptomatic, but diarrhea, abdominal discomfort, vomiting and weight loss have been reported. In some cases, intestinal obstruction can occur. It is estimated that in the early 1970s nine million humans were affected by diphyllobothriasis. More recent data[40] indicates that about 20 million individuals are infected worldwide but no estimation of the global prevalence is available. The overall incidence of human infections has decreased particularly in North America and Europe but a reemergence has occurred in Russia, South Korea, Japan and South America.

6.4 Nematoda

The Nematoda are roundworms. When the eggs open the larval stages undergo a series of developmental stages until they differentiate into male and female adults.

The larvae migrate through various organs and tissues of the definitive host before they settle down in a permanent location, which is usually the intestines.

6.4.1 *Trichinella*

Trichinosis can be acquired by ingesting meat containing tissue cysts (encysted larvae) of *Trichinella*. During the process of digestion, a larva excysts and takes residence in the small bowel mucosa where it differentiates into an adult worm. The female worm releases larvae that migrate to other tissues and encyst (Fig. 6.3(a)) where they can remain for many years. This cycle repeats when omnivorous or carnivorous animals eat meat from infected animals containing the encysted larvae. Humans become accidental hosts when meat improperly cooked containing the encysted larvae is consumed. *Trichinella spiralis* is transmitted by the consumption of pork. Other species of *Trichinella*, particularly *T. murrelli*, have been identified in humans. These have been associated with consumption of horse, bear, cougar, deer, walrus, seal and wild game.[41,42]

Trichinosis initially can present with nausea, vomiting and gastrointestinal illness including diarrhea and mild fever.[41] In some cases infected individuals are asymptomatic. When a larva migrates to the muscle, the most common manifestations are eosinophilia, myositis with myalgia, stiffness, fever, rash or headaches.

Trichinosis is a notifiable disease in the US, and has shown a decreasing trend over time. In 2008–2010, about 20 cases a year were reported, whereas the number of cases associated with raw or undercooked wild game meat remained the same. Since 1997–2007, 40% of 126 cases and 60% of 15 outbreaks were attributed to consumption of raw or undercooked bear meat. In 2008, in northern California, 30 attendees of an event acquired trichinosis after consumption of black bear meat. That same year, five cases were reported in another county in California.[41]

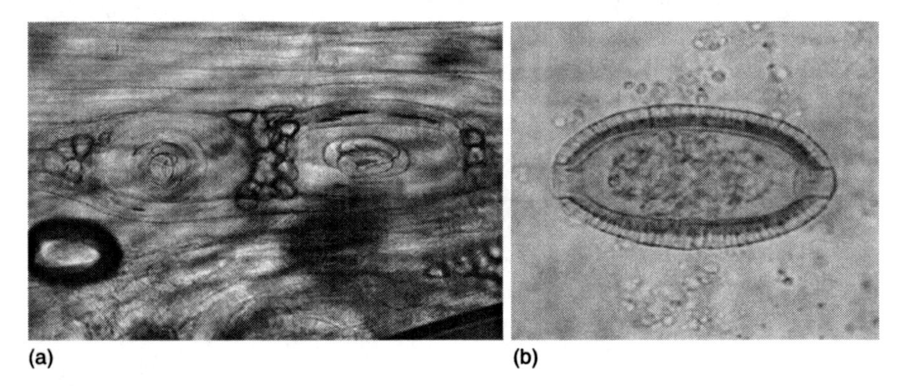

Fig. 6.3 Nematoda. (a) Encysted larvae of *Trichinella* spp. (b) Egg of *Capillaria philippinensis*. Source: www.dpd.cdc.gov/dpdx/HTML/Image_Library.htm.

6.4.2 *Angiostrongylus*

This nematode can infect the gastrointestinal tissues or migrate and infect the CNS. In rats, the adult nematodes live in the pulmonary arteries and produce larvae that migrate to the pharynx and are swallowed and excreted in the feces and can parasitize an intermediary host (snail or slug). *Angiostrongylus* will molt to other larval stages. The natural definitive hosts are rats. When the definitive host ingests the infected mollusks, the third-stage larva will migrate to the brain where it will develop into an adult. *A. cantonensis* localizes preferentially in the CNS whereas *A. costaricensis* resides in the ileocecal arterioles. Larvae will be produced but they will degenerate and cause intense local inflammation. Infection can also occur when humans ingest paratenic hosts containing the larval stages. Humans acquire the infection by ingestion of snails or slugs containing the larval stages or by ingestion of paratenic hosts such as crabs or freshwater shrimps.[42]

People can be considered aberrant hosts, and their symptoms may resemble bacterial meningitis, including nausea, vomiting, stiffness and headaches. Eosinophilia is observed and in some cases infection can lead to severe neurological dysfunction and death. *A. costaricensis* can cause fever, nausea, vomiting and abdominal pain. It can be confused with appendicitis if the larvae migrate to the mesenteric arteries resulting in thrombosis and gastrointestinal hemorrhage. The eggs can also cause a localized inflammatory reaction. Massive eosinophilic inflammatory reaction, intestinal obstruction and perforation can occur.[43]

Angiostrongylus cantonensis has been reported in individuals in South-East Asia and the Pacific Basin. Cases have also been reported in some parts of Africa, Australia, the Caribbean, Hawaii and Louisiana. *A. costaricensis* have been reported in Latin America and the Caribbean. During 2001–2005, 18 cases were reported in Hawaii. Headaches were the most frequently reported symptom.[44]

6.4.3 *Anisakis* and *Pseudoterranova*

Both of these nematodes reside in the stomach of various marine mammals. Eggs are passed in the feces, are fully embryonated and hatch while in the water. The larvae are ingested by crustaceans where they morph to third-stage larvae. These larvae are infectious to fish and squid. The larvae mature in the peritoneal cavity and when the fish dies, the larvae migrate into the muscle and the cycle will continue when these fish are eaten by other fish. If a marine mammal ingests an infected fish, the larvae will molt and form adult worms, which will in turn multiply. Worms will produce eggs and mature to form the larvae. If humans ingest fish containing the larvae, the worms will penetrate the gastric and intestinal mucosa.[45]

Individuals with anisakiasis can present with abdominal pain, mild fever, nausea, vomiting, diarrhea, abdominal distension, blood and mucus in feces. Allergic reactions have also being reported in cases with *Anisakis* infection. *Anisakis* and *Pseudoterranova* are common in Japan. However, more cases are being reported in other areas of the world including in the US, Europe and South America.[46]

6.4.4 Other nematodes: *Capillaria* and *Gnathostoma*

Capillaria philippinensis and *C. hepatica* are frequently reported in humans. Human and bird (from birds that eat fish) feces containing *Capillaria philippinensis* eggs can contaminate water (Fig. 6.3(b)). Humans acquire the infection by ingestion of fish infected with larvae.

Capillaria hepatica can infect rodents, pigs, carnivores and primates. An adult takes residence in the liver and produces eggs, which will remain in the parenchyma of the infected host. When the infected host dies or a predator eats the infected animal, the unembryonated eggs are released into the environment. Infective eggs require 30 days in the environment to become infectious. When an egg is ingested, it hatches in the intestine and the larva reaches the liver via the portal vein. In humans this migration can be to other organs including the lungs and kidneys. Infection in humans can occur by ingestion of contaminated water, soil or vegetables.[47]

Infections are frequently asymptomatic but patients can present with abdominal pain, diarrhea, nausea, vomiting, weight loss and even death. *Capillaria philippinensis* is endemic in the Philippines and Thailand whereas *C. hepatica* and *C. aerophila* have been reported worldwide.[47]

The definitive hosts of *Gnathostoma* are pigs, cats, dogs and wild fish-eating animals. The worm resides in an induced tumor in the intestinal wall. The eggs are excreted into the environment in the feces of an infected animal. The eggs embryonate in water and if ingested by crustaceans (such as *Cyclops*), the larvae will molt to form a second larval stage. If the infected crustacean is ingested by a fish, frog or snake, the larva will molt to a third-stage larva. When another host eats the second host, the larva will develop into an adult parasite in the stomach wall. Humans acquire infection by ingestion of fish or poultry containing the third larval stage or by drinking water containing infected *Cyclops*.[48]

Larvae migrating through the subcutaneous tissues can be painful and cause pruritic swelling. They can also cause coughing, hematuria and ocular involvement resulting in eosinophilia. Complications can result in eosinophilic meningitis. *Gnathostoma* are found in Thailand and Japan and most recently in Mexico.[48]

6.5 Trematoda

Trematodes, also known as flukes, have been identified as infecting the lungs, the liver and the intestines. Infection occurs through ingesting freshwater plants and aquatic fauna. Trematodes are characterized by a dorsoventral, flat, symmetrical body. Trematoda are hermaphrodites and digenean. They have oral and ventral suckers, which are used to attach to the epithelial layers of the host's intestine, bile ducts or lung parenchyma. They require an aquatic snail to develop. Most of the foodborne trematodes have up to six larval stages, which infect mollusks and vertebrate hosts. The life cycle starts when eggs are excreted into water and embryonation occurs. A miracidium is released from an egg and infects a snail where asexual reproduction occurs. The miracidium develops into a cercaria, which is motile. The cercaria exits the snail and infects the second intermediary host (an aquatic plant, crustacean or fish) and forms the metacercaria. When the definitive host, an animal including humans, ingests a metacercaria, the metacercaria excysts during the process of digestion, migrates and localizes in the lungs or intestine and develops into an adult fluke.

6.5.1 *Paragonimus*

Paragonimus westermani (Fig. 6.4(a)) is the species most commonly reported as infecting humans. The eggs are excreted in the sputum of infected individuals. When the eggs mature in water, the miracidia hatch and infect snails, which are the first intermediary hosts. Multiplication occurs in the snail and after going through several developmental stages, the cercariae emerge from the snail. A cercaria infects a crustacean, which is the second intermediary host, where it will encyst and become a metacercaria. Humans can acquire paragonimiasis by ingestion of infected freshwater crabs, crayfish or wild boar meat. The larval stages migrate to the lungs. Eggs are excreted in the sputum and are occasionally found in the feces of an infected individual.[49] Sometimes *Paragonimus* migrates to the central nervous system.

Infections can range from asymptomatic to acute and are characterized by coughing, abdominal pain, discomfort, fever and bloody sputum. Long-term infections can have a productive cough with brown sputum and can be confused with bronchitis or tuberculosis.[49]

In the US, paragonimiasis is mostly caused by *Paragonimus kellicotti*. Most infections have been reported in Missouri and were associated with ingestion of raw crayfish and crabs. Other animals, including cats, dogs, mink and muskrats, can also acquire the infection.[50]

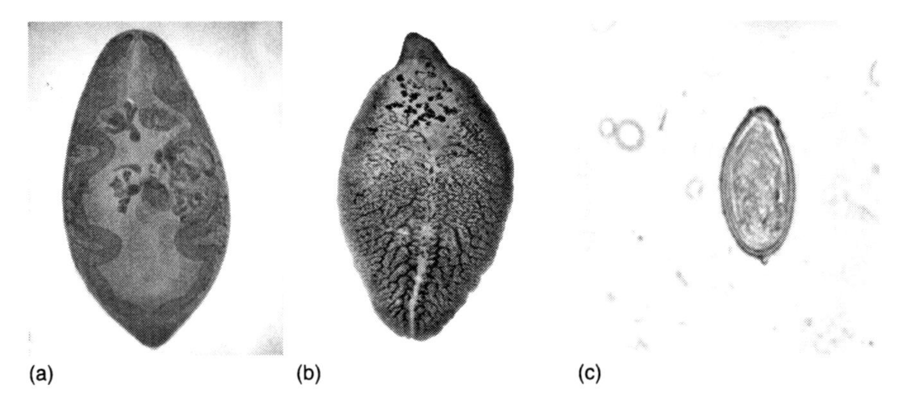

Fig. 6.4 Trematoda. (a) Adult *Paragonimus westermani*. (b) Adult *Fasciola hepatica*. (c) Egg of *Opisthorchis* spp. Source: www.dpd.cdc.gov/dpdx/HTML/Image_Library.htm.

6.5.2 *Fasciola* and *Fasciolopsis*

It is estimated that 2.4–7 million people are infected with *Fasciola* worldwide. Eggs are excreted in the feces of an infected individual. Infection occurs when a susceptible host ingests an encysted cercaria that is attached to the surface of an aquatic plant (watercress, water chestnut, etc.). *Fasciola hepatica* can also be acquired by ingestion of either a contaminated animal liver or water. A metacercaria excysts, migrates and localizes in a biliary duct where it will grow and develop into an adult worm. *F. hepatica* (Fig. 6.4(b)) can measure up to 3 cm long.[49]

Dyspepsia, fever, hepatomegaly, splenomegaly and jaundice are frequent symptoms of infection. Fatty food intolerance and cholecystitis are also characteristic of this infection. *Fasciolopsis buski* is acquired by ingestion of aquatic plants. Gastrointestinal symptoms including diarrhea, constipation, headaches and anorexia are characteristic of *Fasciolopsis* infection.[49]

Fasciola hepatica can be found worldwide but most frequently in South America, the Caribbean, Europe, the Middle East, Africa, Asia and Oceania. Highly endemic areas include the highlands of Peru and Bolivia.[49]

6.5.3 Other trematodes: *Opisthorchis*, *Clonorchis*, *Echinostoma*

Opisthorchis and *Clonorchis* (liver flukes) infection can be acquired by ingestion of small freshwater fish. *C sinensis* is the most frequently identified and is found worldwide. The global estimate for the number of people infected with *C. sinensis* is 35 million. Infected individuals usually have diarrhea, fever, enlargement of the liver, biliary obstruction, cholecystitis, and biliary and liver abscesses.

Opisthorchis can also parasitize the liver (Fig. 6.4(c)). The two species most commonly infecting humans are *O. viverrini* and *O. felineus*: it is estimated that 10 million people are infected with *O. viverrini* and 1.2 million with *O. felineus*. The symptomology for the former frequently ranges from asymptomatic to flatulence, fatigue and hepatomegaly, whereas people with the latter present with fever and hepatic involvement. Abdominal pain, nausea and vomiting have also been reported. The chronic phase is characterized by obstruction of the biliary tree, liver abscesses, pancreatitis and suppurative cholangitis.[49,50] *Clonorchis* can be found in Asia including Korea, China, Vietnam, Japan and Russia. In the US and non-endemic areas, cases have been reported in immigrants from endemic regions.[49]

Echinostoma (intestinal flukes) infection is acquired by ingestion of raw snails or frogs, which act as intermediary hosts for this parasite. A fluke requires two intermediary hosts to fully develop. The first host is a snail or other mollusk and the second could be a snail or tadpole. Its oral sucker has a head collar of spines. Of the seven species that infect humans, *E. revolutum* has been the most frequently reported. *E. hortense* can cause mucosal ulcerations and bleeding of the upper part of the duodenum and the distal part of the stomach. It can also infect ducks, geese and muskrats. Nausea, vomiting, diarrhea, fever and abdominal pain are characteristic in patients infected with *Echinostoma*. Most of the cases of *Echinostoma* spp. have been described in Asia, South-East Asia and Europe.[51]

6.6 Epidemiology, prevention and control

6.6.1 Epidemiology

As described for each parasite earlier, cases have been reported in various parts of the world including the US. The distribution and prevalence of these parasites in the food supply varies based on whether the environmental conditions favor the multiplication of reservoir populations, including snails, crustaceans, mollusks and fish. These diseases are prevalent in humans where the reservoirs are present and if food or water is consumed raw. *Cyclospora* is highly seasonal and the factors influencing it are unknown. *Cryptosporidium, Giardia* and *Toxoplasma* are ubiquitous and because of their zoonotic potential, any interaction between animals (and their products) and humans may result in human infection. Infections with cestodes are usually a result of ingestion of contaminated game meat, vegetables or water. The role of food handlers has been evident, particularly for cryptosporidiosis and cysticercosis. Other parasitic infections could well be acquired by contamination of foods by food handlers. Most trematodes are infrequent in the US; however, a rise in the number of *Paragonimus* cases has been attributed to ingestion of raw crayfish.[52]

6.6.2 Prevention and control

Meat

Toxoplasma, *Sarcocystis*, platyhelminths and nematodes can be rendered non-infectious in meat if the meat is frozen for 48 h at −20°C or cooked at 60°C, 70°C or 100°C for 20, 15 or 5 min, respectively.[28] To inactivate trematodes and nematodes, it is recommended that fish intended for raw consumption is frozen for 7 days at −20°C or for 15 h at −35°C.[49] *Cryptosporidium* oocysts in oysters (shucked or unshucked) can be inactivated by a dose of 2 kGy of gamma irradiation.[53] High temperatures are very effective at eliminating all parasites; however, consumption of raw meat or fish poses a risk of infection by foodborne parasites. *Toxoplasma* can be rendered non-infectious when meats containing tissue cysts are treated with 50 krad (using cesium-137 or cobalt-60).[54]

Vegetables

Contamination of vegetables can occur via irrigation or application of pesticides using water contaminated with parasites. Prevention can be achieved by not using water from where feces could be present or from near where there is soil that has not been properly composted. Wild animals can also be reservoirs of these parasites (as is the case for *Toxoplasma* and *Cryptosporidium*). Produce contamination can be reduced by preventing traffic and defecation of intermediary and definitive hosts in or close to agricultural fields. Good hygienic practices by field workers and harvesters can also reduce and prevent fresh produce contamination. Gamma irradiation of 0.5 kGy rendered *Toxoplasma* oocysts non-infectious, suggesting that this could also be used on fruits and vegetables for other coccidian oocysts.

Water

Filtration, ozonation, UV radiation and, to a lesser extent, chlorination can eliminate parasites in drinking water. Boiling water can effectively eliminate parasites. Microfiltration and ultrafiltration have been used to remove *Cryptosporidium* and *Giardia*. More than 7 log units could be removed using ultrafiltration membranes with a nominal cutoff of 13 000 Da. In another challenge study, a 5 log removal was obtained using membrane filters. UV radiation has also been used to inactivate protozoan parasites. Up to 99% of *Cryptosporidium* are inactivated when exposed to UV light (0.2 mW/cm^2, 254 nm) for 36 min, but in the presence of titanium dioxide this inactivation could be reached in 22 min. Gamma irradiation at 50 kGy was effective at killing *C. parvum* oocysts resuspended in water whereas for *Giardia* only 7 kGy was needed. High-pressure inactivation has also been tried in a variety of foods. In apple and orange juice, 3.4 and 4.1 log units of *Cryptosporidium* were inactivated when treated at 80 000 psi for 30 s.[53]

Chlorine is not very effective at inactivating protozoa. Cryptosporidium Oocysts exposed to 5.25% for 120 min were still viable and infectious to mice. For a 2 log reduction a Ct value of 7 200 mg·min/lt was needed, whereas for *Giardia* a Ct value of 41 was needed to achieve 2 log inactivation.[53]

6.7 Conclusions

Parasites and their effect on public health have been known for centuries. In developed countries the incidence of parasitic infections has decreased by controlling sanitary conditions in households and by improvements in the quality of municipal water and the sewers servicing cities. Most of the cases with exotic and neglected parasitic infections have been reported in returning travelers or expatriates.

However, the increase in the number of these parasitic infections is attributed to ingestion of ready-to-eat products that are imported from endemic countries where sanitary conditions and good agricultural practices may not be followed. In addition, an influx of food handlers working in the food industry who may be carriers of parasitic diseases may be contributing to the increase in the number of cases. The consumption of meat that has not reached an adequate inactivation temperature, or was not frozen as recommended, is also contributing to the increased number of cases in developed countries. Inspection of meat, fish and mollusks is crucial for identifying infected products. Game meat and their products should be cooked adequately. Sausages or fermented meats can harbor viable parasites and proper disinfection or kill steps should be used during preparation of these foods.

Promoting awareness and education on food safety and parasites should be considered by producers, exporters and importing companies as part of the globalization trend for the food supply.

6.8 References

1. ORTEGA, Y. R., R. NAGLE, R. H. GILMAN, J. WATANABE, J. MIYAGUI *et al.* 1997. Pathologic and clinical findings in patients with cyclosporiasis and a description of intracellular parasite life-cycle stages. *J. Infect. Dis.* 176:1584–1589.
2. ORTEGA, Y. R., C. R. STERLING, R. H. GILMAN, V. A. CAMA and F. DIAZ. 1993. *Cyclospora* species – A new protozoan pathogen of humans. *N. Engl. J. Med.* 328:1308–1312.
3. ORTEGA, Y. R. and R. SANCHEZ. 2010. Update on *Cyclospora cayetanensis*, a food-borne and waterborne parasite. *Clin. Microbiol. Rev.* 23:218–234.
4. BERN, C., Y. ORTEGA, W. CHECKLEY, J. M. ROBERTS, A. G. LESCANO, L. CABRERA, M. VERASTEGUI, R. E. BLACK, C. STERLING and R. H. GILMAN. 2002. Epidemiologic differences between cyclosporiasis and cryptosporidiosis in Peruvian children. *Emerg. Infect. Dis.* 8:581–585.
5. EBERHARD, M. L., E. K. NACE, A. R. FREEMAN, T. G. STREIT, A. J. DA SILVA *et al.* 1999. *Cyclospora cayetanensis* infections in Haiti: A common occurrence in the absence of watery diarrhea. *Am. J. Trop. Med. Hyg.* 60:584–586.
6. PAPE, J. W., R. I. VERDIER, M. BONCY, J. BONCY and W. D. JOHNSON, Jr. 1994. *Cyclospora* infection in adults infected with HIV. Clinical manifestations, treatment, and prophylaxis. *Ann. Intern. Med.* 121:654–657.
7. BERN, C., B. HERNANDEZ, M. B. LOPEZ, M. J. ARROWOOD, M. A. DE MEJIA, A. M. DE MERIDA, A. W. HIGHTOWER, L. VENCZEL, B. L. HERWALDT and R. E. KLEIN. 1999. Epidemiologic studies of *Cyclospora cayetanensis* in Guatemala. *Emerg. Infect. Dis.* 5:766–774.
8. HOGE, C. W., P. ECHEVERRIA, R. RAJAH, J. JACOBS, S. MALTHOUSE *et al.* 1995. Prevalence of *Cyclospora* species and other enteric pathogens among children less than 5 years of age in Nepal. *J. Clin. Microbiol.* 33:3058–3060.
9. SHERCHAND, J. B. and J. H. CROSS. 2001. Emerging pathogen *Cyclospora cayetanensis* infection in Nepal. *SE Asian J. Trop. Med. Public Health* 32 Suppl. 2:143–150.
10. HERWALDT, B. L. and M. L. ACKERS. 1997. An outbreak in 1996 of cyclosporiasis associated with imported raspberries. The *Cyclospora* Working Group. *N. Engl. J. Med.* 336:1548–1556.
11. HERWALDT, B. L. and M. J. BEACH. 1999. The return of *Cyclospora* in 1997: Another outbreak of cyclosporiasis in North America associated with imported raspberries. *Cyclospora* Working Group. *Ann. Intern. Med.* 130:210–220.
12. SCALLAN, E., R. M. HOEKSTRA, F. J. ANGULO, R. V. TAUXE, M. A. WIDDOWSON *et al.* 2011. Foodborne illness acquired in the United States – Major pathogens. *Emerg. Infect. Dis.* 17:7–15.
13. BUZBY, J. C. and T. ROBERTS. 1997. Economic costs and trade impacts of microbial foodborne illness. *World Health Stat. Q.* 50:57–66.
14. WEISS, L. M. and J. P. DUBEY. 2009. Toxoplasmosis: A history of clinical observations. *Int. J. Parasitol.* 39:895–901.
15. PEREIRA, K. S., R. M. FRANCO and D. A. LEAL. 2010. Transmission of toxoplasmosis (*Toxoplasma gondii*) by foods. *Adv. Food Nutr. Res.* 60:1–19.
16. BOWIE, W. R., A. S. KING, D. H. WERKER, J. L. ISAAC-RENTON, A. BELL, S. B. ENG and S. A. MARION. 1997. Outbreak of toxoplasmosis associated with municipal drinking water. The BC Toxoplasma Investigation Team. *Lancet* 350:173–177.
17. DE MOURA, L., L. M. BAHIA-OLIVEIRA, M. Y. WADA, J. L. JONES, S. H. TUBOI *et al.* 2006. Waterborne toxoplasmosis, Brazil, from field to gene. *Emerg. Infect. Dis.* 12:326–329.
18. BALASUNDARAM, M. B., R. ANDAVAR, M. PALANISWAMY, and N. VENKATAPATHY. 2010. Outbreak of acquired ocular toxoplasmosis involving 248 patients. *Arch. Ophthalmol.* 128:28–32.
19. THOMPSON, R. C. and P. MONIS. 2012. *Giardia* – From genome to proteome. *Adv. Parasitol.* 78:57–95.
20. XIAO, L. 2010. Molecular epidemiology of cryptosporidiosis: An update. *Exp. Parasitol.* 124:80–89.

21. FAYER, R., E. J. LEWIS, J. M. TROUT, T. K. GRACZYK, M. C. JENKINS *et al.* 1999. *Cryptosporidium parvum* in oysters from commercial harvesting sites in the Chesapeake Bay. *Emerg. Infect. Dis.* 5:706–710.
22. CHECKLEY, W., L. D. EPSTEIN, R. H. GILMAN, R. E. BLACK, L. CABRERA and C. R. STERLING. 1998. Effects of *Cryptosporidium parvum* infection in Peruvian children: Growth faltering and subsequent catch-up growth. *Am. J. Epidemiol.* 148:497–506.
23. CHECKLEY, W., R. H. GILMAN, L. D. EPSTEIN, M. SUAREZ, J. F. DIAZ, L. CABRERA, R. E. BLACK and C. R. STERLING. 1997. Asymptomatic and symptomatic cryptosporidiosis: Their acute effect on weight gain in Peruvian children. *Am. J. Epidemiol.* 145:156–163.
24. FAYER, R., J. M. TROUT, E. J. LEWIS, M. SANTIN, L. ZHOU *et al.* 2003. Contamination of Atlantic coast commercial shellfish with *Cryptosporidium*. *Parasitol. Res.* 89:141–145.
25. GRACZYK, T. K., R. FAYER, E. J. LEWIS, J. M. TROUT and C. A. FARLEY. 1999. *Cryptosporidium* oocysts in Bent mussels (*Ischadium recurvum*) in the Chesapeake Bay. *Parasitol. Res.* 85:518–521.
26. BUDU-AMOAKO, E., S. J. GREENWOOD, B. R. DIXON, H. W. BARKEMA and J. T. MCCLURE. 2011. Foodborne illness associated with *Cryptosporidium* and *Giardia* from livestock. *J. Food Prot.* 74:1944–1955.
27. QUIROZ, E. S., C. BERN, J. R. MACARTHUR, L. XIAO, M. FLETCHER *et al.* 2000. An outbreak of cryptosporidiosis linked to a foodhandler. *J. Infect. Dis* 181:695–700.
28. FAYER, R. 2004. *Sarcocystis* spp. in human infections. *Clin. Microbiol. Rev.* 17:894–902.
29. CDC. 2012. Notes from the field: Acute muscular sarcocystosis among returning travelers – Tioman Island, Malaysia, 2011. *Morb. Mortal. Wkly. Rep.* 61:37–38.
30. TOSO, M. A., U. F. VIAL and N. GALANTI. 2011. Oral transmission of Chagas' disease. *Rev Med. Chil.* 139:258–266.
32. RASSI, A., Jr., A. RASS, and R. J. MARCONDES DE. 2012. American trypanosomiasis (Chagas' disease). *Infect. Dis. Clin. North Am.* 26:275–291.
33. BERN, C., S. KJOS, M. J. YABSLEY and S. P. MONTGOMERY. 2011. *Trypanosoma cruzi* and Chagas' disease in the United States. *Clin. Microbiol. Rev.* 24:655–681.
34. SORVILLO, F., P. WILKINS, S. SHAFIR and M. EBERHARD. 2011. Public health implications of cysticercosis acquired in the United States. *Emerg. Infect. Dis.* 17:1–6.
35. DEL BRUTTO, O. H. 2012. Neurocysticercosis: A review. *Sci. World J.* 2012:159821.
36. SORVILLO, F. J., C. DEGIORGIO and S. H. WATERMAN. 2007. Deaths from cysticercosis, United States. *Emerg. Infect. Dis.* 13:230–235.
37. SCHANTZ, P. M., A. C. MOORE, J. L. MUNOZ, B. J. HARTMAN, J. A. SCHAEFE *et al.* 1992. Neurocysticercosis in an orthodox Jewish community in New York City. *N. Engl. J. Med.* 327:692–695.
38. BRUNETTI, E. and A. C. WHITE, Jr. 2012. Cestode infestations: Hydatid disease and cysticercosis. *Infect. Dis. Clin. North Am.* 26:421–435.
39. GROSSO, G., S. GRUTTADAURIA, A. BIONDI, S. MARVENTANO and A. MISTRETTA. 2012. Worldwide epidemiology of liver hydatidosis including the Mediterranean area. *World J. Gastroenterol.* 18:1425–1437.
40. SCHOLZ, T., H. H. GARCIA, R. KUCHTA and B. WICHT. 2009. Update on the human broad tapeworm (genus *Diphyllobothrium*), including clinical relevance. *Clin. Microbiol. Rev.* 22:146–60.
41. HALL, R. L., A. LINDSAY, C. HAMMOND, S. P. MONTGOMERY, P. P. WILKINS *et al.* 2012. Outbreak of human trichinellosis in northern California caused by *Trichinella murrelli*. *Am. J. Trop. Med. Hyg.* 87:297–302.
42. KENNEDY, E. D., R. L. HALL, S. P. MONTGOMERY, D. G. PYBURN and J. L. JONES. 2009. Trichinellosis surveillance – United States, 2002–2007. *MMWR Surveill Summ.* 58:1–7.
43. WANG, Q. P., Z. D. WU, J. WEI, R. L. OWEN and Z. R. LUN. 2012. Human *Angiostrongylus cantonensis*: An update. *Eur. J. Clin. Microbiol. Infect. Dis.* 31:389–395.
44. HOCHBERG, N. S., B. G. BLACKBURN, S. Y. PARK, J. J. SEJVAR, P. V. EFFLER *et al.* 2011. Eosinophilic meningitis attributable to *Angiostrongylus cantonensis* infection in

Hawaii: Clinical characteristics and potential exposures. *Am. J. Trop. Med. Hyg.* 85:685–690.

45. HOCHBERG, N. S. and D. H. HAMER. 2010. Anisakidosis: Perils of the deep. *Clin. Infect. Dis.* 51:806–812.
46. CHAI, J-Y., K.D. MURRELL and A.J. LYMBERY. 2012. Fish-borne parasitic zoonoses: Status and issues. *International J. Parasitol.* 35:1233–1254.
47. FUEHRER, H. P., P. IGEL and H. AUER. 2011. *Capillaria hepatica* in man – An overview of hepatic capillariosis and spurious infections. *Parasitol. Res.* 109:969–979.
48. HERMAN, J. S. and P. L. CHIODINI. 2009. Gnathostomiasis, another emerging imported disease. *Clin. Microbiol. Rev.* 22:484–492.
49. KEISER, J. and J. UTZINGER. 2009. Food-borne trematodiases. *Clin. Microbiol. Rev.* 22:466–483.
50. LANE, M. A., L. A. MARCOS, N. F. ONEN, L. M. DEMERTZIS, E. V. HAYES *et al.* 2012. *Paragonimus kellicotti* flukes in Missouri, USA. *Emerg. Infect. Dis.* 18:1263–1267.
50. SCHUSTER, R. K. 2010. Opisthorchiidosis – A review. *Infect. Disord. Drug Targets.* 10:402–415.
51. CHAI, J.-Y. 2009. Echinostomes in humans, p. 147–183. In R. Toledo and B. Fried (ed.), *The Biology of Echinostomes*. Springer, N.Y.
52. CDC. 2010. Human paragonimiasis after eating raw or undercooked crayfish – Missouri, July 2006–September 2010. *Morb. Mortal. Wkly. Rep.* 59:1573–1576.
53. ERICKSON, M. C. and Y. R. ORTEGA. 2006. Inactivation of protozoan parasites in food, water, and environmental systems. *J. Food Prot.* 69:2786–2808.
54. DUBEY, J. P., R. J. BRAKE, K. D. MURRELL and R. FAYER. 1986. Effect of irradiation on the viability of *Toxoplasma gondii* cysts in tissues of mice and pigs. *Am. J. Vet. Res.* 47:518–522.

7

New research on antimicrobial resistance in foodborne pathogens

J. Threlfall, HPA, UK

DOI: 10.1533/9780857098740.2.134

Abstract: There are three major aspects of antimicrobial resistance in foodborne pathogens: (1) the resistance to quinolones and cephalosporins in *Salmonella enterica*, and to quinolones and macrolides in *Campylobacter*, (2) multidrug-resistant clones of monophasic *Salmonella* Typhimurium-like organisms, (3) strains of Enterobacteriaceae, in particular *Escherichia coli*, either exhibiting resistance to beta (*β*)-lactam antibiotics or producing AmpC *β*-lactamases.

Controlling antibiotic usage in food animals is necessary in order to reduce the incidence of resistant organisms. A single control measure is unlikely to limit the transmission of resistance through the food chain. Surveillance of resistance in humans and food animals is vital for measuring the long-term effectiveness of any control measure.

Key words: antimicrobial drug resistance, food chain, *Salmonella*, *Campylobacter*, fluoroquinolone, cephalosporin, extended-spectrum-β-lactamase (ESBL), AmpC, control measures.

7.1 Introduction

Resistance to antimicrobials in foodborne bacterial pathogens is of concern because if treatment is warranted, first-line antimicrobials may no longer be effective and treatment options are limited. In an effort to mitigate the risk of antimicrobial drug resistance (AMR) to human health arising from the use of antimicrobials in veterinary medicine, in 2007 the World Health Organization (WHO) developed a list of antimicrobials according to how important they are in the treatment of human illness (Collignon *et al.*, 2009; WHO, 2007). Critically important antimicrobials (CIAs) in this ranking include, among others, quinolones, macrolides and third- and fourth-generation cephalosporins, which are the antimicrobials of choice in treating *Salmonella* and *Campylobacter* infections. The World Health Organization, the Food and Agriculture Organization of the

United Nations (FAO) and the World Organisation for Animal Health (OIE) consider AMR in zoonotic bacteria as a public health threat and recognise that resistance may be the consequence of the use of antimicrobials in food animals and may be transmitted to humans.

There are currently three overarching major aspects of antimicrobial resistance in foodborne pathogens. These are: (1) the increasing occurrence of resistance to quinolones and cephalosporins in *Salmonella enterica*, and to quinolones and macrolides in *Campylobacter*; (2) the recent emergence of new multidrug-resistant clones of monophasic *Salmonella* Typhimurium-like organisms, which have caused outbreaks of infection in many European Union Member States (MSs) and also worldwide since 2000; and (3) the rapid emergence and spread of strains of Enterobacteriaceae, and in particular *Escherichia coli*, either exhibiting resistance to a variety of beta (β)-lactam antibiotics including penicillins, second-, third- and fourth-generation cephalosporins (ESBLs) and monobactams, or producing AmpC β-lactamases conferring resistance to penicillins, second- and third-generation cephalosporins, including β-lactam/inhibitor combinations, and cephamycins.

The emergence and spread of the above organisms, all exhibiting resistance to therapeutically important antibiotics, have promoted research into their origin, the mechanisms of resistance and into meaningful control measures.

7.2 Increasing occurrence of resistance to quinolones and cephalosporins in *Salmonella*

7.2.1 *Salmonella*: quinolone resistance

Mechanisms of resistance

Chromosomal resistance

Chromosomal resistance to quinolones arises spontaneously under antimicrobial pressure due to point mutations that result in: (i) amino acid substitutions within the topoisomerase II (DNA gyrase) and IV subunits *gyr*A, *gyr*B, *par*C or *par*E, (ii) decreased expression of outer membrane porins or alteration of lipopolysaccharide, or (iii) overexpression of multidrug efflux pumps. Mutations in the *gyr*A, *gyr*B, *par*C or *par*E genes in regions that form the fluoroquinolone binding site (termed the quinolone resistance-determining region, QRDR) change the topoisomerase structure so that fluoroquinolones are unable to bind to these target sites. Single mutations affect firstly only older quinolones such as nalidixic acid in their inhibitory action. The minimum inhibitory concentration (MIC) for nalidixic acid is in the range of 64–128 mg/l, whereas the MICs for fluoroquinolones are generally in the range of 0.25–1.0 mg/l. This level of resistance is generally regarded as epidemiological and is becoming increasingly used in many countries worldwide, particularly for food and veterinary isolates, but may be of limited clinical relevance. Additional mutations are required to decrease the susceptibility to later and more recently introduced fluoroquinolones (ciprofloxacin, danofloxacin, difloxacin, enrofloxacin, levofloxacin and marbofloxacin). These additional mutations result in the development of clinical resistance, with MICs greater than 2 mg/l.

Plasmid-mediated quinolone resistance (PMQR)

Plasmid-mediated quinolone resistance (PMQR) is mediated by genes (*qnr*) encoding proteins that protect DNA gyrase from inhibition by ciprofloxacin. One such gene, *qnrA*, confers resistance to nalidixic acid (MIC: 8–16 mg/l) and epidemiological resistance to fluoroquinolones (ciprofloxacin, MIC: 0.25–1.0 mg/l). The basal level of quinolone resistance provided by *qnr* genes is low and strains can appear susceptible to quinolones according to Clinical and Laboratory Standards Institute (CLSI) guidelines. Their clinical importance lies in increasing the MIC of quinolone-resistant, and in particular nalidixic acid-resistant, strains of *Salmonella* to levels that are clinically relevant.

Salmonella and *Campylobacter* resistance to quinolones, cephalosporins and macrolides, classified as critically important antimicrobials (CIAs) by the WHO (WHO, 2007), was considered in depth in Europe in the *Joint Scientific Opinion on antimicrobial resistance (AMR) focused on zoonotic infections*, as compiled by a working group of experts from the European Centre for Disease Prevention and Control (ECDC), the European Food Safety Authority's (EFSA) Panel on Biological Hazards, the Committee for Medicinal Products for Veterinary Use (now replaced by the European Medicines Agency, EMA), and the Scientific Committee on Emerging and Newly Identified Health Risks (SCENIHR) (EFSA/ECDC/EMA/SCENIHR, 2009).

Occurrence of resistance to quinolones

Isolates from humans

In an Enter-Net study of 135 591 isolates of cases of human infection in ten European countries over the five-year period 2000–4, the occurrence of resistance to ciprofloxacin (MIC: > 1.0 mg/l) remained constant at approximately 0.8% whereas resistance to nalidixic acid increased from 14% to 20% (Meakins *et al.*, 2008). Although resistance to ciprofloxacin remained constant in most serotypes, for nalidixic acid considerable variation between serovars was observed. For example, in *S.* Enteritidis, the most commonly isolated serotype, resistance increased from 10% to 26% over the four-year period but remained constant at approximately 6% in *S.* Typhimurium. The highest incidence of resistance to both ciprofloxacin and nalidixic acid was seen in *S.* Virchow, with 68% of isolates resistant to nalidixic acid in 2002 and approximately 4–5% of isolates exhibiting resistance to ciprofloxacin. This study possibly represents the most comprehensive baseline investigation of resistance in human isolates of *Salmonella* undertaken, and remains the gold standard for determination of changes in resistance levels over a defined time period.

More recently, in a joint study between the EFSA and ECDC, which brought together studies on isolates from humans, food and food animals (EFSA/ECDC, 2012), 15% of 22 117 *Salmonella* isolates from cases of human infection in 19 EU

MSs in 2010 exhibited resistance to quinolone antibiotics (specifically nalidixic acid) and 9% to ciprofloxacin. There was considerable variation between MSs, ranging from 7% to 29% for nalidixic acid and 0% to 15% for ciprofloxacin. These differences are accentuated for ciprofloxacin, as some MSs report only clinical levels of resistance, whereas others report epidemiological cut-off (ECOFF) levels. Resistance to quinolone antibiotics and ciprofloxacin was particularly common in *S.* Enteritidis with an overall level of 19% for nalidixic acid in 6904 isolates studied, and 8% to ciprofloxacin. Again, there was considerable variation between MSs, varying from 5% to 72% for resistance to nalidixic acid, and 0% to 19% for ciprofloxacin. For *S.* Typhimurium, the second most common serovar, 9% of 5734 isolates were resistant to nalidixic acid and 5% of 6412 isolates to ciprofloxacin, again with considerable variance between MSs, compounded by some countries reporting resistance at clinical rather than epidemiological levels, and vice versa.

Isolates from food

Ten MSs provided data on the occurrence of resistance to nalidixic acid in *Salmonella* from pig meat in 2010. Of 527 isolates, 4% were resistant to nalidixic acid and 5% to ciprofloxacin. The majority of resistance to these antimicrobials was found in isolates of *S.* Typhimurium.

For broiler meat 24% of 548 isolates from seven MSs were resistant to nalidixic acid and ciprofloxacin. Data for individual serovars were not provided and recording of resistance was at ECOFF levels for both antimicrobials. There was a high incidence of resistance to nalidixic acid, ranging from 13% to 90%. Resistance to ciprofloxacin was variable, with most MSs not reporting such resistance but with two MSs reporting high levels (13% and 81%, respectively).

Overall little data are available on the occurrence of resistance in isolates from food. The most comprehensive data in this respect comes from a series of UK studies in the mid-2000s. In these studies (Little *et al.*, 2007, 2008a, 2008b) quinolone-resistant *S.* Typhimurium was isolated from lamb and pork on retail sale and quinolone-resistant *S.* Enteritidis from imported shell eggs and poultry meat on retail sale. In the Netherlands a high incidence (>40%) of *S.* Java and other serovars with ciprofloxacin and nalidixic acid resistance in poultry meat was reported in 2007 (MARAN, 2007).

Isolates from food-producing animals

Figures for 934 *S.* Enteritidis isolates from poultry in 12 MSs in 2010 indicated levels of resistance to nalidixic acid to be 23% and for ciprofloxacin to be 25%. Again, there were considerable differences between MSs, ranging from 0% to 97% for nalidixic acid, and 05 to 95% for ciprofloxacin. For *S.* Typhimurium only 2% of 124 isolates from pigs from 10 MSs were resistant to nalidixic acid, and 3% to ciprofloxacin. Again, there was considerable variance in the occurrence of resistance to these antimicrobials between MSs. In 2007 there was an incidence of 42% resistance to nalidixic acid/ciprofloxacin in *S.* Enteritidis from Dutch poultry, and of over 50% in *S.* Java from poultry (MARAN, 2007).

Resistance to quinolones was also monitored in a EFSA-funded study undertaken by the Danish Technical University in 2009, investigating the occurrence of resistance to various antimicrobials in *Salmonella* from food-production animals in MSs from 2004–2007. For isolates from food-producing animals, epidemiological resistance to ciprofloxacin was common in *S.* Enteritidis from broiler meat and hens, particularly in isolates from countries in southern Europe but also from certain countries in northern Europe. With the exception of certain new MSs, such resistance was relatively uncommon in isolates of *S.* Typhimurium from pork, pigs and cattle. With the exception of one northern European country, there was a high incidence of quinolone resistance in *Salmonella* from turkeys.

Comparison of isolation frequencies: human, animal and food

Direct comparison of quinolone resistance data between the three categories – human, animal and food – is difficult in that certain countries, in particular Denmark and the Netherlands, use epidemiological levels (MIC: >0.125 mg/L) to define resistance to ciprofloxacin for *Salmonella* isolations from food animals and foods (MARAN, 2007, 2009; DANMAP, 2007), whilst other countries use epidemiological values to define decreased susceptibility, but clinical levels (MIC: > 1 mg/L) for defining resistance. This is particularly the case with isolations from cases of human infection in the study by Enter-Net laboratories referred to above (Meakins *et al.*, 2008). Nevertheless, certain trends are apparent, particularly the increasing occurrence of nalidixic acid resistance and decreased susceptibility to fluoroquinolones in cases of human infection with *S.* Enteritidis mirrored by similar nalidixic acid-resistant strains in isolates from poultry meat and shell eggs (MARAN, 2007; Little *et al.*, 2007, 2008b). The studies quoted above do underline the necessity for agreed levels of significance for resistance reporting for human, food and veterinary isolates at an international level. Without such an agreement, meaningful comparisons between isolates from the different sources are extremely difficult, if not impossible.

Attribution studies

Microbial sub-typing can provide useful information during outbreak investigations but also at a population level. Isolates derived from humans, animals or food are obtained, analysed using a discriminatory method and are finally compared. The attribution of antimicrobial-resistant *Salmonella*-related cases has also been investigated by Hald *et al.* (2008). In this study, from a consideration of the attribution of resistant, multiresistant and quinolone-resistant strains it was concluded that imported poultry and Danish eggs were important sources for quinolone-resistant *Salmonella*, and that travel was also associated with the acquisition of quinolone-resistant *Salmonella* strains.

Temporal studies

In the UK fluoroquinolones were licensed for veterinary use in 2003. Subsequent studies of the occurrence of resistance to quinolones in an *S.* Typhimurium definitive phage type (DT104) showed a temporal increase of quinolone-resistant strains in isolations of multidrug-resistant DT104 from humans, cattle, poultry and pigs (Threlfall, 2000).

Outbreak investigations
The transfer of quinolone-resistant *S.* Typhimurium through the food chain from food-production animals to humans has been documented in two outbreaks, in Denmark (Mølbak *et al.*, 1999) and the UK (Walker *et al.*, 2000). In each outbreak, the transfer of a multiresistant strain of *S.* Typhimurium DT104 with additional resistance to nalidixic acid coupled with decreased susceptibility to ciprofloxacin from food animal to humans was conclusively demonstrated.

Plasmid-mediated quinolone resistance
Plasmid-mediated quinolone resistance (PMQR) has now been reported in isolations of *Salmonella* from many countries (EFSA/ECDC/EMA/SCENIHR, 2009). A variety of serotypes and *qnr* genes (A1, B1, B2, B5 and S1) has been reported, frequently associated with resistance to several unrelated antimicrobials including third-generation cephalosporin antibiotics. The majority of isolations were from cases of human infection, and were mostly associated with travel to countries outside the EU. Unusually, PMQR (*qnrS*) has been identified in a single isolate of *S.* Infantis from poultry in Germany, a single isolate of *S.* Bredeney (*qnrS1*) from a Dutch broiler and in Denmark from serovars *S.* Saintpaul (*qnrS1*), *S.* Newport (*qnrB51*) and *S.* Hadar (*qnrB5*) from imported turkey meat (for a review, see EFSA/ECDC/EMA/SCENIHR, 2009).

7.2.2 *Salmonella*: cephalosporin resistance

Mechanisms of resistance
The main mechanism of resistance to cephalosporins is through production of β-lactamase enzymes, which hydrolyse the β-lactam ring, thereby inactivating cephalosporin: this is called the enzymatic barrier. The genes coding for these enzymes, of which there are a large number of different types, are acquired by horizontal transmission from other bacteria since they are invariably absent from naturally occurring *Salmonella* strains.

There are two broad types of β-lactamase enzyme that have been reported most frequently in *Salmonella* and which confer resistance to third-generation cephalosporins. These are:

(i) Extended-spectrum β-lactamases (ESBLs) (e.g. the TEM and SHV variants and the CTX-M enzymes).
(ii) AmpC β-lactamases, which hydrolyse oxyimino-cephalosporins and cephamycins and are also resistant to clavulanate; they are class C enzymes in Ambler's molecular classification (Ambler, 1980).

In addition to these types of β-lactamase, other types have also been reported, for example OXA enzymes, which are assigned to a different molecular class (class

D) and the KPC enzymes (carbapenemases), which also confer resistance to cephalosporins.

Occurrence of resistance

Isolates from humans

Among *Salmonella*, the global spread of multidrug-resistant clones (resistant to chloramphenicol, florfenicol, streptomycin, spectinomycin, sulphonamides, tetracyclines and trimethoprim) belonging to serovars *S.* Agona with *Salmonella* genomic island 1 (SGI-1) and *S.* Typhimurium DT104 carrying SGI-1, has been linked to TEM-52 production. Particular clones of *S.* Infantis seem to be widespread among poultry in some EU countries, for example, in France and Belgium (Cloeckaert *et al.*, 2007) and also in Japan and China (Dahshan *et al.*, 2010; Yang *et al.*, 2010).

CTX-M-9, CTX-M-15 and CTX-17 to -18 enzymes have also been identified in different serovars from humans in the UK (Batchelor *et al.*, 2005), but at low frequency.

Isolates from animals and food

Different national surveys performed in Germany, Denmark, Norway and Spain and collected data from EU MSs and from a EFSA-funded study have demonstrated that resistance to broad-spectrum cephalosporins is low among *Salmonella* from animals and food products in northern MSs although higher in southern and eastern European countries. Nevertheless, a continuous increase in this prevalence is observed in several countries, such as the Netherlands, Denmark and France, mainly linked to the spread of clonal lines, namely *S.* Typhimurium, *S.* Paratyphi B variant Java (i.e., *S.* Java) and *S.* Agona. Plasmid-mediated CTX-M-like enzymes have been increasingly reported in a range of serovars in food-production animals in several European MSs, notably *S.* Virchow in Spain, Belgium and France (Carattoli, 2009). The use of third-generation cephalosporins in poultry has been suggested as major contributory factor in this respect. A range of *Salmonella* serovars (*S.* Agona, *S.* Virchow, *S.* Infantis and *S.* Typhimurium) has also been associated with the dissemination of ESBLs in poultry, cattle and pigs.

A very different situation has been reported in Canada, where ceftiofur-resistant, AmpC-producing *S.* Heidelberg has been regularly isolated from retail chicken since the mid-2000s. A strong correlation between such strains and ceftiofur-resistant *S.* Heidelberg from cases of human infection in that country has been demonstrated. The widespread use of ceftiofur in poultry in Canada is considered to be a significant factor in the appearance of such strains (Dutil *et al.*, 2010). Epidemics of AmpC-producing *S.* Newport have also been reported in North America and particularly in the USA since the early 2000s, often associated with cattle (Zansky *et al.*, 2002; Harbottle *et al.*, 2006; Varma *et al.*, 2006).

Conclusions

The prevalence of resistance to third-generation cephalosporins in *Salmonella* from animals is currently low in all MSs. Nevertheless, the prevalence in humans, animals and food and reports of linkages between epidemiological groups show that transfer along the food chain can occur. The EU picture is also affected by global food imports and human travel-associated exposure to *Salmonella*. Notwithstanding these considerations, it seems safe to conclude from published reports and from the data submitted to the EFSA, that the overall prevalence of resistance to third-generation cephalosporins in *Salmonella* in EU MSs is low. This is contrary to the situation in Canada and the USA, where AmpC-producing strains of *S.* Heidelberg with resistance to third-generation cephalosporins have spread to humans from poultry and caused several serious infections. Similarly AmpC-producing strains of *S.* Newport have caused extensive outbreaks in cattle and humans in the USA since 2000 (Zansky *et al.*, 2002; Varma *et al.* 2006).

7.2.3 *Salmonella*: overall conclusions

Resistance to quinolone antimicrobials has increased substantially in *Salmonella* from both food-production animals and from cases of human infection in several EU MSs since the early 2000s. Resistance to third- and fourth-generation cephalosporins has remained low and when encountered in human infections, has generally been related to foreign travel. An exception to this has been recently observed in the Netherlands, where isolates of *S.* Java with resistance to third-generation cephalosporins has increased in recent years (MARAN, 2009). Plasmid-mediated resistance to quinolones, although increasing in incidence in several serovars, is not yet a significant public health problem, although of concern is the frequent association of plasmid-mediated quinolone resistance with resistance to third-generation cephalosporins in such strains.

7.3 Increasing occurrences of resistance to quinolones and macrolides in *Campylobacter*

7.3.1 *Campylobacter*: quinolone resistance

Mechanisms of resistance

Quinolone resistance in *Campylobacter* is principally due to single mutations in *gyr*A and occasionally in topoisomerase IV (*par*C). The resultant MICs are in the range 64–128 mg/l for nalidixic acid and 16–64 mg/l for ciprofloxacin. There is also evidence, albeit rarely, of resistance by efflux, with consequent cross-resistance to a range of therapeutic antimicrobials.

Occurrence of resistance

Human isolates

The first comprehensive report demonstrating a link between the increasing occurrence of resistance to quinolone antibiotics in poultry and in cases of human infection was that of Endtz *et al.* (1991), when approximately 11% of *Campylobacter* from human infections in the Netherlands were resistant to quinolones. Further studies by Engberg and colleagues (2001) demonstrated an overall increase in quinolone resistance in *Campylobacter* worldwide, thereby reducing the efficacy of quinolones for treatment, should such therapy be required. By 2005, 37% of *C. jejuni* and 48% of *C. coli* from cases of human infection in EU MSs were resistant to ciprofloxacin (EFSA, 2007), and by 2010, 52% of *C. jejuni* and 66% of *C. coli* from 11 MSs were resistant to ciprofloxacin (EFSA/ECDC, 2012).

Isolates from animals

The ability for *C. jejuni* to rapidly develop mutational resistance to fluoroquinolones in poultry treated with fluoroquinolone-containing compounds has been clearly demonstrated in a series of experiments by McDermott and colleagues (McDermott *et al.*, 2002). In 2010, 47% of 638 isolates *of C. jejuni* from poultry in nine MSs were resistant to ciprofloxacin; 84% of 243 isolates of *C. jejuni* from poultry also exhibited resistance to this antimicrobial, as did 40% of 537 isolates from pigs (EFSA/ECDC, 2012).

7.3.2 *Campylobacter*: macrolide resistance

Mechanisms of resistance

Modification of the macrolide ribosomal targets is the most common macrolide resistance mechanism encountered in *Campylobacter* spp. and this develops due to a mutation. Two nucleotides close to each other are target sites for modification. Mutation of A2075G results in a high-level erythromycin resistance (MIC: >128 mg/ml) in clinical strains of *C. jejuni* and *C. coli*. Mutations affecting the ribosomal proteins L4 and L22 have also been identified. These were associated with both *C. jejuni* and *C. coli* that possessed an A2075G polymorphism in the 23S rRNA gene. A number of different mutations have been described in both ribosomal protein-encoding genes. In addition, the RND pump CmeABC is known to contribute to intrinsic and acquired resistance to fluoroquinolones and macrolides in *C. jejuni* and *C. coli*.

Occurrence of resistance

In human isolates of *Campylobacter*, only 2.8% of 34 838 isolates from 11 EU MSs exhibited resistance to macrolides (erythromycin) in 2010 (EFSA/ECDC, 2012). As with quinolone resistance in *Campylobacter*, resistance varies between serotypes, with only 1.7% of 8969 isolates of *C. jejuni* exhibiting resistance compared to 11% of 1099 isolates *of C. coli*. Similarly, only 2% of isolates of *C. jejuni* from meat from broilers from seven MSs exhibited resistance, as did 12% of 374 isolates of *C. coli*.

7.3.3 *Campylobacter*: overall conclusions

Resistance to quinolones in *Campylobacter* spp. taken from cases of human infection is increasing rapidly in many countries worldwide. The predominant mechanism of resistance is by mutation in *gyrA*. A temporal association between the emergence of quinolone resistance and its increase in isolates both from animals and humans following the use of a fluoroquinolone antibiotic in poultry production has been demonstrated in many countries. In contrast, resistance to macrolides such as erythromycin has not significantly increased over the last decade.

7.4 The recent emergence and spread of new multidrug-resistant clones of monophasic *Salmonella* Typhimurium-like organisms

7.4.1 Background

S. enterica serovar 4,[5],12:i:- is considered a monophasic variant of serovar *S.* Typhimurium (which has the antigenic structure 4,[5],12:i:1,2) due to antigenic and genotypic similarities between the two serovars (Echeita *et al.*, 2001). The emergence, in many different countries worldwide, of a variety of apparently different monophasic *S.* Typhimurium-like strains of different phage types, genotypes and antimicrobial resistance profiles, has been well documented over the past two decades (Echeita *et al.*, 1999; Amavisit *et al.*, 2005; Mossong *et al.*, 2007; Switt *et al.*, 2009; Soyer *et al.*, 2009). Such studies indicate that certain strains of monophasic *S.* Typhimurium represent multiple clones or clonal lines, which can only be further differentiated by highly sensitive molecular methods and which have emerged through independent deletion events (Soyer *et al.*, 2009). Some of these lines appear to be ecologically successful since they have spread rapidly in certain animal populations. In this respect within EU MSs, two major clonal lines of monophasic *S.* Typhimurium-like strains have emerged. One clonal line emerged in Spain in the late 1990s and exhibits plasmid-mediated resistance to a range of antimicrobials. The second clonal line has become particularly common in several MSs since 2000 and is characterised by chromosomally encoded resistance to ampicillin (A), streptomycin (S), sulphonamides (Su) and tetracyclines (T) (known as R-type ASSuT). Such strains appear to have rapidly increased in prevalence in cases of human illness in the EU over a relatively short time period and appear to be derived from the *S.* Typhimurium genetic lineages. They therefore behave exactly like normal pathogenic strains of *S.* Typhimurium in terms of their ability to infect and cause disease in both animals and the human population.

Salmonella Typhimurium (antigenic structure 4,[5],12:i:1,2) is the second most common serovar associated with human cases of *Salmonella* infection in the EU (EFSA/ECDC, 2012). In contrast, isolates of serovar 4,[5],12:i:- were rarely identified before the mid-1990s but now occur in the top ten most common serovars from humans in several countries. Cases of infection with serovar 4,[5],12:i:- have reportedly been severe, with a 70% hospitalisation rate during an outbreak in New York City in 1998 (Agasan *et al.*, 2002), although a much lower rate of 21% was observed during an outbreak in Luxembourg in 2006 (Mossong *et al.*, 2007). Infections have also been particularly associated with cases of septicaemia in Thailand (Pornruangwong *et al.*, 2008) and Brazil (Tavechio *et al.*, 2004). Overall, cases of infection have been linked to a number of sources, including poultry (Zamperini *et al.*, 2007) and cattle, but particularly pigs and pork products (de la Torre *et al.*, 2003; Vieira-Pinto *et al.*, 2005; Hopkins *et al.*, 2010; Hauser *et al.*, 2010).

7.4.2 Occurrence and epidemiology

A marked increase in prevalence of *S. enterica* serovar 4,[5],12:i:- of R-type ASSuT and with a characteristic pulsed field gel electrophoresis (PFGE) and variable number of tandem repeats (VNTR) profile has been noted in both foodborne infections and in pigs and pork meat in several European countries over the last ten years (Mossong *et al.* 2007; van Pelt *et al.*, 2008; Dionisi *et al.*, 2009; EFSA, 2010; EFSA/ECDC, 2012). In a baseline study from fattening pigs (Commission Decision 2006/668/EC), Spanish strains of *S. enterica* serovar 4,[5],12:i:- represented 14% of the isolates, 53% of which exhibited the ASSuT R-type. A recent study described emergence of a clonal group of serovar Typhimurium and 4,[5],12:i:- R-type ASSuT strains in Italy, Denmark and the UK (Lucarelli *et al.*, 2010). In 2008, the monophasic variant represented 42.2% of all *S.* Typhimurium isolates from humans analysed at the National Reference Centre in Germany, and were responsible for several large, diffuse outbreaks often involving increased hospitalisation in comparison with classic *S.* Typhimurium (Trüpschuch *et al.*, 2010).

Epidemiological studies in several European countries have indicated that closely-related *S. enterica* 4,[5],12:i:- clones of R-type ASSuT and with characteristic PFGE and VNTR profiles have spread within several European countries, with pigs as a likely reservoir of infection. Because of differences in nomenclature, the true incidence of such strains is difficult to determine. Over 10 000 infections with such clones have now been reported in different European countries since the early 2000s (EFSA, 2010).

The diversity of PFGE and VNTR profiles within *S. enterica* serovar 4,[5],12:i:- R-type ASSuT isolates, and the differences between these isolates and those previously described in Spain (Guerra *et al.*, 2000, 2001), suggests that *S. enterica* serovar 4,[5],12:i:- is likely to represent several clones/strains that have emerged independently from serovar Typhimurium. Recent genotypic studies have shown that in addition to the Spanish 4,[5],12:i- clone, other 4,[5],12:i:- lineages exist (Soyer *et al.*, 2009).

7.4.3 Molecular studies

Isolates of *S. enterica* serovar 4,[5],12:i:- are characterised by lack of the *flj*B flagellar gene found in isolates of classic *S.* Typhimurium with the antigenic structure 4,[5],12:i:1,2, and by the ASSuT R-type encoded by the resistance genes $bla_{\text{TEM-1}}$, *strA-strB, sul2* and *tetB.* These genes are localised on the bacterial chromosome in a resistance island of approximately 16 kb, which differs from the characteristic SGI-1 resistance island present in epidemic *S.* Typhimurium DT104 (Hopkins *et al.*, 2010).

Studies in Germany have shown that isolates of *S. enterica* serovar 4,[5],12:i:- harbour *S.* Typhimurium specific Gifsy-1, Gifsy-2 and ST64B prophages. Searches for insertions in tRNA sites resulted in the detection of an 18.4-kb fragment adjacent to the thrW tRNA locus, exhibiting a lower G+C content compared to the *S.* Typhimurium LT2 genome. Sequence analysis identified 17 potential open reading frames, some with a high similarity to enterobacterial phage sequences and sequences from *Shigella boydii, S. dysenteriae*, avian pathogenic *Escherichia coli* and other *Escherichia* spp. (Trüpschuch *et al.*, 2010).

More recently, DNA sequencing has demonstrated that the genomic resistance region in *S. enterica* serovar 4,[5],12:i:- coding for ASSuT consists of two regions, region 1 (RR1), conferring resistance to ampicillin, streptomycin and sulphonamides, and resistance region 2 (RR2), conferring tetracycline resistance. These resistance regions are both surrounded by IS26 elements. Sequence comparative analysis showed 99% sequence identity with a region of plasmid pO111_1 from an *E. coli* strain. All monophasic strains examined were positive for the four resistance genes and the left and right junctions and the internal regions of RR1 and RR2. All the strains lacked the STM1053-1997 and STM2694 genes and showed deletion of the *flj*A-*flj*B operon (Lucarelli *et al.*, 2012).

7.4.4 Public health measures

In April 2011, the European Commission stated that: 'Monophasic strains of *S.* Typhimurium have rapidly become one of the most commonly found *Salmonella* serotypes in several species of animals and in clinical isolates of humans.' According to the Scientific Opinion on monitoring and assessment of the public health risk of '*Salmonella* Typhimurium-like strains', adopted by the Panel on Biological Hazards of EFSA on 22 September 2010 (EFSA, 2010), monophasic *S.* Typhimurium strains with the antigenic formula 1,4,[5],12:i:- are considered as variants of *S.* Typhimurium and pose a public health risk comparable to that of other *Salmonella* Typhimurium strains.

Accordingly, for the purposes of clarity of Union legislation, it is appropriate to amend Regulation (EC) No 2160/2003 and Commission Regulation (EU) No 200/2010 of 10 March 2010 implementing Regulation (EC) No 2160/2003 of the European Parliament and of the Council as regards a Union target for the reduction of the prevalence of *Salmonella* serotypes in adult breeding flocks of *Gallus gallus* in order to provide that *S.* Typhimurium include monophasic strains with the antigenic formula 1,4,[5],12:i:-.' (European Commission, 2011).

7.4.5 Conclusions

Multidrug-resistant strains of *S. enterica* serovar 4,[5],12:i:-, now regarded as monophasic *S.* Typhimurium-like strains, are an increasing problem worldwide, and particularly in the European Union, where widespread outbreaks have been reported over the past decade. In order to prevent a global epidemic of these newly emerging clones/strains, as occurred with *S.* Typhimurium DT104, appropriate intervention strategies need to be put in place as soon as possible, particularly in pig husbandry throughout the EU. In this respect the move by the European Commission, to place *S. enterica* serovar 4,[5],12:i:- under the same controls as *S.* Enteritidis and *S.* Typhimurium in poultry is particularly welcome.

7.5 Rapid emergence and spread of strains of Enterobacteriaceae

7.5.1 Background

In the last decade a variety of β-lactamases have emerged in Gram-negative bacteria, resulting in reduced susceptibility to broad spectrum β-lactams (Arlet *et al.* 2006; ECDC, 2011). These β-lactamases include both extended spectrum β-lactamases (ESBLs) and AmpC β-lactamases (AmpC). ESBL-producing organisms are frequently co-, or multiresistant, exhibiting resistance to other antimicrobial classes such as fluoroquinolones, aminoglycosides and trimethoprim-sulphamethoxazole due to associated resistance mechanisms, which may be either chromosomally or plasmid-encoded. The most frequently encountered ESBLs in Enterobacteriaceae belong to the TEM, SHV and CTX-M families (Paterson and Bonomo, 2005).

AmpC β-lactamases are intrinsic cephalosporinases found on the chromosomal DNA of many Gram-negative bacteria, including many members of the Enterobacteriaceae (but, notably, not in *Klebsiella* or *Salmonella)*, and opportunistic pathogens such as *Pseudomonas* and *Acinetobacte*r. These enzymes confer resistance to penicillins, second- and third-generation cephalosporins including β-lactam/inhibitor combinations, cephamycins (cefoxitin), but usually not to fourth-generation cephalosporins (cefepime, cefquinome) and carbapenems. A growing number of AmpC enzymes have now escaped on to plasmids. These are the so-called acquired, or plasmidic, AmpCs.

7.5.2 Public health concerns

ESBL enzymes are capable of breaking down cephalosporin antibiotics, which are commonly used in hospitals as front-line treatments. Recently the CTX-M group of ESBLs, so called after their ability to hydrolyse the third-generation (extended spectrum) cephalosporin cefotaxime, have become prominent and are common in *Escherichia coli*, with many infections in community patients, often resulting in deaths in vulnerable individuals (Pitout and Laupland, 2008). Such infections have increased both throughout Europe and worldwide in the last 10 years (Livermore *et al.*, 2007). CTX-M-15 *E. coli* O25:H4-ST131 has emerged as the dominant clone in human infections in many countries (Lau *et al.*, 2008), but different clones have emerged in food animals, particularly broilers. For example the dominant isolate in broilers in the UK, the Netherlands and France is a CTX-M-1 ESBL. Similarly CTX-M-14 is not common in human isolates but frequently found in turkeys and cattle in the UK. A similar observation has been made in the Netherlands where CTX-M-1 ESBLs have been found in poultry, chicken meat at retail and in humans (MARAN, 2009).

The increase of ESBL bacteria, which has been observed in both healthcare and community settings as well as among animals in recent years, suggests that food and the environment could be sources of these resistant bacteria. In this respect recent studies at the UK Health Protection Agency found cephalosporin-resistant *E. coli* in over 30% of 210 batches of imported raw chicken sampled. Of these cephalosporin-resistant isolates 30% produced group 2 CTX-M ESBLs, 27% produced group 8 CTX-M ESBLs, 42% produced CMY-type AmpC enzymes and 1% produced a group 2 CTX-M. In contrast, none produced CTX-M-15 ESBL nor belonged to the ST131 clone (Dhanji *et al.*, 2010).

7.5.3 Epidemiology of acquired resistance with ESBLs and AmpC in food-producing animals and food

Organisms producing ESBLs (e.g. TEM, SHV, CTX-M and PER) and AmpCs (e.g. CMY, DHA-1 and ACT-1) have been detected in food-producing animals (poultry, swine, bovines, horses, rabbits, ostriches and wild boar), marine aquaculture systems and foods of animal origin (for a review, see EFSA, 2011). The most common ESBL genes are those encoding CTX-M enzymes, $bla_{TEM\text{-}52}$ and $bla_{SHV\text{-}12}$, while the most common AmpC gene is $bla_{CMY\text{-}2}$. It is of note that ESBL producers are mostly found in Europe, while AmpC producers are particularly common in North America, mirroring the trends observed for human isolates. ESBL and AmpC genes have been most frequently detected in *E. coli*, non-typhoidal *Salmonella* and to a lesser extent *Klebsiella, Citrobacter* and *Enterobacter* in samples from terrestrial food-producing animals. (For a review, see EFSA, 2011).

7.5.4 Transmission

Transmission of ESBL and AmpC genes in both humans and animals is mainly driven by the IncF, IncI, IncN, IncA/C, IncL/M and IncK plasmid families (Carattoli, 2009). IncA/C plasmids have been associated with the emergence of $bla_{CMY\text{-}2}$ in both the USA and the UK (Hopkins *et al.*, 2006) and IncN, IncI and IncL/M plasmids with the spread of $bla_{CTX\text{-}M\text{-}1}$, $bla_{CTX\text{-}M\text{-}3}$ and $bla_{TEM\text{-}52}$, respectively in Europe (Cloeckaert *et al.*, 2007). IncK plasmids carrying the $bla_{CTX\text{-}M\text{-}14}$ gene have become diffused in both Spain and the UK (Cottell *et al.*, 2011).

Recent reports documenting outbreaks of ESBL producers of Shiga toxin-producing *E. coli* (STEC, also called verotoxin-producing *E. coli*, VTEC) clones O111:H8 and O104:H4 linked to food or food-producing animals are of particular concern (Valat *et al.*, 2012; Buchholz *et al.*, 2011; King *et al.*, 2012). It is noteworthy that the causative strain of *E. coli* responsible for the 2011 outbreak of STEC/VTEC O104:H4, centred in northern Germany but with over 1500 cases in at least eight countries, exhibits resistance to a range of antibiotics including several second- and third-generation cephalosporins. This strain produces an

ESBL as a result of possession of a *bla*$_{CTX\text{-}M\text{-}15}$ gene. The source of the outbreak train has not been determined, although different salad vegetables, including organic fenugreek sprouts (Buchholz *et al.*, 2011; King *et al.*, 2012) have been implicated as possible vehicles of infection.

7.5.5 Possible reservoirs

Although person-to-person spread is recognised as the main method for spreading ESBL/AmpC-*β*-lactamase-producing *E. coli* both in hospitals and the community, the primary reservoirs of such organisms are contentious. ESBL/AmpC-producing *E. coli* have been isolated from food animals in many European countries, particularly poultry and cattle. Farm animals, particularly poultry, are now recognised as important carriers (Dierikx *et al.*, 2010; Hasman *et al.*, 2006; Randall *et al.*, 2011). Similarly there have been an increasing number of reports of isolations from foods of animal origin (Bergenholtz *et al.*, 2009; Carattoli, 2008; Mora *et al.*, 2010; Leverstein-van Hall *et al.*, 2011; EFSA, 2011). These reports raise questions about the possible role of animal- and food-related reservoirs in the spread of ESBL/AmpC-producing microorganisms.

7.5.6 Resistance to carbapenems

The occurrence of Enterobacteriaceae resistant to carbapenems is a growing threat in human medicine. The presence of such resistance in bacteria from animals is largely unknown, although *E. coli* producing VIM-1 carbapenemase have been recently recorded in pigs in one European country (Fischer *et al.*, 2012). As yet there are no indications of the zoonotic transfer of such resistance to humans.

7.5.7 Conclusions

The rapid emergence and spread of strains of Enterobacteriaceae, and in particular *E. coli*, either exhibiting plasmid-mediated resistance to third-generation cephalosporin antibiotics or producing AmpC *β*-lactamases, is now a major public health problem worldwide. Much work has gone into establishing a link between such strains and the widespread use of *β*-lactam antibiotics in food-production animals, particularly poultry. Nevertheless it is important that the contribution of corresponding antibiotics in human medicine in both hospitals and the community is not overlooked, particularly as certain epidemic clones – e.g. CTX-M-15 *E. coli* O25:H4-ST131 – have a human-to-human epidemiology, with only very tenuous links to food animals.

7.6 Control measures

7.6.1 Background

As antibiotics are not recommended for the treatment of mild to moderate *Salmonella*- or *Campylobacter*-induced enteritis in humans, it may be argued that drug resistance in such organisms is of little consequence for public health. Nevertheless antibiotics are used for the treatment of gastroenteritis in immunocompromised patients and sometimes for treating particularly vulnerable patients; in such cases treatment with an appropriate antibiotic is often essential and may be life-saving. The increased occurrence, both nationally and internationally, of strains of *S. enterica* and *Campylobacter* with decreased susceptibility or resistance to quinolone and fluoroquinolones antibiotics is therefore an unwelcome development.

For Enterobacteriaceae, the broad resistance profile in bacteria that follows the production of ESBL/AmpC β-lactamases is highly significant in human infections (Rodriguez-Bano *et al.*, 2010). This is primarily because many community infections, and also infections treated or transmitted within hospitals, are caused by bacteria that are no longer sensitive to second-, third- or fourth-generation cephalosporins (Ben-Ami *et al.*, 2006). The multiresistant nature of bacteria that produce ESBLs and AmpCs can affect the selection and timely administration of appropriate antimicrobials for community-acquired and healthcare-associated infections, since many first-line antimicrobials are no longer active against them.

Over the last 40 years many outbreaks and cases of infection in developed countries, particularly those caused by such drug-resistant non-typhoidal *Salmonella* and *Campylobacter* strains have been linked to foods of animal origin. In turn this has led to speculation about the role of antimicrobials in animals bred for food in contributing to the development and spread of such strains.

7.6.2 Use of growth promoters (feed additives)

Concern about the use of therapeutic antibiotics as growth promoters was expressed in the United Kingdom as long ago as 1969. At this time a joint committee on the use of antibiotics in animal husbandry and veterinary medicine (the Swann Committee) published a report, which resulted in the banning of the use of certain compounds (penicillins and tetracyclines) as growth promoters in the UK, and also made other therapeutic antibiotics available for use in animals by prescription only (Joint Committee, 1969). In the EU the use of growth-promoting antibiotics in food-production animals was banned by Sweden in 1999. Avoparcin was banned as a feed additive in food animals in EU countries in 1997, and bacitracin, spiramycin, tylosin and virginiamycin in 1999. On 1 January 2006, the use of all growth-promoting antibiotics was banned throughout the EU following an EC decision to this effect. The results of these actions are controversial. Substantive declines in the incidence of vancomycin-resistant enterococci in both humans and food animals have been reported in some countries (van den Bogaard *et al.*, 2000), but increases in the overall usage of therapeutic antibiotics in food animals have also been reported (Casewell *et al.*, 2003).

7.6.3 Prophylactic usage and control options

Of major concern is the continued prophylactic use, in animal husbandry, of CIAs for human health (WHO, 2007), particularly as such usage has undoubtedly contributed to the appearance and spread of drug-resistant organisms in the food chain. These concerns were recognised in the USA in 2000, and resulted in the banning of fluoroquinolones for use in poultry in 2005. This was as regarded as a major success for public health (Nelson *et al.*, 2007). Further research is now required to evaluate the effect of this ban on the incidence of fluoroquinolone-resistant *Campylobacter* in that country.

In the UK there has been considerable speculation about the contribution of the prophylactic use in food-production animals of antimicrobials such as third-generation cephalosporins and quinolones to the development of resistance in a range of organisms, including *Salmonella.* This has culminated in a recent report from the Chief Medical Officer (Chief Medical Officer, 2009), who stated that in addition to the substantive use of antibiotics in human medicine, antibiotics are also used in large quantities on animals, thereby adding to the threat of resistance. Following on from this observation, he recommended that there should be a ban on the use of certain types of antibiotics (quinolones and cephalosporins) in animals, in order to protect their activity in humans. Whether such a ban can be implemented unilaterally for the UK is debatable, but the comments in the report do highlight increasing concern about the use of certain key therapeutic animals in livestock.

Elimination or reducing the incidence of drug-resistant epidemic strains is an option which, to a certain extent, has been overlooked. In this respect the recent move by the European Commission to place *S. enterica* serovar 4,[5],12:i:- under the same controls as *S.* Enteritidis and *S.* Typhimurium in poultry is particularly welcome.

The WHO Regional Office for Europe has recently published a booklet, *Tackling Antibiotic Resistance from a Food Safety Perspective in Europe* (WHO, 2011), in which a range of options for combating antibiotic resistance have been listed, including regulation, prudent usage, integrated surveillance, advocacy and communication, and training and capacity building. The need for continued research into the problem is also highlighted.

There are no data on the comparative efficiency of individual control options in reducing public health risks caused by ESBL- and/or AmpC-producing bacteria related to food-producing animals. In a recent EFSA Opinion (EFSA, 2011) it was considered that a highly effective control option to reduce selection of ESBL/AmpC-producing bacteria at an EU level would be to stop all uses of cephalosporins or systemically active third- and fourth-generation cephalosporins, or to restrict their use (i.e. only allowed under specific circumstances). The more comprehensive the restriction, the more prominent the effect on selection pressure would be, although it was agreed that a very restrictive policy might have unintended consequences on animal health and welfare if effective antimicrobials were not readily available for treatment.

7.6.4 Conclusion

The consequences for public health resulting from the use of antibiotics in food-production animals are now well recognised. Antibiotics are generally recognised as being important for animal health and welfare, and complete withdrawal of such a vital armamentarium is unrealistic. Nevertheless controls are necessary to reduce the incidence of resistant organisms, which pose a threat to human health. Whether such controls should be enforced by legislation is debatable, particularly as changes in the incidence of resistance in certain foodborne zoonotic pathogens do not of necessity correlate to the veterinary use of antibiotics in a particular country (Threlfall *et al.*, 2006). The long-term effects of the European ban on antibiotic-containing growth promoters on overall antibiotic use in the EU have not yet been fully quantified. Likewise the effects of the ban of fluoroquinolone antibiotics in poultry in the USA on the occurrence of fluoroquinolone-resistant *Campylobacter* in human infections have not yet been elucidated. Similarly, although severe restrictions on the use of cephalosporins or systemically active third- and fourth-generation cephalosporins in poultry in the EU have been advocated (EFSA, 2011), because of the ubiquity of ESBL- and AmpC-producing bacterial strains and of the associated genetic determinants, it is unlikely that any single control measure will be sufficient to limit their transmission through the food chain.

In conclusion, there is widespread agreement that controls on the use of antibiotics in food-production animals are essential in controlling the spread of resistant organisms through the food chain. Nevertheless continued surveillance of resistance in both the human population and in food animals is vital in measuring the long-term effectiveness of any such control measures.

7.7 References

AGASAN A, KORNBLUM J, WILLIAMS G, PRATT C C, FLECKENSTEIN P *et al.* (2002), Profile of *Salmonella enterica* subsp. *enterica* (subspecies I) serotype 4,5,12:i:- strains causing food-borne infections in New York City, *J Clin Microbiol*, 40, 1924–1929.

AMAVISIT P, BOONYAWIWAT W and BANGTRAKULNONT A (2005), characterization of *Salmonella enterica* serovar Typhimurium and monophasic *Salmonella* serovar 1,4,[5],12:i:- isolates in Thailand, *J Clin Microbiol*, 43, 2736–2740.

AMBLER R P (1980), The structure of beta-lactamases, *Philos Trans R Soc Lon B Biol Sci*, 289, 321–331.

ARLET G, BARRETT T J, BUTAYE P, CLOECKAERT A, MULVEY M R *et al.* (2006), *Salmonella* resistant to extended-spectrum cephalosporins: Prevalence and epidemiology, *Microbes Infect*, 8, 1945–1954.

BATCHELOR M, CLIFTON-HADLEY F A., STALLWOOD, A D, PAIBA, G A, DAVIES RH *et al.* (2005), Detection of multiple cephalosporin-resistant *Escherichia coli* from a cattle faecal sample in Great Britain, *Microb Drug Resist*, 11, 58–61.

BEN-AMI R, SCHWABER M J, NAVON-VENEZIA S, SCHWARTZ D *et al.* (2006), Influx of extended-spectrum beta-lactamase-producing Enterobacteriaceae into the hospital, *Clin Infect Dis*, 42, 925–934.

BERGENHOLTZ R D, JORGENSEN M S, HANSEN L H, JENSEN LB and HASMAN H (2009), Characterization of genetic determinants of extended-spectrum cephalosporinases (ESCs) in *Escherichia coli* isolates from Danish and imported poultry meat, *J Antimicrob Chemother*, 64, 207–209.

BUCHHOLZ U, BERNARD H, WERBER D *et al.* (2011), German outbreak of *Escherichia coli* O104:H4 associated with sprouts, *N Engl J Med*, 365, 1763–1770.

CARATTOLI A (2008), Animal reservoirs for extended spectrum beta-lactamase producers. *Clin Microbiol Infect*, 14, Suppl 1, 1171–1123.

CARATTOLI A (2009), Resistance plasmid families in Enterobacteriaceae, *Antimicrob Agents Chemother*, 53, 2227–2238.

CASEWELL M, FRISS C, MARCO E, MCMULLEN P and PHILLIPS I (2003), The European ban on growth-promoting antibiotics and emerging consequences for human and animal health, *J Antimicrob Chemother*, 52, 157–161.

CHIEF MEDICAL OFFICER (2009), Annual Report 2009: On the state of public health, Department of Health. Available at: www.dh.gov.uk/publications.

CLOECKAERT A, PRAUD A, DOUBLET B, BERTINI A, CARATTOLI A *et al.* (2007), Dissemination of an extended-spectrum-β-lactamase bla_{TEM-52} gene-carrying IncI1 plasmid in various *Salmonella enterica* serovars isolated from poultry and humans in Belgium and France between 2001 and 2005, *Antimicrob Agents Chemother*, 51, 1872–1875.

COLLIGNON P, POWERS J H, CHILLER T M, AIDARA-KANE A and AARESTRUP F M (2009), World Health Organization ranking of antimicrobials according to their importance in human medicine: A critical step for developing risk management strategies for the use of antimicrobials in food production animals, *Clin Infect Dis*, 49, 132–141.

COTTELL J L, WEBBER M A, COLDHAM N G *et al.* (2011), Complete sequence and molecular epidemiology of IncK epidemic plasmid encoding blaCTX-M-14, *Emerg Infect Dis*, 17, 645–652.

DAHSHAN H, CHUMA T, SHAHADA F *et al.* (2010), Characterization of antibiotic resistance and the emergence of AmpC-producing *Salmonella* Infantis from pigs, *J Vet Med Sci*, 72, 1437–1442.

DANMAP (2007), Use of antimicrobial agents and occurrence of antimicrobial resistance in bacteria from food animals, foods and humans in Denmark. Available at: www.danmap.org/pdfFILES/DANMAP_2007.pdf.

DE LA TORRE E, ZAPATA D, TELLO M, MEJÍA W, FRÍAS N *et al.* (2003), Several *Salmonella enterica* subsp. *enterica* serotype 4,5,12:i:- phage types isolated from swine samples originate from serotype Typhimurium DT U302, *J Clin Microbiol*, 41, 2395–2400.

DHANJI H, MURPHY N M, DOUMITH M *et al.* (2010), Cephalosporin resistance mechanism in *Escherichia coli* isolated from raw chicken imported into the UK. *J Antimicrob Chemother*, 65, 2534–2539.

DIERIKX C, VAN ESSEN-ZANDBERGEN A, VELDMAN K, SMITH H and MEVIUS D (2010), Increased detection of extended spectrum beta-lactamase producing *Salmonella enterica* and *Escherichia coli* isolates from poultry, *Vet Microbiol*, 145, 273–278.

DIONISI A M, GRAZIANI C, LUCARELLI C, FILETICI E, VILLA L *et al.* (2009), Molecular characterization of multidrug-resistant strains of *Salmonella enterica* serotype Typhimurium and monophasic variant (*S.* 4,[5],12:i:-) isolated from human infections in Italy, *Foodborne Pathog Dis*, 6, 711–717.

DUTIL L, IRWIN R, FINLEY R, NG L K, AVERY B *et al.* (2010) Ceftiofur resistance in *Salmonella enterica* serovar Heidelberg from chicken meat and humans, Canada, *Emerg Infect Dis*, 16, 48–54.

ECDC (2011), Antimicrobial resistance surveillance in Europe 2010. Annual Report of the European Antimicrobial Resistance Surveillance Network (EARS-Net), 2011. Stockholm, ECDC, 2011.

ECHEITA M A, ALADUEÑA A, CRUCHAGA S and USERA M A (1999), Emergence and spread of an atypical *Salmonella enterica* subsp. *enterica* serotype 4,5,12:i:- strain in Spain, *J Clin Microbiol*, 37, 3425–3428.

ECHEITA M A, HERRERA S and USERA M A (2001), Atypical, *flj*B-negative *Salmonella enterica* subsp. *enterica* strain of serovar 4,5,12:i:- appears to be a monophasic variant of serovar Typhimurium, *J Clin Microbiol*, 39, 2981–2983.

EFSA (2007), The community summary report of trends and sources of zoonoses, zoonotic agents, antimicrobial resistance and foodborne outbreaks in the European Union in 2005, *EFSA Journal 2006*, 94, 3–288.

EFSA (2010), Scientific Opinion on monitoring and assessment of the public health risk of '*Salmonella* Typhimurium-like' strains, *EFSA Journal*, 8, 1826. doi: 10.2903/j.efsa.2010.1826.

EFSA (2011), The public health risks of bacterial strains producing ESBLs and/or AmpC β-lactamases in food and food-producing animals, *EFSA Journal*, 9, 2322.

EFSA/ECDC (2012), The European Union Summary Report on antimicrobial resistance in zoonotic and indicator bacteria from humans, animals and food in 2010, *EFSA Journal*, 10, 2598.

EFSA/ECDC/EMA/SCENIHR (2009), *Joint Opinion on antimicrobial resistance (AMR) focused on zoonotic infections* – Scientific opinion of the European Centre for Disease Prevention and Control; Scientific opinion of the Panel on Biological Hazards; Opinion of the Committee for Medicinal Products for Veterinary Use; Scientific opinion of the Scientific Committee on Emerging and Newly Identified Health Risks, *EFSA Journal*, 7, 1372. Available at: www.ema.europa.eu/docs/en_GB/document_library/Other/2009/11/WC500015452.pdf. Also: European Medicines Agency Reference EMEA/CVMP/447259/2009.

ENDTZ H P, RUIS G J, VAN KLINGEREN B, JANSEN W H *et al* (1991), Quinolone resistance in *Campylobacter* isolates from man and poultry following the introduction of fluoroquinolones in veterinary medicine, *J Antimicrob Chemother*, 27, 199–208.

ENGBERG J, AARESTRUP F M, TAYLOR D E, GERNER-SMIDT P and NACHAMKIN I (2001), Quinolone and macrolide resistance in *Campylobacter jejuni* and *C. coli*: Resistance mechanisms and trends in human isolates, *Emerg Infect Dis*, 7, 24–34.

EUROPEAN COMMISSION (2011), Commission Regulation (EU) No 517/2011 of 25 May 2011, *Official J European Union* L/45 2011.

FISCHER J, RODRIGUEZ I, SCHMOGER S *et al.* (2012), *Escherichia coli* producing VIM-1 carbapenemase isolated on a pig farm, *J Antimicrob Chemother*, 67, 1793–1795.

GUERRA B, LACONCHA I, SOTO S M, GONZALEZ-HEVIA M A and MENDOZA M C (2000), Molecular characterisation of emergent multiresistant *Salmonella enterica* serotype [4,5,12:i:-] organisms causing human salmonellosis, *FEMS Microbiol Lett*, 190, 341–347.

GUERRA B, SOTO S M, ARGÜELLES J M and MENDOZA M C (2001), Multidrug resistance is mediated by large plasmids carrying a class 1 integron in the emergent *Salmonella enterica* serotype [4,5,12:i:-], *Antimicrob Agents Chemother*, 45, 1305–1308.

HALD T, DANILO, M A, WONG L F and AARESTRUP F M (2008), The attribution of human infections with antimicrobial resistant *Salmonella* bacteria in Denmark to sources of animal origin, *Microb Drug Resist*, 14, 31–35.

HARBOTTLE H, WHITE DG, MCDERMOTT PF, WALKER R D and ZHAO S (2006), Comparison of multilocus sequence typing, pulsed-field gel electrophoresis, and antimicrobial susceptibility typing for characterization of *Salmonella enterica* serotype Newport isolates, *J Clin Microbiol*, 44, 2449–2457.

HASMAN H, MEVIUS D, VELDMAN K, OLESEN I and AARESTRUP F M (2006), β-lactamases among extended-spectrum β-lactamase (ESBL)-resistant *Salmonella* from poultry, poultry products and human patients in the Netherlands, *J Antimicrob Chemother*, 56, 115–121.

HAUSER E, TIETZE E, HELMUTH R, JUNKER E, BLANK K *et al.* (2010), Pork contaminated with *Salmonella enterica* serovar 4,[5],12:i:-, an emerging health risk for humans, *Appl Environ Microbiol*, 76, 4601–4610.

HOPKINS K L, LIEBANA E, VILLA L, BATCHELOR M, THRELFALL E J *et al.* (2006), Replicon typing of plasmids carrying CTX-M or CMY beta-lactamases circulating among *Salmonella* and *Escherichia coli* isolates, *Antimicrob Agents Chemother*, 50, 3203–3206.

HOPKINS K L, KIRCHNER M, GUERRA B, GRANIER S A, LUCARELLI C *et al.* (2010), Multiresistant *Salmonella enterica* serovar 4,[5],12:i:- in Europe: a new pandemic strain?, *Euro Surveill*, 15, 19580–19583.

JOINT COMMITTEE (1969), On the use of antibiotics in animal husbandry and veterinary medicine. London, HMSO.

KING L A, NOGAREDA F, WEILL F X *et al.* (2012), Outbreak of Shiga toxin-producing *Escherichia coli* O104:H4 associated with organic fenugreek sprouts, France, June 2011, *Clin Infect Dis*, 54, 1588–1594.

LAU S H, KAUFFMAN M E, LIVERMORE D M, WOODFORD N *et al.* (2008), UK epidemic *Escherichia coli* strains A-E, with CTX-M-15 beta-lactamase, all belong to the international O25:H4-ST131 clone, *J Antimicrob Chemother*, 62, 1241–1244.

LEVERSTEIN-VAN HALL M A, DIERIKX C M, COHEN STUART J *et al.* (2011), Dutch patients, retail chicken meat and poultry share the same ESBL genes, plasmids and strains, *Clin Microbiol Infect*, 17, 873–880.

LITTLE C L, WALSH S, HUCKLESBY L, SURMAN-LEE S, PATHAK K *et al.* (2007), Survey of *Salmonella* contamination of non-United Kingdom-produced raw shell eggs on retail sale in the northwest of England and London, 2005 to 2006, *J Food Prot*, 70, 2259–2265.

LITTLE C L, RICHARDSON J F, OWEN R J, DE PINNA E and THRELFALL E J (2008a), Prevalence, characterisation and antimicrobial resistance of *Campylobacter* and *Salmonella* in raw poultrymeat in the UK, 2003–2005, *Int J Environ Health Res*, 18, 403–414.

LITTLE C L, RICHARDSON J F, OWEN R J, DE PINNA E and THRELFALL E J (2008b), *Campylobacter* and *Salmonella* in raw red meats in the United Kingdom: Prevalence, characterization and antimicrobial resistance pattern, 2003–2005, *Food Microbiol*, 25, 538–543.

LIVERMORE D M, CANTON R, GNIADOWSKI M, NORDMANN P *et al.* (2007), CTX-M: The changing face of ESBLs in Europe, *J Antimicrob Chemother*, 59, 165–174.

LUCARELLI C, DIONISI A M, TORPDAHL M, VILLA L, GRAZIANI C *et al.* (2010), Evidence for a second genomic island conferring resistance in a clonal group of *Salmonella* Typhimurium and its monophasic variant circulating in Italy, Denmark and United Kingdom. *J Clin Microbiol*, 48, 2103–2109.

LUCARELLI C, DIONISI A M, FILETICI E, OWCZAREK S, LUZZI I and VILLA L (2012), Nucleotide sequencing of the chromosomal region conferring multidriug resistance (R-type ASSuT) in *Salmonella* Typhimurium and monophasic *Salmonella* Typhimurium strains, *J Antimicrob Chemother*, 67, 111–114.

MARAN (2007), Monitoring of antimicrobial resistance and antibiotic usage in animals in the Netherlands in 2006/2007. Available at: www.cvi.wur.nl.

MARAN (2009), Monitoring of antimicrobial resistance and antibiotic usage in animals in the Netherlands in 2009. Available at: www.lei.wur.nl/NR/rdonlyres/4ED137F2-5A0D-4449-B84E-61A3A8AC4A42/135753/MARAN_2009.pdf.

MCDERMOTT P F, BODELS S M, ENGLISH L J, WHITE D G *et al.* (2002), Ciprofloxacin resistance in *Campylobacter jejuni* evolves rapidly in chickens treated with fluoroquinolones, *J Inf Dis*, 185, 837–840.

MEAKINS S, FISHER I S, BERGHOLD C, GERNER-SMIDT P, TSCHÄPE H *et al.* (2008), Antimicrobial drug resistance in human nontyphoidal *Salmonella* isolates in Europe 2000–2004: A report from the Enter-Net International Surveillance Network, *Microb Drug Resist*, 14, 31–35.

MØLBAK K, BAGGESEN D L, AARESTRUP F M, EBBESEN J M, ENGBERG J *et al.* (1999), An outbreak of multidrug-resistant, quinolone-resistant *Salmonella enterica* serotype Typhimurium DT104, *N Engl J Med*, 341, 1420–1425.

MORA A, HERRERA A, MAMANI R *et al* (2010), Recent emergence of clonal group O25b:K1:H4-B2-ST131 ibeA strains among *Escherichia coli* poultry isolates, including CTX-M-9-producing strains, and comparison with clinical human isolates, *Appl Environ Microbiol*, 76, 6991–6697.

MOSSONG J, MARQUES P, RAGIMBEAU C, HUBERTY-KRAU P, LOSCH S *et al.* (2007), Outbreaks of monophasic *Salmonella enterica* serovar 4,[5],12:i:- in Luxembourg, 2006, *Euro Surveill*, 12(6), E11–E12.

NELSON J M, CHILLER T M, POWERS J H and ANGULO F J (2007), Fluoroquinolone-resistant *Campylobacter* species and the withdrawal of fluoroquinolones for use in poultry: A public health success story, *Clin Inf Dis*, 44, 977–980.

PATERSON D L and BONOMO R A (2005), Extended-spectrum beta-lactamases: A clinical update, *Clin Microbiol Rev*, 18, 657–686.

PITOUT J D and LAUPLAND K B (2008), Extended-spectrum beta-lactamase-producing Enterobacteriaceae: An emerging public-health concern, *Lancet Infect Dis*, 8, 159–166.

PORNRUANGWONG S, SRIYAPAI T, PULSRIKARN C, SAWANPANYALERT P *et al.* (2008), The epidemiological relationship between *Salmonella enterica* serovar Typhimurium and *Salmonella enterica* serovar 4,[5],12:i:- isolates from humans and swine in Thailand, *Southeast Asian J Trop Med Public Health*, 39, 288–296.

RANDALL L P, CLOUTING C, HORTON R A *et al.* (2011), Prevalence of *Escherichia coli* carrying extended-spectrum beta-lactamases (CTX-M and TEM-52) from broiler chickens and turkeys in Great Britain between 2006 and 2009, *J Antimicrob Chemother*, 66, 86–95.

RODRIGUEZ-BANO J, PICON E, GIJON P, HERNANDEZ J R, CISNEROS J M *et al.* (2010), Risk factors and prognosis of nosocomial bloodstream infections caused by extended-spectrum-beta-lactamase-producing *Escherichia* coli, *J Clin Microbiol*, 48, 1726–1731.

SOYER Y, MORENO S A, DAVIS M A, MAURER J, MCDONOUGH P L *et al.* (2009), *Salmonella* 4,5,12:i:-: An emerging *Salmonella* serotype that represents multiple distinct clones, *J Clin Microbiol*, 47, 3546–3556.

SWITT A I, SOYER Y, WARNICK L D and WIEDMANN M (2009), Emergence, distribution, and molecular and phenotypic characteristics of *Salmonella enterica* serotype 4,5,12:i:-, *Foodborne Pathog Dis*, 6, 407–415.

TAVECHIO A T, GHILARDI A C and FERNANDES S A (2004), 'Multiplex PCR' identification of the atypical and monophasic *Salmonella enterica* subsp. *enterica* serotype 1,4,[5],12:i:- in São Paulo State, Brazil: Frequency and antibiotic resistance patterns, *Rev Inst Med Trop Sao Paulo*, 46, 115–117.

THRELFALL E J (2000), Epidemic *Salmonella* Typhimurium DT 104 – A truly international epidemic clone, *J Antimicrob Chemother*, 46, 7–10.

THRELFALL E J, DAY M, DE PINNA E, CHARLET A and GOODYEAR K (2006), Assessment of factors contributing to changes in the incidence of antimicrobial drug resistance in *Salmonella enterica* serotypes Enteritidis and Typhimurium from humans in England and Wales in 2000, 2002 and 2004, *Int J Antimicrob Agents*, 5, 389–395.

TRÜPSCHUCH S, LAVERDE GOMEZ J A, EDIBERIDZE I A *et al.* (2010), Characterisation of multidrug-resistant *Salmonella* Typhimurium 4,[5],12:i:- strains carrying a novel genomic island adjacent to the thrW TRNA locus, *Int J Med Microbiol*, 300, 279–288.

VALAT C, HAENNI M, SARAS E *et al.* (2012), CTX-M-15 extended-spectrum beta-lactamase in a Shiga toxin-producing *Escherichia coli* isolate of serotype O111:H8, *Appl Environ Microbiol*, 78, 1308–1309.

VAN DEN BOGAARD A C, BRUINSMA N and STOBBERING E E (2000), The effect of banning avoparcin on VRE in the Netherlands, *J Antimicrob Chemother*, 46, 146–148.

VAN PELT W, NOTERMANS D, MEVIUS D, VENNEMA H, KOOPMANS M P G *et al.* (2008), Trends in gastro-enteritis van 1996–2006: verdere toename van ziekenhuisopnames, maar stabiliserende sterfte, *Infect Bull*, 19, 24–31. In Dutch.

VARMA J K, MARCUS R, STENZEL S A *et al.* (2006), Highly resistant *Salmonella* Newport-MDRAmpC transmitted through the domestic US food supply: A FoodNet case-control study of sporadic *Salmonella* Newport infections, 2002–2003, *J Infect Dis*, 194, 222–230.

VIEIRA-PINTO M, TEMUDO P and MARTIN C (2005), Occurrence of *Salmonella* in the ileum, ileocolic lymph nodes, tonsils, mandibular lymph nodes and carcasses of pigs slaughtered for consumption, *J Vet Med B Infect Dis Vet Publ Health*, 52, 476–481.

WALKER R A, LAWSON A J, LINDSAY E A, WARD L R *et al.* (2000), Decreased susceptibility to ciprofloxacin in outbreak-associated multiresistant *Salmonella* Typhimurium DT104, *Vet Rec*, 147, 395–396.

WHO (2007), *Critically Important Antimicrobials for Human Medicine: Categorization for the Development of Risk Management Strategies to Contain Antimicrobial Resistance due to Non-Human Antimicrobial Use*, ISBN 978 92 4 159574 2.

WHO (2011), *Tackling Antibiotic Resistance from a Food Safety Perspective in Europe*, ISBN 978 92 890 1421 2 (print); ISBN 978 92 890 1422 9 (ebook).

YANG B, QU D, ZHANG X *et al.* (2010), Prevalence and characterization of *Salmonella* serovars in retail meats of marketplace in Shaanxi, China, *Int J Food Microbiol*, 141, 63–72.

ZAMPERINI K, SONI V, WALTMAN D, SANCHEZ S, THERIAULT E C *et al.* (2007), Molecular characterization reveals *Salmonella enterica* serovar 4,[5],12:i:- from poultry is a variant Typhimurium serovar, *Avian Dis*, 51, 958–964.

ZANSKY S, WALLACE B, SCHOONMAKER-BOPP D *et al.* (2002), From the Centers for Disease Control and Prevention. Outbreak of multi-drug resistant *Salmonella* Newport, United States, January–April 2002, JAMA, 288, 951–953.

8

Antibiotic resistance development and identification of response measures

B. ter Kuile and S. Brul, University of Amsterdam, The Netherlands

DOI: 10.1533/9780857098740.2.157

Abstract: Antibiotic resistance in human pathogens is becoming a major threat to human health. This is partly due to agriculture, because large amounts of antibiotics are used to treat animals, which encourages resistant genes to appear and transfer to humans through foodstuffs. This is a food safety issue.

Measures against selection of resistance in agriculture have long been limited to monitoring. The scientific knowledge acquired in recent years can be used to introduce cost-effective measures, such as harmonized monitoring and prudent usage, which should be enforced by the competent national authorities. Some antibiotics should be reserved for human use only and this list should be reviewed regularly.

Key words: antibiotic resistance, agriculture, cost-effective measures, pathogen.

8.1 Introduction

The resistance of pathogenic microorganisms against antibiotics has recently developed into a major challenge to human healthcare, as therapy failure is becoming more and more common. While deaths attributable to resistant strains were rare in the past and are still rare in Northern Europe, they have become commonplace in Southern Europe and the rest of the world. The European Centre for Disease Prevention and Control (ECDC) and the European Medicines Agency (EMEA) estimated in a joint technical report that within the EU, drug-resistant infections cause more than 25 000 deaths and cost €1.5 billion in extra healthcare costs every year (ECDC/EMEA, 2009). The extra costs to hospitals in treating infections caused by resistant pathogens, compared to susceptible variants of the same strains, are between €5000 and €35 000 per case (Maragakis *et al.*, 2008; Slama, 2008). Methicillin-resistant *Staphylococcus aureus* (MRSA) infections within the healthcare setting in the EU are estimated to cost €380 million annually (Kock *et al.*, 2010).

Part of the resistance encountered in human healthcare has its origin in the agricultural sector (Heymann, 2006; Hald *et al.*, 2007; Geenen *et al.*, 2010). The Committee for Medicinal Products for Veterinary Use (CVMP) of the EMEA stated that animals used for human food are a source of resistant microorganisms in humans (EMEA/CVMP, 2006; ECDC/EMEA, 2009). Preventing the incorrect use of antibiotics in the treatment of human disease is critical in avoiding the development of resistance among human pathogens, but the agricultural sector also contributes indirectly. Exactly how much of the resistance encountered in human healthcare originates from selection in the agricultural sector is unclear (Miller *et al.*, 2006), but estimates of between one third and one half are often suggested, the remainder being due to usage in human healthcare. To prevent both human and veterinary infectious diseases from becoming untreatable, the selection of resistant variants should be reduced to a minimum. Therefore risk control strategies should be developed and subsequently applied to control foodborne and animal-borne antimicrobial resistance and the transfer of that resistance to the human healthcare system.

Since 1960, approximately 200 000 scientific articles have appeared on the subject of antibiotic resistance and a large number of risk assessments and scientific opinions have been devoted to this problem. By and large these reflect a consensus on what should be done in principle and in great detail they describe which antibiotics should not be used in certain situations. However, practical, implementable risk control options that can be enforced by competent authorities are rarely formulated as such. One of the reasons may be that the attribution of resistance encountered in human healthcare is complex (Miller *et al.*, 2006). The aim of this review is to extract from the scientific literature and various other documents, control options for the competent authorities that can be implemented straightaway and to discuss and consider these with the possibilities for immediate application in mind. Given the rapid increase of resistance it seems imperative not to postpone taking measures that can be implemented now while the research recommended in many of the documents is being performed. Therefore, the recommendations in the various reports and opinions are reviewed with the aim of implementation on farms and elsewhere in the chain and the possibilities for effective control.

Food, meat included, is transported all over the world (Havelaar *et al.*, 2010). Microorganisms, pathogenic or not, are unintentionally distributed along the same lines (Pires *et al.*, 2009). Antibiotic resistance is therefore a global concern that should be addressed globally (Heymann, 2006). The stakeholder consultation on the transatlantic task force on antimicrobial resistance clearly demonstrated that all respondents desire a worldwide approach. If different practices are implemented in different countries, meat-producing businesses in countries that try to enforce effective measures may be outcompeted by those that are not required to make these investments. As a result, the spread of resistance might increase rather than decrease. This is not meant to imply that no measures should be taken unless worldwide consensus can be achieved.

The total amount of antibiotics used for the treatment of individual animals is small in comparison to the quantities used for flock or herd treatment and it is likely that individual animals that do not need treatment will receive it. Therefore, there is probably no need to regulate the application of antibiotics to treat infections in individual animals beyond adherence to the prudent-use principles. Flock/herd treatment is often not limited to situations when the whole group have clinical symptoms, but is also applied when only a few animals actually have infections (Hill *et al.*, 2009). The total amount of antibiotics administered to flocks has not gone down by much after the ban on the use of antibiotics as growth promoters in 2006, indicating that prescribed antibiotics are in fact being used as growth promoters (Mevius *et al.*, 2010). Hence it seems useful to examine possibilities for better supervision of flock treatment, aiming at preventing the stealth use of prescribed antibiotics as growth promoters.

8.2 Existing international risk assessments

A large variety of international organizations and government agencies have performed risk assessments of antibiotic resistance. The European Food Safety Authority (EFSA) has issued opinions on antimicrobial resistance, both independently and in cooperation with other European organizations (EFSA, 2008a, 2009). Both opinions provide a comprehensive review of the scientific knowledge available at the time of writing. One of the conclusions was that for some combinations of microorganisms and antibiotics, the resistance from clinical isolates from humans reflected the trends in resistance of animal isolates. The joint opinion of ECDC, EFSA, EMEA and the Scientific Committee on Emerging and Newly Identified Health Risks (SCENIHR) suggested research options for the development of practical implementable measures (EFSA, 2009). For risk control, prevention and control options, the EFSA opinion of 2008 refers to WHO statements of 1997–2000 and provides detailed recommendations focussed on specific subjects, such as MRSA, quinolones and third- and fourth-generation cephalosporins. An expert report by the Institute of Food Technologists explored the implications of antimicrobial resistance for the food system and suggested control measures (Doyle *et al.*, 2006). The more general recommendations in *Foodborne Antimicrobial Resistance as a Biological Hazard* require a transformation step to be implementable at the farm level (EFSA, 2008a).

The Codex Alimentarius Commission has adopted guidelines for the risk analysis of foodborne antimicrobial resistance (Codex Alimentarius, 2011). These guidelines incorporate the earlier *Code of Practice to Minimize and Contain Antimicrobial Resistance* (Codex Alimentarius, 2005). It also quoted *Guidelines for the Design and Implementation of National Regulatory Food Safety Assurance Programmes Associated With the Use of Veterinary Drugs in food producing animals* (Codex Alimentarius, WHO/FOA, 2012), which are aimed at preventing undesirable levels of antibiotics in meat and are not specifically designed to prevent development of antibiotic resistance. The guidelines of the code of practice, CAC/RCP 61-2005, put the full responsibility for minimizing the development of antibiotic resistance on veterinarians and producers. The role of the competent authorities is limited to supervision. The draft Codex guidelines, in addition, emphasize the need for permanent monitoring and reviews of risk management options, without at this stage formulating such options.

8.3 Control targets

Selection for and spread of antibiotic resistance should be counteracted at the source, since measures at stages downstream in the production chain can never be completely effective at preventing the spreading of resistance genes (Cassone and Giordano, 2009; Alexander *et al.*, 2010). The reason for this is that resistant microbes can spread from the farm into the environment and reach the human population in many ways, not only through the food production chain (Friesema *et al.*, 2010). Measures to reduce the microbial load applied in later production stages therefore remove only some of the resistance genes selected for at the farm.

8.4 Monitoring and reporting usage and resistance

Monitoring antimicrobial resistance is not so much a scientific aim in itself, instead it is used to perform risk assessments and advise policymakers on potentially effective measures. This is reflected in the design of the various government-sponsored monitoring programmes. These are mostly carried out at a relatively high level of integration and cannot always be linked with data on exposure to antibiotics. For an epidemiological approach towards elucidating the quantitative relationship between exposure to antibiotics and the development of and selection for resistance, data on resistance and the usage of antibiotics must be coupled.

Veterinary practice differs strongly between countries and even within states, certainly with respect to prescribing antibiotics (Mathew *et al.*, 2007). A major difficulty in policy development at the moment is the lack of straightforward comparable data on the usage of antibiotics in the different EU member states. Comparing usage in units of tons of antibiotics (e.g. Table 2 of EFSA, 2009) is not informative, because the specific weight of active ingredients and the number of animals involved are not known. The only valid comparison would be daily doses per animal per year (DD/AY). Easy-to-compare data on usage are crucial for implementing measures to reduce usage and for monitoring their effects. Similarly a standard methodology for monitoring resistance should be agreed upon as well.

For monitoring antibiotic resistance several options exist. It is essential that the information gathered in the different sectors is compatible, to enable tracking of the origin of resistance to source and preferably to determine quantitative attributions. High-throughput whole genome sequencing (WGS) is a powerful tool and is rapidly becoming so cheap that it is likely to become the general diagnostic tool in clinical microbiology (Pallen *et al.*, 2010). Using WGS as a monitoring tool for antimicrobial resistance has major advantages, because all of the resistance genes are detected in one procedure and the data can subsequently also be used for other types of analysis. Reconstructing the development and spread of resistance using an epidemiological methodology is feasible if a large number of randomly obtained sequences of resistant strains are available. As well as finding resistance genes, WGS data can also be used to detect diffuse outbreaks of foodborne pathogens (Boxrud, 2010).

An obvious question concerning the monitoring of antimicrobial resistance is whether this will provide enough new and useful information to justify the effort. After all, that exposure to antibiotics leads to resistance is by now well accepted in the scientific literature (Gould, 2009). However, the exact relationship between the amount of antibiotic used and the development of resistance has not yet been determined quantitatively (Roberts *et al.*, 2008). Another very important number is the percentage of resistance encountered in the human sector that was originally selected for in the agricultural sector (Alexander *et al.*, 2009). Sophisticated monitoring that yields more than overall information at a high level of integration can definitely contribute to a reliable estimate of the attribution. Therefore the most likely answer to the above question is that monitoring antibiotic resistance is useful and informative if it meets a number of requirements:

- Antibiotic resistance must be determined for a representative number of isolates of predetermined species for a fixed set of antibiotics using pre-set cut-off values for each antibiotic (Hald *et al.*, 2007).
- To trace strains backwards and forwards between the agricultural and the human sectors, detailed strain typing is needed, complemented with precise timing and location of isolation.
- To quantify the resistance transferred from the agricultural to the human sector, resistance genes must be characterized for as many strains as feasible (Withee and Dearfield, 2007).
- The amount of antibiotics used should be monitored as well, to ascertain the selection pressure. Integration of all of the data can be used to assess the relationship between usage and development of antimicrobial resistance (Miller *et al.*, 2006).

Even if these requirements are met and the information from the monitoring programme meets expectations, it could still be argued that it would not influence the policies of national authorities, which regulate antibiotic usage in animal husbandry. These could simply demand that all antibiotic use is reduced as much as possible, to reduce the selective pressure. A simple reduction of the amount of antibiotics used, however, is unlikely to be the most effective and efficient way to combat the proliferation of antimicrobial resistance (van der Horst *et al.*, 2011). To design cost-effective measures and to create acceptance for these among all the parties concerned, the effect of the measures must be predictable. To achieve this, a well thought out antibiotic resistance monitoring programme is crucial.

8.5 Reserving antibiotics for human healthcare

One practical risk control option that is already in place in various countries is reserving several antibiotics exclusively for treating humans. Within the EU the following antibiotics are reserved solely for human healthcare: carbapenems and other penems, glycopeptides, synergistins, streptogramins, glycylcyclines, oxazolidinones and lipopeptides. Policymakers often debate whether to reserve third- and fourth-generation cephalosporins solely for human usage. The alternative option is to allow their use as a treatment of last resort in agriculture (Use, 2009). Such measures are most effective if they are adopted and enforced worldwide. Other points of consideration are cross-resistance and co-selection. Antibiotics that can cause resistance to a structurally related antibiotic that is reserved for human use, should also not be used in farms. In the past avoparcin use in chickens has contributed to vancomycin resistance in human pathogens (van den Bogaard *et al.*, 2000; van den Bogaard and Stobberingh, 2000). Resistance to unrelated antibiotics can be coded for by genes that are located on a single plasmid (Canton and Ruiz-Garbajosa, 2011). If, for instance, a resistance gene against any of the antibiotics reserved for humans is located on a plasmid together with a gene for resistance against tetracycline, the application of tetracycline will select for these other resistance genes as well.

8.6 Prudent use

The prudent use of antibiotics, often mentioned in official documents, means choosing the right antibiotic and applying it correctly (Allerberger *et al.*, 2008). However, the definition and implications of this expression are not always the same in each document. The list given below attempts to summarize the most commonly mentioned guidelines and mostly adheres to those of the Federation of Veterinarians of Europe (www.fve.org/news/publications/pdf/antibioen.pdf):

- The animals for which antibiotics are prescribed must be under the care of the prescribing veterinarian.
- All antibiotics must be supplied on prescription and records of the prescription, supply and administration of the antibiotics must be kept.
- The basis for a prescription is an accurate diagnosis and an antibiotic that is approved for the species and indication.
- The infectious agent must be susceptible to the drug applied. Therefore known or predictable sensitivities of the microorganism must be ascertained, if possible before therapy is started.
- The dosage regimen should be based on known pharmacokinetics to avoid sub-therapeutic doses and the duration of the treatment should be limited to that required for sufficient therapeutic effect.
- The application of multiple antibiotics to provide a broad coverage should be limited to those cases when failure of the therapy would be a major risk for the patient.

- Group medication can be prescribed only when necessary to prevent the spread of disease within a flock or herd.

These guidelines imply that the prescribing veterinarian has examined the animals in person and preferably has a long-term relationship with the farmer.

8.7 Role of the competent authorities

Regulatory authorities have several responsibilities concerning antibiotic usage in the agricultural sector (Turnidge, 2004):

- approve new drugs, control quality of those in use and regulate their marketing;
- monitor and report the occurrence of antibiotic resistance and antibiotic usage;
- promote prudent use of antibiotics in cooperation with the veterinary and agricultural sectors;
- supervise the conditions on farms and other animal holding facilities in order to prevent unnecessary usage of antibiotics;
- supervise adherence to approved practices and enforce laws and regulations on application of antibiotics;
- perform risk assessments in particular on the threat for human healthcare resulting from antibiotic usage in the agricultural sector.

The procedures for approving new drugs are outside the scope of this chapter. Possible actions concerning the other responsibilities will be discussed below.

The promotion of prudent use is most effectively done by the professional medical and veterinary organizations, as these can directly inform their members of guidelines for prescription and best practices (Earnshaw *et al.*, 2009). Given the continuous flow of new data and novel insights, regular adjustments of the recommendations will be needed. The competent authorities will therefore have to maintain a close relationship with medical and veterinary organizations in order to coordinate the respective efforts of both sides.

Continuous monitoring of antibiotic resistance and usage in the agricultural sector is essential for development and readjustment of sound policies to prevent the spread of resistance, selected for in animal husbandry, to human healthcare (Mathew *et al.*, 2007). Responsibility for this monitoring rests with national authorities. Standardization of the methodologies for monitoring and reporting may be crucial for achieving the consistency needed to facilitate comparisons of the data from different countries. The purpose of monitoring antibiotic usage is to provide risk managers with feedback on the effectiveness of their measures.

New scientific data and insights on antibiotic resistance are becoming available constantly. In combination with the permanently increasing levels of resistance encountered in the agricultural sector, these data and insights necessitate the updating of risk assessments on a regular basis. Each time a new set of monitoring data becomes available the need for an adjusted risk assessment should be examined. The relationships between usage, resistance and the transfer of resistance genes from the agricultural sector to human healthcare should be critically reviewed. Whenever relevant, control measures should be adapted based on the new insights. An example is the demonstration that sensitive *Staphylococcus aureus*, transferred from human healthcare to the agricultural sector, became resistant due to the ample usage of antibiotics and transferred back as MRSA (Price *et al.*, 2012). The new insight derived from this is that the agricultural sector may function as an amplifier of resistance in ways that are different from previously realized.

8.8 Responsibilities for veterinarians

On the one hand, veterinary surgeons have a professional responsibility towards their customers and the animals, and the efforts of governments and their agencies to reduce antibiotic usage should not interfere with this (Earnshaw *et al.*, 2009). On the other hand, the experiences in Denmark show that monitoring the prescription behaviour of veterinarians and calling to order those who prescribe much more antibiotics than their colleagues is a very effective way of reducing antibiotic usage (Vieira *et al.*, 2011). One of the reasons that supervision of prescribing is effective is that it prevents farmers from shopping around for veterinarians who are willing to prescribe more antibiotics than strictly needed to treat infectious diseases. Given the possibilities of information technology, it should be simple to automatically register a prescription together with the relevant data, as the veterinarian will enter these into an information system as a standard procedure.

8.9 Farmers' mission

Infections occur more often in poor hygiene situations than when animals are kept in clean facilities (EFSA, 2008a). Therefore clean production environments and adherence to good agricultural practice (GAP) guidelines by farmers can reduce the total amount of antibiotics needed to maintain animal health. Right now, there are no standards by which to judge the cleanliness of farms. So if it is decided to enforce hygiene rules at the farm level, such rules will have to be developed and agreed upon first. The GAP guidelines could be a good starting point. A long-term exclusive relationship between a farmer and a veterinarian will also be advantageous in this respect, as the veterinarian will be a steadfast advisor on all animal health and welfare matters. Timely vaccination, isolating sick animals and rapid removal of dead animals are examples of other measures that can be taken on farms to reduce the need for antibiotics. Manure management can be an effective measure in reducing the load of resistant microbes emanating from the farm into the environment and possibly indirectly into the human healthcare system (Chee-Sanford *et al.*, 2009). In addition, manure might contain antibiotics excreted by the animals. Application of hazard analysis and critical control point (HACCP) principles, not only for manure management but for all farm procedures, should be considered, because then all controls that can be applied will be reviewed.

8.10 Conclusion

Several risk control options can be envisioned, which will implement recommendations that have been made by EFSA, Codex, WHO and other organizations in the past few years.

8.10.1 Further harmonize reporting

Reporting both antibiotic usage and resistance levels in a way that allows policymakers to draw conclusions about the effectiveness of measures aimed at reducing the development and spread of resistance, is crucial. Since measures must be implemented EU-wide for maximal effect, reporting should be harmonized as well. Therefore guidelines for reporting usage and resistance levels should be agreed so that comparisons between member states' data can be made without the need for conversion.

8.10.2 Application of the prudent-use guidelines

The prudent-use guidelines of the Federation of Veterinarians of Europe are an effective means for reducing the development and spread of antibiotic resistance (Earnshaw *et al.*, 2009). Their application should be strongly encouraged and whenever possible enforced by the competent authorities.

8.10.3 Improper incentives to prescribe antibiotics

Veterinary surgeons sometimes obtain part of their income from the sale of drugs (Aarestrup *et al.*, 2008b). Even if this arrangement does not influence prescription behaviour, it still creates the semblance of a conflict of interest. Monitoring individual prescribing behaviour combined with preventing veterinarians from receiving financial benefits from prescribing antibiotics is a risk control option that deserves to be urgently explored.

8.10.4 Exclusive relationship between farmer and veterinary surgeon

The option for farmers to pressure veterinary surgeons into prescribing antibiotics by threatening to move their business to another practice should be prevented. One way to achieve this is to demand that an exclusive long-term relationship exists between a farmer or production facility and a veterinarian or veterinary practice. For example, farmers should be forbidden from switching practice more than once a year. Only the home vet should be allowed to prescribe antibiotics to livestock.

8.10.5 Enforce good agricultural practice

In well-managed production facilities, a lower amount of antibiotics are needed (EFSA, 2008a). Careful cleaning during operations and between batches and the constant removal of refuse would reduce the number of infections and consequently the need to use antibiotics for treatment (Bao *et al.*, 2009). Reducing the density of animals on farms is likely to reduce the chance of spreading infections and thus the number of animals that need treatment (Aarestrup *et al.*, 2008a). Another option to be considered is encouraging the use of breeds of animals that have a good immune system and are less prone to infections.

8.10.6 Carry-over of antibiotics added to feed

When flocks are treated with antibiotics, the drug is often mixed into the feed. After batches of medicated feed have been produced, the following batch of feed, which shoud be free of antibiotics, often contains considerable levels that can contribute to antibiotic resistance. Depending on the combination of the drug and the feed, good manufacturing practice (GMP) can reduce this carry-over to levels below 5% (EFSA, 2008b). Additional measures to further reduce the carry-over of antibiotics from medicated batches into regular batches should be considered.

8.10.7 Different antibiotics for agricultural and human use

Several classes of antibiotics should be reserved for use in human healthcare (Miller *et al.*, 2006). At this moment these include carbapenems and other penems, glycopeptides, synergistins, streptogramins, glycylcyclines, oxazolidinones and lipopeptides. Possibly the use of cephalosporins and enrofloxacin in the agricultural sector should be reconsidered, as the former are essential to human healthcare and resistance to the latter develops very quickly upon exposure (Schuurmans *et al.*, 2010). Given the swiftness of the changes in resistance patterns, it may be worthwhile reviewing the list of antibiotics that are reserved for human use every other year.

8.10.8 Selection for multiple resistance

Antibiotic resistance genes are often clustered on plasmids or other genetic constructs that carry several genes for resistance against different antibiotics (Partridge, 2011). As a result, exposure to one antibiotic may result in resistance against others, not necessarily chemically or structurally related. Monitoring programmes for antibiotic resistance therefore need to explore resistance patterns to investigate the presence of multi-resistance and verify the occurrence of plasmids or other genetic constructs that will be co-selected upon exposure to an antibiotic that is part of the cluster. Whole genome sequencing seems an ideal instrument for this goal if it can be applied to a large set of isolates. The results of monitoring programmes need to be evaluated regularly and existing risk assessments readjusted based on these results. Guidelines for veterinarians for the prescription of antibiotics will have to be amended accordingly.

8.11 Control measures that can be implemented immediately

This review has yielded a number of measures that can be taken on the basis of the present scientific knowledge, such as:

- Ensure that the methodologies for monitoring the usage of antibiotics and occurrence of resistance are harmonized and suitable for explicating the relationships between usage and resistance quantitatively.
- Encourage and when possible enforce the application of the prudent-use guidelines.
- Prevent veterinary surgeons from receiving any financial benefit from prescribing antibiotics.
- Insist on a long-term relationship between the veterinary surgeon prescribing antibiotics and the farmer or producer.
- Monitor the prescribing behaviour of veterinarians on an individual basis.
- Ensure that veterinary organizations incorporate EFSA and WHO recommendations into prescription guidelines. If necessary for enforcing adherence, the incorporation of guidelines into EU regulations should be considered.
- Reduce the need for antibiotics on farms by insisting on hygienic measures, GAP and where applicable HACCP principles as well as optimizing overall management.
- Where available consider the use of appropriate vaccines to prevent disease before it starts, thereby reducing the need for antimicrobial treatment.
- Reduce carry-over from medicated batches of feed to regular lots to a level as low as reasonably achievable.
- Implement measures to prevent the spread of antibiotic-resistance genes from farms to other environments both through the foodborne pathway and otherwise.
- Using an iterative process, identify those antibiotics that cause most resistance and adjust the prescription guidelines for veterinarians accordingly.
- Review the list of antibiotics reserved for human healthcare every other year.

8.12 References

AARESTRUP, F. M., OLIVER DURAN, C. and BURCH, D. G. 2008a. Antimicrobial resistance in swine production. *Anim Health Res Rev*, 9, 135–48.

AARESTRUP, F. M., WEGENER, H. C. and COLLIGNON, P. 2008b. Resistance in bacteria of the food chain: Epidemiology and control strategies. *Expert Rev Anti Infect Ther*, 6, 733–50.

ALEXANDER, T. W., INGLIS, G. D., YANKE, L. J., TOPP, E., READ, R. R. *et al.* 2010. Farm-to-fork characterization of *Escherichia coli* associated with feedlot cattle with a known history of antimicrobial use. *Int J Food Microbiol*, 137, 40–8.

ALEXANDER, T. W., REUTER, T., SHARMA, R., YANKE, L. J., TOPP, E. *et al.* 2009. Longitudinal characterization of resistant *Escherichia coli* in fecal deposits from cattle fed subtherapeutic levels of antimicrobials. *Appl Environ Microbiol*, 75, 7125–34.

ALLERBERGER, F., LECHNER, A., WECHSLER-FORDOS, A. and GAREIS, R. 2008. Optimization of antibiotic use in hospitals – Antimicrobial stewardship and the EU project ABS international. *Chemotherapy*, 54, 260–7.

BAO, Y., ZHOU, Q., GUAN, L. and WANG, Y. 2009. Depletion of chlortetracycline during composting of aged and spiked manures. *Waste Manag*, 29, 1416–23.

BOXRUD, D. 2010. Advances in subtyping methods of foodborne disease pathogens. *Curr Opin Biotechnol*, 21, 137–41.

CANTON, R. and RUIZ-GARBAJOSA, P. 2011. Co-resistance: An opportunity for the bacteria and resistance genes. *Curr Opin Pharmacol*, 11, 477–85.

CASSONE, M. and GIORDANO, A. 2009. Resistance genes traveling the microbial internet: Down the drain, up the food chain? *Expert Rev Anti Infect Ther*, 7, 637–9.

CHEE-SANFORD, J. C., MACKIE, R. I., KOIKE, S., KRAPAC, I. G., LIN, Y. F. *et al.* 2009. Fate and transport of antibiotic residues and antibiotic resistance genes following land application of manure waste. *J Environ Qual*, 38, 1086–108.

CODEX ALIMENTARIUS 2005. *Code of Practice to Minimize and Contain Antimicrobial Resistance*. CAC/RCP 61–2005.

CODEX ALIMENTARIUS 2011. *Guidelines for Risk Analysis of Foodborne Antimicrobial Resistance.*

CODEX ALIMENTARIUS, WHO/FOA 2012. *Guidelines for the Design and Implementation of National Regulatory Food Safety Assurance Programme Associated With the Use of Veterinary Drugs in Food Producing Animals*. CAC/GL 71-2009.

DOYLE, M.P., BUSTA, F.F., CORDS, B.R., DAVIDSON, P.M., HAWKE, J. *et al.* 2006. Antimicrobial Resistance: Implications for the Food System. Comprehensive Reviews in Food Science and Food Safety, Institute of Food Technologists, Chicago, IL. pp. 71–122

EARNSHAW, S., MONNET, D. L., DUNCAN, B., O'TOOLE, J., EKDAHL, K. *et al.* 2009. European Antibiotic Awareness Day, 2008 – The first Europe-wide public information campaign on prudent antibiotic use: Methods and survey of activities in participating countries. *Euro Surveill*, 14, 19280.

ECDC/EMEA 2009. The Bacterial Challenge: Time to react. Technical report. Stockholm: European Centre for Disease Prevention and Control.

EFSA 2008a. *Foodborne Antimicrobial Resistance as a Biological Hazard.* Scientific Opinion of the Panel on Biological Hazards (Question No EFSA-Q-2007–089). *EFSA Journal*, 765, 1–87.

EFSA 2008b. Opinion of the Scientific Panel on Contaminants in the Food Chain on a request from the European Commission on cross-contamination of non-target feedingstuffs by maduramicin authorised for use as a feed additive. *EFSA Journal* 549, 1–30.

EFSA 2009. EMEA/CVMP/447259/2009 Joint Opinion on antimicrobial resistance (AMR) focused on zoonotic infections. *EFSA Journal*, 7.

EMEA/CVMP 2006. CVMP Strategy on antimicrobials 2006–2010 and status report on activities on antimicrobials. In: EMEA/CVMP (ed.). London: EMEA.

FRIESEMA, I. H., VAN DE KASSTEELE, J., DE JAGER, C. M., HEUVELINK, A. E. and VAN PELT, W. 2010. Geographical association between livestock density and human Shiga toxin-producing *Escherichia coli* O157 infections. *Epidemiol Infect*, 1–7.

GEENEN, P. L., KOENE, M. G. J., BLAAK, H., HAVELAAR, A. H. and VAN DE GIESSEN, A. W. 2010. *Risk Profile on Antimicrobial Resistance Transmissible from Food Animals to Humans.* Bilthoven: RIVM.

GOULD, I. M. 2009. Antibiotic resistance: The perfect storm. *Int J Antimicrob Agents*, 34 Suppl 3, S2–5.

HALD, T., LO FO WONG, D. M. and AARESTRUP, F. M. 2007. The attribution of human infections with antimicrobial resistant *Salmonella* bacteria in Denmark to sources of animal origin. *Foodborne Pathog Dis*, 4, 313–26.

HAVELAAR, A. H., BRUL, S., DE JONG, A., DE JONGE, R., ZWIETERING, M. H. *et al.* 2010. Future challenges to microbial food safety. *Int J Food Microbiol*, 139 Suppl 1, S79–94.

HEYMANN, D. L. 2006. Resistance to anti-infective drugs and the threat to public health. *Cell*, 124, 671–5.

HILL, A. E., GREEN, A. L., WAGNER, B. A. and DARGATZ, D. A. 2009. Relationship between herd size and annual prevalence of and primary antimicrobial treatments for common diseases on dairy operations in the United States. *Prev Vet Med*, 88, 264–77.

KOCK, R., BECKER, K., COOKSON, B., VAN GEMERT-PIJNEN, J. E., HARBARTH, S. *et al.* 2010. Methicillin-resistant *Staphylococcus aureus* (MRSA): Burden of disease and control challenges in Europe. *Euro Surveill*, 15, 19688.

MARAGAKIS, L. L., PERENCEVICH, E. N. and COSGROVE, S. E. 2008. Clinical and economic burden of antimicrobial resistance. *Expert Rev Anti Infect Ther*, 6, 751–63.

MATHEW, A. G., CISSELL, R. and LIAMTHONG, S. 2007. Antibiotic resistance in bacteria associated with food animals: A United States perspective of livestock production. *Foodborne Pathog Dis*, 4, 115–33.

MEVIUS, D. J., WIT, B., VAN PELT, W. and BONDT, N. 2010. *MARAN Monitoring of Antimicrobial Resistance and Antibiotic Usage in Animals in the Netherlands in 2008*. Lelystad: Central Veterinary Institute.

MILLER, G. Y., MCNAMARA, P. E. and SINGER, R. S. 2006. Stakeholder position paper: Economist's perspectives on antibiotic use in animals. *Prev Vet Med*, 73, 163–8.

PALLEN, M. J., LOMAN, N. J. and PENN, C. W. 2010. High-throughput sequencing and clinical microbiology: Progress, opportunities and challenges. *Curr Opin Microbiol*, 13, 625–31.

PARTRIDGE, S. R. 2011. Analysis of antibiotic resistance regions in Gram-negative bacteria. *FEMS Microbiol Rev*, 35, 820–55.

PIRES, S. M., EVERS, E. G., VAN PELT, W., AYERS, T., SCALLAN, E. *et al.* 2009. Attributing the human disease burden of foodborne infections to specific sources. *Foodborne Pathog Dis*, 6, 417–24.

PRICE, L. B., STEGGER, M., HASMAN, H., AZIZ, M., LARSEN, J. *et al.* 2012. *Staphylococcus aureus* CC398: Host adaptation and emergence of methicillin resistance in livestock. *MBio*, 3, e00305–11.

ROBERTS, J. A., KRUGER, P., PATERSON, D. L. and LIPMAN, J. 2008. Antibiotic resistance – What's dosing got to do with it? *Crit Care Med*, 36, 2433–40.

SCHUURMANS, J. M., KOENDERS, B. B., TER KUILE, B. H. and HAYALI, A. S. 2010. Rapid induction of resistance to the fluoroquinolone enrofloxacin in *Pseudomonas putida* NCTC 10936. *Int J Antimicrob Agents*, 35, 612–13.

SLAMA, T. G. 2008. Gram-negative antibiotic resistance: There is a price to pay. *Crit Care*, 12 Suppl 4, S4.

TURNIDGE, J. 2004. Antibiotic use in animals – Prejudices, perceptions and realities. *J Antimicrob Chemother*, 53, 26–7.

USE, S., GREKO, C., BADIOLA, J. I., CATRY, B., VAN DUIJKEREN, E. *et al.* 2009. Reflection paper on the use of third and fourth generation cephalosporins in food producing animals in the European Union: Development of resistance and impact on human and animal health. *J Vet Pharmacol Ther*, 32, 515–33.

VAN DEN BOGAARD, A. E. and STOBBERINGH, E. E. 2000. Epidemiology of resistance to antibiotics. Links between animals and humans. *Int J Antimicrob Agents*, 14, 327–35.

VAN DEN BOGAARD, A. E., BRUINSMA, N. and STOBBERINGH, E. E. 2000. The effect of banning avoparcin on VRE carriage in the Netherlands. *J Antimicrob Chemother*, 46, 146–8.

VAN DER HORST, M. A., SCHUURMANS, J. M., SMID, M. C., KOENDERS, B. B. and TER KUILE, B. H. 2011. *De novo* acquisition of resistance to three antibiotics by *Escherichia coli*. *Microb Drug Resist*, 17, 141–47.

VIEIRA, A. R., PIRES, S. M., HOUE, H. and EMBORG, H. D. 2011. Trends in slaughter pig production and antimicrobial consumption in Danish slaughter pig herds, 2002–2008. *Epidemiol Infect*, 139, 1601–9.
WITHEE, J. and DEARFIELD, K. L. 2007. Genomics-based food-borne pathogen testing and diagnostics: Possibilities for the US Department of Agriculture's Food Safety and Inspection Service. *Environ Mol Mutagen*, 48, 363–8.

Part III

Pathogen surveillance, detection and identification

9

Advances in separation and concentration of microorganisms from food samples

S. H. Suh and L.-A. Jaykus, North Carolina State University, USA and B. Brehm-Stecher, Iowa State University, USA

DOI: 10.1533/9780857098740.3.173

Abstract: Current methods for detecting bacterial pathogens from foods and the environment are slow. Although newer, more rapid detection techniques have been developed, they have insufficient sensitivity and produce residual matrix-associated inhibitory compounds. The detection of foodborne pathogens could be more rapid if the target cells were separated and concentrated before detection, a procedure known as pre-analytical sample processing. We review sample preparation techniques, focusing on target-specific methods. Unlike non-target-specific separation/concentration methods, these approaches use bioaffinity ligands, preventing co-precipitation of target pathogens with residual matrix components. Bioaffinity ligands such as bacteriophage, phage-derived biomolecules, nucleic acid/peptide aptamers, carbohydrate ligands, antimicrobial peptides and synthetic ligands are described. The use of bioaffinity ligands in pathogen detection is new and requires further development. Such ligands have a future in foodborne pathogen detection and may allow us to substitute culture-based methods with more rapid, yet sensitive non-cultural methods.

Key words: pathogen detection, pre-analytical sample processing, pathogen concentration, pathogen separation, aptamer, bacteriophage, lectin, vancomycin, immunomagnetic separation, antimicrobial peptide, nanotechnology.

9.1 Introduction

Foodborne disease is an important public health problem worldwide. In the United States alone, it is estimated that there are 9.4 million episodes of foodborne illness annually, resulting in 55 961 hospitalizations and 1351 deaths (Scallan *et al.*, 2011).

In addition to other interventions, rapid and accurate identification of pathogens can be an important tool for prevention, control or mitigation of foodborne disease. Historically, methods for detecting foodborne bacterial pathogens have relied upon cultural enrichment and selective/differential plating, followed by biochemical identification and, sometimes, serological characterization. These methods are laborious and it usually takes days to weeks to confirm a positive result, a time frame that is out of sync with today's rapid food production and distribution networks (Donnelly, 2002; Yang *et al.*, 2007). Standard culture-based pathogen detection methods have been refined in efforts to reduce time to detection, and incremental improvements have been made. Newer detection techniques, including enzyme-linked immunosorbent assay (ELISA), DNA/RNA probes and the polymerase chain reaction (PCR) have also been developed, but the truly rapid use of these methods has been limited by their relatively high detection limits, necessitating the use of some degree of cultural enrichment prior to their use in detection. These methods can also be limited due to their sensitivity to potential interference from various inhibitory compounds present in food matrices (Yu *et al.*, 2001).

9.2 The need for pre-analytical sample processing

It has been suggested that detection of foodborne pathogens could be made more rapid, and larger sample sizes analyzed, if the target cells were separated, concentrated, and purified from the sample matrix before detection (Brehm-Stecher *et al.*, 2009). Such so-called pre-analytical sample processing could facilitate the removal of residual food matrix components while simultaneously concentrating target pathogen cells. An ideal sample preparation method would achieve at least six distinct goals, as listed in Fig. 9.1. In practice, however, few if any methods embody all six goals. Along with removal of inhibitors, effective pre-analytical sample preparation methods may reduce the initial sample size by 10–1000 fold (from liter or milliliter volumes down to microliters), making the use of molecular-based methods more realistic with respect to sample size. Ultimately, this could improve our ability to detect low levels of pathogens or sporadic contamination, reducing the time or even the need for cultural enrichment prior to detection. This chapter will focus primarily on cell (or analyte) separation and concentration from food matrices, as these are the key steps from which the other elements (i.e. volume reduction, purification, exclusion of inhibitory substances and homogenization) are realized.

9.2.1 Non-specific methods for pre-analytical sample processing

Pre-analytical sample processing methods can be classified into two major categories: non-specific and specific methods. Non-specific methods can concentrate a wide variety of microorganisms, regardless of any unique features associated with their antigenicity, surface and/or genetic characteristics. Traditional non-specific methods include filtration, centrifugation, using cationic/anionic exchange resins, aqueous two-phase partitioning and using metal hydroxides. These methods promote separation based on particle size, density

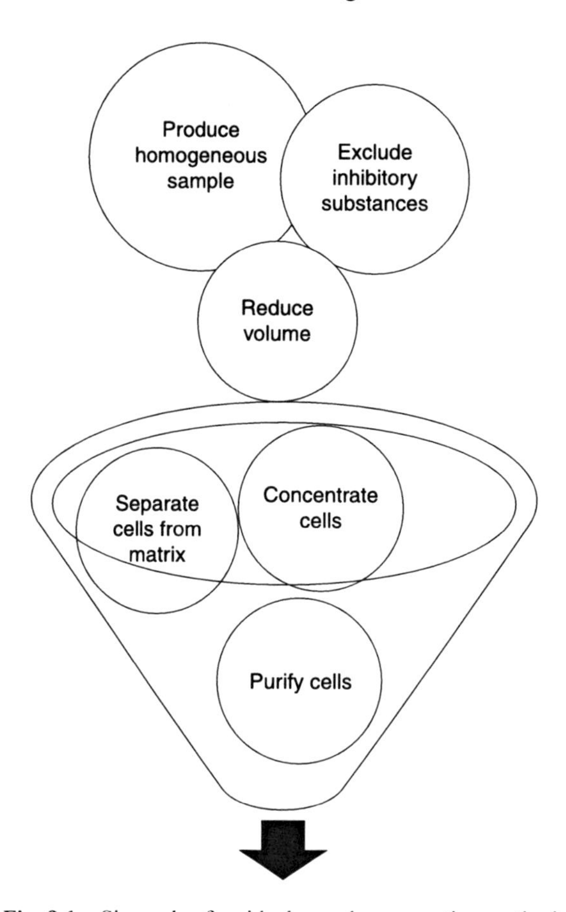

Fig. 9.1 Six goals of an ideal sample preparation method.

and/or charge. They are frequently used in conjunction with one another. Overall, the major advantages of non-specific methods are their cost, simplicity and ability to handle fairly large sample volumes. The major disadvantage is the tendency to co-concentrate non-target microflora or residual matrix components that are of approximately the same size, density or charge as the target microbe(s). For this reason, these methods are reasonable choices as a first-stage preparatory step or for concentration of pathogens from relatively simple sample matrices (such as water or juice), but are not always ideal for single-step pre-analytical sample processing of more complex foods. These methods are described in greater detail elsewhere (Stevens and Jaykus, 2004; Dwivedi and Jaykus, 2011).

9.2.2 Target-specific methods for pre-analytical sample processing

Separation and concentration methods based on naturally occurring biological interactions between bioaffinity ligands and specific cell surface receptor(s) have the advantages of high selectivity and binding affinity. Antibodies are the most commonly used ligand in this regard, and the method of coupling of antibodies to magnetic beads to facilitate pathogen concentration (called immunomagnetic separation or IMS) was introduced two decades ago (Skjerve *et al.*, 1990). Since that time, this approach has been incorporated into a number of standard food microbiology methods (FDA, 2011; FSIS, 2012). As a research tool, IMS has been used for the separation and concentration of foodborne pathogens such as *Salmonella enterica, Listeria monocytogenes, Escherichia coli* O157:H7, *Vibrio parahaemolyticus, Cryptosporidium parvum* oocysts and enteric viruses in a wide variety of food and environmental samples (Stevens and Jaykus, 2004).

Recently, there has been interest in bypassing, or at least shortening, cultural enrichment, and antibodies are a logical ligand for use in upstream sample processing steps designed to help realize this goal. For example, Favrin *et al.* (2001) used IMS to isolate cells of *S. enterica*, followed by infecting the captured cells with bacteriophage SJ2 and endpoint detection of the released phage progeny by measuring the optical density. The assay detection limit was less than 10^4 CFU/ml with a time-to-detection of 4 to 5 h. A similar IMS-coupled bacteriophage amplification assay was reported by Madonna *et al.* (2001, 2003) using matrix-assisted laser desorption/ionization mass time-of-flight (MALDI-TOF) spectrometry to detect generic *E. coli* with a detection limit of $\sim 5.0 \times 10^4$ cells/ml.

Although IMS is widely used in sample preparation, antibodies have some limitations, including the requirement for a living host for their production, relatively high cost and somewhat limited shelf life. It can also be difficult to produce broadly reactive antibodies for some target pathogens. Perhaps the best examples are *Salmonella enterica*, a species comprising over 2500 serovars, and human noroviruses. Additionally, antibodies can be sensitive to extremes of pH and can bind non-specifically to food matrix components. Finally, antigen expression can be affected by properties of the sample matrix, the enrichment medium used, incubation temperature and bacterial growth phase, making IMS sensitive to these factors as well (Geng *et al.*, 2006; Hahm and Bhunia, 2006; Stancik *et al.*, 2002; Dwivedi and Jaykus, 2011).

9.3 Emerging approaches

Nanotechnology is the manipulation of matter on a molecular scale, an approach that may allow scientists to develop a new generation of portable devices to simplify sample preparation and reduce pathogen detection time. Nanomaterials, which are the foundation of nanotechnology, are minute materials, structures or devices ranging in size from 1–100 nm. These include nanowires, nanofibers, liposomes, polymeric nanoparticles and magnetic nanoparticles (Gu *et al.*, 2006). Due to their high surface-to-volume ratios, nanomaterials can theoretically improve target capture efficiency because they have a greater surface area for attachment of bioaffinity ligands. Because of their small size, nanoparticles can move more rapidly than larger particles, penetrate inaccessible matrix interstices and interact more effectively with their relatively large cellular targets, resulting in increased capture efficiency, decreased capture time and faster, more sensitive detection (El-Boubbou *et al.*, 2007; Yang *et al.*, 2008). For example, Varshney *et al.* (2005) used magnetic nanoparticles conjugated with polyclonal antibodies to detect *E. coli* O157:H7 in 0.5 ml samples of homogenized ground beef, achieving a minimum capture efficiency of 94% in immunoreactions with starting inoculum levels ranging from 1.6×10^1 to 7.2×10^7 CFU/ml. These experiments were done without prior cultural enrichment and results were obtained in as little as 15 min.

Antibodies are not the only bioaffinity ligands that have potential for use in magnetic particle-based pre-analytical sample processing. Alternative ligands include bacteriophages and their tail fiber proteins, aptamers and carbohydrate moieties, among others. Indeed, magnetic nanoparticles have been functionalized with many different types of biorecognition molecules (Torres-Chavolla and Alocilja, 2009; Johnson *et al.*, 2008; Lin *et al.*, 2002; El-Boubbou *et al.*, 2007). Several of these novel binders are described below.

9.3.1 Bacteriophages

Bacteriophages (bacteria-infecting viruses or phages) have long been used in microbial diagnostics, in amplification-based assays, directly as cell capture reagents and even as selective agents in pathogen-specific enrichment media. These applications hinge on the unique binding affinities of phages for their host(s), resulting in highly specific assays (Wolber and Green, 1990; Tanji *et al.*, 2004). For example, the *Salmonella*-specific lytic phage *Sapphire* has been immobilized on polystyrene surfaces and used to capture *Salmonella* from mixed bacterial populations, followed by detection using PCR (Bennett *et al.*, 1997). This approach was reportedly able to detect 10^5 CFU *Salmonella* per milliliter in suspensions containing other Enterobacteriaceae and the capture efficiency was around 1%. In another application, magnetic particles conjugated to bacteriophage engineered to encode the *lux* operon were used for the selective concentration of *S. enterica* serovar Enteritidis, with subsequent detection based on the degree of bioluminescence; this assay demonstrated 20% capture efficiency (Sun *et al.*, 2001).

Although such studies illustrate proof-of-concept, target capture remains relatively inefficient. To address this, investigators have sought to develop methods for phage immobilization that promote optimal target binding (Tolba *et al.*, 2010; Gervais *et al.*, 2007; Singh *et al.*, 2009; Minikh *et al.*, 2010). Most of this work has been done using bacteriophage T4, which binds tail first to host cells. For example, this phage has been genetically modified by fusing a biotin carboxyl carrier protein gene (*bccp*) or the cellulose-binding module gene (*cbm*) with the small outer phage capsid protein gene (*soc*), resulting in expression of the ligand on the phage head (Tolba *et al.*, 2010). The recombinant phages, when immobilized on solid surfaces such as streptavidin-labeled magnetic beads or microcrystalline cellulose beads, were properly oriented to facilitate capture.

Tolba *et al.* (2010) used this approach to capture *E. coli* B cells at initial cell concentrations of 10^1–10^5 CFU/ml with an efficiency ranging from 72–90%. The same group immobilized bioengineered phages on nano-aluminum fiber-based filters (Disruptor™), which had a higher density, achieving detection (based on an adenosine triphosphate (ATP) bioluminescence assay) of as few as 10^3 *E. coli* cells/ml within 2 h (Minikh *et al.*, 2010). Improved efficiency of bacteriophage immobilization can also be achieved by chemical modification of solid surfaces. For example, Singh and colleagues (2009) immobilized *E. coli* host (EC12)-specific bacteriophages on gold surfaces chemically modified using sugars and amine groups, demonstrating up to a 37-fold improvement in immobilization efficiency with a corresponding ninefold improvement in *E. coli* capture efficiency.

Biomolecules derived from bacteriophage have also been used for the selective separation of bacterial cells. For example, Kretzer and colleagues (2007) expressed cell wall binding domains (CBDs) of bacteriophage-encoded peptidoglycan hydrolases (endolysins). The CBDs from a *L. monocytogenes* phage endolysin specifically bound to unique peptidoglycans on the surface of *Listeria* cells. Using paramagnetic beads conjugated with the recombinant phage endolysin-derived CBD molecules, members of the *Listeria* genus could be captured and recovered at an efficiency ranging from 86–99% in buffer systems containing target cells at 10^3–10^5 CFU/ml.

In addition to cell capture applications, bacteriophages can also be used for target-specific detection, with candidate detection platforms including MALDI-TOF (Madonna *et al.*, 2001, 2003), competitive ELISA (Guan *et al.*, 2006) and quantitative PCR (qPCR) (Tolba *et al.*, 2010). One particularly interesting method is the phage amplification technique (Favrin *et al.*, 2003). Adapted for *Salmonella*, pre-enriched samples are processed for pathogen-specific capture using antibody-conjugated magnetic beads. The captured cells are then infected with the *Salmonella*-specific phage SJ2, and the efficiency of infection determined using freshly cultured *Salmonella* cells. The optical density (OD) at 600 nm is measured before and after phage infection and lysis, and the ratio between the two OD values is used to determine sample positivity.

Labeling bacteriophages with reporter molecules (e.g., *lux* and *lacZ*) can also be used to facilitate detection of foodborne pathogens (Goodridge and Griffiths, 2002). For example, Goodridge *et al.* (1999) were able to detect *E. coli* O157:H7 by flow cytometry using a fluorescently labeled bacteriophage, reporting assay detection limits (without prior cultural enrichment) of 10^4 cells/ml. More recently, Oda *et al.* (2004) engineered phage PP01 (specific to *E. coli* O157:H7) to express green fluorescent protein (GFP), with positive results visualized by fluorescence microscopy. A similar study was done by Tanji *et al.* (2004) using GFP-tagged T4 for detection of generic *E. coli.* Chen and Griffiths (1996) developed a reporter containing the *lux* gene and used this for the detection of *Salmonella* in eggs, achieving detection limits as low as 10 CFU/ml after only a 6 h pre-enrichment.

Bacteriophages and their proteins have several advantages over other recognition elements, such as the fact that they are environmentally ubiquitous, easy and inexpensive to culture and highly stable. However, they may be less desirable for applications to pathogens having many serotypes or phage types (e.g. *S. enterica*), or for bacterial species for which phage profiles are poorly characterized (e.g., *Campylobacter* spp.). There are also concerns about undesired consequences of the interactions with target cells, such as the potential for target cell lysis and degradation of DNA (Dwivedi and Jaykus, 2011).

9.3.2 Nucleic acid ligands

Nucleic acid aptamers are single-stranded RNA or DNA oligonucleotides that fold into three-dimensional structures having binding affinity and specificity to target molecules and/or organisms (Jayasena, 1999; Wilson and Szostak, 1999). Naturally occurring aptamers include nucleic acid sequences capable of binding specific metals or proteins. In the lab, aptamers with binding specificity toward desired biotargets may be selected for using a combinatorial approach called SELEX (systematic evolution of ligands by exponential enrichment) (Tuerk and Gold, 1990). Nucleic acid aptamers have been used in wide-ranging applications, including as therapeutics (Green *et al.*, 1995; Nimjee *et al.*, 2005; Que-Gewirth and Sullenger, 2007; Keefe *et al.*, 2010), as diagnostic tools (Brody and Gold, 2000) and for the development of new drugs (Tombelli *et al.*, 2005). In recent years, DNA aptamers have been applied to detect microbial agents and their metabolites, including foodborne pathogens and mycotoxins (Bruno and Kiel, 1999; Cruz-Aguado and Penner, 2008; Jeffrey and Fischer, 2008). Recent studies in which nucleic acid aptamers have been used for pre-analytical sample processing and detection of foodborne pathogens are summarized in Table 9.1.

As early as 1999, Bruno and Kiel (1999) generated DNA aptamers against spores of the *Bacillus anthracis* Sterne strain, subsequently developing an aptamer-magnetic bead-electrochemiluminescence (AM-ECL) sandwich assay for detection. This assay was able to detect <10 anthrax spores. Jeffrey and Fischer (2008) and Cruz-Aguado and Penner (2008) generated high affinity DNA aptamers specific to the *Clostridium botulinum* neurotoxin and ochratoxin A, respectively. Recently, Joshi *et al.* (2009) immobilized DNA aptamers selected against purified outer membrane proteins of *S. enterica* serovar Typhimurium on magnetic beads, using the beads to capture the organism from whole carcass chicken rinse samples. When their magnetic pull-down assay was combined with qPCR, *S.* Typhimurium could be detected at concentrations as low as 10^1–10^2 CFU/9 ml rinsate, while 10^2–10^3 CFU/25 ml was the lower detection limit using a recirculating magnetic capture method combined with qPCR. In a similar study, DNA aptamers specific to surface proteins of *Campylobacter jejuni* were linked to magnetic beads for capture and subsequent detection using a quantum dot-based fluorescent sandwich assay in which endpoint detection was achieved using fluorometry (Bruno *et al.*, 2009). This assay was able to detect 2.5 CFU equivalents of *C. jejuni* in buffer and 10^1–2.5×10^2 CFU/ml in various seeded food matrices without prior cultural enrichment.

There are two different strategies for selecting nucleic acid aptamers. In the first, illustrated by the studies described above, the SELEX process is carried out

Table 9.1 Aptamer-based methods for capture and detection of foodborne pathogens

Target	Type of aptamer	Affinity target	Detection	Limit of detection (without cultural enrichment)	Reference
Salmonella enterica serovar Typhi	RNA	Structural protein of type IVB pili	N/A	N/A	Pan *et al.* (2005)
Salmonella enterica serovar Typhimurium	RNA	Outer membrane protein	Real-time PCR	10^1–10^2 CFU/9 ml and 10^2–10^3 CFU/25 ml of chicken rinsate	Joshi *et al.* (2009)
Campylobacter jejuni	ssDNA[a]	Membrane protein	Quantum dot sandwich assay	10^1–2.5×10^2 CFU/ml in food samples	Bruno *et al.* (2009)
Campylobacter jejuni	ssDNA	Whole cell	Flow cytometry	N/A	Dwivedi *et al.* (2010)
Listeria monocytogenes	ssDNA	Internalin A (InlA)	Optic aptamer sensor	10^3 CFU/ml in pure culture	Ohk *et al.* (2010)
Escherichia coli	RNA	Whole cell	Real-time PCR	10^1 CFU/ml in 1 ml buffer	Lee *et al.* (2009)
Staphylococcus aureus	ssDNA	Whole cell	Flow cytometry	N/A	Cao *et al.* (2009)
Vibrio alginolyticus	ssDNA	Whole cell	N/A	N/A	Zheng *et al.* (2010)
Francisella tularensis	ssDNA	Protein lysate	ELISA	1.7×10^3 CFU/ml buffer	Vivekananda and Kiel (2006)
Escherichia coli	RNA	N/A	MPN	N/A	So *et al.* (2008)

Note: [a] single-stranded DNA.

against a purified target, such as a cell surface protein. Because purified proteins may no longer retain their native conformations, these aptamers may have less than optimal binding affinities for their targets when presented in cell-associated form. To circumvent this problem, whole cells can be used as the target for aptamer development (i.e., whole-cell SELEX). The selection of aptamers using whole cells or whole-cell lysates may have some inherent advantages. For example, different aptamers having binding affinity to different cell surface targets can be produced, and using them in combination (aptamer 'cocktails') could theoretically increase assay sensitivity, specificity and capture efficiency, and perhaps even help discriminate between different cellular states (Dwivedi *et al.*, 2010; Shamah *et al.*, 2008). Production of nucleic acid aptamers using whole-cell SELEX has been an area of active investigation, as illustrated by their use for capture and/or detection of *Campylobacter jejuni* (Dwivedi *et al.*, 2010), *E. coli* (Lee *et al.*, 2009), *Staphylococcus aureus* (Cao *et al.*, 2009) and *Vibrio* spp. (Zheng *et al.*, 2010).

Aptamers, like antibodies and bacteriophages, can be used for capture, detection or for both simultaneously. For example, a DNA aptamer specific for the internalin A protein of *L. monocytogenes* was conjugated with a fluorescent dye and used as a reporter in a fiber-optic sensor. The aptamer biosensor was able to detect *L. monocytogenes* at a concentration of approximately 10^3 CFU/ml in a mixed culture without enrichment, and in artificially contaminated ready-to-eat meat products having initial concentrations of 10^2 CFU/25 g if preceded by an 18 h enrichment (Ohk *et al.*, 2010). Vivekananda and Kiel (2006) used two different aptamers, one for capture and the other for detection, in a two-site binding sandwich assay that they called aptamer-linked immobilized sorbent assay (ALISA). The capture ligands, which were specific to *Francisella tularensis*, were immobilized on a 96-well microtiter plate; detection was achieved using a biotinylated secondary aptamer, with visualization of a positive binding reaction achieved using streptavidin-conjugated horseradish peroxidase (HRP) with 2,2′-azino-bis (3-ethylbenzothiazoline-6-sulfonic acid) enhancer (ABTS) as the substrate. Detection of *F. tularensis* was achieved at concentrations as low as 1.7×10^3 CFU/ml without prior enrichment.

Nucleic acid aptamers have several advantages over other ligands, especially antibodies. Like antibodies, aptamers specifically bind to a selected target; however, because they are composed of nucleic acid rather than protein, the user has more control over their manipulation. Additionally, aptamers are resistant to protease degradation, inexpensive to manufacture, highly stable, easy to modify chemically and more consistently produced. The primary limitation of nucleic acid aptamers (particularly RNA aptamers) is their nuclease sensitivity. However, chemical modification of the ribose ring can enhance their stability (Pieken *et al.*, 1991). Another potential drawback is the relatively high dissociation constants of some aptamers, which can limit assay sensitivity (Shlyahovsky *et al.*, 2007). Cross reactivity with non-target substances can also be problematic, particularly for aptamers targeting complex analytes such as intact bacterial cells.

9.3.3 Peptide ligands

Like nucleic acid aptamers, peptide aptamers are combinatorial binders selected for their affinities to target proteins or small molecules. They consist of a variable peptide loop attached at both ends to a protein scaffold, a structure that results in a ligand with binding affinity comparable to that of an antibody (i.e., in the nM range) (Hoppe-Seyler and Butz, 2000; Johnson *et al.*, 2008). The variable loop usually consists of 10–20 amino acids, while the scaffold can consist of any protein that is both compact and soluble. Peptide aptamers have mostly been developed for use in therapeutics (Hoppe-Seyler and Butz, 2000), drug delivery and drug discovery (Crawford *et al.*, 2003; Baines and Colas, 2006). However, a few antiviral (Butz *et al.*, 2001; Trahtenherts *et al.*, 2008) and antibacterial (Blum *et al.*, 2000) peptide aptamers have been reported in the literature.

Peptide aptamers can be selected in a variety of ways, including the phage display technique, first introduced by Smith (1985). In phage display, random DNA sequences are inserted into gene III or gene VIII, corresponding to coat proteins located on the surface of a filamentous phage, generating a large library (~10^9 clones) of fusion phages. A binding-based selection technique (biopanning) is then used to screen for and purify phage clones capable of binding to the target analyte.

Over the last decade, peptides produced by phage display technology have been used as biorecognition molecules for selective separation and concentration of target proteins and microorganisms, including some foodborne pathogens. For example, a peptide-mediated magnetic separation technique has been developed for the selective isolation of *Mycobacterium avium* subsp. *paratuberculosis* from bulk milk. The candidate peptide was chemically synthesized, biotinylated and conjugated to streptavidin-coated paramagnetic beads. In artificially contaminated milk, the combined peptide-mediated magnetic separation-PCR method was able to detect *M. avium* subsp. *paratuberculosis* at concentrations as low as 10 cells/ml. When applied to milk samples derived from naturally infected herds, seven of nine samples classified as serologically positive for *M. avium* subsp. *paratuberculosis* were also positive by this method (Stratmann *et al.*, 2002). More recently, Foddai *et al.* (2010, 2011) used magnetic beads coated with a chemically synthesized biotinylated polypeptide specific for *M. avium* subsp. *paratuberculosis*. When applied to broth containing 10^3–10^4 CFU/ml, this method demonstrated capture efficiencies ranging from 85% to100%.

Due to their complex structures and amphipathic nature, which provides many differently charged regions for potential target binding, peptide aptamers may actually have stronger affinity and selectivity when compared to other ligands. Unlike nucleic acid aptamers, peptide aptamers may have limited abilities to penetrate bacterial cell walls, and may therefore be most appropriate as surface-binding reagents. Drawbacks include the expense of peptide aptamer synthesis and the potential for denaturation from matrix-associated proteases or the extremes of heat and pH that may be encountered in some assays.

Antimicrobial peptides (AMPs) are part of the innate defense system found in higher eukaryotes and are widely distributed in nature. Hundreds of AMPs have been discovered, and they are classified primarily on the basis of their secondary structures (Boman, 1995). A key step in the action of any antimicrobial is the ability to bind to microbial surfaces, and most AMPs exhibit a relatively broad spectrum of molecular recognition for various Gram-negative and Gram-positive bacteria, fungi and viruses (Nicolas and Mor, 1995; Brogden, 2005). The linear cationic AMPs are of particular interest due to their smaller size and intrinsic stability. These AMPs are attracted to the net negative charges on the outer envelope of Gram-negative bacteria, including anionic phospholipids and phosphate groups on the lipopolysaccharide.

The ease of synthesis and intrinsic stability of AMPs render them good candidates for use as molecular recognition elements in pathogen separation, concentration and detection (Arcidiacono *et al.*, 2008) (Table 9.2). For example, cecropin P1, a porcine homolog of the AMP originally isolated from the moth species *Hyalophora cecropia*, was used to capture pathogenic and non-pathogenic *E. coli* strains when covalently immobilized on maleic anhydride microplates (Gregory and Mello, 2005). Similarly, synthetic magainin I, originally extracted from the skin of the African clawed frog *Xenopus laevis*, was immobilized on glass slides and used as a biosensor for foodborne pathogen detection. In this work, magainin I was used as a capture reagent for fluorescently labeled target microorganisms, producing detection limits of 1.6×10^5 and 6.5×10^4 cells/ml of *E. coli* O157:H7 and *Salmonella* Typhimurium, respectively (Kulagina *et al.*, 2005).

The primary limitation of AMPs for use in pre-analytical sampling is the potential for their degradation by proteolytic enzymes, especially trypsin-like proteases. Another consideration is that, because AMPs bind to net negative charges that exist on the outer envelope of bacteria, their cell recognition activity tends to be relatively non-specific with respect to genus or species. Some of these drawbacks can be addressed through the use of AMP-like biomimetic polymers, discussed further in Section 9.3.5 below.

9.3.4 Carbohydrate ligands

Non-covalent interactions between proteins and carbohydrates also occur widely in nature. Prominent examples are carbohydrate-specific enzymes, antibodies directed against carbohydrates and lectins. This latter group has been used rather extensively in biological research. Lectins are proteins that reversibly bind to mono- and oligosaccharides with relatively high specificity. They are found in most living organisms and even in some viruses. Historically, lectins have been isolated and purified using affinity chromatography; more recently, they have been generated using recombinant DNA techniques (Lis and Sharon, 1998).

The binding efficiency of lectins varies as a function of their source, the target microorganism and sample matrix. For example, Payne *et al.* (1992) showed that lectins isolated from wheat (*Triticum vulgaris*) showed high binding affinity for *L. monocytogenes* (87–100%) and *S. aureus* (80–100%), and more modest binding efficiency for *Salmonella* spp. (33–45%) and *E. coli* (42–77%). In comparison,

Table 9.2 Antimicrobial peptides for capture of foodborne pathogens

Target	Antimicrobial peptide	Original source of AMP	Immobilization and detection	Reference
E. coli O157:H7	Cecropin P1	*Hyalophora cecropia* (moth), pig intestine	Fiber-optic biosensor detection	Arcidiacono *et al.* (2008)
	Cathelicidin (SMAP29)	Lysosomes in macrophages and polymorphonuclear leukocytes (PMNs)		
	PGQ	*Xenopus laevis* (African clawed frog)		
E. coli	Cecropin P1	*Hyalophora cecropia* (moth), pig intestine	Fluorescent microscopic observation	Gregory and Mello (2005)
E. coli O157:H7, *Salmonella* Typhimurium	Magainin I	*Xenopus laevis* (African clawed frog)	Biosensor detection	Kulagina *et al.* (2005)
E. coli, Salmonella	Magainin I	Chemical synthesis	Biosensor detection (impedance spectroscopy)	Mannor *et al.* (2010)

lectins isolated from the mushroom species *Agaricus bisporus* bound 31–63% of *L. monocytogenes* cells, 83% of *S. aureus* cells, but only 3–5% of representative *Salmonella* cells. Lectins derived from the Burgundy snail (*Helix pomatia*) bound >92% of *S. aureus* and 64% of *L. monocytogenes*, but showed poor binding affinity for Gram-negative organisms.

Lectins such as these have been used as pathogen recognition agents in microarray-based detection methods. For example, two biotinylated lectins [agglutinin I, derived from the castor oil plant (*Ricinus communis*), and concanavalin A, extracted from the jack bean (*Canavalia ensiformis*)] were conjugated to gold nanoparticles and used to detect *E. coli* and *B. cereus*, detecting as few as 10^3 cells per assay (Gao *et al.*, 2010). Similarly, lectin-based screen-printed gold electrodes have been used for the impedimetric label-free detection of *E. coli* at cell concentrations ranging from 5.0×10^3 to 5.0×10^7 CFU/ml (Gamella *et al.*, 2009).

A major limitation of lectins is the difficulty in isolating them from their natural biological sources, making them both expensive and in short supply. In addition, the elution of target cells from lectins can be inefficient unless stringent elution conditions are used, which can impact cell viability and/or interfere with downstream detection methods like PCR.

9.3.5 Additional binders: vancomycin and biomimetic polymers

Vancomycin is a glycopeptide antibiotic capable of recognizing cell surface moieties of Gram-positive bacteria. Recognition and binding occurs via hydrogen bonding between vancomycin's heptapeptide backbone and the D-alanyl-D-alanine dipeptide displayed on the bacterial cell wall. Gram-negative bacteria are intrinsically resistant to vancomycin due to the permeability barrier posed by the outer membrane, which restricts access of the antibiotic to its target in the periplasm (Walsh *et al.*, 1996).

Because of its ability to bind bacterial cells, several research groups have used vancomycin-conjugated surfaces for pre-analytical sample processing. For example, vancomycin-modified magnetic nanoparticles have been employed for selective isolation of Gram-positive pathogens from pure cultures (Lin *et al.*, 2005). In this case, the isolated cells were further identified by matrix-assisted laser desorption ionization mass spectroscopy (MALDI-MS), with a detection limit of $\sim 7 \times 10^4$ CFU/ml obtained for both *Staphylococcus saprophyticus* and *S. aureus* in urine. In another study, vancomycin-modified nanoparticles were used to isolate *S. aureus* and *Enterococcus faecalis* from aqueous solutions (Kell *et al.*, 2008). These investigators found that the size of the vancomycin-nanoparticle linker molecule and the size of the nanoparticle were critical variables impacting the efficiency of magnetic capture.

The primary limitation of using vancomycin as a bioaffinity ligand is the relative ease with which important pathogens such as *S. aureus* develop resistance to the antibiotic. A common mode for the development of resistance involves chemical modification of the cell wall, which also results in the loss of affinity between vancomycin and the target cell. Apart from natural antimicrobials such as vancomycin, biomimetic antimicrobial polymers, which are fully synthetic compounds with AMP-like functionalities, may also have promise as alternative binders (Tew *et al.*, 2002). Computer-aided modeling and design can be used to guide development of polymers that are at least broadly selective for specific cell types (e.g. Gram-negative bacteria or yeasts), followed by *in vitro* screening for desired binding (or antimicrobial) activities. The benefits of biomimetic AMPs over their natural counterparts include lower synthesis costs, inherent stability and resistance to proteases, the ability to bind under various salt or cationic conditions, and computer-aided rational design of mimics having desirable performance characteristics. Figure 9.2 demonstrates proof of concept for whole-cell capture of *E. coli* on an atomic force microscope (AFM)-patterned biomimetic polymer ultramicroarray. Similar strategies for micro-functionalization of surfaces could be used to create binding or recognition elements within biosensors.

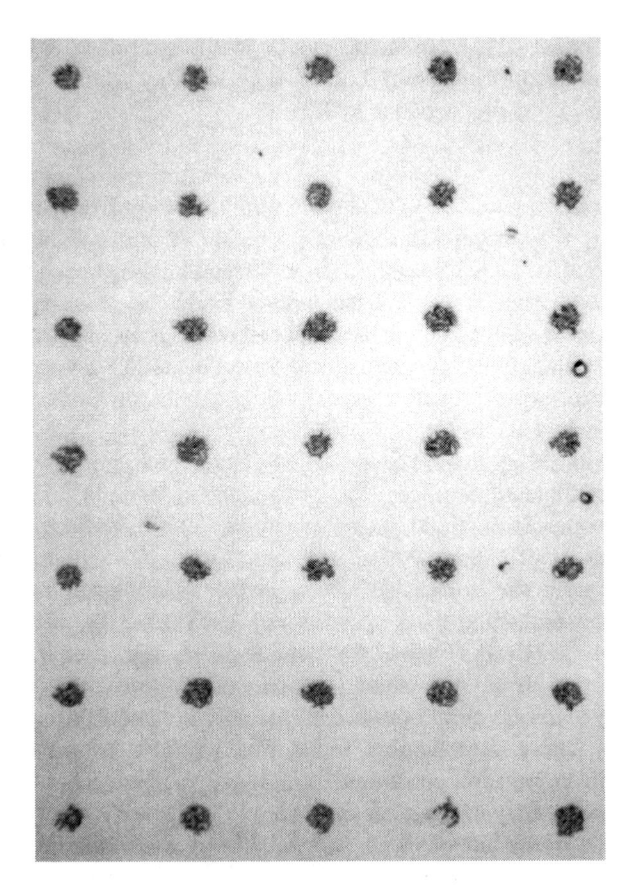

Fig. 9.2 Whole *E. coli* cells captured on an AFM-patterned biomimetic polymer ultramicroarray.

9.3.6 Green chemistry and novel solvents

As a universal medium for sample manipulations such as solvation, extraction, partitioning or washing, water has its limitations, as not all food matrices, cell surfaces or cell constituents are hydrophilic in nature. While the combined use of organic solvents, chaotropic agents and/or detergents is effective for direct extraction of target molecules such as RNA from complex, high fat matrices such as cheese (Monnet *et al.*, 2008), such inherently toxic or disruptive processes are not compatible with the extraction of living cells from these matrices, and cannot be used if cell enumeration or subsequent cultural studies are the goals of the analysis (Brehm-Stecher *et al.*, 2009; Mayrl *et al.*, 2009; Rossmanith *et al.*, 2007). The availability of culture-compatible materials able to bridge the hydrophilic–hydrophobic divide is a critical limitation when dealing with multiphasic matrices or with certain target cells or diagnostic molecules where neither aqueous nor organic extractants provide an ideal solution. Possible approaches include leveraging the power of aqueous interfacial chemistry to partition target analytes (cells or molecules) between immiscible biocompatible polymer phases, where they can be recovered (Pedersen *et al.*, 1998; Cabral, 2007). Alternatively, amphipathic molecules may also act as intermediates between hydrophilic and hydrophobic phases.

In accord with the general trends within the greater chemistry community, green or ecologically friendlier alternatives to hazardous organic solvents are also in demand (Han and Armstrong, 2007; Roosen *et al.*, 2008). These approaches may ultimately find applications in food analyses or separations, including in food microbiology (Mester *et al.*, 2010). Possible approaches for addressing the chemical limitations of current solvent or separation systems, while simultaneously achieving a greener chemical footprint, include the use of next-generation solvents such as ionic liquids (Mester *et al.*, 2010; Han and Armstrong, 2007) and stimulus-responsive materials such as peptide surfactants (pepfactants) (Malcolm *et al.*, 2007) and switchable solvents that can reverse their character (i.e. from hydrophilic to hydrophobic) in response to pH or other external stimuli (Mercer and Jessop, 2010).

9.4 Conclusions

Pre-analytical sample processing is a critical cornerstone in effective microbiological analysis and there is significant interest in this topic. Despite advances, the identification of sample preparation methods that are both reliable and universally applicable has remained elusive and unfortunately even the emerging techniques described in this paper are still not considered a substitute for classic methods that are based on cultural enrichment. Additional effort and resources will be needed to develop methods that can ultimately eliminate the need for an enrichment culture, the key rate-limiting step in microbiological detection.

9.5 References

ARCIDIACONO, S., PIVARNIK, P., MELLO, C. M. and SENECAL, A. 2008. Cy5 labeled antimicrobial peptides for enhanced detection of *Escherichia coli* O157:H7. *Biosensors and Bioelectronics*, 23, 1721–1727.

BAINES, I. C. and COLAS, P. 2006. Peptide aptamers as guides for small-molecule drug discovery. *Drug Discovery Today*, 11, 334–341.

BENNETT, A. R., DAVIDS, F. G. C., VLAHODIMOU, S., BANKS, J. G. and BETTS, R. P. 1997. The use of bacteriophage-based systems for the separation and concentration of Salmonella. *Journal of Applied Microbiology*, 83, 259–265.

BLUM, J. H., DOVE, S. L., HOCHSCHILD, A. and MEKALANOS, J. J. 2000. Isolation of peptide aptamers that inhibit intracellular processes. *Proceedings of the National Academy of Sciences*, 97, 2241.

BOMAN, H. G. 1995. Peptide antibiotics and their role in innate immunity. *Annual Review of Immunology*, 13, 61–92.

BREHM-STECHER, B., YOUNG, C., JAYKUS, L.-A. and TORTORELLO, M. L. 2009. Sample preparation: The forgotten beginning. *Journal of Food Protection*, 72, 1774–1789.

BRODY, E. N. and GOLD, L. 2000. Aptamers as therapeutic and diagnostic agents. *Reviews in Molecular Biotechnology*, 74, 5–13.

BROGDEN, K. A. 2005. Antimicrobial peptides: Pore formers or metabolic inhibitors in bacteria? *Nature Reviews Microbiology*, 3, 238–250.

BRUNO, J. G. and KIEL, J. L. 1999. *In vitro* selection of DNA aptamers to anthrax spores with electrochemiluminescence detection. *Biosensors and Bioelectronics*, 14, 457–464.

BRUNO, J. G., PHILLIPS, T., CARRILLO, M. P. and CROWELL, R. 2009. Plastic-adherent DNA aptamer-magnetic bead and quantum dot sandwich assay for *Campylobacter* detection. *Journal of Fluorescence*, 19, 427–435.

BUTZ, K., DENK, C., FITSCHER, B., CRNKOVIC-MERTENS, I., ULLMANN, A. *et al.* 2001. Peptide aptamers targeting the hepatitis B virus core protein: A new class of molecules with antiviral activity. *Oncogene*, 20, 6579.

CABRAL, J. 2007. Cell partitioning in aqueous two-phase polymer systems. *Cell Separation*, 106, 151–171.

CAO, X., LI, S., CHEN, L., DING, H., XU, H. *et al.* 2009. Combining use of a panel of ssDNA aptamers in the detection of *Staphylococcus aureus*. *Nucleic Acids Research*, 37, 4621–4628.

CHEN, J. and GRIFFITHS, M. W. 1996. *Salmonella* detection in eggs using *Lux*+ bacteriophages. *Journal of Food Protection*, 59, 908–914.

CRAWFORD, M., WOODMAN, R. and FERRIGNO, P. K. 2003. Peptide aptamers: Tools for biology and drug discovery. *Briefings in Functional Genomics and Proteomics*, 2, 72–79.

CRUZ-AGUADO, J. A. and PENNER, G. 2008. Determination of ochratoxin a with a DNA aptamer. *Journal of Agricultural and Food Chemistry*, 56, 10456–10461.

DONNELLY, C. W. 2002. Detection and isolation of *Listeria monocytogenes* from food samples: Implications of sublethal injury. *Journal of AOAC International*, 85, 495–500.

DWIVEDI, H. P. and JAYKUS, L.-A. 2011. Detection of pathogens in foods: The current state-of-the-art and future directions. *Critical Reviews in Microbiology*, 37, 40–63.

DWIVEDI, H. P., SMILEY, R. D. and JAYKUS, L. A. 2010. Selection and characterization of DNA aptamers with binding selectivity to *Campylobacter jejuni* using whole-cell SELEX. *Applied Microbiology and Biotechnology*, 87, 2323–2334.

EL-BOUBBOU, K., GRUDEN, C. and HUANG, X. 2007. Magnetic glyco-nanoparticles: A unique tool for rapid pathogen detection, decontamination, and strain differentiation. *Journal of the American Chemical Society*, 129, 13392.

FAVRIN, S. J., JASSIM, S. A. and GRIFFITHS, M. W. 2001. Development and optimization of a novel immunomagnetic separation-bacteriophage assay for detection of *Salmonella enterica* serovar Enteritidis in broth. *Applied and Environmental Microbiology*, 67, 217–224.

FAVRIN, S. J., JASSIM, S. A. and GRIFFITHS, M. W. 2003. Application of a novel immunomagnetic separation-bacteriophage assay for the detection of *Salmonella enteritidis* and *Escherichia coli* O157: H7 in food. *International Journal of Food Microbiology*, 85, 63–71.

FDA 2011. Diarrheagenic *Escherichia coli*, Chapter 4A in *Bacteriological Analytical Manual*. FDA.

FODDAI, A., ELLIOTT, C. T. and GRANT, I. R. 2010. Maximizing capture efficiency and specificity of magnetic separation for *Mycobacterium avium* subsp. *paratuberculosis* cells. *Applied and Environmental Microbiology*, 76, 7550.

FODDAI, A., STRAIN, S., WHITLOCK, R. H., ELLIOTT, C. T. and GRANT, I. R. 2011. Application of a peptide-mediated magnetic separation-phage assay for detection of viable *Mycobacterium avium* subsp. *paratuberculosis* to bovine bulk tank milk and feces samples. *Journal of Clinical Microbiology*, 49, 2017–2019.

FSIS 2012. *Microbiology Laboratory Guide Book*, MLG 5B.03. USDA FSIS.

GAMELLA, M., CAMPUZANO, S., PARRADO, C., REVIEJO, A. J. and PINGARRÓN, J. M. 2009. Microorganisms recognition and quantification by lectin adsorptive affinity impedance. *Talanta*, 78, 1303–1309.

GAO, J., LIU, C., LIU, D., WANG, Z. and DONG, S. 2010. Antibody microarray-based strategies for detection of bacteria by lectin-conjugated gold nanoparticle probes. *Talanta*, 81, 1816–1820.

GENG, T., HAHM, B. K. and BHUNIA, A. K. 2006. Selective enrichment media affect the antibody-based detection of stress-exposed *Listeria monocytogenes* due to differential expression of antibody-reactive antigens identified by protein sequencing. *Journal of Food Protection*, 69, 1879–1886.

GERVAIS, L., GEL, M., ALLAIN, B., TOLBA, M., BROVKO, L. *et al.* 2007. Immobilization of biotinylated bacteriophages on biosensor surfaces. *Sensors and Actuators B: Chemical*, 125, 615–621.

GOODRIDGE, L. and GRIFFITHS, M. 2002. Reporter bacteriophage assays as a means to detect foodborne pathogenic bacteria. *Food Research International*, 35, 863–870.

GOODRIDGE, L., CHEN, J. R. and GRIFFITHS, M. 1999. Development and characterization of a fluorescent-bacteriophage assay for detection of *Escherichia coli* O157:H7. *Applied and Environmental Microbiology*, 65, 1397–1404.

GREEN, L. S., JELLINEK, D., BELL, C., BEEBE, L. A., FEISTNER, B. D. *et al.* 1995. Nuclease-resistant nucleic acid ligands to vascular permeability factor/vascular endothelial growth factor. *Chemistry and Biology*, 2, 683–695.

GREGORY, K. and MELLO, C. M. 2005. Immobilization of *Escherichia coli* cells by use of the antimicrobial peptide cecropin P1. *Applied and Environmental Microbiology*, 71, 1130–1134.

GU, H. W., XU, K. M., XU, C. J. and XU, B. 2006. Biofunctional magnetic nanoparticles for protein separation and pathogen detection. *Chemical Communications*, 941–949.

GUAN, J., CHAN, M., ALLAIN, B., MANDEVILLE, R. and BROOKS, B. W. 2006. Detection of multiple antibiotic-resistant *Salmonella enterica* serovar Typhimurium DT104 by phage replication competitive enzyme-linked immunosorbent assay. *Journal of Food Protection*, 69, 739–742.

HAHM, B. K. and BHUNIA, A. K. 2006. Effect of environmental stresses on antibody-based detection of *Escherichia coli* O157: H7, *Salmonella enterica* serotype Enteritidis and *Listeria monocytogenes*. *Journal of Applied Microbiology*, 100, 1017–1027.

HAN, X. and ARMSTRONG, D. W. 2007. Ionic liquids in separations. *Accounts of Chemical Research*, 40, 1079–1086.

HOPPE-SEYLER, F. and BUTZ, K. 2000. Peptide aptamers: Powerful new tools for molecular medicine. *Journal of Molecular Medicine*, 78, 426–430.

JAYASENA, S. D. 1999. Aptamers: An emerging class of molecules that rival antibodies in diagnostics. *Clinical Chemistry*, 45, 1628.

JEFFREY, B. H. T. and FISCHER, N. O. 2008. Single microbead SELEX for efficient ssDNA aptamer generation against botulinum neurotoxin. *Chemical Communications*, 1883–1885.

JOHNSON, S., EVANS, D., LAURENSON, S., PAUL, D., DAVIES, A. G. *et al.* 2008. Surface-immobilized peptide aptamers as probe molecules for protein detection. *Analytical Chemistry*, 80, 978–983.

JOSHI, R., JANAGAMA, H., DWIVEDI, H. P., KUMAR, T. M. A. S., JAYKUS, L.-A. *et al.* 2009. Selection, characterization, and application of DNA aptamers for the capture and detection of *Salmonella enterica* serovars. *Molecular and Cellular Probes*, 23, 20–28.

KEEFE, A. D., PAI, S. and ELLINGTON, A. 2010. Aptamers as therapeutics. *Nature Reviews Drug Discovery*, 9, 537–550.

KELL, A. J., STEWART, G., RYAN, S., PEYTAVI, R., BOISSINOT, M. *et al.* 2008. Vancomycin-modified nanoparticles for efficient targeting and preconcentration of Gram-positive and Gram-negative bacteria. *Acs Nano*, 2, 1777–1788.

KRETZER, J. W., LEHMANN, R., SCHMELCHER, M., BANZ, M., KIM, K. P. *et al.* 2007. Use of high-affinity cell wall-binding domains of bacteriophage endolysins for immobilization and separation of bacterial cells. *Applied and Environmental Microbiology*, 73, 1992–2000.

KULAGINA, N. V., LASSMAN, M. E., LIGLER, F. S. and TAITT, C. R. 2005. Antimicrobial peptides for detection of bacteria in biosensor assays. *Analytical Chemistry*, 77, 6504–6508.

LEE, H.-J., KIM, B. C., KIM, K.-W., KIM, Y. K., KIM, J. *et al.* 2009. A sensitive method to detect *Escherichia coli* based on immunomagnetic separation and real-time PCR amplification of aptamers. *Biosensors and Bioelectronics*, 24, 3550–3555.

LIN, C. C., YEH, Y. C., YANG, C. Y., CHEN, C. L., CHEN, G. F. *et al.* 2002. Selective binding of mannose-encapsulated gold nanoparticles to type 1 pili in *Escherichia coli*. *Journal of the American Chemical Society*, 124, 3508–3509.

LIN, Y. S., TSAI, P. J., WENG, M. F. and CHEN, Y. C. 2005. Affinity capture using vancomycin-bound magnetic nanoparticles for the MALDI-MS analysis of bacteria. *Analytical Chemistry*, 77, 1753–1760.

LIS, H. and SHARON, N. 1998. Lectins: Carbohydrate-specific proteins that mediate cellular recognition. *Chemical Reviews*, 98, 637–674.

MADONNA, A. J., BASILE, F., FURLONG, E. and VOORHEES, K. J. 2001. Detection of bacteria from biological mixtures using immunomagnetic separation combined with matrix-assisted laser desorption/ionization time-of-flight mass spectrometry. *Rapid Communications in Mass Spectrometry*, 15, 1068–1074.

MADONNA, A. J., VAN CUYK, S. and VOORHEES, K. J. 2003. Detection of *Escherichia coli* using immunomagnetic separation and bacteriophage amplification coupled with matrix-assisted laser desorption/ionization time-of-flight mass spectrometry. *Rapid Communications in Mass Spectrometry*, 17, 257–263.

MALCOLM, A. S., DEXTER, A. F. and MIDDELBERG, A. P. J. 2007. Peptide surfactants (pepfactants) for switchable foams and emulsions. *Asia-Pacific Journal of Chemical Engineering*, 2, 362–367.

MANNOOR, M.S., ZHANG, S., LINK, A. J. and MCALPINE, M. C. 2010. Electrical detection of pathogenic bacteria via immobilized antimicrobial peptides. *Proceedings of the National Academy of Sciences*, 107: 19207–19212.

MAYRL, E., ROEDER, B., MESTER, P., WAGNER, M. and ROSSMANITH, P. 2009. Broad range evaluation of the matrix solubilization (matrix lysis) strategy for direct enumeration of foodborne pathogens by nucleic acids technologies. *Journal of Food Protection*, 72, 1225–1233.

MERCER, S. M. and JESSOP, P. G. 2010. "Switchable water": Aqueous solutions of switchable ionic strength. *ChemSusChem*, 3, 467–470.

MESTER, P., WAGNER, M. and ROSSMANITH, P. 2010. Use of ionic liquid-based extraction for recovery of *Salmonella* Typhimurium and *Listeria monocytogenes* from food matrices. *Journal of Food Protection*, 73, 680–687.

MINIKH, O., TOLBA, M., BROVKO, L. Y. and GRIFFITHS, M. W. 2010. Bacteriophage-based biosorbents coupled with bioluminescent ATP assay for rapid concentration and detection of *Escherichia coli*. *Journal of Microbiological Methods*, 82, 177–183.

MONNET, C., ULVÉ, V., SARTHOU, A. S. and IRLINGER, F. 2008. Extraction of RNA from cheese without prior separation of microbial cells. *Applied and Environmental Microbiology*, 74, 5724–5730.

NICOLAS, P. and MOR, A. 1995. Peptides as weapons against microorganisms in the chemical defense system of vertebrates. *Annual Reviews in Microbiology*, 49, 277–304.

NIMJEE, S. M., RUSCONI, C. P. and SULLENGER, B. A. 2005. Aptamers: An emerging class of therapeutics. *Annual Review of Medicine*, 56, 555–583.

ODA, M., MORITA, M., UNNO, H. and TANJI, Y. 2004. Rapid detection of *Escherichia coli* O157:H7 by using green fluorescent protein-labeled PP01 bacteriophage. *Applied and Environmental Microbiology*, 70, 527.

OHK, S. H., KOO, O. K., SEN, T., YAMAMOTO, C. M. and BHUNIA, A. K. 2010. Antibody-aptamer functionalized fibre-optic biosensor for specific detection of *Listeria monocytogenes* from food. *Journal of Applied Microbiology*, 109, 808–817.

PAN, Q., ZHANG, X. L., WU, H. Y., HE, P. W., WANG, F. *et al.* 2005. Aptamers that preferentially bind type IVB pili and inhibit human monocytic-cell invasion by *Salmonella enterica* serovar Typhi. *Antimicrobial Agents and Chemotherapy*, 49(10): 4052–4060.

PAYNE, M. J., CAMPBELL, S., PATCHETT, R. A. and KROLL, R. G. 1992. The use of immobilized lectins in the separation of *Staphylococcus aureus*, *Escherichia coli*, *Listeria* and *Salmonella* spp. from pure cultures and foods. *Journal of Applied Microbiology*, 73, 41–52.

PEDERSEN, L., SKOUBOE, P., ROSSEN, L. and RASMUSSEN, O. 1998. Separation of *Listeria monocytogenes* and *Salmonella berta* from a complex food matrix by aqueous polymer two-phase partitioning. *Letters in Applied Microbiology*, 26, 47–50.

PIEKEN, W. A., OLSEN, D. B., BENSELER, F., AURUP, H. and ECKSTEIN, F. 1991. Kinetic characterization of ribonuclease-resistant 2′-modified hammerhead ribozymes. *Science*, 253, 314.

QUE-GEWIRTH, N. S. and SULLENGER, B. A. 2007. Gene therapy progress and prospects: RNA aptamers. *Gene Therapy*, 14, 283–291.

ROOSEN, C., MÜLLER, P. and GREINER, L. 2008. Ionic liquids in biotechnology: Applications and perspectives for biotransformations. *Applied Microbiology and Biotechnology*, 81, 607–614.

ROSSMANITH, P., WAGNER, M. and HEIN, I. 2007. Development of matrix lysis for concentration of Gram positive bacteria from food and blood. *Journal of Microbiological Methods*, 69, 504–511.

SCALLAN, E., HOEKSTRA, R. M., WIDDOWSON, M.-A., HALL, A. J. and GRIFFIN, P. M. 2011. Foodborne illness acquired in the United States response. *Emerging Infectious Diseases*, 17, 1339–1340.

SHAMAH, S. M., HEALY, J. M. and CLOAD, S. T. 2008. Complex target SELEX. *Accounts of Chemical Research*, 41, 130–138.

SHLYAHOVSKY, B., DI, L., WEIZMANN, Y., NOWARSKI, R., KOTLER, M. *et al.* 2007. Spotlighting of cocaine by an autonomous aptamer-based machine. *Journal of the American Chemical Society*, 129, 3814–3815.

SINGH, A., GLASS, N., TOLBA, M., BROVKO, L., GRIFFITHS, M. *et al.* 2009. Immobilization of bacteriophages on gold surfaces for the specific capture of pathogens. *Biosensors and Bioelectronics*, 24, 3645–3651.

SKJERVE, E., RORVIK, L. M. and OLSVIK, O. 1990. Detection of *Listeria monocytogenes* in foods by immunomagnetic separation. *Applied and Environmental Microbiology*, 56, 3478–3481.

SMITH, G. P. 1985. Filamentous fusion phage: Novel expression vectors that display cloned antigens on the virion surface. *Science*, 228, 1315.

SO, H. M., PARK, D. W., JEON, E. K., KIM, Y. H., KIM, B. S. *et al.* 2008. Detection and titer estimation of *Escherichia coli* using aptamer functionalized single walled carbon nanotube field effect transistors. *Small* 4(2), 197–201.

STANCIK, L. M., STANCIK, D. M., SCHMIDT, B., BARNHART, D. M., YONCHEVA, Y. N. *et al.* 2002. pH-dependent expression of periplasmic proteins and amino acid catabolism in *Escherichia coli*. *Journal of Bacteriology*, 184, 4246.

STEVENS, K. A. and JAYKUS, L.-A. 2004. Bacterial separation and concentration from complex sample matrices: A review. *Critical Reviews in Microbiology*, 30, 7–24.

STRATMANN, J., STROMMENGER, B., STEVENSON, K. and GERLACH, G. F. 2002. Development of a peptide-mediated capture PCR for detection of *Mycobacterium avium* subsp. *paratuberculosis* in milk. *Journal of Clinical Microbiology*, 40, 4244.

SUN, W., BROVKO, L. and GRIFFITHS, M. 2001. Use of bioluminescent *Salmonella* for assessing the efficiency of constructed phage-based biosorbent. *Journal of Industrial Microbiology and Biotechnology*, 27, 126–128.

TANJI, Y., FURUKAWA, C., NA, S. H., HIJIKATA, T., MIYANAGA, K. and UNNO, H. 2004. *Escherichia coli* detection by GFP-labeled lysozyme-inactivated T4 bacteriophage. *Journal of Biotechnology*, 114, 11–20.

TEW, G. N., LIU, D., CHEN, B., DOERKSEN, R. J., KAPLAN, J. *et al.* 2002. *De novo* design of biomimetic antimicrobial polymers. *Proceedings of the National Academy of Sciences*, 99, 5110.

TOLBA, M., MINIKH, O., BROVKO, L. Y., EVOY, S. and GRIFFITHS, M. W. 2010. Oriented immobilization of bacteriophages for biosensor applications. *Applied and Environmental Microbiology*, 76, 528–535.

TOMBELLI, S., MINUNNI, M. and MASCINI, M. 2005. Analytical applications of aptamers. *Biosensors and Bioelectronics*, 20, 2424–2434.

TORRES-CHAVOLLA, E. and ALOCILJA, E. C. 2009. Aptasensors for detection of microbial and viral pathogens. *Biosensors and Bioelectronics*, 24, 3175–3182.

TRAHTENHERTS, A., GAL-TANAMY, M., ZEMEL, R., BACHMATOV, L., LOEWENSTEIN, S. *et al.* 2008. Inhibition of hepatitis C virus RNA replicons by peptide aptamers. *Antiviral Research*, 77, 195–205.

TUERK, C. and GOLD, L. 1990. Systematic evolution of ligands by exponential enrichment: RNA ligands to bacteriophage T4 DNA polymerase. *Science*, 249, 505.

VARSHNEY, M., YANG, L. J., SU, X. L. and LI, Y. B. 2005. Magnetic nanoparticle-antibody conjugates for the separation of *Escherichia coli* O157:H7 in ground beef. *Journal of Food Protection*, 68, 1804–1811.

VIVEKANANDA, J. and KIEL, J. L. 2006. Anti-*Francisella tularensis* DNA aptamers detect tularemia antigen from different subspecies by aptamer-linked immobilized sorbent assay. *Laboratory Investigation*, 86, 610–618.

WALSH, C., FISHER, S., PARK, I. S., PRAHALAD, M. and WU, Z. 1996. Bacterial resistance to vancomycin: Five genes and one missing hydrogen bond tell the story. *Chemistry and Biology*, 3, 21–28.

WILSON, D. S. and SZOSTAK, J. W. 1999. *In vitro* selection of functional nucleic acids. *Annual Review of Biochemistry*, 68, 611–647.

WOLBER, P. K. and GREEN, R. L. 1990. Detection of bacteria by transduction of ice nucleation genes. *Trends in Biotechnology*, 8, 276.

YANG, H., LI, H. and JIANG, X. 2008. Detection of foodborne pathogens using bioconjugated nanomaterials. *Microfluidics and Nanofluidics*, 5, 571–583.

YANG, H., QU, L., WIMBROW, A. N., JIANG, X. and SUN, Y. 2007. Rapid detection of *Listeria monocytogenes* by nanoparticle-based immunomagnetic separation and real-time PCR. *International Journal of Food Microbiology*, 118, 132–138.

YU, L. S. L., UKNALIS, J. and TU, S. I. 2001. Immunomagnetic separation methods for the isolation of *Campylobacter jejuni* from ground poultry meats. *Journal of Immunological Methods*, 256, 11–18.

ZHENG, J., LI, Y. B., LI, J. X., WANG, J. and SU, Y. Q. 2010. *In vitro* selection of oligonucleotide acid aptamers against pathogenic *Vibrio alginolyticus* by SELEX. *Key Engineering Materials*, 439, 1456–1462.

10

Second-generation polymerase chain reaction (PCR) and DNA microarrays for *in vitro* and *in situ* study of foodborne pathogens

K. Rantsiou and L. Cocolin, University of Turin, Italy

DOI: 10.1533/9780857098740.3.193

Abstract: Since the invention of the polymerase chain reaction (PCR) in the 1980s, applied molecular biology has progressed enormously. Following technological advancements, PCR analysis has become quantitative rather than qualitative. Quantitative PCR (qPCR) revolutionized molecular approaches to food safety, due to the ability to quantify specific pathogens in food. New methods for studying the transcription process, such as microarrays, can be used to investigate the behavior of pathogens in the food environment. Indeed, the effects of food-related conditions on gene expression in pathogenic microorganisms have already been demonstrated.

Key words: foodborne pathogens, quantitative polymerase chain reaction (qPCR), microarrays, microbial behavior, risk assessment.

10.1 Introduction

In the last 50 years, consumer demand for fresh, minimally processed food has provoked food business operators (FBOs) to change their approach to food safety. From traditional means of controlling foodborne pathogens, such as freezing, blanching, sterilization, curing and the use of preservatives, the food industry has recently come under pressure to introduce mild technologies such as modified atmosphere and vacuum packaging, low heating, high hydrostatic pressure and the use of natural antimicrobial compounds such as bacteriocins and essential oils (Abee *et al.*, 2004). These changes have had a direct impact on the ecology of foodborne pathogens and their behavior in foods, because pathogenic microorganisms have specifically adapted to overcome new food-processing technologies. A relevant example is *Listeria monocytogenes*, a psychrotrophic bacterium that emerged as a foodborne pathogen after the establishment of refrigeration throughout the food chain, in order to meet consumer demand for freshness and longer shelf life (Rantsiou *et al.*, 2011).

Hurdle technology (Leistner, 1995) was developed in response to consumer demand for foods that retained the characteristics of freshness without the use of extreme preservation techniques. However, hurdle technology creates conditions that favor the viable but non-culturable (VBNC) state in microorganisms. As defined by Oliver (1993), a bacterium in the VBNC state is 'a cell which is metabolically active, while being incapable of undergoing the cellular division required for growth in or on a medium normally supporting growth of that cell'. It is believed that cells enter into a VBNC state as a survival strategy, in response to adverse environmental conditions, such as starvation or acid stress (Rowan, 2004). Cells in a VBNC state are a great risk for the food industry. Although there is a general lack of knowledge concerning the potential pathogenicity of VBNC cells, the possibility that such cells will change into a culturable state when entering the human body and reaching a target organ (such as the gastrointestinal tract) cannot be excluded. VBNC cells often have reduced capacity for gene expression and so a slower rate of protein synthesis than cells growing under optimal conditions. Nevertheless, in the VBNC state, the normal mechanisms of transcription of DNA into RNA should still exist. VBNC states have been identified for several foodborne pathogens, such as *Escherichia coli, L. monocytogenes, Campylobacter jejuni* and *Salmonella enterica* serovar Typhimurium (Rowan, 2004), and recent reports highlight the possible presence of VBNC pathogenic bacteria in food-processing lines in the meat and dairy sectors (Alessandria *et al.*, 2010; Melero *et al.*, 2011; Osés *et al.*, 2010).

Despite the significant growth of knowledge regarding food safety, the control of foodborne pathogens throughout the food chain remains a challenge for food producers, authorities and consumers. Worldwide, various efforts have been made to reduce the incidence of diseases related to food consumption, but with limited success (Rantsiou *et al.*, 2011). The lack of reliable detection methods for foodborne pathogens is one reason for this failure. Indeed, in the last 20 years it has been repeatedly demonstrated that traditional microbiological methods, based on the cultivation of microorganisms in synthetic media, have failed to correctly describe the prevalence of foodborne pathogens in foodstuffs. This is mainly due to the fact that stressed or injured cells are not able to recover in laboratory media that are often characterized by the presence of selective agents, including antibiotics, which are used to select the target microorganism. False-negative results can also be ascribed to the presence of VBNC cells (Cocolin *et al.*, 2011).

New requirements for food microbiological testing have driven research into new tools that are able to detect pathogenic bacteria in food properly. The polymerase chain reaction (PCR) is now a standard method, used within the food industry as well as research laboratories. An impressive number of papers have been published in the field of food microbiology and safety, and PCR methods have been developed for the detection and identification of almost all foodborne pathogens. Instruments able to detect PCR product synthesis in real time became available in the late 1990s. This technique, called quantitative PCR (qPCR), revolutionized molecular approaches to food safety by making it possible to quantify specific pathogens in food.

Methods for studying the transcription process, such as microarrays, have allowed the investigation of environmental influences on foodborne pathogen behavior, including food composition, temperature, pH and oxygen content. This is particularly relevant in terms of pathogenic potential, since it has already been demonstrated that food-related conditions can change gene expression in pathogenic microorganisms (Rantsiou *et al.*, 2012a, 2012b; Bae *et al.*, 2011; Olesen *et al.*, 2010).

The focus of this chapter is on modern molecular approaches, mainly qPCR and microarrays, for the quantification and study of gene expression in foodborne pathogens. Two important aspects are addressed: (i) the necessity of applying a culture-independent approach, to avoid the problems inherent to cultivation; and, (ii) the need to switch from *in vitro* to *in situ* analysis, in order to better comprehend the physiology of pathogenic microorganisms and to design more effective preventive measures for their control in the food chain.

10.2 The quantitative polymerase chain reaction (qPCR)

The applications of qPCR in food microbiology and safety have grown exponentially in the last three years, and a remarkable number of review papers were published in 2011 alone (Cocolin *et al.*, 2011; Martinez *et al.*, 2011; Postollec *et al.*, 2011). Briefly, qPCR assays measure reaction progress during each amplification cycle, rather than after reaching a plateau. Amplification increases the amount of double-stranded DNA resulting in an increase in fluorescence. By plotting the increase in fluorescence versus cycle number, the system produces amplification plots that provide a more complete picture of the PCR process than assaying product accumulation after a fixed number of cycles. Molecules produce fluorescence in correlation with the amount of PCR product synthesized, and the PCR reaction progress is measured using the amount of fluorescence at regular intervals. A number of qPCR chemistries are currently commercially available for use in PCR. These can be divided into those that are not sequence specific, such as DNA minor groove-binding dyes, and those that are sequence specific, and which might even afford simultaneous detection and confirmation of the target amplicon during the PCR reaction (McKillip and Drake, 2004).

Undoubtedly, qPCR possesses some features that make it extremely useful in the field of food microbiology. The target microorganism is detected and quantified regardless of its physiological state, and stressed, injured or VBNC cells are thereby identified, avoiding the false-negative results that occur in traditional microbiological methods. The time required for analysis is much shorter compared to cultural methods; responses to the presence of specific microorganisms can be obtained within 3–24 h, depending on whether an enrichment step is carried out. The method is also highly specific to the target microorganism, avoiding false-positive results due to misidentification.

On the other hand, qPCR methods have some disadvantages, which limit its application in the food sector. Since the target for amplification is mainly represented by DNA, a molecule characterized by high stability even after cell death, a positive result from PCR does not necessarily indicate the presence of living microorganisms. This aspect is extremely important for processed foods, particularly thermally processed ones, where production kills foodborne pathogens. This weakness of PCR has been addressed in different ways by researchers, such as using RNA analysis to quantify living populations (Rantsiou *et al.*, 2008), and the use of ethidium monoazide (EMA) for the selective quantification of live populations (Nogva *et al.*, 2003). Furthermore, it is generally accepted that if positive results are obtained after an enrichment step, this implies the presence of living populations that were able to multiply and their DNA was subsequently amplified by PCR (Rossen *et al.*, 1992). Lastly, there is a need for a significant investment in equipment and trained personnel, notwithstanding that in recent years the market has seen a significant decrease in the prices of thermal cyclers for DNA amplification (Cocolin and Rantsiou, 2012).

The most critical aspects for the correct implementation of qPCR protocols in food microbiology are the extraction of DNA and RNA from complex food matrices and the quantification procedure.

10.2.1 Extraction of nucleic acids from food matrices

Studies in the 1990s clearly highlighted that fats, proteins, culture medium components and specific classes of compounds due to the origin of the food (polyphenols in vegetables or blood in meat and meat products) can dramatically influence the efficiency of the amplification reaction, thereby increasing the risk of false-negative results (Rossen *et al.*, 1992). The problem of inhibitors in foods is particularly relevant to qPCR. The presence of compounds able to interfere with DNA polymerase activity can prevent amplification and thereby cause the number of target microorganisms present to be underestimated. Thus, there is a need for dedicated (often labor-intensive) sample preparation to recover cells from foods and acquire good quality DNA for amplification. Currently, sample preparation is considered the bottleneck in the quantification of microorganisms using qPCR. The cell separation and concentration methods often used in extraction procedures may increase the amount of inhibitory substances from the food matrix (Brehm-Stecher *et al.*, 2009). This can jeopardize the efficient functioning of qPCR, which is prone to inhibition, and in turn might lead to false-negative results. Choosing an appropriate sample preparation approach is a critical issue prior to qPCR detection. Sample preparation methods should be selected with care depending upon the food matrices under analysis and the sensitivity of the test required. Choosing the best protocol is always a compromise between cost, rapidity and universal application with regard to the food matrix.

10.2.2 Quantification of the target microorganism in food

A prerequisite for qPCR in quantifying pathogens in foods is the generation of calibration curves. Calibration curves correlate colony-forming units (cfu) per gram or per milliliter, which is commonly used in traditional microbiological analysis, with the $C(t)$ value, the threshold cycle in qPCR. In this way, qPCR results can be translated into a value relevant to microbiological criteria. It is important to underline that calibration curves should be created in food matrices. One should avoid creating standard curves to quantify a microorganism in a specific food, by diluting a pure culture, extracting the DNA from each dilution and then performing qPCR. The influence of the food matrices on DNA extraction and during the amplification step is not taken into account in this approach, thereby resulting in underestimation of the microbial load. Moreover, according to the internationally accepted method of reporting microbial counts, standard curves cannot be constructed from DNA dilutions, as is the usual approach in molecular biology applications, and the results should not be reported as genome equivalents. Generally the limit of quantification for foodborne pathogens has been reported to be 10^3–10^4 cfu/g or ml. These quantification limits of qPCR are too high for most practical applications, since for most samples taken along the food chain, the contamination is low (usually less than 100 cfu/g). For this reason, in food safety the main application of qPCR is for qualitative, presence/absence testing, performed after an enrichment step. Only in a few cases (based on the number of papers published in the last five years) have authors aimed to develop qPCR protocols that could be used for quantification purposes without enrichment (Cocolin *et al.*, 2011).

10.3 Transcriptomics

Recent advances in nucleic acid analysis have created a number of new tools that can be exploited for food microbiology. One aspect that can be explored is the behavior of foodborne pathogens in food matrices, which takes into account the expression of specific traits *in situ*, namely virulence and stress response. The virulence expression of pathogenic bacteria is a complex process and its elucidation is still a challenge. The ability to rapidly determine the virulence potential of foodborne pathogens is of great importance for controlling and preventing the presence of pathogenic bacteria in foods. As a consequence, it is important to understand the causes and mechanisms of virulence expression in order to provide a safe food supply (Rantsiou *et al.*, 2011).

Transcriptomics analyse the RNA transcripts produced by the genotype at a given time and provide a link between the genome, the proteome and the cellular phenotype. This approach gives a better understanding of the molecular basis of virulence expression and further insight into the complex expression events involved.

Transcriptomics can be used to investigate and explain microbial behavior under different environmental conditions. Parameters such as chemical composition, physicochemical parameters such as pH and water activity, the presence of other microorganisms, conditions of storage and transport, and the preservation techniques employed to improve food safety, influence the behavior of foodborne pathogens. It is hypothesized that the physiological state of a microorganism when found in food will play a role in its ability to cause disease to humans. Pathogens are known to possess mechanisms that allow them to respond to various stresses and it has been demonstrated that stress response is connected to virulence. The molecular networks activated to respond to various stresses are interconnected with pathways that play a role in virulence (Chaturongakui and Boor, 2006). In food, microorganisms are subjected to various stresses and, through transcriptomics, it is possible to study the effect of food on microbial stress response and virulence. So far, the application of transcriptomics in studying the behavior of pathogens in food has shown that a high degree of intraspecies biodiversity exists, suggesting that not all representatives of one pathogenic species behave in the same way (Rantsiou *et al.*, 2012b). This finding implies that in the future it will be necessary to shift from the enumeration of pathogenic microorganisms per unit of food to determine acceptability, towards understanding the behavior or the virulence potential of these microorganisms in a particular type of food.

The most frequently used technologies in transcriptomics are microarrays, which evaluate global gene expression events and provide information on the differentially expressed genes, and reverse transcription qPCR (RT-qPCR), which quantifies (and confirms) the differential expression of selected genes. Generally, it is recognized that qPCR is the appropriate method for studying a moderate number of genes in a number of samples. In contrast, microarrays provide the possibility for whole genome discovery experiments in a small number of samples (VanGuilder *et al.*, 2008). Recently, more function-focused sub-arrays have been developed that target specific cellular functions, such as virulence regulons for pathogens (Rantsiou *et al.*, 2012a). This trend allows application to a larger number of samples and facilitates interpretation of the data obtained.

10.3.1 Challenges in transcriptomic studies in food matrices

Moving from the study of gene expression in pure cultures (*in vitro* studies) to the study of gene expression of microorganisms in real food samples, or any other complex environmental sample (*in situ* studies), presents considerable difficulties. Most foods encompass microbial communities with representatives of different species of bacteria, yeasts and molds. So far, the common approach used to study the effect of the food matrix on one species specifically, has been the use of model matrices, which are either sterile or contaminated at very low levels. It is possible to investigate the behavior of the microorganism of interest inoculated in these matrices. This approach has primarily been employed in liquid foods, such as wine and milk, which are easily heat or filter sterilized. Solid food matrices, such as meat-based products, remain a challenge, due to the difficulties in eliminating the autochthonous microbiota before conducting gene expression experiments.

Perhaps the most critical parameter still requiring improvement is the extraction of high quality RNA. This issue is being addressed by different scientists (Monnet *et al.*, 2008; Ulvé *et al.*, 2008; Olesen *et al.*, 2010) and the common outcome of these studies is that RNA extraction has to be optimized for the combination of food matrix and microorganism being studied, and no universal protocol exists. It is also important to recognize that certain microbial groups, primarily foodborne pathogens, when present in a food, constitute the minority of the total microbial community. This means that to mimic real contamination levels and study gene expression, foodborne species should be inoculated at very low levels, in most cases not above 10^2 cfu/g or ml of food. This has not yet been achieved and so far, studies concerning pathogenic species have been conducted with significantly higher inocula (Olesen *et al.*, 2010; Duodu *et al.*, 2010; Rieu *et al.*, 2009; Rantsiou *et al.*, 2012a, 2012b).

10.4 Conclusions and future trends

The availability of new, molecular biology-based methodological approaches has enabled the development of tools for food microbiology and food safety, which can be used to accurately identify the presence of pathogenic microorganisms in the food chain. Although their behavior in terms of the expression of genes related to stress and virulence in foods is now comprehensively understood, so far these new possibilities have only been applied on a limited scale. This is partly due to the technical difficulty of extracting good quality DNA and RNA from food matrices, and partly owing to the sensitivity and specificity requirements described above. The detection, and more importantly, the quantification achieved by qPCR, is still not adequate to guarantee the safety of foodstuffs. Because the amplification process can also pick up dead cells, the need for an enrichment step remains. An enrichment step negates the most important advantage of qPCR, that is, quantification. Nevertheless, it should be underlined that future efforts by scientists to optimize protocols for the detection and quantification of low numbers of pathogenic bacteria are very necessary. Only with robust and sensitive quantification methods can the prevalence of foodborne pathogens in the food chain be investigated and understood.

Interestingly, apart from the application of qPCR in foods, in the last couple of years its potential to quantify pathogenic bacteria in meat and dairy processing plants has been explored (Alessandria *et al.*, 2010; Melero *et al.*, 2011; Osés *et al.*, 2010). The authors of these studies underlined that the application of qPCR could allow FBOs to better comprehend the spread of specific pathogens in processing plants, thereby enabling the implementation of corrective actions to eliminate or decrease the risk associated with their presence in the final product (Alessandria *et al.*, 2010).

The study of microbial behavior has too often been performed in standardized laboratory conditions, without taking into consideration that pathogenic bacteria in foods face a number of environmental stresses, which can have a significant influence on their virulence potential. This issue has been addressed in the last couple of years by several researchers and the results from *in situ* studies are sometimes in disagreement with previous information acquired *in vitro*. The scientific community should be ready to accept that well-established knowledge may have to be reconsidered in light of the new evidence from *in situ* approaches. RT-qPCR is a method, which can be applied relatively easily to food matrices. However, due to the limited number of genes that can be studied, a restricted view of the microorganism's physiology is obtained. On the other hand, microarray technology, which allows whole genome analysis, still suffers from important drawbacks, mainly due to the lack of specificity, preventing its application in most food matrices unless proper preparation is carried out to eliminate the natural microbiota. From results available so far, there seems to be significant intraspecies heterogeneity concerning stress response and virulence gene expression. This heterogeneity emphasizes the need to move away from a determination of the number of pathogens in food towards quantifying their behavior. Providing information to the risk assessment evaluation regarding survival and/or virulence potential should be considered. Lastly, the strain and its provenience and/or its history during food production need to be taken into consideration in risk assessment, since they influence the stress response and virulence potential. Methods that determine gene expression *in situ* will provide the necessary information for risk assessment, which considers not only the numbers of microorganisms but also their behavior.

10.5 References

ABEE, T., VAN SCHAIK, W. and SIEZEN, R.J. (2004) Impact of genomics on microbial food safety, *Trends Biotechnol*, 22, 653–660.

ALESSANDRIA, V., RANTSIOU, K., DOLCI, P. and COCOLIN, L. (2010) Molecular methods to assess *Listeria monocytogenes* route of contamination in a dairy processing plant, *Int J Food Microbiol*, 141, S156–S162.

BAE, D., CROWLEY, M.R. and WANG, C. (2011) Transcriptome analysis of *Listeria monocytogenes* grown on a ready-to-eat meat matrix, *J Food Prot*, 74, 1104–1111.

BREHM-STECHER, B., YOUNG, C., JAYKUS, L.A. and TORTORELLO, M.L. (2009) Sample preparation: The forgotten beginning, *J Food Prot*, 72, 1774–1789.

CHATURONGAKUI, S. and BOOR, K.J. (2006) σ^B activation under environmental and energy stress conditions in *Listeria monocytogenes*, *Appl Environ Microbiol*, 72, 5197–5203.

COCOLIN, L. and RANTSIOU, K. (2012) Quantitative polymerase chain reaction in food microbiology. In M. Filion (Ed), *Quantitative Real Time PCR in Applied Microbiology*, pp. 149–160, Caister Academic Press, Norfolk, UK.

COCOLIN, L., RAJKOVIC, A., RANTSIOU, K. and UYTTENDAELE, M. (2011) The challenge of merging food safety diagnostics needs with quantitative PCR platforms, *Trends Food Sci Tech*, 22, S30–S38.

DUODU, S., HOLST-JENSEN, A., SKJERDAL, T., CAPPELIER J.M., PILET, M.F. *et al.* (2010) Influence of storage temperature on gene expression and virulence potential of *Listeria monocytogenes* strains grown in salmon matrix, *Food Microbiol*, 27, 795–801.

LEISTNER, L. (1995) Principles and applications of hurdle technology. In G.W. Gould (Ed), *New Methods of Food Preservation*, pp. 1–21, Blackie Academic and Professional, Glasgow, UK.

MARTINEZ, N., MARTIN, M.C., HERRERO, A., FERNANDEZ, M., ALVAREZ, M.A. *et al.* (2011) qPCR as a powerful tool for microbial food spoilage quantification: Significance for food quality, *Trends Food Sci Tech*, 22, 367–376.

MCKILLIP, J.L. and DRAKE, M. (2004) Real-time nucleic acid based detection methods for pathogenic bacteria in food, *J Food Prot*, *67*, 823–832.

MELERO, B., COCOLIN, L., RANTSIOU, K., JAIME, I. and ROVIRA, J. (2011) Comparison between conventional and qPCR methods for enumerating *Campylobacter jejuni* in a poultry processing plant, *Food Microbiol*, 28, 1353–1358.

MONNET, C., ULVÉ, V, SARTHOU, A.S. and IRLINGER, F. (2008) Extraction of RNA from cheese without prior separation of microbial cells, *Appl Environ Microbiol*, 74, 5724–5730.

NOGVA, H., DROMTORP, S.M., NISSEN, H. and RUDI, K. (2003) Ethidium monoazide for DNA-based differentiation of viable and dead bacteria by 5′-nuclease PCR, *Biotechniques*, 34, 804–813.

OLESEN, I., THORSEN, L. and JESPERSEN, L. (2010) Relative transcription of *Listeria monocytogenes* virulence genes in liver pates with varying NaCl content, *Int J Food Microbiol*, 141, S60–S68.

OLIVER, J.D. (1993) Formation of viable but nonculturable cells. In S. Kjelleberg (Ed.), *Starvation in Bacteria*, pp. 239–272, Plenum Press, New York, NY, USA.

OSÉS, S.M., RANTSIOU, K., COCOLIN, L., JAIME, I. and ROVIRA, J. (2010) Prevalence and quantification of Shiga-toxin producing *Escherichia coli* along the lamb food chain by quantitative PCR, *Int J Food Microbiol*, 141, S163–S169.

POSTOLLEC, F., FALENTIN, H., PAVAN, S., COMBRISSON, J. and SOHIER, D. (2011) Recent advances in quantitative PCR (qPCR) applications in food microbiology, *Food Microbiol*, 28, 848–861.

RANTSIOU, K., ALESSANDRIA, V., URSO, R., DOLCI, P. and COCOLIN, L. (2008) Detection, quantification and vitality of *Listeria monocytogenes* in food as determined by quantitative PCR, *Int J Food Microbiol*, 121, 99–105.

RANTSIOU, K., GREPPI, A., GAROSI, M., ACQUADRO, A., MATARAGAS, M. *et al.* (2012a) Strain dependent expression of stress response and virulence genes of *Listeria monocytogenes* in meat juices as determined by microarrays, *Int J Food Microbiol*, 152, 116–122

RANTSIOU, K., MATARAGAS, M., ALESSANDRIA, V. and COCOLIN, L. (2012b) Expression of virulence genes of *Listeria monocytogenes* in food, *J Food Safety*, 32, 161–168.

RANTSIOU, K., MATARAGAS, M., JESPERSEN, L. and COCOLIN, L. (2011) Understanding the behavior of foodborne pathogens in the food chain: New information for risk assessment analysis, *Trends Food Sci Tech*, 22, S21–S29.

RIEU, A., GUZZO, J. and PIVETEAU, P. (2009) Sensitivity to acetic acid, ability to colonize abiotic surfaces and virulence potential of *Listeria monocytogenes* EGD-e after incubation on parsley leaves, *J Appl Microbiol*, 108, 560–570.

ROSSEN, L., NORSKOV, P., HOLMSTROM, K. and RASMUSSEN, O.F. (1992) Inhibition of PCR by components of food samples, microbial diagnostic assays and DNA-extraction solutions, *Int J Food Microbiol*, 17, 37–45.

ROWAN, N.J. (2004) Viable but non-culturable forms of food and waterborne bacteria: *Quo vadis?*, *Trends Food Sci Tech*, 15, 462–467.

ULVÉ, V.M., MONNET, C., VALENCE, F., FAUQUANT, J., FALENTIN, H. *et al.* (2008) RNA extraction from cheese for analysis of *in situ* gene expression of *Lactococcus lactis*, *J Appl Microbiol*, 105, 1327–1333.

VANGUILDER, H.D., VRANA, K.E. and FREEMAN, W.M. (2008) Twenty-five years of quantitative PCR for gene expression analysis, *Biotechniques*, 44, 619–626.

11

New approaches in microbial pathogen detection

L. N. Kahyaoglu and J. Irudayaraj, Purdue University, USA

DOI: 10.1533/9780857098740.3.202

Abstract: Viruses are common causes of foodborne outbreaks. Viral diseases have low fatality rates but transmission to humans via food is important due to the high probability of consuming fecally contaminated food or water because of poor food handling. Because of the low infectious doses of some foodborne viruses, there is a need for standardization and the development of new sensitive methods for detecting viruses. The focus is on molecular and non-molecular approaches, and emerging methods for the detection of foodborne viruses. The detection of noroviruses, hepatitis A and E viruses, rotaviruses and adenoviruses will be discussed. The chapter will conclude with insights into future research directions.

Key words: foodborne outbreak, virus, detection, food safety, nanotechnology.

11.1 Introduction

An estimated 40 million illnesses each year are caused by foodborne pathogens, including 5.2 million (13%) due to bacteria, 2.5 million (7%) due to parasites and 30.9 million (80%) due to viruses (Mead *et al.*, 1999). In the last decade, the occurrence of gastroenteritis in humans as a result of consumption of foods contaminated by viruses has increased (WHO, 2000). The transmission of viruses has been predominantly associated with the consumption of shellfish, mainly, raw oysters (Koopmans and Duizer, 2004; Widdowson *et al.*, 2005), which have been contaminated by polluted water or virus-infected food handlers (Bosch *et al.*, 2011). Common symptoms of viral gastroenteritis include vomiting and diarrhea (FAO/WHO, 2008). Foodborne viruses can be divided into three categories based on disease symptoms: those that cause gastroenteritis (noroviruses, rotaviruses and adenoviruses), those that cause fecal-orally transmitted hepatitis (hepatitis A and E viruses), and those that cause other illnesses after they migrate to other organs, such as the central nervous system or the liver (enteroviruses). Of these, noroviruses and the hepatitis A virus have been recognized as the most important human foodborne pathogens in terms of the number of outbreaks reported and the number of people affected in the world (Koopmans and Duizer, 2004; Cook and Rzezutka, 2006; Verhoef *et al.*, 2008). Even though these viruses are the major cause of foodborne outbreaks, there is still an urgent need for standardization and the validation of detection methods at the national and international levels (FAO/WHO, 2008). In this chapter, an overview of existing detection methods for foodborne viruses are presented including non-molecular and molecular approaches as well as promising emerging methods for virus detection. The detection of noroviruses, hepatitis A and E viruses, rotaviruses and adenoviruses will be discussed as these viral agents have a high potential for foodborne outbreaks. The rest of this section briefly gives the essential characteristic features of these viruses.

Noroviruses (NVs) comprise a genus in the family Caliciviridae. They are mostly linked to non-bacterial gastroenteritis in humans and are estimated to cause 93% of the food-related outbreaks of gastroenteritis in the United States (Widdowson *et al.*, 2005; Fankhauser *et al.*, 2002). An NV is a small round virus, 27 to 35 nm in diameter, with a single-stranded, positive-sense, polyadenylated RNA genome of 7400 to 7500 nucleotides (Atmar and Estes, 2001). These small viruses show high genomic diversity and antigenic variation within five genogroups (GI, GII, GIII, GIV and GV) based on the genome sequence of the RNA-dependent RNA polymerase (RdRp) and the capsid regions (Vinjé *et al.*, 2004).

Hepatoviruses (or hepatitis A viruses or HAVs) comprise a genus in the family Picornaviridae with a diameter in the range between 27 and 32 nm. They were first identified by electron microscopy in 1973 (Lemon and Robertson, 1993; Koopmans *et al.*, 2002). They are small, non-enveloped, spherical viruses with a single (positive-) stranded RNA genome of approximately 7.5 kb in length (Koopmans *et al.*, 2002). Each year, approximately 1.4 million people become ill with HAV costing $1.5–3.0 billion worldwide (WHO, 2000).

The hepatitis E virus (HEV), one of the major causes of viral hepatitis other than HAV, causes infection with a high rate of mortality particularly in pregnant women (Widen *et al.*, 2011; Ahn *et al.*, 2005). A HEV, similar to a HAV, is a small, non-enveloped, positive sense, single-stranded RNA virus, reclassified in the genus *Hepevirus* of the family Hepeviridae although previously classified as a member of Caliciviridae family (Berke and Matson, 2000; Gyarmati *et al.*, 2007). The primary source of infection is the consumption of fecally contaminated drinking water in developing countries (Koopmans *et al.*, 2002). Based on geographical origin, all isolated HEV strains can be identified into four genotypes as Asian/African, Chinese, Mexican or US/European (Emerson and Purcell, 2001). Among these genotypes, genotypes 1 and 2 only infect humans, while genotypes 3 and 4 appear to infect other hosts, particularly pigs, and are associated with zoonotic transmission (Gyarmati *et al.*, 2007).

Rotaviruses comprise a genus in the family *Reoviridae* and are one of the major causes of acute diarrhea in infants and young children with high morbidity and mortality, especially in developing countries (Kapikian *et al.*, 1996). These round, non-enveloped viruses are estimated to cause the death of more than 600 000 children worldwide every year and infect almost all children under five years of age (Pineda *et al.*, 2009; Gutierrez-Aguirre *et al.*, 2008).

Adenoviruses are non-enveloped viruses with double-stranded DNA (Vasickova *et al.*, 2005). Human adenoviruses are the only human enteric DNA viruses and are often detected in association with other human enteroviruses or hepatitis A viruses (Rigotto *et al.*, 2005). Outbreaks have caused gastroenteritis in children (Vasickova *et al.*, 2005). Adenovirus types 40 and 41 cause gastroenteritis when transmitted through a fecal–oral cycle (Koopmans *et al.*, 2002).

11.2 Detection methods

HAVs and other enteric viruses may be found in large numbers in clinical samples ($\geq 10^6$ virus particles per gram of stool); however, they are usually found in much lower numbers in food, e.g. 0.2–224 particles per 100 g shellfish meat (Sanchez *et al.*, 2007). The infectious dose of HAVs and NVs is estimated to be as low as 10–100 infectious viral particles even though the ingestion of thousands of cells is required for bacterial infection to occur with the same probability (Sair *et al.*, 2002; Gerba, 2006; Guevremont *et al.*, 2006). Unlike bacterial pathogens, viruses cannot multiply in foods, making the traditional food microbiological techniques of cultural enrichment and selective plating inapplicable (D'Souza and Jaykus, 2006). Therefore, methods with high reliability and sensitivity are required for viral detection. In the sections below we discuss some of the currents methods.

11.2.1 General approaches

Conventional assay systems to detect enteric viruses in clinical specimens cannot be directly used for food (Rodriguez-Lazaro *et al.*, 2007). In general, electron microscopy, tissue cultures and immunological and molecular methods are used to detect viruses in food. Viruses were diagnosed historically by scanning a stool suspension under an electron microscope (EM) (Koopmans and Duizer, 2004). Many of the small round viruses, including HAVs, astroviruses, noroviruses, sapovirus and parvoviruses, were first discovered through the use of EM (Greening, 2006). EM is fairly insensitive, labor intensive and requires a minimum of 10^6 virus particles per milliliter of sample for detection in patient fecal samples, thus, using this method, detecting viruses at low levels in contaminated food, water and environmental samples is not possible (Koopmans and Duizer, 2004; Seymour and Appleton, 2001).

Detection by cell culture depends on cytopathic effects, and virus quantification is performed by plaque assay, the most probable number or 50% tissue culture infectious dose ($TCID_{50}$) (Bosch *et al.*, 2011). Cell-culture-based assay can differentiate between infectious and non-infectious viruses; nevertheless it is limited and not practical, mainly due to the lack of sensitivity, the long analysis time and the lack of susceptible cell lines for many epidemiologically important enteric viruses (Casas and Sunen, 2001; Verhoef *et al.*, 2008). Even though these assays are commonly used to enumerate levels of viable polioviruses and adenoviruses, they are inadequate for the detection of the two most important foodborne viruses, HAVs and NVs, since neither of these replicate or express themselves efficiently in cell cultures (Goyal, 2006; Jiang *et al.*, 2004; Koopmans and Duizer, 2004). Thus, HAVs and NVs have been detected conventionally using EM and enzyme-linked immunosorbent assay (ELISA) but even these methods are insensitive, lengthy and expensive (Morales-Rayas *et al.*, 2010).

Non-culture-based detection methods, such as immunoassays, have been developed to detect viruses over the years (Lees, 2000). Although immunoassays, such as ELISA, have been used to detect viruses in water and HAVs in shellfish, reports are very limited and not always successful (Lees, 2000). The limited success of this approach is probably due to the lack of sensitivity of the immunoassay and like EM requires a thousand or more virus particles for a positive result (Kogawa *et al.*, 1996). Therefore, new approaches have focused on molecular methods as these techniques for detecting enteric viruses are faster and more sensitive compared to infectivity tests performed with *in vitro* cell cultures or with immunological methods, even though molecular methods cannot discriminate between infectious and non-infectious particles (Green and Lewis, 1999; Morales-Rayas *et al.*, 2010).

11.2.2 Molecular approaches

Several molecular methods using nucleic acid amplification have been developed for virus detection in food (Jean *et al.*, 2003). In recent years, polymerase chain reaction (PCR)-based methods in particular, have become the gold standard for virus detection in food due to their high sensitivity, specificity and potential to detect even a single virus particle (Bosch *et al.*, 2011; Martinez-Martinez *et al.*, 2011; Richards *et al.*, 2003; Cook and Rzezutka, 2006). Selected examples with detection limits are listed in Table 11.1.

Reverse transcription PCR (RT-PCR)

Reverse transcription PCR (RT-PCR), a modified form of PCR that allows the amplification of viral RNA, is currently the most sensitive and widely used method for foodborne virus detection (Casas and Sunen, 2001; Morales-Rayas *et al.*, 2010). However, the application of this technique for routine analysis of food matrices is elaborate due to the need for sample concentration and the presence of residual food-related PCR inhibitors (Sair *et al.*, 2002). Since only low numbers of viruses are present in food, inhibition is a more serious issue (Morales-Rayas *et al.*, 2010). Therefore, several methods have been developed to concentrate and purify viruses and remove inhibitors from food samples before RT-PCR (Dubois *et al.*, 2002; Croci *et al.*, 2008).

The sample preparation procedures for detecting viruses in food typically involve one or more of the following: (i) elution of the virus particles from the

Table 11.1 Selected examples of molecular approaches for detection of foodborne viruses

Type of virus	Method of detection	Detection limit	Samples tested	References
Norovirus	TaqMan qRT-PCR	0.01 PDU	Clinical	Lamhoujeb *et al.*, 2009
	Real-time NASBA	0.01 PDU		
	RT-PCR	1 RT PCRU/25 g	Green onion	Guevremont *et al.*, 2006
	RT-PCR	1–10 PCRU/mL	Ham	Kim *et al.*, 2008b
	Nested PCR			
Hepatitis A	Duplex qRT-PCR	10 PFU/1.5 L	Bottled water	Blaise-Boisseau *et al.*, 2010
		100 PFU/1.5 L	Tap water	
		50 PFU/25 g	Fresh raspberries	
		100 PFU/25 g	Frozen raspberries	
	TaqMan RT-PCR	14 PFU/g	Tomato sauces	Love *et al.*, 2008
		33 PFU/g	Blended strawberries	
	Nested RT-PCR	1 $TCID_{50}$/10 g	Mollusks	Croci *et al.*, 1999
Hepatitis E	TaqMan qRT-PCR	1.2 PID_{50}	Water	Jothikumar *et al.*, 2006
Rotavirus	qRT-PCR	125 PFU/g	Oyster	Kittigul *et al.*, 2008
	NASBA-ELISA	0.2 PFU	Water	Jean *et al.*, 2002a, 2002b
		15 PFU	Sewage effluent	
Adenovirus	Nested PCR	1.2 PFU/g	Oysters	Rigotto *et al.*, 2005
	Conventional PCR	1.2×10^2 PFU/g		
	ICC-PCR	1.2×10^2 PFU/g		
	Nested mPCR	1 copy of adenovirus DNA/ PCR reaction	Sewage Shellfish	Formiga-Cruz *et al.*, 2005

Note: PFU: plaque-forming unit; PCRU: RT-PCR amplifiable unit; $TCID_{50}$: 50% tissue culture infective dose; PID: pig infectious dose.

food using a variety of buffers and solutions including solutions of glycine and sodium chloride, borate and beef extract, saline and beef extract, and beef extract alone; (ii) extraction with an organic solvent, most commonly with Freon to remove insoluble or poorly soluble organic compounds in the water; (iii) concentration of the viruses using sedimentation by antibody or ligand capture, flocculation, ultra-centrifugation or precipitation (commonly polyethylene glycol precipitation); and (iv) extraction of viral nucleic acids (there are two main approaches using phenol: chloroform extraction and guanidinium isothiocyanate extraction) (Cook and Rzezutka, 2006; Goyal, 2006; Rodriguez-Lazaro *et al.*, 2007). Various strategies have been proposed to improve the performance of each step over the years.

There are several commercial kits for nucleic acid purification, which are reliable, produce reproducible results and are easy to use. Most of these kits are based on guanidinium lysis and the capture of nucleic acids on a column or bead of silica (Bosch *et al.*, 2011). However, sample preparation methods still require improvement to isolate viral particles from diverse food matrices without decreasing the sensitivity of the molecular method used for detection (Morales-Rayas *et al.*, 2010).

The sensitivity and specificity of RT-PCR assays depends mainly on primer selection (Atmar and Estes, 2001). The major obstacle in NV detection with PCR arises from the very high genomic diversity of NV since new variants continue to evolve constantly (Widen *et al.*, 2011). Therefore, it is difficult to select a single or even a small number of probes that can detect all possible NV variants (Atmar and Estes, 2001). Although ORF1 of the RdRp gene has been targeted in most of the assays (Nakayama *et al.*, 1996; Jiang *et al.*, 1999), the ORF1-ORF2 region has also been shown to be well conserved and is used in several assays (Katayama *et al.*, 2002; Hohne and Schreier, 2004; Jothikumar *et al.*, 2005b). One of the first enteric viruses detected by RT-PCR was HAV (Jansen *et al.*, 1990). The VP1 capsid region was previously commonly targeted by primers in HAV detection; however, nowadays the 5′ non-coding region is highly preferred for targeting. It has similar performance as VP1, approximately 1 RNA copy per reaction (Sanchez *et al.*, 2007). For HEV detection, various specific sets of primers have been developed to amplify conserved regions within ORF1, ORF2 and ORF3 (Enouf *et al.*, 2006). Most of the RT-PCR assays developed for rotaviruses target the structural genes VP4, VP6 and VP7 (Atmar, 2006). The hexon gene in adenoviruses is most commonly used as the target in PCR assays; it has been shown to be reactive in all adenovirus species (Jothikumar *et al.*, 2005a; Atmar, 2006). More recently, a FRET-based real-time assay, which amplifies the adenovirus fiber gene, was described. It showed slightly better performance in terms of detection limits of AdV40 and AdV41 compared to TaqMan assays (Jothikumar *et al.*, 2005a).

The major limitation of RT-PCR is its inability to distinguish between infectious and non-infectious viruses (Richards, 1999). Integrated cell culture PCR (ICC-PCR) and ICC/strand-specific RT-PCR have been proposed to compensate for this problem (Atmar, 2006; Jiang *et al.*, 2004). ICC/strand-specific RT-PCR is a combination of cell culture and molecular biology-based methods, which requires initial propagation of infectious virus particles in a cell culture and the detection of a negative-strand RNA replicative intermediate as an indicator of viral replication (Jiang *et al.*, 2004). The limitations of RT-PCR were eliminated in environmental samples by increasing the equivalent sample volume and thereby reducing the effects of inhibitory compounds (Reynolds *et al.*, 1996). ICC-PCR and ICC/strand-specific RT-PCR assays targeting the VP3 genes, which code for a major HAV capsid protein, have been developed to detect viruses in water (Jiang *et al.*, 2004). The ICC/strand-specific RT-PCR used in this study was demonstrated to be a novel, rapid, sensitive and reliable method, since it can detect infectious HAVs at inoculation level of 100 $TCID_{50}$ per flask within four days in water samples.

Even though RT-PCR is a rapid and sensitive method and can detect viruses that are difficult or impossible to culture (Casas and Sunen, 2001), several different types of RT-PCR have been developed to improve the specificity and sensitivity of the standard method for foodborne virus detection such as nested RT-PCR (Love *et al.*, 2008; Croci *et al.*, 1999) and multiplex RT-PCR (Rosenfield and Jaykus, 1999; Formiga-Cruz *et al.*, 2005; Coelho *et al.*, 2003).

Nested PCR

In nested PCR, two different primer pairs are used successively to amplify a target sequence (Haqqi *et al.*, 1988). Nested PCR was developed to ensure detection specificity, to minimize false-positive results and to enhance the amplification signal (Rigotto *et al.*, 2005). It has been widely used in the performance evaluation and verification of different PCR-based methods as well as viral extraction, concentration and purification (Kim *et al.*, 2008a, 2008b; Di Pinto *et al.*, 2003; Jothikumar *et al.*, 2005b).

The superior sensitivity of nested PCR over other methods has been demonstrated in several studies (Croci *et al.*, 1999; Rigotto *et al.*, 2005; Love *et al.*, 2008). Nested PCR gives a more sensitive and specific identification of HAV at concentrations as low as 1 $TCID_{50}$/10 g of mollusk compared to 10^3–10^4 TCID/10 g of mollusk after one round of PCR (Croci *et al.*, 1999). It had a higher level of sensitivity in shellfish compared to conventional PCR and ICC-PCR when detecting adenoviruses (Rigotto *et al.*, 2005) (Table 11.1). Recently, TaqMan RT-PCR has been used to detect HAV RNA from artificially inoculated tomato sauce and blended strawberries (Love *et al.*, 2008). The lower limits of HAV detection were reported as 14 PFU/g (plaque-forming units per gram) of tomato sauce and 33 PFU/g of blended strawberries at initial seeding levels. Moreover, the nested RT-PCR was not inhibited by undiluted final RNA extracts of tomato sauce or blended strawberries unlike TaqMan RT-PCR.

The sensitivity of standard RT-PCR was further increased when combined with semi-nested or nested PCR by using an aliquot of the product from the primary RT-PCR as a template for the second round of amplification (O'Connell, 2002; Abad *et al.*, 1997). Nested multiplex real-time PCR (mRT-PCR) has also been developed to provide a highly sensitive, rapid and cost-efficient approach for HAV, adenovirus and enterovirus detection in urban sewage and shellfish (Formiga-Cruz *et al.*, 2005). This method was able to detect as little as one copy of adenovirus DNA, and ten copies of both enterovirus and HAV RNA, which was shown to be similar to the previously determined sensitivities of monoplex PCR with 1–10 viral particles for adenoviruses, and 5–10 viral particles for enteroviruses both in sewage and shellfish samples (Formiga-Cruz *et al.*, 2005). Most recently, RT nested PCR targeting the VP7 gene of rotaviruses in naturally contaminated oyster samples was shown to give the highest sensitivity and the lowest detection limit of 125 PFU/g of oyster with acid adsorption–alkaline elution (Kittigul *et al.*, 2008).

Multiplex PCR

In multiplex PCR, two or more primer sets are used simultaneously in the amplification of different target sequences in a single tube (Chamberlain *et al.*, 1988). Thus, this method could be used for the detection of more than one virus in a single reaction tube (Rosenfield and Jaykus, 1999; Coelho *et al.*, 2003; Beuret, 2004). A multiplex reverse transcription polymerase chain reaction (mRT-PCR) method has been described for the simultaneous detection of the human enteroviruses, HAV and NV (Rosenfield and Jaykus, 1999). Detection limits lower than 1 infectious unit (poliovirus type 1 (PV1) and HAV) or RT-PCR-amplifiable unit (NV) for all viruses were obtained by the multiplex method. In a similar vein, mRT-PCR has been developed to concentrate and purify HAV, PV1 and simian rotaviruses (RV-SA11) simultaneously from experimentally seeded oysters (Coelho *et al.*, 2003). However, this method could not detect the three viruses simultaneously when tested on experimentally contaminated raw oysters. This was attributed to the low concentration of viral RNA present in the oyster extract as a result of an ineffective extraction method.

Quantitative real-time PCR

Quantitative real-time PCR (qRT-PCR) is used to amplify and quantify simultaneously a targeted DNA molecule by using DNA-binding fluorophores or, commonly, by specific fluorescently labeled oligoprobes (Atmar, 2006). In recent years, qRT-PCR has been widely used in food virology as the most promising nucleic acid detection method, since it offers several advantages over conventional RT-PCR, including high sensitivity, the possibility of simultaneous amplification, detection and quantification of the target nucleic acids in a single step, and with minimum risk of carry-over contamination through the use of a closed system (Mackay *et al.*, 2002; Bosch *et al.*, 2011; Houde *et al.*, 2007). Sensitive and specific detection with real-time PCR is achieved using novel fluorescent technology probes (Espy *et al.*, 2006). In qPCR assays, three types of fluorescently labeled target-specific probes have been used most often: TaqMan probes, molecular beacons and fluorescence resonance energy transfer (FRET) hybridization probes (Sanchez *et al.*, 2007). These detection methods all depend on the transfer of light energy between two adjacent dye molecules, a process known as fluorescence resonance energy transfer (Espy *et al.*, 2006).

TaqMan-based assays have been widely used to detect HAVs using qPCR in recent years (Sanchez *et al.*, 2007). These assays combine a specific linear dual-labeled oligoprobe in the TaqMan master mix to eliminate the need for post amplification steps and also offer the opportunity of multiplexing amplification reactions (Houde *et al.*, 2007). Several studies targeting the 5′ non-coding region (NCR) have been performed with TaqMan qRT-PCR to detect HAVs (Costa-Mattioli *et al.*, 2002; El Galil *et al.*, 2005; Jothikumar *et al.*, 2005b, Costafreda *et al.*, 2006). The detection limits ranged from one to five copies per reaction. In NV detection, real-time TaqMan-PCR targeting in the well conserved ORF1-ORF2 region has been developed (Hohne and Schreier, 2004; Jothikumar *et al.*, 2005b). These TaqMan RT-PCR assays were able to detect as few as 100 genomic equivalents of different NV strains, including subtypes of GI and GII, rapidly, sensitively and reliably. A TaqMan RT-PCR assay targeting a conserved region in ORF3 has also been developed to detect HEVs in clinical and environmental samples (Jothikumar *et al.*, 2006). This assay was shown to be sensitive and specific for detecting HEV genotypes 1–4 with the detection limit as few as four genome equivalent copies of HEV plasmid DNA and as low as 0.12 50% pig infectious dose (PID_{50}) of swine HEV. Moreover, the detection of different concentrations of swine HEVs (120–1.2 PID_{50}) in a surface water concentrate was performed successfully.

Molecular beacons (MBs) are single-stranded fluorescent probes and have a stem-loop structure that is labeled both with a fluorescent dye and a universal quencher at the 5′ and 3′ ends, respectively (El Galil *et al.*, 2005). MBs undergo a fluorogenic conformational change upon binding to their target, which allows the progress of the reaction to be followed in real-time PCR (El Galil *et al.*, 2004; Valdivia-Granda *et al.*, 2005). A qRT-PCR based on the amplification of 5′-NCR was used to detect genome copies of HAVs using TaqMan and MB probes in clinical and shellfish samples (Costafreda *et al.*, 2006). MB had a lower sensitivity and reproducibility compared to TaqMan probes, which was able to detect as little as 0.05 infectious unit and 10 copies of a single-stranded RNA (ssRNA) synthetic transcript.

Two FRET hybridization probes, made from DNA, are used: one with a fluorescent dye on the 3′ end and the other with an acceptor dye on the 5′ end. They are intended to anneal next to each other in a head-to-tail configuration on the PCR product (Espy *et al.*, 2006). These probes are also referred to as LightCycler probes and are commercially available (Espy *et al.*, 2006; Sanchez *et al.*, 2006). A commercial qRT-PCR assay, the LightCycler HAV quantification kit (Roche Diagnostics), coupled with immunomagnetic separation (IMS) pretreatment, has been shown to be sensitive and specific in the detection of HAVs in fresh produce (Shan *et al.*, 2005). IMS is based on the isolation of an antigen from the sample with a monoclonal antibody against HAV (anti-HAV 1009) combined with streptavidin-coated magnetic beads to recover low levels of viruses and to remove PCR inhibitors. In this assay, 5′ NCR was chosen as the highly conserved target region and a detection limit as low as 1 PFU was obtained. In a similar study, two commercial qRT-PCR HAV assays, the LightCycler HAV quantification kit (Roche Diagnostics) and the RealArt HAV LC RT PCR kit (artus GmbH), were compared in terms of precision, accuracy, linearity and detection limits (Sanchez *et al.*, 2006). The results showed that both kits were suitable for detecting and quantifying HAVs; however, the Roche kit had a slightly better detection limit with the capability of differentiating between different HAV strains and it was also able to detect HAVs in spiked water and food samples.

Several commercial kits for detecting and quantifying NVs have been developed due to the high incidence of NV outbreaks (Butot *et al.*, 2010). The NV qRT-PCR Kit (AnDiaTec GmbH and Co. KG, Kornwestheim, Germany) and the NV Type I and Type II kits (Generon S.r.l., Castelnuovo, Italy) were evaluated and compared with the assay designed by the CEN/TC/WG6/TAG4 research group in the specific detection and quantification of 59 NV samples, including different subtypes of NV genogroups I and II (Butot *et al.*, 2010). The commercial kits failed to detect the vast majority of NV strains, showing poor performance.

The challenges associated with the detection of foodborne viruses, such as PCR inhibitors and low virus concentrations in foods, affect the efficiency of real-time assay adversely, therefore, for process control (PC) an internal amplification control (IAC), which is extracted and amplified with the target sequence, is crucial in the evaluation of PCR and to prevent false negatives (Di Pasquale *et al.*, 2010). A real-time PCR IAC has been developed recently for the simultaneous detection of GI and GII NVs, which may also reduce the cost of the assay (Stals *et al.*, 2009). Likewise, the use of non-pathogenic viruses, such as the mutant mengovirus MC_0 strain, the MS2 bacteriophage and feline calicivirus (FCV), as sample process controls has been proposed in detecting HAVs in different food matrices (e.g. shellfish, raspberries and strawberries) (Costafreda *et al.*, 2006; Blaise-Boisseau *et al.*, 2010; Di Pasquale *et al.*, 2010). In these studies, no loss of HAV detection sensitivity was observed after the addition of controls.

Nucleic acid sequence-based amplification (NASBA)

NASBA is an alternative approach to PCR-based molecular methods. In this method, an RNA template is amplified under isothermal conditions using three enzymes (avian myeloblastosis virus reverse transcriptase, RNase H and T7 RNA polymerase) in the reaction tube (Compton, 1991). NASBA is particularly suitable for detecting RNA viruses since the direct amplification of RNA targets is possible without a separate reverse transcription step (Jean *et al.*, 2001, 2004). It has also been shown to be less susceptible to environmental PCR inhibitors (Rutjes *et al.*, 2006). Even though the amplification power and the sensitivity of NASBA assays are comparable or even better than that of RT-PCR (Jean *et al.*, 2001), NASBA assays have been used in relatively few studies for detecting enteric viruses compared to RT-PCR.

NASBA assays have been multiplexed or coupled to RT-PCR and ELISA assays to achieve lower detection limits, high sensitivity and specificity in virus detection (Jean *et al.*, 2002a, 2002b, 2003, 2004). Amplification of viral RNA from HAVs and human rotaviruses with selected primers in the multiplex NASBA mixture had detection limits of 40 and 400 PFU/ml for rotaviruses and HAVs, respectively (Jean *et al.*, 2002b). In this study, highly conserved regions in rotavirus gene 9 and in the HAV VP2 gene encoding a major capsid protein were targeted for amplification. Accordingly, multiplex NASBA has been used to detect HAVs, GI and GII noroviruses from representative ready-to-eat foods (Jean *et al.*, 2004). All three viruses were detected in the food matrix simultaneously through targeting relatively conserved genomic regions for each of these, with detection limits ranging from 2×10^2 to 2×10^3 PFU/9 cm^2. These results show that NASBA is a promising alternative to RT-PCR as it offers rapid and simultaneous detection in a single reaction tube.

A semi-quantitative form of real-time NASBA estimated a viral load in less than half an hour (Patterson *et al.*, 2006). Molecular beacons can be used with NASBA coupled with RNA amplification to produce a specific fluorescent signal, which can be monitored in real time. The measurable fluorescence is directly proportional to the concentration of the target sequence (Leone *et al.*, 1998). More recently, real-time NASBA using a MB probe has been demonstrated to be a sensitive and specific assay for NV detection in clinical and environmental samples (Lamhoujeb *et al.*, 2009). Molecular methods, despite being sensitive and specific, cannot differentiate between infectious and non-infectious viruses. Hence, an enzymatic treatment followed by molecular beacon NASBA targeting of the highly conserved ORF1-ORF2 junction has been developed to distinguish infectious from non-infectious NVs in ready-to-eat food (Lamhoujeb *et al.*, 2008). The proposed enzymatic pretreatment utilized proteinase K and RNase at the same time to digest non-infectious virus particles (Nuanualsuwan and Cliver, 2002).

11.3 Emerging methods

In general, current detection methods have poor sensitivity and selectivity at low virus concentrations. In the main, PCR-based methods have been used to overcome the challenges associated with virus detection; however, these methods also have limitations in terms of complexity in sample preparation and amplification. Thus, the following section is an overview of emerging detection methods.

11.3.1 Spectroscopic approaches

Spectroscopic techniques to detect and identify viral infections are promising owing to their sensitivity, speed, cost and simplicity (Erukhimovitch *et al.*, 2011).

Surface enhanced Raman spectroscopy

Surface enhanced Raman spectroscopy (SERS) and electrochemical impedance spectroscopy are the most commonly used spectroscopic approaches in virus detection.

Even though Raman spectroscopy has been used previously to characterize virus structures, it lacks sensitivity due to the extremely small cross section of

Raman scattering, which is about 12–14 orders of magnitude less than fluorescence cross sections (Porter *et al.*, 2008; Shanmukh *et al.*, 2006; Kneipp *et al.*, 2002). With the help of metallic nanostructures, SERS amplifies low-level Raman signals within highly localized optical fields on metallic surfaces. It overcomes the limitations of conventional Raman spectroscopy because of the electromagnetic field or chemical enhancement (Kneipp *et al.*, 2002). SERS spectral fingerprints have been used to discriminate between different types of viruses (Fan *et al.*, 2010; Shanmukh *et al.*, 2006). Recently, several food and waterborne viruses, namely noroviruses, adenoviruses, parvoviruses, rotaviruses, coronaviruses, paramyxoviruses and herpesviruses, were detected and identified using a gold substrate (Fan *et al.*, 2010). Viruses with or without an envelope were differentiated using multivariate statistical analyses (SIMCA) with more than 95% classification accuracy. For SERS, the detection limit was a titer of 10^2, demonstrating promise for the rapid detection and identification of viruses in food and water samples.

In addition to discriminating between different virus types, SERS has also been used to detect different strains of a single virus type (Tripp *et al.*, 2008). Using silver nanorod arrays fabricated by oblique angle deposition (OAD), SERS was able to detect trace levels of DNA viruses (adenoviruses) and RNA viruses (rhinoviruses and human immunodeficiency viruses (HIVs)) in real time. Moreover, it was able to discriminate between respiratory viruses, virus strains and viruses with gene deletions in biological media (Shanmukh *et al.*, 2006). Further studies indicated that SERS spectra could be used to differentiate between respiratory syncytial virus (RSV) strains and detect viruses with gene deletions using partial least squares (Shanmukh *et al.*, 2008). In a similar study, SERS-active silver nanorod arrays prepared by OAD detected and differentiated between the molecular fingerprints of several important human pathogens, including RSV, HIV and rotavirus (Driskell *et al.*, 2008). SERS also showed high sensitivity and specificity in the identification and classification of rotavirus strains (Driskell *et al.*, 2010). Even though the spectra were similar for each strain, the relative intensities were different (Fig. 11.1). Besides being determined as rotavirus positive or rotavirus negative, samples could be classified by the difference in spectral shapes.

Recently, tip-enhanced Raman spectroscopy (TERS), a combination of optical spectroscopy with SERS, was used to obtain a representative virus spectrum in the identification of different virus strains of avipoxvirus and adeno-associated virus (Hermann *et al.*, 2011). In recent years, SERS substrates and probes have been developed to detect viral genes from HIVs, West Nile viruses and RSVs (Liang *et al.*, 2007; Malvadkar *et al.*, 2010; Zhang *et al.*, 2011a). These studies indicate that the specificity, speed and sensitivity may make SERS-based virus detection a competitive alternative to current detection methods used for food matrices.

Electrochemical impedance spectroscopy

Electrochemical impedance spectroscopy (EIS) monitors the electrical response of a system when a periodic, small amplitude AC signal is applied (Hassen *et al.*,

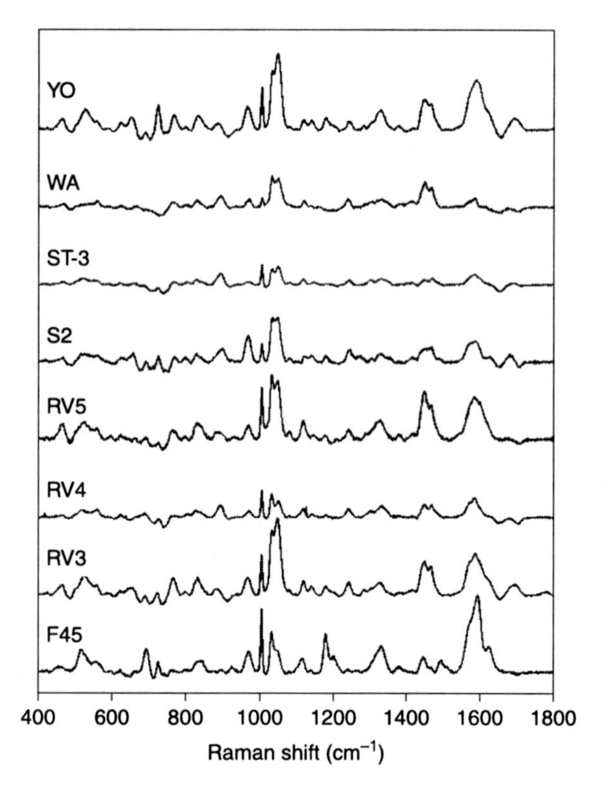

Fig. 11.1 Average SERS spectra of eight strains of rotavirus and the negative control (MA104 cell lysate). Taken from Driskell *et al.* (2010) with permission.

2008). EIS has been used to detect several viruses including the influenza virus, the rabies virus, the hepatitis B virus and HIV (Kukol *et al.*, 2008; Hassen *et al.*, 2008, 2011; Hnaien *et al.*, 2008). Many types of biosensor are based on EIS (Hassen *et al.*, 2008). Recently, the influenza A virus was detected using EIS with an antibody-neutravidin-thiol structure immobilized on the surface of an Au electrode in solutions of phosphate buffer saline with large amounts of non-target protein, which showed the detection sensitivity and selectivity (Hassen *et al.*, 2011). The detection limit was as low as 8 ng/ml, which shows the efficiency of this approach for virus detection. A biosensor based on EIS has been used to detect the label-free viral DNA hybridization of avian influenza virus (Kukol *et al.*, 2008). Even though EIS has not been used for foodborne virus detection, it is a promising approach in terms of sensitivity, selectivity and response.

11.3.2 Immunoassays

Immunoassays are analytical methods that produce a sensitive, selective and measurable response based on highly specific antibody and antigen interactions

(Li *et al.*, 2011a; Lee *et al.*, 2011). Until recently, ELISA and enzyme immunoassays were widely used in foodborne virus detection. Even though these methods are reliable, they are time-consuming and labor intensive. An immunoassay using microsphere technology can overcome the limitations associated with traditional ELISA (Go *et al.*, 2008). The well-known microsphere assay system, the xMap system (Luminex Corp., Austin, TX), combines three well-established technologies: bioassays, solution phase microspheres and flow cytometry (Go *et al.*, 2008). A liquid suspension array consisting of unique color-coded microsphere polystyrene beads is coupled to antigens and antibody reactions, and the emissions are then measured by a flow-based detector (Deregt *et al.*, 2006). Microsphere immunoassays offer several advantages, including accuracy, high sensitivity, specificity, reproducibility, high-throughput sample analysis and multiplexing capability, over traditional ELISAs (Go *et al.*, 2008). In particular, the multiplexing capability enables the detection of a multiplex analyte in a single reaction tube based on individually identifiable, fluorescently coded sets of polystyrene microbeads (Binnicker *et al.*, 2011; Khan *et al.*, 2006). In the last decade, a number of microsphere-based immunoassays have been described for the antigen and antibody detection of several viruses including HIV (Bellisario *et al.*, 2001), non-human primate viruses (Khan *et al.*, 2006), avian influenza virus (Deregt *et al.*, 2006), West Nile virus (Johnson *et al.*, 2007), Epstein–Barr virus (Binnicker *et al.*, 2008) and hepatitis C virus (Fonseca *et al.*, 2011).

Immuno-PCR (IPCR) is a method similar to ELISA. Reporter DNA is used instead of an enzyme in IPCR, which may have a 10^2 to 10^5 increase in sensitivity as a result of the amplification of the reporter DNA (Deng *et al.*, 2011b). More recently, this method has been used in rapid screening for trace levels of avian influenza viruses (Deng *et al.*, 2011b), Newcastle disease viruses (Deng *et al.*, 2011a), RSVs (Perez *et al.*, 2011) and foot and mouth disease viruses (Ding *et al.*, 2011). IPCR had an approximately 1000-fold improvement over conventional ELISA, and a 100-fold enhancement over RT-PCR. The detection limit was as low as 10^{-4} EID_{50} (50% egg infective dose) for the H5 subtype avian influenza virus (Deng *et al.*, 2011b).

11.3.3 Microelectromechanical systems and microfluidics

Microelectromechanical systems (MEMSs) can act as transducers for sensing and actuation in various engineering applications. They can be used to integrate micron-sized mechanical parts with electronics and they can be batch fabricated in large quantities (Gau *et al.*, 2001). MEMS-based and microfluidic-based biosensing approaches have received considerable interest in recent years owing to their advantages over conventional methods including low cost and sample volume, portability, disposability, parallel processing and automation (Wang *et al.*, 2011). More recently, a MEMS biosensor has been developed to detect hepatitis A and hepatitis C viruses (HCVs) in serum using dynamic-mode microcantilevers without any labels or preamplification (Timurdogan *et al.*, 2011). Electroplated nickel MEMS cantilevers functionalized with HAV or HCV

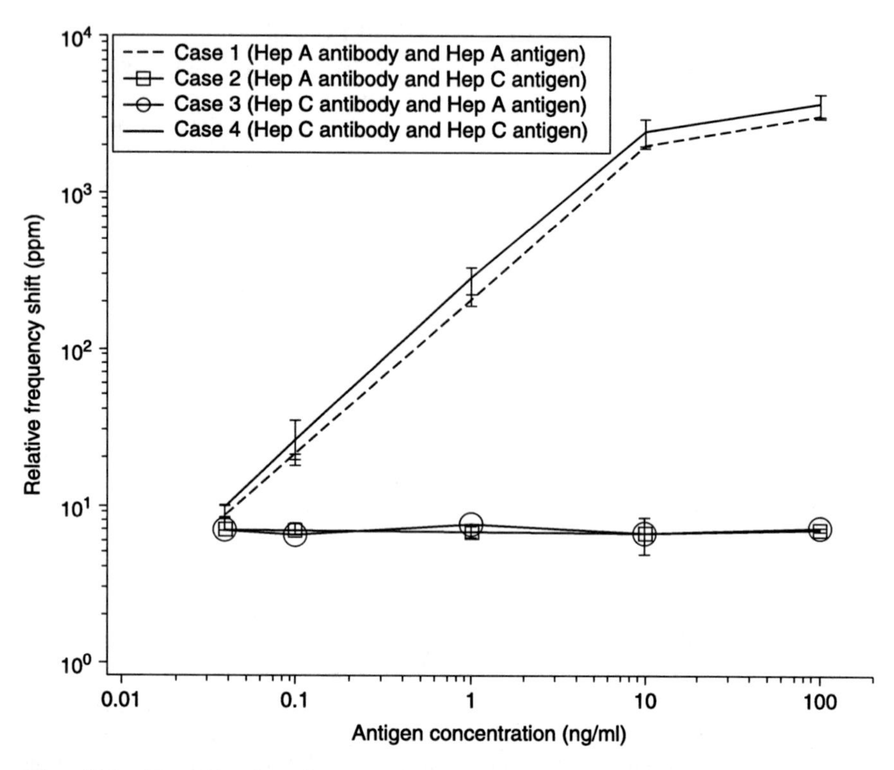

Fig. 11.2 Hepatitis detection measurement results using two biosensor chips. Measurements were taken at different concentrations ranging from 0.04 to 100 ng/ml for negative and positive controls. Different HAV and HCV concentrations were introduced into fetal bovine serum. Taken from Timurdogan *et al.* (2011) with permission.

antibodies were exposed to either HAV antigens (Case 1 and Case 3) or HCV antigens (Case 2 and Case 4), in increasing concentrations in an undiluted serum (Fig. 11.2). The minimum detection limit concentration was 0.1 ng/ml for both HAVs and HCVs, which is comparable with labeled sensing detection methods such as ELISA. Moreover, it was shown that the dynamic range of this biosensor was in excess of 1000:1 for the specific type of hepatitis antibody used.

MEMS technology enables PCR using microfluidics and consequently the synthesis of complementary DNA (cDNA) on microfluidic devices (Li *et al.*, 2011c). This microfluidic-based PCR method has several advantages including lower thermal capacitance giving rapid thermal cycling, reduced analysis times, low consumption of sample and reagent, portability and the potential for high automation and integration of various analytical procedures (Li *et al.*, 2011b). Microfluidic-based RT-PCR has been developed to detect foodborne viruses (Li *et al.*, 2011b, 2011c). An integrated microfluidic system for continuous-flow RT-PCR reactions with online fluorescence detection has been developed for the rapid identification of NVs and rotaviruses; the limit of detection (LOD) is

6.4×10^4 copies per μl using a one-step RT-PCR process (Li *et al.*, 2011b). This restricted LOD was mainly attributed to the inhibition effect of the channel surface. Detection of the amplified products was carried out online using fluorescence microscopy with SybrGreen I. This method did not require the time-consuming and labor-intensive agarose gel electrophoresis and ethidium bromide staining and had much faster reaction times compared to conventional RT-PCR.

11.3.4 Nanostructures

Spherical (quantum dots) and linear particles (nanowires, nanotubes or nanorods) with specific optical, electrical, mechanical, thermal and magnetic properties can be fabricated by combining different metals, semiconductors and carbon (Valdivia-Granda *et al.*, 2005). Nanoparticles (NPs) can be used to provide additional functional properties, including signal enhancement or purification, in virus detection (Fournier-Wirth and Coste, 2010).

Quantum dots (QDs), clusters of a few hundred to a few thousand atoms, are synthesized from metallic materials such as gold, silver or cobalt and semiconductor materials such as cadmium sulfite, cadmium selenide and cadmium telluride (Valdivia-Granda *et al.*, 2005). QDs have often been used to label biomolecules owing to their outstanding properties such as negligible photobleaching, fairly high quantum yield, stability, narrow emission spectrum and broad excitation spectrum (Zhang *et al.*, 2011b). These particles have been conjugated to antibodies and nucleic acids and used as a label in the detection of several viruses including RSV (Agrawal *et al.*, 2005), porcine reproductive virus (Stringer *et al.*, 2008), cauliflower mosaic virus (Huang *et al.*, 2009), Newcastle disease virus and avian virus arthritis virus (Wang *et al.*, 2010) and the Epstein–Barr virus (Chen *et al.*, 2010); however, QDs have never been used to detect foodborne viruses.

Carbon nanotubes (CNTs) are widely used in novel nanostructures and devices due to their large surface area per unit mass and excellent mechanical and electrical properties (Bhattacharya *et al.*, 2011). Moreover, the functionalization of CNTs through the alteration of the surface chemistry increases their potential for use as biosensing markers (Valdivia-Granda *et al.*, 2005). Using the surface functionalization feature, CNTs can be used to immobilize antibodies or nucleic acid that target a type of virus. The process can be monitored using a change in the mechanical or electrical property of the CNTs (Bhattacharya *et al.*, 2011). This concept has been recently used to detect hepatitis C viruses (Dastagir *et al.*, 2007), avian influenza viruses (Zhu *et al.*, 2009; Tam *et al.*, 2009) and swine influenza viruses (Lee *et al.*, 2011) other than foodborne viruses. As CNT-based biosensors are easy to produce, have reproducible results and are inexpensive, and since they have better sensitivity and time responses than current techniques, they are very promising for detecting viruses.

One other promising approach for detecting biomolecules is the use of a semiconducting nanowire where the conductance is proportional to the viral load. The change in conductance is in response to binding between the target and the probe, which is attached to the nanowire (Patolsky *et al.*, 2004; Valdivia-Granda *et al.*, 2005). Nanowires act as a capture agent on the sensor surface and selectively bind target biomolecules much like CNTs (Ishikawa *et al.*, 2009). Nanowires have several attractive features for the real-time detection of a single virus with high selectivity (Valdivia-Granda *et al.*, 2005). Silicon nanowires have been used in label-free field effect transistor (FET)-based biosensors to detect influenza A viruses (Zheng *et al.*, 2005) and dengue viruses (Zhang *et al.*, 2010). The results showed that silicon nanowire-based sensors are more sensitive and have a more rapid response compared to traditional methods. Recently, an alternative nanomaterial to silicon nanowire, a metal oxide nanowire, has been used to detect a protein related to severe acute respiratory syndrome (SARS) at a subnanomolar concentration in a background of 44 μM bovine serum albumin (Ishikawa *et al.*, 2009).

11.4 Future trends

Research into the detection of foodborne viruses has grown in recent years due to the high incidence of outbreaks. Currently, immunological and PCR-based methods are commonly used to detect viruses in food samples. Despite their reliability, most of these methods have limitations in terms of speed and sensitivity owing to low viral concentrations and inhibitory substances present in food. Even though methods for concentrating and purifying viruses in food samples have been widely investigated and developed, the inhibitory substances can remain and cause false-negative results. Therefore, new detection methods that are rapid and sensitive are necessary for direct detection in food samples.

Some of the approaches described for detecting viruses are relatively new and some are still in their infancy. It is expected that electrochemical-based detection techniques will become more prominent, while spectroscopic and microfluidic assays will be developed in parallel. It is anticipated that research using microfluidics will focus on combining the pretreatment of a viral sample and multiplex detection into a biochip. Thus, the microfluidic approach could be a promising platform for rapid detection of viruses. Additionally, the conjugation of antibodies, antigens and nucleic acids with quantum dots, nanowires and carbon nanotubes offers several advantages over current detection methods in terms of sensitivity and speed. These systems may also be used for the label-free detection of very low concentration of viral particles, or even for detecting a single virus without amplification. These approaches can be further improved with the advent of novel nanostructures. Even though most of these approaches have not been used to detect foodborne viruses, all of them are promising and can complement the existing methods.

11.5 References

ABAD, F., PINTO, R., VILLENA, C., GAJARDO, R. and BOSCH, A. 1997. Astrovirus survival in drinking water. *Applied and Environmental Microbiology*, 63, 3119–3122.

AGRAWAL, A., TRIPP, R., ANDERSON, L. and NIE, S. 2005. Real-time detection of virus particles and viral protein expression with two-color nanoparticle probes. *Journal of Virology*, 79, 8625–8628.
AHN, J., KANG, S., LEE, D., SHIN, S. and YOO, H. 2005. Identification of novel human hepatitis E virus (HEV) isolates and determination of the seroprevalence of HEV in Korea. *Journal of Clinical Microbiology*, 43, 3042–3048.
ATMAR, R. L. 2006. Molecular methods of virus detection in foods. In: Goyal, S. M. (ed.) *Viruses in Foods*. New York, USA: Springer.
ATMAR, R. and ESTES, M. 2001. Diagnosis of noncultivatable gastroenteritis viruses, the human caliciviruses. *Clinical Microbiology Reviews*, 14, 15.
BELLISARIO, R., COLINAS, R. and PASS, K. 2001. Simultaneous measurement of antibodies to three HIV-1 antigens in newborn dried blood-spot specimens using a multiplexed microsphere-based immunoassay. *Early Human Development*, 64, 21–25.
BERKE, T. and MATSON, D. 2000. Reclassification of the Caliciviridae into distinct genera and exclusion of hepatitis E virus from the family on the basis of comparative phylogenetic analysis. *Archives of Virology*, 145, 1421–1436.
BEURET, C. 2004. Simultaneous detection of enteric viruses by multiplex real-time RT-PCR. *Journal of Virological Methods*, 115, 1–8.
BHATTACHARYA, M., HONG, S., LEE, D., CUI, T. and GOYAL, S. 2011. Carbon nanotube based sensors for the detection of viruses. *Sensors and Actuators B – Chemical*, 155, 67–74.
BINNICKER, M., JESPERSEN, D., HARRING, J., ROLLINS, L. and BEITO, E. 2008. Evaluation of a multiplex flow immunoassay for detection of Epstein–Barr virus-specific antibodies. *Clinical and Vaccine Immunology*, 15, 1410–1413.
BINNICKER, M., JESPERSEN, D. and ROLLINS, L. 2011. Evaluation of the Bio-Rad BioPlex measles, mumps, rubella, and Varicella-Zoster virus IgG multiplex bead immunoassay. *Clinical and Vaccine Immunology*, 18, 1524–1526.
BLAISE-BOISSEAU, S., HENNECHART-COLLETTE, C., GUILLIER, L. and PERELLE, S. 2010. Duplex real-time qRT-PCR for the detection of hepatitis A virus in water and raspberries using the MS2 bacteriophage as a process control. *Journal of Virological Methods*, 166, 48–53.
BOSCH, A., SANCHEZ, G., ABBASZADEGAN, M., CARDUCCI, A., GUIX, S. *et al.* 2011. Analytical methods for virus detection in water and food. *Food Analytical Methods*, 4, 4–12.
BUTOT, S., LE GUYADER, F., KROL, J., PUTALLAZ, T., AMOROSO, R. *et al.* 2010. Evaluation of various real-time RT-PCR assays for the detection and quantitation of human norovirus. *Journal of Virological Methods*, 167, 90–94.
CASAS, N. and SUNEN, E. 2001. Detection of enterovirus and hepatitis A virus RNA in mussels (*Mytilus* spp.) by reverse transcriptase-polymerase chain reaction. *Journal of Applied Microbiology*, 90, 89–95.
CHAMBERLAIN, J., GIBBS, R., RANIER, J., NGUYEN, P. and CASKEY, C. 1988. Deletion screening of the Dushenne muscular-dystrophy locus via multiplex DNA amplification. *Nucleic Acids Research*, 16, 11141–11156.
CHEN, L., QI, Z., CHEN, R., LI, Y. and LIU, S. 2010. Sensitive detection of Epstein–Barr virus-derived latent membrane protein 1 based on CdTe quantum dots-capped silica nanoparticle labels. *Clinica Chimica Acta*, 411, 1969–1975.
COELHO, C., VINATEA, C., HEINERT, A., SIMOES, C. and BARARDI, C. 2003. Comparison between specific and multiplex reverse transcription-polymerase chain reaction for detection of hepatitis A virus, poliovirus and rotavirus in experimentally seeded oysters. *Memorias Do Instituto Oswaldo Cruz*, 98, 465–468.
COMPTON, J. 1991. Nucleic acid sequence based amplification. *Nature*, 350, 91–92.
COOK, N. and RZEZUTKA, A. 2006. Hepatitis viruses. In: Motarjemi, Y. and Adams, M. (eds.) *Emerging Foodborne Pathogens*. Woodhead Publishing.
COSTA-MATTIOLI, M., MONPOEHO, S., NICAND, E., ALEMAN, M., BILLAUDEL, S. *et al.* 2002. Quantification and duration of viraemia during hepatitis A infection as determined by real-time RT-PCR. *Journal of Viral Hepatitis*, 9, 101–106.

COSTAFREDA, M., BOSCH, A. and PINTO, R. 2006. Development, evaluation, and standardization of a real-time TaqMan reverse transcription-PCR assay for quantification of hepatitis A virus in clinical and shellfish samples. *Applied and Environmental Microbiology*, 72, 3846–3855.

CROCI, L., DE MEDICI, D., MORACE, G., FIORE, A., SCALFARO, C. *et al.* 1999. Detection of hepatitis A virus in shellfish by nested reverse transcription-PCR. *International Journal of Food Microbiology*, 48, 67–71.

CROCI, L., DUBOIS, E., COOK, N., DE MEDICI, D., SCHULTZ, A. *et al.* 2008. Current methods for extraction and concentration of enteric viruses from fresh fruit and vegetables: Towards international standards. *Food Analytical Methods*, 1, 73–84.

DASTAGIR, T., FORZANI, E., ZHANG, R., AMLANI, I., NAGAHARA, L. *et al.* 2007. Electrical detection of hepatitis C virus RNA on single wall carbon nanotube-field effect transistors. *Analyst*, 132, 738–740.

DENG, M., LONG, L., XIAO, X., WU, Z., ZHANG, F., *et al.* 2011a. Immuno-PCR for one step detection of H5N1 avian influenza virus and Newcastle disease virus using magnetic gold particles as carriers. *Veterinary Immunology and Immunopathology*, 141, 183–189.

DENG, M., XIAO, X., ZHANG, Y., WU, X., ZHU, L. *et al.* 2011b. A highly sensitive immuno-PCR assay for detection of H5N1 avian influenza virus. *Molecular Biology Reports*, 38, 1941–1948.

DEREGT, D., FURUKAWA-STOFFER, T., TOKARYK, K., PASICK, J., HUGHES, K. *et al.* 2006. A microsphere immunoassay for detection of antibodies to avian influenza virus. *Journal of Virological Methods*, 137, 88–94.

DI PASQUALE, S., PANICONI, M., DE MEDICI, D., SUFFREDINI, E. and CROCI, L. 2010. Duplex real time PCR for the detection of hepatitis A virus in shellfish using feline calicivirus as a process control. *Journal of Virological Methods*, 163, 96–100.

DI PINTO, A., FORTE, V., TANTILLO, G., TERIO, V. and BUONAVOGLIA, C. 2003. Detection of hepatitis A virus in shellfish (*Mytilus galloprovincialis*) with RT-PCR. *Journal of Food Protection*, 66, 1681–1685.

DING, Y., LIU, Y., ZHOU, J., CHEN, H., WEI, G. *et al.* 2011. A highly sensitive detection for foot-and-mouth disease virus by gold nanoparticle improved immuno-PCR. *Virology Journal*, 8.

DRISKELL, J., SHANMUKH, S., LIU, Y., HENNIGAN, S., JONES, L. *et al.* 2008. Infectious agent detection with SERS-active silver nanorod arrays prepared by oblique angle deposition. *IEEE Sensors Journal*, 8, 863–870.

DRISKELL, J., ZHU, Y., KIRKWOOD, C., ZHAO, Y., DLUHY, R. *et al.* 2010. Rapid and sensitive detection of rotavirus molecular signatures using surface enhanced Raman spectroscopy. *Plos One*, 5, e10222.

D'SOUZA, D. H. and JAYKUS, L. A. 2006. Molecular approaches for the detection of foodborne viral pathogens. In: Maurer, J. (ed.) *PCR Methods in Foods. Food Microbiology and Food Safety Series*. New York, USA: Springer.

DUBOIS, E., AGIER, C., TRAORE, O., HENNECHART, C., MERLE, G. *et al.* 2002. Modified concentration method for the detection of enteric viruses on fruits and vegetables by reverse transcriptase-polymerase chain reaction or cell culture. *Journal of Food Protection*, 65, 1962–1969.

EL GALIL, K., EL SOKKARY, M., KHEIRA, S., SALAZAR, A., YATES, M. *et al.* 2004. Combined immunomagnetic separation-molecular beacon-reverse transcription-PCR assay for detection of hepatitis A virus from environmental samples. *Applied and Environmental Microbiology*, 70, 4371–4374.

EL GALIL, K., EL SOKKARY, M., KHEIRA, S., SALAZAR, A., YATES, M. *et al.* 2005. Real-time nucleic acid sequence-based amplification assay for detection of hepatitis A virus. *Applied and Environmental Microbiology*, 71, 7113–7116.

EMERSON, S. and PURCELL, R. 2001. Recombinant vaccines for hepatitis E. *Trends in Molecular Medicine*, 7, 462–466.

ENOUF, V., DOS REIS, G., GUTHMANN, J., GUERIN, P., CARON, M. *et al.* 2006. Validation of single real-time TaqMan (R) PCR assay for the detection and quantitation of four major genotypes of hepatitis E virus in clinical specimens. *Journal of Medical Virology*, 78, 1076–1082.

ERUKHIMOVITCH, V., BOGOMOLNY, E., HULEIHIL, M. and HULEIHEL, M. 2011. Infrared spectral changes identified during different stages of herpes viruses infection *in vitro*. *Analyst*, 136, 2818–2824.

ESPY, M., UHL, J., SLOAN, L., BUCKWALTER, S., JONES, M. *et al.* 2006. Real-time PCR in clinical microbiology: Applications for a routine laboratory testing. *Clinical Microbiology Reviews*, 19, 165–256.

FAN, C., HU, Z., RILEY, L., PURDY, G., MUSTAPHA, A. *et al.* 2010. Detecting food- and waterborne viruses by surface-enhanced Raman spectroscopy. *Journal of Food Science*, 75, M302–M307.

FANKHAUSER, R. L., MONROE, S. S., NOEL, J. S., HUMPHREY, C.D., BRESEE, J. S. *et al.* 2002. Epidemiologic and molecular trends of 'Norwalk-like viruses' associated with outbreaks of gastroenteritis in the United States. *Journal of Infectious Diseases*, 186, 1–7.

FAO/WHO 2008. Viruses in food: Scientific advice to support risk management activities. meeting report. *Microbiological Risk Assessment Series*. Rome.

FONSECA, B., MARQUES, C., NASCIMENTO, L., MELLO, M., SILVA, L. *et al.* 2011. Development of a multiplex bead-based assay for detection of hepatitis C virus. *Clinical and Vaccine Immunology*, 18, 802–806.

FORMIGA-CRUZ, M., HUNDESA, A., CLEMENTE-CASARES, P., ALBINANA-GIMENEZ, N., ALLARD, A. *et al.* 2005. Nested multiplex PCR assay for detection of human enteric viruses in shellfish and sewage. *Journal of Virological Methods*, 125, 111–118.

FOURNIER-WIRTH, C. and COSTE, J. 2010. Nanotechnologies for pathogen detection: Future alternatives? *Biologicals*, 38, 9–13.

GAU, J., LAN, E., DUNN, B., HO, C. and WOO, J. 2001. A MEMS based amperometric detector for *E. coli* bacteria using self-assembled monolayers. *Biosensors and Bioelectronics*, 16, 745–755.

GERBA, C. P. 2006. Food virology: Past, present, and future. In: Goyal, S. M. (ed.) *Viruses in Foods*. New York, USA: Springer.

GO, Y., WONG, S., BRANSCUM, A., DEMAREST, V., SHUCK, K. *et al.* 2008. Development of a fluorescent-microsphere immunoassay for detection of antibodies specific to equine arteritis virus and comparison with the virus neutralization test. *Clinical and Vaccine Immunology*, 15, 76–87.

GOYAL, S. M. 2006. Methods of virus detection in foods. In: Goyal, S. M. (ed.) *Viruses in Foods*. New York, USA: Springer.

GREEN, D. and LEWIS, G. 1999. Comparative detection of enteric viruses in wastewaters, sediments and oysters by reverse transcription PCR and cell culture. *Water Research*, 33, 1195–1200.

GREENING, G. E. 2006. Molecular methods of virus detection in foods. In: Goyal, S. M. (ed.) *Viruses in Foods*. New York, USA: Springer.

GUEVREMONT, E., BRASSARD, J., HOUDE, A., SIMARD, C. and TROTTIER, Y. 2006. Development of an extraction and concentration procedure and comparison of RT-PCR primer systems for the detection of hepatitis A virus and norovirus GII in green onions. *Journal of Virological Methods*, 134, 130–135.

GUTIERREZ-AGUIRRE, I., STEYER, A., BOBEN, J., GRUDEN, K., POLJSAK-PRIJATELJ, M. *et al.* 2008. Sensitive detection of multiple rotavirus genotypes with a single reverse transcription-real-time quantitative PCR assay. *Journal of Clinical Microbiology*, 46, 2547–2554.

GYARMATI, P., MOHAMMED, N., NORDER, H., BLOMBERG, J., BELAK, S. *et al.* 2007. Universal detection of hepatitis E virus by two real-time PCR assays: TaqMan (R) and primer-probe energy transfer. *Journal of Virological Methods*, 146, 226–235.

HAQQI, T., SARKAR, G., DAVID, C. and SOMMER, S. 1988. Specific amplification with PCR of a refractory segment of genomic DNA. *Nucleic Acids Research*, 16, 11844.

HASSEN, W., CHAIX, C., ABDELGHANI, A., BESSUEILLE, F., LEONARD, D. *et al.* 2008. An impedimetric DNA sensor based on functionalized magnetic nanoparticles for HIV and HBV detection. *Sensors and Actuators B – Chemical*, 134, 755–760.

HASSEN, W., DUPLAN, V., FROST, E. and DUBOWSKI, J. 2011. Quantitation of influenza A virus in the presence of extraneous protein using electrochemical impedance spectroscopy. *Electrochimica Acta*, 56, 8325–8328.

HERMANN, P., HERMELINK, A., LAUSCH, V., HOLLAND, G., MOLLER, L. *et al.* 2011. Evaluation of tip-enhanced Raman spectroscopy for characterizing different virus strains. *Analyst*, 136, 1148–1152.

HNAIEN, M., DIOUANI, M., HELALI, S., HAFAID, I., HASSEN, W. *et al.* 2008. Immobilization of specific antibody on SAM functionalized gold electrode for rabies virus detection by electrochemical impedance spectroscopy. *Biochemical Engineering Journal*, 39, 443–449.

HOHNE, M. and SCHREIER, E. 2004. Detection and characterization of norovirus outbreaks in Germany: Application of a one-tube RT-PCR using a fluorogenic real-time detection system. *Journal of Medical Virology*, 72, 312–319.

HOUDE, A., GUEVREMONT, E., POITRAS, E., LEBLANC, D., WARD, P. *et al.* 2007. Comparative evaluation of new TaqMan real-time assays for the detection of hepatitis A virus. *Journal of Virological Methods*, 140, 80–89.

HUANG, D., LIU, H., ZHANG, B., JIAO, K. and FU, X. 2009. Highly sensitive electrochemical detection of sequence-specific DNA of 35S promoter of cauliflower mosaic virus gene using CdSe quantum dots and gold nanoparticles. *Microchimica Acta*, 165, 243–248.

ISHIKAWA, F., CHANG, H., CURRELI, M., LIAO, H., OLSON, C. *et al.* 2009. Label-free, electrical detection of the SARS virus N-protein with nanowire biosensors utilizing antibody mimics as capture probes. *ACS Nano*, 3, 1219–1224.

JANSEN, R., SIEGL, G. and LEMON, S. 1990. Molecular epidemiology of human hepatitis A virus defined by an antigen-capture polymerase chain reaction method. *Proceedings of the National Academy of Sciences of the United States of America*, 87, 2867–2871.

JEAN, J., BLAIS, B., DARVEAU, A. and FLISS, I. 2001. Detection of hepatitis A virus by the nucleic acid sequence-based amplification technique and comparison with reverse transcription-PCR. *Applied and Environmental Microbiology*, 67, 5593–5600.

JEAN, J., BLAIS, B., DARVEAU, A. and FLISS, I. 2002a. Rapid detection of human rotavirus using colorimetric nucleic acid sequence-based amplification (NASBA)-enzyme-linked immunosorbent assay in sewage treatment effluent. *FEMS Microbiology Letters*, 210, 143–147.

JEAN, J., BLAIS, B., DARVEAU, A. and FLISS, I. 2002b. Simultaneous detection and identification of hepatitis A virus and rotavirus by multiplex nucleic acid sequence-based amplification (NASBA) and microtiter plate hybridization system. *Journal of Virological Methods*, 105, 123–132.

JEAN, J., D'SOUZA, D. and JAYKUS, L. 2003. Transcriptional enhancement of RT-PCR for rapid and sensitive detection of noroviruses. *FEMS Microbiology Letters*, 226, 339–345.

JEAN, J., D'SOUZA, D. and JAYKUS, L. 2004. Multiplex nucleic acid sequence-based amplification for simultaneous detection of several enteric viruses in model ready-to-eat foods. *Applied and Environmental Microbiology*, 70, 6603–6610.

JIANG, X., HUANG, P., ZHONG, W., FARKAS, T., CUBITT, D. *et al.* 1999. Design and evaluation of a primer pair that detects both Norwalk- and Sapporo-like caliciviruses by RT-PCR. *Journal of Virological Methods*, 83, 145–154.

JIANG, Y., LIAO, G., ZHAO, W., SUN, M., QIAN, Y. *et al.* 2004. Detection of infectious hepatitis A virus by integrated cell culture/strand-specific reverse transcriptase-polymerase chain reaction. *Journal of Applied Microbiology*, 97, 1105–1112.

JOHNSON, A., CHESHIER, R., COSENTINO, G., MASRI, H., MOCK, V. *et al.* 2007. Validation of a microsphere-based immunoassay for detection of anti-West Nile virus and anti-St. Louis encephalitis virus immunoglobulin M antibodies. *Clinical and Vaccine Immunology*, 14, 1084–1093.

JOTHIKUMAR, N., CROMEANS, T., HILL, V., LU, X., SOBSEY, M. *et al.* 2005a. Quantitative real-time PCR assays for detection of human adenoviruses and identification of serotypes 40 and 41. *Applied and Environmental Microbiology*, 71, 3131–3136.

JOTHIKUMAR, N., LOWTHER, J., HENSHILWOOD, K., LEES, D., HILL, V. *et al.* 2005b. Rapid and sensitive detection of noroviruses by using TaqMan-based one-step reverse transcription-PCR assays and application to naturally contaminated shellfish samples. *Applied and Environmental Microbiology*, 71, 1870–1875.

JOTHIKUMAR, N., CROMEANS, T., ROBERTSON, B., MENG, X. and HILL, V. 2006. A broadly reactive one-step real-time RT-PCR assay for rapid and sensitive detection of hepatitis E virus. *Journal of Virological Methods*, 131, 65–71.

KAPIKIAN, A., HOSHINO, Y., CHANOCK, R. and PEREZSCHAEL, I. 1996. Efficacy of a quadrivalent rhesus rotavirus-based human rotavirus vaccine aimed at preventing severe rotavirus diarrhea in infants and young children. *Journal of Infectious Diseases*, 174, S65–S72.

KATAYAMA, K., SHIRATO-HORIKOSHI, H., KOJIMA, S., KAGEYAMA, T., OKA, T. *et al.* 2002. Phylogenetic analysis of the complete genome of 18 Norwalk-like viruses. *Virology*, 299, 225–239.

KHAN, I., MENDOZA, S., YEE, J., DEANE, M., VENKATESWARAN, K. *et al.* 2006. Simultaneous detection of antibodies to six nonhuman-primate viruses by multiplex microbead immunoassay. *Clinical and Vaccine Immunology*, 13, 45–52.

KIM, D., KIM, S., KWON, K., LEE, J. and OH, M. 2008a. Detection of hepatitis A virus from oyster by nested PCR using efficient extraction and concentration method. *Journal of Microbiology*, 46, 436–440.

KIM, S., KIM, D., KWON, K., HWANG, I. and OH, M. 2008b. Detection of norovirus in contaminated ham by reverse transcriptase-PCR and nested PCR. *Food Science and Biotechnology*, 17, 651–654.

KITTIGUL, L., POMBUBPA, K., RATTANATHAM, T., DIRAPHAT, P., UTRARACHKIJ, F. *et al.* 2008. Development of a method for concentrating and detecting rotavirus in oysters. *International Journal of Food Microbiology*, 122, 204–210.

KNEIPP, K., KNEIPP, H., ITZKAN, I., DASARI, R. and FELD, M. 2002. Surface-enhanced Raman scattering and biophysics. *Journal of Physics – Condensed Matter*, 14, R597–R624.

KOGAWA, K., NAKATA, S., UKAE, S., ADACHI, N., NUMATA, K. *et al.* 1996. Dot blot hybridization with a cDNA probe derived from the human calicivirus Sapporo 1982 strain. *Archives of Virology*, 141, 1949–1959.

KOOPMANS, M. and DUIZER, E. 2004. Foodborne viruses: An emerging problem. *International Journal of Food Microbiology*, 90, 23–41.

KOOPMANS, M., VON BONSDORFF, C., VINJE, J., DE MEDICI, D. and MONROE, S. 2002. Foodborne viruses. *FEMS Microbiology Reviews*, 26, 187–205.

KUKOL, A., LI, P., ESTRELA, P., KO-FERRIGNO, P. and MIGLIORATO, P. 2008. Label-free electrical detection of DNA hybridization for the example of influenza virus gene sequences. *Analytical Biochemistry*, 374, 143–153.

LAMHOUJEB, S., FLISS, I., NGAZOA, S. and JEAN, J. 2008. Evaluation of the persistence of infectious human noroviruses on food surfaces by using real-time nucleic acid sequence-based amplification. *Applied and Environmental Microbiology*, 74, 3349–3355.

LAMHOUJEB, S., CHAREST, H., FLISS, I., NGAZOA, S. and JEAN, J. 2009. Real-time molecular beacon NASBA for rapid and sensitive detection of norovirus GII in clinical samples. *Canadian Journal of Microbiology*, 55, 1375–1380.

LEE, D., CHANDER, Y., GOYAL, S. and CUI, T. 2011. Carbon nanotube electric immunoassay for the detection of swine influenza virus H1N1. *Biosensors and Bioelectronics*, 26, 3482–3487.

LEES, D. 2000. Viruses and bivalve shellfish. *International Journal of Food Microbiology*, 59, 81–116.

LEMON, S. and ROBERTSON, B. 1993. Current perspectives in the virology and molecular biology of hepatitis A virus. *Seminars in Virology*, 4, 285–295.

LEONE, G., VAN GEMEN, B. and SCHOEN, C. D. 1998. Molecular beacon probes combined with amplification by NASBA enable homogeneous, real-time detection of RNA. *Nucleic Acids Research*, 26, 2150–2155.

LI, F., MEI, L., LI, Y., ZHAO, K., CHEN, H. *et al.* 2011a. Facile fabrication of magnetic gold electrode for magnetic beads-based electrochemical immunoassay: Application to the diagnosis of Japanese encephalitis virus. *Biosensors and Bioelectronics*, 26, 4253–4256.

LI, Y., ZHANG, C. and XING, D. 2011b. Fast identification of foodborne pathogenic viruses using continuous-flow reverse transcription-PCR with fluorescence detection. *Microfluidics and Nanofluidics*, 10, 367–380.

LI, Y., ZHANG, C. and XING, D. 2011c. Integrated microfluidic reverse transcription-polymerase chain reaction for rapid detection of food- or waterborne pathogenic rotavirus. *Analytical Biochemistry*, 415, 87–96.

LIANG, Y., GONG, J., HUANG, Y., ZHENG, Y., JIANG, J. *et al.* 2007. Biocompatible core-shell nanoparticle-based surface-enhanced Raman scattering probes for detection of DNA related to HIV gene using silica-coated magnetic nanoparticles as separation tools. *Talanta*, 72, 443–449.

LOVE, D., CASTEEL, M., MESCHKE, J. and SOBSEY, M. 2008. Methods for recovery of hepatitis A virus (HAV) and other viruses from processed foods and detection of HAV by nested RT-PCR and TaqMan RT-PCR. *International Journal of Food Microbiology*, 126, 221–226.

MACKAY, I., ARDEN, K. and NITSCHE, A. 2002. Real-time PCR in virology. *Nucleic Acids Research*, 30, 1292–1305.

MALVADKAR, N., DEMIREL, G., POSS, M., JAVED, A., DRESSICK, W. *et al.* 2010. Fabrication and use of electroless plated polymer surface-enhanced Raman spectroscopy substrates for viral gene detection. *Journal of Physical Chemistry C*, 114, 10730–10738.

MARTINEZ-MARTINEZ, M., DIEZ-VALCARCE, M., HERNANDEZ, M. and RODRIGUEZ-LAZARO, D. 2011. Design and application of nucleic acid standards for quantitative detection of enteric viruses by real-time PCR. *Food and Environmental Virology*, 3, 92–98.

MEAD, P., SLUTSKER, L., DIETZ, V., MCCAIG, L., BRESEE, J. *et al.* 1999. Food-related illness and death in the United States. *Emerging Infectious Diseases*, 5, 607–625.

MORALES-RAYAS, R., WOLFFS, P. and GRIFFITHS, M. 2010. Simultaneous separation and detection of hepatitis A virus and norovirus in produce. *International Journal of Food Microbiology*, 139, 48–55.

NAKAYAMA, M., UEDA, Y., KAWAMOTO, H., HANJUN, Y., SAITO, K. *et al.* 1996. Detection and sequencing of Norwalk-like viruses from stool samples in Japan using reverse transcription-polymerase chain reaction amplification. *Microbiology and Immunology*, 40, 317–320.

NUANUALSUWAN, S. and CLIVER, D. 2002. Pretreatment to avoid positive RT-PCR results with inactivated viruses. *Journal of Virological Methods*, 104, 217–225.

O'CONNELL, J. 2002. *RT-PCR Protocols*. Totowa, NJ: Humana Press.

PATOLSKY, F., ZHENG, G., HAYDEN, O., LAKADAMYALI, M., ZHUANG, X. *et al.* 2004. Electrical detection of single viruses. *Proceedings of the National Academy of Sciences of the United States of America*, 101, 14017–14022.

PATTERSON, S., SMITH, M., CASPER, E., HUFFMAN, D., STARK, L. *et al.* 2006. A nucleic acid sequence-based amplification assay for real-time detection of norovirus genogroup II. *Journal of Applied Microbiology*, 101, 956–963.

PEREZ, J., VARGIS, E., RUSS, P., HASELTON, F. and WRIGHT, D. 2011. Detection of respiratory syncytial virus using nanoparticle amplified immuno-polymerase chain reaction. *Analytical Biochemistry*, 410, 141–148.

PINEDA, M., CHAN, L., KUHLENSCHMIDT, T., CHOI, C., KUHLENSCHMIDT, M. *et al.* 2009. Rapid specific and label-free detection of porcine rotavirus using photonic crystal biosensors. *IEEE Sensors Journal*, 9, 470–477.

PORTER, M., LIPERT, R., SIPERKO, L., WANG, G. and NARAYANANA, R. 2008. SERS as a bioassay platform: Fundamentals, design, and applications. *Chemical Society Reviews*, 37, 1001–1011.

REYNOLDS, K., GERBA, C. and PEPPER, I. 1996. Detection of infectious enteroviruses by an integrated cell culture PCR procedure. *Applied and Environmental Microbiology*, 62, 1424–1427.

RICHARDS, G. 1999. Limitations of molecular biological techniques for assessing the virological safety of foods. *Journal of Food Protection*, 62, 691–697.

RICHARDS, A., LOPMAN, B., GUNN, A., CURRY, A., ELLIS, D. *et al.* 2003. Evaluation of a commercial ELISA for detecting Norwalk-like virus antigen in faeces. *Journal of Clinical Virology*, 26, 109–115.

RIGOTTO, C., SINCERO, T., SIMOES, C. and BARARDI, C. 2005. Detection of adenoviruses in shellfish by means of conventional-PCR, nested-PCR, and integrated cell culture PCR (ICC/PCR). *Water Research*, 39, 297–304.

RODRIGUEZ-LAZARO, D., LOMBARD, B., SMITH, H., RZEZUTKA, A., D'AGOSTINO, M. *et al.* 2007. Trends in analytical methodology in food safety and quality: Monitoring microorganisms and genetically modified organisms. *Trends in Food Science and Technology*, 18, 306–319.

ROSENFIELD, S. and JAYKUS, L. 1999. A multiplex reverse transcription polymerase chain reaction method for the detection of foodborne viruses. *Journal of Food Protection*, 62, 1210–1214.

RUTJES, S., HAROLD, H. J. L., VAN DEN BERG, H., LODDER, W. J. and DE RODA HUSMAN, A. M. 2006. Real-time detection of noroviruses in surface water by use of a broadly reactive nucleic acid sequence-based amplification assay. *Applied and Environmental Microbiology*, 72, 5349–5358.

SAIR, A., D'SOUZA, D., MOE, C. and JAYKUS, L. 2002. Improved detection of human enteric viruses in foods by RT-PCR. *Journal of Virological Methods*, 100, 57–69.

SANCHEZ, G., POPULAIRE, S., BUTOT, S., PUTALLAZ, T. and JOOSTEN, H. 2006. Detection and differentiation of human hepatitis A strains by commercial quantitative real-time RT-PCR tests. *Journal of Virological Methods*, 132, 160–165.

SANCHEZ, G., BOSCH, A. and PINTO, R. M. 2007. Hepatitis A virus detection in food: Current and future prospects. *Letters in Applied Microbiology*, 45, 1–5.

SEYMOUR, I. and APPLETON, H. 2001. Foodborne viruses and fresh produce. *Journal of Applied Microbiology*, 91, 759–773.

SHAN, X., WOLFFS, P. and GRIFFITHS, M. 2005. Rapid and quantitative detection of hepatitis A virus from green onion and strawberry rinses by use of real-time reverse transcription-PCR. *Applied and Environmental Microbiology*, 71, 5624–5626.

SHANMUKH, S., JONES, L., DRISKELL, J., ZHAO, Y., DLUHY, R. *et al.* 2006. Rapid and sensitive detection of respiratory virus molecular signatures using a silver nanorod array SERS substrate. *Nano Letters*, 6, 2630–2636.

SHANMUKH, S., JONES, L., ZHAO, Y., DRISKELL, J., TRIPP, R. *et al.* 2008. Identification and classification of respiratory syncytial virus (RSV) strains by surface-enhanced Raman spectroscopy and multivariate statistical techniques. *Analytical and Bioanalytical Chemistry*, 390, 1551–1555.

STALS, A., BAERT, L., BOTTELDOORN, N., WERBROUCK, H., HERMAN, L. *et al.* 2009. Multiplex real-time RT-PCR for simultaneous detection of GI/GII noroviruses and murine norovirus 1. *Journal of Virological Methods*, 161, 247–253.

STRINGER, R., SCHOMMER, S., HOEHN, D. and GRANT, S. 2008. Development of an optical biosensor using gold nanoparticles and quantum dots for the detection of porcine reproductive and respiratory syndrome virus. *Sensors and Actuators B – Chemical*, 134, 427–431.

TAM, P., HIEU, V., CHIEN, N., LE, A. and TUAN, M. 2009. DNA sensor development based on multi-wall carbon nanotubes for label-free influenza virus (type A) detection. *Journal of Immunological Methods*, 350, 118–124.

TIMURDOGAN, E., ALACA, B., KAVAKLI, I. and UREY, H. 2011. MEMS biosensor for detection of hepatitis A and C viruses in serum. *Biosensors and Bioelectronics*, 28, 189–194.

TRIPP, R., DLUHY, R. and ZHAO, Y. 2008. Novel nanostructures for SERS biosensing. *Nano Today*, 3, 31–37.

VALDIVIA-GRANDA, W., KEATING, C., KANN, M., BERESFORD, R., KELLEY, S. *et al.* 2005. Detection of encephalic and hemorrhagic viruses: Integration of micro- and nano-fabrication with computational tools. *2005 International Conference on MEMS, NANO and Smart Systems, Proceedings*, 411–417.

VASICKOVA, P., DVORSKA, L., LORENCOVA, A. and PAVLIK, I. 2005. Viruses as a cause of foodborne diseases: A review of the literature. *Veterinarni Medicina*, 50, 89–104.

VERHOEF, L., BOXMAN, I. L. and KOOPMANS, M. 2008. Viruses transmitted through the food chain: A review of the latest developments. *CAB Reviews: Perspectives in Agriculture, Veterinary Science, Nutrition and Natural Resources*, 78, 3–15.

VINJÉ, J., HAMIDJAJA, R. A. and SOBSEY, M.D. 2004. Development and application of a capsid VP1 (region D) based reverse transcription PCR assay for genotyping of genogroup I and II noroviruses. *Journal of Virological Methods*, 116, 109–117.

WANG, G., XIE, P., XIAO, C., YUAN, P. and SU, X. 2010. Magnetic fluorescent composite nanoparticles for the fluoroimmunoassays of Newcastle disease virus and avian virus arthritis virus. *Journal of Fluorescence*, 20, 499–506.

WANG, C., LIEN, K., HUNG, L., LEI, H. and LEE, G. 2011. An integrated microfluidic system for diagnosis and multiple subtyping of influenza virus. *2011 IEEE 24th International Conference on Micro Electro Mechanical Systems (MEMS)*, 841–844.

WHO 2000. Hepatitis A. *WHO Recommended Surveillance Standards*, 55–59.

WIDDOWSON, M., SULKA, A., BULENS, S., BEARD, R., CHAVES, S. *et al.* 2005. Norovirus and foodborne disease, United States, 1991–2000. *Emerging Infectious Diseases*, 11, 95–102.

WIDEN, F., VAGSHOLM, I., BELAK, S. and MURADRASOLI, S. 2011. Achievement V – Methods for breaking the transmission of pathogens along the food chain: Detection of viruses in food. *Trends in Food Science and Technology*, 22, S49–S57.

ZHANG, G., ZHANG, L., HUANG, M., LUO, Z., TAY, G. *et al.* 2010. Silicon nanowire biosensor for highly sensitive and rapid detection of dengue virus. *Sensors and Actuators B – Chemical*, 146, 138–144.

ZHANG, H., HARPSTER, M., PARK, H. and JOHNSON, P. 2011a. Surface-enhanced Raman scattering detection of DNA derived from the West Nile virus genome using magnetic capture of Raman-active gold nanoparticles. *Analytical Chemistry*, 83, 254–260.

ZHANG, H., LIU, L., LI, C., FU, H., CHEN, Y. *et al.* 2011b. Multienzyme-nanoparticles amplification for sensitive virus genotyping in microfluidic microbeads array using Au nanoparticle probes and quantum dots as labels. *Biosensors and Bioelectronics*, 29, 89–96.

ZHENG, G., PATOLSKY, F., LIEBER, C., SEAL, S., BARATON, M. *et al.* 2005. Multiplexed electrical detection of single viruses. *Semiconductor Materials for Sensing*, 828, 79–84.

ZHU, X., AI, S., CHEN, Q., YIN, H. and XU, J. 2009. Label-free electrochemical detection of avian influenza virus genotype utilizing multi-walled carbon nanotubes-cobalt phthalocyanine-PAMAM nanocomposite modified glassy carbon electrode. *Electrochemistry Communications*, 11, 1543–1546.

12

Tracking of pathogens via virulence factors: Shiga toxin-producing *Escherichia coli* in cattle and potential risks for human disease

J. Elder, Washington State University, USA and K. Nightingale, Texas Tech University, USA

DOI: 10.1533/9780857098740.3.227

Abstract: Enterohemorrhagic *Escherichia coli* (EHEC) belong to a group of emerging foodborne pathogens that are a significant public health threat throughout the world. Much is unknown about these organisms: their prevalence, their burden on public health and how they cause disease. The acquisition of virulence factors is responsible for transforming commensal *E. coli* into pathogens, as virulence factors allow for intimate adherence to and subsequent colonization of the gastrointestinal tract, evasion of the immune system and/or the ability to induce disease. Identifying virulence factors to serve as definitive markers of pathogenicity across EHEC seropathotypes, for molecular detection assays, is a focus of current research.

Key words: Shiga toxin-producing *Escherichia coli*, enterohemorrhagic *Escherichia coli*, virulence factors, cattle.

12.1 Introduction

Escherichia coli is normally a commensal microorganism found in the intestinal tracts of mammals; however, several pathotypes that are capable of causing intestinal disease in humans have evolved. These pathotypes include enterotoxigenic *E. coli* (ETEC), enteroinvasive *E. coli* (EIEC), enteropathogenic *E. coli* (EPEC), enteroaggregative *E. coli* (EAggEC) and diffusely adherent *E. coli* (DAEC). Enterohemorrhagic *E. coli*, the focus of this chapter, is a pathotype that has recently emerged and forms a subgroup of the Shiga toxin-producing *E. coli* (STEC), which are defined by the presence of one or more of the Shiga toxin genes (*stx*). STEC is a significant public health concern and economic burden in the US. These organisms exhibit a wide range of virulence characteristics, from non-pathogenic STEC that are carried asymptomatically to pathogenic EHEC, which have the potential to cause severe life-threatening disease, especially in immune-compromised individuals (Karmali, 1989). EHEC infections are characterized by hemolytic colitis (HC) and can lead to the severe complication, hemolytic uremic syndrome (HUS), the most common cause of renal failure in children under the age of five in the US (Scheiring *et al.*, 2008; Tarr *et al.*, 2005). The disparity in pathogenic ability of STEC is due to the combination of virulence factors located within the bacterium's genetic material (Reid *et al.*, 2000). Virulence factors allow pathogens to invade and colonize the host, avoid or escape the host's immune system, and/or induce disease (Peterson, 1996). The ability to identify STEC that harbor the virulence factors required to induce severe disease in humans has been identified as a global health priority by the World Health Organization (WHO) (WHO, 1998). Farmer and Davis (1985) discovered that EHEC belonging to serotype O157:H7 exhibited a unique phenotype that provided a relatively easy method for their identification. These *E. coli* exhibited delayed sorbitol fermentation and thereby could be differentiated from non-pathogenic *E. coli* on sorbitol MacConkey agar. The ability to easily detect and differentiate *E. coli* O157:H7 has likely led to disease incidence and outbreak investigation reporting bias, which has influenced the development of regulations that focus solely on this serogroup. However, it has become clear that there are other serotypes of *E. coli* that have the ability to cause severe disease equivalent to that of O157:H7 (Hughes *et al.*, 2006). These so-called non-O157 STEC include the serotypes other than O157 that carry *stx* gene(s) (i.e. O26, O45, O103, O111, O121 and O145) and so far no phenotype has been identified that can be used to differentiate them from non-pathogenic *E. coli*. Due to the difficulty in the detection and isolation of non-O157 STEC, current efforts are focused on the development of assays to detect the virulence factors responsible for transforming *E. coli* from benign commensals to potent pathogens. Selecting virulence factors for these assays is a challenge in itself as the processes by which STEC cause disease are not well understood, especially which virulence factors are required for full pathogenicity of STEC.

The central dogma of biology states that DNA provides the template for the transcription of RNA, RNA is then translated by the ribosome into protein, and proteins act as the catalytic machines that drive the cellular processes. Virulence factors are produced by pathogens following this process and therefore there are several points at which one can attempt to detect these virulence factors. Detection of the end product, a protein, is one option. Assays that depend on the recognition of a particular protein often rely on the use of antibodies. Antibodies are complex proteins produced by the immune systems of vertebrates, which bind strongly to specific molecules called antigens. Antigens are a wide array of molecules from simple sugars to large, complex proteins. There are methods for developing antibodies against specific antigens of interest. For example, antibodies can be raised against Shiga toxin and those antibodies may be used in assays to detect the presence of this toxin. However, there are drawbacks associated with using antibodies. The specificity of antibodies can be imperfect: cross reactions with molecules other than the target are not uncommon, particularly when a high load of non-target microorganisms is present in a sample. These assays may also be costly, difficult to perform and difficult to interpret. Bacteria do not express all of their proteins at all times and if the bacteria fail to express the targeted protein or antigen the detection kit will fail to detect them. On the other hand, molecular-based assays can be extremely sensitive and specific, and virulence factors can be detected regardless of the expression of the protein. Due to these factors, molecular-based assays for the detection of virulence factors, especially in the food industry, have become the method of choice.

The rest of the chapter further discusses the burden of EHEC and virulence factors associated with the most virulent strains. The chapter also includes a review of studies investigating the prevalence of STEC in cattle and beef products along with a review of current methods and assays used for the detection of STEC.

12.2 The impact of Shiga toxin-producing *E. coli* (STEC) on public health

12.2.1 Prevalence and epidemiology of EHEC infections

EHEC are recognized as primarily a public health concern of industrialized nations. In the US, *E. coli* O157:H7 are widely accepted as the etiological agents of most outbreaks and cases of severe disease attributed to EHEC since 1994 (Bell *et al.*, 1994); however, as surveillance of non-O157 STEC improved after O157 was declared an adulterant in ground beef in 1994, the estimated incidence of disease caused by this serotype has decreased, while the number of non-O157 STEC infections has increased (CDC, 2011). The most recent estimates of domestically acquired STEC infections in the US attribute 63 153 cases annually to *E. coli* O157 with 68% of these infections being foodborne. The number of domestically acquired non-O157 STEC cases was estimated to be 112 752 each year, 82% of which were foodborne (Scallan *et al.*, 2011). In preliminary data collected by the Foodborne Diseases Active Surveillance Network (FoodNet), of the non-O157 STEC isolates that were successfully serotyped, O26, O103 and O111 accounted for 78% of the infections (CDC, 2011). It was recognized by Karmali *et al.* (2003) that although many *E. coli* serogroups have been isolated from human clinical cases, only a small collection of serogroups accounted for the majority of human disease, which is reflected in the data presented above. Karmali and others organized STEC into a seropathotype classification system that separated the serogroups into seropathotypes A through E based on their association with outbreaks of foodborne illness and the severity of clinical manifestations (Karmali *et al.*, 2003).

E. coli O157:H7, which with O157:H- makes up seropathotype A, is the most commonly isolated EHEC from patients suffering from HUS in the US (Banatvala *et al.*, 2001), the British Isles (Lynn *et al.*, 2005), France (Espié *et al.*, 2008),

Table 12.1 *E. coli* pathotypes and STEC seropathotypes

Virotype	Seropathotype	Serotype	Virulence factors
Enteroinvasive *E. coli*	N/A	O28:NM, O124:H30, O124:NM, O136:NM, O143:NM, O144:NM, O152:NM, O164:NM	Large virulence plasmid similar to that of *Shigella* spp.
Diffusely adherent *E. coli*	N/A	N/A	Fimbrial adhesin F1845 or homologue
Adherent invasive *E. coli*	N/A	N/A	Type 1 pili, flagella, outer membrane vesicles, outer membrane protein C (OmpC)
Enterotoxigenic *E. coli*	N/A	O6:H16, O8:H19, O15:H11, O78:H11, O78:H12	Colonization factor antigen (CFA), heat-labile and heat-stable enterotoxins
Enteropathogenic *E. coli*	N/A	O55:H6, O55:NM, O111:H2, O111:NM, O119:H6, O119:NM, O125:H21, O126:H27, O127:H6, O127:NM, O128:H2, O128:H12, O142:H6	LEE, partial to complete OI-122
Enterohemorrhagic *E. coli*	A	O157:H7, O157:NM	*stx*, LEE, complete OI-122
	B	O26:H11, O103:H2, O111:NM, O121:H19, O145:NM	*stx*, LEE, partial to complete OI-122
	C	O91:H21, O104:H21, O113:H21, others	*stx, saa, subAB*, usually partial OI-122
Shiga toxin-producing *E. coli*	D	Multiple	*stx*
	E	Multiple	*stx*
Enterohemorrhagic *E. coli* with aggregative adherence	N/A	O104:H4, O111:H2	*stx*, AAF
Enteroaggregative *E. coli* (EAggEC)	N/A	O3:H2, O15:H18, O77:H18, O86:NM, O111:H21, O126:H27, O127:H2	AAF, EAST

N/A, not applicable. HUS, hemolytic uremic syndrome.

Germany (Gerber *et al.*, 2002) and Japan (Takeda, 1999) (Table 12.1). For a review of *E. coli* pathotypes see Kaper *et al.*, 2004. Studies in other regions have isolated an alarming number of non-O157 STEC, those that make up seropathotypes B and C, from patients suffering from HUS. A significant proportion (43%) of STEC-related pediatric HUS cases in Germany and Austria from 1997 to 2000 were attributed to non-O157:H7 STEC (Gerber *et al.*, 2002). Non-O157 STEC were isolated from the majority of patients suffering from diarrhea and slightly less than half of the patients with HUS in Denmark (Ethelberg *et al.*, 2004). Non-O157 STEC were found to be responsible for the majority of STEC associated HUS cases in Italy (Tozzi *et al.*, 2003). In a surveillance of HUS cases in Australian children from 1994 to 1998, serotype O157:H7 was not isolated and serotype O111:H- was the most commonly isolated (Elliott *et al.*, 2001). Brazil has an unusually low rate of HUS and HC associated with STEC compared to Argentina, which has the highest rate of HUS in children in the world (López, 1998). The first reported case of STEC-related HUS in Brazil was described in 2001 and this case was attributed to serotype O26:H11 (Guth *et al.*, 2002). *E. coli* O157:H7 has been isolated from human disease cases in Brazil but does not appear to be as significant as non-O157 STEC (Irino *et al.*, 2002; Vaz *et al.*, 2004). Studies in Argentina have found conflicting results regarding the significance of non-O157 STEC in disease. Rivas *et al.* (2006) found that a majority (59%) of STEC strains isolated from children were *E. coli* O157; however, in a study conducted 13 years earlier Lopez *et al.* (1989) isolated O157:H7 from only one patient out of 51 suffering from HUS and two of 44 with diarrhea. Current detection and isolation methods for *E. coli* O157 typically rely on the picking of a few morphologically typical colonies from some selective and differential agar. Despite the lack of sensitive diagnostic methods, the data presented here indicate that non-O157 STEC are a significant cause of severe disease and the development of more sensitive detection methods and increased surveillance will aid in the determination of the true public health impact of non-O157 STEC. Non-O157 STEC probably represents a public health threat that is equivalent to that represented by *E. coli* O157:H7, based on disease frequencies and clinical manifestations of disease (Brooks *et al.*, 2005) (see Table 12.2).

12.2.2 Risk characterization and regulatory definition of STEC

As a result of the public health threat represented by non-O157 STEC, the US Department of Agriculture's (USDA) Food Safety Inspection Service (FSIS) declared on 5 March 2012 that six non-O157 serotypes (i.e., O26, O45, O103, O111, O121 and O145) are to be considered adulterants in raw, non-intact beef products after the agency implemented routine testing for these organisms. The FSIS has determined that a sample will be deemed potential positive when it produces positive polymerase chain reaction (PCR) results for *stx* (either stx_1, stx_2 or both), *eae* and one of the O-antigen-specific genes from serotypes O26, O45, O103, O111, O121 and O145 (USDA, 2011b). The FSIS expects that most samples will be found negative by the initial screening and results will be available

Table 12.2 Non-O157 STEC outbreaks since 2005 (US)

Year	Serogroup	Total ill	Confirmed	Vehicle	Hospitalizations	HUS	Deaths
2005	O45:NM	52	16	Food worker	3	0	0
2006	O121:H19	73	4	Iceberg lettuce	3	3	N/A
2006	O121	5	N/A	Person to person (daycare)	5	Yes	N/A
2006	O26	5	N/A	Strawberries or blueberries	1	0	N/A
2006	O45	11	N/A	Animal contact (goats)	0	0	0
2007	O45	5	N/A	Animal contact	0	0	0
2007	O111	23	N/A	Ground beef	0	0	0
2007	O121, O26, O84	135	N/A	Cheese or margarine	10	N/A	0
2007	O157:NM	7	5	Animal contact	2	N/A	0
2007	O111	6	N/A	Person to person (elementary school)	1	N/A	N/A
2007	O111	8	N/A	Person to person (daycare)	0	0	0
2007	O111	8	N/A	Unknown	0	N/A	0
2008	O111	3	N/A	Person to person (daycare)	N/A	N/A	N/A
2008	O111	344	60	Unknown (restaurant)	71	26	1
2008	O157:NM	14	7	Raw milk	5	N/A	0
2008	O157:NM	9	N/A	Unknown (restaurant)	0	N/A	0
2010	O145	7	N/A	Smoked game	1	N/A	N/A
2010	O26	10	10	Unknown (daycare)	0	0	0
2010	N/A	3	N/A	Unknown (fair)	N/A	N/A	N/A
2010	O26	3	N/A	Ground beef	0	0	0
2010	N/A	24	N/A	Well water	N/A	N/A	N/A
2010	O26	14	13	Person to person (daycare)	N/A	N/A	N/A
2010	O145	33	26	Romaine lettuce	13	N/A	0
2010	O111:NM	10	10	Unknown (correctional facility)	N/A	N/A	N/A
2011	O26	8	N/A	Unknown	0	0	0

N/A, not applicable. HUS, hemolytic uremic syndrome.

within 48 hours of the samples arriving in the laboratory. By targeting these serotypes, the FSIS hopes to reduce the illness associated with the serotypes implicated in the majority of human disease, but it should be understood that these six serotypes do not account for all the non-O157 STEC that have the ability to cause severe disease and outbreaks, and defining a positive by these metrics will exclude some EHEC. Although less common, there are documented cases of HC and HUS attributed to strains lacking the *eae* gene (Bonnet *et al.*, 1998; Elliott *et al.*, 2001; Paton *et al.*, 1999). Using a PCR screen for these genes from an enriched culture is not without drawbacks. The enriched samples will undoubtedly be mixed cultures and it is not unlikely that some of the organisms will harbor the *eae* gene and others may harbor *stx* and/or the O-antigen-specific genes. This problem is addressed in confirmation steps that would follow a positive finding by the initial screening; however, a food product may be held unnecessarily during the confirmatory steps, which is undesirable, especially for raw non-intact beef products that have a short shelf life. A test result is considered as presumptive positive when there are one or more colonies that agglutinate beads specific for the O-antigen are detected by the second RT-PCR reaction. The latex positive colonies are confirmed if PCR finds them to be *eae* and *stx* positive and positive for the O-antigen-specific gene detected previously. A sample is determined to be a confirmed positive when the pure culture isolated carries these genes and is biochemically confirmed as *E. coli* (USDA, 2011a). In the future it may be prudent to include more genes within some of the PCR reactions. As the virulence potential of STEC is better understood, tests could be tailored to more accurately assess the pathogenicity of isolates found in food; however, until then the best approach may be to target those serotypes that account for the majority of disease incidence.

12.3 STEC virulence factors

Shiga toxin-producing *E. coli* are a group of pathogens that have emerged recently. Although the evolution of this group of pathogens has not been entirely elucidated, the evidence points to the parallel evolution of virulence through horizontally acquired factors (Ogura *et al.*, 2009; Reid *et al.*, 2000). These virulence genes are found on mobile elements such as prophages, integrative elements and plasmids (Ogura *et al.*, 2009) (see Table 12.3). When Ogura *et al.* (2009) compared the genome sequences of serotypes O26, O103, O111 and O157 they found that, although these pathogens have a similar repertoire of virulence genes, the structure of the mobile elements and location of virulence genes within the genomes were quite different. There seems to be a selective pressure for STEC to acquire mobile elements that contain homologous genes or genes that fulfill a similar role in pathogenesis.

Prophages are responsible for shuttling the genes encoding the Shiga toxins, stx_1 and stx_2, into STEC. Numerous pathogenicity islands (PAIs), a type of integrative genetic element, are found in STEC including the locus of enterocyte effacement (LEE). The LEE, which is absent in non-pathogenic *E. coli* but found

Table 12.3 EHEC virulence factors

Name	Gene	Location	Putative function	References
Intimin	*eae*	LEE	Adherence (LEE positive)	McKee *et al.*, 1995
Translocated intimin receptor	*tir*	LEE	Adherence (LEE positive)	Frankel *et al.*, 2001
Tir-cytoskeleton coupling protein	*tccP/espF*$_u$	Prophage CP-933U of O157:H7	Actin polymerization (LEE positive)	Campellone *et al.*, 2004; Garmendia *et al.*, 2004
STEC autoagglutinating adhesin	*saa*	pO113	Adherence (LEE negative)	Paton *et al.*, 2001
Sorbitol fermenting EHEC O157 fimbriae	*sfp*	pSFO157	Adherence (O157:NM)	Brunder *et al.*, 2001
IrgA homologue adhesion	*iha*	OI43, OI48, pO113	Adherence (non-O157 STEC)	Tarr *et al.*, 2000
EHEC factor for adherence or lymphostatin	*efa-1/lifA*	OI122	Adherence (LEE positive)	Nicholls *et al.*, 2000
E. coli common pilus	*ecpA*	Chromosome	Adherence (all *E. coli*)	Rendón *et al.*, 2007
Long polar fimbriae	*lpfA*	OI141, OI154	Adherence (all seropathotypes)	Doughty *et al.*, 2002; Toma *et al.*, 2004; Torres *et al.*, 2002
Hemorrhagic coli pilus	*hcpA*	Chromosome	Adherence	Xicohtencatl-Cortes *et al.*, 2007
Non-LEE encoded effectors	*nleA/espI*	OI71	Disruption of host intestinal barrier	Gruenheid *et al.*, 2004; Thanabalasuriar *et al.*, 2010
	nleB	OI122	Anti-inflammatory, inhibits NF-kappaB	Newton *et al.*, 2010
	nleC	OI36	Anti-inflammatory, inhibits NF-kappaB	Baruch *et al.*, 2011; Pearson *et al.*, 2011
	nleD	OI36	Anti-inflammatory, inhibits NF-kappaB	Baruch *et al.*, 2011
	nleE	OI122	Anti-inflammatory, inhibits NF-kappaB	Nadler *et al.*, 2010; Newton *et al.*, 2010; Vossenkämper *et al.*, 2010
	nleF	OI71	Unknown	Echtenkamp *et al.*, 2008

	nleG	OI57, OI71	Modulation of host ubiquitination	Wu *et al.*, 2010
	nleH	OI36, OI71	Inhibition of enterocyte exfoliation	Hemrajani *et al.*, 2010
	nleL/espX7	Prophage Sp6	Regulation of pedestal formation	Piscatelli *et al.*, 2011
E. coli secreted proteins	*espA*	LEE	T3SS component that forms a bridge between bacterium and host	Knutton *et al.*, 1998
	espF	LEE	Induction of host cell apoptosis and disruption of host intestinal barrier	Crane *et al.*, 2001; McNamara *et al.*, 2001
	espJ	Prophage CP933U	Inhibition of phagocytosis	Dahan *et al.*, 2005; Marchès *et al.*, 2008
	espK	Prophage CP933N	Unknown	Vlisidou *et al.*, 2006
	espO	Prophage	Inhibition of enterocyte exfoliation	Kim *et al.*, 2009
Serine protease	*espP*	pO157	Adherence	Brunder *et al.*, 1997
	espZ	LEE	Inhibition of enterocyte exfoliation	Shames *et al.*, 2010
Cycle inhibiting factor	*cif*	Prophage	Inhibition of enterocyte exfoliation by suspending mitosis	Marchès *et al.*, 2003
Subtilase cytotoxin	*subAB*	pO113	Cytotoxin	Paton *et al.*, 2004
	toxB	pO157	Inhibition of host immune activation and adherence	Klapproth *et al.*, 2000; Stevens *et al.*, 2004; Tatsuno *et al.*, 2001
Enterohemolysin	*ehx*	pO157, pO113, pO26	Cytotoxin, unknown	Schmidt *et al.*, 1994
Catalase-peroxidase	*katP*	pO157	Protection from oxidative damage generated by host immune system	Brunder *et al.*, 1996
STEC autotransporter mediating biofilm formation	*sab*	pO113	Biofilm formation	Herold *et al.*, 2009
Autotransporter protein	*epeA*	pO113	Proteolysis and destruction of host mucus barrier	Leyton *et al.*, 2003

in EHEC and EPEC, contains a cassette of genes for the formation of a type III secretion system (T3SS) as well as effector molecules required for the formation of the characteristic attaching and effacing (A/E) lesions (McDaniel *et al.*, 1995). Other putative virulence genes are found on virulence plasmids such as pO113, pO26 and pO157.

These A/E lesions in the human gastrointestinal tract are pathognomonic of EHEC and EPEC infections. These lesions are characterized by the intimate attachment of the bacterium with the intestinal epithelium, effacement of the microvilli and host actin polymerization to form pedestals on which the bacterium appears to rest. As described above, the majority of the genes required to produce A/E lesions are encoded within the LEE, a 35 kb PAI (McDaniel *et al.*, 1995). In LEE positive STEC, intimate adherence to host mucosa is mediated by a bacterial outer membrane bound protein, intimin (Eae), which binds the translocated intimin receptor (Tir) (Frankel *et al.*, 2001). The cloning of an EHEC LEE from O157:H7 into *E. coli* K12 was not sufficient to cause the production of A/E lesions on HEp-2 cells (Elliott *et al.*, 1999), which led to the discovery of an additional virulence-associated factor located outside the LEE, which is required by EHEC in the formation of A/E lesions. This factor, Tir-cytoskeleton coupling protein (TccP), was determined to be required for efficient actin polymerization in pedestal formation (Campellone *et al.*, 2004; Garmendia *et al.*, 2004). Although relatively rare, cases of human disease have been associated with LEE-negative STEC, primarily from seropathotype C (Mellmann, 2009; Paton *et al.*, 1999; Vidal *et al.*, 2008). LEE-negative strains such as O113:H21 adhere differently and do not produce A/E lesions; this serotype, as well as serotypes O91:H21 and O104:H21, has acquired alternative adhesion factors such as the STEC autoagglutinating adhesion (Saa) (Bugarel *et al.*, 2010; Paton *et al.*, 2001). Saa allows the bacteria to adhere to epithelial cells in a semilocalized adherence pattern. Although the mechanism by which these serotypes, as well as other LEE-negative strains, attach is not well understood, it is likely that they rely on multiple factors and mechanisms for full adherence (Toma *et al.*, 2008). Other putative adherence factors in EHEC, proposed to either complement Eae or fulfill its role in pathogenesis, include: sorbitol fermenting EHEC O157 fimbriae (Sfp) (Brunder *et al.*, 2001), IrgA homologue adhesion (Iha) (Tarr *et al.*, 2000), EHEC factor for adherence (Efa1) (Nicholls *et al.*, 2000), *E. coli* common pilus (ECP) (Rendón *et al.*, 2007), long polar fimbriae in EHEC O113:H21 ($LpfA_{O113}$) (Doughty *et al.*, 2002), $LpfA_{O26}$ (Toma *et al.*, 2004), $LpfA_{O157/O1-141}$ (Torres *et al.*, 2002), $LpfA_{O157/O1-154}$ (Toma *et al.*, 2004) and hemorrhagic coli pilus (HCP) (Xicohtencatl-Cortes *et al.*, 2007).

In addition to virulence factors aiding the attachment, T3SS effector molecules may modulate the virulence of EHEC. These molecules are transported into the host cells by the T3SS, carried within the LEE, where they are responsible for the manipulation of host cellular processes (Bhavsar *et al.*, 2007; Galán and Collmer, 1999). A homology search of the genome sequence of the O157:H7 Sakai strain identified 49 effectors that were potentially functional and 39 of these were confirmed to be functional (Tobe *et al.*, 2006). The majority of the effectors found in this search were located in lambdoid prophages (Tobe *et al.*, 2006). In addition to prophages, O islands (OIs) 122, 57 and 71 harbor these effectors, termed non-LEE effectors (Nle's) and *E. coli* secreted proteins (Esp's). These OIs have been linked to strains of EHEC that belong to seropathotypes A, B and C, which are responsible for outbreaks and severe disease, with the strains carrying the most complete OIs associated with the most severe disease and outbreaks (Coombes *et al.*, 2008; Karmali *et al.*, 2003; Konczy *et al.*, 2008). Of particular interest are the genes *nleA, nleB, nleF, nleH1-2* and *ent/espL2*. The genes *nleB* and *nleE*, of OI-122, have been studied in the mouse model using the LEE-positive bacterium, *Citrobacter rodentium*, which forms A/E lesions in the colons of mice. Deletion mutants of either gene prevented mortality in mice (Gruenheid *et al.*, 2004; Wickham *et al.*, 2006), suggesting they could play a pivotal role in the virulence of EHEC. The protein NleF may also play a role in virulence, as *nleF* deletion mutants show significantly reduced colonization of *C. rodentium* when co-infected with the wild type parent strain in a competitive colonization assay in mice (Echtenkamp *et al.*, 2008). Several NleG proteins have been found to share homology with eukaryote proteins involved in the process of ubiquitination of host proteins and therefore have the potential to modulate host cell activity in a way that is advantageous to EHEC (Wu *et al.*, 2010). NleH has been shown to inhibit apoptosis and exfoliation of host enterocytes, which could prevent the detachment of the bacteria (Hemrajani *et al.*, 2010). NleC has been shown to be an inhibitor of the innate immune system (Pearson *et al.*, 2011). *E. coli* secreted protein K (EspK), located on the prophage CP-933N, is another type III effector molecule that has been shown to affect colonization of calves (Vlisidou *et al.*, 2006). A homologue of EspO, OspE in *Shigella flexneri*, has been shown to prevent the shedding of host gut epithelial cells, which could aid in prolonging infection (Kim *et al.*, 2009). The cycle inhibiting factor (Cif) found in both EHEC and EPEC, prevents the progression of the host cell cycle, which prevents exfoliation of enterocytes (Marchès *et al.*, 2003). It has been suggested that *nle* genes may have an additive effect on virulence, as strains of the most virulent seropathotypes (i.e., A, B and C) have been found to carry the most complete sets of *nle* genes (Coombes *et al.*, 2008; Karmali *et al.*, 2003). However, with the large number of putative effectors, the variability in the number of and alleles of effectors in disease-causing strains, and the redundancy in effector functions make it difficult to conclusively determine which T3SS effectors contribute to virulence. However, with the large number of effectors and the variability observed between strains, it is difficult to conclusively determine which T3SS effectors contribute to virulence. Knockout studies, where gene function is prevented through mutations, cannot account for the synergy of multiple factors: the process of pathogenesis is too complex to target only a single gene and conclusively determine its role.

In addition to the O islands, the majority of EHEC, both LEE-positive and LEE-negative strains, carry a variable repertoire of putative virulence factors on plasmids (Beutin *et al.*, 1989, 1994; Sandhu *et al.*, 1997; Schmidt *et al.*, 1995). The *ehx* gene cluster, encoding EHEC hemolysin (*ehx*), is a characteristic feature of STEC virulence plasmids and is found on the *E. coli* O157:H7 plasmid, pO157.

In addition to carrying the genes for *ehx* (*ehxCABD*), pO157 also has the genes for a catalase-peroxidase (*katP*), serine protease (*espP*) and a type II secretion system (*etpC-O*) (Makino *et al.*, 1998). KatP, identified by Brunder *et al.* (1996), is a catalase-peroxidase that may contribute to the protection of STEC from oxidative damage (Uhlich, 2009). ToxB was found to be required for the full adherence of O157 strains to epithelial cells in *in vitro* experiments and *toxB* deletion mutants exhibited reduced expression of T3SS effector proteins (Stevens *et al.*, 2004; Tatsuno *et al.*, 2001). ToxB was also found to inhibit the activation of host immune cells (Klapproth *et al.*, 2000). *toxB* was found in 50% of serotype O26 isolates but was not found in serotype O103 or O111 isolates (Tozzoli *et al.*, 2005). ToxB is carried on the pO113 virulence plasmid of LEE-negative STEC (Newton, 2009). ToxB may thus play a role in virulence in a subset of EHEC. EspP, a secreted protein, has the ability to cleave human coagulation factor V and has been found in seropathotypes A and B (Brockmeyer *et al.*, 2007; Brunder *et al.*, 1997). In addition to cleaving human coagulation factor V, EspP shows homology to PssA, a toxin found in *E. coli* serotype O26:NM (Djafari *et al.*, 1997), and was found to be related to the adherence of STEC to the gut mucosa of calves (Dziva *et al.*, 2007). Plasmids pO113 and pO26-Vir have also been shown to carry *espP* (Fratamico *et al.*, 2011; Leyton *et al.*, 2003). In addition to sharing the *ehx* operon and *espP*, pO113 contains the putative adhesion *iha*, which is also found on pO157 (Newton, 2009). This plasmid was shown to have evolved independently of pO157 despite sharing these genes (Boerlin *et al.*, 1998; Newton, 2009). Putative virulence factors unique to pO113 include; an autotransporter protein (EpeA), Saa, subtilase cytotoxin (SubAB), STEC autotransporter mediating biofilm formation (Sab) and LH0189. EpeA may contribute to virulence with its proteolytic and mucinase activities (Leyton *et al.*, 2003). The *saa* gene mentioned above, is responsible for the diffuse attachment phenotype of some LEE-negative EHEC (Paton *et al.*, 2001a). SubAB is a potent cytotoxin found only in *eae*-negative strains (Irino *et al.*, 2010; Paton *et al.*, 2004). Sab contributes to the adherence of STEC as well as biofilm formation (Herold *et al.*, 2009). The plasmid pSFO157 of *E. coli* O157:H- shares a common ancestor with pO157 but lacks the genes *espP* and *katP* but contains the aforementioned *sfp*, which may contribute to adherence (Brunder *et al.*, 2006). The plasmid pO26-Vir contains homology with pO157 but also contains the genes necessary for the formation of a type IV pilus, which may contribute to virulence (Fratamico *et al.*, 2011). A previous study identified this locus on pO113 but did not find it in LEE-positive STEC of serotype O26 (Srimanote *et al.*, 2002), in contrast to the findings of Fratamico *et al.* (2011). Many of the virulence genes found on these plasmids are associated with insertion elements and DNA recombinases (Newton, 2009), which allow these genes to jump in and out of the plasmids, resulting in the observed variability. Collectively, these previous studies indicate that the properties of these virulence plasmids are highly variable which, like the OI genes, makes it difficult to identify conclusively the full cassette of genes and networking among them required for the full virulence of EHEC.

The Shiga toxins (Stx's) are the defining feature of the STEC and are essential in pathogenesis and the progression of the severe complications associated with some EHEC infections. These toxins are found on temperate, lambdoid bacteriophages that lysogenize the *E. coli* host and incorporate their viral DNA into the *E. coli* chromosome. After lysogeny, the *E. coli* hosts express the Shiga toxins and are thereby transformed into STEC. Stx binds to the Gb3 receptor found on some mammalian cells and induces cell death by blocking translation. In the case of severe human disease the toxin enters the bloodstream and causes endothelial damage that leads to HUS or thrombotic thrombocytopenic purpura (Ray and Liu, 2001). It has been suggested that the toxin binds to polymorphonuclear leukocytes (PMNs), which inadvertently deliver the toxin to organs beyond the gastrointestinal tract (te Loo *et al.*, 2000); however, the role of PMNs in toxin delivery remains controversial. Although both Stx_1 and Stx_2 are capable of inducing toxic effects in cells bearing the Gb3 receptor, strains carrying the stx_2 gene are more apt to cause severe disease compared to strains expressing stx_1 (Boerlin *et al.*, 1999). A standard nomenclature and typing method for both stx_1 and stx_2 was recently proposed by Scheutz and others (2012), which uses the sequence of the toxin genes to classify stx_1 into three subtypes (a, c and d) and stx_2 into subtypes a through g. While the sequence of stx_1 is fairly conserved, with only three known variants, stx_{1a}, stx_{1c} (Zhang *et al.*, 2002) and stx_{1d} (Burk *et al.*, 2003), many subtypes and subtype variants of stx_2 have been identified including stx_{2a} of *E. coli* O157:H7 EDL933 (Datz *et al.*, 1996), stx_{2c} (Schmitt *et al.*, 1991), stx_{2d} (Jelacic *et al.*, 2003; Melton-Celsa *et al.*, 1996; Pierard *et al.*, 1998), stx_{2e} (Weinstein *et al.*, 1988), the highly divergent stx_{2f} (Schmidt *et al.*, 2000) and stx_{2g} (Leung *et al.*, 2003). Strains expressing stx_{2a} alone or with stx_{2c} are the strains most frequently isolated from patients suffering HUS or TTP (Eklund *et al.*, 2002; Friedrich *et al.*, 2002; Persson *et al.*, 2007). Stx_{2d} is also associated with severe disease in humans caused by LEE-negative strains (Bielaszewska *et al.*, 2006; Jelacic *et al.*, 2003; Ito *et al.*, 1990; Melton-Celsa *et al.*, 1996). Several studies have also found that the vast majority of STEC infections that lead to HUS are those that carry stx_2 and *eae* (Boerlin *et al.*, 1999; Werber *et al.*, 2003). Supporting the epidemiological results, the toxicity of Stx_2 was found to be much higher than that of Stx_1 (Siegler *et al.*, 2003; Tesh *et al.*, 1993).

A recent outbreak of HC and HUS in Germany, associated with bean sprouts, was attributed to a unique strain of *E. coli* belonging to serotype O104:H4 (Struelens *et al.*, 2011). This strain carried a combination of virulence factors from two *E. coli* pathotypes, including EAggEC and EHEC. EAggEC are associated with infantile diarrhea in developing countries (Bhan *et al.*, 1989a, 1989b; Cravioto *et al.*, 1991; Wanke *et al.*, 1991) but has rarely been associated with HC and HUS in industrialized nations (Morabito *et al.*, 1998). The stx_2 gene this outbreak strain carried was nearly identical to the stx_{2a} found in *E. coli* O157 EDL933, differing by only one nucleotide (Struelens *et al.*, 2011). Phylogenetic analysis of the genome found that the O104:H4 outbreak strain was more closely related to EAggEC compared to the EHEC of serotypes O26:H11, O111:H- and O103:H2 (Mellmann *et al.*, 2011). A strain of *E. coli* with the aggregative adherence phenotype and which also carried stx_2 has been reported from an outbreak of HUS in France (Morabito *et al.*, 1998); additionally one of the groups that analyzed the genome of the recent O104:H4 outbreak strain, sequenced the genome of another EAggEC O104:H4 strain, which carried stx_2 and was the etiological agent of a case of HUS in Germany ten years prior to the sprout outbreak. Comparison of the genomes from the 2011 outbreak strain, the 2001 HUS case and that of another typical EAggEC O104:H4 strain, which lacked stx_2, led the authors to conclude that the two strains carrying stx_2 share a common EAggEC progenitor, which carried stx_2. Since diverging from this progenitor the strains have acquired and lost different plasmids. The 2011 EAggEC O104:H4 strain acquired a plasmid that contains aggregative adherence fimbriae type I (AAF/I) as well as a 90 kb resistance plasmid and lost the plasmid with the aggregative adherence fimbriae type III (AAF/III) and the gene for EAggEC heat-stable enterotoxin (*astA*) found in the proposed progenitor. The 2001 strain gained a 95 kb resistance plasmid, while retaining the plasmid for AAF/III and *astA*. Both the 2001 and 2011 strains had the adherence factor *iha* associated with LEE-negative STEC that was listed above (Mellmann *et al.*, 2011). The high rate of HUS associated with the O104:H4 outbreak strain could be explained by this unique combination of virulence factors. Besides the virulence factors required for adherence, the 2011 strain harbored many resistance genes including those for tellurite, mercury, extended-spectrum beta-lactams, chloramphenicol, tetracycline and streptomycin resistance (Brzuszkiewicz *et al.*, 2011), although antibiotics are not used in the treatment of EHEC infections. The aggregative adherence displayed by this strain was equivalent to, if not more effective, than that of typical EHEC that form A/E lesions in the colonization of the host gut mucosa.

12.4 Prevalence of STEC in cattle and other reservoirs

12.4.1 Prevalence of STEC in cattle

Ruminants, especially cattle, have been recognized as the major reservoir of STEC (Beutin *et al.*, 1993; Montenegro *et al.*, 1990; Wells *et al.*, 1991). The STEC carried by ruminants may be transmitted to humans through several routes, including direct contact (Caprioli *et al.*, 2005; Heuvelink *et al.*, 2002), insufficiently pasteurized or raw milk products (Allerberger *et al.*, 2003; Keene *et al.*, 1997a), ground beef (Caprioli *et al.*, 2005), or fecal contamination of vegetables (Caprioli *et al.*, 2005; Söderström *et al.*, 2008), fruit (Caprioli *et al.*, 2005; Cody *et al.*, 1999), and water (Hrudey *et al.*, 2003).

Cattle worldwide have been found to carry STEC; however, a wide range in the prevalence of STEC carriage among cattle has been reported. A review examining the worldwide STEC prevalence in cattle found that non-O157 STEC prevalence rates have been reported to range from 4.6% to 55.9%, 4.7% to 44.8%, and 2.1% to 70.1% in feedlots, grazing cattle, and at the time of slaughter respectively (Hussein, 2007). The prevalence of both O157 and non-O157 strains vary by season with O157 strains having the highest prevalence on cattle hides during the summer months and non-O157 in the fall months (Barkocy-Gallagher *et al.*, 2003). Multiple studies have described an elevated prevalence of STEC carriage among cattle in South America. For example, 62.7% of feedlot cattle in Argentina were found to carry STEC at some point in the duration of a six-month study (Padola *et al.*, 2004). A study looking at seasonal variation of STEC shedding found overall that 37.5% and 43% of Argentinean milking cows and calves were positive for *stx* (Fernández *et al.*, 2009). Another study of beef and dairy cattle in Brazil found the *stx* genes in 71% of the fecal samples collected; this same study was the first to isolate *E. coli* O157:H7 in Brazil (Cerqueira *et al.*, 1999). Another surveillance study of calves in Brazil found 12.7% of fecal samples to be *stx* positive. Sanz *et al.* (1998) isolated STEC from 33% of the Brazilian cattle and calves included in their study. STEC were isolated from 49% of dairy cattle samples in Brazil (Moreira *et al.*, 2003).

A lower prevalence of STEC in cattle was found in several Asian countries. A study of cattle in Japan prior to slaughter found 37.2% of the cattle were STEC positive, defined by a positive PCR result for *stx* (Fukushima and Seki, 2004) another study of calves and pre-slaughter cattle found an incidence rate of 40.8% (Shinagawa *et al.*, 2000). A study of healthy cattle in Thailand found 18.7% of the feces collected were positive for stx_1 and/or stx_2 (Panutdaporn *et al.*, 2004). STEC were cultured and isolated from 23% of cattle in Vietnam (Vu-Khac and Cornick, 2008).

The US and Europe appear to have a more moderate carriage rate in cattle, similar to the findings in Asia. A study of Swiss cattle found that 57.6% of fecal samples were positive for either *stx* gene (Kuhnert *et al.*, 2005). The genes for either *stx* were found in 18.1% of fecal samples from healthy cattle in France (Rogerie *et al.*, 2001). In a study of German cattle, the authors were able to isolate STEC from 10.8% of the animals (Montenegro *et al.*, 1990). In a study involving a cattle herd in Scotland, 25% of fecal samples were positive for STEC; however, the authors were only able to recover STEC isolates from 13% of the samples (Jenkins *et al.*, 2002). In a study of dairy cattle on US farms in the states of Wisconsin and Washington, 8.4% of dairy cows and 19.0% of heifers and calves harbored non-O157 STEC, which the authors were able to isolate (Wells *et al.*, 1991).

It is important to note that it is difficult to compare the results of the various prevalence studies presented here. A standard isolation method for non-O157 STEC has not been established and as a result the authors of the studies cited here have employed a variety of methodologies. One should also recognize that detecting *stx* using enrichment with a rapid screening method (i.e., PCR) is not equivalent to the isolation and subsequent confirmation of a STEC strain from a sample. STEC are isolated at a much lower rate than PCR positive screens for *stx* genes. A PCR screen will detect a few copies of the *stx* gene if the gene is found in dead or injured cells, organisms other than *E. coli* or bacteriophages – none of which would yield a viable STEC isolate that would be culturally confirmed as STEC. The factors affecting STEC shedding are not well understood; variables such as season, feed and animal management practices may add to the variability observed between different studies.

Studies of STEC prevalence in beef products have also been conducted to probe the food attribution of STEC infections to beef. A study examining carcasses in a slaughter plant in Mexico found 20.5% of carcasses tested positive for non-O157 STEC and 7.7% of carcasses were positive for either *E. coli* O157:H7 or *E. coli* O157:NM (Varela-Hernández *et al.*, 2007). In a recent study conducted by Bosilevac and Koohmaraie (2011), the *stx* genes were detected in 24.3% of ground beef samples collected in the US. In a study characterizing STEC strains isolated from food, strains in beef products carried the *stx* variants most commonly attributed to severe human disease (Beutin *et al.*, 2007). Overall, these results indicate that cattle are a major reservoir of STEC worldwide and STEC carried in cattle have the potential to enter the human food chain and cause disease.

12.4.2 Prevalence of STEC in other reservoirs

Ruminants other than cattle have been shown to carry STEC including wild deer (Asakura *et al.*, 1998; Bardiau *et al.*, 2010; Martin and Beutin, 2011; Sánchez *et al.*, 2009), sheep (Beutin *et al.*, 1993; Martin and Beutin, 2011; Sekse *et al.*, 2011), goats (Beutin *et al.*, 1993; Martin and Beutin, 2011) and water buffalo (Oliveira *et al.*, 2007; Vu-Khac and Cornick, 2008). A study of STEC in Spain found that the majority of the strains carried by sheep belonged to serotypes that have also been isolated from human disease; 51% of the serotypes were previously associated with HUS (Blanco *et al.*, 2003). Despite the fact that serotypes associated with human disease have been isolated from sheep, the strains associated with these animals typically carry the Shiga toxin variant stx_{1c}, which is not associated with human disease (Brett *et al.*, 2003; Koch *et al.*, 2001; Urdahl *et al.*, 2002a), and although cases of HUS and HC have been linked to goats and sheep (Bielaszewska *et al.*, 1997; Espié *et al.*, 2006; Urdahl *et al.*, 2002b) these animals are not assumed to be a significant source of outbreaks of human disease.

Ruminants are not the only reservoirs of STEC. The first isolation of an stx_2 variant, stx_{2f}, was from feral pigeons, which are considered to be reservoirs of STEC bearing this gene (Schmidt *et al.*, 2000). The *stx* genes were detected in 10.8% of pigeon fecal enrichments (Morabito *et al.*, 2001) and other studies have detected or isolated STEC from bird feces (Farooq *et al.*, 2009; Kobayashi *et al.*, 2009; Makino *et al.*, 2000; Wallace *et al.*, 1997). It has been suggested that pigeons and other birds may serve as reservoirs of STEC that are potentially pathogenic to humans as STEC carrying stx_{2f} have been isolated from cases of diarrhea (Gannon *et al.*, 1990; Isobe *et al.*, 2004; Prager *et al.*, 2009; Sonntag *et al.*, 2005); however, cases associated with stx_{2f} are rare. Swine are recognized as an important reservoir for STEC carrying the stx_{2e} variant (Fratamico *et al.*, 2004, 2008), and STEC are found in pork products (Bouvet *et al.*, 2002; Brooks *et al.*, 2001; Read *et al.*, 1990). However, these STEC most commonly carry the stx_2 variant stx_{2e}, which causes edema disease in weaned pigs (MacLeod *et al.*, 1991; Moxley, 2000) and, on occasion, has been associated with human disease (Pierard *et al.*, 1991; Thomas *et al.*, 1994), but this variant does not seem to be significantly linked to severe illness in humans (Beutin *et al.*, 2004, 2008; Franke *et al.*, 1995; Friedrich *et al.*,

2002). STEC isolated from pork products and swine often lack virulence factors required to cause disease in humans (Bouvet *et al.*, 2002; Desrosiers *et al.*, 2001).

Other animals from which STEC have been isolated include dogs (Beutin *et al.*, 1993), cats (Beutin *et al.*, 1993), wild boar (Wacheck *et al.*, 2009), rats (Nielsen *et al.*, 2004), wild rabbits and hares (Martínez *et al.*, 2011), and domesticated rabbits (García and Fox, 2003). Although only sporadic cases of EHEC infections have been associated with consumption of wild game (Keene *et al.*, 1997b; Rabatsky-Ehr *et al.*, 2002), wildlife could act as vehicles for the transmission of EHEC to and from domesticated animals as STEC with identical virulence factors and pulsed-field gel electrophoresis patterns have been isolated from wild animals and cattle living on the same farm (Nielsen *et al.*, 2004). Characterization of STEC isolated from game meat found strains that are potentially pathogenic as 54 of 140 of the STEC isolates belonged to serotypes that have been previously reported in human disease (Miko *et al.*, 2009). Wild deer carrying *E. coli* O157:H7 were considered to be the source of an outbreak caused by unpasteurized apple juice (Cody *et al.*, 1999). Despite the prevalence of STEC in other ruminant and non-ruminant animals, cattle are considered the most significant reservoir for STEC that cause human disease (Hussein, 2007).

12.5 Challenges and considerations for the detection of STEC

E. coli O157:H7 isolates have a phenotype (i.e., delayed fermentation of sorbitol) that allows differentiation from other *E. coli.* There are standardized methods to detect and isolate *E. coli* O157:H7 (e.g., USDA, 2010). In addition, several commercial real-time PCR assays have been validated to screen for *E. coli* O157:H7 in sample enrichments, and there are standardized methods that include PCR screening followed by using a microbiological culture to confirm screen positives (e.g., USDA, 2010). The greatest challenges that face researchers developing assays for the detection of non-O157 STEC are the lack of distinguishing phenotypes conserved among this diverse set of microorganisms and selecting the genes that are most indicative of high virulence and which should be targeted by the assays. The combination of genes should be common to all EHEC (ideally O157 and non-O157 serotypes capable of causing HUS) but absent from non-pathogenic *E. coli.* The *FSIS Microbiology Laboratory Guidebook* method 5B.03 (USDA, 2011a) delineates a standardized method for the detection of O26, O45, O103, O111, O121 and O145 STEC in meat products. This method utilizes custom PCR primers and probes (the sequences are provided in the protocol) to amplify and thereby detect *stx, eae* and the O-antigen-specific genes. Custom PCR reactions may be designed at a relatively low cost but these require optimization and validation before implementation. Commercial PCR assays (see Table 12.4) are an alternative to custom PCR assays that have been optimized and validated for the user; however, the exact sequences targeted by the assay are not typically made available. The majority of PCR-based assays for the detection of non-O157 STEC are still in development or in the process of being

Table 12.4 Comparison of commercial kits for the detection of non-O157 STEC

Assay name	Serogroups detected by assay	Targets	Total time	Sensitivity	Chemistry	Samples per run	Open or closed platform	Number of PCR reactions per sample	Description of platform
GeneDisc STEC & *Salmonella* (Pall)[a]	O157, *Salmonella*	stx_1, stx_2, *eae*, $rfbE_{O157}$	9 h min.	Less than 1 cfu/25 g	Probe-based RT (Taqman)	6–12 samples per disc, 96 samples per platform	Closed	5 plus 1 inhibition control	Tubes are arranged in a disc that rotates continuously though a circular thermocycler block
GeneDisc STEC (Pall)[a]	O157	stx_1, stx_2, *eae*, $rfbE_{O157}$	9 h min.	Less than 1 cfu/25 g	Probe-based RT (Taqman)	6–12 samples per disc, 96 samples per platform	Closed	4 plus 1 inhibition control	Same as above
GeneDisc EHEC 5 ID (Pall)[a]	O26, O103, O111, O145, H7	wzx_{O26}, wzx_{O103}, $wbdL_{O111}$, $ihpL_{O145}$, $fliC_{H7}$	9 h min.	Less than 1 cfu/25 g	Probe-based RT (Taqman)	6 samples per disc, 48 samples per platform	Closed	5 plus 1 inhibition control	Same as above
GeneDisc EHEC 6 ID (Pall)[a]	O26, O45, O103, O111, O121, O145	wzx_{O26}, wzx_{O103}, $wbdL_{O111}$, $ihpL_{O145}$, wzx_{O45}, wzx_{O121}	9 h min.	Less than 1 cfu/25 g	Probe-based RT (Taqman)	6 samples per disc, 48 samples per platform	Closed	6 plus 1 inhibition control	Same as above

Rapid Finder™ STEC Screening Kit and Confirmation Assay (Life Technologies)[a]	O45, O121, O103, O145, O111, O26, O157:H7	stx_1, stx_2 (all alleles including stx_{2f} and stx_{2g}), *eae* (all alleles), O-antigen specific genes	8–9 h	1–3 cfu/g	Probe-based RT (Taqman)	96 samples on 7500 Fast platform	Open	2	Five-color channel platform with traditional block with 96 wells
Assurance GDS for Top STEC (BioControl)	O26, O45, O103, O111, O121, O145	*eae*, stx_1, stx_2	14 h min.	Not available	Probe-based RT (MGB Eclipse ssDNA probe)	Not available	Not available	2	Rotary format within a chamber, tubes are heated and cooled with air, not contact with a thermocycler block
iQ-Check STEC (BIORAD)	O26, O45, O103, O111, O121, O145, O157	*eae*, stx_1, stx_2, serogroup-specific genes	Not available	Not available	Probe-based RT (dsDNA probe)	96 reactions per run	Not available	Not available	96-well heating block, 5 color (CFX96) or 2 color channels (Miniopticon)
BAX(R) System Real-Time PCR Assay for STEC (DuPont Qualicon)[a]	O26, O45, O103, O111, O121, O145, O157	(*stx*, *eae*, O157:H7 initial screen) (serogroup-specific genes, Big 6 secondary screen)	10 h min.	10^4 cfu/mL	Probe-based RT (Scorpion)	96 reactions per run	Closed	3	Traditional thermocycling block with 12 strips and 8 tubes per strip

(*Continued*)

Table 12.4 Continued

Assay name	Serogroups detected by assay	Targets	Total time	Sensitivity	Chemistry	Samples per run	Open or closed platform	Number of PCR reactions per sample	Description of platform
NeoSEEK (Neogen)[a]	O26, O45, O103, O111, O121, O145, O157	Two targets for each of the serotypes: O26, O45, O103, O111, O121, O145, O157; two serotype SNP specific markers for: O26, O45, O103, O111, O121, O145 Stx_1, Stx_2 Stx_{2c}, *eae*, eae_{α}, eae_{β}, $eae_{\gamma1}$, $eae_{\gamma2/}\theta$, $tir_{\alpha1}$, $tir_{\beta1}$, tir_{γ}, $tir_{\gamma2/\theta,}$ $espA_{\alpha1}$, $espA_{\beta1}$, $espA_{\gamma1}$, $espB_{\alpha1}$, $espB_{\beta1}$, $espB_{\gamma1}$, $espD_{\alpha1}$, $espD_{\beta1}$, $espD_{\gamma1}$, *nleB*, *nleE*, *nleH*, *ent/espL*, $fliC_{H2}$, $fliC_{H6}$, $fliC_{H7}$, $fliC_{H8}$, $fliC_{H9}$, $fliC_{H10}$, $fliC_{H11}$, $fliC_{H16}$, $fliC_{H19}$, $fliC_{H21}$, $fliC_{H25}$, $fliC_{H28}$	Less than 24 hours	10^4 cfu/mL	Multiplex PCR/ Sequenom iPLEX	8–88, each sample is run in duplicate	Open	4 PCR reactions per sample	PCR is coupled with mass spectrometry and MALDI-TOF analysis

Note:[a] confirmed by company representatives.

validated for screening. In general, these assays include multiple phases of testing where the initial test phase targets the virulence factors *eae* and *stx*. *stx* is essential for the development of HC and HUS but is not a good indicator of EHEC, since effective colonization of the host is required prior to delivery of the Shiga toxins. As discussed above some *stx* subtypes are rarely associated with human disease and other microorganisms can carry *stx*. The study by Bosilevac and Koohmaraie (2011) illustrates the prevalence of organisms carrying *stx* in ground beef; however, despite the fact that nearly one fourth of ground beef samples tested positive for *stx*, after conducting a molecular risk assessment of the isolated STEC only 0.24% of the samples produced STEC that had the OI and *nle* gene repertoire necessary to be considered pathogenic. The combination of *stx* and *eae* is a better choice for detecting EHEC but this criterion may include non-pathogenic *E. coli* as well. The addition of O-antigen-specific genes in follow-up phases of testing in a detection assay reduces positives to only the serotypes that account for the majority of non-O157 STEC infections, which excludes EHEC of seropathotype C. Since identification of the Big Six (the six most frequently isolated STEC in the US) requires, at a minimum, inclusion of six O-antigen-specific genes detection assays typically encompass two to three PCR reactions dedicated solely to these targets. It is unlikely that an assay will be developed that detects all EHEC while excluding all non-pathogenic *E. coli*. The most feasible approach may be to target the EHEC that have been found to cause the majority of human disease while continuing to characterize clinical isolates to monitor these rapidly evolving pathogens. Full genome sequencing, as was performed on isolates from the German sprouts outbreak, is likely to prove to be a great tool in the rapid characterization of EHEC outbreak strains. Array-based methods, such as the GeneDisc® array employed by Bugarel *et al.* (2010) or NeoSEEK®, are of value when characterizing larger collections of isolates and looking for putative virulence factors.

12.6 Summary and discussion

The true burden of disease caused by non-O157 STEC is not entirely clear; however, routine screening for *E. coli* O157 in cases of HC has been recommended by the Council of State and Territorial Epidemiologists since 1993 (CSTE, 1993). The detection bias and focus on *E. coli* O157:H7 has led to a significant knowledge gap in the literature on non-O157 STEC of clinical importance. The development of more sensitive and high throughput detection methods for non-O157 STEC as well as increased surveillance are necessary for bridging this knowledge gap. Some of the assays presented here are likely to aid in understanding the prevalence of these organisms in food products, animal reservoirs and human patients. Besides determining the public health burden of EHEC, investigating the process by which they cause disease and determining which virulence factors are critical for this process is paramount. Genome sequencing of outbreak strains in the future will aid in the determination of virulence factors. As the FSIS moves to declaring the Big Six as adulterants in raw ground beef, developing a better understanding of the virulence potential, fitness and prevalence of these pathogens could have a huge impact on the meat industry. It is clear that non-O157 STEC are present in many animal reservoirs, they have the ability to be translocated from farm to fork and they have the ability to cause severe disease equivalent to that observed with *E. coli* O157:H7. The recent emergence of EHEC as well as the devastating outbreak associated with the novel O104:H4 strain illustrates the ability of these organisms to acquire virulence factors rapidly through the horizontal transfer of genes. It is the genomic plasticity of these organisms that makes them a challenge to understand and a threat to public health.

12.7 References and further reading

ALLERBERGER F, FRIEDRICH A W, GRIF K, DIERICH M P, DORNBUSCH H-R *et al.* (2003), Hemolytic-uremic syndrome associated with enterohemorrhagic *Escherichia coli* O26:H infection and consumption of unpasteurized cow's milk, *Int J of Infect Dis*, 7, 42–45.

ASAKURA H, MAKINO S I, SHIRAHATA T, TSUKAMOTO, KURAZONO *et al.* (1998), Detection and genetical characterization of Shiga toxin-producing *Escherichia coli* from wild deer, *Microbiol Immunol*, 42, 815–822.

BANATVALA N, GRIFFIN P M, GREENE K D, BARRETT T J, BIBB W F *et al.* (2001), The United States National Prospective Hemolytic Uremic Syndrome Study: Microbiologic, serologic, clinical, and epidemiologic findings, *J Infect Dis*, 183, 1063–1070.

BARDIAU M, GRÉGOIRE F, MUYLAERT A, NAHAYO A, DUPREZ J N *et al.* (2010), Enteropathogenic (EPEC), enterohaemorragic (EHEC) and verotoxigenic (VTEC) *Escherichia coli* in wild cervids, *J of Appl Micro*, 109, 2214–2222.

BARKOCY-GALLAGHER G A, ARTHUR T M, RIVERA-BETANCOURT M, NOU X, SHACKELFORD S D *et al.* (2003), Seasonal prevalence of Shiga toxin-producing *Escherichia coli*, including O157:H7 and non-O157 serotypes, and *Salmonella* in commercial beef processing plants, *J of Food Prot*, 66, 1978–1986.

BARUCH K, GUR-ARIE L, NADLER C, KOBY S, YERUSHALMI G *et al.* (2011), Metalloprotease type III effectors that specifically cleave JNK and NF-κB, *EMBO J*, 30, 221–231.

BELL B, GOLDOFT M, GRIFFIN P, DAVIS M, GORDON D *et al.* (1994), A multistate outbreak of *Escherichia coli* O157:H7-associated with bloody diarrhea and hemolytic uremic syndrome from hamburgers: The Washington experience, *JAMA*, 117, 1229–1231.

BEUTIN L, ALEKSIC S, ZIMMERMANN S and GLEIER K (1994), Virulence factors and phenotypical traits of verotoxigenic strains of *Escherichia coli* isolated from human patients in Germany, *Med Microbiol and Immunol*, 183, 13–21.

BEUTIN L, GEIER D, STEINRUCK H, ZIMMERMANN S and SCHEUTZ F (1993), Prevalence and some properties of verotoxin (Shiga-like toxin)-producing *Escherichia coli* in seven different species of healthy domestic animals, *J Clin Microbiol*, 31, 2483–2488.

BEUTIN L, KRAUSE G, ZIMMERMANN S, KAULFUSS S and GLEIER K (2004), Characterization of Shiga toxin-producing *Escherichia coli* strains isolated from human patients in Germany over a 3-year period, *J Clin Microbiol*, 42, 1099–1108.

BEUTIN L, KRUGER U, KRAUSE G, MIKO A, MARTIN A *et al.* (2008), Evaluation of major types of Shiga toxin 2e-producing *Escherichia coli* bacteria present in food, pigs, and the environment as potential pathogens for humans, *Appl Environ Microbiol*, 74, 4806–4816.

BEUTIN L, MIKO A, KRAUSE G, PRIES K, HABY S *et al.* (2007), Identification of human-pathogenic strains of Shiga toxin-producing *Escherichia coli* from food by a combination of serotyping and molecular typing of Shiga toxin genes, *Appl Environ Microbiol*, 73, 4769–4775.

BEUTIN L, MONTENEGRO M A, ORSKOV I, ORSKOV F, PRADA J *et al.* (1989), Close association of verotoxin (Shiga-like toxin) production with enterohemolysin production in strains of *Escherichia coli*, *J Clin Microbiol*, 27, 2559–2564.

BHAN M K, KHOSHOO V, SOMMERFELT H, RAJ P, SAZAWAL S *et al.* (1989a), Enteroaggregative *Escherichia coli* and *Salmonella* associated with nondysenteric persistent diarrhea, *Pediatric Infect Dis J*, 8, 499–501.

BHAN M K, RAJ P, LEVINE M M, KAPER J B, BHANDARI N *et al.* (1989b), Enteroaggregative *Escherichia coli* associated with persistent diarrhea in a cohort of rural children in India, *J Infect Dis*, 159, 1061–1064.

BHAVSAR A P, GUTTMAN J A and FINLAY B B (2007), Manipulation of host-cell pathways by bacterial pathogens, *Nature*, 449, 827–834.

BIELASZEWSKA M, FRIEDRICH A W, ALDICK T, SCHÜRK-BULGRIN R and KARCH H (2006), Shiga toxin activatable by intestinal mucus in *Escherichia coli* isolated from humans: Predictor for a severe clinical outcome, *Clin Infect Dis*, 43, 1160–1167.

BIELASZEWSKA M, JANDA J, BLÁHOVÁ K, MINARÍKOVÁ H, JÍKOVÁ E, KARMALI M A *et al.* (1997), Human *Escherichia coli* O157:H7 infection associated with the consumption of unpasteurized goat's milk, *Epidemiol and Infect*, 119, 299–305.

BLANCO M, BLANCO J E, MORA A, REY J, ALONSO J M *et al.* (2003), Serotypes, virulence genes, and intimin types of Shiga toxin (verotoxin)-producing *Escherichia coli* isolates from healthy sheep in Spain, *J Clin Microiol*, 41, 1351–1356.

BOERLIN P, CHEN S, COLBOURNE J K, JOHNSON R, DE GRANDIS S *et al.* (1998), Evolution of enterohemorrhagic *Escherichia coli* hemolysin plasmids and the locus for enterocyte effacement in Shiga toxin-producing *E. coli*, *Infect Immun*, 66, 2553–2561.

BOERLIN P, MCEWEN S A, BOERLIN-PETZOLD F, WILSON J B, JOHNSON R P *et al.* (1999), Associations between virulence factors of Shiga toxin-producing *Escherichia coli* and disease in humans, *J Clin Microbiol*, 37, 497–503.

BONNET R, SOUWEINE B, GAUTHIER G, RICH C, LIVRELLI V *et al.* (1998), Non-O157:H7 Stx2-producing *Escherichia coli* strains associated with sporadic cases of hemolytic-uremic syndrome in adults, *J Clin Microbiol*, 36, 1777–1780.

BOSILEVAC J M and KOOHMARAIE M (2011), Prevalence and characterization of non-O157 Shiga toxin-producing *Escherichia coli* isolates from commercial ground beef in the United States, *Appl and Environ Microbiol*, 77, 2103–2112.

BOUVET J, MONTET M P, ROSSEL R, LE ROUX A, BAVAI C *et al.* (2002), Prevalence of verotoxin-producing *Escherichia coli* (VTEC) and *E. coli* O157:H7 in French pork, *J Appl Microbiol*, 93, 7–14.

BRETT K N, HORNITZKY M A, BETTELHEIM K A, WALKER M J and DJORDJEVIC S P (2003), Bovine non-O157 Shiga toxin 2-containing *Escherichia coli* isolates commonly possess *stx2*-EDL933 and/or *stx2vhb* subtypes, *J Clin Microbiol*, 41, 2716–2722.

BROCKMEYER J, BIELASZEWSKA M, FRUTH A, BONN M L, MELLMANN A *et al.* (2007), Subtypes of the plasmid-encoded serine protease EspP in Shiga toxin-producing *Escherichia coli*: distribution, secretion, and proteolytic activity, *Appl Environ Microbiol*, 73, 6351–6359.

BROOKS H J L, MOLLISON B D, BETTELHEIM K A, MATEJKA K, PATERSON K A *et al.* (2001), Occurrence and virulence factors of non-O157 Shiga toxin-producing *Escherichia coli* in retail meat in Dunedin, New Zealand, *Lett Appl Microbiol*, 32, 118–122.

BROOKS J T, SOWERS E G, WELLS J G, GREENE K D, GRIFFIN P M *et al.* (2005), Non-O157 Shiga toxin-producing *Escherichia coli* infections in the United States, 1983–2002, *J Infect Dis*, 192, 1422–1429.

BRUNDER W, KARCH H and SCHMIDT H (2006), Complete sequence of the large virulence plasmid pSFO157 of the sorbitol-fermenting enterohemorrhagic *Escherichia coli* O157:H- strain 3072/96, *Int J Med Microbiol*, 296, 467–474.

BRUNDER W, KHAN A S, HACKER J and KARCH H (2001), Novel type of fimbriae encoded by the large plasmid of sorbitol-fermenting enterohemorrhagic *Escherichia coli* O157:H–, *Infect Immun*, 69, 4447–4457.

BRUNDER W, SCHMIDT H and KARCH H (1996), KatP, a novel catalase-peroxidase encoded by the large plasmid of enterohaemorrhagic *Escherichia coli* O157:H7, *Microbiology*, 142, 3305–3315.

BRUNDER W, SCHMIDT H and KARCH H (1997), EspP, a novel extracellular serine protease of enterohaemorrhagic *Escherichia coli* O157:H7 cleaves human coagulation factor V, *Mol Microbiol*, 24, 767–778.

BRZUSZKIEWICZ E, THÜRMER A, SCHULDES J, LEIMBACH A, LIESEGANG H *et al.* (2011), Genome sequence analyses of two isolates from the recent *Escherichia coli* outbreak in Germany reveal the emergence of a new pathotype: Entero-aggregative-haemorrhagic *Escherichia coli* (EAHEC), *Arch Microbiol*, 193, 883–891.

BUGAREL M, BEUTIN L, MARTIN A, GILL A and FACH P (2010), Micro-array for the identification of Shiga toxin-producing *Escherichia coli* (STEC) seropathotypes associated with hemorrhagic colitis and hemolytic uremic syndrome in humans, *Int J Food Microbiol*, 142, 318–329.

BURK C, DIETRICH R, ACAR G, MORAVEK M, BULTE M *et al.* (2003), Identification and characterization of a new variant of Shiga toxin 1 in *Escherichia coli* ONT:H19 of bovine origin, *J Clin Microbiol*, 41, 2106–2112.

CAMPELLONE K G, ROBBINS D and LEONG J M (2004), EspFU is a translocated EHEC effector that interacts with Tir and N-WASP and promotes Nck-independent actin assembly, *Dev Cell*, 7, 217–228.

CAPRIOLI A, MORABITO S, BRUGÈRE H and OSWALD E (2005), Enterohaemorrhagic *Escherichia coli*: Emerging issues on virulence and modes of transmission, *Vet Res*, 36, 289–311.

CDC (2011), Vital signs: Incidence and trends of infection with pathogens transmitted commonly through food – Foodborne diseases active surveillance network, 10 US sites, 1996–2010, *MMWR Morb Mortal Wkly Rep*, 60, 749–755.

CERQUEIRA A M F, GUTH B E C, JOAQUIM R M and ANDRADE J R C (1999), High occurrence of Shiga toxin-producing *Escherichia coli* (STEC) in healthy cattle in Rio de Janeiro State, Brazil, *Vet Microbiol*, 70, 111–121.

CODY S H, GLYNN M K, FARRAR J A, CAIRNS K L, GRIFFIN P M *et al.* (1999), An outbreak of *Escherichia coli* O157:H7 infection from unpasteurized commercial apple juice, *Ann Intern Med*, 130, 202–209.

COOMBES B K, WICKHAM M E, MASCARENHAS M, GRUENHEID S, FINLAY B B *et al.* (2008), Molecular analysis as an aid to assess the public health risk of non-O157 Shiga toxin-producing *Escherichia coli* strains, *Appl Environ Microbiol*, 74, 2153–2160.

CRANE J K, MCNAMARA B P and DONNENBERG M S (2001), Role of EspF in host cell death induced by enteropathogenic *Escherichia coli*, *Cell Microbiol*, 3, 197–211.

CRAVIOTO A, TELLO A, NAVARRO A, RUIZ J, VILLAFÁN H *et al.* (1991), Association of *Escherichia coli* HEp-2 adherence patterns with type and duration of diarrhoea, *Lancet*, 337, 262–264.

CSTE (1993), CSTE position statement #4: National surveillance of *Escherichia coli* O157:H7. Formerly available from: www.cste.org/dnn/AnnualConference/PositionStatements/tabid/191/Default.aspx [Accessed on 16 November 2011].

DAHAN S, WILES S, LA RAGIONE R M, BEST A, WOODWARD M J *et al.* (2005), EspJ Is a prophage-carried type III effector protein of attaching and effacing pathogens that modulates infection dynamics, *Infect Immun*, 73, 679–686.

DATZ M, JANETZKI-MITTMANN C, FRANKE S, GUNZER F, SCHMIDT H *et al.* (1996), Analysis of the enterohemorrhagic *Escherichia coli* O157 DNA region containing lambdoid phage gene p and Shiga-like toxin structural genes, *Appl Environ Microbiol*, 62, 791–797.

DESROSIERS A, FAIRBROTHER J M, JOHNSON R P, DESAUTELS C, LETELLIER A *et al.* (2001), Phenotypic and genotypic characterization of *Escherichia coli* verotoxin-producing isolates from humans and pigs, *J Food Prot*, 64, 1904–1911.

DJAFARI S, EBEL F, DEIBEL C, KRÄMER S, HUDEL M *et al.* (1997), Characterization of an exported protease from Shiga toxin-producing *Escherichia coli*, *Mol Microbiol*, 25, 771–784.

DOUGHTY S, SLOAN J, BENNETT-WOOD V, ROBERTSON M, ROBINS-BROWNE R M *et al.* (2002), Identification of a novel fimbrial gene cluster related to long polar fimbriae in locus of enterocyte effacement-negative strains of enterohemorrhagic *Escherichia coli*, *Infect Immun*, 70, 6761–6769.

DZIVA F, MAHAJAN A, CAMERON P, CURRIE C, MCKENDRICK I J *et al.* (2007), EspP, a type V-secreted serine protease of enterohaemorrhagic *Escherichia coli* O157:H7, influences intestinal colonization of calves and adherence to bovine primary intestinal epithelial cells, *FEMS Microbiol Lett*, 271, 258–264.

ECHTENKAMP F, DENG W, WICKHAM M E, VAZQUEZ A, PUENTE J L *et al.* (2008), Characterization of the NleF effector protein from attaching and effacing bacterial pathogens, *FEMS Microbiol Lett*, 281, 98–107.

EKLUND M, LEINO K and SIITONEN A (2002), Clinical *Escherichia coli* strains carrying *stx* genes: *stx* variants and *stx*-positive virulence profiles, *J Clin Microbiol*, 40, 4585–4593.

ELLIOTT E J, ROBINS-BROWNE R M, O'LOUGHLIN E V, BENNETT-WOOD V, BOURKE J *et al.* (2001), Nationwide study of haemolytic uraemic syndrome: Clinical, microbiological, and epidemiological features, *Arch Dis Child*, 85, 125–131.

ELLIOTT S J, YU J and KAPER J B (1999), The cloned locus of enterocyte effacement from enterohemorrhagic *Escherichia coli* O157:H7 is unable to confer the attaching and effacing phenotype upon *E. coli* K-12, *Infect Immun*, 67, 4260–4263.

ESPIÉ E, GRIMONT F, MARIANI-KURKDJIAN P, BOUVET P, HAEGHEBAERT S *et al.* (2008), Surveillance of hemolytic uremic syndrome in children less than 15 years of age, a system to monitor O157 and non-O157 Shiga toxin-producing *Escherichia coli* infections in France, 1996–2006, *Pediatr Infect Dis J*, 27, 595–601.

ESPIÉ E, GRIMONT F, VAILLANT V, MONTET M P, CARLE *et al.* (2006), O148 Shiga toxin-producing *Escherichia coli* outbreak: Microbiological investigation as a useful complement to epidemiological investigation, *Clin Microbiol Infect*, 12, 992–998.

ETHELBERG S, OLSEN K E P, SCHEUTZ F, JENSEN C, SCHIELLERUP P *et al.* (2004), Virulence factors for hemolytic uremic syndrome, Denmark, *Emerg Infect Dis*, 10, 842–847.

FARMER J J, 3rd and DAVIS B R (1985), H7 antiserum-sorbitol fermentation medium: A single tube screening medium for detecting *Escherichia coli* O157:H7 associated with hemorrhagic colitis, *J Clin Microbiol*, 22, 620–625.

FAROOQ S, HUSSAIN I, MIR M A, BHAT M A and WANI S A (2009), Isolation of atypical enteropathogenic *Escherichia coli* and Shiga toxin 1 and 2f-producing *Escherichia coli* from avian species in India, *Lett Appl Microbiol*, 48, 692–697.

FERNÁNDEZ D, RODRÍGUEZ E M, ARROYO G H, PADOLA N L and PARMA A E (2009), Seasonal variation of Shiga toxin-encoding genes (*stx*) and detection of *E. coli* O157 in dairy cattle from Argentina, *J Appl Microbiol*, 106, 1260–1267.

FRANKE S, HARMSEN D, CAPRIOLI A, PIERARD D, WIELER L *et al.* (1995), Clonal relatedness of Shiga-like toxin-producing *Escherichia coli* O101 strains of human and porcine origin, *J Clin Microbiol*, 33, 3174–3178.

FRANKEL G, PHILLIPS A D, TRABULSI L R, KNUTTON S, DOUGAN G *et al.* (2001), Intimin and the host cell – is it bound to end in Tir(s)?, *Trends Microbiol*, 9, 214–218.

FRATAMICO P M, BAGI L K, BUSH E J and SOLOW B T (2004), Prevalence and characterization of Shiga toxin-producing *Escherichia coli* in swine feces recovered in the National Animal Health Monitoring System's Swine 2000 study, *Appl Environ Microbiol*, 70, 7173–7178.

FRATAMICO P M, BHAGWAT A A, INJAIAN L and FEDORKA-CRAY P J (2008), Characterization of Shiga toxin-producing *Escherichia coli* strains isolated from swine feces, *Foodborne Pathog Dis*, 5, 827–838.

FRATAMICO P M, YAN X, CAPRIOLI A, ESPOSITO G, NEEDLEMAN D S *et al.* (2011), The complete DNA sequence and analysis of the virulence plasmid and of five additional plasmids carried by Shiga toxin-producing *Escherichia coli* O26:H11 strain H30, *Int J Med Microbiol*, 301, 192–203.

FRIEDRICH A W, BIELASZEWSKA M, ZHANG W-L, PULZ M, KUCZIUS T *et al.* (2002), *Escherichia coli* harboring Shiga toxin 2 gene variants: Frequency and association with clinical symptoms, *J Infect Dis*, 185, 74–84.

FUKUSHIMA H and SEKI R (2004), High numbers of Shiga toxin-producing *Escherichia coli* found in bovine faeces collected at slaughter in Japan, *FEMS Microbiol Lett*, 238, 189–197.

GALÁN J E and COLLMER A (1999), Type III secretion machines: Bacterial devices for protein delivery into host cells, *Science*, 284, 1322–1328.

GANNON V P J, TEERLING C, MASRI S A and GYLES C L (1990), Molecular cloning and nucleotide sequence of another variant of the *Escherichia coli* Shiga-like toxin II family, *J General Microbiol*, 136, 1125–1135.

GARCÍA A and FOX J G (2003), The rabbit as a new reservoir host of enterohemorrhagic *Escherichia coli*, *Emerg Infect Dis*, 9, 1592–1597.

GARMENDIA J, PHILLIPS A D, CARLIER M-F, CHONG Y, SCHULLER S *et al.* (2004), TccP is an enterohaemorrhagic *Escherichia coli* O157:H7 type III effector protein that couples Tir to the actin-cytoskeleton, *Cell Microbiol*, 6, 1167–1183.

GERBER A, KARCH H, ALLERBERGER F, VERWEYEN H M and ZIMMERHACKL L B (2002), Clinical course and the role of Shiga toxin-producing *Escherichia coli* infection in the hemolytic-uremic syndrome in pediatric patients, 1997–2000, in Germany and Austria: A prospective study, *J Infect Dis*, 186, 493–500.

GRUENHEID S, SEKIROV I, THOMAS N A, DENG W, O'DONNELL P *et al.* (2004), Identification and characterization of NleA, a non-LEE-encoded type III translocated virulence factor of enterohaemorrhagic *Escherichia coli* O157:H7, *Mol Microbiol*, 51, 1233–1249.

GUTH B E C, LOPES DE SOUZA R, VAZ T M I and IRINO K (2002), First Shiga toxin-producing *Escherichia coli* isolate from a patient with hemolytic uremic syndrome, Brazil, *Emerg Infect Dis*, 8, 535–536.

HEMRAJANI C, BERGER C N, ROBINSON K S, MARCHÈS O, MOUSNIER A *et al.* (2010), NleH effectors interact with Bax inhibitor-1 to block apoptosis during enteropathogenic *Escherichia coli* infection, *Proc Nat Acad Sci USA*, 107, 3129–3134.

HEROLD S, PATON J C and PATON A W (2009), Sab, a novel autotransporter of locus of enterocyte effacement-negative Shiga-toxigenic *Escherichia coli* O113:H21, contributes to adherence and biofilm formation, *Infect Immun*, 77, 3234–3243.

HEUVELINK A E, VAN HEERWAARDEN C, ZWARTKRUIS-NAHUIS J T, VAN OOSTEROM R, EDINK K *et al.* (2002), *Escherichia coli* O157 infection associated with a petting zoo, *Epidemiol Infect*, 129, 295–302.

HRUDEY S E, PAYMENT P, HUCK P M, GILLHAM R W and HRUDEY E J (2003), A fatal waterborne disease epidemic in Walkerton, Ontario: Comparison with other waterborne outbreaks in the developed world, *Water Sci Technol*, 47, 7–14.

HUGHES J M, WILSON M E, JOHNSON K E, THORPE C M and SEARS C L (2006), The emerging clinical importance of non-O157 Shiga toxin-producing *Escherichia coli*, *Clin Infect Dis*, 43, 1587–1595.

HUSSEIN H S (2007), Prevalence and pathogenicity of Shiga toxin-producing *Escherichia coli* in beef cattle and their products, *J Anim Sci*, 85, E63–E72.

IRINO K, VAZ T M I, KATO M A M F, NAVES Z V F, LARA R R *et al.* (2002), O157:H7 Shiga toxin-producing *Escherichia coli* strains associated with sporadic cases of diarrhea in São Paulo, Brazil, *Emerging Infectious Diseases*, 8, 446–447.

IRINO K, VIEIRA M A M, GOMES T A T, GUTH B E C, NAVES Z V F *et al.* (2010), Subtilase cytotoxin-encoding *subAB* operon found exclusively among Shiga toxin-producing *Escherichia coli* strains, *J Clin Microbiol*, 48, 988–990.

ISOBE J, KIMATA K, SHIMOJIMA M, HOSOROGI S, TANAKA D *et al.* (2004), Isolation of *Escherichia coli* O128:HNM harboring *stx2f* gene from diarrhea patients, *Kansenshōgaku Zasshi*, 78, 1000–1005.

ITO H, TERAI A, KURAZONO H, TAKEDA Y and NISHIBUCHI M (1990), Cloning and nucleotide sequencing of vero toxin 2 variant genes from *Escherichia coli* O91:H21 isolated from a patient with the hemolytic uremic syndrome, *Microb Pathog*, 8, 47–60.

JELACIC J K, DAMROW T, CHEN G S, JELACIC S, BIELASZEWSKA M *et al.* (2003), Shiga toxin-producing *Escherichia coli* in Montana: Bacterial genotypes and clinical profiles, *J Infect Dis*, 188, 719–729.

JENKINS C, PEARCE M C, CHART H, CHEASTY T, WILLSHAW G A *et al.* (2002), An eight-month study of a population of verocytotoxigenic *Escherichia coli* (VTEC) in a Scottish cattle herd, *J Appl Microbiol*, 93, 944–953.

KAPER J B, NATARO J P and MOBLEY H L T (2004), Pathogenic *Escherichia coli*, *Nat Rev Microbiol*, 2, 123–140.

KARMALI M A (1989), Infection by verocytotoxin-producing *Escherichia coli*, *Clin Microbiol Rev*, 2, 15–38.

KARMALI M A, MASCARENHAS M, SHEN S, ZIEBELL K, JOHNSON S *et al.* (2003), Association of genomic O island 122 of *Escherichia coli* EDL 933 with verocytotoxin-producing *Escherichia coli* seropathotypes that are linked to epidemic and/or serious disease, *J Clin Microbiol*, 41, 4930–4940.

KEENE W E, HEDBERG K, HERRIOTT D E, HANCOCK D D, MCKAY R W *et al.* (1997a), A prolonged outbreak of *Escherichia coli* O157:H7 infections caused by commercially distributed raw milk, *J Infect Dis*, 176, 815–818.

KEENE W E, SAZIE E, KOK J, RICE D H, HANCOCK D D *et al.* (1997b), An outbreak of *Escherichia coli* O157:H7 infections traced to jerky made from deer meat, *JAMA*, 277, 1229–1231.

KIM M, OGAWA M, FUJITA Y, YOSHIKAWA Y, NAGAI T *et al.* (2009), Bacteria hijack integrin-linked kinase to stabilize focal adhesions and block cell detachment, *Nature*, 459, 578–582.

KLAPPROTH J-M A, SCALETSKY I C A, MCNAMARA B P, LAI L-C, MALSTROM C *et al.* (2000), A large toxin from pathogenic *Escherichia coli* strains that inhibits lymphocyte activation, *Infect Immun*, 68, 2148–2155.

KNUTTON S, ROSENSHINE I, PALLEN M J, NISAN I, NEVES B C *et al.* (1998), A novel EspA-associated surface organelle of enteropathogenic *Escherichia coli* involved in protein translocation into epithelial cells, *EMBO J*, 17, 2166–2176.

KOBAYASHI H, KANAZAKI M, HATA E and KUBO M (2009), Prevalence and characteristics of *eae*- and *stx*-positive strains of *Escherichia coli* from wild birds in the immediate environment of Tokyo Bay, *Appl Environ Microbiol*, 75, 292–295.

KOCH C, HERTWIG S, LURZ R, APPEL B and BEUTIN L (2001), Isolation of a lysogenic bacteriophage carrying the *stx1OX3* gene, which is closely associated with Shiga toxin-producing *Escherichia coli* strains from sheep and humans, *J Clin Microbiol*, 39, 3992–3998.

KONCZY P, ZIEBELL K, MASCARENHAS M, CHOI A, MICHAUD C *et al.* (2008), Genomic O island 122, locus for enterocyte effacement, and the evolution of virulent verocytotoxin-producing *Escherichia coli*, *J Bacteriol*, 190, 5832–5840.

KUHNERT P, DUBOSSON C R, ROESCH M, HOMFELD E, DOHERR M G *et al.* (2005), Prevalence and risk-factor analysis of Shiga toxigenic *Escherichia coli* in faecal samples of organically and conventionally farmed dairy cattle, *Vet Microbiol*, 109, 37–45.

LEUNG P H M, PEIRIS J S M, NG W W S, ROBINS-BROWNE R M, BETTELHEIM K A *et al.* (2003), A newly discovered verotoxin variant, vt2g, produced by bovine verocytotoxigenic *Escherichia coli*, *Appl Environ Microbiol*, 69, 7549–7553.

LEYTON D L, SLOAN J, HILL R E, DOUGHTY S and HARTLAND E L (2003), Transfer region of pO113 from enterohemorrhagic *Escherichia coli*: Similarity with R64 and identification of a novel plasmid-encoded autotransporter, EpeA, *Infect Immun*, 71, 6307–6319.

LÓPEZ E L 1998. Epidemiology of Shiga toxin-producing *Escherichia coli* in South America. In KAPER J B and O'BRIEN A D (eds) *Escherichia coli O157:H7 and Other Shiga Toxin-producing Escherichia coli*. Washington DC: American Society of Microbiology Press.

LOPEZ E L, DIAZ M, GRINSTEIN S, DEVOTO S, MENDILAHARZU F *et al.* (1989), Hemolytic uremic syndrome and diarrhea in Argentine children: The role of Shiga-like toxins, *J Infect Dis*, 160, 469–475.

LYNN R M, O'BRIEN S J, TAYLOR C M, ADAK G K, CHART H *et al.* (2005), Childhood hemolytic uremic syndrome, United Kingdom and Ireland, *Emerg Infect Dis*, 11, 590–596.

MACLEOD D L, GYLES C L and WILCOCK B P (1991), Reproduction of edema disease of swine with purified Shiga-like toxin-II variant, *Vet Pathol*, 28, 66–73.

MAKINO K, ISHII K, YASUNAGA T, HATTORI M, YOKOYAMA K *et al.* (1998), Complete nucleotide sequences of 93-kb and 3.3-kb plasmids of an enterohemorrhagic *Escherichia coli* O157:H7 derived from Sakai outbreak, *DNA Res*, 5, 1–9.

MAKINO S, KOBORI H, ASAKURA H, WATARAI M, SHIRAHATA T *et al.* (2000), Detection and characterization of Shiga toxin-producing *Escherichia coli* from seagulls, *Epidemiol Infect*, 125, 55–61.

MARCHÈS O, COVARELLI V, DAHAN S, COUGOULE C, BHATTA P *et al.* (2008), EspJ of enteropathogenic and enterohaemorrhagic *Escherichia coli* inhibits opsono-phagocytosis, *Cell Microbiol*, 10, 1104–1115.

MARCHÈS O, LEDGER T N, BOURY M, OHARA M, TU X *et al.* (2003), Enteropathogenic and enterohaemorrhagic *Escherichia coli* deliver a novel effector called Cif, which blocks cell cycle G2/M transition, *Molr Microbiol*, 50, 1553–1567.

MARTIN A and BEUTIN L (2011), Characteristics of Shiga toxin-producing *Escherichia coli* from meat and milk products of different origins and association with food producing animals as main contamination sources, *Int J Food Microbiol*, 146, 99–104.

MARTÍNEZ R, GARCÍA A, BLANCO J, BLANCO J, REY J *et al.* (2011), Occurrence of verocytotoxin-producing *Escherichia coli* in the faeces of free-ranging wild lagomorphs in southwest Spain, *Eur J Wildl Res*, 57, 187–189.

MCDANIEL T K, JARVIS K G, DONNENBERG M S and KAPER J B (1995), A genetic locus of enterocyte effacement conserved among diverse enterobacterial pathogens, *Proc Natl Acad of Sci USA*, 92, 1664–1668.

MCKEE M, MELTON-CELSA A, MOXLEY R, FRANCIS D and O'BRIEN A (1995), Enterohemorrhagic *Escherichia coli* O157:H7 requires intimin to colonize the gnotobiotic pig intestine and to adhere to HEp-2 cells, *Infect Immun*, 63, 3739–3744.

MCNAMARA B P, KOUTSOURIS A, O'CONNELL C B, NOUGAYRÉDE J-P, DONNENBERG M S *et al.* (2001), Translocated EspF protein from enteropathogenic *Escherichia coli* disrupts host intestinal barrier function, *J Clin Invest,* 107, 621–629.

MELLMANN A (2009), Phylogeny and disease association of Shiga toxin-producing *Escherichia coli* O91, *Emerg Infect Dis*, 15, 1474–1477.

MELLMANN A, HARMSEN D, CUMMINGS C A, ZENTZ E B, LEOPOLD S R *et al.* (2011), Prospective genomic characterization of the German enterohemorrhagic *Escherichia coli* O104:H4 outbreak by rapid next generation sequencing technology, *PLoS ONE*, 6, e22751.

MELTON-CELSA A, DARNELL S and O'BRIEN A (1996), Activation of Shiga-like toxins by mouse and human intestinal mucus correlates with virulence of enterohemorrhagic *Escherichia coli* O91:H21 isolates in orally infected, streptomycin-treated mice, *Infect Immun*, 64, 1569–1576.

MIKO A, PRIES K, HABY S, STEEGE K, ALBRECHT N *et al.* (2009), Assessment of Shiga toxin-producing *Escherichia coli* isolates from wildlife meat as potential pathogens for humans, *Appl Environ Microbiol*, 75, 6462–6470.

MONTENEGRO M A, BULTE M, TRUMPF T, ALEKSIC S, REUTER G *et al.* (1990), Detection and characterization of fecal verotoxin-producing *Escherichia coli* from healthy cattle, *J Clin Microbiol*, 28, 1417–1421.

MORABITO S, DELL'OMO G, AGRIMI U, SCHMIDT H, KARCH H *et al.* (2001), Detection and characterization of Shiga toxin-producing *Escherichia coli* in feral pigeons, *Vet Microbiol*, 82, 275–283.

MORABITO S, KARCH H, MARIANI-KURKDJIAN P, SCHMIDT H, MINELLI F *et al.* (1998), Enteroaggregative, Shiga toxin-producing *Escherichia coli* O111:H2 associated with an outbreak of hemolytic-uremic syndrome, *J Clin Microbiol*, 36, 840–842.

MOREIRA C N, PEREIRA M A, BROD C S, RODRIGUES D P, CARVALHAL J B *et al.* (2003), Shiga toxin-producing *Escherichia coli* (STEC) isolated from healthy dairy cattle in southern Brazil, *Vet Microbiol*, 93, 179–183.

MOXLEY R (2000), Edema disease, *Vet Clin North Am Food Anim Pract*, 16, 175–185.

NADLER C, BARUCH K, KOBI S, MILLS E, HAVIV G *et al.* (2010), The type III secretion effector NleE inhibits NF-κB activation, *PLoS Pathog*, 6, e1000743.

NEWTON H J (2009), Shiga toxin-producing *Escherichia coli* strains negative for locus of enterocyte effacement, *Emerg Infect Dis*, 15, 372–380.

NEWTON H J, PEARSON J S, BADEA L, KELLY M, LUCAS M *et al.* (2010), The type III effectors NleE and NleB from enteropathogenic *E. coli* and OspZ from *Shigella* block nuclear translocation of NF-kappaB p65, *PLoS Pathog*, 6, e1000898.

NICHOLLS L, GRANT T H and ROBINS-BROWNE R M (2000), Identification of a novel genetic locus that is required for *in vitro* adhesion of a clinical isolate of enterohaemorrhagic *Escherichia coli* to epithelial cells, *Mol Microbiol*, 35, 275–288.

NIELSEN E M, SKOV M N, MADSEN J J, LODAL J, JESPERSEN J B *et al.* (2004), Verocytotoxin-producing *Escherichia coli* in wild birds and rodents in close proximity to farms, *Appl Environ Microbiol*, 70, 6944–6947.

OGURA Y, OOKA T, IGUCHI A, TOH H, ASADULGHANI M *et al.* (2009), Comparative genomics reveal the mechanism of the parallel evolution of O157 and non-O157 enterohemorrhagic *Escherichia coli*, *Proc Natl Acad Sci USA*, 106, 17939–17944.

OLIVEIRA M G, BRITO J R, CARVALHO R R, GUTH B E, GOMES T A *et al.* (2007), Water buffaloes (*Bubalus bubalis*) identified as an important reservoir of Shiga toxin-producing *Escherichia coli* in Brazil, *Appl Environ Microbiol*, 73, 5945–5948.

PADOLA N L, SANZ M E, BLANCO J E, BLANCO M *et al.* (2004), Serotypes and virulence genes of bovine Shigatoxigenic *Escherichia coli* (STEC) isolated from a feedlot in Argentina, *Vet Microbiol*, 100, 3–9.

PANUTDAPORN N, CHONGSA-NGUAN M, NAIR G B, RAMAMURTHY T, YAMASAKI S *et al.* (2004), Genotypes and phenotypes of Shiga toxin-producing *Escherichia coli* isolated from healthy cattle in Thailand, *J Infect*, 48, 149–160.

PATON A, SRIMANOTE P, TALBOT U M, WANG H and PATON J C (2004), A new family of potent AB5 cytotoxins produced by Shiga toxigenic *Escherichia coli*, *J Exp Med*, 200, 35–46.

PATON A W, SRIMANOTE P, WOODROW M C and PATON J C (2001), Characterization of Saa, a novel autoagglutinating adhesin produced by locus of enterocyte effacement-negative Shiga-toxigenic *Escherichia coli* strains that are virulent for humans, *Infect Immun*, 69, 6999–7009.

PATON A W, WOODROW M C, DOYLE R M, LANSER J A and PATON J C (1999), Molecular characterization of a Shiga toxigenic *Escherichia coli* O113:H21 strain lacking *eae* responsible for a cluster of cases of hemolytic-uremic syndrome, *J Clin Microbiol*, 37, 3357–3361.

PEARSON J S, RIEDMAIER P, MARCHÈS O, FRANKEL G and HARTLAND E L (2011), A type III effector protease NleC from enteropathogenic *Escherichia coli* targets NF-κB for degradation, *Mol Microbiol*, 80, 219–230.

PERSSON S, OLSEN K E P, ETHELBERG S and SCHEUTZ F (2007), Subtyping method for *Escherichia coli* Shiga toxin (verocytotoxin) 2 variants and correlations to clinical manifestations, *J Clin Microbiol*, 45, 2020–2024.

PETERSON J W (1996), Bacterial pathogenesis. In BARON S (ed.) *Medical Microbiology*. 4th ed. Galveston TX: University of Texas Medical Branch at Galveston.

PIERARD D, HUYGHENS L, LAUWERS S and LIOR H (1991), Diarrhoea associated with *Escherichia coli* producing porcine oedema disease verotoxin, *Lancet*, 338, 762.

PIERARD D, MUYLDERMANS G, MORIAU L, STEVENS D and LAUWERS S (1998), Identification of new verocytotoxin type 2 variant B-subunit genes in human and animal *Escherichia coli* isolates, *J Clin Microbiol*, 36, 3317–3322.

PISCATELLI H, KOTKAR S A, MCBEE M E, MUTHUPALANI S, SCHAUER D B *et al.* (2011), The EHEC type III effector NleL is an E3 ubiquitin ligase that modulates pedestal formation, *PLoS ONE*, 6, e19331.

PRAGER R, FRUTH A, SIEWERT U, STRUTZ U and TSCHÄPE H (2009), *Escherichia coli* encoding Shiga toxin 2f as an emerging human pathogen, *Int J Med Microbiol*, 299, 343–353.

RABATSKY-EHR T, DINGMAN D, MARCUS R, HOWARD R *et al.* (2002), Deer meat as the source for a sporadic case of *Escherichia coli* O157:H7 infection, Connecticut, *Emerg Infect Dis*, 8, 525–527.

RAY P E and LIU X-H (2001), Pathogenesis of Shiga toxin-induced hemolytic uremic syndrome, *Pediatr Nephrol*, 16, 823–839.

READ S C, GYLES C L, CLARKE R C, LIOR H and MCEWEN S (1990), Prevalence of verocytotoxigenic *Escherichia coli* in ground beef, pork, and chicken in southwestern Ontario, *Epidemiol Infect*, 105, 11–20.

REID S D, HERBELIN C J, BUMBAUGH A C, SELANDER R K and WHITTAM T S (2000), Parallel evolution of virulence in pathogenic *Escherichia coli*, *Nature*, 406, 64–67.

RENDÓN M A, SALDAÑA Z, ERDEM A L, MONTEIRO-NETO V, VÁZQUEZ A *et al.* (2007), Commensal and pathogenic *Escherichia* coli use a common pilus adherence factor for epithelial cell colonization, *Proc Nat Acad Sci USA*, 104, 10637–10642.

RIVAS M, MILIWEBSKY E, CHINEN I, ROLDÁN C D, BALBI L *et al.* (2006), Characterization and epidemiologic subtyping of Shiga toxin-producing *Escherichia coli* strains isolated from hemolytic uremic syndrome and diarrhea cases in Argentina, *Foodborne Pathog Dis*, 3, 88–96.

ROGERIE F, MARECAT A, GAMBADE S, DUPOND F, BEAUBOIS P and LANGE M (2001), Characterization of Shiga toxin-producing *E. coli* and O157 serotype *E. coli* isolated in France from healthy domestic cattle, *Int J Food Microbiol*, 63, 217–223.

SÁNCHEZ S, GARCÍA-SÁNCHEZ A, MARTÍNEZ R, BLANCO J, BLANCO J E *et al.* (2009), Detection and characterisation of Shiga toxin-producing *Escherichia coli* other than *Escherichia coli* O157:H7 in wild ruminants, *Vet J*, 180, 384–388.

SANDHU K S, CLARKE R C and GYLES C L (1997), Hemolysin phenotypes and genotypes of *eaeA*-positive and *eaeA*-negative bovine verotoxigenic *Escherichia coli*, *Adv Exp Med Biol*, 412, 295–302.

SANZ M E, VIÑAS M R and PARMA A E (1998), Prevalence of bovine verotoxin-producing *Escherichia coli* in Argentina, *Eur J Epidemiol*, 14, 399–403.

SCALLAN E, HOEKSTRA R M, ANGULO F J, TAUXE R V, WIDDOWSON M A *et al.* (2011), Foodborne illness acquired in the United States – Major pathogens, *Emerg Infect Dis*, 17, 7–15.

SCHEIRING J, ANDREOLI S and ZIMMERHACKL L (2008), Treatment and outcome of Shiga-toxin-associated hemolytic uremic syndrome (HUS), *Pediatr Nephrol*, 23, 1749–1760.

SCHEUTZ F, TEEL L D, BEUTIN L, PIÉRARD D, BUVENS G *et al.* (2012), Multicenter evaluation of a sequence-based protocol for subtyping Shiga toxins and standardizing *stx* nomenclature, *J Clin Microbiol* 50, 2951–2963.

SCHMIDT H, BEUTIN L and KARCH H (1995), Molecular analysis of the plasmid-encoded hemolysin of *Escherichia coli* O157:H7 strain EDL 933, *Infect Immun*, 63, 1055–1061.

SCHMIDT H, KARCH H and BEUTIN L (1994), The large-sized plasmids of enterohemorrhagic *Escherichia coli* O157 strains encode hemolysins which are presumably members of the *E. coli* alpha-hemolysin family, *FEMS Microbiol Lett*, 117, 189–196.

SCHMIDT H, SCHEEF J, MORABITO S, CAPRIOLI A, WIELER L H *et al.* (2000), A new Shiga toxin 2 variant (Stx2f) from *Escherichia coli* isolated from pigeons, *Appl Environ Microbiol*, 66, 1205–1208.

SCHMITT C K, MCKEE M L and O'BRIEN A D (1991), Two copies of Shiga-like toxin II-related genes common in enterohemorrhagic *Escherichia coli* strains are responsible for the antigenic heterogeneity of the O157:H- strain E32511, *Infect Immun*, 59, 1065–1073.

SEKSE C, SUNDE M, LINDSTEDT B-A, HOPP P, BRUHEIM T *et al.* (2011), Potentially human pathogenic *Escherichia coli* O26 in Norwegian sheep flocks, *Appl and Environ Microbiol*, 77, 4949–4958.

SHAMES S R, DENG W, GUTTMAN J A, DE HOOG C L, LI Y *et al.* (2010), The pathogenic *E. coli* type III effector EspZ interacts with host CD98 and facilitates host cell prosurvival signalling, *Cell Microbiol*, 12, 1322–1339.

SHINAGAWA K, KANEHIRA M, OMOE K, MATSUDA I, HU D-L *et al.* (2000), Frequency of Shiga toxin-producing *Escherichia coli* in cattle at a breeding farm and at a slaughterhouse in Japan, *Vet Microbiol*, 76, 305–309.

SIEGLER R L, OBRIG T G, PYSHER T J, TESH V L *et al.* (2003), Response to Shiga toxin 1 and 2 in a baboon model of hemolytic uremic syndrome, *Pediatr Nephrol*, 18, 92–96.

SÖDERSTRÖM A, ÖSTERBERG P, LINDQVIST A, JÖNSSON B *et al.* (2008), A large *Escherichia coli* O157 outbreak in Sweden associated with locally produced lettuce, *Foodborne Pathog Dis*, 5, 339–349.

SONNTAG A K, ZENNER E, KARCH H and BIELASZEWSKA M (2005), Pigeons as a possible reservoir of Shiga toxin 2f-producing *Escherichia coli* pathogenic to humans, *Berl Munch Tierarztl Wochenschr*, 118, 464–470.

SRIMANOTE P, PATON A W and PATON J C (2002), Characterization of a novel type IV pilus locus encoded on the large plasmid of locus of enterocyte effacement-negative Shiga-toxigenic *Escherichia coli* strains that are virulent for humans, *Infect Immun*, 70, 3094–3100.

STEVENS M P, ROE A J, VLISIDOU I, VAN DIEMEN P M, LA RAGIONE R M *et al.* (2004), Mutation of *toxB* and a truncated version of the *efa-1* gene in *Escherichia coli* O157:H7 influences the expression and secretion of locus of enterocyte effacement-encoded proteins but not intestinal colonization in calves or sheep, *Infect Immun*, 72, 5402–5411.

STRUELENS M J, PALM D and TAKKINEN J (2011), Enteroaggregative, Shiga toxin-producing *Escherichia coli* O104:H4 outbreak: New microbiological findings boost coordinated investigations by European public health laboratories, *Euro Surveill*, 24, 2.

TAKEDA Y (1999), Enterohemorrhagic *Escherichia coli* infection in Japan, *Pediatr Int*, 41, 198–201.

TARR P I, BILGE S S, VARY J C, Jr, JELACIC S, HABEEB R L *et al.* (2000), Iha: A novel *Escherichia coli* O157:H7 adherence-conferring molecule encoded on a recently acquired chromosomal island of conserved structure, *Infect Immun*, 68, 1400–1407.

TARR P I, GORDON C A and CHANDLER W L (2005), Shiga toxin-producing *Escherichia coli* and haemolytic uraemic syndrome, *Lancet*, 365, 1073–1086.

TATSUNO I, HORIE M, ABE H, MIKI T, MAKINO K *et al.* (2001), *toxB* Gene on pO157 of enterohemorrhagic *Escherichia coli* O157:H7 is required for full epithelial cell adherence phenotype, *Infect Immun*, 69, 6660–6669.

TE LOO D M W M, MONNENS L A H, VAN DER VELDEN T J A M, VERMEER M A, PREYERS F *et al.* (2000), Binding and transfer of verocytotoxin by polymorphonuclear leukocytes in hemolytic uremic syndrome, *Blood*, 95, 3396–3402.

TESH V L, BURRIS J A, OWENS J W, GORDON V M, WADOLKOWSKI E A *et al.* (1993), Comparison of the relative toxicities of Shiga-like toxins type I and type II for mice, *Infect Immun*, 61, 3392–3402.

THANABALASURIAR A, KOUTSOURIS A, WEFLEN A, MIMEE M *et al.* (2010), The bacterial virulence factor NleA is required for the disruption of intestinal tight junctions by enteropathogenic *Escherichia coli*, *Cell Microbiol*, 12, 31–41.

THOMAS A, CHEASTY T, CHART H and ROWE B (1994), Isolation of verocytotoxin-producing *Escherichia coli* serotypes O9ab:H- and O101:H-carrying VT2 variant gene sequences from a patient with haemolytic uraemic syndrome, *Eur J Clin Microbiol Infect Dis*, 13, 1074–1076.

TOBE T, BEATSON S A, TANIGUCHI H, ABE H, BAILEY C M *et al.* (2006), An extensive repertoire of type III secretion effectors in *Escherichia coli* O157 and the role of lambdoid phages in their dissemination, *Proc Nat Acad Sci USA*, 103, 14941–14946.

TOMA C, MARTINEZ ESPINOSA E, SONG T, MILIWEBSKY E, CHINEN I *et al.* (2004), Distribution of putative adhesins in different seropathotypes of Shiga toxin-producing *Escherichia coli*, *J Clin Microbiol*, 42, 4937–4946.

TOMA C, NAKASONE N, MILIWEBSKY E, HIGA N *et al.* (2008), Differential adherence of Shiga toxin-producing *Escherichia coli* harboring *saa* to epithelial cells, *Int J Med Microbiol*, 298, 571–578.

TORRES A G, GIRON J A, PERNA N T, BURLAND V *et al.* (2002), Identification and characterization of *lpfABCC'DE*, a fimbrial operon of enterohemorrhagic *Escherichia coli* O157:H7, *Infect Immun*, 70, 5416–5427.

TOZZI A E, CAPRIOLI A, MINELLI F, GIANVITI A, DE PETRIS L *et al.* (2003), Shiga toxin-producing *Escherichia coli* infections associated with hemolytic uremic syndrome, Italy, 1988–2000, *Emerg Infect Dis*, 9, 106–108.

TOZZOLI R, CAPRIOLI A and MORABITO S (2005), Detection of *toxB*, a plasmid virulence gene of *Escherichia coli* O157, in enterohemorrhagic and enteropathogenic *E. coli*, *J Clin Micro*, 43, 4052–4056.

UHLICH G A (2009), KatP contributes to OxyR-regulated hydrogen peroxide resistance in *Escherichia coli* serotype O157:H7, *Microbiology*, 155, 3589–3598.

URDAHL A M, BEUTIN L, SKJERVE E and WASTESON Y (2002a), Serotypes and virulence factors of Shiga toxin-producing *Escherichia coli* isolated from healthy Norwegian sheep, *J Appl Microbiol*, 93, 1026–1033.

URDAHL A M, CUDJOE K, WAHL E, HEIR E and WASTESON Y (2002b), Isolation of Shiga toxin-producing *Escherichia coli* O103 from sheep using automated immunomagnetic separation (AIMS) and AIMS-ELISA: Sheep as the source of a clinical *E. coli* O103 case?, *Lett Appl Microbiol*, 35, 218–222.

USDA (2010), Detection, isolation and identification of *Escherichia coli* O157:H7 from meat products, *FSIS Microbiology Laboratory Guidebook* method 5.06, 1–12. Available from: www.fsis.usda.gov/science/microbiological_lab_guidebook [Accessed 16 November 2011].

USDA (2011a), Detection and isolation of non-O157 Shiga-toxin producing *Escherichia coli* (STEC) from meat products, *FSIS Microbiology Laboratory Guidebook* 5B.03, 1–17. Available from: www.fsis.usda.gov/science/microbiological_lab_guidebook [Accessed 16 November 2011].

USDA (2011b), Shiga Toxin-Producing *Escherichia coli* in certain raw beef products, *Federal Register*, 76, 58157–58165. Available from: www.fsis.usda.gov/OPPDE/rdad/FRPubs/2010-0023.pdf [Accessed 15 October 2011].

VARELA-HERNÁNDEZ J J, CABRERA-DIAZ E, CARDONA-LÓPEZ M A, IBARRA-VELÁZQUEZ L M, RANGEL-VILLALOBOS H *et al.* (2007), Isolation and characterization of Shiga toxin-producing *Escherichia coli* O157:H7 and non-O157 from beef carcasses at a slaughter plant in Mexico, *Int J Food Microbiol*, 113, 237–241.

VAZ T M I, IRINO K, KATO M A M F, DIAS A M G, GOMES T A T *et al.* (2004), Virulence properties and characteristics of Shiga toxin-producing *Escherichia coli* in Sao Paulo, Brazil, from 1976 through 1999, *J Clin Microbiol*, 42, 903–905.

VIDAL M, PRADO V, WHITLOCK G C, SOLARI A, TORRES A G *et al.* (2008), Subtractive hybridization and identification of putative adhesins in a Shiga toxin-producing *eae*-negative *Escherichia coli*, *Microbiol*, 154, 3639–3648.

VLISIDOU I, MARCHÉS O, DZIVA F, MUNDY R, FRANKEL G *et al.* (2006), Identification and characterization of EspK, a type III secreted effector protein of enterohaemorrhagic *Escherichia coli* O157:H7, *FEMS Microbiol Lett*, 263, 32–40.

VOSSENKÄMPER A, MARCHÈS O, FAIRCLOUGH P D, WARNES G, STAGG A J *et al.* (2010), Inhibition of NF-κB signaling in human dendritic cells by the enteropathogenic *Escherichia coli* effector protein NleE, *J Immunol*, 185, 4118–4127.

VU-KHAC H and CORNICK N A (2008), Prevalence and genetic profiles of Shiga toxin-producing *Escherichia coli* strains isolated from buffaloes, cattle, and goats in central Vietnam, *Vet Microbiol*, 126, 356–363.

WACHECK S, FREDRIKSSON-AHOMAA M, KÖNIG M, STOLLE A and STEPHAN R (2009), Wild boars as an important reservoir for foodborne pathogens, *Foodborne Pathog Dis*, 7, 307–312.

WALLACE J S, CHEASTY T and JONES K (1997), Isolation of verocytotoxin-producing *Escherichia coli* O157 from wild birds, *J Appl Microbiol*, 82, 399–404.
WANKE C A, SCHORLING J B, BARRETT L J, DESOUZA M A and GUERRANT R L (1991), Potential role of adherence traits of *Escherichia coli* in persistent diarrhea in an urban Brazilian slum, *Pediatr Infect Dis J*, 10, 746–751.
WEINSTEIN D L, JACKSON M P, SAMUEL J E, HOLMES R K and O'BRIEN A D (1988), Cloning and sequencing of a Shiga-like toxin type II variant from *Escherichia coli* strain responsible for edema disease of swine, *J Bacteriol*, 170, 4223–4230.
WELLS J G, SHIPMAN L D, GREENE K D, SOWERS E G, GREEN J H *et al.* (1991), Isolation of *Escherichia coli* serotype O157:H7 and other Shiga-like-toxin-producing *E. coli* from dairy cattle, *J Clin Microbiol*, 29, 985–989.
WERBER D, FRUTH A, BUCHHOLZ U, PRAGER R, KRAMER M H *et al.* (2003), Strong association between Shiga toxin-producing *Escherichia coli* O157 and virulence genes *stx 2* and *eae* as possible explanation for predominance of serogroup O157 in patients with haemolytic uraemic syndrome, *Eur J Clin Microbiol Infect Dis*, 22, 726–730.
WHO (1998), Zoonotic non-O157 Shiga toxin-producing *Escherichia coli* (STEC). *Report of a WHO Scientific Workshop Group Meeting.* Berlin: World Health Organization.
WICKHAM M E, LUPP C, MASCARENHAS M, VÁZQUEZ A, COOMBES B K *et al.* (2006), Bacterial genetic determinants of non-O157 STEC outbreaks and hemolytic-uremic syndrome after infection, *J Infect Dis*, 194, 819–827.
WU B, SKARINA T, YEE A, JOBIN M-C, DILEO R *et al.* (2010), NleG type 3 effectors from enterohaemorrhagic *Escherichia coli* are U-Box E3 ubiquitin ligases, *PLoS Pathog*, 6, e1000960.
XICOHTENCATL-CORTES J, MONTEIRO-NETO V, LEDESMA M A, JORDAN D M, FRANCETIC O *et al.* (2007), Intestinal adherence associated with type IV pili of enterohemorrhagic *Escherichia coli* O157:H7, *J Clin Invest*, 117, 3519–3529.
ZHANG W, BIELASZEWSKA M, KUCZIUS T and KARCH H (2002), Identification, characterization, and distribution of a Shiga toxin 1 gene variant (*stx1c*) in *Escherichia coli* strains isolated from humans, *J Clin Microbiol*, 40, 1441–1446.

13

New research on estimating the global burden of foodborne disease

R. Lake, Institute of Environmental Science and Research, New Zealand, A. H. Havelaar, National Institute for Public Health and the Environment, The Netherlands and T. Kuchenmüller, WHO, Denmark

DOI: 10.1533/9780857098740.3.260

Abstract: Burden of disease estimates inform public health policy by ranking issues, indicating trends, assessing intervention options and supporting risk management. Foodborne disease caused by microbiological, chemical or physical hazards includes acute and chronic conditions and is an important cause of morbidity and mortality worldwide. To characterise the burden of foodborne disease, the WHO Department of Food Safety Zoonoses and Foodborne Diseases launched the Initiative to Estimate the Global Burden of Foodborne Diseases. The Initiative is supported by the Foodborne Disease Burden Epidemiology Reference Group (FERG), which has established task forces to address elements required for the burden estimate. Another task force supports national burden of illness studies.

Key words: burden, foodborne, disease, attribution, disability adjusted life year, metric.

13.1 Introduction

The burden of disease is used to describe the effect of a defined health issue on a specific population. The effect may apply to society as a whole or individuals. The burden metric combines data on the prevalence or incidence of health states with indices of severity. Several severity indices have been developed, which use either a monetary value or assign a scaled value. The assigned values are elicited from patients, lay people or medical professionals.

Burden of disease estimates are used to inform public health policy at the population level in a number of ways:

- ranking of health issues to prioritise actions;
- assessment of trends;

Published by Woodhead Publishing Limited, 2013

- description of the benefit of an intervention as a reduction in burden;
- ranking of alternative interventions to reduce the burden;
- provision of an evidence base to support the development of regulatory instruments (standards and guidelines);
- provision of elements of economic analysis (cost-benefit, cost-effectiveness or cost-utility analysis).

Foodborne diseases are caused by hazards present in food, which acts as a vehicle for exposure. These hazards may be microbiological, chemical or physical contaminants. Foodborne disease includes acute and chronic conditions and is an important cause of morbidity and mortality worldwide. Diarrhoeal diseases alone – a large proportion of which is foodborne – kill 2.2 million people globally every year (WHO, 2008). Beyond its impact on global public health security, foodborne disease jeopardises the achievement of several of the Millennium Development Goals set in 2000, including the overarching goal of poverty reduction (Tauxe *et al.*, 2010).

Estimates of the national burden of foodborne disease have been published for a number of developed countries, although the metrics used are not always consistent or comparable (Buzby *et al.*, 1996; Kemmeren *et al.*, 2006; Lake *et al.*, 2010). However, the magnitude of the burden of foodborne diseases worldwide still remains unknown. The greatest burden of foodborne disease falls on populations in developing countries due to, *inter alia*, adverse environments and reduced capacity to enforce food safety measures. Burden of foodborne disease data from developing countries are scarce.

The World Health Organization (WHO) has recognised the need to address this data gap and increase global and national capacity to measure the size and nature of the burden of foodborne disease at global, regional and national levels. Under the leadership of the WHO's Department of Food Safety Zoonoses and Foodborne Diseases (FOS), the Initiative to Estimate the Global Burden of Foodborne Diseases (hereafter called the Initiative) was launched in September 2006.

The purpose of the Initiative is consistent with:

- The WHO global strategy for food safety, which has three principal goals: to advocate risk-based food safety systems, to develop science-based measures to prevent exposure to hazards through food and to assess and communicate foodborne risks (WHO, 2002).
- The WHO's food safety resolution, 'Advancing food safety initiatives', in which member states asked WHO 'to continue to provide global leadership in providing technical assistance and tools that meet the needs of Member States and the Secretariat for scientific estimations on foodborne risks and foodborne disease burden from all causes' (WHO, 2010).

As described on the website dedicated to this Initiative (WHO, 2012a), the intention is to provide data and tools to support policymakers and other stakeholders to set appropriate, evidence-informed priorities for food safety at the country level.

Through the support of a special external advisory group, the Foodborne Disease Burden Epidemiology Reference Group (FERG), the Initiative aims to:

- strengthen the capacity of countries in conducting burden of foodborne disease assessments and to increase the number of countries that have undertaken a burden of foodborne disease study;
- encourage countries to use burden of foodborne disease estimates to set evidence-informed policies;
- provide estimates on the global burden of foodborne diseases according to age, sex and region for a defined list of causative agents of microbial, parasitic and chemical origin.

This chapter will describe burden of foodborne disease estimation in general terms, and provide a detailed description of the activity of the Initiative to date (December 2012).

13.2 Estimating the burden of foodborne diseases: metrics and attribution

13.2.1 Metrics

Several metrics to quantify the burden of disease as a summary measure of public health have been put forward (Murray *et al.*, 2002; Mangen *et al.*, 2010). Broadly, these provide numerical values for health states in two ways: (1) as a monetary value (the cost of illness) and (2) in units of health.

Monetary value

Assigning a monetary value to a health state involves estimating:

- direct costs (medical and non-medical);
- indirect costs (estimate of the monetary value of the inability of the cases (or a caregiver) to undertake normal activities).

Medical direct costs include expenses required to access healthcare and medication. Non-medical direct costs usually include items such as the transport requirements for cases to access healthcare. Other non-medical costs may include those associated with risk management by governments (surveillance and the development of regulatory instruments) and industry (preventive control measures, recalls and loss of product and brand credibility).

Indirect costs for morbidity usually include the economic value of lost productivity through not being able to work, while costs for mortality may be estimated from lost earnings over the remaining period of normal life expectancy (human capital method, Buzby *et al.*, 1996) or over the (much shorter) period needed to replace a deceased person on the labour market (friction period method, van den Hout *et al.*, 2010).

Alternatively, a willingness-to-pay approach may be taken to provide an economic value for health states. This approach involves a variety of mechanisms by which individuals indicate how much they would be willing to pay to avoid a particular health state. Estimates may be generated by addressing the health state itself, or else developing estimates of the higher prices consumers might pay for safer products.

Unit of health

The quantity of life (mortality) and changes in the quality of life (morbidity) can be amalgamated into a single unit of health metric. The most common of these health adjusted life year metrics are the quality adjusted life year (QALY, usually employed to measure health gains from interventions) and the disability adjusted life year (DALY) (Gold *et al.*, 2002). The DALY expresses the gap between an existing health state and a hypothetical ideal, and thus is a metric suited to the burden of disease.

In order to amalgamate the quantity and quality of life into a unit of health metric, a set of values or weights are developed to express the severity of the health states. The values are assigned on a scale of 0 to 1, with perfect health and the worst imaginable health state representing the opposite extremes. The specific values are elicited from panels of lay people, patients or medical professionals, using a variety of methods (Gold *et al.*, 2002; Haagsma, 2010). Methods involving the exchange of a period of life expectancy to avoid an adverse health state (a time trade-off) are considered the most rigorous.

The DALY metric was originally developed by the WHO for the Global Burden of Disease (GBD) Study (Murray and Acharya, 1997; Murray and Lopez, 1997). National burden of foodborne disease estimates using the DALY metric have been published for the Netherlands and New Zealand (Kemmeren *et al.*, 2006; Lake *et al.*, 2010; Mangen *et al.*, 2004).

The DALY is calculated by adding the number of years of life lost to mortality (YLL) to the number of years lived with disability due to morbidity (YLD):

$$\text{DALY} = \text{YLL} + \text{YLD}$$

The YLL due to a specific disease in a specified population is calculated by summing all fatal cases (n) due to the health outcomes (l) of that specific disease, each case multiplied by the expected individual lifespan (e) at the age of death:

$$\text{YLL} = \sum_{l} n_l \times e_l$$

YLD is calculated by summing over all health outcomes (l), the product of the number of cases (n), the duration of the illness (t) and the severity weight (w) of a specific disease. It should be noted that the calculation for YLL implicitly includes a severity weight factor. The severity weight or disability weight factors are in the range 0 to 1, with the severity weight for death being equal to 1:

$$\text{YLD} = \sum_{l} n_l \times t_l \times w_l$$

DALYs may be calculated using a prevalence approach, which estimates the current burden of disease in a population, considering previous events. However, the more common approach is to use incidence, i.e. both current and future health outcomes are included. Future outcomes include sequelae and mortality resulting from the initial disease within a defined time period.

To define the life expectancy needed for the calculation of YLL, life expectancy tables for the population being studied may be used. Alternatively, life expectancy that reflects an ideal of human potential may be used. These values (based on those for the Japanese population, who have the highest life expectancy globally) are taken from the Coale-Demeny Regional Model Life Tables (WHO, 2102b).

13.2.2 Attribution

Since the hazards that humans are exposed to via food also often occur with other exposure vehicles, the proportion of the overall incidence caused by exposure from food consumption needs to be determined (i.e. a proportion of the overall incidence of disease needs to be attributed to food).

Comparative risk assessments, as undertaken by the GBD Study, assign burden of disease estimates to risk factors based on a counterfactual exposure distribution (Ezzati *et al.*, 2006). Reviews of the literature are used to provide suitable counterfactual exposure distributions, and the stratified estimates for existing exposures.

To estimate the minimum theoretical exposure distribution (i.e. zero exposure) as a counterfactual for foodborne hazards, it is necessary to estimate the current exposure. Such estimates rely on data about food consumption, hazard prevalence and concentration. While such data may be available for some hazards (particularly chemicals), for others alternative approaches must be used to assign the attribution. Alternative approaches often involve a multitude of methods, including an analysis of surveillance and outbreak data, analytical epidemiological studies, subtyping of pathogens from humans and putative sources, and comparative risk assessments (Batz *et al.*, 2005; Pires *et al.*, 2009).

13.3 Foodborne Disease Burden Epidemiology Reference Group (FERG) structure and process

While the WHO's FOS provides leadership, coordination and administration functions, the Initiative is operated through the FERG, an international expert group advising the WHO Director-General on issues related to global, regional and country-level foodborne disease burden estimation. FERG consists of a central reference group and six task forces, which work on the Initiative's two main tracks:

Track 1: Assembly, appraisal and reporting of the burden of foodborne disease estimates.

Track 2: In-depth country studies to supplement the work of FERG and enable countries to conduct their own burden of disease studies.

Of the six task forces established by FERG, three are hazard based:

- chemical hazards
- parasitic hazards
- hazards causing enteric illness.

A fourth task force was established to address source attribution as a fundamental component of assigning the burden of disease.

For the Initiative's second track, a Country Studies Task Force was established in 2009 to advise WHO on the initiation, conduct and completion of national burden of foodborne disease studies.

In 2012, a new Computational Task Force was established to develop and implement a strategy to calculate the DALY estimates of the global and regional burdens of foodborne disease from age and sex stratified incidence data, including methods to complement existing data gaps.

Each task force includes FERG members, representatives from WHO's FOS, as well as *ad hoc* resource advisors recruited for their particular expertise. Some work is undertaken by task force members, while additional work (in particular systematic reviews) is commissioned to various scientific institutions.

13.3.1 Hazard-specific task forces

Each of these task forces has compiled a list of priority hazards to be included in the global burden estimates. Each of these hazards was chosen by considering:

- its global distribution
- a qualitative estimate of its likely contribution to the overall burden
- practical considerations of data availability and quality

For each of the priority hazards, a defined set of health outcomes has been described. For most of the parasitic hazards, systematic reviews have been commissioned to generate agent-specific incidence estimates for the outcomes. This is also the case for some of the hazards causing enteric illness. However, as the most common outcome is acute gastrointestinal disease (AGI), the incidence of illness caused by several of the hazards being considered by the Enteric Disease Task Force will be estimated on the basis of assigning a fraction of the overall incidence of AGI, based on aetiological studies of cases.

The Chemical Hazards Task Force was faced with the issue that most of the relevant health outcomes are multi-factorial, may be caused by a number of agents and studies of cases to determine aetiology are not available. Consequently systematic reviews have been commissioned to determine not only the incidence of the health outcomes but also to provide data to support exposure assessment so that dose response relationships can be used to estimate attribution.

13.3.2 Source Attribution Task Force

To address the issue of source attribution from the global and regional perspectives needed for the FERG estimates, the Source Attribution Task Force has reviewed all available methods (Pires *et al.*, 2009). Inevitably, data to support attribution is extremely sparse in the regions of the world where the burden of foodborne disease is likely to be the greatest.

To overcome this problem, the Source Attribution Task Force commissioned an expert elicitation exercise, to provide estimates for attribution on a regional basis. To provide direction for this exercise, the task force has developed agreed definitions for:

- the regional basis on which the attribution estimates will be made;
- the food, water, human, animal contact and environment pathways for which attribution will be made;
- the categories to be used for attribution to specific foods.

13.3.3 Related burden of disease studies

It is important for the FERG estimates to be consistent and comparable with the data generated by other institutions involved in global burden of disease assessments. The various GBD studies, in particular the Global Burden of Diseases, Injuries, and Risk Factors Study 2010, will be a crucial source of methodological direction for DALY estimates as well as reference data. The GBD 2010 Study is about to publish estimates of the burden of diseases, injuries and risk factors for two time periods, 1990 and 2005, with projections for 2010 (IHME, 2012).

The Child Health Epidemiology Reference Group has provided important estimates of the incidence of diarrhoeal mortality in children less than five years of age (CHERG, 2012; Black *et al.*, 2010). This disease is the second largest cause of child mortality in children (after pneumonia), and is consequently expected to be a major contributor to the foodborne disease burden in developing countries.

13.3.4 DALY calculation

A key input to the DALY calculation is the severity (or disability) weight for each of the health outcomes relevant to foodborne hazards. The GBD 2010 Study has developed disability weights for the illnesses and injuries included in their burden estimates, and some of these will be useful for the foodborne diseases being considered by the FERG. An alternative source of disability weights, which are particularly relevant to foodborne diseases caused by microbial pathogens (i.e. acute gastrointestinal disease), is a Dutch study which used a time trade-off approach (Haagsma *et al.*, 2008). These two sources are not necessarily compatible, and decisions will need to be made about which weights to use.

Additional decisions for calculating DALYs have to be made. These relate to:

- Relevant health outcomes: Foodborne hazards may cause a variety of adverse acute health problems and sequelae, or health problems that occur after chronic exposure. These have been defined for the priority hazards chosen by each task force.

- Life expectancy: For consistency, the ideal lifespan approach will be used.
- Age weighting: Although previous burden of disease estimates have weighted age ranges differently, all years will be weighted equally by the FERG.
- Discounting of future burden: DALYs occurring in the future will not be discounted.
- Comorbidity: If more than one adverse condition occurs in a person at the same time, an adjustment will be made to allocate the burden correctly.

13.3.5 Country Studies Task Force

The Country Studies Task Force was established to initiate and provide guidance to countries conducting national burden of foodborne disease studies. These studies will provide first-hand burden estimates and will be accompanied by: (a) specific capacity building opportunities designed for and offered to developing countries who have successfully applied to the Initiative's country-level track, and (b) knowledge transitions to ensure that the burden data are meaningful to end users and foster research uptake.

The objectives of the Initiative's country-level work are to:

- deliver burden of disease estimates for foodborne diseases;
- contribute to the burden of disease scientific and knowledge translation development within each country;
- provide results that can be translated into food safety policy for each country;

To accomplish the above objectives, two subgroups of the Country Studies Task Force have been established: a Burden of Disease Group to address the burden of disease study, and the Knowledge Translation and Policy Group to promote the use of burden of disease estimates in setting food safety policy within each country.

Although it is intended that the national foodborne disease studies will be conducted as much as possible by local scientists, the Burden of Disease Group has developed protocols and tools to assist with these studies. These tools will also mean that national burden of foodborne disease studies will be comparable with the global and regional estimates being developed by the FERG, by using consistent methodology.

The second subgroup, the Knowledge Translation and Policy Group, has developed a manual to assist with context mapping analyses to better understand the actors, dynamics, structures and processes surrounding priority food safety issues at a national level. These analyses will focus on stakeholders, the political context of food safety, national policy processes and what windows of opportunity might open for the evidence to influence national policymaking.

13.3.6 Annual meetings, stakeholders and communication

From its inception, the FERG has been active in promoting communication to publicise its work and engaging with stakeholders to open new channels for multi-sectoral technical cooperation and ensure two-way communication with the food safety stakeholder community. Annual meetings of the FERG and associated task forces were held each year from 2007 to 2010. In addition to progress and planning discussions, each of these meetings included a stakeholders' day. Representatives from industry, the media, NGOs and governments were invited to attend briefings and provide feedback about the direction and progress of the FERG.

13.4 FERG outputs to date

Scientific publications that have emerged from this project include:

- global estimates of the incidence and aetiology of diarrhoeal morbidity and mortality in older children, adolescents and adults (Fischer Walker and Black, 2010);
- a systematic review of the aetiology of diarrhoea in older children, adolescents and adults (Fischer Walker *et al.*, 2010);
- a systematic review of the parameters needed to estimate the global burden of peanut allergy (Ezendam and van Loveren, 2012);
- a systematic review of the relationship between aflatoxin exposure and growth impairment (Khlangwiset *et al.*, 2011);
- global incidence estimates of human neurocysticercosis (Ndimubanzi *et al.*, 2010);
- a systematic review of clinical manifestations of human neurocysticercosis (Carabin *et al.*, 2011);
- global incidence estimates of human alveolar echinococcosis (Torgerson *et al.*, 2010);
- a report on the worldwide occurrence and burden of trichinellosis (Murrell and Pozio, 2011);
- a report on the global burden of human foodborne trematodiasis (Fürst *et al.*, 2011);
- a systematic review of the disease frequency of brucellosis (Dean *et al.*, 2012).

The FERG website has links to these publications, as well as copies of meeting reports and promotional material.

Throughout 2009 and 2010, the Country Studies Task Force developed protocols and tools to assist individual countries. An application round was conducted and four countries were selected to conduct the pilot studies. As of early 2012, Thailand, Albania, Japan and Uganda have all commenced their national studies.

13.5 Future trends

Currently the FERG foresees the presentation of initial estimates of the global and regional burden of foodborne disease in early 2014. During 2012 and 2013 the programme of work will address:

- the development of a DALY calculator, which will provide burden estimates from age-stratified incidence estimates of the relevant health outcomes and include the methodological choices required;
- the consolidation of the incidence estimates for the chosen health outcomes from the systematic reviews and other sources;
- the completion of the expert elicitation exercise to provide attribution estimates;
- the completion of the country studies.

As with any global health analysis, developing estimates of the burden of foodborne disease is an ambitious undertaking. Inevitably, the first estimates will be subject to much discussion and future refinement. Ultimately, the estimates should inform the setting of priorities for ways to reduce the global burden of foodborne disease, in the same manner as the Disease Control Priorities Project (World Bank, 2006). The participants in this Initiative have been committed and enthusiastic, which is welcome as this type of undertaking relies to a large extent on voluntary contributions.

13.6 References

BATZ MB, DOYLE MP, MORRIS JG, PAINTER J, SINGH R *et al.* (2005) 'Attributing illness to food', *Emerg Infect Dis*, 11, 993–999.

BLACK RE, COUSENS S, JOHNSON HL, LAWN JE, RUDAN I *et al.* (2010) 'Global, regional, and national causes of child mortality in 2008: A systematic analysis', *Lancet*, 375, 1969–1987.

BUZBY J, ROBERTS T, LIN C-T and MACDONALD J (1996) 'Bacterial foodborne disease: Medical costs and productivity losses', *Agricultural Economic Report* No. 741. United States Department of Agriculture Economic Research Service: Washington DC.

CARABIN H, NDIMUBANZI PC, BUDKE CM, NGUYEN H, QIAN Y *et al.* (2011) 'Clinical manifestations associated with neurocysticercosis', *PloS Neglect Trop Dis*, 5, e1152. doi: 10.1371/journal.pntd.0001152.

CHERG (2012) 'Child Health Epidemiology Reference Group', Geneva, World Health Organization. Available from: cherg.org (accessed 12 December 2012).

DEAN AS, CRUMP L, GRETER H, SCHELLING E and ZINSSTAG J (2012) 'Global burden of human brucellosis: A systematic review of disease frequency', *PLoS Negl Trop Dis*, 6: e1865. doi: 10.1371/journal.pntd.0001865.

EZENDAM J and VAN LOVEREN H (2012) 'Parameters needed to estimate the global burden of peanut allergy. Systematic literature review', RIVM report 340007002/2012. Available from: library.wur.nl/WebQuery/clc/1988290 (accessed 5 December 2012).

EZZATI M, VANDER HOORN S, LOPEZ A, DANAEI G, RODGERS A *et al.* (2006). 'Comparative quantification of mortality and burden of disease attributable to selected risk factors', in Lopez AD, Mathers CD, Ezzati M, Jamison DT and Murray CL (edd), *Global Burden of Disease and Risk Factors*, World Bank and Oxford University Press, New York, 241–396.

FISCHER WALKER CL and BLACK RE (2010) 'Diarrhoea morbidity and mortality in older children, adolescents, and adults', *Epidemiol Infect*, 138, 1215–1226.

FISCHER WALKER CL, SACK D and BLACK RE (2010) 'Etiology of diarrhea in older children, adolescents and adults: A systematic review', *PLoS Negl Trop Dis*, 4, e768. Epub 7 August 2010.

FÜRST T, KEISER J and UTZINGER J (2011) 'Global burden of human food-borne trematodiasis: A systematic review and meta-analysis', *Lancet Inf Dis* 21 November, doi: 10.1016/S1473-3099(11)70294–8.

GOLD M, STEVENSON D and FRYBACK D (2002) 'HALYs and QALYs and DALYs, Oh My: similarities and differences in summary measures of population health', *Annu Rev Publ Health*, 23, 115–134.

HAAGSMA J (2010) 'Disability adjusted life years and acute onset disorders. Improving estimates of the non-fatal burden of injuries and infectious intestinal disease', A thesis submitted in partial fulfilment of the degree of Doctor of Philosophy, Erasmus University, Rotterdam. Available from: repub.eur.nl/res/pub/21187/Proefschrift%20Juanita%20A%20Haagsma.pdf (accessed 14 February 2012).

HAAGSMA JA, HAVELAAR AH, JANSSEN BMF and BONSEL GJ (2008) 'Disability adjusted life years and minimal disease: Application of a preference-based relevance criterion to rank enteric pathogens', *Popul Health Metrics*, 7, doi: 10.1186/1478-7954-6-7.

IHME (2012) 'Global Burden of Disease Study', Seattle, Institute for Health Metrics and Evaluation. Available from: www.globalburden.org (accessed 12 December 2012).

KEMMEREN JM, MANGEN M-JJ, VAN DUYNHOVEN YTPH and HAVELAAR AH (2006) 'Priority setting of foodborne pathogens', National Institute for Public Health and the Environment, the Netherlands (RIVM), RIVM report 330080001/2006. Available from: www.rivm.nl/bibliotheek/rapporten/330080001.pdf (accessed 14 February 2012).

KHLANGWISET P, SHEPHERD GS and WU F (2011) 'Aflatoxins and growth impairment: A review', *Crit Rev Toxicol*, 41, 740–755.

LAKE R, CRESSEY P, CAMPBELL D and OAKLEY E (2010) 'Risk ranking for foodborne microbial hazards in New Zealand: Burden of disease estimates', *Risk Anal*, 30, 743–752.

MANGEN M-JJ, BATZ MB, KASBOHRER A, HALD T, MORRIS JG, Jr *et al.* (2010) 'Integrated approaches for the public health prioritization of foodborne and zoonotic pathogens', *Risk Anal*, 30, 782–797.

MANGEN M-JJ, HAVELAAR AH and DE WIT GA (2004) 'Campylobacteriosis and sequelae in the Netherlands. Estimating the disease burden and the cost-of-illness.' RIVM report 250911004/2004. Available from: www.rivm.nl/bibliotheek/rapporten/250911004.pdf. National Institute for Public Health and the Environment (RIVM). Accessed 20 February 2013.

MURRAY CJ and ACHARYA AK (1997) 'Understanding DALYs', *J Health Econ*, 16, 703–730.

MURRAY CJL and LOPEZ AD (1997) 'Global mortality, disability, and the contribution of risk factors: Global burden of disease study', *Lancet*, 349, 1436–1442.

MURRAY CJL, SALOMON JA, MATHERS CD and LOPEZ AD (2002) *Summary Measures of Public Health: Concepts, Ethics, Measurement and Application.* World Health Organization; Geneva.

MURRELL KD and POZIO E (2011) 'Worldwide occurrence and impact of human trichinellosis, 1986–2009', *Emerg Infect Dis*, 17, 2194–2202.

NDIMUBANZI PC, CARABIN H, BUDKE CM, NGUYEN H, QIAN Y-J *et al.* (2010) 'A systematic review of the frequency of neurocyticercosis with a focus on people with epilepsy', *PLoS Negl Trop Dis*, 4(11): e870. doi: 10.1371/journal.pntd.0000870.

PIRES SM, EVERS EG, VAN PELT W, AYERS T, SCALLAN E *et al.* (2009) 'Attributing the human disease burden of foodborne infections to specific sources', *Foodborne Pathog Dis*, 6, 417–424.

TAUXE RV, DOYLE MP, KUCHENMÜLLER T, SCHLUNDT J and STEIN CE (2010) 'Evolving public health approaches to the global challenge of foodborne infections', *Int J Food Microbiol*, 139, Suppl 1,16–28.

TORGERSON PR, KELLER K, MAGNOTTA M and RAGLAND N (2010) 'The global burden of alveolar echinococcosis', *PLoS Negl Trop Dis*, 4(6): e722. doi: 10.1371/journal.pntd.0000722.

VAN DEN HOUT WB (2010) 'The value of productivity: Human-capital versus friction-cost method', *Ann Rheum Dis*, 69, i89–i91.

WHO (2002) 'WHO global strategy for food safety: Safer food for better health', Geneva, World Health Organization. Available from: www.who.int/foodsafety/publications/general/en/strategy_en.pdf (accessed 14 February 2012).

WHO (2008) 'The global burden of disease. 2004 update', Geneva, World Health Organization. Available from: www.who.int/healthinfo/global_burden_disease/GBD_report_2004update_full.pdf (accessed 14 February 2012).

WHO (2010) 'Advancing food safety initiatives', WHA63.3, Geneva, World Health Organization. Available from: apps.who.int/gb/ebwha/pdf_files/WHA63/A63_R3-en.pdf (accessed 14 February 2012).

WHO (2012a) 'Initiative to estimate the global burden of foodborne diseases', Geneva, World health Organization. Available from: www.who.int/foodsafety/foodborne_disease/ferg/en/index.html (accessed 12 December 2012).

WHO (2012b) 'Model life tables', Geneva, World Health Organization. Available from: www.un.org/esa/population/publications/Model_Life_Tables (accessed 14 December 2012).

WORLD BANK (2006) 'Disease control priorities in developing countries', 2nd edition, Washington DC, The International Bank for Reconstruction and Development / The World Bank. Available from: files.dcp2.org/pdf/DCP/DCP.pdf (accessed 12 December 2012).

Part IV

Food preservation techniques

14

Novel methods for pathogen control in livestock pre-harvest: an update*

T. R. Callaway, R. C. Anderson, T. S. Edrington, K. J. Genovese, R. B. Harvey, T. L. Poole and D. J. Nisbet, USDA-ARS, USA

DOI: 10.1533/9780857098740.4.275

Abstract: Pathogenic bacteria are found asymptomatically in food animals, which often results in pathogen entry into the food chain. Processing plants reduce pathogen contamination with intervention strategies, yet foodborne illnesses still occur at an unacceptable frequency. Strategies are needed against pathogenic bacteria before they can enter processing facilities and the food chain. Reducing farm levels of zoonotic pathogens will enhance human health and food safety. Several pre-slaughter intervention strategies are under investigation: (1) direct anti-pathogen strategies, (2) competitive enhancement strategies and (3) animal management strategies. These include methods such as: vaccination against foodborne pathogenic bacteria, probiotics, prebiotics and competitive exclusion, using viruses to reduce pathogen populations, chemical methods and dietary changes.

Key words: interventions, foodborne diseases, pathogens, pre-harvest.

14.1 Introduction

Consumers rightfully expect and demand a safe food supply. However, yearly millions of consumers become ill from consuming foods contaminated with zoonotic pathogenic bacteria in or on foods (Scallan *et al.*, 2011). Human illnesses caused by the most common foodborne pathogens cost the United States economy

* Mandatory Disclaimer: Proprietary or brand names are necessary to report factually on available data; however, the USDA neither guarantees nor warrants the standard of the product, and the use of the name by the USDA implies no approval of the product, or exclusion of others that may be suitable.

alone more than $152 billion each year (Scharff, 2010). Because some outbreaks have been linked to consumption of meat-based products, or to contact with animals or their wastes, this review will focus on issues affecting animal agriculture. The zoonotic pathogens responsible for most of the American human foodborne illnesses include: *Campylobacter* spp., *Salmonella*, enterohemorrhagic *Escherichia coli* (EHEC; including O157:H7) and *Listeria monocytogenes* (USDA-ERS, 2001; Scallan *et al.*, 2011; Scharff, 2010).

Post-harvest pathogen reduction strategies effectively reduce pathogen contamination after animals enter the abattoir (Arthur *et al.*, 2009, 2010; Koohmaraie *et al.*, 2005). However, focusing pathogen reduction efforts only after animals have been harvested has not eliminated human foodborne illnesses and does not address environmental contamination and dissemination or illnesses resulting from direct animal exposure via water or foods of plant origin contaminated by animal waste. Thus recent years have found researchers exploring new avenues to reduce zoonotic pathogens in animals before they enter the food chain (Sargeant *et al.*, 2007; LeJeune and Wetzel, 2007; Loneragan and Brashears, 2005; Oliver *et al.*, 2008). Since fecal pathogen shedding has been correlated with carcass contamination, pathogen populations in the live animal are now viewed as critical to the production of safe and wholesome food, as well as to the assurance of environmental quality.

This chapter presents several potential intervention and management strategies that have been proposed to reduce zoonotic foodborne pathogenic bacteria populations found in live food animals. These strategies can be categorized into three general approaches: (1) competitive enhancement strategies, (2) direct anti-pathogen strategies and (3) animal management strategies. Some of these intervention strategies are available to animal producers today, but others require a great deal of further research before they can be introduced as another hurdle/enhancement in the farm-to-fork food safety continuum.

14.2 Foodborne pathogenic bacteria: human exposure routes

Foods for human consumption naturally contain bacteria acquired during growth, harvesting, preparation, processing and production, and they are indicative of re-environmental contamination throughout the food production and preparation chain (Arthur *et al.*, 2008; Doane *et al.*, 2007). Typically, this bacterial population is harmless to humans, but when animal and vegetable products are not cooked or pasteurized, they can often be sources of human illnesses by carrying pathogenic bacteria (Noal *et al.*, 2010; Jain *et al.*, 2009; LeJeune and Rajala-Schultz, 2009; Cody *et al.*, 1999a, 1999b).

Foodborne pathogenic bacteria can survive in a variety of environments (Semenov *et al.*, 2010; Roesch *et al.*, 2007; Coyne *et al.*, 1997), including animals and have been found on all types of animal production farms and in all stages of animal growth and production (Oliver *et al.*, 2005; Murinda *et al.*, 2004; Alali *et al.*, 2010; LeJeune and Kersting, 2010). Enterohemorrhagic *Escherichia coli* (including *E. coli* O157:H7), *Salmonella, Campylobacter* and *Listeria* are some of the most common foodborne pathogenic bacteria isolated from human outbreaks, and have all been isolated from cattle, swine and poultry (Oliver *et al.*, 2005; Callaway *et al.*, 2006a; Borland, 1975). Foodborne pathogenic bacteria can also live asymptomatically in the gut or on the skin and hide of food animals (Arthur *et al.*, 2007; Reid *et al.*, 2002; Doyle and Erickson, 2006; Porter *et al.*, 1997) where they can be transmitted directly to meat products during processing (Mackey and Derrick, 1979; Arthur *et al.*, 2010). These bacteria can be transmitted to other foods by direct contact with animals or their feces, as well as by contact with vectors such as insects, mice, birds and other mammals (Nielsen *et al.*, 2004; Jay *et al.*, 2007; Talley *et al.*, 2009). Furthermore irrigation water can be a route of fruit or vegetable contamination (Mackey and Derrick, 1979; Natvig *et al.*, 2002; Manshadi *et al.*, 2001). Perhaps more significantly (from a public health perspective), human foodborne illness outbreaks have been linked to indirect human contact with animal feces via water supplies (both drinking and irrigation) as well as direct fecal contact (Anonymous, 2000; Jay *et al.*, 2007; LeJeune and Kersting, 2010). Increasing incidences of human illnesses linked to direct animal contact via petting zoos, fairs and open farms emphasizes the need for strategies to focus on reduction of pathogens on the farm (Chapman *et al.*, 2000; Pritchard *et al.*, 2000; Varma *et al.*, 2003; Keen *et al.*, 2003, 2006; Durso *et al.*, 2005).

Given these multiple routes of human exposure to these critical foodborne pathogenic bacteria, it is apparent that intervening in the processing plant or later can only solve part of the problem. The introduction of pre-harvest pathogen reduction interventions on farms can reduce the environmental contamination load and can potentially reduce human illness from direct contact with animal or feces, and the entry of foodborne pathogens into the water supply. Furthermore, reducing pathogens in live animals will reduce the pathogen burden entering the processing plants, which will allow the already efficacious in-plant treatments to work against a lower incidence or concentration of pathogens, resulting in a synergistic decrease in pathogen entry into the food supply.

14.3 The gastrointestinal tract microflora and ecological inertia

The microbial population of the intestinal tract of mammals is very diverse, with upward of 600 microbial species being found in individual samples (Callaway *et al.*, 2010a; Lu *et al.*, 2003). In newborn animals, the digestive tract population is very low in both numbers and diversity (Falk *et al.*, 1998). Over time maternal and/or environmental exposure begins to populate the intestinal tract with a succession of microbial species, resulting in the development of a mature, complex, microbial ecosystem (Lu *et al.*, 2003). Endogenous gastrointestinal bacteria compete fiercely with one another for available nutrients (Hungate, 1950, 1952). An established gastrointestinal microbial population makes an animal more resistant to transient opportunistic infections (Fuller, 1989; Jayne-Williams and Fuller, 1971). A fully mature ecosystem has all environmental niches occupied, and as a result opportunistic or transient populations have a difficult time establishing a foothold in the environment (the gastrointestinal tract). Thus, the presence of the microbial population acts as a form of inertia or inherent ecological conservatism, preventing drastic changes in the population. However, disturbances in the environment (host animal), such as dietary changes, stress, transport to a new location and contact with new animals, can disrupt the microbial ecology in the intestinal ecosystem, allowing pathogens to colonize the intestinal tract and gain a foothold in the ecosystem. Thus, many of the proposed live animal interventions are aimed at enhancing the environmental inertia of the microbial population or at decreasing the amplitude of changes inflicted by environmental disturbances (i.e., the severity of gastrointestinal disruption caused by stresses on the host animal) and the subsequent carriage of zoonotic foodborne pathogenic bacteria.

In recent years, the role of the microbial ecosystem or microbial organ has been examined in regards to its role in health, as well as for the production parameters (Turnbaugh *et al.*, 2006, 2009; Ley *et al.*, 2006; Xu and Gordon, 2003; Finegold, 2008; Murphy, 2004; Lyte, 2010). Though much of the microbial ecology research has focused on the effects of the microbial communities, an increasing amount of research has delved into the microbial organ of animals (Freestone and Lyte, 2010; Bailey *et al.*, 2011; Pullinger *et al.*, 2010). The cross-talk between the host and the resident microbial flora in the realm of microbial endocrinology offers a new mechanism to researchers and to the animal industry (Lyte, 2010). With the view that the microbial population (or probiotics) can be used as a type of drug delivery mechanism, the modification of animal performance and food safety via alterations in the microbial population is a real possibility (Price *et al.*, 2010).

14.4 Delineation of anti-pathogen versus pro-commensal strategies

In this section we delineate the live animal pathogen reduction strategies into two basic approaches: anti-pathogen and pro-commensal. Anti-pathogen strategies are those that directly kill or reduce the growth rate in the pathogens of interest. The best example of this would be the use of antimicrobials and antibiotics, but given concerns about antibiotic resistance it is expected in the future that prophylactic antibiotic usage will be reduced, causing probiotic/competitive enhancement strategies to become more effective and more widely adopted. Conversely, pro-commensal strategies seek to promote the growth of groups of beneficial bacteria that are competitive with, or antagonistic to, pathogens (Fuller, 1989). Unfortunately, competitive enhancement products (probiotics) have not been widely implemented in the animal industry, often due to the use of antibiotics (Steer *et al.*, 2000).

14.5 Competitive enhancement strategies to reduce foodborne pathogens

Introducing native or novel microflora (fungi or bacteria) to reduce pathogenic bacteria in the gut is a probiotic strategy (Fuller, 1989; Ouwehand *et al.*, 1999). Another pro-commensal approach is to provide a limiting nutrient (prebiotic) to the microbial population in the gastrointestinal tract, which can give an existing commensal microbial population a competitive advantage (Schrezenmeir and de Vrese, 2001; Mosenthin and Bauer, 2000; Steer *et al.*, 2000). No matter which competitive enhancement strategy is used, the overall goal is to fill all ecological niches in the gut and thereby prevent the establishment of, or cause the displacement of, a pathogenic bacterial population in the gut (Endt *et al.*, 2010). This beneficial effect of the natural microbial population has been variously described as 'bacterial antagonism' (Freter *et al.*, 1983), 'bacterial interference' (Dubos, 1963), the 'barrier effect' (Fedorka-Cray *et al.*, 1999) or competitive exclusion (Lloyd *et al.*, 1977).

14.5.1 Probiotics

A probiotic is a 'live microbial feed supplement which beneficially affects the host animal by improving intestinal microbial balance' and they are included in animal rations to enhance performance or to reduce zoonotic pathogens (Collins and Gibson, 1999; Fuller, 1989; Sherman, 2009). Probiotics can be: (1) live cultures of yeast or bacteria, (2) heat-treated (or otherwise inactivated) cultures of yeast or bacteria, or (3) fermentation end products from culturing yeast or bacteria. The most common probiotic bacteria used in food animals remain *Bifidobacterium* and *Lactobacillus* (Gomes and Malcata, 1999).

In general, probiotics are aimed at improving animal growth and performance, but some have been reported to reduce foodborne pathogens in food animals (Ohya *et al.*, 2000, 2001; Etienne-Mesmin *et al.*, 2011; Leatham *et al.*, 2009; Stephens *et al.*, 2007a, 2007b). In swine, the addition of a probiotic culture comprising *Streptococcus (Enterococcus) faecium* reduced enterotoxigenic *E. coli* (ETEC) colonization and subsequent diarrhea in swine (Underdahl *et al.*, 1982; Ushe and Nagy, 1985). Inclusion of a *Saccharomyces* fermentation feed product did not reduce populations of *Salmonella* in experimentally infected pigs, but did reduce the negative effects of *Salmonella* infection in these animals (Price *et al.*, 2010). Other swine studies found that feeding a probiotic coupled with a prebiotic (a synbiotic, discussed below) reduced ETEC shedding and effects (Krause *et al.*, 2010) as well as *Clostridium difficile* colonization (Songer *et al.*, 2007). Other studies found that *Lactobacillus* and *Streptococcus* cultures reduced *Salmonella* populations in poultry (Zhang *et al.*, 2007a, 2007b).

Research studies of ruminants have demonstrated that a probiotic containing *Lactobacillus acidophilus* cultures (single strain, known as a direct fed microbial or DFM) added to the feed of finishing cattle reduced *E. coli* O157:H7 shedding by more than 50% (Brashears *et al.*, 2003a, 2003b; Elam *et al.*, 2003; Brashears

Published by Woodhead Publishing Limited, 2013

and Galyean, 2002). In an independent evaluation, it was found that in cattle fed this DFM, fecal shedding of *E. coli* O157:H7 was 13%, while in the control group it reached 46% (Ransom *et al.*, 2003). Further studies have confirmed that this product can reduce *E. coli* O157 populations in cattle (Stephens *et al.*, 2007a, 2007b; Younts-Dahl *et al.*, 2004; Moxley *et al.*, 2003). Currently, this probiotic is fed to beef cattle to reduce pathogens and improve growth efficiency; this ensures that the pathogen reduction pays for its inclusion in the cattle ration, a critical factor in any pathogen reduction strategy. Other probiotics comprising non-toxigenic *E. coli* strains that specifically target the reduction of *E. coli* O157:H7 in ruminants have been developed as well, and are in the product development stream (Zhao *et al.*, 2003).

Drawbacks/limitations: Probiotics by definition need to be fed daily, and there have often been issues with specific probiotic versus diet interactions, which must be examined for the most common diets.

14.5.2 Prebiotics

Substrates that are unavailable to or indigestible by the host animal, but are digestible by a segment of its microbial population, are generally classified as prebiotics (Crittenden, 1999; Collins and Gibson, 1999; Walker and Duffy, 1998). Prebiotics have been used in humans largely in an effort to promote intestinal health and stimulate native immunity, as well as modifying the microbial ecology of the intestinal tract (Meyer, 2008; Crittenden, 1999; Janczyk *et al.*, 2010). Prebiotics provide energy or other limiting nutrients to the intestinal mucosa as well as substrates for intestinal bacterial fermentation, resulting in enhanced production of vitamins and antioxidants that further directly benefit the host animal (Kim *et al.*, 2011; Janardhana *et al.*, 2009; Meyer, 2008; Bailey, 2009). Some prebiotics can provide a competitive advantage to members of the native microflora, which can help to exclude pathogenic bacteria from the intestine via direct competition for nutrients or binding sites through the production of blocking factors, or via the production of antimicrobial compounds (e.g., bacteriocins and volatile fatty acids) (Zopf and Roth, 1996; Baines *et al.*, 2011; Price *et al.*, 2010; Shoaf *et al.*, 2006).

Generally speaking, feeding prebiotics to food animals has been limited due to the prohibitive expense of these compounds. Recent years have seen an upswing in research into the use of prebiotics for food animals, including efforts that utilize prebiotics to reduce foodborne pathogenic bacteria populations. Prebiotics have been used to improve growth and nutrition in pigs successfully and have been shown to affect *Salmonella* populations in swine (Budiño *et al.*, 2010; Wells *et al.*, 2005; Price *et al.*, 2010; Krause *et al.*, 2010; Van Loo, 2007; Kien *et al.*, 2007). Other research has indicated that specific carbohydrate prebiotics reduced colitis-type symptoms and improved mucosal immunity in swine (Pouillart *et al.*, 2010). In poultry, the use of prebiotics has been found to increase immune measures and performance (Vandeplas *et al.*, 2010; Falaki *et al.*, 2010; Janardhana *et al.*, 2009; Midilli *et al.*, 2008; Piray and Kermanshahi, 2008) and in some cases it has reduced *Salmonella* and *Clostridium perfringens* populations in broilers and models of *in vitro* intestinal infection (Kim *et al.*, 2011; da Silva *et al.*, 2011; Awad *et al.*, 2011).

Drawbacks/limitations: Currently, prebiotics are prohibitively expensive for use in food animals; however, that is changing with the development of molecular technologies. Prebiotics must also be fed daily, and their use in ruminants will always be problematic due to the catabolic potential of the mixed ruminal microbial consortium.

14.5.3 Competitive exclusion

Competitive exclusion (CE) is a specific type of probiotic strategy that involves the addition of a (non-pathogenic) bacterial culture to the intestinal tract of food animals in order to reduce colonization or decrease populations of pathogenic bacteria in the gastrointestinal tract (Fuller, 1989; Nurmi *et al.*, 1992; Callaway and Martin, 2006; Schneitz, 2005). A CE culture may be composed of one or more strains or species of bacteria, but it should be derived from the animal of interest (e.g., a chicken CE culture from a chicken or a swine CE from swine). Thus theoretically, CE cultures attempt to conserve and take advantage of synergies acquired during co-evolution of host and microorganism. Unfortunately, there is considerable confusion over the use of the competitive exclusion terminology based on the exclusionary effect of some probiotic preparations; however, labeling a product as CE is closely regulated by the US Food and Drug Administration (FDA).

Salmonella colonization in young chickens was reduced by administration of a preparation of gut bacteria originating from healthy adult chickens (Nurmi and Rantala, 1973; Snoeyenbos *et al.*, 1978). Other treatments that utilized defined and undefined CE cultures have shown that mixtures of gut bacteria from healthy adult animals that do not contain pathogenic bacteria provide a degree of protection against pathogen colonization in newly hatched poultry (Lloyd *et al.*, 1974, 1977; Reid and Barnum, 1984; Stavric and D'Aoust, 1993). The extensive use of CE cultures to reduce pathogen colonization in poultry around the world has been thoroughly reported (Nava *et al.*, 2005; Bielke *et al.*, 2003; Stavric and D'Aoust, 1993; Stavric, 1992; Mead, 1989, 2000; Schneitz, 2005). The beneficial effects of CE in poultry have led to the development of several successful commercial CE products around the world (Weinack *et al.*, 1982; Snoeyenbos *et al.*, 1978; Schneitz, 2005; Nisbet *et al.*, 1993, 1994; Hofacre *et al.*, 2003). Other researchers have demonstrated that a swine mucosal CE culture could reduce *Salmonella* populations in young pigs (Fedorka-Cray *et al.*, 1999). Recent studies demonstrated that a swine CE culture derived from the cecal contents of healthy pigs reduced the incidence of *Salmonella choleraesuis* (Anderson *et al.*, 1999) and enterotoxigenic *E. coli* (Genovese *et al.*, 2001; Harvey *et al.*, 2003, 2005). The use of true CE products in ruminants has been limited to date because of the complexity of the ruminant gastrointestinal microbial population, and the length of time involved in cattle production (up to 18 months).

Drawbacks/limitations: Competitive exclusion must be used with neonates, but this is difficult for cattle born in pasture. The use of antibiotics is contra-effective to effective CE utilization, and currently the use of antibiotics in the US is more economically feasible, but given rumored impending changes this situation is still fluid.

14.5.4 Synbiotics

The simultaneous application of CE (and some probiotic products) with prebiotics is known as synbiotics, and could have a synergistic effect in the reduction of foodborne pathogenic bacterial populations in food animals prior to slaughter (Nemcova *et al.*, 2007; Bomba *et al.*, 2002). Furthermore, synbiotic approaches increased glucose transport in broilers (Awad *et al.*, 2008) and increased the availability of B vitamins (Branner and Roth-Maier, 2006). Several studies have demonstrated that synbiotic strategies could reduce post-weaning *E. coli* diarrhea in swine (Krause *et al.*, 2010) and necrotic enteritis in poultry (Hofacre *et al.*, 2003), but to date few studies have examined the effects of synbiotics per se. In a new twist on the synbiotic concept, researchers included bacteriophages along with a competitive exclusion culture to successfully reduce *Salmonella* population in poultry (Toro *et al.*, 2005). This result suggests a future where custom synbiotics can be prepared to target foodborne pathogens in food animals where they can be effective against pathogens even during the stressful transport and lairage periods.

14.6 Direct anti-pathogen strategies to reduce foodborne pathogens

Antibiotics are a well-known method of targeting pathogenic bacteria, but their utility in reducing foodborne pathogenic bacteria in food animals is limited. Therefore, several alternative non-antibiotic anti-foodborne pathogen strategies for live food animals have been investigated in recent years: (1) vaccination, (2) sodium chlorate, (3) antimicrobial proteins produced by bacteria or fungi, and (4) bacteriophages.

14.6.1 Antibiotics

Antibiotics have often been used as a direct method to alter the microbial ecology of the intestinal tract and to increase animal growth rate and/or efficiency. Because of concerns over antibiotic resistance, it is likely that the use of antibiotics in food animals will become even more highly regulated. In spite of these potential drawbacks in the use of antibiotics, researchers found that neomycin-treated animals shed fecal *E. coli* O157:H7 at a rate of 0% and reduced hide contamination by *E. coli* O157:H7; untreated animals shed the pathogen at a rate of 46% (Ransom *et al.*, 2003). However, in spite of these potential benefits, the use of antibiotics also used in human medicine is not recommended.

14.6.2 Vaccination

Vaccines have a long history of use in food animal production, primarily for reducing populations of disease-causing bacteria and viruses in food animals, though some vaccines have been considered for use against foodborne pathogenic bacteria (Johansen *et al.*, 2000; Amani *et al.*, 2011; Yekta *et al.*, 2011; McNeilly *et al.*, 2010; Gyles, 1998). For example, vaccines against *Salmonella* strains responsible for animal diseases have been developed for use in swine and dairy cattle (House *et al.*, 2001). Vaccination has also been successfully used to combat post-weaning *E. coli* edema disease in young pigs (Johansen *et al.*, 2000) and to reduce *Salmonella* colonization in poultry and swine (Stabel *et al.*, 1993; Zhang *et al.*, 1999; Zhang-Barber *et al.*, 1999). Because foodborne pathogenic bacteria (especially *E. coli* O157) generally do not cause illness in the host animals, a different approach has had to be taken in harnessing the immune system to reduce the prevalence of foodborne pathogenic bacteria.

A vaccine has been developed for use in feedlot cattle, which significantly reduces fecal *E. coli* O157:H7 shedding (Potter *et al.*, 2004; Finlay, 2003). Preliminary experimental results indicate that this vaccine reduced *E. coli* O157:H7 shedding in feedlot cattle from 23% to less than 9% (Moxley *et al.*, 2003). In an evaluation study, it was demonstrated that vaccinated animals had *E. coli* O157:H7 fecal shedding rates of 14% compared to 46% for untreated control animals (Ransom *et al.*, 2003). This vaccine has undergone repeated evaluation and is currently being marketed, along with other E. *coli* O157 (and other EHEC) vaccines (Moxley *et al.*, 2008, 2009; Smith *et al.*, 2009a, 2009b; Fox *et al.*, 2009; Thomson *et al.*, 2009; Van Donkersgoed *et al.*, 2005). Further vaccines have been developed, which target specific proteins secreted by EHEC, and the future development of these broad vaccines is promising (McNeilly *et al.*, 2010; Amani *et al.*, 2011; Yekta *et al.*, 2011; Khare *et al.*, 2010).

While some technical issues remain to be resolved, the use of vaccination to reduce foodborne pathogens appears to hold promise, and can be included as edible vaccines (Amani *et al.*, 2011; Judge *et al.*, 2004). Because of the nature of vaccination (using the native immunity of the host), it is possible that vaccination could be used in conjunction with other pathogen reduction strategies to produce synergistic results (Moxley *et al.*, 2003).

Drawbacks/limitations: Most animals are vaccinated, especially cattle and swine, so the methods are well understood. Development of a vaccine against EHEC strains and various *Salmonella* serotypes is challenging due to the difficulty in targeting such different organisms. Technical issues surrounding the number of vaccinations needed to achieve full immunity also remain, especially in regards to handling cattle multiple times in order to vaccinate them against this single pathogen, because handling stress affects growth efficiency.

14.6.3 Sodium chlorate

Salmonella and *E. coli* bacteria can respire under anaerobic conditions by reducing nitrate to nitrite via a dissimilatory nitrate reductase (Stouthamer, 1969;

Stewart, 1988). The intracellular bacterial enzyme nitrate reductase does not differentiate between nitrate and its analog, chlorate, which is reduced to chlorite in the cytoplasm; chlorite accumulation kills bacteria equipped with nitrate reductase (Stewart, 1988). Chlorate treatment *in vitro* quickly reduced populations of *E. coli* O157:H7 and *Salmonella* by more than 5 $\log_{10}$ CFU (Anderson *et al.*, 2000). The addition of chlorate to animal rations reduced experimentally inoculated *Salmonella* and *E. coli* O157:H7 populations in swine and sheep intestinal tracts (Anderson *et al.*, 2001a, 2001b; Edrington *et al.*, 2003a; Callaway *et al.*, 2003a; Burkey *et al.*, 2004) and poultry feces and intestinal contents (Moore *et al.*, 2006; Jung *et al.*, 2003; Byrd *et al.*, 2003). Other studies indicated that soluble chlorate administered via drinking water significantly reduced ruminal, cecal and fecal *E. coli* O157:H7 populations in cattle (Anderson *et al.*, 2005; Callaway *et al.*, 2002) and populations of *E. coli* on their hides. *In vitro* and *in vivo* results have indicated that chlorate treatment does not adversely affect the ruminal or the cecal/colonic fermentation in ruminant or monogastric animals (Anderson *et al.*, 2000; Callaway *et al.*, 2002). Additional studies have demonstrated that chlorate does not alter antibiotic resistance or toxin production by *E. coli* O157:H7 (Callaway *et al.*, 2004a, 2004b). Adding chlorate to the last meal (approximately 24 h) before animals are shipped for processing has been suggested as a way of reducing foodborne pathogens; the chlorate would work against these pathogens during transportation. The use of chlorate for food animals to reduce foodborne pathogenic bacteria is presently under review by the FDA.

Drawbacks/limitations: Chlorate is not currently approved, but is likely to be most useful immediately pre-harvest or on open farms and petting zoos to prevent direct contamination.

14.6.4 Antimicrobial proteins

Some bacteria produce antimicrobial proteins as they engage in biochemical warfare against bacteria that occupy the same (or similar) environmental niche in order to sequester nutrients (Klaenhammer, 1988; Jack *et al.*, 1995). These antimicrobial proteins are classified as bacteriocins; those that specifically target *E. coli* species are known as colicins (Jack *et al.*, 1995; Lakey and Slatin, 2001). Intestinal bacteria can produce bacteriocins including colicins (Wells *et al.*, 1997; Laukova and Marckova, 1993; Smarda and Obdrzalek, 2001), and some of these bacterial strains are often incorporated into CE cultures or probiotics (Zhao *et al.*, 1998, 2003; Reissbrodt *et al.*, 2009; Walsh *et al.*, 2008; Otero and Nader-Macías, 2006; Murinda *et al.*, 1996, 1998). The feeding of probiotic bacteria that produce colicins or other bacteriocins has been suggested and successfully utilized against animal and foodborne pathogens in experimental models (Schamberger *et al.*, 2004; Stahl *et al.*, 2004; Walsh *et al.*, 2008; Cutler *et al.*, 2007). Only recently have purified bacteriocins, including colicins, been produced in quantities sufficient for examination as feed additives for reducing foodborne pathogens in food animals (Callaway *et al.*, 2004c; Stahl *et al.*, 2004), and potentially for use on carcasses and ready-to-eat foods to prevent foodborne illness (Patton *et al.*, 2007, 2008). However, for bacteriocins and colicins to be economically feasible as an additive in animal feed for improving food safety, their production must be scaled up.

Drawbacks/limitations: Colicins and bacteriocins are proteins and must be protected from enzymatic degradation by the animal and must also be protected from ruminal degradation, as a bypass protein, for efficacy in the lower gastrointestinal tract.

14.6.5 Bacteriophages

Bacteriophages are viruses that prey solely upon bacteria and have rather narrow target spectra (Lederberg, 1996; Sulakvelidze and Kutter, 2005). Specificity allows phages to be used as smart bombs against specific microorganisms in a mixed microbial population without perturbing the overall ecosystem, and have been used in place of antibiotics for nearly 100 years (d'Herelle, 1917; Summers, 2001; Sulakvelidze and Barrow, 2005; Sulakvelidze and Kutter, 2005). Recently, the FDA approved the use of phages as surface-cleaning agents as well as in live animal hide sprays to reduce *E. coli* O157:H7 on cattle hides before they enter the slaughter plant (FDA, 2006). A commercial product that is an anti-*E. coli* O157:H7 phage spray has been produced and is currently undergoing a commercial field evaluation.

Phages are a normal component of the gastrointestinal (Callaway *et al.*, 2006a, 2010b; Raya *et al.*, 2006; Dhillon *et al.*, 1976) and farm microbial ecosystems (McLaughlin *et al.*, 2006; Niu *et al.*, 2009), and as such interest in using them against disease and foodborne pathogenic bacteria has grown in recent years (Johnson *et al.*, 2008; Sulakvelidze and Barrow, 2005). Phages have reduced the incidence of diseases that impact production efficiency or health in swine, sheep and poultry (Smith and Huggins, 1982, 1983, 1987; Huff *et al.*, 2002, 2005), including ETEC-induced diarrhea and splenic ETEC colonization (Smith and Huggins, 1983, 1987). Bacteriophages have been used experimentally to control foodborne pathogenic bacteria in ruminant animals, especially *E. coli* O157:H7 (Kudva *et al.*, 1999; Bach *et al.*, 2009; Rozema *et al.*, 2009; Niu *et al.*, 2008; Raya *et al.*, 2011; Callaway *et al.*, 2008; Stanford *et al.*, 2010). Further studies have found that *Salmonella* populations in swine and poultry could also be controlled by phage addition (Higgins *et al.*, 2005; Wall *et al.*, 2010; Callaway *et al.*, 2011). In spite of the recent spate of exciting research, more studies must be performed with this re-emerging tool before phages can be considered a viable method to control populations of foodborne pathogenic bacteria in food animals.

Drawbacks/limitations: Bacteria can quickly become resistant to a single phage, therefore phages must be used in cocktails to prevent the development of resistance. Furthermore, phage cocktails must be formulated specifically to target multiple serotypes or species of foodborne pathogenic bacteria. They can be tailored against a wide variety of physiological receptors.

14.7 Animal management intervention strategies

Good management of cattle is critical for efficient animal production, but to date no typical management procedures have been shown to affect colonization or shedding of foodborne pathogenic bacteria. Processing cattle with squeeze chutes has been shown to increase the probability of hide contamination with *E. coli* O157 (Mather *et al.*, 2007). Lairage and handling have also been demonstrated to play a role in *Salmonella* transmission in swine (Rostagno *et al.*, 2003; Hurd *et al.*, 2001; Rostagno and Callaway, 2012). Although no management practices have been shown to directly improve food safety (LeJeune and Wetzel, 2007), some practices may reduce horizontal transmission (Ellis-Iversen and Watson, 2008).

Increasing amounts of research have investigated the effects of stress on animal susceptibility to pathogen colonization and pathogen shedding (Rostagno, 2009; Bailey *et al.*, 2011; Freestone and Lyte, 2010; Salak-Johnson and McGlone, 2007). Results of stress studies have been mixed because the definition of stress in animals is variable between individual animals and is cumulative (Brown-Brandl *et al.*, 2009; Schuehle Pfeiffer *et al.*, 2009; Ghareeb *et al.*, 2008). However, in general it does appear that handling stress (Brown-Brandl *et al.*, 2009; Dowd *et al.*, 2007; Midgley and Desmarchelier, 2001) and new animal contact (social stress) can impact foodborne pathogen populations (Callaway *et al.*, 2006b; Bailey *et al.*, 2011; Bach *et al.*, 2004), but further research in this area is clearly indicated.

Farm hygiene has been shown to have some limited role in fecal shedding and in hide contamination (Dargatz *et al.*, 1997), but only recently have researchers begun to examine the role of pens and feed bunks in the spread of foodborne pathogenic bacteria. Bedding material can harbor bacteria that cause animals disease, as well as foodborne pathogenic bacteria (Wetzel and LeJeune, 2006; Oliver *et al.*, 2005; Richards *et al.*, 2006; Davis *et al.*, 2005). Modeling research has shown that an increase in the frequency of cleaning bedding would increase the death rate of pathogens such as *E. coli* O157:H7 (Vosough Ahmadi *et al.*, 2007). Researchers have shown that sand bedding reduced transmission of *E. coli* O157:H7 compared to sawdust (LeJeune and Kauffman, 2005). Recent molecular studies have indicated that the bacterial communities of feedlot surfaces (and by extension other animal housing environments) are distinct from fecal bacterial populations (Durso *et al.*, 2011). Wet surfaces allow foodborne pathogens to survive longer periods than in dry conditions (Berry and Miller, 2005; Smith *et al.*, 2001). Overall, bedding or pen cleaning or other changes will not eliminate foodborne pathogens from any farm or feedlot environment, but can slow the spread of pathogens. Closed herds prevent the spread of pathogens from one farm to another (Ellis-Iversen and Watson, 2008). Off-site rearing of dairy heifers may be an important solution for reducing foodborne pathogens, as has been shown in regard to *Salmonella* (Hegde *et al.*, 2005), and the risk of transmission back to the farm by heifers returning from an off-site facility was found to be low (Edrington *et al.*, 2008). Animal density may also play a role in the horizontal spread of foodborne pathogens (Vidovic and Korber, 2006). Densely packed animals have a greater chance of contamination via fecal-oral routes. Higher animal density can be linked to an increased risk of carriage of some EHEC, including O157:H7 (Frank *et al.*, 2008; Vidovic and Korber, 2006). The use of split marketing in swine has also been linked to an increased risk of *Salmonella* shedding (Rostagno *et al.*, 2009). Thus some simple steps to improve food safety are reducing stress and novel animal contact/environmental contamination during transportation and lairage. Several research groups are currently working in this area and they are developing some promising animal handling processes that minimize stress and contamination.

14.7.1 Dietary composition, changes and effects

Diet affects the environment of the gastrointestinal tract of food animals by causing changes in the environment at the microbial level. Rapid changes in rations cause stress and a disruption of the intestinal microbial ecology, which can indirectly affect pathogen populations (Callaway *et al.*, 2009; Bailey *et al.*, 2010). When cattle were abruptly switched from a maize-based finishing ration to a 100% hay diet, fecal *E. coli* populations declined significantly within 5 d (Diez-Gonzalez *et al.*, 1998). Other results have indicated that longer-term forage feeding had no effect, or even increased *E. coli* O157:H7 shedding (Kudva *et al.*, 1995; Buchko *et al.*, 2000a, 2000b). Based on a large number of studies from around the world, it appears that an abrupt shift from grain to forage does indeed affect *E. coli* populations (Callaway *et al.*, 2009; Jacob *et al.*, 2009), but the magnitude of this effect is not always consistent, and may be related to forage quality. In spite of the potential benefits (if any) of forage feeding, we must weigh the benefits versus the negative impacts on carcass quality, as well as economic, logistic and other infrastructure-related issues before advocating widespread adoption.

Some animals spend a period of time off-feed (by choice or during feed withdrawal prior to shipment) during their lives, and this has been shown to increase fecal shedding of *Salmonella* and *E. coli* (Grau *et al.*, 1968, 1969; Brownlie and Grau, 1967; Nettelbladt *et al.*, 1997; Rostagno and Callaway, 2012; Holt, 1993; Buchko *et al.*, 2000b). The type of grain used in cattle rations can significantly impact fecal shedding of *E. coli* O157:H7. For instance, barley has been linked to an increased shedding of *E. coli* O157:H7 (Dargatz *et al.*, 1997; Berg *et al.*, 2004; Bach *et al.*, 2005). Other studies have found an increased risk of *E. coli* O157:H7 shedding in cattle fed distiller's grains as part of their concentrate (Jacob *et al.*, 2008a, 2008b, 2008c, 2010; Wells *et al.*, 2009), while other studies have found this effect to be less pronounced (Edrington *et al.*, 2010).

Ionophores (such as monensin) are antimicrobials fed to cattle to improve production efficiency that are unrelated to antibiotics used in human medicine. The use of ionophores in cattle rations is temporally related to the discovery of *E. coli* O157:H7 in cattle, thus it was hypothesized that feeding cattle with ionophores could select for the growth of this important gram-negative pathogen. One study found that when fed with forage including monensin, cattle shed *E. coli* O157:H7 for longer periods (Van Baale *et al.*, 2004). However, because of the physiology of common foodborne pathogenic bacteria, ionophore feeding did not affect foodborne pathogenic bacterial populations *in vitro* or in experimentally infected animals in a majority of studies (Busz *et al.*, 2002; Edrington *et al.*, 2003b, 2003c, 2006; Bach *et al.*, 2002; McAllister *et al.*, 2006; Jacob *et al.*, 2008b). Thus it appears that ionophore feeding is not responsible for the rise of *E. coli* O157:H7 in cattle. Additionally, ionophore resistance is not linked to medically important antibiotic resistance (Callaway *et al.*, 2003b), although an increase in resistance to macrolide antibiotics in enterococci was noted in a single study (Jacob *et al.*, 2008b).

14.8 Future trends

The development of pre-harvest interventions to reduce foodborne pathogenic bacteria is a relatively recent innovation. While much of the work that has been performed thus far has been foundational or exploratory, this young field appears to have reached a critical mass of knowledge where forward progress can be made relatively rapidly. Because there are so many approaches to pre-harvest pathogen reduction, it is likely that several of these strategies can be applied concurrently to produce a synergistic reduction in pathogen populations. As the various interventions described above are utilized in the real world, pragmatic analysis of the results will demonstrate which need refinement (or replacement), and which provide synergistic benefits when applied concurrently (such as vaccination and probiotics) or sequentially (such as phages and sodium chlorate).

Many of the interventions described in this chapter depend to some degree on the native microbial ecosystem or the host immune system. We have only recently been able to quantify the microbial populations of large numbers of genera simultaneously (Callaway *et al.*, 2010a; Dowd *et al.*, 2008). By understanding which members of the microbial ecosystem are correlated with positive production, health or food safety results, we can determine which populations to target with probiotics and prebiotics, and how phages can affect niches that pathogens may occupy. Furthermore, microbial endocrinology is a new and developing field encompassing the two-way interaction between the host and its microbial organ within the gastrointestinal tract (Lyte, 2010). As we begin to understand more of the microbial ecosystem and its interactions, both within the microbial population as well as between the microbes and the host animals (Lyte, 2010; Lyte and Freestone, 2009), new interventions will undoubtedly be developed that are closely tied to these interactions. With such developments in understanding the commensal relationships within and between the microbial ecosystem and the host, it is clear that pre-harvest strategies will undergo radical enhancements in the near future.

14.9 Conclusions

Although North American and European food supplies are generally safe, foodborne illnesses still occur and are too frequently associated with products derived from animal agriculture. Whilst animal agriculture has sought to enhance the safety of its products, until relatively recently pre-slaughter intervention points were not fully explored as potential strategies for improving the microbiological status of animals and consequently the safety of the derived food products. This has changed in recent years with the development of vaccines, prebiotics, probiotics, competitive exclusion, antibiotics, antimicrobials and proper animal management, all of which can potentially reduce the number of foodborne pathogenic bacteria that enter an abattoir. Further research into intervention strategies and management procedures that specifically target the pre-slaughter critical control points is crucial to improving overall food and environmental safety. However, it must be noted that none of the interventions has been shown to be a magic bullet. They must be utilized as part of a coherent, coordinated anti-pathogen strategy (and commitment), which encompasses the continuum from farm to fork. The introduction of multiple hurdles to the entry of pathogenic bacteria into the production of food, which includes barriers that begin near birth, during growth, finishing and in-plant, can produce a synergistic reduction of foodborne illnesses in humans.

It must also be noted that the intervention strategies discussed in this chapter cannot be implemented without imposing an economic cost on animal or food producers. The increased cost of production must be shared by all concerned parties, including consumers, to increase the safety of our food supply. Without an economic incentive to implement these pathogen-reduction procedures, many intervention strategies will remain economically unfeasible.

14.10 References

ALALI, W. Q., THAKUR, S., BERGHAUS, R. D., MARTIN, M. P. and GEBREYES, W. A. 2010. Prevalence and distribution of *Salmonella* in organic and conventional broiler poultry farms. *Foodborne Path. Dis.*, 7, 1363–1371.

AMANI, J., MOUSAVI, S. L., RAFATI, S. and SALMANIAN, A. H. 2011. Immunogenicity of a plant-derived edible chimeric EspA, Intimin and Tir of *Escherichia coli* O157:H7 in mice. *Plant Sci.*, 180, 620–627.

ANDERSON, R. C., BUCKLEY, S. A., CALLAWAY, T. R., GENOVESE, K. J., KUBENA, L. F. *et al.* 2001a. Effect of sodium chlorate on *Salmonella* Typhimurium concentrations in the weaned pig gut. *J. Food Prot.*, 64, 255–259.

ANDERSON, R. C., BUCKLEY, S. A., KUBENA, L. F., STANKER, L. H., HARVEY, R. B. *et al.* 2000. Bactericidal effect of sodium chlorate on *Escherichia coli* O157:H7 and *Salmonella* Typhimurium DT104 in rumen contents *in vitro*. *J. Food Prot.*, 63, 1038–1042.

ANDERSON, R. C., CALLAWAY, T. R., BUCKLEY, S. A., ANDERSON, T. J., GENOVESE, K. J. *et al.* 2001b. Effect of oral sodium chlorate administration on *Escherichia coli* O157:H7 in the gut of experimentally infected pigs. *Int. J. Food Microbiol.*, 71, 125–130.

ANDERSON, R. C., CARR, M. A., MILLER, R. K., KING, D. A., CARSTENS, G. E. *et al.* 2005. Effects of experimental chlorate preparations as feed and water supplements on *Escherichia coli* colonization and contamination of beef cattle and carcasses. *Food Microbiol.*, 22, 439–447.

ANDERSON, R. C., STANKER, L. H., YOUNG, C. R., BUCKLEY, S. A., GENOVESE, K. J. *et al.* 1999. Effect of competitive exclusion treatment on colonization of early-weaned pigs by *Salmonella* serovar Choleraesuis. *Swine Health Prod.*, 7, 155–160.

ANONYMOUS 2000. Waterborne outbreak of gastroenteritis associated with a contaminated municipal water supply, Walkerton, Ontario, May–June 2000. *Can. Commun. Dis. Rep.*, 26, 170–173.

ARTHUR, T. M., BOSILEVAC, J. M., BRICHTA-HARHAY, D. M., KALCHAYANAND, N., KING, D. A. *et al.* 2008. Source tracking of *Escherichia coli* O157:H7 and *Salmonella* contamination in the lairage environment at commercial U.S. beef processing plants and identification of an effective intervention. *J. Food Prot.*, 71, 1752–1760.

ARTHUR, T. M., BOSILEVAC, J. M., NOU, X., SHACKLEFORD, S. D., WHEELER, T. L. *et al.* 2007. Comparison of the molecular genotypes of *Escherichia coli* O157:H7 from the hides of beef cattle in different regions of North America. *J. Food Prot.*, 70, 1622–1626.

ARTHUR, T. M., BRICHTA-HARHAY, D. M., BOSILEVAC, J. M., KALCHAYANAND, N., SHACKELFORD, S. D. *et al.* 2010. Super shedding of *Escherichia coli* O157:H7 by cattle and the impact on beef carcass contamination. *Meat Sci.*, 86, 32–37.

ARTHUR, T. M., KEEN, J. E., BOSILEVAC, J. M., BRICHTA-HARHAY, D. M., KALCHAYANAND, N. *et al.* 2009. Longitudinal study of *Escherichia coli* O157:H7 in a beef cattle feedlot and role of high-level shedders in hide contamination. *Appl. Environ. Microbiol.*, 75, 6515–6523.

AWAD, W. A., GHAREEB, K. and BÖHM, J. 2011. Evaluation of the chicory inulin efficacy on ameliorating the intestinal morphology and modulating the intestinal electrophysiological properties in broiler chickens. *J. Anim. Physiol. Anim. Nutr.*, 95, 65–72.

AWAD, W. A., GHAREEB, K., NITSCH, S., PASTEINER, S., ABDEL-RAHEEM, S. *et al.* 2008. Effects of dietary inclusion of prebiotic, probiotic and synbiotic on the intestinal glucose absorption of broiler chickens. *Int. J. Poult. Sci.*, 7, 686–691.

BACH, S. J., JOHNSON, R. P., STANFORD, K. and MCALLISTER, T. A. 2009. Bacteriophages reduce *Escherichia coli* O157:H7 levels in experimentally inoculated sheep. *Can. J. Anim. Sci.*, 89, 285–293.

BACH, S. J., MCALLISTER, T. A., MEARS, G. J. and SCHWARTZKOPF-GENSWEIN, K. S. 2004. Long-haul transport and lack of preconditioning increases fecal shedding of *Escherichia coli* and *Escherichia coli* O157:H7 by calves. *J. Food Prot.*, 67, 672–678.

BACH, S. J., MCALLISTER, T. A., VEIRA, D. M., GANNON, V. P. and HOLLEY, R. A. 2002. Effect of monensin on survival and growth of *Escherichia coli* O157:H7 *in vitro*. *Can. Vet. J.*, 43, 718–719.

BACH, S. J., SELINGER, L. J., STANFORD, K. and MCALLISTER, T. 2005. Effect of supplementing corn- or barley-based feedlot diets with canola oil on faecal shedding of *Escherichia coli* O157:H7 by steers. *J. Appl. Microbiol.*, 98, 464–475.

BAILEY, M. 2009. The mucosal immune system: Recent developments and future directions in the pig. *Develop. Comp. Immunol.*, 33, 375–383.

BAILEY, M. T., DOWD, S. E., GALLEY, J. D., HUFNAGLE, A. R., ALLEN, R. G. *et al.* 2011. Exposure to a social stressor alters the structure of the intestinal microbiota: Implications for stressor-induced immunomodulation. *Brain Behav. Immun.*, 25, 397–407.

BAILEY, M. T., DOWD, S. E., PARRY, N. M. A., GALLEY, J. D., SCHAUER, D. B. *et al.* 2010. Stressor exposure disrupts commensal microbial populations in the intestines and leads to increased colonization by *Citrobacter rodentium*. *Infect. Immun.*, 78, 1509–1519.

BAINES, D., ERB, S., LOWE, R., TURKINGTON, K., SABAU, E. *et al.* 2011. A prebiotic, Celmanax, decreases *Escherichia coli* O157:H7 colonization of bovine cells and feed-associated cytotoxicity *in vitro*. *BMC Res. Notes*, 4, 110.

BERG, J., MCALLISTER, T., BACH, S., STILBORN, R., HANCOCK, D. *et al.* 2004. *Escherichia coli* O157:H7 excretion by commercial feedlot cattle fed either barley- or corn-based finishing diets. *J. Food Prot.*, 67, 666–71.

BERRY, E. D. and MILLER, D. N. 2005. Cattle feedlot soil moisture and manure content: II. Impact on *Escherichia coli* O157. *J. Environ. Qual.*, 34, 656–663.

BIELKE, L. R., ELWOOD, A. L., DONOGHUE, D. J., DONOGHUE, A. M., NEWBERRY, L. A. *et al.* 2003. Approach for selection of individual enteric bacteria for competitive exclusion in turkey poults. *Poult. Sci.*, 82, 1378–1382.

BOMBA, A., NEMCOVÁ, R., MUDRONOVÁ, D. and GUBA, P. 2002. The possibilities of potentiating the efficacy of probiotics. *Trends Food Sci. Technol.*, 13, 121–126.

BORLAND, E. D. 1975. *Salmonella* infection in poultry. *Vet. Rec.*, 97, 406–408.

BRANNER, G. R. and ROTH-MAIER, D. A. 2006. Influence of pre-, pro-, and synbiotics on the intestinal availability of different B-vitamins. *Arch. Anim. Nutr.*, 60, 191–204.

BRASHEARS, M. M. and GALYEAN, M. L. 2002. Testing of probiotic bacteria for the elimination of *Escherichia coli* O157:H7 in cattle [Online]. Amer. Meat Inst. Found. Available: www.amif.org/PRProbiotics042302.htm [Accessed 24 April 2007].

BRASHEARS, M. M., GALYEAN, M. L., LONERAGAN, G. H., MANN, J. E. and KILLINGER-MANN, K. 2003a. Prevalence of *Escherichia coli* O157:H7 and performance by beef feedlot cattle given *Lactobacillus* direct-fed microbials. *J. Food Prot.*, 66, 748–754.

BRASHEARS, M. M., JARONI, D. and TRIMBLE, J. 2003b. Isolation, selection, and characterization of lactic acid bacteria for a competitive exclusion product to reduce shedding of *Escherichia coli* O157:H7 in cattle. *J. Food Prot.*, 66, 355–363.

BROWN-BRANDL, T. M., BERRY, E. D., WELLS, J. E., ARTHUR, T. M. and NIENABER, J. A. 2009. Impacts of individual animal response to heat and handling stresses on *Escherichia coli* and *E. coli* O157:H7 fecal shedding by feedlot cattle. *Foodborne Path. Dis.*, 6, 855–864.

BROWNLIE, L. E. and GRAU, F. H. 1967. Effect of food intake on growth and survival of *Salmonellas* and *Escherichia coli* in the bovine rumen. *J. Gen. Microbiol.*, 46, 125–134.

BUCHKO, S. J., HOLLEY, R. A., OLSON, W. O., GANNON, V. P. J. and VEIRA, D. M. 2000a. The effect of different grain diets on fecal shedding of *Escherichia coli* O157:H7 by steers. *J. Food Prot.*, 63, 1467–1474.

BUCHKO, S. J., HOLLEY, R. A., OLSON, W. O., GANNON, V. P. J. and VEIRA, D. M. 2000b. The effect of fasting and diet on fecal shedding of *Escherichia coli* O157:H7 by cattle. *Can. J. Anim. Sci*, 80, 741–744.

BUDIÑO, F. E. L., JÚNIOR, F. G. D. C. and OTSUK, I. P. 2010. Frutooligosaccharide addition in diets for weaned pigs: Performance, diarrhea incidence and metabolism. *Rev. Brasil. Zootec.*, 39, 2187–2193.

BURKEY, T. E., JOHNSON, B. J., MINTON, J. E., DRITZ, S. S., NIETFELD, J. C. *et al.* 2004. Effect of dietary mannanoligosaccharide and sodium chlorate on the growth performance, acute-phase response, and bacterial shedding of weaned pigs challenged with *Salmonella enterica* serotype Typhimurium. *J. Anim. Sci.*, 82, 397–404.

BUSZ, H. W., MCALLISTER, T. A., YANKE, L. J., OLSON, M. E., MORCK, D. W. *et al.* 2002. Development of antibiotic resistance among *Escherichia coli* in feedlot cattle. *J. Anim. Sci.*, 80 (Suppl. 1), 102.

BYRD, J. A., ANDERSON, R. C., CALLAWAY, T. R., MOORE, R. W., KNAPE, K. D. *et al.* 2003. Effect of experimental chlorate product administration in the drinking water on *Salmonella* Typhimurium contamination of broilers. *Poult. Sci.*, 82, 1403–1406.

CALLAWAY, T. R. and MARTIN, S. A. 2006. Use of competitive exclusion cultures and oligosaccharides. In *Feedstuffs Direct-fed Microbial, Enzyme and Forage Additive Compendium, 8th Ed.* Minnetonka, MN: Miller Publishing.

CALLAWAY, T. R., ANDERSON, R. C., EDRINGTON, T. S., BISCHOFF, K. M., GENOVESE, K. J. *et al.* 2004a. Effects of sodium chlorate on antibiotic resistance in *Escherichia coli* O157:H7. *Foodborne Path. Dis.*, 1, 59–63.

CALLAWAY, T. R., ANDERSON, R. C., EDRINGTON, T. S., JUNG, Y. S., BISCHOFF, K. M. *et al.* 2004b. Effects of sodium chlorate on toxin production by *Escherichia coli* O157:H7. *Curr. Iss. Intest. Microbiol.*, 5, 19–22.

CALLAWAY, T. R., ANDERSON, R. C., GENOVESE, K. J., POOLE, T. L., ANDERSON, T. J. *et al.* 2002. Sodium chlorate supplementation reduces *E. coli* O157:H7 populations in cattle. *J. Anim. Sci.*, 80, 1683–1689.

CALLAWAY, T. R., CARR, M. A., EDRINGTON, T. S., ANDERSON, R. C. and NISBET, D. J. 2009. Diet, *Escherichia coli* O157:H7, and cattle: A review after 10 years. *Curr. Iss. Mol. Biol.*, 11, 67–80.

CALLAWAY, T. R., DOWD, S. E., EDRINGTON, T. S., ANDERSON, R. C., KRUEGER, N. *et al.* 2010a. Evaluation of bacterial diversity in the rumen and feces of cattle fed different levels of dried distillers grains plus solubles using bacterial tag-encoded FLX amplicon pyrosequencing. *J. Anim. Sci.*, 88, 3977–3983.

CALLAWAY, T. R., EDRINGTON, T. S., ANDERSON, R. C., GENOVESE, K. J., POOLE, T. L. *et al.* 2003a. *Escherichia coli* O157:H7 populations in sheep can be reduced by chlorate supplementation. *J. Food Prot.*, 66, 194–199.

CALLAWAY, T. R., EDRINGTON, T. S., BRABBAN, A. D., ANDERSON, R. C., ROSSMAN, M. L. *et al.* 2008. Bacteriophage isolated from feedlot cattle can reduce *Escherichia coli* O157:H7 populations in ruminant gastrointestinal tracts. *Foodborne Path. Dis.*, 5, 183–192.

CALLAWAY, T. R., EDRINGTON, T. S., BRABBAN, A. D., KEEN, J. E., ANDERSON, R. C. *et al.* 2006a. Fecal prevalence of *Escherichia coli* O157, *Salmonella, Listeria*, and bacteriophage infecting *E. coli* O157:H7 in feedlot cattle in the southern plains region of the United States. *Foodborne Path. Dis.*, 3, 234–244.

CALLAWAY, T. R., EDRINGTON, T. S., BRABBAN, A. D., KUTTER, E., KARRIKER, L. *et al.* 2010b. Isolation of *Salmonella* spp. and bacteriophages active against *Salmonella* spp. from commercial swine feces. *Foodborne Path. Dis.*, 7, 851–856.

CALLAWAY, T. R., EDRINGTON, T. S., BRABBAN, A. D., KUTTER, E. M., KARRIKER, L. *et al.* 2011. Evaluation of phage treatment as a strategy to reduce *Salmonella* populations in growing swine. *Foodborne Path. Dis.*, 8, 261–266.

CALLAWAY, T. R., EDRINGTON, T. S., RYCHLIK, J. L., GENOVESE, K. J., POOLE, T. L. *et al.* 2003b. Ionophores: Their use as ruminant growth promotants and impact on food safety. *Curr. Iss. Intest. Microbiol.*, 4, 43–51.

CALLAWAY, T. R., MORROW, J. L., EDRINGTON, T. S., GENOVESE, K. J., DOWD, S. *et al.* 2006b. Social stress increases fecal shedding of *Salmonella* Typhimurium by early weaned piglets. *Curr. Iss. Intest. Microbiol.*, 7, 65–72.

CALLAWAY, T. R., STAHL, C. H., EDRINGTON, T. S., GENOVESE, K. J., LINCOLN, L. M. *et al.* 2004c. Colicin concentrations inhibit growth of *Escherichia coli* O157:H7 *in vitro*. *J. Food Prot.*, 67, 2603–2607.

CHAPMAN, P. A., CORNELL, J. and GREEN, C. 2000. Infection with verocytotoxin-producing *Escherichia coli* O157 during a visit to an inner city open farm. *Epidemiol. Infect.*, 125, 531–536.

CODY, S. H., ABBOTT, S. L., MARFIN, A. A., SCHULZ, B., WAGNER, P. *et al.* 1999a. Two outbreaks of multidrug-resistant *Salmonella* serotype Typhimurium DT104 infections linked to raw-milk cheese in northern California. *J. Amer. Med. Assoc.*, 281, 1805–1810.

CODY, S. H., GLYNN, K., FARRAR, J. A., CAIRNS, K. L., GRIFFIN, P. M. *et al.* 1999b. An outbreak of *Escherichia coli* O157:H7 infection from unpasteurized commercial apple juice. *Ann. Intern. Med.*, 130, 202–209.

COLLINS, D. M. and GIBSON, G. R. 1999. Probiotics, prebiotics, and synbiotics: Approaches for modulating the microbial ecology of the gut. *Amer. J. Clin. Nutr.*, 69, 1052S–1057S.

COYNE, M. S., HOWELL, J. M. and PHILLIPS, R. E. 1997. How do bacteria move through soil? *Soil Science News and Views*. Lexington, KY: University of Kentucky, Coop. Ext. Ser.

CRITTENDEN, R. G. 1999. Prebiotics. In: Tannock, G. W. (ed.) *Probiotics: A Critical Review*. Wymondham, UK: Horizon Scientific Press.

CUTLER, S. A., LONERGAN, S. M., CORNICK, N., JOHNSON, A. K. and STAHL, C. H. 2007. Dietary inclusion of colicin E1 is effective in preventing postweaning diarrhea caused by F18-positive *Escherichia coli* in pigs. *Antimicrob. Ag. Chemother.*, 51, 3830–3835.

D'HERELLE, F. 1917. Sur un microbe invisible antagoniste des bacilles dysentériques. *Comptes rendus Acad Sci Paris*, 165, 373–375.

DA SILVA, W. T. M., NUNES, R. V., POZZA, P. C., DOS SANTOS POZZA, M. S., APPELT, M. D. *et al.* 2011. Evaluation of inulin and probiotic for broiler chickens. *Acta Scient. Anim. Sci.*, 33, 19–24.

DARGATZ, D. A., WELLS, S. J., THOMAS, L. A., HANCOCK, D. D. and GARBER, L. P. 1997. Factors associated with the presence of *Escherichia coli* O157 in feces of feedlot cattle. *J. Food Prot.*, 60, 466–470.

DAVIS, M. A., CLOUD-HANSEN, K. A., CARPENTER, J. and HOVDE, C. J. 2005. *Escherichia coli* O157:H7 in environments of culture-positive cattle. *Appl. Environ. Microbiol.*, 71, 6816–6822.

DHILLON, T. S., DHILLON, E. K., CHAU, H. C., LI, W. K. and TSANG, A. H. 1976. Studies on bacteriophage distribution: Virulent and temperate bacteriophage content of mammalian feces. *Appl. Environ. Microbiol.*, 32, 68–74.

DIEZ-GONZALEZ, F., CALLAWAY, T. R., KIZOULIS, M. G. and RUSSELL, J. B. 1998. Grain feeding and the dissemination of acid-resistant *Escherichia coli* from cattle. *Science*, 281, 1666–1668.

DOANE, C. A., PANGLOLI, P., RICHARDS, H. A., MOUNT, J. R., GOLDEN, D. A. *et al.* 2007. Occurrence of *Escherichia coli* O157:H7 in diverse farm environments. *J. Food Prot.*, 70, 6–10.

DOWD, S. E., CALLAWAY, T. R. and MORROW-TESCH, J. 2007. Handling may cause increased shedding of *Escherichia coli* and total coliforms in pigs. *Foodborne Path. Dis.*, 4, 99–102.

DOWD, S. E., CALLAWAY, T. R., SUN, Y., MCKEEHAN, T., HAGEVOORT, R. G. *et al.* 2008. Evaluation of the bacterial diversity in the feces of cattle using bacterial tag-encoded FLX amplicon pyrosequencing (bTEFAP). *BMC Microbiol.*, 8, 125–132.

DOYLE, M. P. and ERICKSON, M. C. 2006. Reducing the carriage of foodborne pathogens in livestock and poultry. *Poult. Sci.*, 85, 960–973.

DUBOS, R. J. 1963. Staphylococci and infection immunity. *Am. J. Dis. Child.*, 105, 643–645.

DURSO, L. M., HARHAY, G. P., SMITH, T. P. L., BONO, J. L., DESANTIS, T. Z. *et al.* 2011. Bacterial community analysis of beef cattle feedlots reveals pen surface is distinct from feces. *Foodborne Path. Dis.*, 8, 647–649.

DURSO, L. M., REYNOLDS, K., BAUER, N. and KEEN, J. E. 2005. Shiga-toxigenic *Escherichia coli* O157:H7 infections among livestock exhibitors and visitors at a Texas county fair. *Vector-borne Zoo. Dis.*, 5, 193–201.

EDRINGTON, T. S., CALLAWAY, T. R., ANDERSON, R. C., GENOVESE, K. J., JUNG, Y. S. *et al.* 2003a. Reduction of *E. coli* O157:H7 populations in sheep by supplementation of an experimental sodium chlorate product. *Small Ruminant Res.*, 49, 173–181.

EDRINGTON, T. S., CALLAWAY, T. R., ANDERSON, R. C. and NISBET, D. J. 2008. Prevalence of multidrug-resistant *Salmonella* on commercial dairies utilizing a single heifer raising facility. *J. Food Prot.*, 71, 27–34.

EDRINGTON, T. S., CALLAWAY, T. R., BISCHOFF, K. M., GENOVESE, K. J., ELDER, R. O. *et al.* 2003b. Effect of feeding the ionophores monensin and laidlomycin propionate and the antimicrobial bambermycin to sheep experimentally infected with *E. coli* O157:H7 and *Salmonella* Typhimurium. *J. Anim. Sci.*, 81, 553–560.

EDRINGTON, T. S., CALLAWAY, T. R., VAREY, P. D., JUNG, Y. S., BISCHOFF, K. M. *et al.* 2003c. Effects of the antibiotic ionophores monensin, lasalocid, laidlomycin propionate and bambermycin on *Salmonella* and *E. coli* O157:H7 *in vitro*. *J. Appl. Microbiol.*, 94, 207–213.

EDRINGTON, T. S., LOOPER, M. L., DUKE, S. E., CALLAWAY, T. R., GENOVESE, K. J. *et al.* 2006. Effect of ionophore supplementation on the incidence of *Escherichia coli* O157:H7 and *Salmonella* and antimicrobial susceptibility of fecal coliforms in stocker cattle. *Foodborne Path. Dis.*, 3, 284–291.

EDRINGTON, T. S., MACDONALD, J. C., FARROW, R. L., CALLAWAY, T. R., ANDERSON, R. C. *et al.* 2010. Influence of wet distiller's grains on prevalence of *Escherichia coli* O157:H7 and

Salmonella in feedlot cattle and antimicrobial susceptibility of generic *Escherichia coli* isolates. *Foodborne Path. Dis.*, 7, 605–608.

ELAM, N. A., GLEGHORN, J. F., RIVERA, J. D., GALYEAN, M. L., DEFOOR, P. J. *et al.* 2003. Effects of live cultures of *Lactobacillus acidophilus* (strains NP45 and NP51) and *Propionibacterium freudenreichii* on performance, carcass, and intestinal characteristics, and *Escherichia coli* strain O157 shedding of finishing beef steers. *J. Anim. Sci.*, 81, 2686–2698.

ELLIS-IVERSEN, J. and WATSON, E. 2008. A 7-point plan for control of VTEC O157, *Campylobacter jejuni/coli* and *Salmonella* serovars in young cattle. *Cattle Pract.*, 16, 103–106.

ENDT, K., STECHER, B., CHAFFRON, S., SLACK, E., TCHITCHEK, N. *et al.* 2010. The microbiota mediates pathogen clearance from the gut lumen after non-typhoidal *Salmonella* diarrhea. *PLoS Pathog.*, 6, 1–18.

ETIENNE-MESMIN, L., LIVRELLI, V., PRIVAT, M., DENIS, S., CARDOT, J. M. *et al.* 2011. Effect of a new probiotic *Saccharomyces cerevisiae* strain on survival of *Escherichia coli* O157:H7 in a dynamic gastrointestinal model. *Appl. Environ. Microbiol.*, 77, 1127–1131.

FALAKI, M., SHARGH, M. S., DASTAR, B. and ZREHDARAN, S. 2010. Effects of different levels of probiotic and prebiotic on performance and carcass characteristics of broiler chickens. *J. Anim. Vet. Adv.*, 9, 2390–2395.

FALK, P. G., HOOPER, L. V., MIDTVEDT, T. and GORDON, J. I. 1998. Creating and maintaining the gastrointestinal ecosystem: What we know and need to know from gnotobiology. *Microbiol. Mol. Biol. Rev.*, 62, 1157–1170.

FDA 2006. Food additives permitted for direct addition to food for human consumption; bacteriophage preparation; final rule. Part 172 of Title 21 (Food and Drugs) of the United States Code of Federal Regulations. Washington, DC: Federal Register.

FEDORKA-CRAY, P. J., BAILEY, J. S., STERN, N. J., COX, N. A., LADELY, S. R. *et al.* 1999. Mucosal competitive exclusion to reduce *Salmonella* in swine. *J. Food Prot.*, 62, 1376–1380.

FINEGOLD, S. M. 2008. Therapy and epidemiology of autism-clostridial spores as key elements. *Med. Hypoth.*, 70, 508–511.

FINLAY, B. 2003. Pathogenic *Escherichia coli*: From molecules to vaccine. *Proc. 5th Int. Symp. on Shiga Toxin-Producing Escherichia coli Infections*, 2003 Edinburgh, UK. p. 23.

FOX, J. T., THOMSON, D. U., DROUILLARD, J. S., THORNTON, A. B., BURKHARDT, D. T. *et al.* 2009. Efficacy of *Escherichia coli* O157:H7 siderophore receptor/porin proteins-based vaccine in feedlot cattle naturally shedding *E. coli* O157. *Foodborne Path. Dis.*, 6, 893–899.

FRANK, C., KAPFHAMMER, S., WERBER, D., STARK, K. and HELD, L. 2008. Cattle density and Shiga toxin-producing *Escherichia coli* infection in Germany: Increased risk for most but not all serogroups. *Vector-Borne Zoo. Dis.*, 8, 635–643.

FREESTONE, P. and LYTE, M. 2010. Stress and microbial endocrinology: Prospects for ruminant nutrition. *Animal*, 4, 1248–1257.

FRETER, R., BRICKNER, H., BOTNEY, M., CLEVEN, D. and ARANKI, A. 1983. Mechanisms that control bacterial populations in continuous-flow culture models of mouse large intestinal flora. *Infect. Immun.*, 39, 676–685.

FULLER, R. 1989. Probiotics in man and animals. *J. Appl. Bacteriol.*, 66, 365–378.

GENOVESE, K. J., HARVEY, R. B., ANDERSON, R. C. and NISBET, D. J. 2001. Protection of suckling neonatal pigs against an infection with an enterotoxigenic *Escherichia coli* expressing 987P fimbriae infection by the administration of a bacterial competitive exclusion culture. *Microb. Ecol. Health Dis.*, 13, 223–228.

GHAREEB, K., AWAD, W. A., NITSCH, S., ABDEL-RAHEEM, S. and BÖHM, J. 2008. Effects of transportation on stress and fear responses of growing broilers supplemented with prebiotic or probiotic. *Int. J. Poult. Sci.*, 7, 678–685.

GOMES, A. M. P. and MALCATA, F. X. 1999. *Bifidobacterium* spp. and *Lactobacillus acidophilus*: Biological, biochemical, technological and therapeutical properties relevant for use as probiotics. *Trends Food Sci. Technol.*, 10, 139–157.

GRAU, F. H., BROWNLIE, L. E. and ROBERTS, E. A. 1968. Effect of some preslaughter treatments on the *Salmonella* population in the bovine rumen and faeces. *J. Appl. Bact.*, 31, 157–163.

GRAU, F. H., BROWNLIE, L. E. and SMITH, M. G. 1969. Effects of food intake on numbers of *Salmonellae* and *Escherichia coli* in rumen and faeces of sheep. *J. Appl. Bact.*, 32, 112–117.

GYLES, C. L. 1998. Vaccines and Shiga toxin-producing *Escherichia coli* in animals. In: Kaper, J. B. and O'Brien, A. D. (eds.) *Escherichia coli O157:H7 and Other Shiga Toxin-Producing E. coli Strains*. Washington, DC: Amer. Soc. Microbiol. Press.

HARVEY, R. B., ANDERSON, R. C., GENOVESE, K. J., CALLAWAY, T. R. and NISBET, D. J. 2005. Use of competitive exclusion to control enterotoxigenic strains of *Escherichia coli* in weaned pigs. *J. Anim. Sci.*, 83 (E. Suppl.), E44–E47.

HARVEY, R. B., EBERT, R. C., SCHMITT, C. S., ANDREWS, K., GENOVESE, K. J. *et al.* 2003. Use of a porcine-derived, defined culture of commensal bacteria as an alternative to antibiotics used to control *E. coli* disease in weaned pigs. *9th Intl. Symp. Dig. Physiol. in Pigs*, Banff, AB, Canada. pp. 72–74.

HEGDE, N. V., COOK, M. L., WOLFGANG, D. R., LOVE, B. C., MADDOX, C. C. *et al.* 2005. Dissemination of *Salmonella enterica* subsp. *enterica* serovar Typhimurium var. Copenhagen clonal types through a contract heifer-raising operation. *J. Clin. Microbiol.*, 43, 4208–4211.

HIGGINS, J. P., HIGGINS, K. L., HUFF, H. W., DONOGHUE, A. M., DONOGHUE, D. J. *et al.* 2005. Use of a specific bacteriophage treatment to reduce *Salmonella* in poultry products. *Poult. Sci.*, 84, 1141–1145.

HOFACRE, C. L., BEACORN, T., COLLETT, S. and MATHIS, G. 2003. Using competitive exclusion, mannan-oligosaccharide and other intestinal products to control necrotic enteritis. *J. Appl. Poult. Res.*, 12, 60–64.

HOLT, P. S. 1993. Effect of induced molting on the susceptibility of white leghorn hens to a *Salmonella* Enteritidis infection. *Avian Dis.*, 37, 412–417.

HOUSE, J. K., ONTIVEROS, M. M., BLACKMER, N. M., DUEGER, E. L., FITCHHORN, J. B. *et al.* 2001. Evaluation of an autogenous *Salmonella* bacterin and a modified live *Salmonella* serotype Choleraesuis vaccine on a commercial dairy farm. *Am. J. Vet. Res.*, 62, 1897–1902.

HUFF, W. E., HUFF, G. R., RATH, N. C., BALOG, J. M. and DONOGHUE, A. M. 2005. Alternatives to antibiotics: Utilization of bacteriophage to treat colibacillosis and prevent foodborne pathogens. *Poult. Sci.*, 84, 655–659.

HUFF, W. E., HUFF, G. R., RATH, N. C., BALOG, J. M., XIE, H. *et al.* 2002. Prevention of *Escherichia coli* respiratory infection in broiler chickens with bacteriophage (SPR02). *J. Poult. Sci.*, 81, 437–441.

HUNGATE, R. E. 1950. The anaerobic mesophilic cellulolytic bacteria. *Bacterial Rev.*, 14, 1–49.

HUNGATE, R. E. 1952. Kinds of cellulolytic cocci in the rumen of cattle and sheep. *Bact. Proc.*, 16.

HURD, H. S., WESLEY, I. V. and KARRIKER, L. A. 2001. The effect of lairage on *Salmonella* isolation from market swine. *J. Food Prot.*, 64, 939–944.

JACK, R. W., TAGG, J. R. and RAY, B. 1995. Bacteriocins of gram-positive bacteria. *Microbiol. Rev.*, 59, 171–200.

JACOB, M. E., CALLAWAY, T. R. and NAGARAJA, T. G. 2009. Dietary interactions and interventions affecting *Escherichia coli* O157 colonization and shedding in cattle. *Foodborne Path. Dis.*, 6, 785–792.

JACOB, M. E., FOX, J. T., DROUILLARD, J. S., RENTER, D. G. and NAGARAJA, T. G. 2008a. Effects of dried distillers' grain on fecal prevalence and growth of *Escherichia coli* O157 in batch culture fermentations from cattle. *Appl. Environ. Microbiol.*, 74, 38–43.

JACOB, M. E., FOX, J. T., NARAYANAN, S. K., DROUILLARD, J. S., RENTER, D. G. *et al.* 2008b. Effects of feeding wet corn distiller's grains with solubles with or without monensin and tylosin on the prevalence and antimicrobial susceptibilities of fecal food-borne pathogenic and commensal bacteria in feedlot cattle. *J. Anim Sci.*, 86, 1182–1190.

JACOB, M. E., PADDOCK, Z. D., RENTER, D. G., LECHTENBERG, K. F. and NAGARAJA, T. G. 2010. Inclusion of dried or wet distillers' grains at different levels in diets of feedlot cattle affects fecal shedding of *Escherichia coli* O157:H7. *Appl. Environ. Microbiol.*, 76, 7238–7242.

JACOB, M. E., PARSONS, G. L., SHELOR, M. K., FOX, J. T., DROUILLARD, J. S. *et al.* 2008c. Feeding supplemental dried distiller's grains increases faecal shedding of *Escherichia coli* O157 in experimentally inoculated calves. *Zoon. Pub. Health*, 55, 125–132.

JAIN, S., BIDOL, S. A., AUSTIN, J. L., BERL, E., ELSON, F. *et al.* 2009. Multistate outbreak of *Salmonella* Typhimurium and Saintpaul infections associated with unpasteurized orange juice – United States, 2005. *Clin. Infect. Dis.*, 48, 1065–1071.

JANARDHANA, V., BROADWAY, M. M., BRUCE, M. P., LOWENTHAL, J. W., GEIER, M. S. *et al.* 2009. Prebiotics modulate immune responses in the gut-associated lymphoid tissue of chickens. *J. Nutr.*, 139, 1404–1409.

JANCZYK, P., PIEPER, R., SMIDT, H. and SOUFFRANT, W. B. 2010. Effect of alginate and inulin on intestinal microbial ecology of weanling pigs reared under different husbandry conditions. *FEMS Microbiol. Ecol.*, 72, 132–142.

JAY, M. T., COOLEY, M., CARYCHAO, D., WISCOMB, G. W., SWEITZER, R. A. *et al.* 2007. *Escherichia coli* O157:H7 in feral swine near spinach fields and cattle, central California coast. *Emerg. Infect. Dis.*, 13, 1908–1911.

JAYNE-WILLIAMS, D. J. and FULLER, R. 1971. The influence of the intestinal microflora on nutrition. In: Bell, D. J. and Freeman, B. M. (eds.) *Physiology and Biochemistry of the Domestic Food*. London, UK: Academic Press.

JOHANSEN, M., ANDRESEN, L. O., THOMSEN, L. K., BUSCH, M. E., WACHMANN, H. *et al.* 2000. Prevention of edema disease in pigs by passive immunization. *Can. J. Vet. Res*, 64, 9–14.

JOHNSON, R. P., GYLES, C. L., HUFF, W. E., OJHA, S., HUFF, G. R. *et al.* 2008. Bacteriophages for prophylaxis and therapy in cattle, poultry and pigs. *Anim. Health Res. Rev.*, 9, 201–215.

JUDGE, N. A., MASON, H. S. and O'BRIEN, A. D. 2004. Plant cell-based intimin vaccine given orally to mice primed with intimin reduces time of *Escherichia coli* O157:H7 shedding in feces. *Infect. Immun.*, 72, 168–175.

JUNG, Y. S., ANDERSON, R. C., BYRD, J. A., EDRINGTON, T. S., MOORE, R. W. *et al.* 2003. Reduction of *Salmonella* Typhimurium in experimentally challenged broilers by nitrate adaptation and chlorate supplementation in drinking water. *J. Food Prot.*, 66, 660–663.

KEEN, J. E., WITTUM, T. E., DUNN, J. R., BONO, J. L. and DURSO, L. M. 2006. Shiga-toxigenic *Escherichia coli* O157 in agricultural fair livestock, United States. *Emerg. Infect. Dis.*, 12, 780–786.

KEEN, J. E., WITTUM, T. E., DUNN, J. R., BONO, J. L. and FONTENOT, M. E. 2003. Occurrence of STEC O157, O111, and O26 in livestock at agricultural fairs in the United States. *Proc. 5th Int. Symp. on Shiga Toxin-Producing Escherichia coli Infections*, Edinburgh, UK. p. 22.

KHARE, S., ALALI, W., ZHANG, S., HUNTER, D., PUGH, R. *et al.* 2010. Vaccination with attenuated *Salmonella enterica* Dublin expressing *E. coli* O157:H7 outer membrane protein intimin induces transient reduction of fecal shedding of *E. coli* O157:H7 in cattle. *BMC Vet. Res.*, 6, 35.

KIEN, C. L., BLAUWIEKEL, R., WILLIAMS, C. H., BUNN, J. Y. and BUDDINGTON, R. K. 2007. Lactulose feeding lowers cecal densities of clostridia in piglets. *J. Parent. Enter. Nutr.*, 31, 194–198.

KIM, G. B., SEO, Y. M., KIM, C. H. and PAIK, I. K. 2011. Effect of dietary prebiotic supplementation on the performance, intestinal microflora, and immune response of broilers. *Poult. Sci.*, 90, 75–82.

KLAENHAMMER, T. R. 1988. Bacteriocins of lactic acid bacteria. *Biochimie*, 70, 337–349.

KOOHMARAIE, M., ARTHUR, T. M., BOSILEVAC, J. M., GUERINI, M., SHACKELFORD, S. D. *et al.* 2005. Post-harvest interventions to reduce/eliminate pathogens in beef. *Meat Sci.*, 71, 79–91.

KRAUSE, D. O., BHANDARI, S. K., HOUSE, J. D. and NYACHOTI, C. M. 2010. Response of nursery pigs to a synbiotic preparation of starch and an anti-*Escherichia coli* K88 probiotic. *Appl. Environ. Microbiol.*, 76, 8192–8200.

KUDVA, I. T., HATFIELD, P. G. and HOVDE, C. J. 1995. Effect of diet on the shedding of *Escherichia coli* O157:H7 in a sheep model. *Appl. Environ. Microbiol.*, 61, 1363–1370.

KUDVA, I. T., JELACIC, S., TARR, P. I., YOUDERIAN, P. and HOVDE, C. J. 1999. Biocontrol of *Escherichia coli* O157 with O157-specific bacteriophages. *Appl. Environ. Microbiol.*, 65, 3767–3773.

LAKEY, J. H. and SLATIN, S. L. 2001. Pore-forming colicins and their relatives. In: van der Goot, F. G. (ed.) *Pore-Forming Toxins*. Berlin: Springer-Verlag GmbH and Co. KG.

LAUKOVA, A. and MARCKOVA, M. 1993. Antimicrobial spectrum of bacteriocin-like substances produced by ruminal staphylococci. *Folia Microbiol.*, 38, 74–76.

LEATHAM, M. P., BANERJEE, S., AUTIERI, S. M., MERCADO-LUBO, R., CONWAY, T. *et al.* 2009. Precolonized human commensal *Escherichia coli* strains serve as a barrier to *E. coli* O157:H7 growth in the streptomycin-treated mouse intestine. *Infect. Immun.*, 77, 2876–2886.

LEDERBERG, J. 1996. Smaller Fleas . . . ad infinitum: Therapeutic bacteriophage redux. *Proc. Natl. Acad. Sci. USA*, 93, 3167–3168.

LEJEUNE, J. T. and KAUFFMAN, M. D. 2005. Effect of sand and sawdust bedding materials on the fecal prevalence of *Escherichia coli* O157:H7 in dairy cows. *Appl. Environ. Microbiol.*, 71, 326–330.

LEJEUNE, J. and KERSTING, A. 2010. Zoonoses: An occupational hazard for livestock workers and a public health concern for rural communities. *J. Agric. Safe. Health*, 16, 161–179.

LEJEUNE, J. T. and RAJALA-SCHULTZ, P. J. 2009. Unpasteurized milk: A continued public health threat. *Clinical Infectious Diseases*, 48, 93–100.

LEJEUNE, J. T. and WETZEL, A. N. 2007. Preharvest control of *Escherichia coli* O157 in cattle. *J. Anim. Sci.*, 85–97.

LEY, R. E., TURNBAUGH, P. J., KLEIN, S. and GORDON, J. I. 2006. Human gut microbes associated with obesity. *Nature*, 444, 1022–1023.

LLOYD, A. B., CUMMING, R. B. and KENT, R. D. 1974. Competitive exclusion as exemplified by *Salmonella* Typhimurium. *Australasian Poult. Sci. Conv.*, World Poult. Sci. Assoc. Austral. Br. p. 155.

LLOYD, A. B., CUMMING, R. B. and KENT, R. D. 1977. Prevention of *Salmonella* Typhimurium infection in poultry by pre-treatment of chickens and poults with intestinal extracts. *Aust. Vet. J.*, 53, 82–87.

LONERAGAN, G. H. and BRASHEARS, M. M. 2005. Pre-harvest interventions to reduce carriage of *E. coli* O157 by harvest-ready feedlot cattle. *Meat Sci.*, 71, 72.

LU, J., IDRIS, U., HOFACRE, C., MAURER, J. J., LEE, M. D. *et al.* 2003. Diversity and succession of the intestinal bacterial community of the maturing broiler chicken. *Appl. Environ. Microbiol.*, 69, 6816–6824.

LYTE, M. 2010. The microbial organ in the gut as a driver of homeostasis and disease. *Med. Hypoth.*, 74, 634–638.

LYTE, M. and FREESTONE, P. 2009. Microbial endocrinology comes of age. *Microbe*, 4, 169–175.

MACKEY, B. M. and DERRICK, C. M. 1979. Contamination of the deep tissues of carcasses by bacteria present on the slaughter instruments or in the gut. *J. Appl. Bacteriol.*, 46, 355–366.

MANSHADI, F. D., GORTARES, P., GERBA, C. P., KARPISCAK, M. and FRENTAS, R. J. 2001. Role of irrigation water in contamination of domestic fresh vegetables. *Gen. Mtg. Amer. Soc. Microbiol.*, p. 561.

MATHER, A. E., INNOCENT, G. T., MCEWEN, S. A., REILLY, W. J., TAYLOR, D. J. *et al.* 2007. Risk factors for hide contamination of Scottish cattle at slaughter with *Escherichia coli* O157. *Prev. Vet. Med.*, 80, 257–270.

McALLISTER, T. A., BACH, S. J., STANFORD, K. and CALLAWAY, T. R. 2006. Shedding of *Escherichia coli* O157:H7 by cattle fed diets containing monensin or tylosin. *J. Food Prot.*, 69, 2075–2083.

McLAUGHLIN, M. R., BALAA, M. F., SIMS, J. and KING, R. 2006. Isolation of *Salmonella* bacteriophages from swine effluent lagoons. *J. Environ. Qual.*, 35, 522–528.

McNEILLY, T. N., MITCHELL, M. C., ROSSER, T., MCATEER, S., LOW, J. C. *et al.* 2010. Immunization of cattle with a combination of purified intimin-531, EspA and Tir significantly reduces shedding of *Escherichia coli* O157:H7 following oral challenge. *Vaccine*, 28, 1422–1428.

MEAD, G. C. 2000. Prospects for 'competitive exclusion' treatment to control salmonellas and other foodborne pathogens in poultry. *Vet. J.*, 159, 111–123.

MEAD, G. C., BARROW, P. A., HINTON, M. H., HUMBERT, F., IMPEY, C. S. *et al.* 1989. Recommended assay for treatment of chicks to prevent *Salmonella* colonization by competitive exclusion. *J. Food Prot.*, 52, 500–502.

MEYER, D. 2008. Prebiotic dietary fibres and the immune system. *Agro-Food Ind. Hi-Tech.*, 19, 12–15.

MIDGLEY, J. and DESMARCHELIER, P. 2001. Pre-slaughter handling of cattle and Shiga toxin-producing *Escherichia coli* (STEC). *Lett. Appl. Microbiol.*, 32, 307–311.

MIDILLI, M., ALP, M., KOCABAGLI, N., MUGLALI, O. H., TURAN, N. *et al.* 2008. Effects of dietary probiotic and prebiotic supplementation on growth performance and serum IgG concentration of broilers. *S. African J. Anim. Sci.*, 38, 21–27.

MOORE, R. W., BYRD, J. A., KNAPE, K. D., ANDERSON, R. C., CALLAWAY, T. R. *et al.* 2006. The effect of an experimental chlorate product on *Salmonella* recovery of turkeys when administered prior to feed and water withdrawal. *Poult. Sci.*, 85, 2101–2105.

MOSENTHIN, R. and BAUER, E. 2000. The potential use of prebiotics in pig nutrition. *Asian-Austral. J. Anim. Sci.*, 13, 315–325.

MOXLEY, R. A., SMITH, D., KLOPFENSTEIN, T. J., ERICKSON, G., FOLMER, J. *et al.* 2003. Vaccination and feeding a competitive exclusion product as intervention strategies to reduce the prevalence of *Escherichia coli* O157:H7 in feedlot cattle. *Proc. 5th Int. Symp. on Shiga Toxin-Producing Escherichia coli Infections*, Edinburgh, UK. p. 23.

MOXLEY, R. A., SMITH, D. R., HANSEN, K., LUEBBE, M. K., ERICKSON, G. E. *et al.* 2008. Vaccination for *Escherichia coli* O157:H7 in feedlot cattle. *Animal Science Department, Nebraska Beef Cattle Reports*. University of Nebraska–Lincoln.

MOXLEY, R. A., SMITH, D. R., LUEBBE, M., ERICKSON, G. E., KLOPFENSTEIN, T. J. *et al.* 2009. *Escherichia coli* O157:H7 vaccine dose-effect in feedlot cattle. *Foodborne Path. Dis.*, 6, 879–884.

MURINDA, S. E., LIU, S. M., ROBERTS, R. F. and WILSON, R. A. 1998. Colicinogeny among *Escherichia coli* serotypes, including O157:H7, representing four closely related diarrheagenic clones. *J. Food Prot.*, 61, 1431–1438.

MURINDA, S. E., NGUYEN, L. T., LANDERS, T. L., DRAUGHON, F. A., MATTHEW, A. G. *et al.* 2004. Comparison of *Escherichia coli* isolates from humans, food, and farm and companion animals for presence of Shiga toxin-producing *E. coli* virulence markers. *Foodborne Path. Dis.*, 1, 178–184.

MURINDA, S. E., ROBERTS, R. F. and WILSON, R. A. 1996. Evaluation of colicins for inhibitory activity against diarrheagenic *Escherichia coli* strains, including serotype O157:H7. *Appl. Environ. Microbiol.*, 62, 3192–3202.

MURPHY, M. 2004. Bacteria could treat symptoms of autism. *Chem. Indust. (London)*, 6.

NATVIG, E. E., INGHAM, S. C., INGHAM, B. H., COOPERBAND, L. R. and ROPER, T. R. 2002. *Salmonella enterica* serovar Typhimurium and *Escherichia coli* contamination of root and leaf vegetables grown in soils with incorporated bovine manure. *Appl. Environ. Microbiol.*, 68, 2737–2744.

NAVA, G. M., BIELKE, L. R., CALLAWAY, T. R. and CASTANEDA, M. P. 2005. Probiotic alternatives to reduce gastrointestinal infections: The poultry experience. *Anim. Health Res. Rev.*, 6, 105–118.

NEMCOVA, R., BOMBA, A., GANCARCIKOVA, S., REIFFOVA, K., GUBA, P. *et al.* 2007. Effects of the administration of lactobacilli, maltodextrins and fructooligosaccharides upon the adhesion of *E. coli* O8:K88 to the intestinal mucosa and organic acid levels in the gut contents of piglets. *Vet. Res. Comm.*, 31, 791–800.

NETTELBLADT, C. G., KATOULI, M., VOLPE, A., BARK, T., MURATOV, V. *et al.* 1997. Starvation increases the number of coliform bacteria in the caecum and induces bacterial adherence to caecal epithelium in rats. *Eur. J. Surg.*, 163, 135–142.

NIELSEN, E. M., SKOV, M. N., MADSEN, J. J., LODAL, J., JESPERSEN, J. B. *et al.* 2004. Verocytotoxin-producing *Escherichia coli* in wild birds and rodents in close proximity to farms. *Appl. Environ. Microbiol.*, 70, 6944–6947.

NISBET, D. J., CORRIER, D. E., SCANLAN, C. M., HOLLISTER, A. G., BEIER, R. C. *et al.* 1993. Effect of a defined continuous flow derived bacterial culture and dietary lactose on *Salmonella* colonization in broiler chicks. *Avian Dis.*, 37, 1017–1025.

NISBET, D. J., RICKE, S. C., SCANLAN, C. M., CORRIER, D. E., HOLLISTER, A. G. *et al.* 1994. Inoculation of broiler chicks with a continuous-flow derived bacterial culture facilitates early cecal bacterial colonization and increases resistance to *Salmonella* Typhimurium. *J. Food Prot.*, 57, 12–15.

NIU, Y. D., MCALLISTER, T. A., XU, Y., JOHNSON, R. P., STEPHENS, T. P. *et al.* 2009. Prevalence and impact of bacteriophages on the presence of *Escherichia coli* O157:H7 in feedlot cattle and their environment. *Appl. Environ. Microbiol.*, 75, 1271–1278.

NIU, Y. D., XU, Y., MCALLISTER, T. A., ROZEMA, E. A., STEPHENS, T. P. *et al.* 2008. Comparison of fecal versus rectoanal mucosal swab sampling for detecting *Escherichia coli* O157:H7 in experimentally inoculated cattle used in assessing bacteriophage as a mitigation strategy. *J. Food Prot.*, 71, 691–698.

NOAL, H., HOFHUIS, A., DE JONGE, R., HEUVELINK, A. E., DE JONG, A. *et al.* 2010. Consumption of fresh fruit juice: How a healthy food practice caused a national outbreak of *Salmonella* Panama gastroenteritis. *Foodborne Path. Dis.*, 7, 375–381.

NURMI, E. and RANTALA, M. 1973. New aspects of *Salmonella* infection in broiler production. *Nature*, 24, 210–211.

NURMI, E., NUOTIO, L. and SCHNCITZ, C. 1992. The competitive exclusion concept: Development and future. *Int. J. Food Microbiol.*, 15, 237–240.

OHYA, T., AKIBA, M. and ITO, H. 2001. Use of a trial probiotic product in calves experimentally infected with *Escherichia coli* O157. *Japan Agric. Res. Quart.*, 35, 189–194.

OHYA, T., MARUBASHI, T. and ITO, H. 2000. Significance of fecal volatile fatty acids in shedding of *Escherichia coli* O157 from calves: Experimental infection and preliminary use of a probiotic product. *J. Vet. Med. Sci.*, 62, 1151–1155.

OLIVER, S. P., JAYARAO, B. M. and ALMEIDA, R. A. 2005. Foodborne pathogens in milk and the dairy farm environment: Food safety and public health implications. *Foodborne Path. Dis.*, 2, 115–129.

OLIVER, S. P., PATEL, D. A., CALLAWAY, T. R. and TORRENCE, M. E. 2008. ASAS Centennial Paper: Developments and future outlook for preharvest food safety. *J. Anim. Sci.*, 87, 419–437.

OTERO, M. C. and NADER-MACÍAS, M. E. 2006. Inhibition of *Staphylococcus aureus* by H_2O_2-producing *Lactobacillus gasseri* isolated from the vaginal tract of cattle. *Anim. Reprod. Sci.*, 96, 35–46.

OUWEHAND, A. C., KIRJAVAINEN, P. V., SHORTT, C. and SALMINEN, S. 1999. Probiotics: Mechanisms and established effects. *Int. Dairy J.*, 9, 43–52.

PATTON, B. S., DICKSON, J. S., LONERGAN, S. M., CUTLER, S. A. and STAHL, C. H. 2007. Inhibitory activity of colicin E1 against *Listeria monocytogenes*. *J. Food Prot.*, 70, 1256–1262.

PATTON, B. S., LONERGAN, S. M., CUTLER, S. A., STAHL, C. H. and DICKSON, J. S. 2008. Application of colicin E1 as a prefabrication intervention strategy. *J. Food Prot.*, 71, 2519–2522.

PIRAY, A. H. and KERMANSHAHI, H. 2008. Effects of diet supplementation of *Aspergillus* meal prebiotic (Fermacto®) on efficiency, serum lipids and immunity responses of broiler chickens. *J. Biol. Sci.*, 8, 818–821.

PORTER, J., MOBBS, K., HART, C. A., SAUNDERS, J. R., PICKUP, R. W. *et al.* 1997. Detection, distribution, and probable fate of *Escherichia coli* O157 from asymptomatic cattle on a dairy farm. *J. Appl. Microbiol.*, 83, 297–306.

POTTER, A. A., KLASHINSKY, S., LI, Y., FREY, E., TOWNSEND, H. *et al.* 2004. Decreased shedding of *Escherichia coli* O157:H7 by cattle following vaccination with type III secreted proteins. *Vaccine*, 22, 362–369.

POUILLART, P. R., DÉPEINT, F., ABDELNOUR, A., DEREMAUX, L., VINCENT, O. *et al.* 2010. Nutrióse, a prebiotic low-digestible carbohydrate, stimulates gut mucosal immunity and prevents TNBS-induced colitis in piglets. *Inflamm. Bowel Dis.*, 16, 783–794.

PRICE, K. L., TOTTY, H. R., LEE, H. B., UTT, M. D., FITZNER, G. E. *et al.* 2010. Use of *Saccharomyces cerevisiae* fermentation product on growth performance and microbiota of weaned pigs during *Salmonella* infection. *J. Anim. Sci.*, 88, 3896–3908.

PRITCHARD, G. C., WILLSHAW, G. A., BAILEY, J. R., CARSON, T. and CHEASTY, T. 2000. Verocytotoxin-producing *Escherichia coli* O157 on a farm open to the public: Outbreak investigation and longitudinal bacteriological study. *Vet. Rec.*, 147, 259–264.

PULLINGER, G. D., VAN DIEMEN, P. M., CARNELL, S. C., DAVIES, H., LYTE, M. *et al.* 2010. 6-hydroxydopamine-mediated release of norepinephrine increases faecal excretion of *Salmonella enterica* serovar Typhimurium in pigs. *Vet. Res.*, 41, 68.

RANSOM, J. R., BELK, K. E., SOFOS, J. N., SCANGA, J. A., ROSSMAN, M. L. *et al.* 2003. Investigation of on-farm management practices as pre-harvest beef microbiological interventions. Centennial, CO National Cattlemen's Beef Association Research Fact Sheet.

RAYA, R. P., OOT, R., MALEY, M., DYEN, M., WIELAND, J. *et al.* 2011. Naturally resident and exogenously applied bacteriophages can reduce *Escherichia coli* O157:H7 levels in ruminant guts. *Bacteriophage*, 1, 15–24.

RAYA, R. R., VAREY, P., OOT, R. A., DYEN, M. R., CALLAWAY, T. R. *et al.* 2006. Isolation and characterization of a new T-even bacteriophage, CEV1, and determination of its potential to reduce *Escherichia coli* O157:H7 levels in sheep. *Appl. Environ. Microbiol.*, 72, 6405–6410.

REID, C. R. and BARNUM, D. A. 1984. The effects of treatments of cecal contents on the protective properties against *Salmonella* in poults. *Avian Dis.*, 29, 1–11.

REID, C. A., SMALL, A., AVERY, S. M. and BUNCIC, S. 2002. Presence of foodborne pathogens on cattle hides. *Food Control*, 13, 411–415.

REISSBRODT, R., HAMMES, W. P., DAL BELLO, F., PRAGER, R., FRUTH, A. *et al.* 2009. Inhibition of growth of Shiga toxin-producing *Escherichia coli* by nonpathogenic *Escherichia coli*. *FEMS Microbiol. Lett.*, 290, 62–69.

RICHARDS, H. A., PEREZ-CONESA, D., DOANE, C. A., GILLESPIE, B. E., MOUNT, J. R. *et al.* 2006. Genetic characterization of a diverse *Escherichia coli* O157:H7 population from a variety of farm environments. *Foodborne Path. Dis.*, 3, 259–265.

ROESCH, L. F., FULTHORPE, R. R., RIVA, A., CASELLA, G., HADWIN, A. K. *et al.* 2007. Pyrosequencing enumerates and contrasts soil microbial diversity. *ISME J.*, 1, 283–290.

ROSTAGNO, M. H. 2009. Can stress in farm animals increase food safety risk? *Foodborne Path. Dis.*, 6, 767–776.

ROSTAGNO, M. H. and CALLAWAY, T. R. 2012. Pre-harvest risk factors for *Salmonella enterica* in pork production. *Food Res. Int.*, 45, in press.

ROSTAGNO, M. H., HURD, H. S. and MCKEAN, J. D. 2009. Split marketing as a risk factor for *Salmonella enterica* infection in swine. *Foodborne Path. Dis.*, 6, 865–869.

ROSTAGNO, M. H., HURD, H. S., MCKEAN, J. D., ZIEMER, C. J., GAILEY, J. K. *et al.* 2003. Preslaughter holding environment in pork plants is highly contaminated with *Salmonella enterica*. *Appl. Environ. Microbiol.*, 69, 4489–4494.

ROZEMA, E. A., STEPHENS, T. P., BACH, S. J., OKINE, E. K., JOHNSON, R. P. *et al.* 2009. Oral and rectal administration of bacteriophages for control of *Escherichia coli* O157.-H7 in feedlot cattle. *J. Food Prot.*, 72, 241–250.

SALAK-JOHNSON, J. L. and MCGLONE, J. J. 2007. Making sense of apparently conflicting data: Stress and immunity in swine and cattle. *J. Anim. Sci.*, 85, E81–88.

SARGEANT, J. M., AMEZCUA, M. R., RAJIC, A. and WADDELL, L. 2007. Pre-harvest interventions to reduce the shedding of *E. coli* O157 in the faeces of weaned domestic ruminants: A systematic review. *Zoonos. Pub. Health*, 54, 260–277.

SCALLAN, E., HOEKSTRA, R. M., ANGULO, F. J., TAUXE, R. V., WIDDOWSON, M.-A. *et al.* 2011. Foodborne illness acquired in the United States – Major pathogens. *Emerg. Infect. Dis.*, 17, 7–15.

SCHAMBERGER, G. P., PHILLIPS, R. L., JACOBS, J. L. and DIEZ-GONZALEZ, F. 2004. Reduction of *Escherichia coli* O157:H7 populations in cattle by addition of colicin E7-producing *E. coli* to feed. *Appl. Environ. Microbiol.*, 70, 6053–6060.

SCHARFF, R. L. 2010. Health-related costs from foodborne illness in the United States [Online]. Washington, DC: Georgetown University. Available: www.producesafetyproject.org/admin/assets/files/Health-Related-Foodborne-Illness-Costs-Report.pdf-1.pdf [Accessed 3 May 2010].

SCHNEITZ, C. 2005. Competitive exclusion in poultry – 30 years of research. *Food Cont.*, 16, 657–667.

SCHREZENMEIR, J. and DE VRESE, M. 2001. Probiotics, prebiotics, and synbiotics – Approaching a definition. *Am. J. Clin. Nutr.*, 73 (Suppl.), 354s–361s.

SCHUEHLE PFEIFFER, C. E., KING, D. A., LUCIA, L. M., CABRERA-DIAZ, E., ACUFF, G. R. *et al.* 2009. Influence of transportation stress and animal temperament on fecal shedding of *Escherichia coli* O157:H7 in feedlot cattle. *Meat Sci.*, 81, 300–306.

SEMENOV, A. M., KUPRIANOV, A. A. and VAN BRUGGEN, A. H. C. 2010. Transfer of enteric pathogens to successive habitats as part of microbial cycles. *Microb. Ecol.*, 60, 239–249.

SHERMAN, M. 2009. Probiotics and microflora. *US Pharmacist*, 34, 42–44.

SHOAF, K., MULVEY, G. L., ARMSTRONG, G. D. and HUTKINS, R. W. 2006. Prebiotic galactooligosaccharides reduce adherence of enteropathogenic *Escherichia coli* to tissue culture cells. *Infect. Immun.*, 74, 6920–6928.

SMARDA, J. and OBDRZALEK, V. 2001. Incidence of colicinogenic strains among human *Escherichia coli*. *J. Basic. Microbiol.*, 41, 367–374.

SMITH, D., BLACKFORD, M., YOUNTS, S., MOXLEY, R., GRAY, J. *et al.* 2001. Ecological relationships between the prevalence of cattle shedding *Escherichia coli* O157:H7 and characteristics of the cattle or conditions of the feedlot pen. *J. Food Prot.*, 64, 1899–1903.

SMITH, D. R., MOXLEY, R. A., KLOPFENSTEIN, T. J. and ERICKSON, G. E. 2009a. A randomized longitudinal trial to test the effect of regional vaccination within a cattle feedyard on *Escherichia coli* O157:H7 rectal colonization, fecal shedding, and hide contamination. *Foodborne Path. Dis.*, 6, 885–892.

SMITH, D. R., MOXLEY, R. A., PETERSON, R. E., KLOPFENSTEIN, T. J., ERICKSON, G. E. *et al.* 2009b. A two-dose regimen of a vaccine against type III secreted proteins reduced *Escherichia coli* O157:H7 colonization of the terminal rectum in beef cattle in commercial feedlots. *Foodborne Path. Dis.*, 6, 155–161.

SMITH, H. W. and HUGGINS, R. B. 1982. Successful treatment of experimental *E. coli* infections in mice using phage: Its general superiority over antibiotics. *J. Gen. Microbiol.*, 128, 307–318.

SMITH, H. W. and HUGGINS, R. B. 1983. Effectiveness of phages in treating experimental *Escherichia coli* diarrhoea in calves, piglets and lambs. *J. Gen. Microbiol.*, 129, 2659–2675.

SMITH, H. W. and HUGGINS, R. B. 1987. The control of experimental *E. coli* diarrhea in calves by means of bacteriophage. *J. Gen. Microbiol.*, 133, 1111–1126.

SNOEYENBOS, G. H., WEINACK, O. M. and SMYSER, C. F. 1978. Protecting chicks and poults from salmonellae by oral administration of 'normal gut microflora'. *Avian Dis.*, 22, 273–285.

SONGER, J. G., JONES, R., ANDERSON, M. A., BARBARA, A. J., POST, K. W. *et al.* 2007. Prevention of porcine *Clostridium difficile*-associated disease by competitive exclusion with nontoxigenic organisms. *Vet. Microbiol.*, 124, 358–361.

STABEL, T. J., MAYFIELD, J. E., MORFITT, D. C. and WANNEMUEHLER, M. J. 1993. Oral immunization of mice and swine with an attenuated *Salmonella choleraesuis* (DELTA-cya-12 DELTA(crp-cdt1)19) mutant containing a recombinant plasmid. *Infect. Immun.*, 61, 610–618.

STAHL, C. H., CALLAWAY, T. R., LINCOLN, L. M., LONERGAN, S. M. and GENOVESE, K. J. 2004. Inhibitory activities of colicins against *Escherichia coli* strains responsible for postweaning diarrhea and edema disease in swine. *Antimicrob. Ag. Chemother.*, 48, 3119–3121.

STANFORD, K., MCALLISTER, T. A., NIU, Y. D., STEPHENS, T. P., MAZZOCCO, A. *et al.* 2010. Oral delivery systems for encapsulated bacteriophages targeted *Escherichia coli* O157:H7 in feedlot cattle. *J. Food Prot.*, 73, 1304–1312.

STAVRIC, S. 1992. Defined cultures and prospects. *Int. J. Food Microbiol.*, 55, 245–263.

STAVRIC, S. and D'AOUST, J.-Y. 1993. Undefined and defined bacterial preparations for competitive exclusion of *Salmonella* in poultry. *J. Food Prot.*, 56, 173–180.

STEER, T., CARPENTER, H., TUOHY, K. and GIBSON, G. R. 2000. Perspectives on the role of the human gut microbiota and its modulation by pro and prebiotics. *Nutr. Res. Rev.*, 13, 229–254.

STEPHENS, T. P., LONERAGAN, G. H., CHICHESTER, L. M. and BRASHEARS, M. M. 2007a. Prevalence and enumeration of *Escherichia coli* O157 in steers receiving various strains of *Lactobacillus*-based direct-fed microbials. *J. Food Prot.*, 70, 1252–1255.

STEPHENS, T. P., LONERAGAN, G. H., KARUNASENA, E. and BRASHEARS, M. M. 2007b. Reduction of *Escherichia coli* O157 and *Salmonella* in feces and on hides of feedlot cattle using various doses of a direct-fed microbial. *J. Food Prot.*, 70, 2386–2391.

STEWART, V. J. 1988. Nitrate respiration in relation to facultative metabolism in enterobacteria. *Microbiol. Rev.*, 52, 190–232.

STOUTHAMER, A. H. 1969. A genetical and biochemical study of chlorate-resistant mutants of *Salmonella* Typhimurium. *Antoine van Leeuwenhoek*, 35, 505–521.

SULAKVELIDZE, A. and BARROW, P. A. 2005. Phage therapy in animals and agribusiness. In: Sulakvelidze, A. and Kutter, E. (eds) *Bacteriophages: Biology and Applications.* New York: CRC Press.

SULAKVELIDZE, A. and KUTTER, E. 2005. Bacteriophage therapy in humans. In: Sulakvelidze, A. and Kutter, E. (eds) *Bacteriophages: Biology and Applications.* New York: CRC Press.

SUMMERS, W. C. 2001. Bacteriophage therapy. *Ann. Rev. Microbiol.*, 55, 437–451.

TALLEY, J. L., WAYADANDE, A. C., WASALA, L. P., GERRY, A. C., FLETCHER, J. *et al.* 2009. Association of *Escherichia coli* O157:H7 with filth flies (Muscidae and Calliphoridae) captured in leafy greens fields and experimental transmission of *E. coli* O157:H7 to spinach leaves by house flies (diptera: Muscidae). *J. Food Prot.*, 72, 1547–1552.

THOMSON, D. U., LONERAGAN, G. H., THORNTON, A. B., LECHTENBERG, K. F., EMERY, D. A. *et al.* 2009. Use of a siderophore receptor and porin proteins-based vaccine to control the burden of *Escherichia coli* O157:H7 in feedlot cattle. *Foodborne Path. Dis.*, 6, 871–877.

TORO, H., PRICE, S. B., HOERR, F. J., KREHLING, J., PERDUE, M. *et al.* 2005. Use of bacteriophages in combination with competitive exclusion to reduce *Salmonella* from infected chickens. *Avian Dis.*, 49, 118–124.

TURNBAUGH, P. J., HAMADY, M., YATSUNENKO, T., CANTAREL, B. L., DUNCAN, A. *et al.* 2009. A core gut microbiome in obese and lean twins. *Nature*, 457, 480–484.

TURNBAUGH, P. J., LEY, R. E., MAHOWALD, M. A., MAGRINI, V., MARDIS, E. R. *et al.* 2006. An obesity-associated gut microbiome with increased capacity for energy harvest. *Nature*, 444, 1027–1031.

UNDERDAHL, R., TORRES-MEDINA, A. and DOSTER, A. R. 1982. Effect of *Streptococcus faecium* C-68 in control of *Escherichia coli* induced diarrhea in gnotobiotic pigs. *Am. J. Vet. Res.*, 43, 2227–2232.

USDA-ERS 2001. ERS estimates foodborne disease costs at $6.9 billion per year [Online]. Economic Research Service, United States Department of Agriculture. Available: www.ers.usda.gov/publications/aer741/aer741.pdf [Accessed 16 October 2009].

USHE, T. C. and NAGY, B. 1985. Inhibition of small intestinal colonization of enterotoxigenic *Escherichia coli* by *Streptococcus faecium* M74 in pigs. *Zbl. Bakt. Hyg. I. Abr. Orig. B.*, 181, 374–382.

VAN BAALE, M. J., SARGEANT, J. M., GNAD, D. P., DEBEY, B. M., LECHTENBERG, K. F. *et al.* 2004. Effect of forage or grain diets with or without monensin on ruminal persistence and fecal *Escherichia coli* O157:H7 in cattle. *Appl. Environ. Microbiol.*, 70, 5336–5342.

VAN DONKERSGOED, J., HANCOCK, D., ROGAN, D. and POTTER, A. A. 2005. *Escherichia coli* O157:H7 vaccine field trial in 9 feedlots in Alberta and Saskatchewan. *Can. Vet. J.*, 46, 724–728.

VAN LOO, J. 2007. How chicory fructans contribute to zootechnical performance and well-being in livestock and companion animals. *J. Nutr.*, 137, 2594S–2597S.

VANDEPLAS, S., DUBOIS DAUPHIN, R., BECKERS, Y., THONART, P. and THEWIS, A. 2010. *Salmonella* in chicken: Current and developing strategies to reduce contamination at farm level. *J. Food Prot.*, 73, 774–785.

VARMA, J. K., GREENE, K. D., RELLER, M. E., DELONG, S. M., TROTTIER, J. *et al.* 2003. An outbreak of *Escherichia coli* O157 infection following exposure to a contaminated building. *J. Amer. Med. Assoc.*, 290, 2709–2712.

VIDOVIC, S. and KORBER, D. R. 2006. Prevalence of *Escherichia coli* O157 in Saskatchewan cattle: Characterization of isolates by using random amplified polymorphic DNA PCR, antibiotic resistance profiles, and pathogenicity determinants. *Appl. Environ. Microbiol.*, 72, 4347–4355.

VOSOUGH AHMADI, B., FRANKENA, K., TURNER, J., VELTHUIS, A. G. J., HOGEVEEN, H. *et al.* 2007. Effectiveness of simulated interventions in reducing the estimated prevalence of *E. coli* O157:H7 in lactating cows in dairy herds. *Vet. Res.*, 38, 755–771.

WALKER, W. A. and DUFFY, L. C. 1998. Diet and bacterial colonization: Role of probiotics and prebiotics. *J. Nutr. Biochem.*, 9, 668–675.

WALL, S. K., ZHANG, J., ROSTAGNO, M. H. and EBNER, P. D. 2010. Phage therapy to reduce preprocessing *Salmonella* infections in market-weight swine. *Appl. Environ. Microbiol.*, 76, 48–53.

WALSH, M. C., GARDINER, G. E., HART, O. M., LAWLOR, P. G., DALY, M. *et al.* 2008. Predominance of a bacteriocin-producing *Lactobacillus salivarius* component of a five-strain probiotic in the porcine ileum and effects on host immune phenotype. *FEMS Microbiol. Ecol.*, 64, 317–327.

WEINACK, O. M., SNOEYENBOS, G. H., SMYSER, C. F. and SOERJADI, A. S. 1982. Reciprocal competitive exclusion of *Salmonella* and *Escherichia coli* by native intestinal microflora of the chicken and turkey. *Avian Dis.*, 26, 585–595.

WELLS, J. E., KRAUSE, D. O., CALLAWAY, T. R. and RUSSELL, J. B. 1997. A bacteriocin-mediated antagonism by ruminal lactobacilli against *Streptococcus bovis*. *FEMS Microbiol. Ecol.*, 22, 237–243.

WELLS, J. E., SHACKELFORD, S. D., BERRY, E. D., KALCHAYANAND, N., GUERINI, M. N. *et al.* 2009. Prevalence and level of *Escherichia coli* O157:H7 in feces and on hides of feed lot steers fed diets with or without wet distillers grains with solubles. *J. Food Prot.*, 72, 1624–1633.

WELLS, J. E., YEN, J. T. and MILLER, D. N. 2005. Impact of dried skim milk in production diets on *Lactobacillus* and pathogenic bacterial shedding in growing-finishing swine. *J. Appl. Microbiol.*, 99, 400–407.

WETZEL, A. N. and LEJEUNE, J. T. 2006. Clonal dissemination of *Escherichia coli* O157:H7 subtypes among dairy farms in Northeast Ohio. *Appl. Environ. Microbiol.*, 72, 2621–2626.

XU, J. and GORDON, J. I. 2003. Honor thy symbionts. *Proc. Nat. Acad. Sci. (USA)*, 100, 10452–10459.

YEKTA, M. A., GODDEERIS, B. M., VANROMPAY, D. and COX, E. 2011. Immunization of sheep with a combination of intimin?, EspA and EspB decreases *Escherichia coli* O157:H7 shedding. *Vet. Immunol. Immunopathol.*, 140, 42–46.

YOUNTS-DAHL, S. M., GALYEAN, M. L., LONERAGAN, G. H., ELAM, N. A. and BRASHEARS, M. M. 2004. Dietary supplementation with *Lactobacillus*- and *Propionibacterium*-based direct-fed microbials and prevalence of *Escherichia coli* O157 in beef feedlot cattle and on hides at harvest. *J. Food Prot.*, 67, 889–893.

ZHANG, G., MA, L. and DOYLE, M. P. 2007a. Potential competitive exclusion bacteria from poultry inhibitory to *Campylobacter jejuni* and *Salmonella*. *J. Food Prot.*, 70, 867–873.

ZHANG, G., MA, L. and DOYLE, M. P. 2007b. Salmonellae reduction in poultry by competitive exclusion bacteria *Lactobacillus salivarius* and *Streptococcus cristatus*. *J. Food Prot.*, 70, 874–878.

ZHANG, X., KELLY, S. M., BOLLEN, W. and CURTISS, R. 1999. Protection and immune responses induced by attenuated *Salmonella* Typhimurium UK-1 strains. *Microb. Pathogenesis*, 26, 121–130.

ZHANG-BARBER, L., TURNER, A. K. and BARROW, P. A. 1999. Vaccination for control of *Salmonella* in poultry. *Vaccine*, 17, 2538–2545.

ZHAO, T., DOYLE, M. P., HARMON, B. G., BROWN, C. A., MUELLER, P. O. E. *et al.* 1998. Reduction of carriage of enterohemorrhagic *Escherichia coli* O157:H7 in cattle by inoculation with probiotic bacteria. *J. Clin. Microbiol.*, 36, 641–647.

ZHAO, T., TKALCIC, S., DOYLE, M. P., HARMON, B. G., BROWN, C. A. *et al.* 2003. Pathogenicity of enterohemorrhagic *Escherichia coli* in neonatal calves and evaluation of fecal shedding by treatment with probiotic *Escherichia coli*. *J. Food Prot.*, 66, 924–930.

ZOPF, D. and ROTH, S. 1996. Oligosaccharide anti-infective agents. *The Lancet (North America)*, 347, 1017–1021.

15

New research on ensuring safety in dry processing environments

J.-L. Cordier, Nestlé Quality Assurance Center, Switzerland

DOI: 10.1533/9780857098740.4.305

Abstract: This chapter is a short overview of the current knowledge regarding pathogens in dry processing environments. The main control measures to ensure the safety of food products are then discussed, focusing on the behavior of *Salmonella* and control measures such as cleaning. Since most of the literature related to contamination of low-moisture foods is related to *Salmonella*, the majority of the topics discussed in the following sections are about this pathogen. Where available and relevant, additional information on other pathogens is also included in the discussion.

Key words: *Salmonella*, dry-cleaning, hygiene control measures, processing environment, verification.

15.1 Introduction

Outbreaks caused by low-moisture products contaminated with *Salmonella* have been reported on several occasions during the last 30–40 years. Examples are chocolate and confectionery products (Werber *et al.*, 2005; Unicomb *et al.*, 2005), raw almonds (Isaacs *et al.*, 2005), breakfast cereals (Sobel *et al.*, 2002), peanut butter (CDC, 2009), paprika-seasoned potato chips (Lehmacher *et al.*, 1995), spices such as black and red pepper (Myers *et al.*, 2010; Lienau *et al.*, 2011), herbal tea (Koch *et al.*, 2005), powdered dairy products and infant formulae (Forsyth *et al.*, 2003; Brouard *et al.*, 2007) and dry dog food (CDC, 2012). Detailed reviews of salmonellosis outbreaks linked to low-moisture foods have been published by the Grocery Manufacturers Association (GMA, 2009a) and Podolak *et al.* (2010). To our knowledge, only a few other vegetative pathogens have been associated with foodborne outbreaks caused by low-moisture foods. *Cronobacter* spp. (*Enterobacter sakazakii*) has been linked to rare but severe outbreaks for a total of about 150–170 cases over a period of more than

40 years (FAO/WHO, 2004, 2006). Recently, a particularly virulent strain of Shiga-toxin producing *Escherichia coli* O104:H4 was identified as the cause of a major outbreak in Germany, a total of 3816 cases (including 54 deaths) being reported (Frank *et al.*, 2011). The outbreak was finally traced back to sprouts prepared from contaminated dry fenugreek seeds (Buchholz *et al.*, 2011). The investigation also showed that the seeds were imported two years earlier from Egypt, indicating survival of the pathogen for an extended period of time.

Published case studies of foodborne outbreaks frequently reveal post-process contamination during manufacture. The use of contaminated ingredients in dry-mixing operations or their addition at processing steps located after heat treatment are among the identified root causes. The presence of pathogens in the processing environment or on processing equipment leading to post-process contamination, has been shown to be the origin of some outbreaks (Reij, *et al.*, 2004). For low-moisture products, the causes of contamination can be attributed to poor design of manufacturing facilities and equipment with regard to hygiene, inadequate maintenance, ineffective or insufficient prerequisite programs for cleaning, sanitation and pest management, and insufficient control over the ingredients used in dry-mixing operations. Detailed reviews of the sources and elements contributing to the contamination of low-moisture foods have been published by the Grocery Manufacturers Association (GMA, 2009a, 2009b) and Podolak *et al.* (2010).

Understanding the occurrence, behavior and fate of vegetative pathogens in premises manufacturing low-moisture ingredients or finished goods is important. This understanding is the basis for appropriate and effective prerequisite programs to prevent contamination. Details of the occurrence and fate of *Cronobacter* spp. and *Salmonella* in processing environments manufacturing infant formulae have been described by Cordier (2008). Recent work has confirmed potential routes of contamination by these organisms during the manufacture of milk proteins and milk powders (Mullane *et al.*, 2008; Jacobs *et al.*, 2011).

15.2 Control measures applied during the manufacture of low-moisture products

Economic considerations – reducing the water content of a product significantly reduces the costs of storage and transport – are certainly an important driver in the manufacture of low-moisture foods and, in particular, powders.

The main target in food preservation is the development of processes and procedures to ensure the safety and stability of the manufactured foods. Food preservation is based on designing foods able to inhibit and prevent the development of pathogens, the formation of toxins and the growth of spoilage microorganisms. The reason for manufacturing low-moisture ingredients or food products is therefore certainly linked to the desire to prolong their shelf life by preventing the growth of microorganisms present in the product. Low levels of pathogens have caused outbreaks and the mere absence of growth will, however, not prevent their occurrence – preventive measures need to ensure the absence of contamination.

Factors contributing to the preservation of foods include several physico-chemical parameters. They have been classified into four groups: (i) intrinsic factors related to the characteristics of the food itself, (ii) extrinsic factors, which are applied to the food and are effective during storage and distribution, (iii) processing factors related to the processing conditions and (iv) implicit factors related to the nature and behavior of the microorganisms themselves.

The manufacture of low-moisture products is therefore based on the application and combination of these factors. These products are manufactured with very different technologies. For some of them, the manufacture of a finished product is limited to blending and mixing individual ingredients followed by intermediate processing steps such as storage and filling. For processes encompassing a bactericidal step, the control of heat-sensitive vegetative pathogens such as *Salmonella* starts with a validated kill step, usually heat treatment. Traditional processes such as pasteurization or direct steam injection (DSI), followed by evaporation and drying, are essential steps in the manufacture of low-moisture products such as milk powders and infant formulae. Such processing conditions have been shown to provide a reduction of vegetative cells ranging from 6 to 7 log units at the lower end to an excess of 50 log units in the case of DSI (FAO/WHO, 2004, 2006), which ensures the safety of the products.

For dry pet foods or dry feeds, the usual biocidal steps are extrusion cooking and pelleting (Jones, 2010; Bianchini *et al.*, 2011). For ingredients such as cocoa beans, nuts or almonds, used in the manufacture of chocolate or confectionery products, roasting with hot air or in oil are common, which has been studied and documented by several authors (Shachar and Yaron, 2006; Ma *et al.*, 2009; Bari *et al.*, 2009; Chang *et al.*, 2010, Du *et al.*, 2010; Beuchat and Mann, 2011).

The presence of vegetative pathogens in low-moisture products is therefore not due to their survival. The main cause of their presence in finished goods is (re-) contamination occurring during further processing. This is true for products manufactured by dry-mixing operations only without any inactivation step but also for products that undergo an inactivation step before further handling. The presence of vegetative pathogens in low-moisture foods can be, as indicated in the introduction, traced back to one of the following causes: (i) recontamination from the processing environment or (ii) addition of contaminated ingredients, alone or in combination.

Contamination of low-moisture products can be minimized or even prevented through the application of appropriate prerequisite programs and hygiene control measures. Hygiene control measures to prevent contamination with pathogens such as *Salmonella* or *Cronobacter* spp., typically associated with low-moisture foods, have been discussed in great detail by Cordier (2008) for infant formulae and follow-up formulae. Prevention is based on the application of prerequisite programs aiming at:

- Preventing the ingress of *Salmonella* in areas where processing lines are installed.
- Preventing their establishment in these processing areas in case of ingress.
- Preventing their multiplication to high levels, thus minimizing the risk of spread throughout the processing areas or into processing equipment, which would then significantly increase the probability of contamination.
- Detecting their presence as soon as possible and then taking appropriate measures to ensure their elimination from the processing areas.

Processing in high hygiene areas is usually limited to the manufacture of infant formula and milk powders. However, the underlying principles of the preventive measures to control *Salmonella* outlined above are valid as well for other low-moisture foods. A similar list of control measures, aimed at minimizing the occurrence of *Salmonella* contamination in low-moisture foods, has been described and discussed by the GMA (2009a). The seven elements considered are outlined as follows:

- Prevent ingress or spread of *Salmonella* in the processing facility.
- Enhance the stringency of hygiene practices and controls in the primary *Salmonella* control area (in the close vicinity of the product).
- Apply hygienic design principles to building and equipment design.
- Prevent or minimize growth of *Salmonella* within the facility.
- Establish a raw material and ingredients control program.
- Validate control measures to inactivate *Salmonella.*
- Establish procedures for the verification of the control measures and apply corrective actions in case of deviations.

Prevention of contamination during manufacturing, in particular at processing steps located after a kill step such as heat treatment, is essential in ensuring the safety of finished goods. This is achieved through the design and the effective implementation of hygiene control measures. Numerous publications have been published describing such control measures and the reader is referred to the Codex Alimentarius (CAC, 1969, 1979), Cordier (2008), the GMA (2009a) and to the extensive handbooks on hygiene control measures such as the ones edited by Lelieveld *et al.* (2003, 2005).

Such control measures have been developed over time to address pathogens such as *Salmonella* (Forsyth *et al.*, 2003). They are not necessarily based on research but represent pragmatic approaches incorporating knowledge of the fate and behavior of pathogens and how this can be used to control them in a processing environment. The aim is to continuously improve and enhance the control measures and a typical example is the management of *Cronobacter* spp. in plants manufacturing infant formulae. In this case the control measures developed over a period of 40 years to address *Salmonella* have been further strengthened to comply with the microbiological requirements for Enterobacteriaceae and *Cronobacter* spp. They are 100–1000 times more strict than a few years ago (CAC, 2008; Cordier, 2008).

15.3 Survival of pathogens in low-moisture products and environments

Salmonella has been shown to adapt very well to extreme conditions of temperature and pH as well as desiccation and it is able to survive for prolonged periods of time. Although growth of *Salmonella* is not possible at water activities below 0.93, it has been well established and documented that *Salmonella* is able to survive for prolonged periods of time at water activities below 0.4–0.6, characteristic for low-moisture foods. As a response to several outbreaks of salmonellosis in the early 1970s and 1980s, several investigations considered the behavior of *Salmonella* in chocolate. Results obtained showed the survival of several serotypes for weeks or even months (e.g. Barille *et al.*, 1970; D'Aoust, 1977). Similar behavior was demonstrated for confectionery products with a very low water activity of 0.18 (Kotzekidou, 1998) and spices such as paprika (Lehmacher *et al.*, 1995). More recent studies concerning foods associated with outbreaks such as peanut butter and peanut-butter-containing products (Burnett *et al.*, 2000) and almonds (Uesugi and Harris, 2006), as well as for different ingredients used in the manufacture of confectionery have confirmed the ability of *Salmonella* to survive (Komitopoulou and Peñaloza, 2009). Studies performed by Breeuwer *et al.* (2003), Edelson-Mammel *et al.* (2005) and Caubilla-Barron and Forsythe (2007) for *Enterobacter sakazakii* (known today as *Cronobacter* spp.) in powdered infant formulae, showed similar behavior. Survival over periods of time ranging from several weeks up to several years has been reported in these studies demonstrating that this pathogen can remain viable under such conditions. Trials performed with Shiga-toxin forming *Escherichia coli* showed similar survival rates as *Salmonella*, while shigellae and other (nonpathogenic) strains of *E. coli* did not survive (Hiramatsu *et al.*, 2005).

Survival of *Salmonella* has also been shown to occur on dry inert matrices such as particles of silica gel (Janning *et al.*, 1994) or polypropylene surfaces (Iibuchi *et al.*, 2009). The behavior of different vegetative pathogens has been investigated by several authors. Kusumaningrum *et al.* (2003), for example, showed that the number of cells of *Salmonella* serovar Enteritidis, *Staphylococcus aureus* and *Campylobacter jejuni* inoculated onto stainless steel surfaces in the form of aqueous suspensions and aerosols and subsequently dried, decreased rapidly. The die-off of the different microorganisms tested seemed to be more marked when low-level inocula were used and *C. jejuni* was the most susceptible to air drying. Trials carried out with suspensions containing food residues such as milk or meat showed varying results. However, in general, the presence of such residues improved the survival of cells on the surface compared to those prepared in saline solutions before being applied onto the surfaces investigated. Studies performed by the same authors to determine the rate of transfer of cells from the stainless steel surfaces to foods such as sliced cucumber or roasted chicken showed significant transfer rates and hence an increase in the level of cross-contamination.

While studies of transfer rates give estimates of the probability of (re-) contamination, no studies have, to our knowledge, been performed on the transfer of cells from dry surfaces to a food matrix.

The mechanisms allowing the prolonged periods of survival of vegetative pathogens in dry environments have been investigated by several authors. In the case of *Cronobacter* spp., which also exhibit an important resistance to desiccation, Riedel and Lehner (2007) investigated the impact of osmotic stress on the synthesis of proteins. The most striking effect was considered to be a down-regulation of the motility apparatus and the formation of filamentous cells. White *et al.* (2006) showed that aggregative fimbriae and cellulose enhanced the resistance of *Salmonella* to desiccation and survival on plastic surfaces over prolonged periods of time. Further information can be obtained in Podolak *et al.*, (2010) and a recent detailed review of the molecular effect of environmental stresses, including desiccation, on *Salmonella enterica* has been published by Spector and Kenyon (2012).

Important research has been performed in the medical field to understand the fate and behavior of nosocomial pathogens on inert surfaces. The data have been compiled into a review by Kramer *et al.* (2006) and more recently by Otter *et al.* (2011). They showed that while survival was observed on dry surfaces, the presence of moisture generally contributed to an increased survival for hours and days. Robine *et al.* (2002) investigated the survival of *Enterococcus faecalis* deposited through aerosols onto different types of metallic surfaces. They observed significant differences in the survival rate depending on the environmental conditions. The relative humidity as well as the type of surface and the presence of traces of nutrients had an effect as well. Their studies illustrate the differences in the ability of cells to generate biofilms on the surfaces tested, comparing those suspended in aqueous suspensions and those deposited through aerosolization. The latter results may be more relevant for food-processing environments and food contact surfaces in areas where limited wet cleaning is carried out. This certainly warrants further investigation using alternatives to generate adhering cells.

Several studies have also considered the effect of food contact surfaces containing antibacterial molecules such as copper or silver and showing bactericidal or bacteriostatic properties (e.g. Wilks *et al.*, 2005). Most if not all these studies, have, however, been performed using aqueous suspensions of the various microorganisms studied. Whether they are relevant for the conditions prevailing in dry processing environments, namely dry food contact surfaces, remains an open question.

15.4 The fate of pathogens in low-moisture processing environments

Studies on the survival of pathogens in foods and on surfaces of materials used in processing environments and processing lines can provide some insight into the establishment of *Salmonella* in harborage niches.

Transmission of pathogens (and microorganisms in general) from the food-processing environment and from food contact surfaces in equipment to the food manufactured plays an important role in the occurrence of foodborne outbreaks. The adhesion and survival of pathogens on such surfaces increases the risk of cross-contamination and this risk is certainly significantly increased in wet or humid environments where growth can take place. Growth will lead to an increased number of cells adhering to surfaces and enable them to generate biofilms. The probability of adhesion and subsequent biofilm formation is, however, negligible or even impossible on dry surfaces as they certainly do not favor adhesion of cells. Should cells nevertheless stick to those surfaces or be entrapped in food residues, the low water activity will not allow for growth.

Case studies performed after the occurrence of outbreaks have shown that poor cleaning and sanitation practices are a significant factor in cross-contamination during the manufacture of food products. The risk of cross-contamination is magnified in situations where microbial growth is possible. The presence of low-moisture food residues cannot completely be avoided in food-processing environments or in food-processing lines. However, as long as these residues remain dry, they will not support the growth of microorganisms including specific pathogens. The presence of water, whether from condensation, the infiltration of rainwater, the activation of a sprinkler system after a fire alarm or from wet cleaning, will lead to significant and rapid growth, especially if temperatures are optimal. Moist residues in spots and niches such as cracks, crevices, interfaces and hollow structures and which cannot be rapidly and thoroughly dried, have a particular risk of growth and hence the build-up of a reservoir of microorganisms. The effect of the presence of water on Enterobacteriaceae, which are used as a hygiene indicator in environmental samples, has been illustrated by Cordier (2008). These data show a rapid increase of counts following the presence of water in the environment. A thorough investigation carried out following an outbreak of *Cronobacter* spp. (Coignard and Vaillant, 2006) showed that incriminated lots were manufactured over a period of about six months during which time wet cleaning had taken place on several occasions. Although no results from environmental samples are available, the levels in finished products (50 g and 100 g samples) and because the same molecular type was found in all isolates are an indication of the persistence of the epidemic strain and a link with the wet cleaning performed.

The occurrence of *Salmonella* in processing environments has been demonstrated in several outbreak investigations. For example, in cases of salmonellosis due to toasted oat cereals, *Salmonella* Agona was found at low levels in the manufacturing plant during the investigation (Sobel *et al.*, 2002). It was found in samples taken from floors, equipment and the air exhaust system. Other weaknesses identified during the investigation, such as poor employee practices, poor design and control of the vitamin supply and the addition system, contributed to the contamination. Although no further details are available for the time being, the recent outbreak due to puffed rice contaminated with *Salmonella* Agona (CDC, 2008) may be linked to this earlier outbreak. The reoccurrence of the same serotype may be indicative of the prolonged survival of a specific strain in the manufacturing premises followed by release into the environment due to as yet unknown reasons, causing the contamination of the finished goods.

In a *Salmonella* Senftenberg outbreak traced to infant cereals, the investigation found the pathogen in residues from a mill (Rushdy *et al.*, 1998). In an investigation into the spread of *Salmonella* in a plant manufacturing oil meal, Morita *et al.* (2006) demonstrated its presence on the operators' footwear and gloves, in different processing areas, in dust in the air as well as in rodents present in the premises. A large number (65%) of positive samples were found in residues on the floor of the processing area, indicating a widespread occurrence. Measures to control the movement of operators between different areas helped to prevent its spread and hence contributed to a reduction in the number of positive samples.

The water found in processing environments and processing lines for low-moisture products is not only from the water used for cleaning. Condensation and water droplets may also be generated through temperature gradients within a building or on or within equipment. Preventative measures based on knowing the dew-point conditions include insulating cold pipes, cooling tunnels and designing air handling units to take into account the possibility of contamination.

In the feed industry, dry conditions should prevail to limit bacterial growth, including *Salmonella*. However, Møretrø *et al.* (2009) have shown that on certain production surfaces, condensation may provide sufficient moisture to allow the growth of microorganisms and the formation of biofilms. The build-up of organic matter may further contribute to the settling and establishing of microorganisms. The impact of such biofilms on the ability of *Salmonella* to survive and persist in feed and food-processing environments has been demonstrated by Joseph *et al.* (2001) and Vestby *et al.* (2009a, 2009b). Habimana *et al.* (2010) studied the resident flora on surfaces in feed mills and concluded that their elimination would impede *Salmonella* colonization and growth.

As shown above, even small quantities of water may have a significant impact. The cooling air for low-moisture products such as powders comes into direct contact with the food. Significant differences between the air temperature and the temperature (gradient) of the building can cause condensation in ducts or on food contact surfaces if the relative humidity is not correctly maintained. If condensation occurs on surfaces in contact with food, the impact is magnified due to possible localized growth and hence the occurrence of spot contamination.

15.5 Cleaning procedures

Cleaning in food-processing facilities is designed to remove residues and to prevent their accumulation and build-up. Cleaning is normally performed at regular intervals. It often involves thorough and complete standard operational procedures specifically designed to eliminate targeted food residues. Depending on the type of product and processing conditions, shorter and simpler intermediate procedures such as flushing can reduce the level of residues to an acceptable level and thus reduce the frequency of more thorough cleaning procedures. The frequency and type of cleaning also depend on production schedules and the number of product changeovers that requiring intermediate cleaning.

The type of cleaning as well as the time between applications will depend on the characteristics of the product manufactured. The layout and design of the processing equipment will greatly influence how rapidly residues of food products generated at different processing steps will accumulate on product contact surfaces or specific locations in pieces of equipment. The nature and composition of the product, and hence of the residues generated, are important elements to consider when establishing the required cleaning frequency as well as whether simple intermediate cleaning procedures are possible. The initial flora of the raw materials and the impact of processing steps such as heat treatment will determine the levels and composition of the microbiological flora in residues accumulating along the processing line.

The ability of these residues to support the growth of microorganisms is an additional factor in determining the appropriate cleaning procedures and the frequency at which they need to be carried out. Residues of neutral high-moisture products are obviously a much higher microbiological risk and require therefore much more attention than residues of dry or low-moisture products unable to support growth.

Residues on processing lines and equipment, especially on surfaces in direct contact with the manufactured product, may have immediate consequences on the quality and safety of products. Consideration must also be given to cleaning of the processing environment such as external surfaces of equipment, the floor underneath equipment, the floor in the near vicinity of processing lines and adjacent areas, and the walls and ceilings of the manufacturing premises. The cleaning of the food-processing environment is necessary to eliminate product residues and dust originating from processing operations such as weighing or through spillages and leaks. In addition dust and other soil particles can be introduced through the air or through packaging materials, pallets and pieces of equipment such as bins, crates and other types of container transported by forklift. Personnel including manufacturing operators and maintenance personnel also contribute to the ingress of dust and soil particles.

Dry-cleaning of food-processing environments and food-processing equipment encompasses sweeping, brushing, scraping, vacuuming and wiping with cloths – all activities that mechanically remove residues. While dry-cleaning is certainly the preferred option in controlling pathogens such as *Salmonella* in factories manufacturing low-moisture foods, the use of limited amounts of water may be required in certain circumstances. Controlled wet cleaning of specific and very limited areas is a possible option for minimizing the risk of building up moist residues or residual water in equipment and the environment, which would favor the growth of microorganisms (EHEDG, 2001, 2003; GMA, 2009a). Controlled wet cleaning implies the use of limited quantities of water and a rapid and thorough drying of the cleaned surfaces immediately after cleaning.

Wet cleaning is the worst option as it often leads to the uncontrolled use of water and should be avoided as far as possible. Wet cleaning removes fine particles and residues effectively and is widely used in the food industry. With the exception of closed cleaning-in-place systems used to clean processing equipment and in which water can be contained, wet cleaning of the processing equipment and other equipment will result in the presence of water in the premises. Such procedures necessitate rapid drying after the cleaning to avoid the presence of standing water or water residues over prolonged periods of time. If drying is difficult for pieces of equipment, which were not designed for being cleaned and dried, then wet cleaning of the processing environment will favor the growth and establishment of pathogens such as *Salmonella* (Cordier, 2008; GMA, 2009a). The risk is increased if there are wet or humid niches in cracks, crevices, pits, holes and junctions where food residues, food debris and dust can accumulate, which are the ideal conditions for growth.

Whether processing equipment and processing lines are wet or dry-cleaned has a significant impact on the required design criteria. For wet cleaning the design has to fulfill general requirements for equipment exposed to the significant quantities of water used during cleaning. Such requirements are described in guidelines published by the European Hygienic Engineering and Design Group (EHEDG), for example, the general requirements (EHEDG, 2004). For dry-cleaning the hygienic requirements could in principle be less stringent. However, this holds true only if cleaning is exclusively dry-cleaning: the occasional wet cleaning or the accidental presence of water, e.g. the activation of fire sprinkler systems, in the equipment or environment will invariably lead to issues. The accumulation of damp residues or the presence of stagnant water in certain parts of the equipment would allow the growth of microorganisms, including pathogens, and lead to a significant increase in the risk of contamination.

The EHEDG has also investigated the cleaning of dry processing environments and dry processing lines and has published relevant guidelines accordingly (EHEDG, 2001, 2003). As outlined by the EHEDG (2001), dry materials can be described by the particle characteristics such as density, hardness, moisture content and size as well as by the characteristics of the bulk material such as the bulk density, flowability, solubility and wettability. The criteria for hygienic design will therefore very much depend on the type of material and its characteristics as well as its moisture content, which in turn will determine the type of deposits and residues which may form in processing equipment. EHEDG (2001) provides guidance on the most appropriate cleaning procedures taking into account the possible need for controlled wet cleaning or simple wet cleaning as well as the hygiene zone in which cleaning is carried out.

For the dry-cleaning of processing premises, the EHEDG has outlined several elements that need to be considered, amongst which are:

- areas and components of the plant and lines requiring cleaning;
- extent of fouling associated with the processes and the need for cleaning;
- degree of cleanliness required as defined by an assessment taking quality and safety parameters into account;

- determination of the acceptable cleaning procedure (dry, wet or controlled wet) in relation to the hygiene zone.

For a changeover of product, dry-cleaning can encompass a simple rinsing or flushing of the processing line with the subsequent product or with a neutral matrix such as maltodextrin or one with a hard structure such as crystalline sugar or salt. This may require the disposal of part or the totality of the subsequent batch due to quality deviations (e.g. composition, color or flavor). The removal of product residues with a vacuum cleaner followed by scraping or brushing of the food contact surfaces is performed in situations where simple flushing is insufficient.

The equipment and tools used for manual cleaning include brushes and scrapers, which mechanically remove product residues adhering to product contact surfaces. Vacuum cleaners are used to collect loose powder and dust residues as well as residues dislodged by brushing and scraping. The design of the tools must be appropriate for the purpose and allow cleaning of difficult-to-reach spots in the equipment. They need to be of a hygienic design to prevent the risk of generating foreign bodies such as brush hairs or plastic pieces and allow for thorough cleaning and sterilization. The tools used to clean the processing environment must be clearly differentiated from the tools used to clean equipment and in particular product contact surfaces – this is usually done by color-coding tools and ensuring proper segregation during storage.

Compressed air and blasting with dry ice (carbon dioxide), sand or bicarbonate soda have been described as well. While these techniques dislodge food residues from surfaces and difficult-to-reach spots, they still require the collection of the residues and subsequent cleaning with the tools mentioned above. These techniques and cleaning procedures bear an inherent risk of spreading small or larger residue particles in the surrounding environment (Jackson *et al.*, 2007; Röder *et al.*, 2010). The dry-cleaning of surfaces by abrasive blasting using dry ice is proposed by several companies for applications in the food industry. This was first described by Hoenig (1986), who showed that during a changeover it enhanced the removal of particles adhering to surfaces. Since the dry-ice particles sublimate to gaseous CO_2 no deposition of particles onto surfaces occurs. The use of a dry-ice jet for surface cleaning was discussed by Liu *et al.* (2011).

15.6 Verification of control measures

The verification of the effectiveness of the control measures is an important element of the pathogen management system and needs to be implemented. Verification consists of taking samples from the processing environment, the processing lines (product contact surfaces) and semi-finished or finished products. Verification plans have been outlined and discussed in great detail by Cordier (2008) and by the Almond Board of California (2010), which provides illustrated guidelines.

15.7 Disinfection and sanitizers

Wet cleaning is often followed by the application of sanitizers to disinfect the cleaned surfaces and to kill any microorganisms still present. Sanitizers require the presence of water or traces of moisture for the active agents to be effective. With dry-cleaning, the use of sanitizers is not possible due to the prevailing conditions and the absence of water. The use of an aqueous solution of a disinfectant, either 70% alcohol or a commercial product, which have been shown to be effective, will invariably lead to the presence of residual water and hence to the potential for the growth of any surviving microorganisms. In addition, even minute quantities of food residues, for example in the form of particles or dust, are often sufficient to inactivate or reduce the effectiveness of the active agents present in such products.

However, in the absence of growth, the need for regular sanitation can certainly be questioned. It would only be needed when there are deviations in the processing environment, such as increased levels of a hygiene indicator or the presence of a pathogen. In such a situation, localized treatment with a highly concentrated chlorine solution followed by immediate drying has proven to be effective.

Du *et al.* (2010) conducted studies using isopropyl alcohol to reduce *Salmonella* present in almond dust, and showed a reduction over time. Kane (2012) investigated a sanitizing system that used a combination of isopropyl alcohol and quaternary ammonium (IPAQ) in combination with CO_2, which gave a 6 log reduction of *Salmonella*, and concluded that such a system could be applied in dry-processing environments. Consideration should, however, be given to the presence of residues of disinfecting agents, which are not eliminated after rinsing with water. Regulatory requirements for the maximal levels of such residues may therefore limit the broad application of these techniques.

15.8 Conclusion

Recent outbreaks of salmonellosis and large recalls of products contaminated with *Salmonella* have recently drawn attention to low-moisture foods as emerging vehicles. While new matrices are certainly involved on occasion, the association of this pathogen with low-moisture products is, however, not new and historical data on outbreaks and recalls can be traced back over more than 40 years. Effective control measures and appropriate verification tools were therefore developed years ago and have been optimized over time. While further fine-tuning may be required, today these measures and tools can be considered to be well established and effective – if implemented and applied correctly and consistently.

15.9 References

ALMOND BOARD OF CALIFORNIA (2010) Pathogen Environmental Monitoring Program (PEM). Available at: www.almondboard.com/Handlers/Documents/pem%20book.pdf. Accessed 12 February 2012.

BARI, M.L., NEI, D., SOTOME, I., NISHINA, I., ISOBE, S. *et al.* (2009) Effectiveness of sanitizers, dry heat, hot water and gas catalytic infrared heat treatment to inactivate *Salmonella* on almonds. *Foodborne Path. Dis.*, **6**, 953–958.

BARRILE, J.C., CONE, J.F. and KEENEY, P.G. (1970) A study of salmonellae survival in milk chocolate. *Manuf. Conf.*, **50**, 34–39.

BEUCHAT, L.R. and MANN, D.A. (2011) Inactivation of *Salmonella* on pecan nut meats by hot air treatment and oil roasting. *J. Food Prot.*, **74**, 1441–1450.

BIANCHINI, A., STRATTON, J., WEIER, S., HARTTER, T., PLATTNER, B. *et al.* (2011) Validation of extrusion as a killing step for *Enterococcus faecium* in a balanced carbohydrate protein meal by using a response surface design. *J. Food Prot.*, **75**, 1646–1653.

BREEUWER, P., LARDEAU, M., PETERZ, M. and JOOSTEN, H.M. (2003) Desiccation and heat tolerance of *Enterobacter sakazakii*. *J. Appl. Microbiol.*, **95**, 967–973.

BROUARD, C., ESPIÉ, E., WEILL, F.X., KÉROUANTON, A., BRISABOIS, A. *et al.* (2007) Two consecutive large outbreaks of *Salmonella enterica* serotype Agona infections in infants linked to the consumption of powdered infant formula. *Pediatric Inf. Dis. J.*, **26**, 148–152.

BUCHHOLZ, U., BERNARD, H., WERBER, D., BÖHMER, M.M., REMSCHMIDT *et al.* (2011) German outbreak of *Escherichia coli* O104:H4 associated with sprouts. *New Engl. J. Med.*, **365**, 1763–1770.

BURNETT, S.L., GEHM, E.R., WEISSINGER, W.R. and BEUCHAT, R. (2000) Survival of *Salmonella* in peanut butter and peanut butter spread. *J. Appl. Microbiol.*, **89**, 472–477.

CAC (1969) Recommended International Code of Practice – General Principles of Food Hygiene. CAC/RCP1-1969, Rev. (1985). FAO/WHO, Rome.

CAC (1979) Recommended International Code of Hygienic Practice for Foods for Infants and Children. CAC/RCP 21-1979. FAO/WHO, Rome.

CAC (2008) Code of Hygienic Practice for Powdered Formulae for Infants and Young Children. CAC/RCP 66-2008. FAO/WHO, Rome.

CAUBILLA-BARRON, J. and FORSYTHE, S.J. (2007) Dry stress and survival time of *Enterobacter sakazakii* and other Enterobacteriaceae in dehydrated powdered infant formula. *J. Food Prot.*, **70**, 2111–2117.

CDC (2008) Investigation of outbreak of infections caused by *Salmonella* Agona. Available at: www.cdc.gov/salmonella/agona. Accessed 10 February 2012.

CDC (2009) Multistate outbreak of *Salmonella* infections associated with peanut butter and peanut butter-containing products – United States, 2008–2009. *MMWR Morb. Mortal. Wkly Rep.*, **58**, 85–90.

CDC (2012) Multistate outbreak of human *Salmonella* Infantis infections linked to dry dog food (final update). Available at: www.cdc.gov/salmonella/dog-food-05-12/index.html. Accessed 20 July 2012.

CHANG, S.S., HAN, A.R., REYES DE CORCUERA, J.I., POWERS, J.R. and KANG, D.H. (2010) Evaluation of steam pasteurization in controlling *Salmonella* serotype Enteritidis on raw almond surfaces. *L. Appl. Microbiol.*, **50**, 393–398.

COIGNARD, B. and VAILLANT, V. (2006) Infections à *Enterobacter sakazakii* associées à la consommation d'une préparation en poudre pour nourrissons, p. 88. Rapport d'Investigation. Institut de Veille sanitaire, Saint-Maurice, France.

CORDIER, J.L. (2008) Production of powdered infant formulae and microbiological control measures. Chapter 6, pp. 145–185 in *Enterobacter sakazakii*, edited by FARBER, J.M. and FORSYTHE, S.J., ASM Press Washington.

D'AOUST, J.Y. (1977) *Salmonella* and the chocolate industry: a review. *J. Food Prot.*, **40**, 718–727.

DU, W.X., ABD, S.I., MCCARTHY, K.L. and HARRIS, L.I. (2010) Reduction of *Salmonella* on inoculated almonds exposed to hot oil. *J. Food Prot.*, **73**, 1238–1246.

EDELSON-MAMMEL, S.G., PORTEOUS, M.K. and BUCHANAN, R.I. (2005) Survival of *Enterobacter sakazakii* in a dehydrated powdered infant formula. *J. Food Prot.*, **68**, 1900–1902.

EHEDG (2001) Design criteria for the safe processing of dry particulate materials. Document 22, 23 pp. European Hygienic Equipment Design Group Secretariat, Brussels.

EHEDG (2003) Hygienic engineering of plants for the processing of dry particulate materials. Document 26, 38 pp. European Hygienic Equipment Design Group Secretariat, Brussels.

EHEDG (2004) Hygienic equipment design criteria. Document 8, second edition, 16 pp. European Hygienic Equipment Design Group Secretariat, Brussels.

FAO/WHO (2004) *Enterobacter sakazakii* and other microorganisms in powdered infant formula: meeting report. Microbiological risk assessment series 6. WHO Press, Geneva, Switzerland. Available at: www.who.int/foodsafety/publications/micro/mra6/en/index.html.

FAO/WHO (2006) *Enterobacter sakazakii* and *Salmonella* in powdered infant formula: meeting report. Microbiological risk assessment series 10. WHO Press, Geneva, Switzerland. Available at: www.who.int/foodsafety/publications/micro/mra10/en/index.html.

FORSYTH, J.R., BENNETT, N.M., HOGBEN, S., HUTCHISON, E.M., ROUCH, G. *et al.* (2003) The year of the *Salmonella* seekers – 1977. *Aust. NZ J. Public Health*, **27**, 385–389.

FRANK, C., WERBER, D., CRAMER, J.P., ASKAR, M., FABER, M. *et al.* (2011) Epidemic profile of Shiga-toxin producing *Escherichia coli* O104:H4 outbreak in Germany. *New Engl. J. Med.*, **365**, 1771–1780.

GMA (2009a) Control of *Salmonella* in low-moisture foods. Available at: www.gmaonline.org/downloads/technical-guidance-and-tools/SalmonellaControlGuidance.pdf. Accessed 12 January 2012.

GMA (2009b) Annex to Control of *Salmonella* in low-moisture foods. Available at: www.gmaonline.org/downloads/wygwam/Salmonellaguidanceannex.pdf. Accessed 22 February 2013.

HABIMANA, O., MØRETRØ, T., LANGSRUD, S., VESTBY, L.K., NESSE, L.L. *et al.* (2010) Micro ecosystems form feed industry surfaces: a survival and biofilm study of *Salmonella* versus host resident flora strains. *BMC Vet. Res.*, **6**, 48–56.

HIRAMATSU, R., MATSUMOTO, M., SAKAE, K. and MIYAZAKI, Y. (2005) Ability of Shiga-toxin-producing *Escherichia coli* and *Salmonella* spp. to survive in a desiccation model system and in dry foods. *Appl. Environ. Microbiol.*, **71**, 6657–6663.

HOENIG, S.A. (1986) Cleaning surfaces with dry ice. *Compressed Air Mag.*, **91**, 22–25.

IIBUCHI, R., HARA-KUDO, Y., HASEGAWA, A. and KUMAGAI, S. (2009) Survival of *Salmonella* on a polypropylene surface under dry conditions in relation to biofilm formation capability. *J. Food Prot.*, **73**, 1506–1510.

ISAACS, S., ARAMINI, J., CIEBIN, B., FARRAR, J.A., AHMED; R. *et al.* and *Salmonella* ENTERITIDIS PT30 OUTBREAK INVESTIGATION WORKING GROUP (2005) An international outbreak of salmonellosis associated with raw almonds contaminated with a rare phage type of *Salmonella* Enteritidis. *J. Food Prot.*, **68**, 191–198.

JACKSON, L.S., AL-TAHER, F.M., MOORMAN, M., DEVRIES, J.W., TIPETT, R. *et al.* (2007) Cleaning and other control and validation strategies to prevent allergen cross-contact in food processing operations. *J. Food Prot.*, **71**, 445–458.

JACOBS, C., BRAUN, P. and HAMMER, P. (2011) Reservoir and routes of transmission of *Enterobacter sakazakii* (*Cronobacter* spp.) in a milk powder producing plant. *J. Dairy Sci.*, **94**, 3801–3810.

JANNING, B., IN'T VELD, P.H., NOTERMANS, S. and KRAMER, J. (1994) Resistance of bacterial strains to dry conditions: use of anhydrous silica gel in a desiccation model system. *J. Appl. Microbiol.*, **77**, 319–324.

JONES, F.T. (2010) A review of practical *Salmonella* control measures in animal feed. *J. Appl. Poultry Res.*, **20**, 102–113.

JOSEPH, B., OTTA, S.K. and KARUNASAGAR, I. (2001) Biofilm formation by *Salmonella* spp. on food contact surfaces and their sensitivity to sanitizers. *Int. J. Food Microbiol.*, **64**, 367–372.

KANE, D.M. (2012) Evaluation of a sanitizing system using isopropyl alcohol quaternary ammonium formula and carbon dioxide for dry-processing environments. Master of Science Kansas State University. Available at: krex.k-state.edu/

dspace/bitstream/handle/2097/14175/DeborahKane2012.pdf?sequence=1. Accessed 22 February 2013.

KOCH, J., SCHRAUDER, A., ALPERS, K., WERBER, D., FRANK, R. *et al.* (2005) *Salmonella* Agona outbreak from contaminated aniseed, Germany. *Emerg. Infect. Dis.*, **11**, 1124–1128.

KOMITOPOULOU, E. and PEÑALOZA, W. (2009) Fate of *Salmonella* in dry confectionery raw materials. *J. Appl. Microbiol.*, **106**, 1892–1900.

KOTZEKIDOU, P. (1998) Microbial stability and fate of *Salmonella* Enteritidis in halva, a low-moisture confection. *J. Food Prot.*, **61**, 181–185.

KRAMER, A., SCHWEBKE, I. and KAMPF, G. (2006) How long do nosocomial pathogens persist on inanimate surfaces? A systematic review. *BMC Infect. Dis.*, **6**, 130–137.

KUSUMANINGRUM, H.D., RIBOLDI, G., HAZELEGER, W.C. and BEUMER, R.R. (2003) Survival of foodborne pathogens on stainless steel surfaces and cross-contamination to foods. *Int. J. Food Microbiol.*, **85**, 227–236.

LEHMACHER, A., BOCKEMÜHL, J. and ALEKSIC, S. (1995) Nationwide outbreak of human salmonellosis in Germany due to contaminated paprika and paprika-powdered potato chips. *Epidemiol. Infect.*, **115**, 501–511.

LELIEVELD, H.L.M., MOSTERT, M.A. and HOLAH, J. (eds) (2003) *Hygiene in Food Processing*. Woodhead Publishing Limited, Cambridge, England.

LELIEVELD, H.L.M., MOSTERT, M.A. and HOLAH, J. (eds) (2005) *Handbook of Hygiene Control in the Food Industry*. Woodhead Publishing Limited, Cambridge, England.

LIENAU, E.K., STRAIN, E., WANG, C., ZHENG, J., OTTESEN, A.R. *et al.* (2011) Identification of salmonellosis outbreak by means of molecular sequencing. *N. Engl. J. Med.*, **364**, 981–982.

LIU, Y.H., MARUYAMA, H. and MATSUSAKA, S. (2011) Effect of particle impact on surface cleaning using dry ice jet. *Aerosol. Sci. Technol.*, **45**, 1519–1527.

MA, L., ZHANG, G., GERNER-SMIDT, P., MANTRIPRAGADA, V., EZEOKE, I. *et al.* (2009) Thermal inactivation of *Salmonella* in peanut butter. *J. Food Prot.*, **72**, 1596–1601.

MØRETRØ, T., VESTBY, L.K., NESSE, L.L., STORHEIM, S.E., KOTLARZ, K. *et al.* (2009) Evaluation of efficacy of disinfectants against *Salmonella* from the feed industry. *J Appl. Microbiol.*, **106**, 1005–1012.

MORITA, T., KITAZAWA, H., IIDA, T. and KAMATA, S. (2006) Prevention of *Salmonella* cross-contamination in an oil meal manufacturing plant. *J. Appl. Microbiol.*, **101**, 464–473.

MULLANE, N., HEALY, B., MEADE, J., WHYTE, P., WALL, P.G. *et al.* (2008) Dissemination of *Cronobacter* spp. (*Enterobacter sakazakii*) in powdered milk protein manufacturing facility. *Appl. Environ. Microbiol.*, **74**, 5913–5919.

MYERS, C., HERNANDEZ, M.F. and KENNELLY, P. (2010) Investigation of Union International Food Company *Salmonella* Rissen outbreak associated with white pepper. California Department of Public Health. Available at: www.cdph.ca.gov/pubsforms/Documents/fdbEIRUFIC2009.pdf. Accessed 12 February 2012.

OTTER, J.A., YEZLI, S. and FRENCH, G.L. (2011) The role played by contaminated surfaces in the transmission of nosocomial pathogens. *Infect. Control Hosp. Epidem.*, **32**, 687–699.

PODOLAK, R., ENACHE, E., STONE, W., BLACK, D.G. and ELLIOTT, P.H. (2010) Sources and risk factors for contamination, survival, persistence and heat resistance of *Salmonella* in low moisture food. *J. Food Prot.*, **73**, 1919–1936.

REIJ, M.W., DEN AANTREKKER, E.D. and THE ILSI EUROPE RISK ANALYSIS IN MICROBIOLOGY TASK FORCE (2004) Recontamination as a source of pathogens in processed foods. *Int. J. Food Microbiol.*, **91**, 1–11.

RIEDEL, K. and LEHNER, A. (2007) Identification of proteins involved in osmotic stress response in *Enterobacter sakazakii* by proteomics. *Proteomics*, **7**, 1217–1231.

ROBINE, E., BOULANGÉ-PETERMANN, L. and DERANGÈRE, D. (2002) Assessing bactericidal properties of materials: the case of metallic surfaces in contact with air. *J. Microbiol. Methods*, **49**, 225–234.

RÖDER, M.R., BALTRUWEIT, I., GRUYTERS, H., IBACH, A., MÜCKE, I. *et al.* (2010) Allergen sanitation in the food industry: a systematic industrial scale approach to reduce hazelnut cross-contamination of cookies. *J. Food Prot.*, **73**, 1671–1679.

RUSHDY, A.A., STUART, J.M., WARD, L.R., BRUCE, J., THRELFALL, E.J. *et al.* (1998) National outbreak of *Salmonella* Senftenberg associated with infant food. *Epidemiol. Infect.*, **120**, 125–128.

SHACHAR, D. and YARON, S. (2006) Heat tolerance of *Salmonella enteric* serovars Agona, Enteritidis and Typhimurium in peanut butter. *J. Food Prot.*, **69**, 2687–2691.

SOBEL, J., GRIFFIN, P.M., SLUTSKER, L., SWERDLOW, D.L. and TAUXE, R.V. (2002) Investigation of multistate foodborne disease outbreak. *Public Health Rep.*, **117**, 8–19.

SPECTOR, M.P. and KENYON, W.J. (2012) Resistance and survival strategies of *Salmonella enterica* to environmental stresses. *Food Res. Int.*, **45**, 455–481.

UESUGI, A.R. and HARRIS, L.J. (2006) Growth of *Salmonella* Enteritidis phage type 30 in almond hull and shell slurries and survival in drying almond hulls. *J. Food Prot.*, **69**, 712–718.

UNICOMB, L.E., SIMMONS, G., MERRITT, T., GREGORY, J., NICOL, C. *et al.* (2005) Sesame seed products contaminated with *Salmonella*: three outbreaks associated with tahini. *Epidemiol. Infect.*, **133**, 1065–1072.

VESTBY, L.K., MØRETRØ, T., LANGSRUD, S., HEIR, E. and NESSE, L.L. (2009a) Biofilm forming abilities of *Salmonella* correlated with persistence in fish meal- and feed factories. *BMC Vet. Res.*, **5**, 43–48.

VESTBY, L.K., MØRETRØ, T., BALANCE, S., LANGSRUD, S. and NESSE, L.L. (2009b) Survival potential of wild type cellulose deficient *Salmonella* from the feed industry. *BMC Vet. Res.*, **5**, 20–25.

WERBER, D., DREESMAN, J., FEIL, F., VAN TREECK, U., FELL, G. *et al.* (2005) International outbreak of *Salmonella* Oranienburg due to German chocolate. *BMC Infect. Dis.*, **5**, 7–17.

WHITE, A.P., GIBSON, D.L., KIM, W., KAY, W.W. and SURETTE, M.G. (2006) Thin aggregative fimbriae and cellulose enhance long-term survival and persistence of *Salmonella*. *J. Bacteriol.*, **188**, 3219–3227.

WILKS, S.A., MICHELS, H. and KEEVIL, C.W. (2005) The survival of *Escherichia coli* O157 on a range of metal surfaces. *Int. J. Food Microbiol.*, **105**, 445–454.

16

New research on bacteriophages and food safety

J. Klumpp and M. J. Loessner, ETH Zurich, Switzerland

DOI: 10.1533/9780857098740.4.321

Abstract: Foodborne bacterial infections are a major healthcare concern worldwide. Bacteriophages, the natural enemies of bacteria, are an ideal means to detect and control foodborne pathogens. In this chapter, bacteriophages for use in food are introduced and general considerations regarding phage characteristics, application-specific parameters and potential problems are presented. Bacteriophage lytic enzymes are discussed as potent novel antimicrobials. The use of bacteriophage preparations in the detection of foodborne pathogens is illustrated. Recent regulatory approvals and scientific advances in bacteriophage-based pathogen detection and control are described.

Key words: biocontrol, pathogen detection, endolysin, infectious dose, post-harvest intervention, pre-harvest intervention, lytic cycle, temperate, virulent, transduction.

16.1 Introduction

Bacterial pathogens are a massive threat to food safety. Amongst the top five foodborne diseases worldwide, four are caused by bacteria, namely *Salmonella*, *Clostridium*, *Campylobacter* and *Staphylococcus*. These are estimated to have caused a total of more than three million infections and intoxications in the US alone in 2011 (CDC, 2011). In the EU, foodborne infections by *Salmonella*, *Campylobacter* and *Listeria* account for over 300 000 human cases annually (EFSA, 2011). Increasing numbers of foodborne pathogens are becoming antibiotic resistant, which can partly be attributed to the extensive use of antibiotics as growth promoters in livestock production. Bacteriophages are the most abundant biological entity on earth and influence most of earth's biogeochemistry through their bacterial hosts (Abedon, 2009; Rohwer and Edwards, 2002). As the natural enemies of bacteria, they are the ideal means to detect and control foodborne pathogens. Phages are highly specific and efficient killing machines, lacking their own metabolism and being totally dependent on the bacterial cell for replication. The concept of opposing bacteria with bacteriophages is nearly a century old: they were used as human therapeutics shortly after being discovered by d'Herelle and Twort early in the nineteenth century (Summers, 2005). In recent years, the use of bacteriophage preparations in food has attracted much research interest (Callewaert *et al.*, 2011; Fenton *et al.*, 2010).

Every step of bacteriophage infection, from phage adsorption to cell lysis can be harnessed for detection and control purposes. Both, detection and biocontrol measures must be fast, simple, robust, inexpensive and ideally exhibit high specificity and sensitivity. Bacteriophage-based procedures can satisfy these criteria. A overview of phage biology in Section 16.2. General considerations are discussed in Section 16.3 and specific applications of bacteriophages and bacteriophage components/enzymes are introduced in Sections 16.4. and 16.5. Section 16.6 details phage-based approaches to pathogen detection.

16.2 Bacteriophages of foodborne pathogens

Bacteriophages of foodborne pathogens have been extensively studied in a number of model organisms, such as *Salmonella*, *Escherichia coli* and *Listeria*. It is important to note that bacteriophages infecting Gram-negative bacteria (e.g. *Salmonella*) face different challenges than bacteriophages infecting Gram-positive hosts, such as *Listeria*. The cell wall composition is very different, which has a number of implications for phage biology. In the Gram-positive case, phage adsorption as the initial step in the infection process occurs mainly at the sugar side chains of wall teichoic acids and the peptidoglycan backbone, whereas in Gram-negative bacteria, which are enveloped by an outer membrane, membrane-anchored proteins and lipopolysaccharides serve as bacteriophage binding ligands. Lysis from without, the destruction of the bacterial cell by phage-encoded peptidoglycan hydrolase enzymes from the outside, without the need for a complete infection cycle, is mostly observed in Gram-positive bacteria. The Gram-negative peptidoglycan layer is not directly accessible to peptidoglycan hydrolases from the outside, due to the presence of the outer membrane (Fischetti, 2010). However, lysis from without can also occur through the action of the tail-associated phage enzyme, which penetrates the cell envelope prior to DNA translocation (Moak and Molineux, 2004). Recent technological advances have made it possible to identify and produce these enzymes for use in pathogen biocontrol (Section 16.5).

16.3 General considerations for bacteriophage application

Bacteriophage application in pre- and post-harvest control of foodborne pathogens features a unique combination of advantages: bacteriophages are designed to bind bacterial cells with high affinity and their killing activity is strictly limited to bacteria. They exhibit high specificity and do not usually cross species borders, thereby allowing selective pathogen removal whilst food fermentation biota is not affected. Unlike antibiotics, phages kill specifically, causing no collateral damage to beneficial microorganisms. Phages are self-replicating and self-limiting, easy to produce, natural and non-toxic. They are abundant in the environment, can be isolated with ease and meet regulatory approval faster than other antibacterial substances, since they are part of the natural flora of the product to be treated. Most importantly, this biocontrol agent co-evolves with the bacterial target, which counters resistance development (in contrast to static antibiotics). Bacteriophage-mediated biocontrol can be part of the hurdle concept, a combination of intrinsic and extrinsic factors to preserve foods.

Because of these advantages, much research activity is focused on pathogen detection and control by bacteriophages. Numerous publications report the use of bacteriophages alone or in combination with other agents, e.g. against *E. coli*, *Salmonella* and *Listeria* in food, food-processing environments and livestock (Guenther *et al.*, 2009; Callaway *et al.*, 2011; Hooton *et al.*, 2011; Anany *et al.*, 2011; Coffey *et al.*, 2011; Patel *et al.*, 2011).

A number of bacteriophage applications are in the development process or have already been commercialized. Among them are the following products:

Listex™ P100, produced by Micreos (former EBI Food Safety), the Netherlands – this product is active against *Listeria monocytogenes* (Carlton *et al.*, 2005) and has received generally-recognized-as-safe status by the FDA and USDA for use in cheese products (2006) and consequently for use in all food materials (2007). The biological status of Listex has been approved EU-wide, allowing application in conventional and organically produced products. Listex has recently been approved as a food-processing aid in Canada and the USA. Micreos has also just launched Salmonellex™, an anti-*Salmonella* phage product that has received a temporary use exemption for large-scale field trials.

ListShield™, produced by Intralytics, USA – ListShield™ has been approved by the FDA as a food additive active against *L. monocytogenes* for ready-to-eat meats (21 CFR §172.785), and by the Environmental Protection Agency as an environmental decontaminant for use in food processing (EPA registration no. 74234-1) (Mai *et al.*, 2010). The company is also currently developing EcoShield™ and SalmoFresh™, two bacteriophage products active against enterohaemorrhagic *E. coli* and *Salmonella*. Both products have recieved regulatory approval.

Agriphage™ (EPA Registration # 67986-1), produced by Omnilytics, USA – this is a bacteriophage-based pesticide specific for *Xanthomonas campestris* pv. *vesicatoria* and *Pseudomonas syringae* pv. *tomato*. Agriphage is used pre-harvest, to protect against bacterial disease in tomato and pepper plants (Balogh *et al.*, 2010; Jones *et al.*, 2007).

16.3.1 Possible intervention points for bacteriophage preparations in food production

Pre-harvest interventions

Bacteriophages or their components can be employed in pre- and post-harvest treatment of raw food materials. It has been postulated that pre-slaughter and pre-harvest interventions may be more effective than post-slaughter/post-harvest interventions (Jordan *et al.*, 1999; Goodridge, 2010). This finding is illustrated by a risk assessment for human *Campylobacter* infections from poultry meat. A 2 log reduction in *Campylobacter* pre-slaughter is estimated to decrease consumer risk by 75% (or 90% if *Campylobacter* in chicken feces is targeted simultaneously). Such reductions in bacterial counts can be achieved by phage biocontrol (Havelaar *et al.*, 2005; Loc Carrillo *et al.*, 2005). Bacteriophages can be used to decontaminate the production environment, to de-colonize slaughter animals or to prevent the animal-to-animal or field-to-field spread of pathogenic bacteria. Animal manure is the main source of *E. coli* and *Salmonella* contamination in the environment and consequently in food products, especially since manure is often used as a fertilizer. Strains can be stable in fields, greenhouses and orchards over a long period of time (Uesugi *et al.*, 2007). A reduction in shedding of these pathogens by animals is highly desirable.

Pre-harvest or pre-slaughter biocontrol using bacteriophages can be very effective if the source of the pathogen in the production environment is targeted or removed, and reintroduction of the pathogen is prevented. Bacteriophages do not necessarily have to exhibit high-log reductions in living animals or plants, as a 1–2 log reduction in pathogen levels still considerably lowers the risk to the consumer (Havelaar *et al.*, 2005). Such reductions have, for example, been achieved by treating *Salmonella* colonization in swine (Callaway *et al.*, 2011). In general, the spread of bacterial strains in large flocks or herds is fast, and the bacteriophages used for pre-harvest interventions in livestock must exhibit an exceptionally broad host range. High doses are not necessarily required, if there is time for multiple rounds of phage replication and if sufficient bacterial cells are present. Not all of these approaches have been found to be effective. The removal of pathogenic bacteria from the ruminant gastrointestinal tract is a promising approach, but several problems, such as the low concentration of target bacteria, the spread of phage to control animals or phage inactivation in the gastrointestinal (GI) tract still have to be solved (Callaway *et al.*, 2008; Rozema *et al.*, 2009; Stanford *et al.*, 2010). The GI system of ruminants is one of the most complex environments for biocontrol measures. The various compartments have different pH, are inhabited by a strong rumen microbiota, feature a large surface area, which inactivates phages through adsorption, and enzymes, gastric acids and bile salts may also inactivate therapeutic phage preparations.

Poultry meat is the main meat source consumed worldwide, with the number of livestock poultry exceeding that of cattle by tenfold (FAO, 2009, 2010). Undercooked poultry meat is a significant threat to human health, due to the presence of *Salmonella* and *Campylobacter*. Reductions of 1–2 log in these pathogens have been achieved by spraying bacteriophages onto chicken flocks

(Borie *et al.*, 2009), or administering phage tail spike proteins to chickens (Waseh *et al.*, 2010). Much research has been dedicated to eradication of *Campylobacter* in poultry flocks or to prevent infection of the animals. Wagenaar *et al.* reported a 1–2 log reduction of *Campylobacter* in broiler chicken with fluctuating levels of bacteria in 2005 (Wagenaar *et al.*, 2005). More recently, El-Shibiny *et al.* and Carvalho *et al.* achieved a 2 log reduction of *Campylobacter* in live chickens, which they claimed corresponds to an extrapolated 30-fold risk reduction of human infection (El-Shibiny *et al.*, 2009; Carvalho *et al.*, 2010).

One of the biggest problems in pre-slaughter biocontrol of livestock is the route of administration of the bacteriophage preparations. Recent results for *Campylobacter* demonstrated that administration via feed leads to a more sustainable reduction of *Campylobacter* than administration by oral gavage (Carvalho *et al.*, 2010). Stanford *et al.* report some success compared to traditional feeding when administering encapsulated phage to feedlot cattle. However, the stability of encapsulated phages, and the precise location of phage liberation from the capsule remain to be researched (Stanford *et al.*, 2010). A similar encapsulation approach has recently been proposed for the *Salmonella*-specific bacteriophage FO1 (Ma *et al.*, 2008).

The situation is different in pre-harvest control of pathogens on fresh produce, e.g. leafy greens. As well as the strict control of irrigation water quality, phage treatment can enhance product safety. However, for technical reasons, phage treatments of produce are mostly conducted after harvesting and initial washing. Bacteriophages are not very stable against UV irradiation and desiccation and quickly become inactivated if applied to growing plants. Successful strategies target potential sources of contamination, such as compost, rather than the pathogen on the plant (Heringa *et al.*, 2010).

One exception is the biocontrol of *Erwinia amylovora*, the fireblight pathogen, which is the subject of considerable research focusing on phage control for apple and pear trees. A number of potential biocontrol phages have been described (Born *et al.*, 2011; Lehman *et al.*, 2009). Spraying phage preparations onto apple or pear blossom is a promising strategy for control of *Erwinia* infections, which cause substantial losses in fruit production worldwide.

An important advantage of pre-harvest intervention strategies is that contaminated raw materials will not enter the production facility and dissemination of the pathogen in the facility can be avoided. However, pre-harvest interventions seem limited, because the plant or animal system they are supposed to free from a certain pathogen is usually very complex, and composed of several distinct compartments with different pH values, temperatures and nutrient availability.

Interventions post-harvest, during production and in the final product

Few industrial food-processing steps permit the conditions needed for application of biocontrol organisms and the relatively long incubation times required for effective treatment. However, machine parts, conveyor belts or other hard-to-clean equipment can be treated with bacteriophage preparations or phage-derived enzymes (Abuladze *et al.*, 2008; Soni and Nannapaneni, 2010; Viazis *et al.*, 2011; Callewaert *et al.*,

2011). Spraying with phages prior to food packaging is the most widely used method to enhance food safety and shelf life. Most commercially available phage preparations are designed to be applied during this step, e.g. spraying sliced ham with Listex P100™. Phage-based products encounter fewer hurdles when used on the final product, compared to the treatment of live animals, such as their active immune system and compartmentalization. Recent advances in the post-harvest use of whole-phage preparations against pathogenic bacteria include the reduction of *E. coli* O157:H7 on tomatoes, spinach, broccoli, ground beef and hard surfaces (Abuladze *et al.*, 2008), *Listeria* in ready-to-eat foods (Guenther *et al.*, 2009), *Salmonella* in pigs (Saez *et al.*, 2011; Callaway *et al.*, 2011) and vegetables (Ye *et al.*, 2010).

16.3.2 Resistance development

Bacterial cells are prone to spontaneous mutation or adaption processes, which may result in the development of phage resistance. Although many studies have reported the isolation of bacteria that are resistant to a phage preparation, the risk of failure of the biocontrol measures remains relatively low. A 2 log reduction, as commonly achieved in pre-harvest pathogen control, equates to about 1% of the bacterial population surviving the disinfection process (Hagens and Loessner, 2010). A similar rate was reported after treatment of *Campylobacter* on live chicken (El-Shibiny *et al.*, 2009). A mutation rate of 10^{-6} during phage treatment has been estimated (O'Flynn *et al.*, 2004). Thus, in phage-treated foods that leave the production facility immediately after treatment, the occurrence of phage-resistant mutants in the surviving bacterial population may be assumed to be a rare event (Hagens and Loessner, 2010).

Any such mutant bacteria are usually not reintroduced into the food-production environment, and thus are not able to multiply or transfer the resistance traits to non-resistant strains. Furthermore, bacteriophage resistance is generally associated with high fitness costs, as it requires e.g. the rearrangement of the bacterial cell wall or the production of enzymatic components. It is expected that such resistance will be quickly lost in the bacterial population, once the selective pressure (the phage) is removed and given that no niches in the production equipment allowing bacterial growth under high phage pressure are maintained (Hagens and Loessner, 2007a, 2010). Bacteriophages co-evolve with the bacteria to counter resistance development and the production of new phages, which are able to overcome the bacterial resistance, can be expected. In order to minimize possible resistance development, the use of a cocktail containing phages with different specificities or the rotation of the treatment preparations is recommended.

16.4 Phage preparations for pathogen detection and control

16.4.1 Pathogen biocontrol

Pathogen biocontrol is defined as eradication of bacteria, which are components of the normal or environmental flora of organisms or products, but act as pathogens towards the consumer. Biocontrol encompasses all approaches for phage-mediated bacterial reduction and has no immediate healthcare connection as opposed to therapy (this concept has been modified from Abedon, 2009). However, the term 'therapy' has been used in conjunction with biocontrol experiments on foodborne pathogens in their natural reservoir, e.g. livestock before slaughter (Johnson *et al.*, 2008). Bacteriophage preparations are safe and are often a very efficient means of controlling bacterial pathogens. However, translating the application of bacteriophages in laboratory-scale experiments to applications in the food industry has an inherent list of potential problems. Amongst them are general questions regarding phage characteristics, propagation and safety, as well as specificity and efficacy questions, and potential complications during application, e.g. resistance development.

Preparations of intact, active and unmodified bacteriophages must satisfy the following criteria for their successful application in food materials and food production:

1. The bacteriophage host range should be very broad. In the ideal case, 100% of a representative set of pathogen strains/serovars/isolates should be sensitive to as few phages as possible. Inherent for success is the choice of such a representative set of target organisms, against which to test biocontrol candidates. In under-researched bacterial species, often with a nonexistent typing scheme, the assembly of such a test set can prove difficult. An alternative would be to assemble a comprehensive set of clinical or food isolates. The use of as few phages as possible is advised, as each additional phage strain in the cocktail adds a multiplicity of potential problems, such as recombination between phages, the potential for antigenicity in the consumer and the need for different phage production procedures.

2. The use of temperate phages must be avoided. Temperate phages can lysogenize their host, i.e. form stable integrations in the host genome (prophage), and may not immediately lyse the target cell. Lysogenized host cells are immune to infection by the same phage, and killing rates decrease during treatment. Temperate phages are also often able to transduce genetic material between bacterial hosts, thereby altering the host phenotype. Temperate phage integration can alter host characteristics (Canchaya *et al.*, 2003). For example, prophages enable *Clostridium botulinum and C. difficile* toxin production (Sekulovic *et al.*, 2011; Sakaguchi *et al.*, 2005), increase *Salmonella* virulence (Zou *et al.*, 2010), influence *Staphylococcus* pathogenicity islands (Tormo-Mas *et al.*, 2010) and have many more effects (reviewed in Waldor and Friedman, 2005). Additionally, temperate phages often have a quite narrow host range, severely diminishing their usefulness as biocontrol agents (Hagens and Loessner, 2010).

3. If a phage cocktail is used, the exact composition must be known and allow perfect reproducibility. Morphological characterization by electron microscopy is the most widely used technology to gain insight into the composition of phage preparations. Often, there are surprises. A bacterial strain, used to amplify the virus, may release a temperate phage and contaminate the preparation. The majority of natural bacteriophage isolates are siphoviruses, which are often

temperate and problematic for industrial use (Canchaya *et al.*, 2003). Electron microscopy can reveal the cocktail composition very early in the development process and is relatively fast and inexpensive. However, morphologically highly similar phages are indistinguishable under the electron microscope and characterization should be combined with genetic methods, such as restriction profiling, pulsed-field gel electrophoresis or genome sequencing.

4. The genome sequence of the bacteriophage must be known and analyzed for genes that encode for unwanted properties, e.g. toxins or antibiotic resistance. Temperate phages can be recognized by the presence of a lysogeny control region in the genome, usually located downstream of the genes encoding structural proteins.

5. An apathogenic host strain should be available for propagation of phages. Large-scale industrial bacteriophage propagation poses a significant security risk, and will most likely not meet regulatory approval. If possible, a closely related, non-pathogenic bacterium should be used. However, most bacteriophages are able to modify their receptor recognition apparatus according to the propagation host, which might confer an altered host range. In the worst case, a phage propagated on the non-pathogen host is no longer able to infect the target pathogen. Possible solutions include constant monitoring of activity against the target organism and propagation on a genetically modified pathogen lacking pathogenicity determinants. The latter may be difficult to achieve for phages infecting Gram-negative enteric bacteria, as most phages require the pyrogenic lipopolysaccharide layer of the cell wall as receptors. Recently, a self-cycling process for bacteriophage production was suggested (Sauvageau and Cooper, 2010).

Regulatory issues

Because of their ubiquitous nature, phages are consumed in high doses every day by every consumer, and are thus generally considered safe (Johnson *et al.*, 2008). Numerous animal and human studies have found no adverse effect of phage ingestion on health (for example, see the review by Housby and Mann, 2009). Recently, several bacteriophage products have been awarded generally-recognized-as-safe or food-additive status (Section 16.3). However, genetically engineered phages, e.g. with a broader host range or increased virulence, are likely to be problematic since the product label would need to indicate that the product contains genetically modified viruses.

Other important considerations for bacteriophage applications in food

Food matrix

Historically, most research on bacteriophages has been performed in liquid media or on agar plates, using high-density bacterial cultures. Although such model systems might be useful for mimicking the conditions in industrial fermentation, most foods are a complex environment, with large, uneven, highly charged, hydrophilic or hydrophobic surfaces with variable moisture content. Bacteriophages can rapidly become adsorbed and thereby inactivated on such surfaces. Niches and cavities provide safe hiding places for bacteria during phage treatment. Thus, the initial dose of phages must be high enough to take into account the inactivation of a large number of phages and still be effective in immediate killing of the target cells (Hagens and Loessner, 2010).

Host cell density and phage titer

Due to modern hygiene regiments, bacterial numbers in industrially produced food are relatively low. It is important to understand that sufficiently high bacteriophage titers need to be present to affect the target bacteria population. Bacteriophages are non-motile and rely on particle diffusion to find and reach their hosts. Even seemingly small surface areas or liquid volumes can present insurmountable hurdles for successful phage infection and the resulting biocontrol. It would take a thousand years for one bacterial cell and one phage particle to meet in 1 mL of liquid. If there were 10^6 plaque-forming units of phage in 1 mL of liquid along with a single bacterial cell, then it would only take on average 12 hours for one of the phages to encounter the bacterium (Hagens and Loessner, 2010). To make matters even more complex, it is noteworthy that phage-bacteria population dynamics differ significantly between high-density and low-density cultures and over the time allowed for co-incubation (Abedon, 2009), making predictions from lab-scale experiments unreliable in industrial settings. Successful biocontrol relies on the initial killing of bacteria by a critical threshold number of phage (passive biocontrol), as well as a build-up of killing efficiency by exponentially increasing phage numbers through release of progeny phage from infected cells (active biocontrol). The latter, however, requires a critical amount of time and high numbers of bacterial target cells to allow completion of the phage lytic cycle and release of progeny particles (Hagens and Loessner, 2010; Abedon, 2009).

Incubation and storage temperatures

Lysis of the bacterial cell can occur from within the cell triggered by bacteriophage-encoded endolysins (see Section 16.5), or from without, by the action of phage-tail associated lysins, which are normally used to penetrate the cell wall in the initial stages of infection (Moak and Molineux, 2004). Low incubation and storage temperatures often do not permit active replication and lysis from within. However, phages can still adsorb to these cells and start to inject DNA. Once the temperature permits the onset of growth (either in the human host or in a culture), infection proceeds, rapidly killing the infected cell. The outcome is the same as with phage treatment at higher incubation temperatures and there is no reason to be concerned regarding biocontrol efficiency.

16.5 Bacteriophage lytic enzymes and their application in food

Bacteriophages employ lytic enzymes to release progeny particles at the end of the lytic cycle. These cell-wall hydrolases degrade the peptidoglycan sacculus, which protects Gram-positive or Gram-negative cells against osmotic pressure and has a barrier function. Tailed phages often have a dual-lysis system, consisting of a small holin protein, which penetrates the cytoplasmic membrane, and the endolysin, which subsequently cleaves the peptidoglycan backbone. This lysis event is precisely regulated to prevent early cell lysis and loss of progeny particles (Loessner, 2005; Young, 2002). Endolysins are usually composed of a C-terminal cell-wall binding domain (CBD), an N-terminal enzymatic active domain (EAD) and a short linker region, allowing for flexible action of the EAD after binding of the CBD to the target region (Korndorfer *et al.*, 2006, 2008; Hermoso *et al.*, 2003). The considerable advantage of endolysin proteins is their specificity, which does not usually extend across species borders, allowing for selective enzymatic biocontrol.

Bacteriophage endolysins have been researched for decades and there have been important technological advances for their use in food (recently reviewed in Callewaert *et al.*, 2011, and Fenton *et al.*, 2010), two of which will selectively be highlighted here.

It was recently demonstrated that lytic domains and binding domains of phage lysins can be shuffled and combined to generate proteins with enhanced catalytic activity and/or altered specificity against different serovars of *Listeria.* Such modified enzymes exhibit up to threefold higher lytic activity than the parental molecule (Schmelcher *et al.*, 2011). The cell-wall binding domains of these endolysins are highly efficient detection tools, able to bind *Listeria* cells with high affinity in the pico- to nanomolar range (Schmelcher *et al.*, 2010) (Section 16.6.4).

Bacteriophage endolysins can be employed to bind and attack a Gram-positive cell from the outside. Gram-negative cells are *per se* not susceptible to endolysin action, as their peptidoglycan backbone is protected by the cell membrane and not accessible. Gram-negative pseudomonads could be targeted by using a phage endolysin combined with cell permeabilizers, opening entirely new ways for the biocontrol of Gram-negative bacteria (Briers *et al.*, 2011).

16.6 Pathogen detection

Detection of bacterial pathogens is traditionally performed by culture-dependent methods combined with biochemical characterization, which is both time- and labor-intensive and oftentimes prone to false-positive or false-negative identification. Molecular methods aid the development of rapid, automatable and reliable pathogen detection but require expensive equipment and skilled operators. Because of the high costs associated, only a limited number of samples can be analyzed in this way.

Bacteriophages exhibit high specificity towards a bacterial genus, species or even towards a specific strain or serovar. Cross-genus infections are uncommon and might only occur (if at all) in very closely related genera, e.g. *E. coli* and *Salmonella.* This specificity can be harnessed for highly efficient pathogen detection assays. Bacteriophages multiply in the infected host, thereby amplifying the detectable signal and increasing the sensitivity. Furthermore, a signal can only be detected from living cells, overcoming the serious limitation of PCR-based assays, which detect the presence of nucleic acids, regardless of the metabolic state of the bacterial cell. Pathogens are detected by observing bacteriophage plaques on agar plates or bacteriophage biomarker molecules, by monitoring the release of molecules from lysed host cells or by using phage-based reporter systems, which activate upon virus multiplication.

Traditionally, bacteriophages have been used to differentiate bacterial strains (phage typing). A panel of bacteriophages with different but complementary host ranges is used to type bacterial isolates by means of lysis patterns. The method is fast, cheap, reliable and widely used. However, with the development of more rapid and precise genome-based typing methods, such as pulsed-field gel electrophoresis (PFGE), randomly amplified polymorphic DNA (RAPD), and amplified fragment length polymorphism (AFLP) (reviewed in Gürtler and Mayall, 2001), phage typing is less and less used and will not be discussed here.

The simplest method for bacterial detection is the direct assessment of the phage particle number, which is directly proportional to the initial population and the number of multiplication cycles on susceptible bacterial hosts. Such phage-amplification assays have been developed for a number of bacterial pathogens, such as *Mycobacterium* (Foddai *et al.*, 2010).

16.6.1 Phage-amplification assay

The replication of bacteriophages results in the formation of plaques, cleared zones of lysis, on a bacterial lawn on plate media. Every plaque can be traced back to one successful infection by one phage particle, thus permitting a direct correlation between plaque number and bacterial counts. Phage amplification can be improved by combining it with modern, indirect detection methods, thereby omitting the labor-intensive and time-consuming plaque-counting steps. In particular, the coupling of phage amplification with mass spectrometry detection is a promising approach (Pierce *et al.*, 2011; Rees and Voorhees, 2005). Other groups have used real-time PCR to detect the increasing number of phage genomes during infection (Kutin *et al.*, 2009).

Such assays are not available for slow-growing organisms, which do not have sufficiently fast phage amplification. A more rapid amplification assay was developed for these cases (Stewart *et al.*, 1998), basically consisting of mixing the test sample with infective phages and allowing time for the establishment of an initial infection. A virucide is then added to eliminate the remaining extracellular phages. Phage genomes, which have been injected into the host cytosol during the first infection round, remain unharmed and successfully complete the infection cycle. By mixing with phage-susceptible indicator bacteria and plating on agar plates, the number of plaques can be counted, which directly correlates with the number of cells present in the initial culture. Phage-amplification assays are very useful for detecting slow-growing bacteria, such as *Mycobacterium* (Stanley *et al.*, 2007). A number of assay optimizations have recently been published

(Foddai *et al.*, 2009, 2010), and the assay has also been evaluated for detection of *M. avium* in milk and cheese (Botsaris *et al.*, 2010).

16.6.2 Detection by lysis

Infection by bacteriophages completing the lytic cycle inevitably leads to cell lysis by bacteriophage lytic proteins. Holin proteins disintegrate the inner membrane, allowing endolysin proteins to access and hydrolyze the rigid cell-wall peptidoglycan layer. Various methods for monitoring the lysis process itself or its by-products (i.e. the substances released from the cell) have been developed (see the review by Griffiths, 2010).

The simplest approach is the direct measurement of cell lysis using turbidity. A more advanced proof-of-principle method used bioluminescent bacteria to monitor the lysis process through a decrease in luminescence, which correlates to the progress of lysis of the bacteria (Kim *et al.*, 2009).

The level of the amount of cytoplasmic markers or ATP released can be measured by a variety of methods. Kannan *et al.* monitored the release of adenylate kinase from lysed cells using a luciferin-luciferase assay (Kannan *et al.*, 2010). The concentration of ATP is directly proportional to the light emission and consequently to the number of bacteria present (Griffiths, 2010). Other assays measure the activity of beta-galactosidase released from lysed cells (Stanek and Falkinham, 2001). Successful completion of the infection cycle results in drastic amplification of the biomarker molecule, thus allowing reliable detection.

Although such methods are fast, easy to perform and high-throughput compatible, they are totally dependent on the ability of the phage (or lysin protein) to specifically lyse the pathogen in question and therefore fail to signal for bacterial strains that are not sensitive to the phage used (Kannan *et al.*, 2010).

16.6.3 Detection using reporter bacteriophages

Genetically engineered bacteriophages can be used to detect bacterial pathogens by transducing a reporter gene into the cell. The activity of the reporter can be monitored, giving rapid and reliable pathogen detection. Widely used reporter systems are the *luxAB* luciferase system (Hagens and Loessner, 2007b), beta-galactosidase (Willford and Goodridge, 2008) or the ice nucleation gene (*ina*) (Wolber and Green, 1990). Regardless of the reporter system, the underlying principle remains the same: the reporter gene of choice is inserted into the phage genome, preferably behind a strong promoter that is expressed during infection, and remains inactive in the phage particle. The reporter gene becomes activated once the phage genome is injected into a susceptible host, transcribed and translated by the host cell machinery. The advantage of this approach is that only viable target cells are detected, as the expression of the reporter gene relies on an active metabolism in the infected cell. Reporter bacteriophages have been constructed for a number of bacterial pathogens, e.g. *Mycobacterium* (Sarkis *et al.*, 1995), *Listeria* (Loessner *et al.*, 1996, 1997), *Bacillus* (Schofield and Westwater, 2009) and *Salmonella* (Kuhn, 2007) (see the review in Smartt and Ripp, 2011). Recently, a hyperthermostable glycosidase reporter gene was used to detect viable *Listeria* cells using phage A511, with flexibility in the use of fluorescent, chemiluminescent or chromogenic detection assay substrates (Hagens *et al.*, 2011). Reporter bacteriophages have also recently been proposed for the detection of Class A bacterial pathogens such as *Yersinia pestis* and *Bacillus anthracis* (Schofield *et al.*, 2009, 2011; Schofield and Westwater, 2009).

A major drawback of the reporter phage detection method is the strict dependency on phage gene expression by the bacterial host, requiring an active metabolism. Viable but non-culturable or dormant cells cannot be detected by this assay. A dependency of reporter protein folding and the resulting signal strength on the incubation temperature was also reported (Hagens *et al.*, 2011).

16.6.4 Detection by receptor binding and adsorption

Bacteriophages use highly specific receptor-recognition and receptor-binding protein domains for attaching to their specific host and targeting their endolysin proteins on the bacterial cell wall. Both whole-phage preparations and purified receptor-binding proteins have been employed in pathogen detection assays. The immobilization of phage particles on an inert carrier material has proven to be challenging but research reports highlight promising results using physical adsorption on quartz crystal surfaces (Nanduri *et al.*, 2007), streptavidin-mediated attachment of biotin-coated phages (Gervais *et al.*, 2007) and the chemical attachment of phages on gold surfaces (Singh *et al.*, 2009). Recently, Li *et al.* reported the coupling of whole filamentous phages to magnetoelastic biosensors, thereby enabling the wireless, on-site detection of *Salmonella* from tomatoes in concentrations exceeding 5×10^2 cfu/mL (Horikawa *et al.*, 2011; Li *et al.*, 2010).

Instead of immobilizing the phages on a solid surface, bacteriophage particles or phage-derived proteins can also be coated onto the surface of paramagnetic beads. As in immunomagnetic separation, the phage/protein coated beads have high recovery rates and enable short assay times (Kretzer *et al.*, 2007). In the case of the foodborne pathogen *Listeria*, endolysin CBD-based paramagnetic separation is superior to immunomagnetic separation because of its high specificity and selectivity compared to commercially available antibodies (Kretzer *et al.*, 2007; Schmelcher *et al.*, 2010). Paramagnetic separation techniques have also been developed for other pathogens, such as *M. avium* (Foddai *et al.*, 2011). Phage proteins have been demonstrated to be superior to antibodies in the electrochemical impedance spectroscopy detection of bacteria, mainly because the binding to phage proteins generates a dual signal, which can be used to distinguish between unspecific and specific binding (Mejri *et al.*, 2010).

16.7 Conclusions

Foodborne infections and intoxications are a major problem worldwide. Bacteriophages and their components can be efficiently harnessed for the sensitive detection of bacteria in food and for the selective removal of such from food matter (biocontrol). As with every antimicrobial agent, certain restrictions apply on the use of bacteriophage preparations in the food industry. Important considerations regarding characterization of the agent, treatment dose, incubation time, target cell number, influence of storage temperature and food composition have to be taken into account. These criteria are even more important in the pre-harvest treatment of plants and livestock. These complex biological systems require tightly controlled intervention treatments, which have to be adapted to the specific conditions of the treated microenvironment and are often prone to failure.

The detection of foodborne pathogens by bacteriophages or bacteriophage components has proven to be a valuable asset in ensuring food safety. The advantages of this type of detection system are its high specificity, high affinity to the target molecule and sensitive detection of low contamination levels or selective amplification of the detection signal with high signal-to-noise ratios.

16.8 References

ABEDON, S. T. (2009), Kinetics of phage-mediated biocontrol of bacteria, *Foodborne Pathog Dis*, 6, 807–15.

ABULADZE, T., LI, M., MENETREZ, M. Y., DEAN, T., SENECAL, A. *et al.* (2008), Bacteriophages reduce experimental contamination of hard surfaces, tomato, spinach, broccoli, and ground beef by *Escherichia coli* O157:H7, *Appl Environ Microbiol*, 74, 6230–8.

ANANY, H., CHEN, W., PELTON, R. and GRIFFITHS, M. W. (2011), Biocontrol of *Listeria monocytogenes* and *Escherichia coli* O157:H7 in meat by using phages immobilized on modified cellulose membranes, *Appl Environ Microbiol*, 77, 6379–87.

BALOGH, B., JONES, J. B., IRIARTE, F. B. and MOMOL, M. T. (2010), Phage therapy for plant disease control, *Curr Pharm Biotechnol*, 11, 48–57.

BORIE, C., SANCHEZ, M. L., NAVARRO, C., RAMIREZ, S., MORALES, M. A. *et al.* (2009), Aerosol spray treatment with bacteriophages and competitive exclusion reduces *Salmonella* Enteritidis infection in chickens, *Avian Dis*, 53, 250–4.

BORN, Y., FIESELER, L., MARAZZI, J., LURZ, R., DUFFY, B. *et al.* (2011), Novel virulent and broad-host-range *Erwinia amylovora* bacteriophages reveal a high degree of mosaicism and a relationship to Enterobacteriaceae phages, *Appl Environ Microbiol*, 77, 5945–54.

BOTSARIS, G., SLANA, I., LIAPI, M., DODD, C., ECONOMIDES, C. *et al.* (2010), Rapid detection methods for viable *Mycobacterium avium* subspecies *paratuberculosis* in milk and cheese, *Int J Food Microbiol*, 141 Suppl 1, S87–90.

BRIERS, Y., WALMAGH, M. and LAVIGNE, R. (2011), Use of bacteriophage endolysin EL188 and outer membrane permeabilizers against *Pseudomonas aeruginosa*, *J Appl Microbiol*, 110, 778–85.

CALLAWAY, T. R., EDRINGTON, T. S., BRABBAN, A. D., ANDERSON, R. C., ROSSMAN, M. L. *et al.* (2008), Bacteriophage isolated from feedlot cattle can reduce *Escherichia coli* O157:H7 populations in ruminant gastrointestinal tracts, *Foodborne Pathog Dis*, 5, 183–91.

CALLAWAY, T. R., EDRINGTON, T. S., BRABBAN, A., KUTTER, B., KARRIKER, L. *et al.* (2011), Evaluation of phage treatment as a strategy to reduce *Salmonella* populations in growing swine, *Foodborne Pathog Dis*, 8, 261–6.

CALLEWAERT, L., WALMAGH, M., MICHIELS, C. W. and LAVIGNE, R. (2011), Food applications of bacterial cell wall hydrolases, *Curr Opin Biotechnol*, 22, 164–71.

CANCHAYA, C., PROUX, C., FOURNOUS, G., BRUTTIN, A. and BRUSSOW, H. (2003), Prophage genomics, *Microbiol Mol Biol Rev*, 67, 238–76.

CARLTON, R. M., NOORDMAN, W. H., BISWAS, B., DE MEESTER, E. D. and LOESSNER, M. J. (2005), Bacteriophage P100 for control of *Listeria monocytogenes* in foods: genome sequence, bioinformatic analyses, oral toxicity study, and application, *Regul Toxicol Pharmacol*, 43, 301–12.

CARVALHO, C. M., GANNON, B. W., HALFHIDE, D. E., SANTOS, S. B., HAYES, C. M. *et al.* (2010), The *in vivo* efficacy of two administration routes of a phage cocktail to reduce numbers of *Campylobacter coli* and *Campylobacter jejuni* in chickens, *BMC Microbiol*, 10, 232.

CDC (2011), *CDC Estimates of Foodborne Illness in the United States*. Available at: www.cdc.gov/foodborneburden/2011-foodborne-estimates.html.

COFFEY, B., RIVAS, L., DUFFY, G., COFFEY, A., ROSS, R. P. *et al.* (2011), Assessment of *Escherichia coli* O157:H7-specific bacteriophages e11/2 and e4/1c in model broth and hide environments, *Int J Food Microbiol*, 147, 188–94.

EFSA (2011), The European Union Summary Report on Trends and Sources of Zoonoses, Zoonotic Agents and Food-borne Outbreaks in 2009, *EFSA Journal*, 9.

EL-SHIBINY, A., SCOTT, A., TIMMS, A., METAWEA, Y., CONNERTON, P. *et al.* (2009), Application of a group II *Campylobacter* bacteriophage to reduce strains of *Campylobacter jejuni* and *Campylobacter coli* colonizing broiler chickens, *J Food Prot*, 72, 733–40.

FAO (2009), FAOSTAT Database. Available at: faostat.fao.org.

FAO (2010), *Agribusiness Handbook: Poultry Meat and Eggs*. Available at: www.responsibleagroinvestment.org/rai/sites/responsibleagroinvestment.org/files/FAO_Agbiz%20handbook_Poultry_Meat.pdf.

FENTON, M., ROSS, P., MCAULIFFE, O., O'MAHONY, J. and COFFEY, A. (2010), Recombinant bacteriophage lysins as antibacterials, *Bioeng Bugs*, 1, 9–16.

FISCHETTI, V. A. (2010), Bacteriophage endolysins: a novel anti-infective to control Gram-positive pathogens, *Int J Med Microbiol*, 300, 357–62.

FODDAI, A., ELLIOTT, C. T. and GRANT, I. R. (2009), Optimization of a phage amplification assay to permit accurate enumeration of viable *Mycobacterium avium* subsp. *paratuberculosis* cells, *Appl Environ Microbiol*, 75, 3896–902.

FODDAI, A., ELLIOTT, C. T. and GRANT, I. R. (2010), Rapid assessment of the viability of *Mycobacterium avium* subsp.s *paratuberculosis* cells after heat treatment, using an optimized phage amplification assay, *Appl Environ Microbiol*, 76, 1777–82.

FODDAI, A., STRAIN, S., WHITLOCK, R. H., ELLIOTT, C. T. and GRANT, I. R. (2011), Application of a peptide-mediated magnetic separation-phage assay for detection of viable *Mycobacterium avium* subsp. *paratuberculosis* to bovine bulk tank milk and feces samples, *J Clin Microbiol*, 49, 2017–19.

GERVAIS, L., GEL, M., ALLAIN, B., TOLBA, M., BROVKO, L. *et al.* (2007), Immobilization of biotinylated bacteriophages on biosensor surfaces, *Sensors Actuators B: Chem*, 125, 615–21.

GOODRIDGE, L. (2010), Application of bacteriophages to control pathogens in food animal production, In: Sabour, P. M. and Griffiths, M. W. (eds.) *Bacteriophages in the Control of Food- and Waterborne Pathogens*, Washington DC, ASM Press.

GRIFFITHS, M. W. (2010), Phage-based methods for the detection of bacterial pathogens, In: Sabour, P. M. and Griffiths, M. W. (eds.) *Bacteriophage in the Control of Food- and Waterborne Pathogens*, Washington, DC, ASM Press.

GUENTHER, S., HUWYLER, D., RICHARD, S. and LOESSNER, M. J. (2009), Virulent bacteriophage for efficient biocontrol of *Listeria monocytogenes* in ready-to-eat foods, *Appl Environ Microbiol*, 75, 93–100.

GÜRTLER, V. and MAYALL, B. C. (2001), Genomic approaches to typing, taxonomy and evolution of bacterial isolates, *Int J System Evol Microbiol*, 51, 3–16.

HAGENS, S. and LOESSNER, M. J. (2007a), Application of bacteriophages for detection and control of foodborne pathogens, *Appl Microbiol Biotechnol*, 76, 513–19.

HAGENS, S. and LOESSNER, M. J. (2007b), Luciferase reporter bacteriophages, In: Marks, R. S., Cullen, D. C., Karube, I., Lowe, C. R. and Weetall, H. H. (eds.) *Handbook of Biosensors and Biochips*, Hoboken, Wiley and Sons.

HAGENS, S. and LOESSNER, M. J. (2010), Bacteriophage for biocontrol of foodborne pathogens: calculations and considerations, *Curr Pharma Biotechnol*, 11, 58–68.

HAGENS, S., DE WOUTERS, T., VOLLENWEIDER, P. and LOESSNER, M. J. (2011), Reporter bacteriophage A511::*celB* transduces a hyperthermostable glycosidase from *Pyrococcus furiosus* for rapid and simple detection of viable *Listeria* cells, *Bacteriophage*, 1, 143–51.

HAVELAAR, A. H., NAUTA, M. J., MANGEN, M.-J. J., DE KOEIJER A. G., BOGAARDT, M.-J. *et al.* (2005), Costs and benefits of controlling *Campylobacter* in the Netherlands, *RIVM report 250911009/2005*, Bilthoven, NL, RIVM.

HERINGA, S. D., KIM, J., JIANG, X., DOYLE, M. P. and ERICKSON, M. C. (2010), Use of a mixture of bacteriophages for biological control of *Salmonella enterica* strains in compost, *Appl Environ Microbiol*, 76, 5327–32.

HERMOSO, J. A., MONTERROSO, B., ALBERT, A., GALAN, B., AHRAZEM, O. *et al.* (2003), Structural basis for selective recognition of pneumococcal cell wall by modular endolysin from phage Cp-1, *Structure*, 11, 1239–49.

HOOTON, S. P., ATTERBURY, R. J. and CONNERTON, I. F. (2011), Application of a bacteriophage cocktail to reduce *Salmonella* Typhimurium U288 contamination on pig skin, *Int J Food Microbiol*, 151, 157–63.

HORIKAWA, S., BEDI, D., LI, S., SHEN, W., HUANG, S. *et al.* (2011), Effects of surface functionalization on the surface phage coverage and the subsequent performance of phage-immobilized magnetoelastic biosensors, *Biosens Bioelectron*, 26, 2361–7.

HOUSBY, J. N. and MANN, N. H. (2009), Phage therapy, *Drug Discov Today*, 14, 536–40.

JOHNSON, R. P., GYLES, C. L., HUFF, W. E., OJHA, S., HUFF, G. R. *et al.* (2008), Bacteriophages for prophylaxis and therapy in cattle, poultry and pigs, *Anim Health Res Rev*, 9, 201–15.

JONES, J. B., JACKSON, L. E., BALOGH, B., OBRADOVIC, A., IRIARTE, F. B. *et al.* (2007), Bacteriophages for plant disease control, *Annu Rev Phytopathol*, 45, 245–62.

JORDAN, D., MCEWEN, S. A., LAMMERDING, A. M., MCNAB, W. B. and WILSON, J. B. (1999), Pre-slaughter control of *Escherichia coli* O157 in beef cattle: a simulation study, *Prev Vet Med*, 41, 55–74.

KANNAN, P., YONG, H. Y., REIMAN, L., CLEAVER, C., PATEL, P. *et al.* (2010), Bacteriophage-based rapid and sensitive detection of *Escherichia coli* O157:H7 isolates from ground beef, *Foodborne Pathog Dis*, 7, 1551–8.

KIM, S., SCHULER, B., TEREKHOV, A., AUER, J., MAUER, L. J. *et al.* (2009), A bioluminescence-based assay for enumeration of lytic bacteriophage, *J Microbiol Methods*, 79, 18–22.

KORNDORFER, I. P., DANZER, J., SCHMELCHER, M., ZIMMER, M., SKERRA, A. *et al.* (2006), The crystal structure of the bacteriophage PSA endolysin reveals a unique fold responsible for specific recognition of *Listeria* cell walls, *J Mol Biol*, 364, 678–89.

KORNDORFER, I. P., KANITZ, A., DANZER, J., ZIMMER, M., LOESSNER, M. J. *et al.* (2008), Structural analysis of the L-alanoyl-D-glutamate endopeptidase domain of *Listeria* bacteriophage endolysin Ply500 reveals a new member of the LAS peptidase family, *Acta Crystallogr D Biol Crystallogr*, 64, 644–50.

KRETZER, J. W., LEHMANN, R., SCHMELCHER, M., BANZ, M., KIM, K. P. *et al.* (2007), Use of high-affinity cell wall-binding domains of bacteriophage endolysins for immobilization and separation of bacterial cells, *Appl Environ Microbiol*, 73, 1992–2000.

KUHN, J. C. (2007), Detection of *Salmonella* by bacteriophage Felix 01, *Methods Mol Biol*, 394, 21–37.

KUTIN, R. K., ALVAREZ, A. and JENKINS, D. M. (2009), Detection of *Ralstonia solanacearum* in natural substrates using phage amplification integrated with real-time PCR assay, *J Microbiol Methods*, 76, 241–6.

LEHMAN, S. M., KROPINSKI, A. M., CASTLE, A. J. and SVIRCEV, A. M. (2009), Complete genome of the broad-host-range *Erwinia amylovora* phage phiEa21-4 and its relationship to *Salmonella* phage felix O1, *Appl Environ Microbiol*, 75, 2139–47.

LI, S., LI, Y., CHEN, H., HORIKAWA, S., SHEN, W. *et al.* (2010), Direct detection of *Salmonella* Typhimurium on fresh produce using phage-based magnetoelastic biosensors, *Biosens Bioelectron*, 26, 1313–19.

LOC CARRILLO, C., ATTERBURY, R. J., EL-SHIBINY, A., CONNERTON, P. L., DILLON, E., *et al.* (2005), Bacteriophage therapy to reduce *Campylobacter jejuni* colonization of broiler chickens, *Appl Environ Microbiol*, 71, 6554–63.

LOESSNER, M. J. (2005), Bacteriophage endolysins – current state of research and applications, *Current Opinion in Microbiology*, 8, 480–7.

LOESSNER, M. J., REES, C. E., STEWART, G. S. and SCHERER, S. (1996), Construction of luciferase reporter bacteriophage A511::luxAB for rapid and sensitive detection of viable *Listeria* cells, *Appl Environ Microbiol*, 62, 1133–40.

LOESSNER, M. J., RUDOLF, M. and SCHERER, S. (1997), Evaluation of luciferase reporter bacteriophage A511::luxAB for detection of *Listeria monocytogenes* in contaminated foods, *Appl Environ Microbiol*, 63, 2961–5.

MA, Y., PACAN, J. C., WANG, Q., XU, Y., HUANG, X. *et al.* (2008), Microencapsulation of bacteriophage felix O1 into chitosan-alginate microspheres for oral delivery, *Appl Environ Microbiol*, 74, 4799–805.

MAI, V., UKHANOVA, M., VISONE, L., ABULADZE, T. and SULAKVELIDZE, A. (2010), Bacteriophage administration reduces the concentration of *Listeria monocytogenes* in the gastrointestinal tract and its translocation to spleen and liver in experimentally infected mice, *Int J Microbiol*, 2010, 624234.

MEJRI, M. B., BACCAR, H., BALDRICH, E., DEL CAMPO, F. J., HELALI, S. *et al.* (2010), Impedance biosensing using phages for bacteria detection: generation of dual signals as the clue for in-chip assay confirmation, *Biosens Bioelectron*, 26, 1261–7.

MOAK, M. and MOLINEUX, I. J. (2004), Peptidoglycan hydrolytic activities associated with bacteriophage virions, *Mol Microbiol*, 51, 1169–83.

NANDURI, V., SOROKULOVA, I. B., SAMOYLOV, A. M., SIMONIAN, A. L., PETRENKO, V. A. *et al.* (2007), Phage as a molecular recognition element in biosensors immobilized by physical adsorption, *Biosens Bioelectron*, 22, 986–92.

O'FLYNN, G., ROSS, R. P., FITZGERALD, G. F. and COFFEY, A. (2004), Evaluation of a cocktail of three bacteriophages for biocontrol of *Escherichia coli* O157:H7, *Appl Environ Microbiol*, 70, 3417–24.

PATEL, J., SHARMA, M., MILLNER, P., CALAWAY, T. and SINGH, M. (2011), Inactivation of *Escherichia coli* O157:H7 attached to spinach harvester blade using bacteriophage, *Foodborne Pathog Dis*, 8, 541–6.

PIERCE, C. L., REES, J. C., FERNANDEZ, F. M. and BARR, J. R. (2011), Detection of *Staphylococcus aureus* using 15N-labeled bacteriophage amplification coupled with matrix-assisted laser desorption/ionization-time-of-flight mass spectrometry, *Anal Chem*, 83, 2286–93.

REES, J. C. and VOORHEES, K. J. (2005), Simultaneous detection of two bacterial pathogens using bacteriophage amplification coupled with matrix-assisted laser desorption/ionization time-of-flight mass spectrometry, *Rapid Commun Mass Spectrom*, 19, 2757–61.

ROHWER, F. and EDWARDS, R. (2002), The phage proteomic tree: a genome-based taxonomy for phage, *J Bacteriol*, 184, 4529–35.

ROZEMA, E. A., STEPHENS, T. P., BACH, S. J., OKINE, E. K., JOHNSON, R. P. *et al.* (2009), Oral and rectal administration of bacteriophages for control of *Escherichia coli* O157:H7 in feedlot cattle, *J Food Prot*, 72, 241–50.

SAEZ, A. C., ZHANG, J., ROSTAGNO, M. H. and EBNER, P. D. (2011), Direct feeding of microencapsulated bacteriophages to reduce *Salmonella* colonization in pigs, *Foodborne Pathog Dis*, 8, 1269–74.

SAKAGUCHI, Y., HAYASHI, T., KUROKAWA, K., NAKAYAMA, K., OSHIMA, K., *et al.* (2005), The genome sequence of *Clostridium botulinum* type C neurotoxin-converting phage and the molecular mechanisms of unstable lysogeny, *Proc Natl Acad Sci USA*, 102, 17472–7.

SARKIS, G. J., JACOBS, W. R., JR and HATFULL, G. F. (1995), L5 luciferase reporter mycobacteriophages: a sensitive tool for the detection and assay of live mycobacteria, *Mol Microbiol*, 15, 1055–67.

SAUVAGEAU, D. and COOPER, D. G. (2010), Two-stage, self-cycling process for the production of bacteriophages, *Microb Cell Fact*, 9, 81.

SCHMELCHER, M., SHABAROVA, T., EUGSTER, M., EICHENSEHER, F., TCHANG, V. S. *et al.* (2010), Rapid multiplex detection and differentiation of *Listeria* cells using fluorescent phage endolysin cell wall binding domains, *Appl Environ Microbiol*, 76, 5745–56.

SCHMELCHER, M., TCHANG, V. S. and LOESSNER, M. J. (2011), Domain shuffling and module engineering of *Listeria* phage endolysins for enhanced lytic activity and binding affinity, *Microb Biotechnol*, 4, 651–62.

SCHOFIELD, D. A. and WESTWATER, C. (2009), Phage-mediated bioluminescent detection of *Bacillus anthracis*, *J Appl Microbiol*, 107, 1468–78.

SCHOFIELD, D. A., MOLINEUX, I. J. and WESTWATER, C. (2009), Diagnostic bioluminescent phage for detection of *Yersinia pestis*, *J Clin Microbiol*, 47, 3887–94.

SCHOFIELD, D. A., MOLINEUX, I. J. and WESTWATER, C. (2011), 'Bioluminescent' reporter phage for the detection of category A bacterial pathogens, *J Vis Exp*, doi: 10.3791/2740.

SEKULOVIC, O., MEESSEN-PINARD, M. and FORTIER, L. C. (2011), Prophage-stimulated toxin production in *Clostridium difficile* NAP1/027 lysogens, *J Bacteriol*, 193, 2726–34.

SINGH, A., GLASS, N., TOLBA, M., BROVKO, L., GRIFFITHS, M. *et al.* (2009), Immobilization of bacteriophages on gold surfaces for the specific capture of pathogens, *Biosens Bioelectron*, 24, 3645–51.

SMARTT, A. E. and RIPP, S. (2011), Bacteriophage reporter technology for sensing and detecting microbial targets, *Anal Bioanal Chem*, 400, 991–1007.

SONI, K. A. and NANNAPANENI, R. (2010), Removal of *Listeria monocytogenes* biofilms with bacteriophage P100, *J Food Prot*, 73, 1519–24.

STANEK, J. E. and FALKINHAM, J. O., III (2001), Rapid coliphage detection assay, *J Virol Methods*, 91, 93–8.

STANFORD, K., MCALLISTER, T. A., NIU, Y. D., STEPHENS, T. P., MAZZOCCO, A. *et al.* (2010), Oral delivery systems for encapsulated bacteriophages targeted at *Escherichia coli* O157:H7 in feedlot cattle, *J Food Prot*, 73, 1304–12.

STANLEY, E. C., MOLE, R. J., SMITH, R. J., GLENN, S. M., BARER, M. R. *et al.* (2007), Development of a new, combined rapid method using phage and PCR for detection and identification of viable *Mycobacterium paratuberculosis* bacteria within 48 hours, *Appl Environ Microbiol*, 73, 1851–7.

STEWART, G. S., JASSIM, S. A., DENYER, S. P., NEWBY, P., LINLEY, K. *et al.* (1998), The specific and sensitive detection of bacterial pathogens within 4 h using bacteriophage amplification, *J Appl Microbiol*, 84, 777–83.

SUMMERS, W. (2005), Bacteriophage research: early history, In: Kutter, E. and SULAKVELIDZE, A. (eds.) *Bacteriophages: Biology and Applications*, Boca Raton, FL, CRC Press.

TORMO-MAS, M. A., MIR, I., SHRESTHA, A., TALLENT, S. M., CAMPOY, S. *et al.* (2010), Moonlighting bacteriophage proteins derepress staphylococcal pathogenicity islands, *Nature*, 465, 779–82.

UESUGI, A. R., DANYLUK, M. D., MANDRELL, R. E. and HARRIS, L. J. (2007), Isolation of *Salmonella* Enteritidis phage type 30 from a single almond orchard over a 5-year period, *J Food Prot*, 70, 1784–9.

VIAZIS, S., AKHTAR, M., FEIRTAG, J. and DIEZ-GONZALEZ, F. (2011), Reduction of *Escherichia coli* O157:H7 viability on hard surfaces by treatment with a bacteriophage mixture, *Int J Food Microbiol*, 145, 37–42.

WAGENAAR, J. A., VAN BERGEN, M. A., MUELLER, M. A., WASSENAAR, T. M. and CARLTON, R. M. (2005), Phage therapy reduces *Campylobacter jejuni* colonization in broilers, *Vet Microbiol*, 109, 275–83.

WALDOR, M. K. and FRIEDMAN, D. I. (2005), Phage regulatory circuits and virulence gene expression, *Curr Opin Microbiol*, 8, 459–65.

WASEH, S., HANIFI-MOGHADDAM, P., COLEMAN, R., MASOTTI, M., RYAN, S. *et al.* (2010), Orally administered P22 phage tailspike protein reduces *Salmonella* colonization in chickens: prospects of a novel therapy against bacterial infections, *PLoS One*, 5, e13904.

WILLFORD, J. and GOODRIDGE, L. D. (2008), An integrated assay for rapid detection of *Escherichia coli* O157:H7 on beef samples, *Food Prot Trends*, 28, 468–72.

WOLBER, P. K. and GREEN, R. L. (1990), Detection of bacteria by transduction of ice nucleation genes, *Trends Biotechnol*, 8, 276–9.

YE, J., KOSTRZYNSKA, M., DUNFIELD, K. and WARRINER, K. (2010), Control of *Salmonella* on sprouting mung bean and alfalfa seeds by using a biocontrol preparation based on antagonistic bacteria and lytic bacteriophages, *J Food Prot*, 73, 9–17.

YOUNG, R. (2002), Bacteriophage holins: deadly diversity, *J Mol Microbiol Biotechnol*, 4, 21–36.

ZOU, Q. H., LI, Q. H., ZHU, H. Y., FENG, Y., LI, Y. G. *et al.* (2010), SPC-P1: a pathogenicity-associated prophage of *Salmonella* Paratyphi C, *BMC Genomics*, 11, 729.

17

New research on modified-atmosphere packaging and pathogen behaviour

A. Vermeulen, P. Ragaert, A. Rajkovic, S. Samapundo, F. Lopez-Galvez and F. Devlieghere, Ghent University, Belgium

DOI: 10.1533/9780857098740.4.340

Abstract: Consumers are switching to fresh, minimally processed foods, creating challenges in terms of ensuring food safety. The shift in food production from local to global has led to a complex logistics chain. These trends and challenges have lead to the development of packaging materials with better barrier properties, and active and intelligent packaging. A recent trend is the increasing sustainability of food packaging. Modified atmosphere or vacuum packaging gives a longer shelf life by reducing the growth of spoilage microorganisms and/or oxidation processes. This chapter focuses on modified-atmosphere packaging (MAP). The effects of high and low O_2, elevated CO_2 concentrations and equilibrium modified-atmosphere packaging (EMAP) are considered. The influence on food infectants, toxin-producing bacteria and mycotoxins is discussed. Recent studies on MAP have had contradictory results, mostly owing to differences in experimental design and materials.

Key words: modified-atmosphere packaging, oxygen, carbon dioxide, pathogens, toxins.

17.1 Introduction

Modified-atmosphere packaging (MAP), which includes vacuum packaging, is a well-known technique used to prolong the shelf life of perishable food products. By eliminating or replacing the air surrounding a packed food product, various deteriorative processes that can occur in food can be restricted. As it is important to maintain the modified atmosphere in the packaging during the shelf life of the product, there has been an increased interest in the production of different packaging concepts. The first section of this chapter deals with the latest developments and trends with regard to packaging materials. Special attention is paid to the various barrier materials, sustainable packaging materials (renewable or biobased), as well as active and intelligent packaging. In the following section the influence of different packaging properties on the microbial safety of packed food products is discussed. A distinction is made between (i) high and low O_2 MAP, (ii) the influence of elevated CO_2 levels in MAP and (iii) the use of equilibrium MAP (EMAP) for the packaging of fruit and vegetables. Finally, a critical evaluation of current research on MAP and the determination of food safety is discussed.

17.2 Trends in packaging configuration

Food packaging is a fast-moving and innovative segment of the food sector, and is influenced by various drivers in today's society (Fig. 17.1). In the past, packaging materials have had one primary function, namely the containment of food. In recent years, however, packaging has evolved to make use of materials that have different protective functions and that offer high levels of convenience to consumers using innovative packaging designs. Besides this higher level of functionality, there is also an increased awareness of sustainability in the food-packaging market. This makes research into food packaging a very multidisciplinary field, as illustrated in Fig. 17.1.

The use of vacuum packaging and MAP in the food industry has resulted in the increased use of different packaging materials with appropriate gas and water barrier properties. Excellent barrier properties can be achieved using multilayer packaging materials containing aluminium, EVOH (ethylene vinyl alcohol) or PVdC (polyvinylidene chloride) as a barrier material, as well as coatings such as SiO_X, AlO_x or carbon (Robertson, 2006; Galić *et al.*, 2009; Yam, 2009). Apart from aluminium-containing materials of sufficient thickness, the most commonly

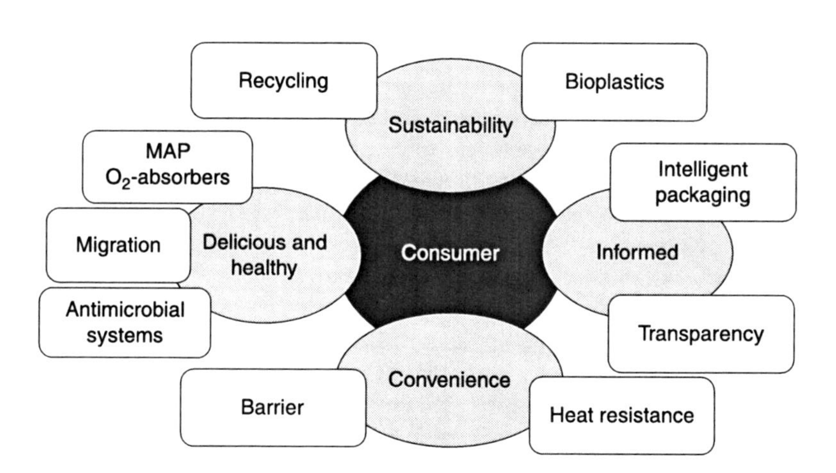

Fig. 17.1 Drivers in food packaging research Pack 4 Food 2011.

used barrier materials in food packaging still have a certain level of gas and water permeability, which allows O_2 to enter the package (Massey, 2003). This eventually leads to a certain amount of O_2 ingress, which can influence the microbial as well as the chemical characteristics of the food product. Furthermore, the filling system used during the packaging process influences the amount of residual O_2 in the headspace after packaging. When flushing a package with a gas combination without a preceding vacuum step, as in the case of many flowpack applications, a higher residual O_2 concentration will usually be present in the headspace. The speed at which the filling system runs can also influence the residual O_2 concentration. Taking the above factors into account, industrial packaging applications with MAP result in many cases in a residual O_2 concentration between 0.5% and 2%, although higher levels can sometimes be present (Galić *et al.*, 2009; Singh *et al.*, 2011b).

New trends in packaging materials focus on two main topics: sustainability and functionality. Regarding the former, increasing interest in lightweight packaging materials has resulted in research on the use of thinner barrier materials, such as different types of plasma coatings (Lee, 2010). Inks, sleeves or injection-moulded labels with integrated barrier functions are also currently important topics in packaging research. An increased focus on recyclability has led to research on the shift from multilayer to monolayer materials whilst not having a significant effect on the gas and/or water barrier properties. As well as research into various coatings, the addition of nanoparticles (e.g. nanoclay) to packaging materials is also being developed, with the aim of increasing the gas and water barrier properties of monolayer materials (Arora and Padua, 2010; Lee, 2010; Duncan, 2011). Another research field that will further expand in the coming years is the use of bioplastics for MAP or vacuum applications. Bioplastics are materials that are compostable and/or originate from renewable resources (Siracusa *et al.*, 2008; Nampoothiri *et al.*, 2010). Materials such as cellulose and starch, possessing good intrinsic gas barrier properties (Dole *et al.*, 2004; Minelli *et al.*, 2010), show promising results as gas barrier materials in multilayer bioplastics. Other bioplastics with poorer barrier properties, such as polylactide (PLA), could be modified by the laminating, blending or coating techniques used nowadays for conventional plastics. One of the main research areas is the use of bioplastic nanocomposites, where bioplastics such as PLA or starch are combined with nanoclays such as montmorillonite or kaolinite (Arora and Padua, 2010). Furthermore, interesting results have come from research into polyhydroxyalkanoate (PHA) biopolymers directly produced by microorganisms (Chen, 2009). It is also to be expected that bioplastics such as bio-polyethylene or bio-polyethylene terephthalate – i.e. plastics already used in many MAP applications and (partly) originating from renewable resources – will be increasingly used as packaging materials.

Regarding research into the increased functionality of packaging materials, the combination of gas and water barrier properties with elevated heat resistance is an important research topic (Belcher, 2006). This is due to the higher demand for in-pack pasteurization or sterilization, where higher oxygen ingress through the packaging material takes place due to the elevated temperatures, especially for plastic packaging materials. It is to be expected that the above mentioned topics will be major areas of study for many years to come. The primary aims of these research topics will also be to improve the heat resistance of different types of materials (e.g. recycled materials and bioplastics), as well as to study the effect of heat treatment on the migration of packaging material components into food products, and to minimize additional oxygen ingress during heat treatment.

One of the research areas related to the latter aim is the incorporation of oxygen-scavenging components in packaging materials. This is a type of active packaging, which interacts with the product to prolong shelf life, enhance product safety and maintain the quality of the product (Restuccia *et al.*, 2010; Singh, 2011a). In the case of O_2 scavengers, this can imply the removal of residual O_2 after the package has been closed and/or avoid O_2 entering the package from the outside atmosphere. In the early days of active packaging, sachets were used to scavenge O_2 or absorb moisture. More recently, research has focused on the incorporation of these active components into the packaging material itself (Ozdemir and Floros, 2004; Xiao-e *et al.*, 2004; Lee, 2010). Active packaging systems can consist of either scavenging (e.g. O_2, moisture or ethylene) or releasing (e.g. CO_2, antimicrobials or antioxidants) systems (Kerry *et al.*, 2006; Galić *et al.*, 2009). Another new trend in scavenging is the use of biological systems (e.g. enzymes or microorganisms) instead of chemical ones (Altieri *et al.*, 2004; Fernández *et al.*, 2008; Anthierens *et al.*, 2011). Commercial O_2 scavengers are based on the oxidation of iron powder, ascorbic acid, photosensitive dyes, enzymes, rice extracts or yeasts, with an iron-based scavenger being most frequently applied. CO_2 emitters are mostly based on the reaction between bicarbonate and water vapour.

Related to this concept, there is increased interest in detecting residual O_2 concentrations in food packaging, which has resulted in research into O_2 sensors, mostly based on components that change colour depending on the absence or presence of O_2 in the headspace of the packaging (Mills, 2005; Roberts *et al.*, 2011). This type of technology is called an intelligent packaging system, which can be defined as packaging that responds to changes in the packed product and/ or storage conditions. It can also include communicating the status of the product to the consumer or end user. Common examples are time-temperature indicators, O_2 sensors, ripeness indicators and radio frequency identification (RFID). Intelligent packaging systems also include biosensors indicating food safety, e.g. by detecting toxins produced by *Staphylococcus aureus* or the growth of *Escherichia coli*, *Salmonella* and *Listeria monocytogenes* (Liu *et al.*, 2007).

17.3 Effect of packaging configuration on food safety

MAP is a widely used preservation technology for many different food products, as it can extend shelf life without modification of the food product itself. Moreover, trends such as salt, sugar and fat reduction, in combination with the desire to use less additives in food products and only mild heat treatment during food production, has led to an increased use of MAP to prolong the shelf stability and safety of food products. Oxygen, carbon dioxide and nitrogen are the most common gases used in MAP, whilst ozone, carbon monoxide, and argon are also used in some countries. The change in packaging atmosphere can be made actively during the packaging of the food or passively for respiring food products (e.g. fresh fruits and vegetables) by using packaging materials with selective permeability. In the former case, a high-barrier packaging material should be used to avoid the exchange of gasses between the internal atmosphere and the surrounding atmosphere of the package (usually air). Although the primary incentive to use MAP for the food industry is the extension of the shelf life of the products (with regard to microbial, visual and sensorial aspects), the use of MAP also has major consequences on the food safety of products with an extended shelf life.

In this section the effects of packaging configuration on the growth of pathogens is subdivided into: (i) the effect of high and low O_2 MAP, (ii) the influence of different CO_2 concentrations and the effect of storage temperature and (iii) equilibrium MAP (EMAP) and controlled atmospheres for fresh produce.

17.3.1 High and low oxygen

High oxygen MAP is particularly effective in inhibiting enzymatic discolouration, so its primary use is to extend the shelf life of fresh produce (e.g. red meat, vegetables and fruit). It is, however, essential to know whether the shelf life extension has also introduced an increased microbial safety risk. Brooks *et al.* (2008) demonstrated that the use of high O_2 MAP (70% O_2, 30% CO_2) for meat resulted in a lower pathogen load (of *Salmonella* and *E. coli* O157) after extended storage compared with air-packed meat in a permeable overwrap foil. For fruits, the same conclusion can be drawn from the study of Siro *et al.* (2006), who proved that the prolonged shelf life of fruits (strawberries and raspberries) using high O_2 did not result in an increased microbial risk from the growth of *E. coli, L. monocytogenes* or *Salmonella*.

The fact that high O_2 MAP is not fully commercialized is mainly due to the lack of information on other important food characteristics, such as the influence of high O_2 on nutritional quality. Furthermore, this technique has higher investment costs because of the precautions that have to be taken, as O_2 levels above 25% are considered to be explosive.

MAP without O_2 is a much more frequently used packaging technique for many food products with an extended shelf life, such as ready-to-eat (RTE) foods, processed meat, cheese, etc. It should be noted that the goal is to achieve an O_2 concentration as low as possible, but that depending on the filling technique the amount of residual O_2 in the headspace can be enough to allow the growth of pathogens. The spore-forming toxin producers, *Bacillus* and *Clostridium* species, behave differently. The *Bacillus* genus includes some strict aerobes, but also facultative anaerobes, which accounts for their ability to grow in food packed in different configurations (MAP, vacuum, active packaging, etc.). Their ability to sporulate makes the *Bacillus* species of most concern in heated, cooked and pasteurized foods because current food production practices rely on low temperatures and MAP for safety and quality during the shelf life. *Bacillus cereus* is the main foodborne pathogen in the genus *Bacillus* and the virulence arsenal of this versatile foodborne pathogen has a number of different toxins. One should look at the effects of MAP on these toxins individually to see whether common conclusions can be drawn, which will influence the use of MAP for foods where *B. cereus* is an important foodborne microbial hazard. Cereulide production is severely impaired by reduced atmospheric oxygen levels (below 2% O_2), suggesting that oxygen is an essential and stimulating factor in cereulide production (Jaaskelainen *et al.*, 2004; Finlay *et al.*, 2002; Rajkovic *et al.*, 2006). Under anaerobic conditions with less than 1–2% O_2, cereulide production does not take place (Rajkovic *et al.*, 2006; Jaaskelainen *et al.*, 2004). *B. weihenstephanensis* strains, belonging to the *B. cereus* group, show similar sensitivity to MAP (Thorsen *et al.*, 2009). In general, it can be concluded that the lower the residual O_2 concentration and the lower the O_2 transfer rate of the packaging materials, the lower the emetic toxin concentrations.

The *Clostridium* genus is strictly anaerobic, which implies that these pathogens can become a risk for MAP products with low (residual) oxygen concentrations. After the (residual) O_2 has been consumed by other microorganisms, respiration of the product or oxidation of food components, *C. botulinum* is able to grow and produce toxins. However, this growth will be very slow and the product may spoil before the toxin is produced. As the results of several studies are not conclusive, additional hurdles must be applied to ensure the safety of products with an extended shelf life, particularly those products where the cold chain is not completely controllable (Sivertsvik *et al.*, 2002).

Next to the growth of and toxin production by bacteria, mould growth and mycotoxin production can also be influenced by the oxygen concentration in MAP. Morales *et al.* (2007) reported that no patulin was detected under low (3% O_2/1% CO_2, rest N_2) and ultralow (1% O_2/1% CO_2, rest N_2) controlled atmosphere (CA) conditions on apples inoculated with *Penicillium expansum* and kept at 1 °C for up to 2.5 months. Patulin was, however, detected after only three days when the apples were stored at 20 °C. Baert *et al.* (2007) reported that the effect of reducing the headspace O_2 level from 20% to 3% on patulin production by *P. expansum* is strain dependent, with either a stimulation or suppression being observed. A further decrease in O_2 level from 3% to 1% resulted in a decrease of patulin production by all strains of *P. expansum* evaluated in that study. With regards to the effect of the headspace O_2 level, it was determined by Samapundo *et al.* (2007b) that the level of O_2 at which the most fumonisin B_1 was produced decreased from 15% to 5% when the water activity (a_w) was decreased from 0.976 to 0.930 for *Fusarium verticillioides*. For *F. proliferatum* the optimum conditions for fumonisin B_1 production shifted from 20% O_2 at a_w 0.976 to 10% O_2 at a_w values of 0.951 and 0.930. Vacuum packaging completely inhibited the growth (and consequently mycotoxin production) of both isolates.

17.3.2 Elevated CO_2 concentration

CO_2 is the most important gas in MAP applications due to its antimicrobial properties. The effectiveness of CO_2 is dependent on the initial CO_2 concentration in the headspace, the storage temperature and the gas/product ratio of the MAP product. The solubility of CO_2 increases with decreasing temperatures, leading to a higher antimicrobial effect. Therefore, most research into MAP with enhanced CO_2 concentrations and food safety focuses on those pathogens that are able to multiply at chilled temperatures (e.g. *Listeria monocytogenes*). After the packaging is opened, CO_2 is slowly released from the product and exerts a preservative effect for a certain period of time (Stammen *et al.*, 1990).

In general, packaging under elevated CO_2 concentrations at chilled temperatures will inhibit the growth of most pathogens, but a modified atmosphere and chilled storage alone is, in some cases, not enough to control the growth of all pathogens. *L. monocytogenes*, for example, will not be affected by elevated CO_2 concentration if there is still >5% O_2 in the package. *Campylobacter jejuni* has been shown to survive better in the presence of higher CO_2 levels, with growth being only weakly inhibited. Recent results from Rajkovic *et al.* (2010) show that the survival of inoculated and naturally present *C. jejuni* on chicken legs was hardly impaired by 80% CO_2, in contrast to the effect of 80% O_2. The opposite was found for the growth of background microflora on the same chicken legs. The inhibition of the background flora by the high CO_2 MAP may also have aided the survival of *C. jejuni*. Other pathogens, such as *E. coli, Salmonella, Staphylococcus aureus, Vibrio cholerae, V. parahaemolyticus* and *Yersinia enterocolitica*, have been reported to be inhibited by enhanced CO_2 concentrations (Sivertsvik *et al.*, 2002). MAP, particularly with high CO_2 content, is most often combined with a cold chain for foods where *B. cereus* control is necessary. In general, *Bacillus* spp. are sensitive to CO_2, but the effect of MAP on the germination and growth of toxin-producing psychrotolerant *Bacillus* spp. is not well understood. Some of the latest results have quantified the effect of MAP with 20% CO_2, concluding that CO_2 MAP can be used to inhibit growth of a psychrotolerant toxin-producing *Bacillus* spp. during chill storage at 8 °C, and can substantially reduce the risk of emetic food poisoning at abuse temperature conditions (Thorsen *et al.*, 2009).

C. botulinum comprises a group of four distinct spore-forming bacteria, which share the common feature of producing a botulinum neurotoxin (BoNT). MAP with a high level of CO_2 can increase the shelf life of a product to such an extent that there is more time for growth and toxin production by *C. botulinum* to occur. These problems might be especially pronounced for individual retail packages where temperature abuse is more likely than at the supplier level. However, the fears for the growth and toxin formation of *C. botulinum* in refrigerated foods packed in modified atmospheres seem to be exaggerated. Among the reasons are the gas distribution in the immediate vicinity of *C. botulinum* (surface vs. inner part of the product), and the fact that *C. botulinum* is not alone in RTE foods, but has to compete with a spoilage flora. Moreover, even though increased partial pressures of carbon dioxide have been shown to stimulate the germination of clostridia spores, a modified atmosphere of 100% CO_2 suppressed toxin production by *C. botulinum*, in comparison with 100% N_2.

In MAP with elevated CO_2 concentrations, mould growth and mycotoxin production are also important. Moulds are facultative aerobes, which are highly sensitive to CO_2 (Smith *et al.*, 1990; Farber, 1991), implying there is considerable potential for modified atmospheres (MAs) and controlled atmospheres to replace the conventional use of chemical agents to protect (bulk) stored products from fungal growth and mycotoxin production. Despite this potential, to date most studies have focused on the ability of MAs to control insect pests in bulk-stored products (Emecki *et al.*, 2002). The importance of such studies should not be underestimated, as the success of MAs and CAs in eradicating insect pests indirectly reduces the possibility of fungal (and consequently mycotoxin) contamination, as most insect pests are fungal vectors. Taniwaki *et al.* (2009) investigated the effect of high CO_2 and low O_2 atmospheres on the amount of aflatoxin, patulin and roquefortine produced by *Aspergillus flavus, Byssochlamys nivea* and *Penicillium roqueforti*, respectively, on artificial media. This study demonstrated that atmospheres with 20%, 40% and 60% CO_2 combined with less than 0.5% O_2 greatly reduced the production of these mycotoxins. In a similar study, Taniwaki *et al.* (2010) observed that despite extensive mycelial growth, *A. flavus* did not produce aflatoxins in an atmosphere with 80% CO_2 and 20% O_2, whilst patulin, roquefortine C and cyclopiazonic acid production by *B. nivea, P. roqueforti* and *Penicillium commune*, respectively, was very low.

Samapundo *et al.* (2007a, 2007b) investigated the effect of CO_2 and O_2 levels on the production of fumonisins by *Fusarium* spp. on yellow dent corn. In Samapundo *et al.* (2007a), it was determined that as little as 10% CO_2 completely inhibited fumonisin B_1 production by *F. verticillioides*, whilst *F. proliferatum* was more resistant and required 40%, 30% and 10% CO_2 at a_w values of 0.984, 0.951 and 0.930, respectively, to completely inhibit fumonisin B_1 production.

Interestingly, the effects of CO_2 on mould growth reported in most of the studies discussed above have been at high storage temperatures (22–30 °C), temperatures at which CO_2 shows little or no activity against bacteria. These observed effects are additional to the effect of the removal of O_2, and could suggest that moulds are far more susceptible to the presence of CO_2 than most bacteria.

17.3.3 Equilibrium modified-atmosphere packaging (EMAP)

Equilibrium modified-atmosphere packaging (EMAP) is a successful technique for preserving minimally processed fresh-cut fruits and vegetables. In current industry practice, packaging films are selected to obtain EMAPs with a low O_2 concentration and low-to-medium CO_2 concentrations. These storage atmospheres lower the metabolic rate of the product, inhibit discoloration and prevent the growth of spoilage microorganisms. As a result, a longer shelf life for the product can be obtained (Allende *et al.*, 2006; Rico *et al.*, 2007). However, it has been suggested that EMAP could increase the health risk associated with these food products, by providing microbial pathogens with a higher chance to grow because of the resultant longer shelf life and reduced competition by the inhibited background microbial flora (Allende *et al.*, 2006; Delaquis *et al.*, 2007; Luo *et al.*, 2010; Rico *et al.*, 2007). The most relevant pathogens associated with these food products (vero-toxigenic *Escherichia coli, Salmonella, Listeria monocytogenes* and noroviruses) are able to survive and, in the case of bacteria, grow, in some cases, under the storage conditions that are usually applied by the fresh-cut produce industry (EMAP combined with refrigeration temperatures) (Francis and O'Beirne, 2006; Tomás-Callejas *et al.*, 2011). However, in spite of the scientific evidence gathered during the last two decades, the effect of EMAP on fresh-cut produce safety (that is, promoting or preventing pathogen survival and growth) remains unclear.

Nevertheless, it is generally acknowledged that the effect of commercially applied storage atmospheres on the survival and growth of pathogens is not important in comparison with the effect of temperature (Delaquis *et al.*, 2007). As a consequence, only time-temperature conditions are taken into account when modelling pathogen growth and survival and performing microbial risk assessments for fresh-cut produce (Koseki and Isobe, 2005; Crepet *et al.*, 2009; Rijgersberg *et al.*, 2010; Tromp *et al.*, 2010; Danyluk and Schaffner, 2011; McKellar and Delaquis, 2011). Furthermore, the outcome of recent studies suggests that the interaction with the background microbial flora could be less relevant than previously assumed (Oliveira *et al.*, 2011; Scifo *et al.*, 2009), and that therefore the inhibition of the indigenous microorganisms by EMAP would not be a very important concern from a safety point of view.

17.4 Critical evaluation of current research on modified-atmosphere packaging and food safety, and some future research trends

For several pathogens, the research into the use of MAP for food safety has produced contradictory results. This could be due to differences in: (i) the filling and the packaging technique, which can influence the residual oxygen levels, (ii) the permeability of the packaging material, (iii) laboratory media versus real food products, (iv) the strain used, (v) the physiological state of the cells, (vi) the parameter evaluated (growth, virulence or toxin production), (vii) the method of analysis or (viii) the inoculation level of the pathogen. In this section, some examples of these discrepancies within the scientific literature will be discussed in depth for various pathogens.

The contradictory effects of oxygen on cereulide production are due to the variable influence of agitation, and thus aeration and oxygen concentration. Somewhat different results have been observed in relation to *B. cereus* enterotoxins. Anaerobic atmospheres result in slower bacterial growth but increased enterotoxin production, because expression of the *nhe* and *hbl* genes, encoding non-hemolytic enterotoxin and hemolysin BL, respectively, is down regulated by atmospheric oxygen (Duport *et al.*, 2006, 2004; van der Voort and Abee, 2009). Maximal enterotoxin production takes place during the exponential phase under anaerobic conditions, but in comparison with aerobiosis it is reported as unaltered (Zigha *et al.*, 2006; Fermanian *et al.*, 1996), delayed (Duport *et al.*, 2004) or accelerated (Van Der Voort and Abee, 2009; Gilois *et al.*, 2007). These discrepancies in the effect of oxygen on toxin expression may be to an extent attributed to the method of mRNA quantification used in different studies.

Similarly, for *C. botulinum* some discrepancies can be found in the literature. Studies have reported that the growth of nonproteolytic *C. botulinum* is slowed by high concentrations of CO_2 (Artin *et al.*, 2008, 2010). This delay results from both an increased lag time and a decreased growth rate, although the reported differences in the maximum specific growth rate (μ_{max}) between the test concentrations of 10% and 35% CO_2 seem not to be statistically and biologically significant, in contrast to the differences between 10% and 35–70% CO_2. For the lag phase, the difference between 10% and 30% was significant, but did not change any more with an increase of CO_2 to 70%. However, from a food safety perspective, it is the production of neurotoxins, not cell growth, which is the important parameter. It was shown by the same authors (Artin *et al.*, 2008) that neurotoxin formation by nonproteolytic *C. botulinum* type E is greater at a higher concentration of CO_2, increasing to 140% and 215% of the production at 10% CO_2 when CO_2 concentration increased to 35% and 70%, respectively. However, it is important to note that these experiments were performed at 30 °C where the inhibitory effect of CO_2 is seriously diminished and at which no refrigerated foods would be stored, and therefore these results require careful interpretation. In the study of Artin *et al.* (2008), both relative gene expression and the amount of neurotoxin produced after 30 h were twofold higher when the headspace CO_2 concentration was 70% than when the CO_2 was 10%. Similar results were observed by Lövenklev *et al.* (2004) with nonproteolytic *C. botulinum* type B.

In RTE foods where MAP follows mild processing, cell injury can occur. In the case of *C. botulinum* and *B. cereus* this mainly refers to the spore injury. Typical symptoms of spore injury include increased germination lag, altered nutritional requirements, sensitivity to recovery on different batches of media, altered optimum temperatures for recovery and sensitivity to sodium chloride and other food preservatives and antibiotics. With MAP, spore recovery is markedly influenced by gas atmosphere and redox potential. The variation in microlag times within populations of spores would explain why the time needed for toxin production becomes much more variable as inoculum size decreases.

For fresh-cut produce, it is important to consider that the information available on the microbiological risk was obtained mainly by determining the counts of pathogens detected by classical culture-based methods. These methods, according to recent studies, could provide a misleading estimate of the microbial population, including pathogens (Scifo *et al.*, 2009). Another trend in fresh-cut produce safety research is the assessment of other aspects besides the numbers of pathogens detectable in the food samples. For example there are studies assessing the effect of EMAP not only on the levels but also on the virulence of the pathogens. Recent studies show that *Salmonella* keeps its virulence in EMAP-packaged lettuce after five days (Oliveira *et al.*, 2011), that the combination of temperature abuse with air storage increased the expression of virulence factors in *E. coli* O157:H7 compared with EMAP storage (Sharma *et al.*, 2011), and that temperature abuse can hamper EMAP and provoke acid resistance in enterohemorrhagic *E. coli*, which could allow the pathogen to overcome the gastric acid barrier (Chua *et al.*, 2008). Additionally, a few recent studies have provided insights into the effect of pre-harvest events on the fitness of pathogens along the food chain (Chua *et al.*, 2008; Tomás-Callejas, 2011).

In the next few years it is expected that integrated models combining the effect of storage atmosphere with that of temperature on the growth and survival of pathogens will become available. Furthermore, more information on the effects of EMAP on virulence, and studies evaluating the effect of pre-harvest conditions on the behaviour of pathogens in post-harvest steps will appear. Additionally, molecular-based detection methods will give us better information on the evolution along the food chain of levels of the pathogens associated with fresh-cut produce (Kisluk *et al.*, 2012). All of these outcomes will provide a more complete knowledge of the effects of EMAP on the safety of fresh-cut produce, and they will improve risk assessments and the selection of safety measures with the aim of mitigating health risks.

17.5 References

ALLENDE, A., TOMAS-BARBERAN, F.A., GIL, M.I. 2006. Minimal processing for healthy traditional foods. *Trends in Food Science and Technology*, 17, 513–519.

ALTIERI, C., SINIGAGLIA, M., CORBO, M.R., BUONOCORE, G.G., FALCONE, P. *et al.* 2004. Use of entrapped microorganisms as biological oxygen scavengers in food packaging applications. *Lebensmittel Wissenschaft und Technologie*, 37, 9–15.

ANTHIERENS, T., RAGAERT, P., VERBRUGGHE, S., OUCHCHEN, A., DE GEEST, B.G. *et al.* 2011. Use of endospore-forming bacteria as an active oxygen scavenger in plastic packaging materials. *Innovative Food Science and Emerging Technologies*, 12, 594–599.

ARORA, A., PADUA, G.W. 2010. Review: nanocomposites in food packaging. *Journal of Food Science*, 75, 1, 43–49.

ARTIN, I., CARTER, A.T., HOLST, E., LOVENKLEV, M., MASON, D.R. *et al.* 2008. Effects of carbon dioxide on neurotoxin gene expression in nonproteolytic *Clostridium botulinum* type E. *Applied and Environmental Microbiology*, 74, 2391–2397.

ARTIN, I., MASON, D.R., PIN, C., SCHELIN, J., PECK, M.W. *et al.* 2010. Effects of carbon dioxide on growth of proteolytic *Clostridium botulinum*, its ability to produce neurotoxin, and its transcriptome. *Applied and Environmental Microbiology*, 76, 1168–1172.

BAERT, K., DEVLIEGHERE, F., FLYPS, H., OOSTERLINCK, M., AHMED, M.M. *et al.* 2007. Influence of storage conditions of apples on growth and patulin production by *Penicillium expansum*. *International Journal of Food Microbiology*, 119, 170–181.

BELCHER, J.N. 2006. Industrial packaging developments for the global meat market. *Meat Science*, 74, 143–148.

BROOKS, J.C., ALVARADO, M., STEPHENS, T.P., KELLERMEIER, J.D., TITTOR, A.W. *et al.* 2008. Spoilage and safety characteristics of ground beef packaged in traditional and modified-atmosphere packages. *Journal of Food Protection*, 71, 293–301.

CHEN, G.Q. 2009. A microbial polyhydroxyalkanoates (PHA) based bio- and materials industry. *Chemical Society Reviews*, 38, 2434–2446.

CHUA, D., GOH, K., SAFTNER, R.A., BHAGWAT, A.A. 2008. Fresh-cut lettuce in modified atmosphere packages stored at improper temperatures supports enterohemorrhagic *E. coli* isolates to survive gastric acid challenge. *Journal of Food Science*, 73, 148–153.

CREPET, A., STAHL, V., CARLIN, F. 2009. Development of a hierarchical Bayesian model to estimate the growth parameters of *Listeria monocytogenes* in minimally processed fresh leafy salads. *International Journal of Food Microbiology*, 131, 112–119.

DANYLUK, M.D., SCHAFFNER, D.W. 2011. Quantitative assessment of the microbial risk of leafy greens from farm to consumption: preliminary framework, data, and risk estimates. *Journal of Food Protection*, 74, 700–708.

DELAQUIS, P., BACH, S., DINU, L.-D. 2007. Behavior of *Escherichia coli* O157:H7 in leafy vegetables. *Journal of Food Protection*, 70, 1966–1974.

DOLE, P., JOLY, C., ESPUCHE, E., ALRIC, I., GONTARD, N. 2004. Gas transport properties of starch based films. *Carbohydrate Polymers*, 58, 335–343.

DUNCAN, T.V. 2011. Applications of nanotechnology in food packaging and food safety: barrier materials, antimicrobials and sensors. *Journal of Colloid and Interface Science*, 363, 1–24.

DUPORT, C., THOMASSIN, S., BOUREL, G., SCHMITT, P. 2004. Anaerobiosis and low specific growth rates enhance hemolysin BL production by *Bacillus cereus* F4430/73. *Archives of Microbiology*, 182, 90–95.

DUPORT, C., ZIGHA, A., ROSENFELD, E., SCHMITT, P. 2006. Control of enterotoxin gene expression in *Bacillus cereus* F4430/73 involves the redox-sensitive ResDE signal transduction system. *Journal of Bacteriology*, 188, 6640–6651.

EMECKI, M., NAVARRO, S., DONAHAYE, E., RINDNER, M., AZRIELI, A., 2002. Respiration of *Tribolium castaneum* (Herbst) at reduced oxygen concentrations. *Journal of Stored Products Research*, 38, 413–425.

FARBER, J.M., 1991. Microbiological aspect of modified atmosphere packaging technology. A review. *Journal of Food Protection*, 54, 58–70.

FERMANIAN, C., LAPEYRE, C., FREMY, J.M., CLAISSE, M. 1996. Production of diarrheal toxin by selected strains of *Bacillus cereus*. *International Journal of Food Microbiology*, 30, 345–358.

FERNÁNDEZ, A., CAVA, D., OCIO, M.J., LAGARÓN, J.M. 2008. Perspectives for biocatalysts in food packaging. *Trends in Food Science and Technology*, 19, 198–206.

FINLAY, W.J.J., LOGAN, N.A., SUTHERLAND, A.D. 2002. *Bacillus cereus* emetic toxin production in relation to dissolved oxygen tension and sporulation. *Food Microbiology*, 19, 423–430.

FRANCIS, G.A., O'BEIRNE, D. 2006. Isolation and pulsed-field gel electrophoresis typing of *Listeria monocytogenes* from modified atmosphere packaged fresh-cut vegetables collected in Ireland. *Journal of Food Protection*, 69, 2525–2528.

GALIĆ, K., CURIĆ, D., GABRIĆ, D. 2009. Shelf life of packaged bakery goods – a review. *Critical Reviews in Food Science and Nutrition*, 49, 405–426.

GILOIS, N., RAMARAO, N., BOUILLAUT, L., PERCHAT, S., AYMERICH S. *et al.* 2007. Growth-related variations in the *Bacillus cereus* secretome. *Proteomics*, 7, 1719–1728.

JAASKELAINEN, E.L., HAGGBLOM, M.M., ANDERSSON, M.A., SALKINOJA-SALONEN, M.S. 2004. Atmospheric oxygen and other conditions affecting the production of cereulide by *Bacillus cereus* in food. *International Journal of Food Microbiology*, 96, 75–83.

KERRY, J.P., O'GRADY, M.N., HOGAN, S.A. 2006. Past, current and potential utilization of active and intelligent packaging systems for meat and muscle-based products: a review. *Meat Science*, 74, 113–130.

KISLUK, G., HOOVER, D.G., KNEIL, K.E., YARON, S. 2012. Quantification of low and high levels of *Salmonella enterica* serovar Typhimurium on leaves. *LWT – Food Science and Technology*, 45, 36–42.

KOSEKI, S., ISOBE, S. 2005. Prediction of pathogen growth on iceberg lettuce under real temperature history during distribution from farm to table. *International Journal of Food Microbiology*, 104, 239–248.

LEE, K.T. 2010. Quality and safety aspects of meat products as affected by various physical manipulations of packaging materials. *Meat Science*, 86, 138–150.

LIU, Y., CHAKRABARTTY, S., ALOCILJA, E.C. 2007. Fundamental building blocks for molecular biowire based forward error-correcting biosensors. *Nanotechnology*, 18, 1–6.

LÖVENKLEV, M., ARTIN, I., HAGBERG, O., BORCH, E., HOLST, E. *et al.* 2004. Quantitative interaction effects of carbon dioxide, sodium chloride, and sodium nitrite on neurotoxin gene expression in nonproteolytic *Clostridium botulinum* type B. *Applied and Environmental Microbiology*, 70, 2928–2934.

LUO, Y., HE, Q., MCEVOY, J.L. 2010. Effect of storage temperature and duration on the behavior of *Escherichia coli* O157:H7 on packaged fresh-cut salad containing romaine and iceberg lettuce. *Journal of Food Science*, 75, 390–397.

MASSEY, L.K. 2003. *Permeability Properties of Plastics and Elastomers: A Guide to Packaging and Barrier Materials*, second edition. Plastics Design Library.

MCKELLAR, R.C., DELAQUIS, P. 2011. Development of a dynamic growth-death model for *Escherichia coli* O157:H7 in minimally processed leafy green vegetables. *International Journal of Food Microbiology*, 151, 7–14.

MILLS, A. 2005. Oxygen indicators and intelligent inks for packaging food. *Chemical Society Reviews*, 34, 1003–1011.

MINELLI, M., BASCHETTI, M.G., DOGHIERI, F., ANKERFORS, M., LINDSTRÖM, T. *et al.* 2010. Investigation of mass transport properties of microfibrillated cellulose (MFC) films. *Journal of Membrane Science*, 358, 67–75.

MORALES, H., SANCHIS, V., ROVIRA, A., RAMOS, A.J., MARIN, S. 2007. Patulin accumulation in apples during postharvest: effect of controlled atmosphere storage and fungicide treatments. *Food Control*, 18, 1443–1448.

NAMPOOTHIRI, K.M., NAIR, N.R., JOHN, R.P. 2010. An overview of the recent developments in polylactide (PLA) research. *Bioresource Technology*, 101, 8493–8501.

OLIVEIRA, M., WIJNANDS, L., ABADIAS, M., AARTS, H., FRANZ, E. 2011. Pathogenic potential of *Salmonella* Typhimurium DT104 following sequential passage through soil, packaged fresh-cut lettuce and a model gastrointestinal tract. *International Journal of Food Microbiology*, 148, 149–155.

OZDEMIR, M., FLOROS, J.D. 2004. Active food packaging technologies. *Critical Reviews in Food Science and Nutrition*, 44, 185–193.

RAJKOVIC, A., TOMIC, N., SMIGIC, N., UYTTENDAELE, M., RAGAERT, P. *et al.* 2010. *International Journal of Food Microbiology*, 140, 201–206.

RAJKOVIC, A., UYTTENDAELE, M., DELEY, W., VAN SOOM, A., RIJSSELAERE, T. *et al.* 2006. Dynamics of boar semen motility inhibition as a semi-quantitative measurement of *Bacillus cereus* emetic toxin (cereulide). *Journal of Microbiological Methods*, 65, 525–534.

RESTUCCIA, D., SPIZZIRRI, U.G., PARISI, O.I., CIRILLO, G., CURCIO, M. *et al.* 2010. New EU regulation aspects and global market of active and intelligent packaging for food industry applications. *Food Control*, 21, 1425–1435.

RICO, D., MARTIN-DIANA, A.B., BARAT, J.M., BARRY-RYAN, C. 2007. Extending and measuring the quality of fresh-cut fruits and vegetables: a review. *Trends in Food Science and Technology*, 18, 373–386.

RIJGERSBERG, H., TROMP, S., JACXSENS, L., UYTTENDAELE, M. 2010. Modeling logistic performance in quantitative microbial risk assessment. *Risk Analysis*, 30, 20–31.

ROBERTS, L., LINES, R., REDDY, S., HAY, J. 2011. Investigation of polyviologens as oxygen indicators in food packaging. *Sensors and Actuators B*, 152, 63–67.

ROBERTSON, G.L. 2006. *Food Packaging: Principles and Practices*, second edition. Taylor & Francis Group.

SAMAPUNDO, S., DE MEULENAER, B., ATUKWASE, A., DEBEVERE, J., DEVLIEGHERE, F. 2007a. The influence of modified atmospheres and their interaction with water activity on the radial growth and fumonisin B_1 production of *Fusarium verticillioides* and *F. proliferatum* on corn. Part I: The effect of initial headspace carbon dioxide concentration. *International Journal of Food Microbiology*, 114, 160–167.

SAMAPUNDO, S., DE MEULENAER, B., ATUKWASE, A., DEBEVERE, J., DEVLIEGHERE, F. 2007b. The influence of modified atmospheres and their interaction with water activity on the radial growth and fumonisin B_1 production of *Fusarium verticillioides* and *F. proliferatum* on corn. Part II: The effect of initial headspace oxygen concentration. *International Journal of Food Microbiology*, 113, 339–345.

SCIFO, G.O., RANDAZZO, C.L., RESTUCCIA, C., FAVA, G., CAGGIA, C. 2009. *Listeria innocua* growth in fresh cut mixed leafy salads packaged in modified atmosphere. *Food Control*, 20, 611–617.

SHARMA, M., LAKSHMAN, S., FERGUSON, S., INGRAM, D.T., LUO, Y. *et al.* 2011. Effect of modified atmosphere packaging on the persistence and expression of virulence factors of *Escherichia coli* O157:H7 on shredded Iceberg lettuce. *Journal of Food Protection*, 74, 718–726.

SINGH, P., WANI, A.A., SAENGERLAUB, S. 2011a. Active packaging of food products: recent trends. *Nutrition and Food Science*, 41, 249–260.

SINGH, P., WANI, A.A., SAENGERLAUB, S., LANGOWSKI, H.C. 2011b. Understanding critical factors for the quality and shelf life of MAP fresh meat: a review. *Critical reviews in Food Science and Nutrition*, 51, 146–177.

SIRACUSA, V., ROCCULI, P., ROMANI, S., DALLA ROSA, M. 2008. Biodegradable polymers for food packaging: a review. *Trends in Food Science and Technology*, 19, 634–643.

SIRO, I., DEVLIEGHERE, F., JACXSENS, L., UYTTENDAELE, M., DEBEVERE, J. 2006. The microbial safety of strawberry and raspberry fruits packaged in high-oxygen and equilibrium-modified atmospheres compared to air storage. *International Journal of Food Science and Technology*, 41, 93–103.

SIVERTSVIK, M., ROSNES, J.T., BERGSLIEN, H. 2002 Modified atmosphere packaging. In *Minimal Processing Technologies in the Food Industry*. Eds. Ohlsson T., Bengtsson N. Woodhead Publishing Limited.

SMITH, J.P. RANSWAMY, H., SIMPSON, B.K. 1990. Developments in food packaging technology. Part 2. storage aspects. *Trends in Food Science and Technology Today*, 1, 112–117.

STAMMEN, K., GERDES, D., CAPORASO F. 1990. Modified atmosphere packaging of seafood. *Critical Reviews in Food Science and Technology*, 29, 301–331.

TANIWAKI, M.H., HOCKING, A.D., PITT, J.I., FLEET, G.H. 2009. Growth and mycotoxin production by food spoilage fungi under high carbon dioxide and low oxygen atmospheres. *International Journal of Food Microbiology*, 132, 100–108.

TANIWAKI, M.H., HOCKING, A.D., PITT, J.I., FLEET, G.H. 2010. Growth and mycotoxin production by fungi in atmospheres containing 80% carbon dioxide and 20% oxygen. *International Journal of Food Microbiology*, 143, 218–225.

THORSEN L., BUDDE, B.B., KOCH, A.G., KLINGBERG, T.D. 2009. Effect of modified atmosphere and temperature abuse on the growth from spores and cereulide production of *Bacillus weihenstephanensis* in a cooked chilled meat sausage. *International Journal of Food Microbiology*, 130, 172–178.

TOMÁS-CALLEJAS, A., LOPEZ-VELASCO, G., CAMACHO, A.B., ARTES, F., ARTES-HERNANDEZ, F. *et al.* 2011. Survival and distribution of *Escherichia coli* on diverse fresh-cut baby leafy greens under preharvest through postharvest conditions. *International Journal of Food Microbiology*, 151, 216–222.

TROMP, S.O., RIJGERSBERG, H., FRANZ, E. 2010. Quantitative microbial risk assessment for *Escherichia coli* O157:H7, *Salmonella enterica*, and *Listeria monocytogenes* in leafy green vegetables consumed at salad bars, based on modeling supply chain logistics. *Journal of Food Protection*, 73, 1830–1840.

VAN DER VOORT, M., ABEE, T. 2009. Transcriptional regulation of metabolic pathways, alternative respiration and enterotoxin genes in anaerobic growth of *Bacillus cereus* ATCC 14579. *Journal of Applied Microbiology*, 107, 795–804.

XIAO-E, L., GREEN, A.N.M., HAQUE, S.A., MILLS, A., DURRANT, J.R. 2004. Light-driven oxygen scavenging by titania/polymer nanocomposite films. *Journal of Photochemistry and Photobiology A: Chemistry*, 162, 253–259.

YAM, K.L. (ed.) 2009. *The Wiley Encyclopedia of Packaging Technology*. John Wiley & Sons.

ZIGHA, A., ROSENFELD, E., SCHMITT, P., DUPORT, C. 2006. Anaerobic cells of *Bacillus cereus* F4430/73 respond to low oxidoreduction potential by metabolic readjustments and activation of enterotoxin expression. *Archives of Microbiology*, 185, 222–233.

18

New research on organic acids and pathogen behaviour

K. Koutsoumanis, Aristotle University of Thessaloniki, Greece and P. Skandamis, Agricultural University of Athens, Greece

DOI: 10.1533/9780857098740.4.355

Abstract: This chapter discusses the research on the use of organic acids as decontamination agents for foods of animal and plant origin. First, the application of organic acids to meat decontamination (i.e., carcasses and trimmings) is presented followed by an overview of research into the use of organic acids for decontaminating vegetables, fruits and fruit juices. Both single and multiple interventions are discussed. Then, the potential concerns of microbial adaptation to acid treatments are examined and the current legislation status at international level is presented. Key points and practical implications for the industry are discussed in the final section of the chapter.

Key words: organic acids, decontamination, fresh produce, meat, natural preservatives.

18.1 Introduction

A total of 5500 foodborne outbreaks involving 48 964 human cases and 46 deaths were reported in Europe in 2009 (EFSA, 2011b). In the United States, 1034 foodborne outbreaks were reported in 2008 (CDC, 2011a). It is estimated that foodborne agents in the US cause 48 million illnesses annually and 3000 deaths, including 9.4 million (20%) illnesses and 1300 (44%) deaths by 31 known pathogens (CDC, 2011a), with the remaining cases attributed to unknown agents (Scallan *et al.*, 2011a, 2011b). The high economic burden (more than $13 billion per year) of foodborne illness, the large number of quality adjusted life years (61 000) (Batz *et al.*, 2012) and the target incidence of certain foodborne illnesses set for 2020 (CDC, 2011b) stress the need for continuous update of pathogen control interventions that aim to improve the safety of foods. Organic acids can be part of such interventions and also meet consumer demands for natural preservation systems. They are indigenous constituents of many foods and they are widely used as additives in food preservation and as sanitizers for decontamination of food surfaces.

Primarily lactic acid and to a lesser extent acetic acid have been used since the early 1990s as meat (beef, lamb, pork and poultry) decontamination agents in the USA, Canada and Australia. Their contribution to the improvement of meat safety by reducing the prevalence and the number of pathogens is well established (Samelis and Sofos, 2003; Skandamis *et al.*, 2010). Currently, lactic acid is permitted for surface application in solutions up to 5% and at temperatures not exceeding 55°C, or 0.2–0.5% of product formulation, depending on the meat part and the dressing stage of the meat, e.g., it is allowed for heads, tongues, ground beef, cured and comminuted sausages, sub-primal beef cuts, trimmings and carcasses prior to fabrication (FSIS, 2012). It is also allowed for the surface treatment of ready-to-eat (RTE) meat products and beef jerky in a mixed solution with other acids and curing salts of pH 1.0–2.0 for 20–30 s before packaging (FSIS, 2012). It can also be used as a flavour enhancer of pork products at levels not exceeding 0.367%. Finally, acetic acid and citric acid are allowed at levels of 2.5% on carcass surfaces as a processing aid (FSIS, 2012).

Lactic, acetic, citric, tartaric, fumaric, levulinic and peroxyacetic acids (an equilibrium mixture of the peroxy radical, acetic acid and hydrogen peroxide) also have recognized potential as disinfectants in fruits and vegetables, especially those intended for preparation of RTE salads (Sapers, 2001; Gil *et al.*, 2009; Ölmez and Kretzschmar, 2009; Raybaudi-Massilia *et al.*, 2009b). Their industrial use as decontaminants of fresh produce is permitted provided that they function as processing aids (EC, 1988; Gil *et al.*, 2009). Some of them, alone or in combination with other disinfectants, have been included in the formulation of commercial sanitizers, such as Purac (80% lactic acid), Tsunami 100 (peroxyacetic acid), Sanoxol 20 (peroxyacetic acid) and Citrox 14W (citric acid and flavonoids) (Allende *et al.*, 2008). However, they should be applied under conditions that maximize effectiveness in microbial reduction and without adverse effects on sensory or nutrient quality.

The decontamination efficacy of organic acids on both meat and plant (vegetables and fruits) surfaces depends on the application conditions (e.g. temperature and pH), whether rinsing is applied after sanitation, the concentration of acid, the characteristics of the food surface, the strength of bacterial attachment and the internalization of cells in the food (Ölmez and Kretzschmar, 2009; Raybaudi-Massilia *et al.*, 2009b; Skandamis *et al.*, 2010). The following paragraphs give an update on the research relevant to the application of organic acids to foods of animal and plant origin over the last 5–10 years. Advantages over classical decontamination methods, potential limitations (e.g. microbial adaptation potential and sensory impact), the existing legislation framework and application in the context of multiple interventions are discussed.

18.2 Use of organic acids for fresh meat decontamination

18.2.1 Carcasses

The majority of the meat decontamination technologies that have been developed and assessed for efficacy in the last three decades involve application of chemical agents at the carcass level, and have been conducted using meat model systems such as artificially inoculated carcasses, meat cuts and meat tissue samples (Loretz *et al.*, 2011a, 2011b; Smulders and Greer, 1998; Theron and Lues, 2007). The application of organic acids, primarily acetic acid and lactic acid, to decontaminate beef and pig carcasses is widespread in the United States and Canada (Loretz *et al.*, 2011a, 2011b). Chemical solutions, including organic acids, are applied either as a pre-evisceration intervention (after hide removal) or after evisceration prior to chilling, during chilling as well as post-chilling and prior to/after fabrication (Byelashov and Sofos, 2009; Edwards and Fung, 2006; Loretz *et al.*, 2011a, 2011b; Skandamis *et al.*, 2010). The application of organic acid solutions to carcasses can be made either by spraying in spraying cabinets, which is the most frequent application at a commercial level, or by immersion (Loretz *et al.*, 2011a, 2011b; Sofos, 2008).

Decontamination treatments are applied during carcass processing at a stage that maximizes contamination reduction by preventing strong bacterial cell attachment (Sofos and Smith, 1998). According to Cabedo *et al.* (1996), the reduction in *Escherichia coli* through chemical decontamination, including acetic acid, on beef carcass tissue samples decreased as the time between the exposure to contamination and the application of the treatment increased. Furthermore, carcass decontamination applied at early stages (i.e., prior to evisceration) of the slaughtering process may reduce bacterial adherence at subsequent processing stages due to changes in the surface physical characteristics of the treated tissue (Dickson, 1995). Nevertheless, the application of chemical interventions such as organic acids both at early and later stages of carcass processing may be needed if the microbiological safety and quality of meat products are to be ensured. There are no research data that demonstrate a significant difference between spraying organic acids onto meat carcasses and immersion. Furthermore, although there are some research data on the relative decontamination efficacy of each method for poultry carcasses, such findings have not been consistent; depending on the organic acid used and the type of microbial contamination, the decontamination efficacy of spraying has been demonstrated as superior, equivalent or inferior to dipping into organic acid solutions (Okolocha and Ellerbroek, 2005; Sakhare *et al.*, 1999; Sinhamahapatra *et al.*, 2004).

The operational parameters of carcass decontamination treatments that may affect the efficacy of organic acids include the temperature and duration of application, and the coverage and contact time of the carcass surface with the solution (Castillo *et al.*, 2001; Cutter and Siragusa, 1994; Hardin *et al.*, 1995; Kotula and Thelappurate, 1994; Sofos and Smith, 1998; Theron and Lues, 2007). According to Cutter *et al.* (1997), temperatures of 30–70°C had no measurable effect on the decontamination efficacy of an acetic acid solution applied by spray washing beef carcasses. However, as supported by the findings of other investigators, the bactericidal activity of organic acids appears to increase with the temperature of the applied solutions (Anderson and Marshall, 1990b; Anderson *et al.*, 1992a; Greer and Dilts, 1992). Although the application of organic acids at elevated temperatures (50–55°C) is generally more effective than at lower temperatures against bacterial contaminants, various levels of decontamination effectiveness may be exhibited at a given temperature by different organic acids. Hardin *et al.* (1995), for instance, reported that lactic acid (2%, 55°C) reduced levels of *E. coli* O157:H7 inoculated on beef carcass surfaces more than acetic acid applied at the same percentage concentration and temperature. The duration of an organic acid treatment of fresh meat has not been consistently positively associated with the efficacy of the treatment against bacterial contaminants. According to some research data, longer application times of organic acid solutions are associated with a higher decontamination effectiveness, while other findings indicate that organic acids' performance is not affected by the exposure time (Kotula and Thelappurate, 1994; Riedel *et al.*, 2009). The length of time for which carcasses are exposed to an organic acid treatment may be changed by varying the processing chain speed and the size (length) of the application equipment used (Sofos and Smith, 1998). When organic acids are applied via spraying, then the spraying pressure is another parameter that needs to be evaluated and chosen. A potential concern is that high spraying pressures may cause bacterial penetration into treated carcass tissues (Anderson *et al.*, 1992b), while lower pressures may result in bacterial redistribution and spreading on the carcass surface. The characteristics of spraying equipment that may have an effect on the extent of contamination removal and/or redistribution include the type, number, distribution and positions of spraying nozzles, as well as the spraying angle and distance (Sofos and Smith, 1998; Yoder *et al.*, 2010).

18.2.2 Trimmings

Trimmings are destined mainly for the production of ground or minced meat, and are also frequently used in the manufacture of sausages, and are the most important source of microbial contamination of these products (Dorsa *et al.*, 1998a). Although several interventions, including chemical decontamination, have been evaluated and shown to be effective in reducing the microbial contamination of carcasses, research into decontamination treatments specifically targeting meat trimmings is relatively limited. However, due to the extensive handling during fabrication, recontamination of meat cuts and trimmings is very likely. Hence, in addition to the implementation of effective decontamination at the carcass level, the application of antimicrobial intervention strategies for meat trimmings is expected to be useful in further improving the microbiological safety and quality of fresh meat products.

Among the antimicrobial treatments that have been evaluated for their efficacy in reducing bacterial pathogens and extending the shelf life of meat trimmings and ground products, hot water treatments, alone or in conjunction with lactic acid, have been found to be promising (Ellebracht *et al.*, 1999; Gill and Badoni, 1997). Nevertheless, hot water washes of trimmings have been associated with negative effects on quality attributes (i.e., colour and emulsion stability) of ground meat (Castelo *et al.*, 2001). In addition to their direct decontamination activity, organic acid treatments applied to fresh meat (carcasses or parts/tissues) may also enhance the safety and quality of further processed final products. For instance, the decontamination of beef or pork meat with organic acids can result in growth suppression of both pathogenic and spoilage bacteria in ground meat or in sausages prepared from such meat (Castillo *et al.*, 2001; Wan *et al.*, 2007). Therefore, the use of organic acids in the processing of meat (mainly beef) trimmings, as a means to provide additional safety of ground meat products, has attracted significant research interest (Conner *et al.*, 1997; Dorsa *et al.*, 1998a, 1998b; Harris *et al.*, 2006; Kang *et al.*, 2001a; Stivarius *et al.*, 2002a, 2002b). The application of organic acid solutions to trimmings is mostly by spraying, while exposure of meat by tumbling (i.e., in a meat tumbler), either aerobically or under vacuum, has also been described (Stivarius *et al.*, 2002a, 2002b).

Conner *et al.* (1997) reported that spraying of beef trim with acetic-lactic acid mixtures (2% or 4%) was only slightly effective against the pathogens *E. coli* O157:H7 and *Listeria monocytogenes*. According to these investigators, processing (grinding and storage) of trim as well as the comminution of adipose tissue with lean tissue may have accounted for the lack of effectiveness of the acid treatments tested (Conner *et al.*, 1997). Research findings regarding the impact of tissue type on the effectiveness of meat decontamination using organic acids have not been conclusive, with some studies reporting higher bacterial reductions on adipose tissue and others on lean beef tissue (Bell *et al.*, 1997; Cutter and Siragusa, 1994; Cutter *et al.*, 1997; Dickson, 1992; Greer and Dilts, 1995). Furthermore, another parameter that needs to be considered is that treating meat trimmings with chemical solutions, including organic acids, may result in increased weight of the trim due to fluid retention. The implementation of a hot air treatment as the final step of trim processing has been proposed as a means of removing the excess fluid gained by chemically treated trimmings (Castelo *et al.*, 2001).

It seems that the greatest challenge in developing chemical decontamination interventions for meat trimmings is quality retention. Due to the increased surface area exposed to antimicrobial treatments, the quality characteristics of ground meat produced from decontaminated trimmings are very susceptible to deterioration during storage and display (Stivarius *et al.*, 2002b). Indeed, despite its important decontamination efficacy against foodborne pathogens, the treatment of beef trimmings with a 5% solution of acetic acid or lactic acid caused changes in the colour (reduced redness) and odour characteristics of the ground beef (Stivarius *et al.*, 2002a, 2002b). Although discolouration is not an important concern for frozen ground beef, which most of the time is cooked directly from the frozen state, it is a considerable quality issue for fresh products. Hence, as is generally the case and particularly for meat trimmings, the objective is to develop decontamination interventions that demonstrate a sufficient reduction in antimicrobial activity with a minimal effect on the resulting quality attributes, and research data indicate that organic acids hold considerable promise for meeting this objective (Harris *et al.*, 2006). Based on findings by Harris *et al.* (2006), acetic or lactic acid solutions (2%), applied by spraying onto beef trim inoculated with *Salmonella enterica* serotype Typhimurium or *E. coli* O157:H7, caused significant reductions in the populations of both pathogens, both in the trim and in the ground beef produced. Moreover, the levels of both organisms were significantly lower in the ground beef produced from the acid-treated trim than in the control (ground beef produced from water-washed trim) during refrigeration and frozen storage, while no adverse sensory changes were associated with the applied decontamination treatments (Harris *et al.*, 2006).

18.2.3 Multiple decontamination interventions for meat

Table 18.1 gives the effectiveness (i.e., reductions in bacterial populations) of single applications of organic acids in the decontamination of carcasses and trimmings. The quantitative data in this table are immediate post-treatment reductions, and refer to differences in bacterial populations between acid-treated and control (untreated or water-treated) samples. As demonstrated by these data, acetic and lactic acids have been extensively evaluated as decontamination agents, and various observations have been made with regard to their effectiveness. Differences in research findings can be attributed to the various decontamination technologies used, as well as to other parameters that may also affect the decontamination efficacy of organic acids such as the type of acid, the type of treated tissue, the bacterial target and the initial bacterial population (Bell *et al.*, 1997; Cutter and Siragusa, 1994; Dickson, 1992; Greer and Dilts, 1992, 1995; Hardin *et al.*, 1995; Rajkovic *et al.*, 2010). In addition to using solutions of organic acids individually, the application of mixtures of organic acids for fresh meat decontamination has also been described (Anderson and Marshall, 1990a; Anderson *et al.*, 1992a; Dubal *et al.*, 2004; Podolak *et al.*, 1996). Furthermore, organic acids or organic acid-containing natural compounds (e.g., vinegar and lemon juice) can be used as marinade ingredients for fresh meat aiming at extending the shelf life and potentially enhancing safety (Kargiotou *et al.*, 2011).

The application of organic acids can also take place simultaneously with or sequentially to other interventions, either physical or chemical (Sofos and Smith, 1998). In the context of this approach, known as multiple hurdle technology, two or more interventions (or hurdles), intelligently applied at suboptimal levels, can demonstrate a synergistic effect on the prevalence or population levels of microbial contaminants, and be more effective than each intervention applied individually at optimal level (Leistner, 2000). In this case, the desired antimicrobial effect is achieved through the metabolic exhaustion of bacterial cells as a result of their exposure to several stress factors, without compromising the quality attributes of the food (Leistner, 2000). Numerous applications combining chemical and physical (e.g., hot-water decontamination and steam pasteurization) decontamination interventions have been described (Geornaras and Sofos, 2005; Skandamis *et al.*, 2010). A characteristic example of the performance of

Table 18.1 Representative research results for fresh meat decontamination using organic acids

Product	Agent (concentration)	Application[a]	Microorganism	Reduction (log CFU)[b]	Reference
Beef carcasses	Acetic acid (1%)	SP, 25°C, 0.5 min	*Escherichia coli*	3.0 cm^{-2}	Bell *et al.*, 1997
			Listeria innocua	2.4 cm^{-2}	
			Salmonella Wentworth	3.2 cm^{-2}	
	Acetic acid (1.5–3.0%)	SP, 32°C, 0.3 min	Aerobic bacteria	1.3–2.0 cm^{-2}	Dorsa *et al.*, 1997
			Escherichia coli O157:H7	>2.7 cm^{-2}	
	Acetic acid (2%); after water washing	SP, 55°C, 0.2 min	*Escherichia coli* O157:H7	2.4–3.7 cm^{-2}	Hardin *et al.*, 1995
			Salmonella Typhimurium	3.2–5.1 cm^{-2}	
	Lactic acid (2%)	SP, ~42°C	Aerobic bacteria	1.6/100 cm^2	Bosilevac *et al.*, 2006
			Enterobacteriaceae	1.0/100 cm^2	
	Lactic acid (1.5–3.0%)	SP, 32°C, 0.3 min	Aerobic bacteria	1.3–2.0 cm^{-2}	Dorsa *et al.*, 1997
			Escherichia coli O157:H7	>2.7 cm^{-2}	
			Listeria innocua	2.8–4.0 cm^{-2}	
	Lactic acid (4%)	SP	Aerobic bacteria	≥2.0 cm^{-2} (distal)	Gill and Badoni, 2004
				≤2.0 cm^{-2} (medial)	
	Lactic acid (2%); after water washing	SP, 55°C, 0.2 min	*Escherichia coli* O157:H7	3.0–4.9 cm^{-2}	Hardin *et al.*, 1995
			Salmonella Typhimurium	3.4–5.0 cm^{-2}	
Beef trimmings	Acetic acid (2%)	SP, 38–54°C, 5.6 sec	Aerobic bacteria	0.7–1.7 cm^{-2}	Graves Delmore *et al.*, 1998
			Escherichia coli	1.3 cm^{-2}	
	Acetic acid (2–4%)	SP, ambient temp.	*Escherichia coli* O157:H7	1.5–2.0 g^{-1}	Harris *et al.*, 2006
			Salmonella Typhimurium	1.5–2.0 g^{-1}	
	Lactic acid (2–4%)	SP, ambient temp.	*Escherichia coli* O157:H7	1.5–2.0 g^{-1}	
			Salmonella Typhimurium	1.5–2.0 g^{-1}	
	Lactic acid (2%)	SP, 15°C	Coliforms (faecal)	0.7–1.1 cm^{-2}	Kang *et al.*, 2001a

(Continued)

Table 18.1 Continued

Product	Agent (concentration)	Application[a]	Microorganism	Reduction (log CFU)[b]	Reference
Pork carcasses	Acetic acid (1.8%)	SP	Aerobic bacteria	1.0–1.5 cm^{-2}	Eggenberger-Solorzano *et al.*, 2002
	Lactic acid (1–5%)	SP, 55°C, 1.5 min	Aerobic bacteria	0.3–1.3 cm^{-2} (mesophilic) 0.1–0.9 cm^{-2} (psychrotrophic)	van Netten *et al.*, 1997
			Enterobacteriaceae	0.2 to >1.1 cm^{-2}	
Pork trimmings	Lactic acid (2%)	SP, 15°C, 0.3–2 min	Coliforms	1.0–2.0 cm^{-2}	Castelo *et al.*, 2001
Sheep carcasses	Lactic acid (1–2%)	SP, 0.5 min	Aerobic bacteria	1.6–1.8 cm^{-2}	Beyaz and Tayar, 2010
			Coliforms	2.7–3.0 cm^{-2}	
			Escherichia coli	2.1–2.2 cm^{-2}	
Sheep/goat carcasses	Lactic acid (2%)	SP, 2–4 min	Aerobic bacteria	0.5 g^{-1}	Dubal *et al.*, 2004
			Escherichia coli	0.4 g^{-1}	

Notes:
[a] SP: spraying.
[b] Bacterial reductions achieved by the applied agent, compared to untreated or water-treated samples, expressed per cm^2 or per g of sample.

interventions that involve simultaneous application of chemical and physical hurdles is the enhanced decontaminating effect of acid solutions under conditions of increased temperature (Cutter *et al.*, 1997). Similarly, the sequential application of chemical and physical treatments has been shown to be more effective than the application of single treatments. For instance, the application of acetic acid (1.8%) to hog carcass tissues considerably increased the decontamination efficacy of hot water (80°C) subsequently applied (Eggenberger-Solorzano *et al.*, 2002). However, an important parameter that needs to be considered for sequential decontamination interventions is the order in which treatments are applied since it may have an impact on the preservation (bacteriostatic) potential of the applied interventions. Based on the findings of Koutsoumanis *et al.* (2004), the application of hot water (75°C) followed by lactic acid (2%, 55°C) was more effective than the reverse-order treatment in limiting *L. monocytogenes* growth during storage (at 4°C, 10°C and 25°C) of vacuum-packaged beef tissues, despite the fact that both treatments caused similar reductions (approximately 2.7 log CFU/cm^2) in the initial populations of the pathogen. In addition to the aforementioned, the sequential application of hurdles for fresh meat decontamination can also involve organic acids in conjunction with other chemical agents (Stopforth *et al.*, 2005a). Given the high susceptibility of meat trimmings to quality deterioration as a result of high-intensity decontamination treatments, as discussed previously in this chapter, the application of organic acids or their salts, such as potassium lactate, as part of a multiple-hurdle process (i.e., a combination of low-intensity hurdles) is expected to be of significant value in the decontamination of this product (Ellebracht *et al.*, 1999; Graves Delmore *et al.*, 1998; Jimenez-Villarreal *et al.*, 2003; Kang *et al.*, 2001a, 2001b; Quilo *et al.*, 2010).

18.3 Use of organic acids for fresh produce decontamination

According to the review by Ilic *et al.* (2012), a total of 269 research studies on decontamination interventions for fresh produce have been published in journals listed in the Science Citation Index between 1990 and 2010. Only 17 of these studies evaluated the effect of organic acids, such as acetic, malic, tartaric, citric, lactic, ascorbic, propionic, succinic, levulinic and caprylic acid or their salts on the reduction of pathogens (Gil *et al.*, 2009; Inatsu *et al.*, 2005; Zhao *et al.*, 2010), whereas the majority of the remaining articles focused on chlorine treatments. Notably, 71.6% of the reports on fresh produce decontamination were for lettuce, followed by cabbage (19.3%), green leafy herbs (15.5%) and spinach (15.2%).

18.3.1 Vegetables: fresh-cut salads

Organic acids may deliver higher reductions of microbial populations on fresh produce than the commonly used disinfectants, such as chorine (Inatsu *et al.*, 2005; Velázquez *et al.*, 2009; Ho *et al.*, 2011); see Table 18.2. Nonetheless, given that they may compromise the sensory characteristics of the tissues treated, optimization of their use is required (Ölmez and Kretzschmar, 2009). The application of organic acids may be by immersing or spraying the fresh-cut tissues for 30 s to 7 min (Allende *et al.*, 2008; Gil *et al.*, 2009; Ho *et al.*, 2011; Rahman *et al.*, 2010). Electrostatic spraying is considered more effective than conventional spraying, because it generates fine disinfectant particles, which are evenly distributed over the whole plant surface and are better retained by the tissue than the bigger droplets which are formed by conventional spraying and which may easily run off (Ganesh *et al.*, 2010). In industrial practice, however, submersion and agitation of vegetables in the sanitizer solution may deliver higher microbial reductions than spraying, apparently due to the continuous contact of the tissue with the antimicrobial compounds (Allende *et al.*, 2008). Organic acids may be used alone or as acidifying agents, e.g., citric acid 1000–10 000 ppm in sodium chlorite solutions (Inatsu *et al.*, 2005). The pH of the solution, usually ranging from 2.7 to 4.0, the use of the surfactant during cleaning, the contact time and the concentration of the organic acid are the most critical factors affecting the decontamination potential of organic acids (Allende *et al.*, 2008; Abadias *et al.*, 2011; Gil *et al.*, 2009; Rahman *et al.*, 2010; Rodgers *et al.*, 2004; Park *et al.*, 2011).

There is marked variability in experimental variables among studies evaluating pathogen reduction by organic acids on fresh produce. Major differences include, temperature (4°C or 25°C), concentration (10–10 000 ppm), contact time (30 s to 10 min) and inoculation method (e.g. dipping, spaying or spot inoculation) (Gil *et al.*, 2009). The common magnitude of the reported pathogen reductions in challenge studies is >3.8 (3.6 to 5.6) log CFU/cm^2 (Table 18.2). In contrast, the respective reductions of natural (epiphytic) aerobic flora, e.g., of white cabbage, grated carrot, leeks and iceberg lettuce, after the application of one of the most effective acid treatments, peroxyacetic acid (80–250 ppm), are limited to the range 1–2.5 log CFU/cm^2 (Allende *et al.*, 2008; Vandekinderen *et al.*, 2008a, 2008b, 2009a, 2009b). This discrepancy, which is also evident in meat decontamination studies, may be associated with the weaker attachment and limited penetration of pathogens inoculated on the surface of the treated food, compared to the long-term attachment of pre-existing epiphytic flora (Skandamis *et al.*, 2010).

The antimicrobial activity of organic acids can be enhanced by increasing the exposure time, mixing acids or by combining acids with other decontamination treatments or agents. Increasing the contact time of 80 ppm of peroxyacetic acid with lettuce, apples, strawberries and cantaloupes from 15 s to 5 min increased the reductions of *E. coli* O157:H7 and *L. monocytogenes* from <1 to 4.4 log CFU/cm^2 (Rodgers *et al.*, 2004). Similarly, increasing the treatment time of apples and lettuce with malic, propionic, acetic, lactic and citric acid for up to 10 min enhanced inactivation of *E. coli* O157:H7, *S.* Typhimurium and *L. monocytogenes* by approximately 1.8–2.2 log CFU/cm^2 (on apples) and 0.5–1.0 log CFU/cm^2 (on lettuce) (Park *et al.*, 2011). Ho *et al.* (2011) demonstrated that mixing lactic acid (2000–2500 ppm) with peroxyacetic acid (70–75 ppm) at 4°C resulted in at least 2.4 log higher reduction of *E. coli* O157:H7, *L. innocua* and *Lactobacillus plantarum* on lettuce and spinach than with the application of each acid alone. Furthermore, the combination of 3% lactic acid with 3% malic acid caused a

Table 18.2 Representative research results for fresh produce decontamination using organic acids

Product	Agent (concentration)	Application	Microorganism	Reduction (log CFU)[a]	Reference
Lettuce and spinach	Single treatment: 1. Lactic acid (0–2.700 ppm) 2. Peroxyacetic acid (0–72 ppm) Combined treatments: Lactic acid + peroxyacetic acid	Dipping and stirring 4.4–7.2°C, 20–30 s	*Escherichia coli* K-12 *Lactobacillus plantarum* *Listeria innocua*	3.5 to >5 cm^{-2}	Ho *et al.*, 2011
Lettuce	Commercial sanitizers Purac (lactic acid) (2500–20000 mg/l)	Dipping and agitation 8°C, 1 min	*Escherichia coli* CECT 471, 516 and 533	0.5–1.5 cm^{-2}	López-Gálvez *et al.*, 2009
	Citrox 14W (citric acid) (1250–5000 mg/l)			1.2–1.5 cm^{-2}	
	Tsunami 100 (peroxyacetic acid) (125–500 mg/l)			2.6–3.0 cm^{-2}	
Escarole	Purac (20 ml/l)	Spray or dipping in water 5°C, 1 min	Mesophilic counts Coliform counts Yeasts and moulds	2.0 cm^{-2} 2.0 cm^{-2} 1.5 cm^{-2}	Allende *et al.*, 2008
	Tsunami 100 (80 μl/l) Citrox 14W (5 ml/l)		Mesophilic counts Coliform counts Yeasts and moulds	2.0 cm^{-2} 1.0 cm^{-2} 0.5 cm^{-2}	
Lettuce	Purac (10–20 ml/l)		Mesophilic counts Coliform counts Yeasts and moulds	0.5–1.0 cm^{-2} 1.5 cm^{-2} 2.0 cm^{-2}	
	Citrox 14W (5 ml/l)		Mesophilic counts Coliform counts Yeasts and moulds	0.5–1.0 cm^{-2} 1.5 cm^{-2} 2.0 cm^{-2}	

(*Continued*)

Table 18.2 Continued

Product	Agent (concentration)	Application	Microorganism	Reduction (log CFU)[a]	Reference
Whole/shredded lettuce Whole/sliced apples Strawberries Cantaloupe	Peroxyacetic acid (80 ppm)	Dipping 21–23°C, 5 min Storage 4°C, 9 days[b]	*Listeria monocytogenes*	After storage: 4.5 to >5 cm^{-2}	Rodgers *et al.*, 2004
			Escherichia coli O157:H7 Mesophilic bacteria Yeasts Moulds	After storage: 4.5 to >5 cm^{-2} 3.0–4.0 cm^{-2} 2.9–3.0 cm^{-2} 1.0–2.0 cm^{-2}	

Notes:
[a] Bacterial reductions achieved by the applied agent, compared to untreated or water-treated samples, expressed per cm^2 of sample.
[b] Bacterial populations increase by 2–3 log CFU/cm^2 after storage.

3 log CFU/cm^2 reduction of *S.* Typhimurium on spinach immediately after treatment and another 3 log decrease was obtained during storage of the treated spinach at 4°C (Ganesh *et al.*, 2010). The use of surfactants, such as sodium dodecyl sulphate (sodium lauryl sulphate), may also enhance the removal of cells by disinfectants, including organic acids (Guan *et al.*, 2010; Ho *et al.*, 2011; Keskinen and Annous, 2011; Ölmez and Kretzschmar, 2009). However, attention should be paid to prevent quality defects being produced during storage of fresh produce treated with a combination of acids and surfactants especially under modified atmosphere packaging (Allende *et al.*, 2008; Guan *et al.*, 2010).

In general, the decontamination of fresh produce by organic acids should optimize any residual effect of the acids on microbial growth during storage and their impact on sensory and nutritional (e.g., flavonoids and vitamins) characteristics of the product (Allende *et al.*, 2008; Vandekinderen *et al.*, 2008a, 2008b, 2009a, 2009b). To achieve this, the concentration and duration of decontamination needs to be strictly applied in accordance with the manufacturer's recommendations, although depending on the acid and the treated vegetable, even compliance with the recommendations might still be insufficient. For instance, the treatment of lettuce or escarole with the recommended dose of lactic acid (20 ml/l) for 3 min, caused the highest reduction in the visual score of treated vegetables after 3 days at 4°C and 5 days at 8°C, compared to other commercial sanitizers such as chlorine (100 mg/l for 1 min), Catalix (40 mg/l for 5 min), Sanova (500 mg/l for 1 min), Sanoxol 20 (20 ml/l for 45 s), Tsunami 100 (80 μl/l for 2 min) and Citrox 14W (5 ml/l for 5 min) (Allende *et al.*, 2008). The microbial reductions of the epiphytic flora (i.e., coliforms, yeasts, moulds and mesophilic flora) caused by these sanitizers ranged from <1.0 to 2.5 log CFU/cm^2, but the differences between lactic acid and other disinfectants in decontamination effectiveness were not as evident as the changes in the sensory properties of the treated vegetables (Allende *et al.*, 2008). The minimum effective dose of the most powerful organic acids, such as acetic and lactic acid, is often above the critical level that causes irreversible sensory damage to plant tissue (Chang and Fang, 2007; López-Gálvez *et al.*, 2009; Ölmez and Kretzschmar, 2009). For example, commercial vinegar (5% acetic acid) has both an immediate and long-term residual antimicrobial effect, but it has an unacceptable sour flavour, and is almost ineffective as a disinfectant if diluted at levels with no sensory impact (Chang and Fang, 2007). However, since organic acids are highly compatible with fresh produce, they remain a promising and natural alternative to chlorine treatments. Their combination with mild decontamination or natural antimicrobials, e.g., ozone, ultrasound, cold atmospheric plasma, pulsed electric fields, plant extracts and bacteriocins, may resolve these sensory issues. Further details are provided in the paragraph on multiple interventions for fresh produce.

18.3.2 Fruits and fruit juices

Raybaudi-Massilia *et al.* (2009b) provided a detailed overview of the conditions of the application of organic acids, such as concentration, method of application (surface, edible films, mixing), target organisms (inoculated pathogen or indigenous microflora) and storage conditions (if applicable) and observed microbial reductions in fruits and fruit juices.

Malic, ascorbic and citric acid as well as lemon juice have been proposed for: (i) the surface decontamination of apple and pears (Raybaudi-Massilia *et al.*, 2007, 2009a, 2009b, 2009c), (ii) enhancing the microbial inactivation during drying (60–62°C) of apples and peaches (DiPersio *et al.*, 2003; Derrickson-Tharrington *et al.*, 2005) and (iii) extending the shelf life of apples and melon cubes coated with edible alginate film containing malic acid (2.5%) or calcium lactate (2%) (Raybaudi-Massilia *et al.*, 2008a, 2008b). In the special case of apples, organic acids added to a solution containing N-acetyl cysteine and glutathione both inhibits microbes and delays enzymatic browning (Raybaudi-Massilia *et al.*, 2007, 2008a, 2009c). Organic acid salts, such as sodium benzoate and potassium sorbate, are well-known preservatives of fruit juices and other acid products, such as ambient-stable sauces (e.g., dressings, ketchup, etc.) (Raybaudi-Massilia *et al.*, 2009b). In addition to these agents, malic acid up to 2.5% is reported to have caused 5 log cycles inactivation of *E. coli* O157:H7, *S.* Enteritidis and *L. monocytogenes* in apple, pear and melon juices (Raybaudi-Massilia *et al.*, 2009a).

18.3.3 Multiple decontamination interventions for fresh produce

The combination of organic acids with other decontamination treatments can markedly enhance microbial inactivation on fresh produce. Malic acid (2%) or 2 ppm ozone alone reduced *Shigella* spp. by less than 3 log CFU/cm^2 on radish and mung bean sprouts, whereas their combination resulted in 4.4 to 4.8 log CFU/cm^2 reductions (Singla *et al.*, 2011). Samples receiving the combined treatment also had suppressed microbial growth and the antioxidant status of the sprouts was maintained during 10 days of storage at 28°C.

Given the increasing popularity of natural antimicrobials, the combination of organic acids with plant antimicrobials or bacteriocins, may give synergistic effects in fresh produce decontamination. Grape seed extract (3%) enhanced the bactericidal activity of various acids, including malic, tartaric and lactic acid against *S.* Typhimurium on spinach (Ganesh *et al.*, 2010). The bactericidal activity of sodium lactate (2%), potassium sorbate (0.02%), phytic acid (0.02%) and citric acid (10 mM) against *L. monocytogenes* was maximized on cabbage, broccoli and mung bean sprouts when combined with nisin (50 μg/ml) or pediocin (100 AU/ml) (Bari *et al.*, 2005). The effective combinations were product dependent.

As with fresh-cut vegetables, in response to consumer demands for foods with less chemicals, mildly processed and with a long shelf life, while maintaining a high safety level, research is driven to consider combinations of organic acids with ultrasound (Sagong *et al.*, 2011), other natural antimicrobials (e.g., nisin) or mild preservation technologies such as a high-intensity pulsed electric field (HIPEF) for increasing the safety of fruit juices (Mosqueda-Melgar *et al.*,

2008a, 2008b; Walker and Phillips, 2008). Combinations of HIPEF (35 kV/cm for 1682 μs or 1709 μs and 4 μs duration) with citric acid (0.5–2%) can cause > 5.0 log CFU/ml reduction of pathogens and complete inactivation of spoilage flora during storage of apple, pear, orange, melon and watermelon juices. However, careful selection of conditions is essential for maintaining quality attributes. Combinations of lactic, malic or citric acid (0.3–2%) with ultrasound (40 kHz) for 5 min increased the reductions of *E. coli* O157:H7, *S.* Typhimurium and *L. monocytogenes* by 0.8 to 1.3 log CFU/cm^2 on organic fresh lettuce compared to individual treatments with organic acids or ultrasound (Sagong *et al.*, 2011). Notably, the decontamination efficiency of the combined treatments containing 0.5% of organic acid was similar to that of a single organic acid treatment with a fourfold higher acid concentration (i.e., 2%) (Sagong *et al.*, 2011). In addition to the immediate microbial reductions, the combined treatments also maintained an acceptable visual appearance of lettuce during storage at 7°C, with no discoloration. Overall, such multiple interventions and combinations of organic acids with mild processing may maximize pathogen reduction and, hence, assist in the achievement of the food safety objectives of the fresh produce industry.

18.4 Risks and concerns: microbial adaptation

Organic acids are effective in controlling microbial contamination and dissemination of foodborne pathogens in post-harvest food production and processing when used as food additives or sanitizers of food and/or equipment surfaces (Dubal *et al.*, 2004; Ricke, 2003; Samara and Koutsoumanis, 2009; Skandamis *et al.*, 2010). However, there are potential risks and concerns associated with decontamination technologies based on organic acids. They include: (i) the induction of acid adaptation in foodborne pathogens, and the resulting enhanced microbial acid resistance, as well as cross-protection from other environmental food-related stresses (Leyer and Johnson, 1993; Kwon *et al.*, 2000; Ryu and Beuchat, 1999; Samelis *et al.*, 2002; Samelis and Sofos, 2003), (ii) the colonization of equipment surfaces and/or recontamination of food surfaces with acid-adapted strains of foodborne pathogens (Samelis and Sofos, 2003; Samelis, 2005), with the former possibly resulting in the development of hardened bacterial biofilms (which are less sensitive to sanitation agents) on equipment surfaces (Stopforth *et al.*, 2003), and (iii) the reduction and/or growth inhibition of natural microbiota by acid decontamination, resulting in reduced microbial competition for pathogenic microorganisms (Ikeda *et al.*, 2003; Nissen *et al.*, 2001; Samelis *et al.*, 2002).

An important concern, specifically for the use of organic acid solutions for decontaminating food, is the induction of acid adaptation or the reselection of acid-resistant strains of foodborne pathogens. The enhanced resistance to low pH after exposure to mildly acidic conditions is called acid habituation (Goodson and Rowbury, 1989) or acid tolerance (Foster and Hall, 1990). Acid tolerance response (ATR) has been observed for several foodborne pathogens including

E. coli O157:H7 (Bearson *et al.*, 1997; Brown *et al.*, 1997; Goodson and Rowbury, 1989), *S.* Typhimurium (Foster, 1995; Foster and Hall, 1990; Bearson *et al.*, 1997) and *L. monocytogenes* (Kroll and Patchett, 1992; Lou and Yousef, 1997; O'Driscoll *et al.*, 1996). The systems underlying ATR include complex biological phenomena associated with the organism, the growth phase, the medium, the type of acid stress (e.g., organic or inorganic acid) and other environmental factors (Bergholz *et al.*, 2009; Foster, 1995; Foster and Hall, 1990; Bearson *et al.*, 1997). For example, ATR is growth-phase specific (Bergholz *et al.*, 2009; Foster and Hall, 1991; Kroll and Patchett, 1992) with distinct responses occurring in both logarithmic and stationary phases, and requires the *de novo* synthesis of acid shock proteins (which promote survival at extremely low pH values), changes in the fatty acid profiles of the cell membrane and associated alterations in membrane permeability (Bergholz *et al.*, 2009; Brown *et al.*, 1997; Foster, 1991, 1993; O'Driscoll *et al.* 1996). The resistance responses of *Salmonella* are different for organic and inorganic acids as reported by Foster (1999), due to the different repair systems that are apparently involved. The majority of the studies conducted to evaluate the development of resistance of pathogenic microorganisms were undertaken under laboratory conditions without taking into account the variable reactions of pathogens in natural systems. However, the observed responses, which have been well studied with pure pathogenic cultures grown under laboratory conditions, could also be a reality in foods (Samelis and Sofos, 2003). In the acid decontamination of meat or vegetables, acid-habituated pathogens are expected to become more resistant to further acid stress if the acid residues are maintained, which has profound implications for food safety. Inadequate cleaning and hygiene after the application of organic acids could enhance this acid-resistant behaviour.

The role of acid adaptation in the microbial resistance to subsequent acid treatments depends on the microorganism and its growth phase – stationary phase cells are more resistant than exponential phase cells (Bergholz *et al.*, 2009; Samelis and Sofos, 2003), the nature of the acidulant and the pH. Lin *et al.* (1995, 1996) demonstrated that depending on the physiological state of the cells (log or stationary phase cells) at the time of the exposure to low pH values, the development of acid resistance can be pH dependent or independent or a combination of both types. Exposing exponentially growing pathogen cells to a pH from neutral to sublethal may confer resistance to the bacteria at extreme acid conditions and, moreover, when the cells enter into the stationary phase they exhibit a pH-independent acid resistance (Arnold and Kaspar, 1995; Lin *et al.*, 1996; O'Driscoll *et al.*, 1996). On the other hand, stationary phase cells have an induced pH-dependent acid resistance that subsequently increases their generalized response stress system, which can vary between different genera or species (Buchanan and Edelson, 1999; Foster 1995).

Koutsoumanis and Sofos (2004) studied stationary phase *L. monocytogenes, E. coli* O157:H7 and *S.* Typhimurium. These pathogens became acid adapted in a tryptic soy broth (TSB), which is glucose free and was acidified with lactic acid at 30°C. The pathogens had increased acid resistance (for pH 3.5), after habituation in the pH ranges 5.0–6.0, 4.0–5.5 and 4.0–5.0, respectively for the three pathogens. The maximum tolerance to acid was induced after habituation at pH 5.5, 5.0 and 4.5, respectively. In addition, in the same research the acid protection was found to be pH dependent.

There is evidence suggesting that the strategies underlying survival under extreme acid conditions are different (i.e., different combinations of ATR) for *E. coli, Shigella flexneri* and *S.* Typhimurium (Lin *et al.*, 1995). Greenacre *et al.* (2003), using TSB acidified with acetic or lactic acid, found that acetic acid was more efficient than lactic acid for inducing ATR in *S.* Typhimurium, while Álvarez-Ordóñez *et al.* (2009) demonstrated that the ability of *S.* Typhimurium to survive extreme pH conditions depended on the growth medium and the type of acidulant used to invoke acid resistance. The order of acids used in inducing ATR in the latter study was: citric>lactic>malic≥HCl>ascorbic. Furthermore, *S.* Typhimurium cells, adapted to exposure to HCl (pH 5.8), had increased resistance to the organic acids commonly present in cheese, including lactic, propionic and acetic acids (Leyer and Johnson, 1992). Moreover, it was shown that acid-adapted *E. coli* O157:H7 cells had a higher ATR to organic acid treatments than non-adapted cells (Buchanan and Edelson, 1996; Deng *et al.*, 1999). In addition, there was enhanced survival on subsequent exposure to organic acids (malic, citric, acetic or lactic) compared with inorganic acids (HCl) and, depending on the strain of *E. coli* O157:H7, the resistance to the organic acids was different (Buchanan and Edelson, 1999). Berry and Cutter (2000), studying the effects of acid adaptation of three *E. coli* O157:H7 strains on the efficacy of acetic acid (2% v/v) spray washes to decontaminate beef carcass tissue, found that the acid-adapted cells of two of the three strains were more resistant against acetic acid sprays than non-adapted cells. These findings (Berry and Cutter, 2000; Buchanan and Edelson, 1999) indicate that the impact of strain variability on ATR is considerable and should not be neglected. Survival was also demonstrated for *L. monocytogenes* in several studies after acid adaptation (Gahan *et al.*, 1996; Kroll and Patchett, 1992; Lou and Yousef, 1997; O'Driscoll *et al.*, 1996). For example, Gahan *et al.* (1996) found that exposing *L. monocytogenes* to pH 5.5 for 60 min enhanced the survival of the pathogen in cottage cheese, orange juice, salad dressing and yogurt. On the other hand, Ikeda *et al.* (2003) showed that, although acid adaptation may increase the survival of *L. monocytogenes* considerably in laboratory media acidified with HCl or organic acids (lactic or acetic), the acid adaptation did not promote the survival or growth of the pathogen on fresh beef, whether the meat was stored at 4°C or 10°C, following acid decontamination with acid or non-acid solutions. Some of the results of studies into acid adaptation are in conflict, and the discrepancies can be attributed to differences in strain, physiological state of the microorganism's cells at the time of the exposure to sublethal and lethal pH values, the substrate (complex or minimal), type of acid, inoculum preparation, pH conditions, time of habituation and other environmental conditions.

A further important consequence of microbial acid adaptation is the induction of cross-protection for another stress. Leyer and Johnson (1993) demonstrated that acid-adapted *S.* Typhimurium cells had cross-protection to various stresses including heat, salt, an activated lactoperoxidase system and the surface-active agents, crystal violet and polymyxin B. Moreover, exposure of *S.* Typhimurium to short chain fatty acids enhanced its survival against various stress conditions such as extremely low pH (pH 3.0), high osmolarity (2.5 M NaCl) and reactive oxygen (20 mM H_2O_2) (Kwon *et al.*, 2000). Acid-adapted strains of *E. coli* also have cross-protection to several adverse conditions, such as salt (Rowe and Kirk, 1999), heat (Ryu and Beuchat, 1999; Shen *et al.*, 2011; Velliou *et al.*, 2012) and other alternative processing techniques, such as multi-frequency Dynashock power ultrasound (Gabriel, 2012). However, microbial adaptation originating from exposure to organic acids does not always have a protective effect for other stresses. For instance, Samara and Koutsoumanis (2009), who investigated the survival of stationary phase cells of *L. monocytogenes* on lettuce that was previously decontaminated with organic acids (lactic acid, acetic acid, propionic acid or citric acid) and then stored at 5°C for 48 h, found that the decontamination treatment did not enhance the acid tolerance of *L. monocytogenes*, which was evaluated by exposing the pathogen to simulated gastric fluid. Furthermore, Greenacre *et al.* (2006) showed that lactic acid-adapted cells of *S.* Typhimurium exhibited sensitivity to H_2O_2 (100 mM). Nevertheless, considering that organic acids can be applied in a multiple-hurdle process for decontaminating food products, it is evident that the cross-protection between stresses has major implications in food preservation (Shen *et al.*, 2011).

A concern for meat processing plants is the long-term consequence of meat decontamination with organic acid solutions when there are acid-adapted pathogens in these plants and products. Washing with organic acids to decontaminate meat carcass surfaces may lead to the development of acid-resistant strains of *E. coli* O157:H7 in the fresh meat environment, when the treatment is not adequately applied in exhausting the pathogen (Samelis and Sofos, 2003). Moreover, there is a potential risk of the extended survival of acid-adapted populations of this pathogen in sublethal pH environments, such as acidic decontamination run-off fluids. Such adverse conditions may increase the potential for the growth of acid-resistant *E. coli* O157:H7 cells, in contrast to cells that have not been exposed to such environments (Samelis *et al.*, 2002, 2004; Skandamis *et al.*, 2007, 2009; Stopforth *et al.*, 2007). In the study by Stopforth *et al.* (2003) it was demonstrated that such situations may also harden bacterial biofilms (which are less sensitive to sanitation agents) that form on equipment surfaces. Furthermore, Skandamis *et al.* (2009) found that bacterial cells that had previously been exposed to diluted organic acid run-off fluids and which could not be detected with the usual enumeration techniques, recovered and restored their acid resistance during a subsequent exposure to fresh meat decontamination run-off fluids. Nevertheless, undiluted or low dilutions of acidic meat washings may minimize the establishment of pathogen biofilm in meat plants (Stopforth *et al.*, 2003). However, it is important to note that the presence of acid-adapted pathogens in meat plants is a result of poor cleaning and sanitation. When the organic acid solutions are properly removed from the plant, the adapted pathogens return to their sensitive stage and the sanitation is more effective.

Another concern in the use of organic acids is that acid decontamination may alter the natural flora of meat (van Netten *et al.*, 1998) as well as the microbial ecology of meat plants (Samelis *et al.*, 2002). The impact of such interventions on the number and types of microorganisms present on meat surfaces is related to competition between the background biota and the pathogens and the potential proliferation of acid-tolerant organisms, which can alter the spoilage organisms in fresh meat (e.g., lactic acid bacteria, moulds and yeasts) (Ikeda *et al.*, 2003; Samelis *et al.*, 2002). Furthermore, spoilage yeasts, such as *Saccharomyces cerevisiae, Zygosaccharomyces bailii* and *Z. rouxii*, can develop adaptive resistance mechanisms when exposed to the weak acids (acetic, propionic, benzoic and sorbic) used for food preservation, potentially compromising food quality (Mira *et al.*, 2010; Piper, 2011).

In addition to the above, there are research data indicating that acid adaptation may increase bacterial virulence (O'Driscoll *et al.*, 1996; Gahan and Hill, 1999). For instance, acid-adapted microorganisms may have enhanced survival during transit through the stomach, resulting in increased likelihood of intestinal colonization and, thus, have enhanced virulence potential (Hill *et al.*, 1995; O'Driscoll *et al.* 1996; Stopforth *et al.*, 2005b). Taking this into account, and considering that acid adaptation may confer cross-resistance to other types of environmental food-related stresses (e.g., heat, osmosis, irradiation or increased concentration of ethanol), it is obvious that this phenomenon impacts on both quantitative microbiology and risk assessment approaches to microbiological food safety (Greenacre *et al.*, 2003). Understanding the adaptive responses of foodborne pathogens to organic acids and identifying the conditions that contribute to or control these responses, could lead to more realistic evaluations of food safety concerns and to a better selection of processes in order to avoid or remove adaptation phenomena and to minimize the potential for food safety risks (Koutsoumanis and Sofos, 2004). Further research into new methods is necessary, in order to develop effective interventions for controlling microbial adaptation in food products. High hydrostatic pressure, irradiation, pulsed electric fields, shock waves, high-intensity light, ultrasound and oscillating magnetic fields are some of the treatments that could be applied as decontamination technologies (Aymerich *et al.*, 2008).

18.5 Legislation

Decontamination treatments, including those that involve the application of organic acid solutions, can be incorporated into hazard analysis and critical control point (HACCP) systems, either as critical control points (CCPs) or as part of the systems' prerequisite programs (Bolton *et al.*, 2001; Hugas and Tsigarida, 2008). Organic acids have been assigned a generally-recognized-as-safe (GRAS) status by the United States Food and Drug Administration (FDA) (CFR, 2011b), and as such, they (primarily lactic and acetic acids) are widely accepted for carcass decontamination in the United States (Smulders and Greer, 1998). Meat processing plants in the United States, and in some cases in Canada and Australia, have integrated chemical decontamination into their HACCP systems, embracing an intervention HACCP approach for the control of microbial contaminants during meat production (Bolton *et al.*, 2001). The United States Department of Agriculture's Food Safety and Inspection Service (FSIS) has recognized that one or more physical or chemical decontamination steps should be included in the slaughter/dressing process as a CCP in the context of HACCP (Bolton *et al.*, 2001), and has approved the application of organic acids (acetic, citric and lactic acids) and other chemical and physical treatments as acceptable interventions to reduce microbial pathogens on meat carcasses, cuts and trimmings (FSIS, 2012). More specifically, organic acids may be applied as aqueous solutions at concentrations of up to 5% and at temperatures of up to 55°C in the form of sprays, rinses or dips (FSIS, 2012).

The importation of beef from acid-treated carcasses from the United States was approved very early by the Canadian regulatory authorities, while the use of organic acids (i.e., lactic and acetic acid sprays) in Canadian abattoirs has been approved as an adjunct to good manufacturing practices (GMPs) during the carcass dressing process (Smulders and Greer, 1998; Theron and Lues, 2007). Lactic acid decontamination has also been adopted by Australian beef plants exporting meat to the United States (Smulders and Greer, 1998).

In produce decontamination, wash water disinfectants in the United States are regulated by the FDA as secondary direct food additives, and there is a list of approved chemicals in the Code of Federal Regulations. Peroxyacetic acid (prepared by reacting acetic acid with hydrogen peroxide) can be used to decontaminate fruits and vegetables at concentrations not exceeding 80 ppm in wash water (CFR, 2011a).

Due to the concern that chemical decontamination compensates for poor hygiene, legislation in the European Union (EU) has traditionally been reluctant towards the implementation of such interventions (Bolton *et al.*, 2001; Hugas and Tsigarida, 2008; Theron and Lues, 2007). Therefore, for a long time meat hygiene regulations within the EU did not allow any method for decontaminating meat carcasses, parts or viscera, other than washing with potable water (Theron and Lues, 2007). Alternatively, the application of strict hygienic practices and GMPs, as well as the adoption of non-chemical decontamination (e.g., hot water and steam pasteurization) as a CCP in HACCP systems in slaughtering operations, has been proposed as a way of providing a sufficient level of protection against carcass bacterial contamination (Skandamis *et al.*, 2010). However, the EU has provided a legal basis for the use of substances other than potable water for decontaminating foods of animal origin (OJEU, 2004), with a draft regulation proposal that specifies the conditions for such decontamination under discussion with Member States and stakeholders (Hugas and Tsigarida, 2008). The European Food Safety Authority (EFSA) is responsible for evaluating the safety and efficacy of substances intended to be used for decontamination. A thorough scientific evaluation of the impact of chemical decontaminants on public health needs to be available before permission is granted for their use

(EFSA, 2008; Hugas and Tsigarida, 2008). According to the EFSA, chemical decontamination should be used within an integrated approach and as a supplementary measure to control the microbial contamination of carcasses (EFSA, 2008). Although several scientific opinions have been issued by the EFSA on the efficacy of chemical agents in decontaminating foods of animal origin, with lactic acid being one of them, their use has not yet received official approval in the EU (Hugas and Tsigarida, 2008; Rajkovic *et al.*, 2010). In a recent scientific opinion on the assessment of the safety and efficacy of lactic acid when used to reduce microbial surface contamination on beef hides, carcasses, cuts and trimmings, it was concluded that: (i) with reference to human toxicological effects, there were no safety concerns with the treatment provided that the substance used complies with the EU specifications for food additives, and (ii) although variable, microbial reductions achieved by the lactic acid treatment of beef are generally significant compared to untreated or water-treated controls (EFSA, 2011a). As is the case for fresh meat decontamination, the use of chemical agents for decontaminating fresh produce is also not authorized in the EU (Rajkovic *et al.*, 2010), and chemicals (e.g., chlorine and chlorine dioxide) can only be used as processing aids (Gil *et al.*, 2009).

18.6 Conclusion

There is continuous interest in research into the application of organic acids in food preservation and food decontamination. Given their natural presence in many foods, organic acids are considered to be natural antimicrobials, especially in fruits and fruit juices, where they are compatible with the sensory properties. Nonetheless, their application has not yet been optimized in terms of time, temperature, acid concentration and processing stage, because, despite their strong antimicrobial character, they may impair the quality (e.g., through discoloration and softening) and nutritional properties of some foods, e.g., fresh-cut vegetables, immediately after application or during storage. In this context, the combination of organic acids with emerging mild non-thermal methods, such as high-intensity pulsed electric fields, ultrasound and ozone may be a promising solution for overcoming these limitations and enhancing the decontamination effectiveness of organic acids.

There is marked improvement in the microbial quality of meat in terms of the prevalence and populations of pathogens in the product, i.e., primal and sub-primal cuts and ground beef. Microbial testing of carcasses in USA beef plants indicates that the adoption of HACCP principles and the associated critical control points when applied as decontamination interventions, including organic acid treatments, caused a significant improvement in hygiene during the slaughtering process (Skandamis *et al.*, 2010). The overall result was a significant increase (>90% of tested samples) in the amount of meat meeting the performance criteria for *E. coli* and the standards for the prevalence of *Salmonella* (Bacon *et al.*, 2000; Rose *et al.*, 2002; FSIS, 2008).

The increasing number of outbreaks associated with fresh-cut packaged (prepared) salads and fruits over the last ten years, has intensified the need for effective decontamination strategies that would reduce the prevalence and growth potential of pathogens in products, especially on the retail side. This also conforms to the aim of substituting chlorine with alternative disinfectants, because of the formation of trihalogens as a result of chlorine coming into contact with organic matter. According to the literature, various decontamination methods are candidates for chlorine substitution; organic acids are the dominant option, alone or in combination with new mild processing technologies, such as pulsed electric fields and ultrasound.

18.7 References and further reading

ABADIAS M, ALEGRE I, USALL J, TORRES R and VIÑAS I (2011), Evaluation of alternative sanitizers to chlorine disinfection for reducing foodborne pathogens in fresh-cut apple. *Postharvest Biol Tec*, 59, 289–297.

ALLENDE A, MCEVOY J, TAO Y and LUO Y (2009), Antimicrobial effect of acidified sodium chlorite, sodium chlorite, sodium hypochlorite, and citric acid on *Escherichia coli* O157:H7 and natural microflora of fresh-cut cilantro. *Food Control*, 20, 230–234.

ALLENDE A, SELMA MV, LÓPEZ-GÁLVEZ F, VILLAESCUSA R and GIL MI (2008), Role of commercial sanitizers and washing systems on epiphytic microorganisms and sensory quality of fresh-cut escarole and lettuce. *Postharvest Biol Technol*, 49, 155–163

ÁLVAREZ-ORDÓÑEZ A, FERNÁNDEZ A, BERNARDO A and LÓPEZ M (2009), Comparison of acids on the induction of an acid tolerance response in *Salmonella* Typhimurium, consequences for food safety. *Meat Sci*, 81, 65–70.

ANDERSON ME and MARSHALL RT (1990a), Reducing microbial populations on beef tissues: concentration and temperature of an acid mixture. *J Food Sci*, 55, 903–905.

ANDERSON ME and MARSHALL RT (1990b), Reducing microbial populations on beef tissues: concentration and temperature of lactic acid. *J Food Saf*, 10, 181–190.

ANDERSON ME, MARSHALL RT and DICKSON JS (1992a), Efficacies of acetic, lactic and two mixed acids in reducing numbers of bacteria on surfaces of lean meat. *J Food Saf*, 12, 139–147.

ANDERSON ME, MARSHALL RT and DICKSON JS (1992b), Estimating depths of bacterial penetration into post-rigor carcass tissue during washing. *J Food Saf*, 12, 191–198.

ARNOLD KW and KASPAR CW (1995), Starvation- and stationary-phase-induced acid tolerance in *Escherichia coli* O157:H7. *Appl Environ Microbiol*, 61, 2037–2039.

AYMERICH T, PICOUET PA and MONFORT JM (2008), Decontamination technologies for meat products. *Meat Sci*, 78, 114–129.

BACON RT, BELK KE, SOFOS JN, CLAYTON RP, REAGAN JO *et al.* (2000), Microbial populations on animal hides and beef carcasses at different stages of slaughter in plants employing multiple-sequential interventions for decontamination. *J. Food Prot*, 63, 1080–1086.

BARI ML, UKUKU DO, KAWASAKI T, INATSU Y, ISSHIKI K *et al.* (2005), Combined efficacy of nisin and pediocin with sodium lactate, citric acid, phytic acid, and potassium sorbate and EDTA in reducing *Listeria monocytogenes* population of inoculated fresh-cut produce. *J. Food Prot*, 68, 1381–1387.

BATZ MB, HOFFMANN S and MORRIS JR, JG (2012), Ranking the disease burden of 14 pathogens in food sources in the United States using attribution data from outbreak investigations and expert elicitation. *J Food Prot*, 75, 1278–1291.

BEARSON S, BEARSON B and FOSTER JW (1997), Acid stress responses in enterobacteria. *FEMS Microbiol Lett*, 147, 173–180.

BELL KY, CUTTER CN and SUMNER SS (1997), Reduction of foodborne micro-organisms on beef carcass tissue using acetic acid, sodium bicarbonate, and hydrogen peroxide spray washes. *Food Microbiol*, 14, 439–448.

BERGHOLZ TM, VANAJA SK and WHITTAM TS (2009), Gene expression induced in *Escherichia coli* O157:H7 upon exposure to model apple juice. *Appl Environ Microbiol*, 75, 3542–3553.

BERRY ED and CUTTER CN (2000), Effects of acid adaptation of *Escherichia coli* O157:H7 on efficacy of acetic acid spray washes to decontaminate beef carcass tissue. *Appl Environ Microbiol*, 66, 1493–1498.

BEYAZ D and TAYAR M (2010), The effect of lactic acid spray application on the microbiological quality of sheep carcasses. *J Anim Vet Adv*, 9, 1858–1863.

BOLTON DJ, DOHERTY AM and SHERIDAN JJ (2001), Beef HACCP: intervention and non-intervention systems. *Int J Food Microbiol*, 66, 119–129.

BOSILEVAC JM, NOU X, BARKOCY-GALLAGHER GA, ARTHUR TM and KOOHMARAIE M (2006), Treatments using hot water instead of lactic acid reduce levels of aerobic bacteria and Enterobacteriaceae and reduce the prevalence of *Escherichia coli* O157:H7 on preevisceration beef carcasses. *J Food Prot*, 69, 1808–1813.

BROWN JL, ROSS T, MCMEEKIN T and NICHOLS PD (1997), Acid habituation of *Escherichia coli* and the potential role of cyclopropane fatty acids in low pH tolerance. *Int J Food Microbiol*, 37, 163–173.

BUCHANAN RL and EDELSON SG (1996), Culturing enterhemorrhagic *Escherichia coli* in the presence and the absence of glucose as a simple means of evaluating the acid tolerance of stationary-phase cells. *Appl Environ Microbiol*, 62, 4009–4013.

BUCHANAN RL and EDELSON SG (1999), pH-dependent stationary-phase acid resistance response of enterohemorrhagic *Escherichia coli* in the presence of various acidulants. *J Food Prot*, 62, 211–218.

BYELASHOV OA and SOFOS JN (2009), Strategies for on-line decontamination of carcasses. In Toldrá F, *Safety of Meat and Processed Meat*, New York, NY, Springer, 149–182.

CABEDO L, SOFOS JN and SMITH GC (1996), Removal of bacteria from beef tissue by spray washing after different times of exposure to fecal material. *J Food Prot*, 59, 1284–1287.

CASTELO MM, KANG D-H, SIRAGUSA GR, KOOHMARAIE M and BERRY ED (2001), Evaluation of combination treatment processes for the microbial decontamination of pork trim. *J Food Prot*, 64, 335–342.

CASTILLO A, LUCIA LM, ROBERSON DB, STEVENSON TH, MERCADO I *et al.* (2001), Lactic acid sprays reduce bacterial pathogens on cold beef carcass surfaces and in subsequently produced ground beef. *J Food Prot*, 64, 58–62.

CDC (2011a), Surveillance for foodborne disease outbreaks – United States, 2008. *Morb Mortal Wkly Rep*, 60, 1197–1202.

CDC (2011b), Vital signs: incidence and trends of infection with pathogens transmitted commonly through food – foodborne diseases active surveillance network, 10 US sites, 1996–2010. *Morb Mortal Wkly Rep*, 60, 749–755.

CFR (2011a), Chemicals used in washing or to assist in the peeling of fruits and vegetables. Code of Federal Regulations 21 CFR Part 173, Section 173.315. Available from: www.accessdata.fda.gov/scripts/cdrh/cfdocs/cfcfr/CFRSearch.cfm?fr=173.315 [Accessed 25 November 2011].

CFR (2011b), Direct food substances affirmed as generally recognized as safe. Code of Federal Regulations 21 CFR Part 184. Available from: www.accessdata.fda.gov/scripts/cdrh/cfdocs/cfcfr/CFRSearch.cfm?CFRPart=184 [Accessed 25 November 2011].

CHANG J-M and FANG TJ (2007), Survival of *Escherichia coli* O157:H7 and *Salmonella enterica* serovars Typhimurium in iceberg lettuce and the antimicrobial effect of rice vinegar against *E. coli* O157:H7. *Food Microbiol*, 24, 745–751.

CONNER DE, KOTROLA JS, MIKEL WB and TAMBLYN KC (1997), Effects of acetic-lactic acid treatments applied to beef trim on populations of *Escherichia coli* O157:H7 and *Listeria monocytogenes* in ground beef. *J Food Prot*, 60, 1560–1563.

CUTTER CN and SIRAGUSA GR (1994), Efficacy of organic acids against *Escherichia coli* O157:H7 attached to beef carcass tissue using a pilot scale model carcass washer. *J Food Prot*, 57, 97–103.

CUTTER CN, DORSA WJ and SIRAGUSA GR (1997), Parameters affecting the efficacy of spray washes against *Escherichia coli* O157:H7 and fecal contamination on beef. *J Food Prot*, 60, 614–618.

DENG Y, RYU J-H and BEUCHAT LR (1999), Tolerance of acid-adapted and non-adapted *Escherichia coli* O157:H7 cells to reduce pH as affected by the type of acidulant. *J Appl Microbiol*, 86, 203–2010.

DERRICKSON-THARRINGTON E, KENDALL PA and SOFOS JN (2005), Inactivation of *Escherichia coli* O157:H7 during storage or drying of apple slices pretreated with acidic solutions. *Int J Food Microbiol*, 99, 79–89.

DICKSON JS (1992), Acetic acid action on beef tissue surfaces contaminated with *Salmonella* Typhimurium. *J Food Sci*, 57, 297–301.

DICKSON JS (1995), Susceptibility of preevisceration washed beef carcasses to contamination by *Escherichia coli* O157:H7 and Salmonellae. *J Food Prot*, 58, 1065–1068.

DIPERSIO PA, KENDAL PA, CALICIOGLU M and SOFOS JN (2003), Inactivation of *Salmonella* during drying and storage of apple slices treated with acidic or sodium metabisulfite solutions. *J Food Prot*, 66, 2245–2251.

DORSA WJ, CUTTER CN and SIRAGUSA GR (1997), Effects of acetic acid, lactic acid and trisodium phosphate on the microflora of refrigerated beef carcass surface tissue inoculated with *Escherichia coli* O157:H7, *Listeria innocua*, and *Clostridium sporogenes*. *J Food Prot*, 60, 619–624.

DORSA WJ, CUTTER CN and SIRAGUSA GR (1998a), Bacterial profile of ground beef made from carcass tissue experimentally contaminated with pathogenic and spoilage bacteria before being washed with hot water, alkaline solution, or organic acid and then stored at 4 or 12°C. *J Food Prot*, 61, 1109–1118.

DORSA WJ, CUTTER CN and SIRAGUSA GR (1998b), Long-term bacterial profile of refrigerated ground beef made from carcass tissue, experimentally contaminated with pathogens and spoilage bacteria after hot water, alkaline, or organic acid washes. *J Food Prot*, 61, 1615–1622.

DUBAL ZB, PATURKAR AM, WASKAR VS, ZENDE RJ, LATHA C *et al.* (2004), Effect of food grade organic acids on inoculated *S. aureus, L. monocytogenes, E. coli* and *S.* Typhimurium in sheep/goat meat stored at refrigeration temperature. *Meat Sci*, 66, 817–821.

EC (1988), On the approximation of the laws of the Member States concerning food additives authorized for use in foodstuffs intended for human consumption. Commission Directive 89/107/EEC. Available from: eur-lex.europa.eu/LexUriServ/LexUriServ.do?uri=OJ:L:1989:040:0027:0033:EN:PDF.

EDWARDS JR and FUNG DYC (2006), Prevention and decontamination of *Escherichia coli* O157:H7 on raw beef carcasses in commercial beef abattoirs. *J Rapid Methods Autom Microbiol*, 14, 1–95.

EFSA (2008), Scientific Opinion of the Panel on Biological Hazards on a request from DG SANCO on the assessment of the possible effect of the four antimicrobial treatment substances on the emergence of antimicrobial resistance. *EFSA J*, 659, 1–26.

EFSA (2011a), Scientific Opinion on the evaluation of the safety and efficacy of lactic acid for the removal of microbial surface contamination of beef carcasses, cuts and trimmings. *EFSA J*, 9, 2317–2351.

EFSA (2011b), Scientific report of EFSA and ECDC: the European Union Summary Report on trends and sources of zoonoses, zoonotic agents and food-borne outbreaks in 2009. *EFSA*, 9, 2090–2468.

EGGENBERGER-SOLORZANO L, NIEBUHR SE, ACUFF GR and DICKSON JS (2002), Hot water and organic acid interventions to control microbiological contamination on hog carcasses during processing. *J Food Prot*, 65, 1248–1252.

ELLEBRACHT EA, CASTILLO A, LUCIA LM, MILLER RK and ACUFF GR (1999), Reduction of pathogens using hot water and lactic acid on beef trimmings. *J Food Sci*, 64, 1094–1099.

FOSTER JW (1991), *Salmonella* acid shock proteins are required for the adaptive acid tolerance response. *J Bacteriol*, 173, 6896–6902.

FOSTER JW (1993), The acid tolerance response of *Salmonella* Typhimurium involves transient synthesis of key acid shock proteins. *J Bacteriol*, 175, 1981–1987.

FOSTER JW (1995), Low pH adaptation and the acid tolerance response of *Salmonella* Typhimurium. *Crit Rev Microbiol*, 21, 215–237.

FOSTER JW (1999), When protons attack: microbial strategies of acid adaptation. *Curr Opin Microbiol*, 2, 170–174.

FOSTER JW and HALL HK (1990), Adaptive acidification tolerance response of *Salmonella* Typhimurium. *J Bacteriol*, 172, 771–778.

FOSTER JW and HALL HK (1991), Inducible pH homeostasis and the acid tolerance response of *Salmonella* Typhimurium. *J Bacteriol*, 173, 5129–5135.

FSIS (2008), Raw ground beef – *E. coli* O157:H7 testing results. Available from: www.fsis.usda.gov/Science/Ground_Beef_E.Coli_Testing_Results/index.asp [Accessed 2 January 2009].

FSIS (2012), Safe and suitable ingredients used in the production of meat, poultry, and egg products, FSIS Directive 7120.1 Revision 13, USDA-FSIS. Available from: www.fsis.usda.gov/OPPDE/rdad/FSISDirectives/7120.1.pdf [Accessed 8 March 2013].

GABRIEL AA (2012), Microbial inactivation in cloudy apple juice by multi-frequency Dynashock power ultrasound. *Ultrason Sonochem*, 19, 346–351.

GAHAN CGM and HILL C (1999), The relationship between acid stress responses and virulence in *Salmonella* Typhimurium and *Listeria monocytogenes*. *Int J Food Microbiol*, 50, 93–100.

GAHAN CGM, O'DRISCOLL B and HILL C (1996), Acid adaptation of *Listeria monocytogenes* can enhance survival in acidic foods and during milk fermentation. *Appl Environ Microbiol*, 62, 3128–3132.

GANESH V, HETTIARACHCHY NS, RAVICHANDRAN M, JOHNSON MG, GRIFFIS CL *et al.* (2010), Electrostatic sprays of food-grade acids and plant extracts are more effective than conventional sprays in decontaminating *Salmonella* Typhimurium on spinach. *J Food Sci*, 75, 574–579.

GEORNARAS I and SOFOS JN (2005), Combining physical and chemical decontamination interventions for meat. In Sofos JN, *Improving the Safety of Fresh Meat*, Cambridge, UK, CRC/Woodhead Publishing Limited, 433–460.

GIL MI, SELMA MV, LÓPEZ-GÁLVEZ F and ALLENDE A (2009), Fresh-cut product sanitation and wash water disinfection: problems and solutions. *Int J Food Microbiol*, 134, 37–45.

GILL CO and BADONI M (1997), The effects of hot water pasteurizing treatments on the appearances of pork and beef. *Meat Sci*, 46, 77–87.

GILL CO and BADONI M (2004), Effects of peroxyacetic acid, acidified sodium chlorite or lactic acid solutions on the microflora of chilled beef carcasses. *Int J Food Microbiol*, 91, 43–50.

GOODSON M and ROWBURY RJ (1989), Habituation to normal lethal acidity by prior growth of *Escherichia coli* at a sublethal acid pH value. *Lett Appl Microbiol*, 8, 77–79.

GRAVES DELMORE LR, SOFOS JN, SCHMIDT GR and SMITH GC (1998), Decontamination of inoculated beef with sequential spraying treatments. *J Food Sci*, 63, 1–4.

GREENACRE EJ, BROCKLEHURST TF, WASPE CR, WILSON DR and WILSON PDG (2003), *Salmonella enterica* serovar Typhimurium and *Listeria monocytogenes* acid tolerance response induced by organic acids at 20°C: optimization and modeling. *Appl Environ Microbiol*, 69, 3945–3951.

GREENACRE EJ, LUCCHINI S, HINTON JCD and BROCKLEHURST TF (2006), The lactic acid-induced acid tolerance response in *Salmonella enterica* serovar Typhimurium induces sensitivity to hydrogen peroxide. *Appl Environ Microbiol*, 72, 5623–5625.

GREER GG and DILTS BD (1992), Factors affecting the susceptibility of meatborne pathogens and spoilage bacteria to organic acids. *Food Res Int*, 25, 355–364.

GREER GG and DILTS BD (1995), Lactic acid inhibition of the growth of spoilage bacteria and cold tolerant pathogens on pork. *Int J Food Microbiol*, 25, 141–151.

GUAN W and HUANG L (2005), Emerging decontamination technologies for meat. In Sofos JN, *Improving the Safety of Fresh Meat*, New York and Washington, CRC Press, 388–417.

GUAN W, HUANG L and FAN X (2010), Acids in combination with sodium dodecyl sulfate caused quality deterioration of fresh-cut iceberg lettuce during storage in modified atmosphere packaging. *J Food Sci*, 75, 435–440.

HARDIN MD, ACUFF GR, LUCIA LM, OMAN JS and SAVELL JW (1995), Comparison of methods for decontamination from beef carcass surfaces. *J Food Prot*, 58, 368–374.

HARRIS K, MILLER MF, LONERAGAN GH and BRASHEARS MM (2006), Validation of the use of organic acids and acidified sodium chlorite to reduce *Escherichia coli* O157 and *Salmonella* Typhimurium in beef trim and ground beef in a simulated processing environment. *J Food Prot*, 69, 1802–1807.

HILL C, O'DRISCOLL B and BOOTH I (1995), Acid adaptation and food poisoning microorganisms. *Int J Food Microbiol*, 28, 245–254.

HO K-LG, LUZURIAGA DA, RODDE KM, TANG S and PHAN C (2011), Efficacy of a novel sanitizer composed of lactic acid peroxyacetic acid against single strains of nonpathogenic *Escherichia coli* K-12, *Listeria innocua*, and *Lactobacillus plantarum* in aqueous solution and on surfaces of romaine lettuce and spinach. *J Food Prot*, 74, 1468–1474.

HUGAS M and TSIGARIDA E (2008), Pros and cons of carcass decontamination: the role of the European Food Safety Authority. *Meat Sci*, 78, 43–52.

IKEDA JS, SAMELIS J, KENDALL PA, SMITH GC and SOFOS JN (2003), Acid adaptation does not promote survival or growth of *Listeria monocytogenes* on fresh beef following acid and nonacid decontamination treatments. *J Food Prot*, 6, 985–992.

ILIC S, RAJIĆ A, BRITTON CJ, GRASSO E, WILKINS W *et al.* (2012), A scoping study characterizing prevalence, risk factor and intervention research, published between 1990 and 2010, for microbial hazards in leafy green vegetables. *Food Control*, 23, 7–19.

INATSU Y, BARI MDL, KAWASAKI S, ISSHIKI K and KAWAMOTO S (2005), Efficacy of acidified sodium chloride treatments in reducing *Escherichia coli* O157:H7 on Chinese cabbage. *J Food Prot*, 68, 251–255.

JIMENEZ-VILLARREAL JR, POHLMAN FW, JOHNSON ZB and BROWN JR AH (2003), Lipid, instrumental color and sensory characteristics of ground beef produced using trisodium phosphate, cetylpypiridinium chloride, chlorine dioxide or lactic acid as multiple antimicrobial interventions. *Meat Sci*, 65, 885–891.

KANG D-H, KOOHMARAIE M, DORSA WJ and SIRAGUSA GR (2001a), Development of a multiple-step process for the microbial decontamination of beef trim. *J Food Prot*, 64, 63–71.

KANG D-H, KOOHMARAIE M and SIRAGUSA GR (2001b), Application of multiple antimicrobial interventions for microbial decontamination of commercial beef trim. *J Food Prot*, 64, 168–171.

KARGIOTOU C, KATSANIDIS E, RHOADES J, KONTOMINAS M and KOUTSOUMANIS K (2011), Efficacies of soy sauce and wine base marinades for controlling spoilage of raw beef. *Food Microbiol*, 28, 158–163.

KESKINEN LA and ANNOUS BA (2011), Efficacy of adding detergents to sanitizer solutions for inactivation of *Escherichia coli* O157:H7 on romaine lettuce. *Int J Food Microbiol*, 147, 157–161

KOTULA KL and THELAPPURATE R (1994), Microbiological and sensory attributes of retail cuts of beef treated with acetic and lactic acid solutions. *J Food Prot*, 57, 665–670.

KOUTSOUMANIS KP and SOFOS JN (2004), Comparative acid stress response of *Listeria monocytogenes, Escherichia coli* O157:H7 and *Salmonella* Typhimurium after habituation at different pH conditions. *Lett Appl Microbiol*, 38, 321–326.

KOUTSOUMANIS KP, ASHTON LV, GEORNARAS I, BELK KE, SCANGA JA *et al.* (2004), Effect of single or sequential hot water and lactic acid decontamination treatments on the survival and growth of *Listeria monocytogenes* and spoilage microflora during aerobic storage of fresh beef at 4, 10, and 25°C. *J Food Prot*, 67, 2703–2711.

KROLL RG and PATCHETT RA (1992), Induced acid tolerance in *Listeria monocytogenes*. *Lett Appl Microbiol*, 14, 224–227.

KWON YM, PARK SY, BIRKHOLD SG and RICKE SC (2000), Induction of resistance of *Salmonella* Typhimurium to environmental stresses by exposure to short-chain fatty acids. *Food Microbiol Saf*, 65, 1037–1040.

LEISTNER L (2000), Basic aspects of food preservation by hurdle technology. *Int J Food Microbiol*, 55, 181–186.

LEYER GJ and JOHNSON EA (1992), Acid adaptation promotes survival of *Salmonella* spp. in cheese. *Appl Environ Microbiol*, 58, 2075–2080.

LEYER GJ and JOHNSON EA (1993), Acid adaptation induces cross-protection against environmental stresses in *Salmonella* Typhimurium. *Appl Environ Microbiol*, 59, 1842–1847.

LIN J, LEE IS, FREY J, SLONCZEWSKI JL and FOSTER JW (1995), Comparative analysis of extreme acid survival in *Salmonella* Typhimurium, *Shigella flexneri*, and *Escherichia coli*. *J Bacteriol*, 177, 4097–4104.

LIN J, SMITH MP, CHAPIN KC, BAIK HS, BENNETT GN *et al.* (1996), Mechanisms of acid resistance in enterohemorrhagic *Escherichia coli*. *Appl Environ Microbiol*, 62, 3094–3100.

LÓPEZ-GÁLVEZ F, ALLENDE A, SELMA MV and GIL MI (2009), Prevention of *Escherichia coli* cross-contamination by different commercial sanitizers during washing of fresh-cut lettuce. *Int J Food Microbiol*, 133, 167–171.

LORETZ M, STEPHAN R and ZWEIFEL C (2011a), Antibacterial activity of decontamination treatments for cattle hides and beef carcasses. *Food Control*, 22, 347–359.

LORETZ M, STEPHAN R and ZWEIFEL C (2011b), Antibacterial activity of decontamination treatments for pig carcasses. *Food Control*, 22, 1121–1125.

LOU Y and YOUSEF AE (1997), Adaptation to sublethal environmental stresses protects *Listeria monocytogenes* against lethal preservation factors. *Appl Environ Microbiol*, 63, 1252–1255.

MIRA NP, TEIXEIRA MC and SÁ-CORREIA I (2010), Adaptive response and tolerance to weak acids in *Saccharomyces cerevisiae*: a genome-wide view. *OMICS J Integr Biol*, 14, 525–540.

MOSQUEDA-MELGAR J, RAYBAUDI-MASSILIA RM and MARTÍN-BELLOSO O (2008a), Non-thermal pasteurization of fruit juices by combining high-intensity pulsed electric fields with natural antimicrobials. *Innov Food Sc Emerg Technol*, 9, 328–340.

MOSQUEDA-MELGAR J, RAYBAUDI-MASSILIA RM and MARTÍN-BELLOSO O (2008b), Combination of high-intensity pulsed electric fields with natural antimicrobials to inactivate pathogenic microorganisms and extend the shelf-life of melon and watermelon juices. *Food Microbiol*, 25, 479–491.

NISSEN H, MAUGESTEN T and LEA P (2001), Survival and growth of *Escherichia coli* O157:H7, *Yersinia enterocolitica* and *Salmonella* Enteritidis on decontaminated and untreated meat. *Meat Sci*, 57, 291–298.

O'DRISCOLL B, GAHAN CGM and HILL C (1996), Adaptive acid tolerance response in *Listeria monocytogenes*: isolation of an acid-tolerant mutant which demonstrates increased virulence. *Appl Environ Microbiol*, 62, 1693–1698.

OJEU (2004), Regulation (EC) No 853/2004 of the European Parliament and of the Council of 29 April 2004 laying down specific hygiene rules for food of animal origin. *Official J European Union*, L226 (25/06/2004), 22–82.

OKOLOCHA EC and ELLERBROEK L (2005), The influence of acid and alkaline treatments on pathogens and the shelf life of poultry meat. *Food Control*, 16, 217–225.

ÖLMEZ H and KRETZSCHMAR U (2009), Potential alternative disinfection methods for organic fresh-cut industry for minimizing water consumption and environment impact. *LWT Food Sci Technol*, 42, 686–693.

PARK S-H, CHOI M-R, PARK J-W, PARK K-H, CHUNG M-S *et al.* (2011), Use of organic acids to inactivate *Escherichia coli* O157:H7, *Salmonella* Typhimurium, and *Listeria monocytogenes* on organic fresh apples and lettuce. *J Food Sci*, 76, 293–298.

PIPER PW (2011), Resistance of yeasts to weak organic acid food preservatives. *Adv Appl Microbiol*, 77, 97–113.

PODOLAK RK, ZAYAS JF, KASTNER CL and FUNG DYC (1996), Inhibition of *Listeria monocytogenes* and *Escherichia coli* O157:H7 on beef by application of organic acids. *J Food Prot*, 59, 370–373.

QUILO SA, POHLMAN FW, DIAS-MORSE PN, BROWN JR AH, CRANDALL PG *et al.* (2010), Microbial, instrumental color and sensory characteristics of inoculated ground beef produced using potassium lactate, sodium metasilicate or peroxyacetic acid as multiple antimicrobial interventions. *Meat Sci*, 84, 470–476.

RAHMAN SME, DING T and OH D-H (2010), Inactivation effect of newly developed low concentration electrolyzed water and other sanitizers against microorganisms on spinach. *Food Control*, 21, 1383–1387.

RAJKOVIC A, SMIGIC N and DEVLIEGHERE F (2010), Contemporary strategies in combating microbial contamination in food chain. *Int J Food Microbiol*, 141, S29–S42.

RAYBAUDI-MASSILIA RM, MOSQUEDA-MELGAR J and MARTÍN-BELLOSO O (2008a), Edible alginate based coating as carrier of antimicrobials to improve shelf-life and safety of fresh-cut melon. *Int J Food Microbiol*, 121, 313–327.

RAYBAUDI-MASSILIA RM, MOSQUEDA-MELGAR J and MARTÍN-BELLOSO O (2009a), Antimicrobial activity of malic acid against *Listeria monocytogenes, Salmonella* Enteritidis and *Escherichia coli* O157:H7 in apple, pear and melon juices. *Food Con*, 20, 105–112.

RAYBAUDI-MASSILIA RM, MOSQUEDA-MELGAR J, SOLIVA-FORTUNY R and MARTÍN-BELLOSO O (2009b), Control of pathogenic and spoilage microorganisms in fresh-cut fruits and fruit juices by traditional and alternative natural antimicrobials. *Compr Rev Food Sci Food Saf*, 8, 157–180.

RAYBAUDI-MASSILIA RM, MOSQUEDA-MELGAR J, SORBINO-LÓPEZ A, SOLIVA-FORTUNY R and MARTÍN-BELLOSO O (2007), Shelf-life extension of fresh-cut 'Fuji' apples at different ripeness stages using natural substances. *Post Harv Biol Technol*, 45, 265–275.

RAYBAUDI-MASSILIA RM, MOSQUEDA-MELGAR J, SORBINO-LÓPEZ A, SOLIVA-FORTUNY R and MARTÍN-BELLOSO O (2009c), Use of malic acid and other quality stabilizing compounds to assure the safety of fresh-cut 'Fuji' apples by inactivation of *Listeria monocytogenes, Salmonella* Enteritidis and *Escherichia coli* O157:H7. *J Food Saf*, 29, 236–252.

RAYBAUDI-MASSILIA RM, ROJAS-GRAÜ MA, MOSQUEDA-MELGAR J and MARTÍN-BELLOSO O (2008b), Comparative study on essential oils incorporated into an alginate-based edible coating to assure the safety and quality of fresh-cut Fuji apples. *J Food Prot*, 71, 1150–1161.

RICKE SC (2003), Perspectives on the use of organic acids and short chain fatty acids as antimicrobials. *Poult Sci*, 82, 632–639.

RIEDEL CT, BRØNDSTED L, ROSENQUIST H, HAXGART SN and CHRISTENSEN BB (2009), Chemical decontamination of *Campylobacter jejuni* on chicken skin and meat. *J Food Prot*, 72, 1173–1180.

RODGERS SL, CASH JN, SIDDIQ M and RYSER ET (2004), A comparison of different chemical sanitizers for inactivating *Escherichia coli* O157:H7 and *Listeria monocytogenes* in solution and on apples, lettuce, strawberries, and cantaloupe. *J Food Prot*, 67, 721–731.

ROSE BE, HILL WE, UMHOLTZ R, RANSOM GM and JAMES WO (2002), Testing for *Salmonella* in raw meat and poultry products collected at federally inspected establishments in the United States, 1998 through 2000. *J. Food Prot*, 65, 937–947.

ROWE MT and KIRK R (1999), An investigation into the phenomenon of cross-protection in *Escherichia coli* O157:H7. *Food Microbiol*, 16, 157–164.

RYU J-H and BEUCHAT LR (1999), Changes in heat tolerance of *Escherichia coli* O157:H7 after exposure to acidic environments. *Food Microbiol*, 16, 317–324.

SAGONG H-G, LEE S-Y, CHANG P-S, HEU S, RYU S *et al.* (2011), Combined effect of ultrasound and organic acids to reduce *Escherichia coli* O157:H7, *Salmonella* Typhimurium, and *Listeria monocytogenes* on organic fresh lettuce. *Int J Food Microbiol*, 145, 287–292.

SAKHARE PZ, SACHINDRA NM, YASHODA KP and NARASIMHA RAO D (1999), Efficacy of intermittent decontamination treatments during processing in reducing the microbial load on broiler chicken carcass. *Food Con*, 10, 189–194.

SAMARA A and KOUTSOUMANIS KP (2009), Effect of treating lettuce surfaces with acidulants on the behaviour of *Listeria monocytogenes* during storage at 5 and 20°C and subsequent exposure to simulated gastric fluid. *Int J Food Microbiol*, 129, 1–7.

SAMELIS J (2005), Meat decontamination and pathogen stress adaptation. In Sofos JN, *Improving the Safety of Fresh Meat*, Cambridge, UK, CRC/Woodhead Publishing Limited, 562–591.

SAMELIS J and SOFOS JN (2003), Strategies to control stress-adapted pathogens. In Yousef AE and Juneja VK, *Microbial Stress Adaptation and Food Safety*, Boca Dalton, FL, CRC Press, 303–351.

SAMELIS J, KENDALL PA, SMITH GC and SOFOS JN (2004), Acid tolerance of acid-adapted and nonadapted *Escherichia coli* O157:H7 following habituation (10°C) in fresh beef decontamination runoff fluids of different pH values. *J Food Prot*, 67, 638–645.

SAMELIS J, SOFOS JN, KENDALL PA and SMITH GC (2002), Effect of acid adaptation on survival of *Escherichia coli* O157:H7 in meat decontamination washing fluids and potential effects of organic acid interventions on the microbial ecology of the meat plant environment. *J Food Prot*, 65, 33–40.

SAPERS GM (2001), Efficacy of washing and sanitizing methods for disinfection of fresh fruit and vegetable products. *Food Technol Biotechnol*, 39, 305–311.

SCALLAN E, GRIFFIN PM, ANGULO FJ, TAUXE RV and HOEKSTRA RM (2011a), Foodborne illness acquired in the United States – unspecified agents. *Emerg Infect Dis*, 17, 16–22.

SCALLAN E, HOEKSTRA RM, ANGULO FJ, TAUXE RV, WIDDOWSON M-A *et al.* (2011b), Foodborne illness acquired in the United States – major pathogens. *Emerg Infect Dis*, 17, 1–15.

SHEN C, GEORNARAS I, BELK KE, SMITH GC and SOFOS JN (2011), Thermal inactivation of acid, cold, heat, starvation, and desiccation stress-adapted *Escherichia coli* O157:H7 in moisture-enhanced nonintact beef. *J Food Prot*, 74, 531–538.

SINGLA R, GANGULI A and GHOSH M (2011), An effective combined treatment using malic acid and ozone inhibits *Shigella* spp. on sprouts. *Food Con*, 22, 1032–1039.

SINHAMAHAPATRA M, BISWAS S, DAS AK and BHATTACHARYYA D (2004), Comparative study of different surface decontaminants on chicken quality. *Br Poult Sci*, 45, 624–630.

SKANDAMIS PN, NYCHAS G-JE and SOFOS JN (2010), Meat decontamination. In Toldrá F, *Handbook of Meat Processing*, Ames, IA, Blackwell Publishing, 43–85.

SKANDAMIS PN, STOPFORTH JD, ASHTON LV, GEORNARAS I, KENDALL PA *et al.* (2009), *Escherichia coli* O157:H7 survival, biofilm formation and acid tolerance under simulated slaughter plant moist and dry conditions. *Food Microbiol*, 26, 112–119.

SKANDAMIS PN, STOPFORTH JD, KENDAL PA, BELK KE, SCANGA JA *et al.* (2007), Modeling the effect of inoculum size and acid adaptation on growth/no growth interface of *Escherichia coli* O157:H7. *Int J Food Microbiol*, 120, 327–237.

SMULDERS FJM and GREER GG (1998), Integrating microbial decontamination with organic acids in HACCP programmes for muscle foods: prospects and controversies. *Int J Food Microbiol*, 44, 149–169.

SOFOS JN (2008), Challenges to meat safety in the 21st century. *Meat Sci*, 78, 3–13.

SOFOS JN and SMITH GC (1998), Nonacid meat decontamination technologies: model studies and commercial applications. *Int J Food Microbiol*, 44, 171–188.

STIVARIUS MR, POHLMAN FW, MCELYEA KS and APPLE JK (2002a), The effects of acetic acid, gluconic acid and trisodium citrate treatment of beef trimmings on microbial, color and odor characteristics of ground beef through simulated retail display. *Meat Sci*, 60, 245–252.

STIVARIUS MR, POHLMAN FW, MCELYEA KS and WALDROUP AL (2002b), Effects of hot water and lactic acid treatment of beef trimmings prior to grinding on microbial, instrumental color and sensory properties of ground beef during display. *Meat Sci*, 60, 327–334.

STOPFORTH JD, ASHTON LV, SKANDAMIS PN, SCANGA JA, SMITH GC *et al.* (2005a), Single and sequential treatment of beef tissue with lactic acid, ammonium hydroxide, sodium metasilicate, and acidic and basic oxidized water to reduce numbers of inoculated *Escherichia coli* O157:H7 and *Salmonella* Typhimurium. *Food Prot Trends*, 25, 14–22.

STOPFORTH JD, O'CONNOR R, LOPES M, KOTTAPALLI B, HILL WE *et al.* (2007), Validation of individual and multiple-sequential innervations for reduction of microbial populations during processing of poultry carcasses and parts. *J Food Prot*, 70, 1393–1401.

STOPFORTH JD, SAMELIS J, SOFOS JN, KENDALL PA and SMITH GC (2003), Influence of organic acid concentration on survival of *Listeria monocytogenes* and *Escherichia coli* O157:H7 in beef carcass wash water and on model equipment surfaces. *Food Microbiol*, 20, 651–660.

STOPFORTH JD, YOON Y, BARMPALIA IM, SAMELIS J, SKANDAMIS PN *et al.* (2005b), Reduction of *Listeria monocytogenes* populations during exposure to a simulated gastric fluid following storage of inoculated frankfurters formulated and treated with preservatives. *Int J Food Microbiol*, 99, 309–319.

THERON MM and LUES JFR (2007), Organic acids and meat preservation: a review. *Food Rev Int*, 23, 141–158.

VAN NETTEN P, MOSSEL DAA and HUIS In 't Veld JHJ (1997), Microbial changes on freshly slaughtered pork carcasses due to 'hot' lactic acid decontamination. *J Food Saf*, 17, 89–111.

VAN NETTEN P, VALENTIJN A, MOSSEL DAA and HUIS In 't Veld JHJ (1998), The survival and growth of acid-adapted mesophilic pathogens that contaminate meat after lactic acid decontamination. *J Appl Microbiol*, 84, 559–567.

VANDEKINDEREN I, DEVLIEGHERE F, DE MEULENAER B, VERAMME K, RAGAERT P *et al.* (2008a), Impact of decontamination agents and a packaging delay on the respiration rate of fresh-cut produce. *Postharvest Biol Tec*, 49, 277–282.

VANDEKINDEREN I, VAN CAMP J, DEVLIEGHERE F, VERAMME K, BERNAERT N *et al.* (2009a), Effect of decontamination on the microbial load, the sensory quality and the nutrient retention of ready-to-eat white cabbage. *Eur Food Res Technol*, 229, 443–455.

VANDEKINDEREN I, VAN CAMP J, DEVLIEGHERE F, RAGAERT P, VERAMME K *et al.* (2009b), Evaluation of the use of decontamination agents during fresh-cut leek processing and quantification of their effect on its total quality by means of a multidisciplinary approach. *Innov Food Sci Emerg*, 10, 363–373.

VANDEKINDEREN I, VAN CAMP J, DEVLIEGHERE F, VERAMME K, DENON Q *et al.* (2008b), Effect of decontamination agents on the microbial population, sensorial quality and nutrient content of grated carrots (*Daucus carota* L.). *Agr Food Sci*, 56, 5723–5731.

VELÁZQUEZ LC, BARBINI NB, ESCUDERO ME, ESTRADA CL and GUZMÁN S (2009), Evaluation of chlorine, benzalkonium chloride and lactic acid as sanitizers for reducing *Escherichia coli* O157:H7 and *Yersinia enterocolitica* on fresh vegetables. *Food Con*, 20, 262–268.

VELLIOU EG, VAN DERLINDEN E, CAPPUYNS AM, GEERAERD AH, DEVLIEGHERE F *et al.* (2012), Heat inactivation of *Escherichia coli* K12 MG1655: effect of microbial metabolites and acids in spent medium. *J Thermal Biol*, 37, 72–78.

WALKER M and PHILLIPS CA (2008), The effect of preservatives on *Alicyclobacillus acidoterrestris* and *Propionibacterium cyclohexanicum* in fruit juice. *Food Con*, 19, 974–981.

WAN T-C, LIN L-C and SAKATA R (2007), Effect of organic acids on the microbial quality of Taiwanese-style sausages. *Anim Sci J*, 78, 407–412.

YODER SF, HENNING WR, MILLIS EW, DOORES S, OSTIGUY N *et al.* (2010), Investigation of water washes suitable for very small meat plants to reduce pathogens on beef surfaces. *J Food Prot*, 73, 907–915.

ZHAO T, ZHAO P and DOYLE MP (2010), Inactivation of *Escherichia coli* O157:H7 and *Salmonella* Typhimurium DT 104 on alfalfa seeds by levulinic acid and sodium dodecyl sulfate. *J Food Prot*, 73, 2010–2017.

19

Progress in intervention programs to eradicate foodborne helminth infections

K. Kniel, University of Delaware, USA

DOI: 10.1533/9780857098740.4.385

Abstract: This chapter discusses the intervention methods used to control two helminthic diseases of public health importance. *Trichinella* and *Taenia* are transmitted by undercooked contaminated meat. Important changes made to animal husbandry and animal feeding have led to a decrease in these diseases in parts of the world. Specific information is provided on disease transmission coupled with the success achieved by intervention programs.

Key words: *Trichinella, Taenia*, cysticercosis, public health.

19.1 Introduction

Foodborne helminths are a varied and complex group of eukaryotic parasitic worms. Many different types of worms are categorized as helminths, including nematodes (roundworms), cestodes (tapeworms) and trematodes (flukes). Each organism has a complicated life cycle involving multiple hosts of mammalian or non-mammalian origin. Worldwide, helminths are an important cause of disease and may be transmitted by water or food. The Disease Control Priorities Project (DCPP, 2012) estimated that the worldwide incidence of intestinal helminth infection, including among people living in developed countries, is three billion cases. Public health eradication programs targeted at helminths have a long history around the globe and are ongoing today. These programs may include vaccination or testing. A crucial issue today is linking eradication programs with the aim of reducing disease in individuals with HIV and other immune-suppressing diseases. Today more people are living relatively long lives with serious immunocompromising illnesses, which in turn may make these same individuals more susceptible to diseases caused by eukaryotic parasites.

Two helminths of agricultural significance are *Trichinella* species and *Taenia* species. Both of these organisms can be transmitted to humans by undercooked meat and spread to other animals via contaminated pasture, food or water. The importance of these organisms and the results of eradication and intervention programs are described in this chapter. The organisms within these two genera are considered in this chapter because there are interesting aspects of the eradication programs and their clear importance to, and impact on, food safety. In particular intensive swine production is an excellent example of how proper intervention programs can yield significant success in pathogen reduction (Davies, 2011). While many social concerns have been raised about intensive livestock production, the elimination of the risk from foodborne helminths like *Taenia solium, Taenia saginata* and *Trichinella spiralis* is a great achievement, as described in this chapter.

19.2 Foodborne helminth infections and food animal contamination

19.2.1 Description of foodborne helminth infections

Trichinella *species*

Trichinae comprise several species of parasitic nematodes, which infect a variety of warm-blooded carnivores and omnivores. While many roundworms have complicated life cycles, trichinae have a direct life cycle, where the organism completes all stages of development in one host (Fig. 19.1). Transmission from one host to another occurs by ingestion of infected muscle tissue containing the encysted larval stage. After ingestion, the larvae excyst within the new host, enter the tissues of the small intestine and develop into adult worms. Adult male and female worms mate and produce many larvae, which are able to break through the intestine, enter the circulatory system and lodge themselves in muscle. Most often this is striated muscle, but in rare and more serious cases this can be heart or diaphragm tissue. The larvae develop into a unique encysted stage called a nurse cell.

The disease trichinellosis, also known as trichinosis, is distributed through temperate, tropical and artic regions (Mawhorter and Kazura, 1993; Bruschi and Murrell, 2006). The disease is still considered endemic in parts of the world where small family-sized outbreaks occur, for example in parts of Eastern Europe, Asia and Latin America (Dupouy-Carnet and Bruschi, 2007). The disease has several phases. The first phase includes symptoms of anorexia, nausea, vomiting, diarrhea and upper abdominal pain. About 1–6 weeks after ingestion of larvae in undercooked contaminated meat, the systemic phase begins, which consists of fever, sweating, facial edema, myalgia, muscle swelling and muscle weakness. Disease confirmation may include correlating symptoms with muscle biopsy or detection of anti-*Trichinella* antibodies in the serum (Eckert, 2005). The disease can be treated with anti-helminthics, like benzimidazoles, and corticosteroids (Schantz and Dietz, 2001).

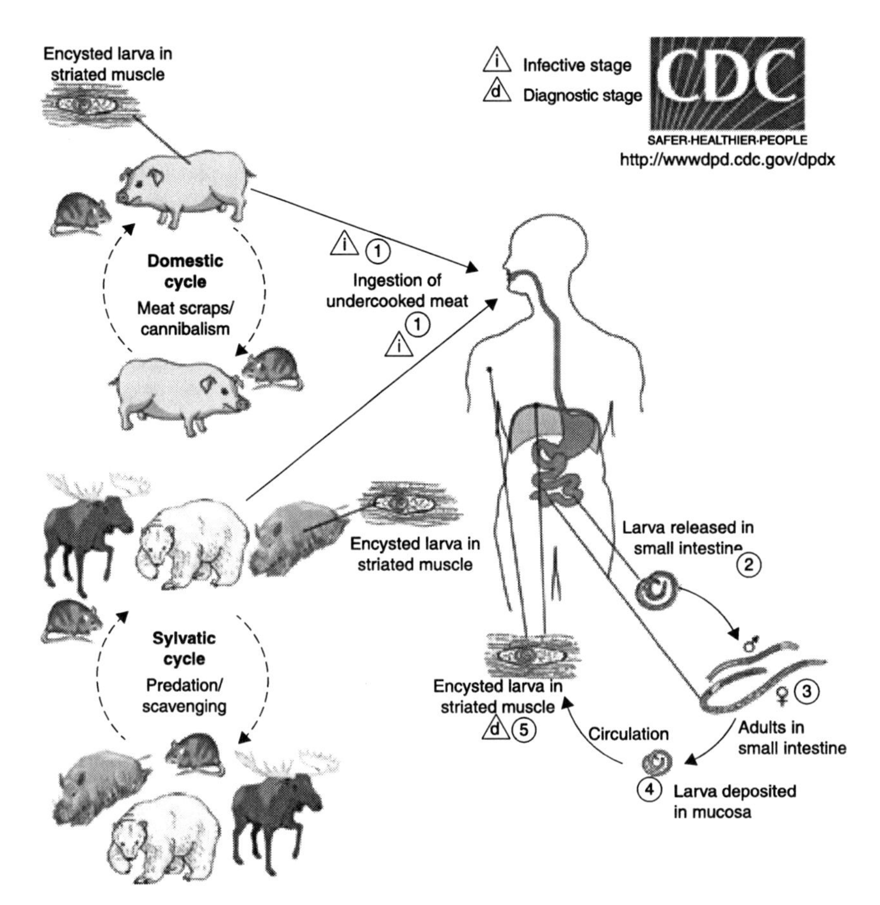

Fig. 19.1 Life cycle of *Trichinella* species. The basic steps of the life cycle for *Trichinella* include (1) release from cysts after exposure to gastric acid and pepsin, (2) invasion of the small intestine where they develop into adult worms, where after 1 week (3–4), the females release larvae that migrate to striated muscles where they encyst (5). Image courtesy of the Centers for Disease Control and Prevention (CDC).

There are seven different species of *Trichinella* and they vary in epidemiological importance. The most common zoonotic species is *T. spiralis*. This organism has spread across five continents. At least two other species are of importance in terms of disease transmission and disease epidemiology. These are the sylvatic species, *T. britovi* and *T. pseudospiralis*, which have caused outbreaks in dogs and pigs across Eastern Europe (Murrell and Pozio, 2000). The sylvatic transmission cycle is a part of the natural transmission cycle of the pathogen, which can survive for years in wildlife.

Taenia *species*

Taenia spp. are long, segmented, parasitic tapeworms. These parasites have an indirect life cycle, cycling between a definitive and an intermediate host

(Fig. 19.2). *Taenia saginata* and *T. solium* are zoonotic, with humans serving as the definitive host, the intermediate host or both. Adult tapeworms live in the intestines of the definitive hosts, and in this case humans are the definitive hosts for *T. solium* (the pork tapeworm) and *T. saginata* (the beef tapeworm). These two species of *Taenia* are important for agricultural livestock, in particular pork and beef: cattle are the intermediate host for *T. saginata* and pigs the intermediate host for *T. solium*. The diseases bovine and porcine cysticercosis are caused by the larval stage. Grazing cattle may ingest pasture or materials contaminated with the tapeworm sections called proglottids. After ingestion these sections release the life stage and the oncospheres penetrate the intestinal wall and travel with the blood supply to striated muscles. Cysts develop within those muscles and may survive for years. The development of this larval stage causes cysticercosis, which is commonly called beef or pork measles.

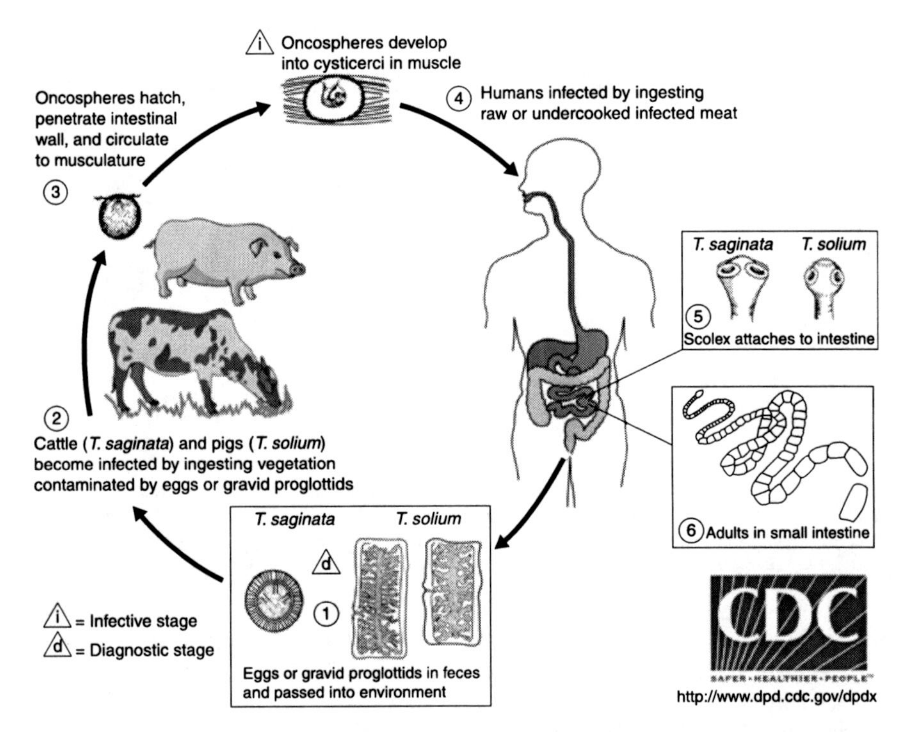

Fig. 19.2 Life cycle of *Taenia* species. The basic steps of the life cycle for *Taenia saginata* and *Taenia solium* are shown. Humans are the only definitive hosts for *T. saginata* and *T. solium*. Eggs or gravid proglottids are passed with feces (1) and cattle (*T. saginata*) or pigs (*T. solium*) become infected by ingesting vegetation contaminated with eggs or gravid proglottids (2). The oncospheres hatch in the animal's intestines (3), invade the intestinal wall and migrate to striated muscles, where they develop into cysticerci. Humans become infected by ingesting raw or undercooked infected meat (4). Over 2 months, a cysticercus develops into an adult tapeworm in the human intestine and adult tapeworms attach to the small intestine by their scolex (5) and live within the small intestine (6). Image courtesy of the Centers for Disease Control and Prevention (CDC).

Humans, as the definitive host, complete the life cycle after ingesting raw or undercooked muscle with cysts containing a viable metacestode. Infected humans can shed gravid proglottids in their feces, which are tapeworm segments full of up to 50 000 eggs. Worldwide, taeniasis and cysticercosis are common parasitic infections: 2–3 million people are thought to be infected with adult *T. solium*, 45 million with adult *T. saginata* and 50 million with *T. solium* cysticerci (OIE, 2005). An estimated 50 000 people die annually from the central nervous system or cardiac complications (OIE, 2005).

19.2.2 Food animal contamination

The measures aimed at reducing the diseases caused by the parasitic worms discussed in this chapter should be widely available and properly used by the majority of animal breeders, hunters and consumers. In some parts of the world, traditions and culinary customs have been correlated with public health conditions. For example, in Romania, the incidence of trichinosis associated with pork decreased markedly after 2001; however, the disease associated with game meat still occurs as does the disease associated with bad habits and poverty (Neghina, 2010). Reductions in disease incidence are attributed to feeding guidelines and regulations and changes in feeding habits along with awareness. There is concern, however, that the considerable degree of success achieved by early control programs for *Trichinella* infection in the 1960s and the disappearance of this pathogen from domestic pigs in industrialized countries may have misled inspectors into a level of complacency (Murrell and Pozio, 2000). This concern is reinforced by the fact that the majority of technicians in charge of the control for *Trichinella* in European slaughterhouses have never even seen a larva of *Trichinella* (Murrell and Pozio, 2000). Certainly, improved detection has played an important role in recognizing the persistence and spread of these zoonoses.

Trichinae

Trichinella spiralis has a long history of association with pork products in the United States and around the globe; however, more recently *Trichinella* contamination of more exotic animals, like cougar, bear and wild boar, has been more common in the United States. In the 1940s, a report from the National Institutes of Health indicated that 16.2% of the population in the United States had been infected with *Trichinella spiralis* (Zimmermann *et al.*, 1973). This information along with the 2.5% infection rate of commercial pigs led to the federal control of methods used to prepare ready-to-eat pork products and publicity about the dangers of eating pork (Zimmermann *et al.*, 1971).

Taenia

Cysticercosis caused by *T. solium* is most common in Latin America, South-east Asia and Africa. It is particularly prevalent in rural areas where domestic pigs are allowed to roam freely. It is diminishing in eastern and southern Europe, and is very rare in Muslim countries. In the United States, *T. solium* may find resurgence in free-range swine.

While *T. saginata* has been virtually eliminated from cattle produced within the United States, spontaneous outbreaks still occur. In 1997 the US Department of Agriculture's (USDA) Food Safety and Inspection Service (FSIS) reported an outbreak in cattle, which was traced back to contaminated cotton burrs (Snowden *et al.*, 1997). This study found that the contamination likely occurred by the overflow of effluent from a municipal sewage lagoon in the Texas Panhandle. The cotton burrs, which are the stemmy portion of cotton left over after ginning, were shipped to many feedlots. At least three feedlots lost carcass value or had carcasses condemned due to *T. saginata* infection. This outbreak was widely investigated since it involved interstate commercial shipments of feed. This outbreak restates the importance of timely reporting by FSIS veterinary medical officers to the appropriate state health departments and regulatory agencies. *T. saginata* is a zoonotic disease of economic importance. The carcasses that are not severely affected enough for condemnation must be frozen in order to kill any parasites that may be present. Like *T. saginata*, *T. solium* can be spread to pigs through contaminated feed.

19.3 Impact of intervention programs

According to FSIS, parasitic worms of public health importance are the beef and pork tapeworms (*Taenia saginata* and *Taenia solium*, respectively) and the roundworm that causes trichinosis (*Trichinella spiralis*). In an animal with a severe tapeworm infection, immature stages (cysts) may be observed. These cysts can be detected during routine inspection by federal or state program personnel or by plant employees in a very small plant. This type of an infection would be condemned and not be further processed for human consumption (FSIS, 1999). The main goal of FSIS is to reduce the risk of initial contamination in the field and also to reduce cross-contamination at processing plants. Many countries have approved methods for the post-mortem inspection of pork and beef (which are not discussed in this chapter) and while these are still important, some inspections now take place in the field and pre-harvest strategies to reduce risk are of increasing importance. The importance of these intervention programs should not be minimized. While helminthic infections may not be a major contributor to deaths in much of the world including tropical regions, they certainly have a significant effect on health, growth and physical fitness, school attendance, worker productivity and the earning potential of people worldwide (Laxminarayan *et al.*, 2006).

19.3.1 *Trichinella*-specific intervention programs

The National Trichinae Certification Program is very important; it was developed through the USDA. The program's standards establish a set of criteria for good swine management practices that minimize the risk of exposure of swine to

T. spiralis. This lets producers market swine that are not considered a risk to public health. The standards in this program were developed over years, and were based on the expertise and input from the National Pork Producers Council (NPPC) and several USDA agencies, including the Animal and Plant Health Inspection Service (APHIS), the Agricultural Research Service (ARS), the former Cooperative States Research, Education and Extension Service (CSREES) and FSIS. The use of risk management rather than testing alone is an important feature of this program.

The program's stages and requirements are outlined below (see also www.aphis.usda.gov). A production site begins initial audit approval by arranging for an audit with the appropriate APHIS administrator. When a Stage I program audit is complete the producer has met all requirements of trichinae good production practices in accordance with the APHIS area office. After five months, a producer may request an audit for Stage II status under which a producer may sell swine as trichinae certified; then 8–10 months later, a producer may seek Stage III status in accordance with the APHIS area office. The success of this program is evident through the careful and specific approach. The flow of events as identified by the USDA is depicted in Fig. 19.3.

Audit completion, review and certification status will be determined on an individual basis for each site. Entry into the program will be determined from the date of the audit and will progress through three stages. A pre-audit package is available for growers (www.aphis.usda.gov/vs/trichinae). When a producer has determined that he/she is ready for the actual audit and program stages, he/she should contact a qualified veterinarian for a Stage I audit. APHIS maintains a list of qualified accredited veterinarians and qualified veterinary medical officers who can perform audits. The auditor will visit the producer and determine if the site has implemented good production practices as deemed appropriate through the trichinae-free program. These are as follows:

- All non-breeding swine entering the site either originate from a certified production site or are less than five weeks old. The source herd trichinae identification number (TIN) must be documented in an animal movement record.
- Sources of feed or feed ingredients meet good manufacturing practices or quality assurance standards recognized by the feed industry.
- Feed prepared on the site must be prepared, stored and delivered in a manner such that the feed has not been contaminated with rodent or wildlife carcasses.
- Exclusion and control of rodents and wildlife is to a level such that fresh signs of activity of these animals are not observed in the swine production or feed preparation and storage areas. The producer maintains a rodent control logbook with a site diagram, or records of a pest control operator, which are updated on at least a monthly basis.
- Wildlife carcasses are not intentionally fed to swine. Swine shall not have access to wildlife harborage or carcasses on the site.
- Feeding of waste containing meat is a major risk factor for swine infection with trichinae. Feeding of waste food to swine requires state licensing and a

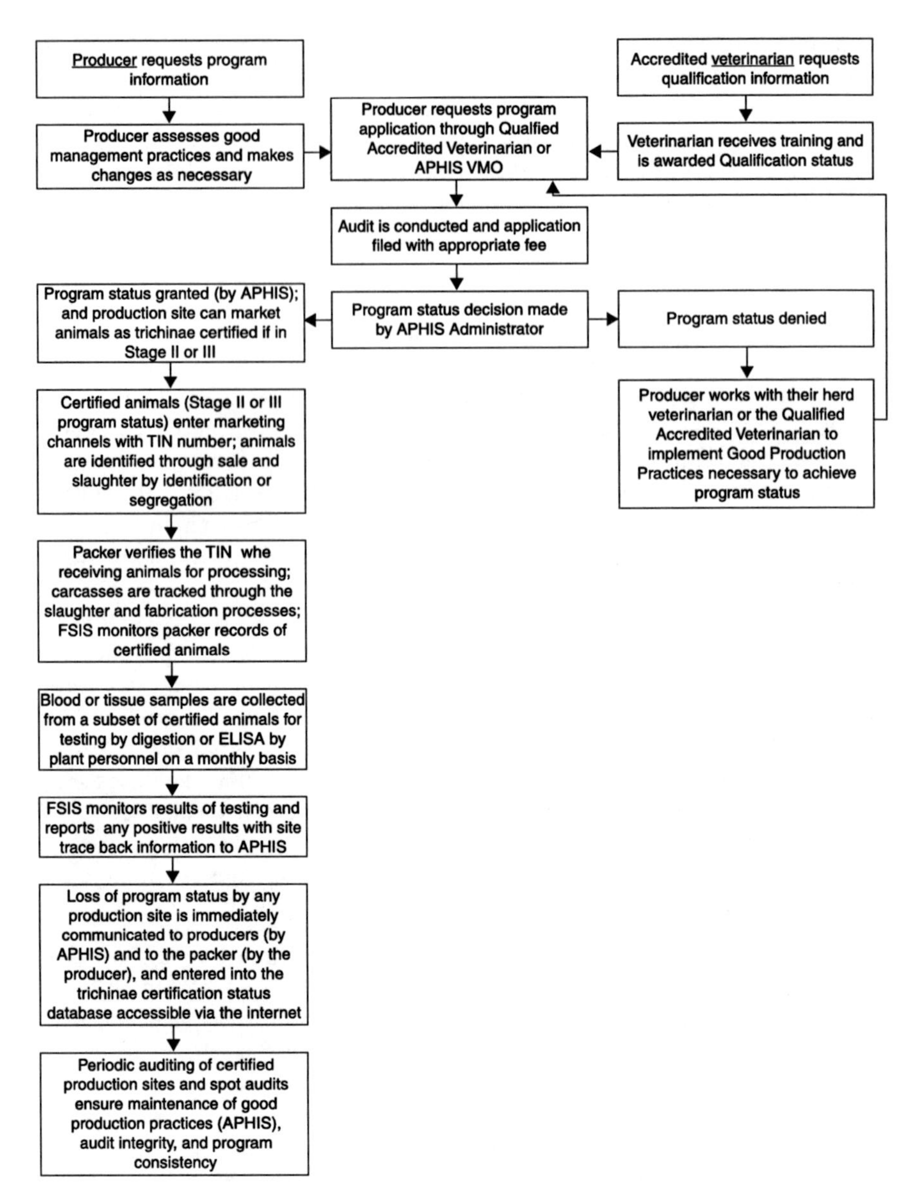

Fig. 19.3 Flow chart for the US National Trichinae Certification Program. The proposed certification process includes the following elements: (1) Veterinarians, trained in good production practices relative to trichinae, work with their producers to ensure that trichinae risk factors are minimized on their farms. (2) The on-farm audit will serve to document the absence of trichinae infection risks. Audits will be done periodically to ensure that good production practices relative to trichinae remain in place. (3) On a regular basis, a statistical sample of the national trichinae certified herd will be tested at slaughter using diaphragm digestion or ELISA to verify the absence of infection. (4) USDA veterinarians will conduct random spot audits of certifications to ensure completeness and the integrity of the program. Flow chart courtesy of FSIS.

cooking time and temperature consistent with state and federal regulations. Storage of waste products to be fed must avoid contaminating cooked material with uncooked material. Uncooked household waste shall not be fed to swine.

- Swine carcasses must be promptly removed from pens to prevent cannibalism. Dead animals should be promptly removed from the site in case they attract rodents or wildlife.
- General hygiene at the facility should be such that rodents or wildlife are not attracted. Solid waste (facility refuse) shall be contained and regularly removed from the site in case it attracts rodents or wildlife. Rodents and wildlife should not have access to solid waste. Spilled feed shall be regularly removed.
- Departures from the site shall be documented on an animal movement record and take place in a manner that ensures the identity of the swine can be traced back to their origin, the certified production site.

After certification, producers must maintain their certification status or start the process from the beginning again. Care must also be used in the buying, selling and moving of swine between producers and sites. When certified pork products are sold commercially as certified pork, the packer/collection points must have procedures in place for obtaining swine from certified sites. Edible pork products from certified swine must be segregated from pork from non-certified sites.

Of course, no program is complete without process verification testing, which in this case can be performed using accepted tissue or blood-based post-mortem tests. Testing is performed by the plant personnel using USDA-approved methods or samples may be sent for testing at a USDA-approved laboratory. Information on how to determine the sample size is available on the APHIS website given above. Information on record keeping, administrative requirements and documentation is also available.

Outside of the United States, the International Commission on Trichinellosis (ICT) has published a comprehensive guide (Gamble *et al.*, 2000). This program offers information on reducing incidence as well as transmission through safe food handling and cooking practices.

There are effective guidelines, which emphasize the fundamental preventive measures that were responsible for the initial decreases in *Trichinella* and other zoonotic agents in the United States. These are (Murrell and Pozio, 2000):

- Strict adherence to garbage-feeding regulations, particularly refuse cooking requirements (bring to a boil for 30 min).
- Stringent rodent control.
- Preventing pigs and other livestock from being exposed to dead animal carcasses of any kind.
- Proper disposal of pig and other animal carcasses (e.g. burial, incineration or rendering). This minimizes the infection risk for commensal wild animals.
- Construction of effective barriers between livestock, wild animals and domestic pets.
- Proper handling and disposal of wild animal carcasses by hunters.

19.3.2 *Taenia*-specific intervention programs

Compared to the trichinae certification program, in general, prevention and eradication programs for *Taenia* have been more on local or regional levels. Reduction programs for trichinae, such as those described above, also reduce *Taenia*, as well as non-helminthic parasites.

Neurocysticercosis is a parasitic disease caused by infection of the central nervous system with *Taenia solium* larval cysts. It is the most common helminthic infection of the central nervous system and a leading cause of acquired epilepsy in Latin America, South-east Asia, and central Africa (Román *et al.*, 2000; Del Brutto *et al.*, 2001). It is generally believed that cysticercosis within the United States is only associated with people who have migrated or traveled abroad (Sorvillo *et al.*, 1992; Townes *et al.*, 2004; del la Garza *et al.*, 2005; O'Neal *et al.*, 2011). The entry of cysticercosis into the United States has been discussed in a study of refugees, who were a large group of approximately 69 000 migrants and travelers, between the years 2000 and 2010. From this study, it was found that seroprevalence was high among all four populations tested: refugees from Burma (23.2%), Lao People's Democratic Republic (18.3%), Bhutan (22.8%) and Burundi (25.8%) (O'Neal *et al.*, 2012).

Intervention programs in the United States rely on education and public health awareness. In particular, educating individuals who may travel frequently to and from cysticercosis-endemic areas of the need to avoid eating undercooked pork and to maintain good hygiene can reduce infection among travelers (O'Neal *et al.*, 2011). As important is the need to increase awareness of *T. solium* infection for public health professionals and in particular to those clinicians who care for Hispanic and other immigrant populations (O'Neal *et al.*, 2011).

Advances in bio-imaging and immunoblot-based testing have led to the recognition of infections across the globe. The majority of infections are in the less developed regions of Mexico, and in parts of South America, Africa and Asia (Schantz, 2002). Cysticercosis in humans often correlates with the disease in pigs that live in close contact with humans (Garcia *et al.*, 2003). In a situation like this in Peru, anti-helminthic drugs were largely but not completely effective at treating the majority of pigs and humans (Garcia, 2002). One potential problem is that diagnosis of the disease in humans is expensive, requiring diagnostic tools like enzyme-linked immunosorbent assays (ELISA) and CT and MRI scans, which are not readily available in less developed areas of the world. In Nepal, where *Taenia* may be endemic, education and training programs were implemented for meat producers and sellers and to improve pig husbandry (Jimba *et al.*, 2003). In this study, meat sellers in Nepal were unaware that they were selling measly pork (meat containing visible *T. solium* cysts).

19.4 Future directions

Reductions in the incidence of illness attributed to zoonotic disease from *Taenia* and *Trichinella* is evident, and is due to an obvious team effort among well-educated and trained producers, excellence in animal husbandry, a strong public health infrastructure, trained veterinary and medical officers, and knowledgeable consumers. *Trichinella* control programs have been active for the past 150 years, and are still making progress and will continue to do so (Murrell and Pozio, 2000). The trichinae-free certification program in the United States is an excellent example. On the global scale, cysticercosis is one of only seven diseases currently considered to be a potential candidate for global eradication (CDC, 2008); however, there is still a long way to go.

The following case study in Romania is an interesting one. Certainly, Romania has made great strides in reducing *Trichinella* parasites and reducing cases of disease (Neghina *et al.*, 2009). There was a small rise in disease following the loss of socialist control and rigorous policies of animal disease control. During the early stages of the post-communist era, the majority of pigs once destined for the large abattoirs were sent to family farms and went untested for disease. Both human and animal trichinosis increased during the 1990s, and reached a peak in 1993 at 15.9 cases per 100 000 inhabitants per year (Neghina *et al.*, 2009). Specific customs including backyard-raised pigs, which are not veterinary tested, and holiday meals overflowing with pork products have been implicated as good sources of infection (Neghina, 2010). *Trichinella* may get a free trip around the world as Romanian pork products are exported or offered as gifts to family and friends. This is certainly important for the US Food and Drug Administration (FDA) and USDA in terms of biosecurity at air terminals and at points of distribution.

Public health education will need to receive higher priority across the globe to aid in the continuing reduction of diseases associated with *Taenia* and *Trichinella*. Another contributing factor is the global increase in the animal and meat trade, which has transferred *Trichinella*, for example, from endemic to non-endemic regions, where the veterinary and medical services are often not familiar with the infection (Murrell and Pozio, 2000). For example, the frequent importation of *Trichinella*-infected horses from eastern European countries and from North America and Mexico has been the source of more than 3300 human infections in France and Italy (Pozio, 2000). Changes in the ways people eat can also affect reductions in both organisms. Free-range animals and increasing populations of wild or feral swine also pose a continually changing obstacle to the reductions of *Taenia* and *Trichinella*.

Another area that may be affected in the future is animal welfare. Many of the methods that have been successful at removing *Trichinella* and *Taenia* from the swine raised in the United States are the same methods that come under fire from animal rights groups. Removal or relaxation of certain containment practices may lead to infections in these animals or worse they may lead to injuries in these animals (Callaway *et al.*, 2005; Rhodes *et al.*, 2005; Davies *et al.*, 1998). While the balance between animal welfare, animal health, human health and food production is a delicate one, tipping the scales in one direction places the others at risk. The public health education mentioned above must be extended to include education of animal welfare and health, which are affected by animal husbandry.

The latter is exceedingly important in maintaining public health, in the US and around the globe, notwithstanding that animal welfare is consistently questioned.

19.5 References

BRUSCHI, F. and MURRELL, K.D. (2006) Trichinellosis. In *Tropical Infectious Diseases Principles, Pathogens & Practice*, Elsevier, New York, 1217–24.

CALLAWAY, T.R., MORROW, J.L., JOHNSON, A.K., DAILEY, J.W., WALLACE, F.M. *et al.* (2005) Environmental prevalence and persistence of *Salmonella* spp. in outdoor swine wallows. *Foodborne Pathog Dis*, 2, 263–73.

CDC (2008) Recommendations of the international task force for disease eradication. *Morbid Mort Weekly Rep*, 42(RR-16):8.

DAVIES, P. (2011) Intensive swine production and pork safety. *Foodborne Pathog Dis*, 8, 189–201.

DAVIES, P.R., MORROW, W.E., DEEN, J., GAMBLE, H.R. and PATTON, S. (1998) Seroprevalence of *Toxoplasma gondii* and *Trichinella spiralis* in finishing swine raised in different production systems in North Carolina, USA. *Prev Vet Med*, 36, 67–76.

DCPP (2012) Disease Control Priorities Project. www.dcp2.org/main/Home.html. [Accessed March 2012.]

DEL BRUTTO, O.H., RAJSHEKHAR, V., WHITE JR, A.C., TSANG, V.C., NASH, T.E. *et al.* (2001) Proposed diagnostic criteria for neurocysticercosis. *Neurology*, 57, 177–83.

DEL LA GARZA, Y., GRAVISS, E.A., DAVER, N.G., GAMBARIN, K.J., SHANDERA, W.X. *et al.* (2005) Epidemiology of neurocysticercosis in Houston, Texas. *Am J Trop Med Hyg*, 73, 766–70.

DUPOUY-CARNET, J. and BRUSCHI, F. (2007) Management and diagnosis of human trichinellosis. In *FAO/WHO/OIW Guidelines for the Surveillance, Management, Prevention and Control of Trichinellosis*, 37–68.

ECKERT, J. (2005) Helminths. In *Medical Microbiology*, Kayser, F.H., Bienz, K.A., Eckert, J. and Zinkermagle, R.M. (eds), Thieme, Stuttgart, 543–602.

FSIS (1999) *Microbiological Hazard Identification Guide for Meat and Poultry Components of Products Produced by Very Small Plants*.

GAMBLE, H.R., BESSONOV, A.S., CUPERLOVIC, K., GAJADHAR, A.A., VAN KNAPEN, F. *et al.* (2000) International Commission on Trichinellosis: recommendations on methods for the control of *Trichinella* in domestic and wild animals intended for human consumption. *Vet Parasitol*, 93, 393–408.

GARCIA, H.H. (2002) Effectiveness of an interventional control program for human and porcine *Taenia solium* cysticercosis in field conditions. In *International Health*, Johns Hopkins University, Baltimore, 250.

GARCIA, H.H., GONZALEZ, A.E., EVANS, C.A. and GILMAN, R.H. (2003) Cysticercosis Working Group in Peru. *Taenia solium* cysticercosis. *Lancet*, 362, 547–56.

JIMBA, M., DURGA, D. JOSHI, A.B. and WAKAI, S. (2003) Health promotion approach for control of *Taenia solium* infection in Nepal. *Lancet*, 362, 1420.

LAXMINARAYAN, R., CHOW, J. and SHAHID-SALLES, S.A. (2006) Intervention cost-effectiveness: overview of main messages. In *Disease Control Priorities in Developing Countries*, Jamison D.T., Breman, J.G., Measham, A.R., Alleyne, G., Claeson, M., *et al.* (eds), The International Bank for Reconstruction and Development /The World Bank, Washington DC, 46.

MAWHORTER, S.D. and KAZURA, J.W. (1993) Trichinosis of the central nervous system. *Semin Neurol*, 13, 148–52.

MURRELL, K.D. and POZIO, E. (2000) Trichinellosis: the zoonosis that won't go quietly. *Int J Parasitol*, 30, 1339–49.

NEGHINA, R. (2010) Trichinellosis, a Romanian never-ending story. An overview of traditions, culinary customs, and public health conditions. *Foodborne Pathog Dis*, 7, 999–1003.

NEGHINA, R., NEGHINA, A.M., MARINCU, I., MOLDOVAN, R. and IACOBICIU, I. (2009) Epidemiology and epizootology of trichinellosis in Romania 1868–2007. *Vector Borne Zoonot Dis*, 10, 323–8.

OIE (2005) *Taenia* Infections. World Organization for Animal Health in collaboration with Iowa State University College of Veterinary Medicine. Available at: www.cfsph.iastate.edu/Factsheets/pdfs/taenia.pdf. [Accessed March 2012.]

O'NEAL, S., NOH, J., WILKINS, P., KEENE, W., ANDERSEN, J. *et al.* (2011) Surveillance and screening for *Taenia solium* infection, Oregon, USA. *Emerg Infect Dis*, 17, 1030–6.

O'NEAL, S.E., TOWNES, J.M., WILKINS, P.P., NOH, J.C., LEE, D. *et al.* (2012) Seroprevalence of antibodies against *Taenia solium* cysticerci among refugees resettled in United States. *Emerg Infect Dis*, 18, 431–8.

POZIO, E. (2000) Is horsemeat trichinellosis an emerging disease in EU? *Parasitol Today*, 16, 266.

RHODES, R.T., APPLEBY, M.C., CHINN, K., DOUGLAS, L., FIRKINS, L.D. *et al.* (2005) A comprehensive review of housing for pregnant sows, *JAVMA*, 227, 1580–90.

ROMÁN, G., SOTELO, J., DEL BRUTTO, O., FLISSER, A., DUMAS, M. *et al.* (2000) A proposal to declare neurocysticercosis an international reportable disease. *Bull World Health Organ*, 78, 399–406.

SCHANTZ, P. (2002) Eradication of *T. solium* cysticercosis, *International Conference on Emerging Infectious Diseases*, CDC. Available at: ftp://ftp.cdc.gov/pub/infectious_diseases/iceid/2002/pdf/schantz.pdf.

SCHANTZ, P.M. and DIETZ, V. (2001) Trichinellosis. In *Principles and Practice of Clinical Parasitology*, Gillepsie, S.H. and Pearson, R.D. (eds), John Wiley and Sons, Chichester, 521–33.

SNOWDEN, K., LATIMER, G.W., JONES, B.L. and BAUER, N.E. (1997) A common source outbreak of *Taenia saginata* in feedlot cattle in Texas, Oklahoma and Colorado. *APHIS Proceedings of the Epidemiology and Economics Symposium*, Ft. Collins, Colorado.

SORVILLO, F.J., WATERMAN, S.H., RICHARDS, F.O. and SCHANTZ, P.M. (1992) Cysticercosis surveillance: locally acquired and travel-related infections and detection of intestinal tapeworm carriers in Los Angeles County. *Am J Trop Med Hyg*, 47, 365–71.

TOWNES, J.M., HOFFMANN, C.J. and KOHN, M.A. (2004) Neurocysticercosis in Oregon, 1995–2000. *Emerg Infect Dis*, 10, 508–10.

ZIMMERMANN, W.J. (1971) The prevalence of trichinosis in swine in the United States, *Health Service Report*, 86, 937–45.

ZIMMERMANN, W.J. (1973) Trichiniasis in the U.S. population, 1966–1970. *Health Service Report*, 88, 606–23.

Part V

Pathogen control management

20

Advances in understanding the impact of personal hygiene and human behaviour on food safety

C. Griffith, Editor of *British Food Journal*, UK

DOI: 10.1533/9780857098740.5.401

Abstract: Food handler behaviour is important for producing safe food with food handler error a factor in many outbreaks. This chapter reviews food handler knowledge, attitudes and practices and the research methods used in their study. Infected food handlers spread pathogens to foods and other workers and the factors influencing this are analysed. Food handlers play a role in cross-contamination, an increasingly reported risk. The mechanisms for this and the components of hand hygiene are discussed. Historically, non-compliance has been assessed at the level of individuals. Recent work on food safety culture is examined and individual behaviour is discussed within the context of an organisation and its leadership. Methods and strategies for effective food hygiene training are reviewed.

Key words: food handler behaviour, infected food handler, cross-contamination and hand hygiene, food safety culture, food hygiene training.

20.1 Introduction

Debate exists as to whether the number of cases of foodborne disease is increasing or not (Newell *et al.*, 2010) although the view of the World Health Organization (WHO) is that it is a growing public health problem. What is not in dispute is that the level of foodborne disease is too high and that 2011 saw, in both the USA and Europe, some of the most significant, large-scale outbreaks ever recorded. This is in spite of libraries containing many volumes of work on food microbiology and the expenditure of large sums of money, around the world, to research and manage the causes of unsafe food. One possible explanation for this is that much of this existing research has focused on the pathogens involved and whilst this work is necessary, it may have been at the expense of work studying food handler behaviour.

A food handler can be defined as anyone involved in the production or preparation of food at any point in the food chain and can include people repairing, maintaining, cleaning or visiting food preparation areas.

Historically microbiological food safety has been based on two basic principles (Griffith, 2000), these are:

- preventing contamination of food with microorganisms;
- preventing the growth or survival of any microorganisms on or in food.

Hygiene practices used during food preparation are based on these two principles. For example, good hand hygiene and cleaning help to minimise food contamination, refrigeration helps to prevent microbial growth and heat processing (e.g. pasteurisation) reduces microbial survival. Except perhaps for completely automated production environments, correct implementation of all of these practices involves food handlers.

Investigations of foodborne disease outbreaks have identified a number of possible risk or contributory factors, e.g. inadequate heating and infected food handlers. Whilst useful this type of approach does have limitations (Griffith and Redmond, 2009). By virtue of the collection methodology some risk factors, especially cross-contamination, may be underreported but more importantly the real or underlying reasons for any errors are not fully explained. For example, inadequate cooking may be due to faulty equipment (not corrected or unreported by a food handler), incorrect setting of oven temperatures, an overloaded oven or food being deliberately or inadvertently removed before cooking is complete. A generalised scheme of error categories has been proposed (Griffith and Redmond, 2009) most of which have a human component or responsibility and it has been suggested up to 97% of general outbreaks may involve human error (Howes *et al.*, 1996). If this is correct it would make food handler error the single most important factor affecting the control of food hazards and in managing food safety risks. However, failure to implement the required food safety practices may not be the only contribution made by food handlers: they may in addition also be the source of the causative organism.

Whilst research into human behaviour and the individual contribution made to food safety by food handlers has been lacking, training to improve food handler safety behaviour has long been used, although this has tended to be generalised food hygiene education delivered in a conventional way (Chapman *et al.*, 2011). Research into the reasons for non-compliance and the strategies and methods for optimum delivery to ensure compliance with hygiene requirements is relatively new. Until very recently research has predominantly been concerned with the role of the individual food handler, their attitudes and practices and why they may not implement known food safety activities. In the last three years attention has also focused on the collective food safety activities within a business. The contribution of supervisors and management and the impact this can have on individual food handler compliance, i.e. the prevailing food safety culture, is increasingly being recognised (Griffith, 2010).

Given the importance of food handler behaviour and personal hygiene, the objectives of this chapter are to:

- Discuss recent approaches and findings concerning food handler behaviour.
- Assess the importance of food handlers as a source of pathogenic microorganisms.
- Analyse the role of food handlers in the spread and transmission of foodborne pathogens.
- Discuss the role of food safety culture, management and leadership as contributory factors in hygiene behaviour.
- Review methods and approaches used in hygiene training.

20.2 Food handler knowledge and practices

20.2.1 Food handler behaviour: knowledge, attitudes and practices

The food safety practices used by food handlers are highly variable and involve the interaction of many different factors (Fig. 20.1). Research over the past 15 years in particular has attempted to understand, quantify and improve food safety behaviour (Griffith and Redmond, 2009). Many of the earlier studies just attempted to determine what food handlers knew about specific food activities (Clayton and Griffith, 2004). The results varied in detail but whilst food handlers' knowledge was deficient in some areas, e.g. the need for cooling foods quickly, the need for hand hygiene was usually well understood and known (Clayton *et al.*, 2002). This type of approach is useful in deciding the messages that need to be communicated. However, provision of this information on its own does not ensure implementation

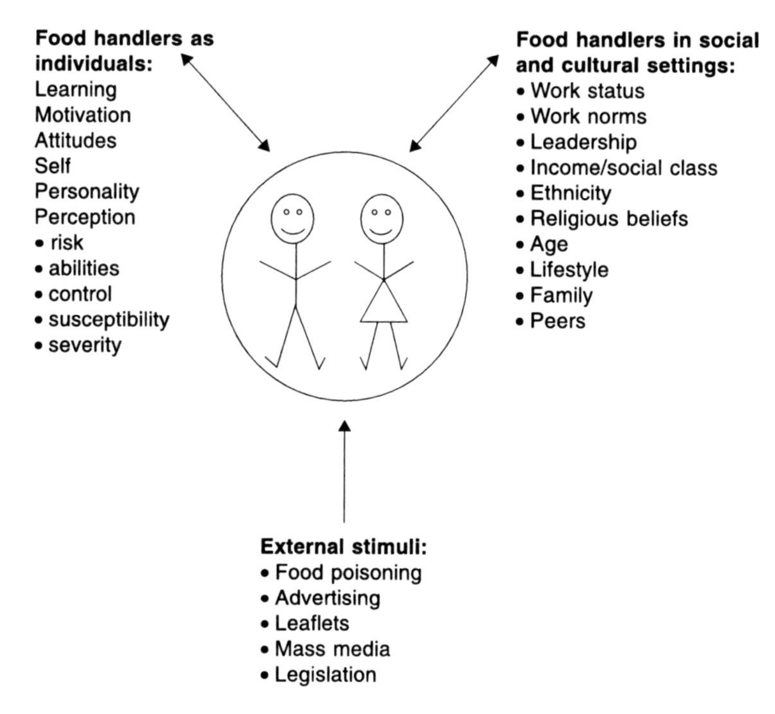

Fig. 20.1 Factors influencing food handler behaviour (Griffith and Redmond, 2009).

of food safety practices and later research has tried to put this data into a social cognitive framework (Clayton and Griffith, 2008), initially using the so-called knowledge attitudes practices (KAP) model. The KAP model works on the assumption that if food handlers are provided with the correct information then their attitudes would change and so would their behaviour. Whilst behavioural and educational models were considered useful it was realised that the KAP model was too simplistic an approach. Subsequently a range of models, including the theory of planned behaviour (TPB), the health belief model (HBM) and others, have been used or proposed (Griffith and Redmond, 2009; Seaman, 2010). The TPB measures behavioural intentions, recognising that certain uncontrollable factors may inhibit their implementation. It also includes the element of subjective norm, which considers the perceived beliefs of others and the individual's desire to comply with these beliefs (effectively like peer pressure). The HBM has been criticised as being more a catalogue of variables rather than a model (Griffith and Redmond, 2014). The health action model (HAM) combines elements of both the TPB and HBM. Whilst these models and theories differ in the detail of their constructs they are all grounded in the assumption that a person's behaviour is determined through an examination of their beliefs, attitudes and norms and that these factors need to be examined within the prevailing social and environmental conditions. This work ultimately gave greater prominence to the concept of food safety culture (Clayton and Griffith, 2008; Griffith, 2010).

An additional element of research has been to try to relate what handlers know to what they practise and do. Initially based on self-reporting, the limitations of this were recognised and it was followed by observation of what people actually did as opposed to what they said they did (Clayton and Griffith, 2004). An important element in some of this work (Clayton *et al.*, 2002) was that food handlers admitted to not carrying out known food safety behaviours. Four per cent admitted to often not carrying out known food safety behaviours whilst 59% admitted they sometimes did not. Due to the methods of data collection this is likely to be an underestimate of non-compliance and implementation but if extrapolated to food handlers in general it indicates that in any one day many known food safety behaviours are not implemented. The main stated barrier to implementation was lack of time, followed by resources and facilities (Clayton *et al.*, 2002). Observations of food handlers have indicated a discrepancy between self-reporting and the actual use of food safety practices and, in particular, the potential for food handlers to cross-contaminate food and surfaces in food preparation areas (Clayton and Griffith, 2004). Staff who had received training were significantly more likely to report they practised good hygiene than those without training.

20.2.2 Food handlers as a source of foodborne pathogens

An infected food handler is a food handler who carries a foodborne pathogen with or without symptoms. This state can be transient or sometimes food handlers can

Table 20.1 Foodborne pathogens most likely to be transmitted by food handlers (adapted from Todd *et al.*, 2008)

Pathogen	Class of microorganism
Salmonella	Bacterial
Shigella	Bacterial
Staphylococcus aureus	Bacterial
Norovirus	Viral
Hepatitis A virus	Viral

be a more permanent source of a range of foodborne pathogens, which are primarily of faecal, nose, throat or skin origin and are summarised in Table 20.1. The infected food handlers themselves may be asymptomatic, suffering or about to suffer an overt infection, convalescent (recovering) or chronic (long-term) carriers. The risk they pose must be considered in relation to a combination of factors:

- the food handling activities they perform and the types of foods touched;
- compliance with hygiene requirements;
- potential to be asymptomatic and carriage rates;
- the numbers of pathogens shed into the environment and their ability to survive and spread;
- duration of colonisation and shedding;
- the dose response curve for the pathogen.

The definition of a food handler is broad and ranges from people who frequently handle large quantities of unwrapped cooked or ready-to-eat food, to people cleaning food premises or handling wrapped food in supermarket checkouts. Thus the degree of food handling activities and hence risk they pose will be highly variable. It has been suggested that food handlers as a source are more significant for multi-ingredient ready-to-eat foods (Todd *et al.*, 2008). The degree of compliance is also highly variable depending on individual beliefs and attitudes (Fig. 20.1) and the prevailing food safety culture. Many of the other characteristics listed have been reviewed recently for a range of pathogens (Todd *et al.*, 2008) although there are still large areas of uncertainty. It would seem, however, that for *Campylobacter* species, Shiga toxin producing *Escherichia coli* (STEC) and norovirus, three of the most important groups of pathogens to have emerged in the last 30–40 years, infection from often a relatively small number of organisms is possible (Todd *et al.*, 2008). For noroviruses, possibly now the world's most common cause of stomach upsets, recent data looking at the latest outbreak strains suggest better and more prolonged environmental survival properties than previously thought (Duizer and Koopmans, 2009). It has been suggested that some genotypes are more likely to be associated with person-to-person spread (Verhoef *et al.*, 2009, 2010). This has implications for the spread and transmission of the virus between food handlers and may be a significant factor in outbreaks of hepatitis A and norovirus, especially when there is prolonged shedding of the virus by colonised people. In general the longer the incubation period and the persistence of symptoms the greater the chance of food becoming contaminated and an outbreak occurring (Todd *et al.*, 2008).

Outbreaks usually involve an interaction of several risk factors. For example, whilst humans can be a good source of *Staphylococcus aureus*, for food poisoning to occur there must usually be product temperature abuse to allow sufficient toxin to be produced to initiate illness.

Hygiene controls to prevent contaminated food handlers from contaminating food rely, to an extent, on good hand hygiene (in its broadest terms see Section 20.3) but this does not necessarily prevent the spread of pathogens to other food handlers. Another strategy is to ensure colonised food handlers do not enter food premises. In that in addition to preventing food contamination this can prevent spread between food handlers. However, it will not exclude asymptomatic food handlers or those who are continuing to shed pathogens after they have recovered from illness.

An area that could be improved is communicating to food workers the importance of not handling open food when ill. One study (Clayton *et al.*, 2002) indicated that whilst food handlers were often aware of the need for hand washing (84%) only 12% thought that reporting and staying away from work when ill were important food safety behaviours.

20.3 Food handlers, cross-contamination and hand hygiene

Cross-contamination is an increasingly reported risk factor in outbreaks of foodborne disease and can be defined as the transfer of microorganisms (or other contaminants) from one surface or food (usually raw) to another surface or food either directly (e.g. direct contact) or indirectly from hands and equipment (Griffith and Redmond, 2005). Post-processing contamination is the contamination of a product after a kill or other critical processing step. Many of these contamination activities involve food handlers.

If food handlers themselves are the source then their hands can spread the pathogen to other staff, foods and surfaces. Studies have shown rapid contamination of the food environment from a contaminated food handler. However, even if food handlers themselves are not the source of the organism they may still cross-contaminate foods and preparation equipment and surfaces. Work on exposure pathways after the routine preparation of food involving the handling of raw chicken (Griffith and Redmond, 2005) indicated the spread of *Campylobacter* species to other previously uncontaminated ready-to-eat foods (8%), cooked chicken (6%) and a range of other surfaces and equipment (12%) at a level that could have been in excess of the organism's infective dose. Pathogens on contaminated equipment or surfaces can remain viable for hours or days and easily spread to other foods (Redmond *et al.*, 2004). The potential for cross-contamination as a factor in outbreaks is much greater than suggested by epidemiological studies, which has been supported by observational studies of food handler activities using a technique known as notational analysis (Clayton and Griffith, 2004). This technique, adapted from sports science, enables the recording of the frequency of observed events in a chronologically ordered sequence. Appropriate decontamination activities, including hand hygiene, equipment cleaning, etc., which were frequently required to be performed, were often poorly or incorrectly implemented leading to the subsequent contamination of foods and surfaces.

Hand hygiene, as a component of personal hygiene, merits particular attention due to: (i) the frequency with which it is required and the levels of non-compliance, (ii) the microbial load that can be found on hands, (iii) the number of times hands touch contaminated objects or surfaces immediately prior to touching ready-to-eat food and (iv) its importance at all steps in the food chain, from initial production to food consumption. It may possibly be the single most important food hygiene activity.

The last few years have seen the subject being considered as much broader than just hand washing. Hand hygiene encompasses: (i) the health of the food handler, (ii) their hand habits and hand health, (iii) the type of food handling activities and practices they use, (iv) the use of gloves and (v) the type, timing and frequency of the hand decontamination procedures they perform.

There are arguments about the advantages and disadvantages of using gloves; in some states and countries, food handlers are not allowed to touch unprotected ready-to-eat food with their bare hands. The potential benefits of this are especially important if the food handler is colonised or infected with a foodborne pathogen, but glove usage is not a replacement for hand decontamination, which is required before wearing and after glove removal. The microbial load on hands can increase significantly during prolonged glove usage (Todd *et al.*, 2010a) and gloves can have micro-punctures in them, allowing transmission of microorganisms between gloved hands and food. A potentially greater problem is that food handlers can develop a false sense of security (Todd *et al.*, 2010a) believing they do not need to change gloves before touching food, even after smoking or emptying waste bins. This is particularly problematic given that during food handling the surfaces that hands touch, including immediately prior to food contact, may be heavily contaminated (Tables 20.2 and 20.3).

Hand decontamination itself is also being viewed differently. Once mostly considered as hand washing, the last ten years has seen much greater consideration and use of antiseptic hand rubs, often based on alcohol as the active agent. Such formulations have many advantages, e.g. there is no need for sinks or other facilities and they can be used virtually anywhere including at the point of hand contamination. They should, however, only be used on clean hands. It is often not appreciated that hand washing and hand rubs are largely based on different principles. Hand washing is mainly a cleaning process (it removes food, other debris and microorganisms) whereas hand rubs are a method of disinfection or antisepsis with the sole aim of destroying any microorganisms present. Heavily soiled hands (with organic or food debris) need to be washed first. As with other

Table 20.2 Surfaces touched by hands during food service prior to touching ready-to-eat (RTE) food in relation to the microbial failure rates

Surface or object touched[a]	% of times touched[b]	Microbiological failure rate[c]
Containers	23%	84%
Ready-to-eat food	15%	–
Knife	12%	58%
Fridge handle	8%	97%
Chopping board	5%	87%
Bin lid	4%	100%
Work surface	4%	88%
Cloth	4%	93%
Raw food	1.5%	100%
Others	25%	56–100%

Notes:
[a] Surfaces touched, without hand washing, immediately prior to touching ready-to-eat (RTE) food.
[b] The percentage number of times the named surface was touched prior to touching RTE food.
[c] The microbiological failure rates of the surfaces, judged in relation to a clean surface, e.g. 84% failure rate means on 84% of occasions this type of surface was not considered microbiologically acceptable.

Source: Adapted from Griffith *et al.* (2002).

Table 20.3 Surfaces most likely to be touched immediately after handling contaminated raw food, and without decontamination, in food service

Surface or object touched	% of times touched
Chopping board	16%
Raw food	11%
Container	10%
Work surfaces	7%
Bin lid	7%
Ready-to-eat food	7%
Tap handles	4%
Others	38%

Source: Adapted from Griffith *et al.* (2002).

disinfectants, organic matter reduces the activity of hand rubs. Additionally, whilst having good efficacy against some pathogens, hand rubs are often limited against others, such as spore formers and some viruses such as norovirus. Their effectiveness is related to many factors, not least their chemical formulation. Their use in the food industry has recently been reviewed (Todd *et al.*, 2010c). Hand washing, in conjunction with the often neglected hand drying, still remains the most efficient overall method for removing soil and microorganisms from the hands and its use in relation to food poisoning outbreaks has been analysed (Todd *et al.*, 2010b). It was found that the time taken to wash the hands and the degree of friction generated during lathering were more important than water temperature and the use of antimicrobial soaps. Paper rather than cloth towels or hot air dryers were considered best for hand drying, which plays an important role in the removal of microorganisms, possibly removing up to an additional 90% of the transient organisms.

Although specific practices designed to prevent cross-contamination are often known and even understood by food handlers they are often not practised and this is a particular problem in the food service and catering industry. Modern food manufacturing plants, especially if externally audited against international standards, meet high standards of design, construction and maintenance, coupled with stringent personal hygiene precautions to minimise cross-contamination; even so widespread *Listeria monocytogenes* contamination can be problematic. Food service establishments are often much smaller and more cramped and less well designed and constructed. Other important differences in food service, which can affect the cross-contamination potential, include the need to make or serve food to order, rather than for stock (as in manufacturing). The need to serve customers to order, who may be standing in front of the food handler, may lead to food handler attitudinal ambivalence. The food handler has to decide what is more important: being hygienic or serving the customer quickly. In food service and retail (e.g. a deli counter in a supermarket), food activities can be subjected to large peaks and troughs in demand and food handler compliance with a normally well-ordered food hygiene management system can collapse at times of peak demand.

Strategies for reducing cross-contamination include good personal hygiene, separation of raw from cooked food, proper cleaning along with waste disposal, pest control and other activities normally considered as prerequisite programmes. These need to be communicated to staff in a way which will ensure and allow compliance. Businesses should be encouraged to take a risk-based approach in assessing the cross-contamination potential within their operations and this should include food handler behaviour (Griffith *et al.*, 2002).

20.4 Food safety culture

Historically food handler behaviour has been considered at the level of the individual and any non-compliance was considered to be the fault of the individuals within the workforce. Typical corrective actions have usually been to train or retrain the miscreants. This approach is based on the often erroneous assumption that the food handlers do not know what is expected of them. Following the studies on organisational culture and non-compliance with systems in other highly regulated industries, research into the food industry has also recently investigated broader aspects of hygiene non-compliance. Initially, studies of the barriers that prevent compliance applied social cognition models to try to understand non-compliance. In turn this research gave rise to studies on organisational food safety culture, which, although mentioned earlier in the literature (Griffith, 2000, 2006), is now receiving much greater attention. In 2009 the first book dedicated to the subject (Yiannas, 2009) was published, followed in 2010 by the first three peer-reviewed papers (Griffith, 2010; Griffith *et al.*, 2010a, 2010b) with in 2011 the first conference devoted to it (the Global Food Safety Initiative Meeting, London, February).

One definition of food safety culture is that it is the aggregation of the prevailing constant learned shared attitudes, values and beliefs contributing to the hygiene behaviours used within a particular food handling environment. It is important to realise that *every* food business has a food safety culture. This may be known and understood by management or, as is more likely, unknown and not understood. Food safety cultures have two dimensions: direction and strength. Direction is a continuum from strongly positive to strongly negative passing through neutral. A positive food safety culture is one which is strongly supportive of food safety and where the business perceives it to be important to their vision of where the company wants to be. Creating a positive food safety culture requires effort, management and leadership. A negative food safety culture is one where food safety has a low priority and where food safety is considered subordinate to other company objectives with often the desire to save money or maximise profits at all costs overriding food safety and being the main driving force. In some cases implementing good food safety practices may be actively discouraged and non-implementation encouraged. Neutral or complacent cultures can be found, often in larger companies, where food safety is taken for granted and there is an air of confidence that foodborne disease will not happen there or that the existing approaches and attitudes to food safety are adequate and unchallenged. The strength of the culture is a measure of the extent to which food safety is practised and embedded throughout the workforce. It reflects how shared and intensely held food safety core values are in the workforce. A problem with larger companies is that a whole variety of sub-cultures can exist sometimes at different levels within the organisation. An additional factor in multi-site companies is the dependence of the food safety culture on the individual unit manager and their beliefs and values.

Outbreaks of foodborne disease have been linked to inappropriate food safety cultures, which often aim to maximise financial returns (Griffith and Redmond, 2009). In one outbreak, the culture of the business was described as 'having little regard for food safety but where making and saving money was the priority' (Pennington, 2009). Individual food handlers learn the food safety culture within the organisation and are more likely to practise what they see rather than what they are told, i.e. they adapt their behaviour to the behavioural patterns of their peers. An example would include the timing, frequency and methods of hand washing. Models of how new staff learn the prevailing culture through a process known as organisational socialisation have been proposed (Buchanan and Huczynski, 2010). Improving the food safety culture within an organisation will improve the hygiene compliance not only of existing staff but also potentially for new employees.

One initial definition of culture was: 'It is the way we do things around here' (Deal and Kennedy, 1982). This implies that a food safety culture could easily be observed; it is now realised that culture is much more complex and what may be observed is the outer layer of the culture, sometimes referred to as the food safety climate (Griffith *et al.*, 2010a). Culture is based upon far more deep-seated assumptions about food safety within the company. Various models of organisational culture (and food safety culture) have been proposed and although their components are often given different titles they reflect the same underlying assumptions and factors (Griffith *et al.*, 2010b). One key factor that contributes to a positive food safety culture is food safety leadership. Whilst food safety management is often discussed, relatively little is spoken about food safety leadership and it has been said that leadership is one of the most observed but least understood phenomena. There are, however, differences between management and leadership and managers and leaders. Food safety leadership can be defined as the ability to engage staff in hygiene and safety performance and compliance with the business's goals, visions and standards. There are many theories of leadership and it is a rapidly changing field of study. It is beyond the scope of this chapter to analyse differences between managers and leaders, discuss different approaches to leadership or examine the other factors that are involved in food safety culture. However, these are topics that will assume greater importance in the future in an attempt to ensure food handlers practise good hygiene and produce safe food. It would probably be reasonable to add a third principle of food safety (in addition to the two outlined in Section 20.1) and that would be: the creation of a strong and positive food safety culture.

20.5 Food hygiene training

In order for food handlers to be hygienic they need to know what to do, with preferably an understanding of why. It is the function of training to provide this information. This does not ensure the knowledge will be used but training does provide a starting point for people, who are able and motivated, to handle food hygienically. In recent years in many countries training (or an equivalent e.g. supervision) has become a legal requirement. This has been achieved to a greater extent in some food sectors than others. It still remains a problem with the food service and some primary production sectors due to high staff turnover and because they have a large number of low paid or part-time workers with limited education and language skills (Griffith and Redmond, 2009). Many of the earlier training initiatives focused on the food handler (as they were the ones doing the food handling) whereas many recent initiatives have concentrated more on the role of the supervisor. This is partly to combat problems of staff turnover and cost issues but also because the thinking is that the person in charge has an important role in assessing the risks within the business and in forming a positive food safety culture. There is some evidence to suggest that businesses with trained supervisors and managers produce food with a better microbiological quality (Little *et al.*, 2002; Sagoo *et al.*, 2003).

Generally there is a lack of information on the efficacy of training and individual studies have produced conflicting results (Egan *et al.*, 2007). This is perhaps not surprising given the wide variety of activities covered under the heading of training. Given the money invested in training, all companies should try to determine its effectiveness and how it could be improved. Unfortunately in many cases food safety training is seen as an end in itself (a certificate to hang on a wall or to meet audit or legislative requirements) rather than a means to an end (the production of safe food).

Different forms of training exist; it can be formal (which is often general, classroom based and externally certificated) or informal (usually in house and more directly work related). Many people feel that competency-based training is more likely to lead to good food hygiene than the classroom-based dissemination of theory. It would be wrong to make generalisations about training efficacy and much of the success of training will depend upon the skill of the trainer. Recent years have seen a much more analytical approach towards training both in how it is delivered and how its efficacy is assessed. The relatively simple KAP model has given way to more sophisticated models for planning and fostering behavioural changes. Other than the previously mentioned models, one which has been advocated for training is the transtheoretical model (TTM) developed by Prochaska and Diclemente, especially within the context of social marketing approaches (Griffith and Redmond, 2009). Social marketing is where a commercial marketing or business approach is used to sell information to the person being trained. TTM suggests that behavioural change is not instantaneous but is a five-step approach consisting of pre-contemplation, contemplation, preparation, action and maintenance.

As part of the approach to make training more effective, two of the more recently studied topics have included refresher training and the types of approaches used in training delivery. Training is usually considered as a requirement for staff when they start a new job, i.e. initial training, but can then be forgotten about; however, research (Worsfold and Griffith, 2009) has shown that refresher and remedial update training (these terms may have slightly different meanings but are often used interchangeably) is valued by food handlers. The methods used in such training are again dominated by traditional approaches to knowledge delivery, rather than orientated towards behavioural change. The strategies and methods used to deliver food safety training have been reviewed (Medeiros *et al.*, 2011). This review indicated that the topic most dealt with was personal hygiene and that interactive training and hands-on activities were the most widely accepted training methods. Recently more innovative approaches to training have been found to be successful in achieving behavioural change. One such strategy used food safety 'infosheets', which were designed to be surprising and generate discussion in the workplace (Chapman *et al.*, 2011). This type of approach could fit well within a social marketing approach to food safety and risk communication involving a mix of training approaches.

The effectiveness of food safety training depends on the dynamic between the trainee, the trainer and the employer. Employers should provide all the necessary facilities and the time for hygiene as well as a positive food safety culture and create a desire in employees to be hygienic. Employers should plan to make training effective and work for them and recommendations have been made for how this can be achieved (Griffith and Redmond, 2014).

20.6 Conclusion and future trends

Food safety continues to be a major public health concern and in spite of much research and expenditure the present reported levels of foodborne disease worldwide are far too high. Food handler behaviour and food safety training are likely to be important elements of attempts to reduce this burden of disease.

Much is known about what food handlers know or do not know; although general research findings on food handler knowledge and attitudes are still likely to be published in the future, they are likely to be of relatively limited value. Specific information on staff food safety knowledge and attitudes within companies is likely to be of more use. Organisational socialisation starts with the careful selection of staff and it is possible that for food safety, personality-style psychometric type tests will be developed for hiring new employees. This may be linked to the further development of social cognition models to better predict likely food safety behaviour.

Food handlers are known sources of some foodborne pathogens but more work is needed on carrier status and the duration of shedding. This in turn may have an impact on legislation and when food handlers can return to work safely after illness. More research is needed on the person-to-person spread of norovirus and the role it plays in outbreaks. Spreading may be linked to trends in environmental survival and the persistence of different pathogens. The role of food handler behaviour in cross-contamination needs greater clarification: currently this is a weak link in risk assessments (Griffith and Redmond, 2005), but given the nature of emerging pathogens and their infective dose this will be a major challenge in the future.

Many papers have been published on hand hygiene but the debate about gloves and their merits along with hand sanitisers continues and further improving hand hygiene is essential. There may not be a need to seek complex strategies since merely giving staff more time to be hygienic may be very effective, providing it is within an appropriate food safety culture. This topic is in its relative infancy and is likely to be a major subject for future research along with food safety leadership. Leadership is required at all levels within a business but senior management set the tone, vision and direction for food safety compliance within a company. Staff will overtly or subliminally pick up on what they perceive to be important business goals. Mechanisms for objectively assessing the food safety culture within a business already exist and inspectors and auditors need to be trained in how this should be undertaken.

Food safety training is likely to be based more on theoretical, psychological and educational models, which will make use of a social marketing approach

(Griffith and Redmond, 2014). Training initiatives should be carefully planned using a broad mix of training strategies and methods to motivate employees. The content of the interventions should be based on a proper training needs analysis, properly evaluated and not just centred on knowledge acquisition.

20.7 Further reading

A particularly useful set of papers to read are the 11 produced by the communicable disease committee of the International Association for Food Protection on outbreaks involving food workers. Some have already been referred to in the text but the full series is outlined below.

GREIG, J. D., TODD, E. C. D., BARTLESON, C. A. and MICHAELS, B. S. (2007). Outbreaks where food workers have been implicated in the spread of foodborne disease. Part 1. Description of the problem, methods, and agents involved. *Journal of Food Protection*, 70 (7), 1752–1761.

TODD, E. C. D., GREIG, J. D., BARTLESON, C. A. and MICHAELS, B. S. (2007a). Outbreaks where food workers have been implicated in the spread of foodborne disease. Part 2. Description of outbreaks by size, severity, and settings. *Journal of Food Protection*, 70 (8), 1975–1993.

TODD, E. C. D., GREIG, J. D., BARTLESON, C. A. and MICHAELS, B. S. (2007b). Outbreaks where food workers have been implicated in the spread of foodborne disease. Part 3. Factors contributing to outbreaks and description of outbreak categories. *Journal of Food Protection*, 70 (9), 2199–2217.

TODD, E. C. D., GREIG, J. D., BARTLESON, C. A. and MICHAELS, B. S. (2008a). Outbreaks where food workers have been implicated in the spread of foodborne disease. Part 4. Infective doses and pathogen carriage. *Journal of Food Protection*, 71 (11), 2339–2373.

TODD, E. C. D., GREIG, J. D., BARTLESON, C. A. and MICHAELS, B. S. (2008b). Outbreaks where food workers have been implicated in the spread of foodborne disease. Part 5. Sources of contamination and pathogen excretion from infected persons. *Journal of Food Protection*, 71 (12), 2582–2595.

TODD, E. C. D., GREIG, J. D., BARTLESON, C. A. and MICHAELS, B. S. (2009). Outbreaks where food workers have been implicated in the spread of foodborne disease. Part 6. Transmission and survival of pathogens in the food processing and preparation environment. *Journal of Food Protection*, 72 (1), 202–219.

TODD, E. C. D., MICHAELS, B. S., GREIG, J. D., SMITH, D., HOLAH, J. *et al.* (2010a). Outbreaks where food workers have been implicated in the spread of foodborne disease. Part 7. Barriers to reduce contamination of food by workers. *Journal of Food Protection*, 73 (8), 1552–1565.

TODD, E. C. D., MICHAELS, B. S., GREIG, J. D., SMITH, D. and BARTLESON, C. A. (2010b). Outbreaks where food workers have been implicated in the spread of foodborne disease. Part 8. Gloves as barriers to prevent contamination of food by workers. *Journal of Food Protection*, 73 (9), 1762–1773.

TODD, E. C. D., MICHAELS, B. S., SMITH, D., GREIG, J. D. and BARTLESON, C. A. (2010c). Outbreaks where food workers have been implicated in the spread of foodborne disease. Part 9. Washing and drying of hands to reduce microbial contamination. *Journal of Food Protection*, 73 (10), 1937–1955.

TODD, E. C. D., MICHAELS, B. S., SMITH, D., GREIG, J. D. and BARTLESON, C. A. (2010d). Outbreaks where food workers have been implicated in the spread of foodborne disease. Part 10. Alcohol-based antiseptics for hand disinfection and a comparison of their effectiveness with soaps. *Journal of Food Protection*, 73 (11), 2128–2140.

TODD, E. C. D., GREIG, J. D., MICHAELS, B. S., BARTLESON, C. A., SMITH, D. *et al.* (2010e). Outbreaks where food workers have been implicated in the spread of foodborne disease. Part 11. Use of antiseptics and sanitizers in community settings and issues of hand hygiene compliance in health care and food industries. *Journal of Food Protection*, 73 (12), 2306–2320.

20.8 References

BUCHANAN, D. A. and HUCZYNSKI, A. A. (2010). *Organizational Behaviour*, 7th ed., Essex, UK, Pearson Education Limited, p. 107.

CHAPMAN, B., MACLAURIN, T. and POWELL, D. (2011). Food safety infosheets: design and refinement of a narrative-based training intervention. *British Food Journal*, 113 (2), 160–186.

CLAYTON, D. and GRIFFITH, C. J. (2004). Observation of food safety practices in catering using notational analysis. *British Food Journal*, 106 (3), 211–227.

CLAYTON, D. A. and GRIFFITH, C. J. (2008). Efficacy of an extended theory of planned behaviour model for predicting caterers' hand hygiene practices. *International Journal of Environmental Health*, 18, 83–98.

CLAYTON, D., GRIFFITH, C. J., PRICE, P. and PETERS, A. C. (2002). Food handlers' beliefs and self-reported practices. *International Journal of Environmental Health*, 12, 25–39.

DEAL, T. and KENNEDY, A. (1982). *Corporate Cultures: The Rights and Rituals of Corporate Life*, Harmondsworth, Penguin.

DUIZER, E. and KOOPMANS, M. (2009). In: Blackburn, C. de W. and McClure, P. J., eds. *Foodborne Pathogens. Hazards, Risk Analysis and Control*, 2nd ed., Cambridge, Woodhead Publishing Ltd, pp. 1161–1192.

EGAN, M. B., RAATS, M. M., GRUBB, S. M., EVES, A., LUMBERS, M. L. *et al.* (2007). A review of food safety and food hygiene training studies in the commercial sector. *Food Control*, 18, 1180–1190.

GRIFFITH, C. J. and REDMOND, E. (2009). Good practices for food handlers and consumers. In: Blackburn, C. de W. and McClure, P. J., eds. *Foodborne Pathogens. Hazards, Risk Analysis and Control*, 2nd ed., Cambridge, Woodhead Publishing Ltd, p. 529, Fig. 15.4.

GRIFFITH, C. J. (2000). 'Food safety in catering establishments'. In: Faber, J. M. and Todd, E. C. D., eds. *Safe Handling of Foods*, New York, Marcel Dekker, pp. 235–56.

GRIFFITH, C. J. (2006). Food safety: where from and where to? *British Food Journal*, 108 (1), 6–15.

GRIFFITH, C. J. (2010). Do businesses get the food poisoning they deserve? The importance of food safety culture. *British Food Journal*, 112 (4), 416–425.

GRIFFITH, C. J. and REDMOND, E. (2005). 'Handling poultry and eggs in the kitchen'. In: Mead, G. C., ed. *Food Safety Control in the Poultry Industry*, Cambridge, UK, Woodhead Publishing Ltd, pp. 524–540.

In: Blackburn, C. de W. and McClure, P. J., eds. *Foodborne Pathogens*.

GRIFFITH, C. J. and REDMOND, E. (2014). In Press. 'Principles of food safety training and health education'. In *Encyclopedia of Food Safety*, Oxford, UK, Elsevier.

GRIFFITH, C. J., DAVIES, C., BREVERTON, J., REDMOND, E. C. and PETERS, A. C. (2002). Assessing and reducing the risk of cross contamination in food stuffs in food handling environments. A report for the Food Standards Agency, B02007, London.

GRIFFITH, C. J., LIVESEY, K. M. and CLAYTON, D. (2010a). Food safety culture: the evolution of an emerging risk factor? *British Food Journal*, 112 (4), 426–438.

GRIFFITH, C. J., LIVESEY, K. M. and CLAYTON, D. (2010b). The assessment of food safety culture. *British Food Journal*, 112 (4), 439–456.

HOWES, M. S., MCEWEN, S., GRIFFITHS, M. and HARRIS, L. (1996). Food handler certification by home study: measuring changes in knowledge and behaviour. *Dairy Food and Environmental Sanitation*, 16, 737–744.

LITTLE, C. L., BARNES, J. and MITCHELL, R. T. (2002). Microbiological quality of take-away cooked rice and chicken sandwiches: effectiveness of food hygiene training of the management. *Communicable Disease and Public Health*, 5 (4), 289–298.

MEDEIROS, C. O., CAVALLI, S. B., SALAY, E. and PROENÇA, R. P. C. (2011). Assessment of the methodological strategies adopted by food safety training programmes for food service workers: a systematic review. *Food Control*, 22, 1136–1144.

NEWELL, D. G., KOOPMANS, M., VERHOEF, L., DUIZER, E., AIDARA-KANE, A. *et al.* (2010). Foodborne diseases – the challenges of 20 years ago still persist while new ones continue to emerge. *International Journal of Food Microbiology*, 139, S3–S15.

PENNINGTON, H. (2009). *Report of the public inquiry into the September 2005 outbreak of E. coli O157 in South Wales*. HMSO, available at wales.gov.uk/ecoliinquiry/?lang=en (accessed December 2009).

SEAMAN, P. (2010). Food hygiene training: introducing the food hygiene training model. *Food Control*, 21, 381–387.

SAGOO, S. K., LITTLE, C. L. and MITCHELL, R. T. (2003). Microbiological quality of open ready-to-eat salad vegetables: effectiveness of food hygiene training of management. *Journal of Food Protection*, 66 (9), 1581–1586.

REDMOND, E., GRIFFITH, C. J., SLADER, J. and HUMPHREY, J. (2004). Microbiological and observational analysis of cross contamination risks during domestic food preparation. *British Food Journal*, 106 (8), 581–597.

TODD, E. C. D., GREIG, J. D., BARTLESON, C. A. and MICHAELS, B. S. (2008). Outbreaks where food workers have been implicated in the spread of foodborne disease. Part 4. Infective doses and pathogen carriage. *Journal of Food Protection*, 71 (11), 2339–2373.

TODD, E. C. D., MICHAELS, B. S., GREIG, J. D., SMITH, D. and BARTLESON, C. A. (2010a). Outbreaks where food workers have been implicated in the spread of foodborne disease. Part 8. Gloves as barriers to prevent contamination of food by workers. *Journal of Food Protection*, 73 (9), 1762–1773.

TODD, E. C. D., MICHAELS, B. S., SMITH, D., GREIG, J. D. and BARTLESON, C. A. (2010b). Outbreaks where food workers have been implicated in the spread of foodborne disease. Part 9. Washing and drying of hands to reduce microbial contamination. *Journal of Food Protection*, 73 (10), 1937–1955.

TODD, E. C. D., MICHAELS, B. S., SMITH, D., GREIG, J. D. and BARTLESON, C. A. (2010c). Outbreaks where food workers have been implicated in the spread of foodborne disease. Part 10. Alcohol-based antiseptics for hand disinfection and a comparison of their effectiveness with soaps. *Journal of Food Protection*, 73 (11), 2128–2140.

VERHOEF, L. P., KRONEMAN, A., VAN DUYNHOVEN, Y., BOSHUIZEN, H., VAN PELT, W. *et al.* (2009). Selection tool for foodborne norovirus outbreaks. *Emerging Infectious Diseases*, 15 (1), 31–8.

VERHOEF, L., VENNEMA, H., VAN PELT, W., LEES, D., BOSHUIZEN, H. *et al.* (2010). Use of norovirus genotypes profiles to differentiate origins of foodborne outbreaks. *Emerging Infectious Diseases*, 16 (4), 617–24.

WORSFOLD, D. and GRIFFITH, C. (2009). Experience and perceptions of secondary food hygiene training: a preliminary study of five larger catering companies in south east Wales. *Perspectives in Public Health*, 129 (2), 1–7.

YIANNAS, F. (2009). *Food Safety Culture: Creating a Behavioural Based Food Safety Management System*, New York, Springer.

21

Expanding the use of HACCP beyond its traditional application areas

W. H. Sperber, Retired, USA

DOI: 10.1533/9780857098740.5.417

Abstract: This chapter outlines the interactions between a hazard analysis and critical control point (HACCP) system and multiple prerequisite programs (PRPs) to produce a food-safety management program that can be applied globally throughout the food supply chain, from farm to table. The HACCP system is supported by three major PRPs: good agricultural practices, good hygienic practices and good consumer practices. While there are many effective food-safety practices and regulations, the current global system of food-safety management is poor. This chapter describes multiple barriers to food-safety progress and proposes new measures for significant improvements in global food-safety management. Such measures are necessary if HACCP is to be applied to non-traditional areas in the supply chain.

Key words: hazard analysis and critical control point (HACCP), prerequisite program (PRP), farm-to-table supply chain, food-safety management program, international efforts.

21.1 Introduction

It is of utmost importance to promote an accurate understanding of hazard analysis and critical control points (HACCPs) and prerequisite programs (PRPs) and their means of application. Development of good consumer practices (GCPs) can plug one of the many holes in the current leaky boat that is food-safety management. Additional progress can be made by developing a single effective food-safety regulatory agency in each country, creating a new international Food Protection Organization and ensuring there is experienced leadership in all food-safety agencies. Such progress will require better collaboration between the regulatory agencies and the food industry to identify and install leaders who have meaningful experience in the production, processing and handling of foods and food ingredients.

21.2 Historical perspective

Modern food-safety management programs and techniques have been initiated and improved in the food industry over the past 40 years. The seeds of these programs were planted during the rapid modernization and expansion of the global food industry after World War II (Sperber, 2006). The application of mechanical refrigeration to the storage and transportation of foods, along with the rapid development of highway, rail, ocean and air transportation systems, enabled the global sourcing of food ingredients and the distribution of consumer foods. These systems replaced home and local production of many foods with the centralized production of foods in large processing facilities. Advances in food science and engineering enabled the production of food ingredients and products with greatly extended shelf life. For example, dried eggs could be used in place of shell eggs and dried milk products could be used in place of fluid milk; this created new categories of convenient consumer products such as fully formulated dessert and beverage products that could be baked, cooked or consumed directly after the addition of water.

Because of their widespread distribution, the new food ingredients and products brought an unexpected hazard into food-processing and food service operations and the home – the presence of *Salmonella* bacteria, which led to illness outbreaks and many recalls of contaminated foods. The food-processing industry attempted to control the new problem of pathogen contamination by extending its quality control (QC) programs to monitor and manage food-safety concerns. QC procedures worked the same way in many industries – food, automotive, manufacturing, etc. – products were produced and placed into storage or quarantine. A prescribed number of samples would be tested; acceptable items or lots would be cleared for shipment.

It soon became apparent that QC procedures were not able to reliably detect the presence of pathogenic microorganisms, particularly if the pathogen were present at a low number or incidence. It was found that *Salmonella*-contaminated foods that had been involved in an outbreak or subjected to recalls often had low numbers and a low incidence of salmonellae (Table 21.1). Several large-scale evaluations of *Salmonella* in North American wheat flour revealed a low and decreasing incidence (Table 21.2). Continued research confirmed that the incidence of salmonellae in many kinds of contaminated foods was often of the order of 0.1%. Applying statistical procedures, if one wanted to detect salmonellae in such a food, one would need to test 3000 25-g samples in order to detect one positive sample at the 95% confidence level (Fig. 21.1) (Sperber, 2010).

Clearly, such a QC sampling plan would be impossibly impractical. Senior managers in the food industry insisted on a better management system to protect their products and businesses from the adverse impacts of pathogen contamination. Fortuitously, during this period (the 1960s) the Pillsbury Company, the US Army and NASA had been developing exactly such a system to better ensure the safety of foods produced for military and space personnel (Sperber, 2006; Sperber and

Table 21.1 *Salmonella* quantification (most probable number or MPN/100 g) in dried food samples taken from known *Salmonella*-positive lots

Product	Number of lots	Mean	Range
Milk chocolate	16	0.78	0.36–2.3
Rosemary spice	8	1.7	0.36–4.3
Dried milk	1	9.3	9.3
Egg pastina	6	36	3.6–93
Gelatin	2	2.6	0.9–4.3
Egg albumin	1	23	23
Powdered trypsin	1	9.1	9.1

Source: Sperber and Deibel, 1969; Fantasia *et al.*, 1969.

Table 21.2 Incidence of *Salmonella* in North American wheat flour

Years	Number of samples	Number positive	% Positive
1989	3040	40	1.32
1984–1991	1170	4	0.34
2003–2005	4358	6	0.14

Source: Sperber and NAMA, 2007.

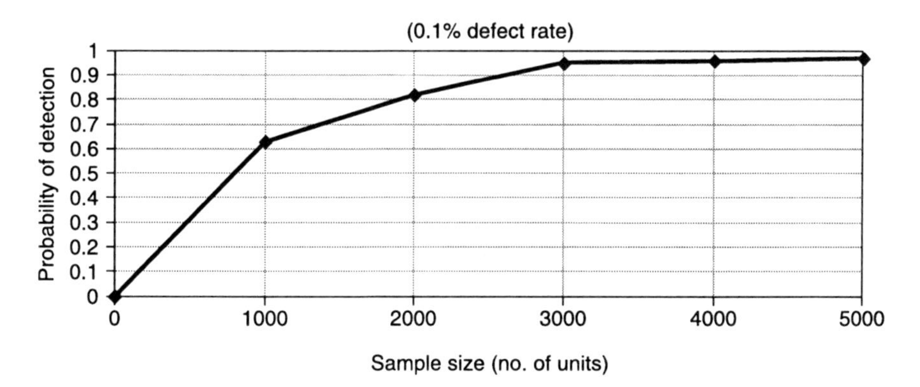

Fig. 21.1 Influence of sample size on the probability of *Salmonella* detection in food.

Stier, 2010). Naming it the hazard analysis and critical control point (HACCP) system, this small team laid the foundation for a food-safety management system, which gradually became the gold standard in the global food industry.

Based on hazard identification and the development of control measures for significant hazards, Pillsbury began in 1972 to apply its HACCP system to the production of its consumer food products. Pillsbury's original HACCP system was based on what are now called HACCP principles 1, 2 and 4. Very quickly, in the mid-1970s, Pillsbury found it necessary to add principles 3 and 5. Principles

6 and 7 were added in 1992 by the US National Advisory Committee on Microbiological Criteria for Foods (NACMCF) and the Codex Committee on Food Hygiene (CCFH):

1. Conduct a hazard analysis.
2. Determine the critical control points (CCPs).
3. Establish critical limit(s).
4. Establish a system to monitor control of the CCP(s).
5. Establish the corrective action to be taken when monitoring indicates that a particular CCP is not under control.
6. Establish procedures for verification to confirm that the HACCP system is working effectively.
7. Establish documentation concerning all procedures and records appropriate to these principles and their application.

The HACCP concept and its potential to enhance food-safety management and public health outcomes were soundly endorsed in 1985 by the US National Academy of Sciences (NAS) (NRC, 1985). In addition, the report suggested that government agencies responsible for food-safety verification would be more effective if inspectors were to audit facility HACCP records, which give a continuous record of food-safety compliance, rather than conducting plant inspections, which in most US food-processing facilities are to this day conducted infrequently, even at intervals of five years or longer. The recommendation for audits rather than inspections has not been accepted by US regulatory agencies, except for the dairy and canned food industries where regulatory compliance based upon audits rather than inspections has been verified since 1923 and 1973, respectively. NAS gave a further boost to HACCP by recommending the formation of NACMCF, which began work in 1988. One of its first major reports was the HACCP document published in 1992; this report was completely harmonized with the CCFH HACCP document in 1997 (CAC, 2003; NACMCF, 1998).

21.3 Contemporary perspective

Today we can reap the full benefit of the HACCP system, whose essential features have remained unchanged for 40 years. It is based upon the identification and control of significant hazards, with key functions embedded in designing food safety into food products and controlling the process by which the products are produced. It does not depend upon product testing and lot acceptance criteria, functioning best as an audit, not as an inspection, tool. Beginning as a voluntary development in the US food-processing industry, HACCP is now globally recognized and used, mainly because of its adoption as a recommended code of practice by the Codex Alimentarius Commission (CAC, 2003). HACCP can be an important component of international trade agreements, as Codex documents such as this have legal status under the World Trade Organization (WTO) and its member nations. The adoption of HACCP principles and procedures by federal governmental agencies and standards organizations gives further credibility to HACCP, facilitating its application in global food commerce.

Despite its obvious appeal and strengths, it did not take the food industry very long to realize that HACCP was not synonymous with food safety. Too often, food-processing facilities with valid HACCP plans experienced failures, including illness outbreaks linked to their products. Evaluation of each incident revealed that the food-safety failures were typically not caused by the failure of an element in the facility's HACCP plan. Rather the failures were usually related to poor cleaning and sanitation practices, which in turn were sometimes related to inadequate sanitary design of the facility or processing equipment. Fortuitously, the US Food and Drug Administration (FDA) had published in 1968 a regulation requiring good manufacturing practices (GMPs) to be used in food-processing facilities (CFR, 2011). The GMP regulation placed significant emphasis on the sanitary design of facilities and processing equipment:

- general provisions
- buildings and facilities
- equipment
- production and process controls
- defect action levels.

In 1969, Codex published its good hygienic practices (GHPs) in its Recommended Code of Practice, as summarized above (CAC, 2003). Because GMPs and GHPs bear many similarities, the more global term GHP will be used in the remainder of this chapter.

For more than 20 years, the food-processing industry has understood that food safety should be implemented through a HACCP system supported by a foundation of prerequisite programs (PRPs) (Sperber *et al.*, 1998). Initially, because HACCP and GHPs were used almost solely in the food-processing industry, the terms PRPs and GHPs were rather synonymous. Since then it has become obvious that HACCP and food-safety management need to be applied to the entire food supply chain, from farm to table. As one particular example, the 1985 British outbreak of bovine spongiform encephalopathy (BSE, or mad cow disease) dramatized the need to develop food-safety protocols that would be effective from farm to table. Because GHPs were developed primarily for use in food-processing facilities at the center of the supply chain, it became obvious that PRPs would also be needed for the farm and table ends of the supply chain (Table 21.3) (Sperber, 2005b). For more than 10 years good agricultural practices (GAPs) have been developed for use at the farm end, while no comparable PRP has yet been developed and promulgated to serve the table end. Therefore, this author is proposing again that GCPs are developed as a PRP to be used at the table end of the food supply chain. GAPs and GCPs will be elaborated in the next section. In the near future we should be able to see food safety assured by HACCP systems that are supported by thee PRPs: GAPs, GHPs and GCPs.

Table 21.3 Farm-to-table use of HACCP and PRPs in major food supply chain areas

Supply chain area		HACCP use	Available PRPs
Farm	Animal production	Minimal	GAP
	Crop production	Minimal	GAP
Processing	Raw products	Minimal	GHP
	Finished products	Major	GHP
Distribution and storage		Minimal	GHP
Table	Food service	Major	GHP and GCP
	Home	Moderate	GCP

21.4 Expansion of HACCP beyond traditional areas of application

It is now obvious to all parties that HACCP can be applied very effectively, with the support of GHPs, in the food-processing industry where hazards can be controlled at CCPs that can reliably kill pathogenic microorganisms or prevent their growth. Such CCPs are often established at process steps including cooking, pasteurization, sterilization, refrigeration, freezing, acidification and drying. Even with this seeming formula for success, it can still be difficult to manage food safety in this traditional area of HACCP application. Employee errors and failures to effectively apply GHPs still occur.

We are now learning that it is even more difficult to apply and maintain HACCP for food-safety assurance in non-traditional areas (Table 21.3). Farmers raise crops and grow animals with essentially no points where CCPs could be established to eliminate contamination with pathogens. Insofar as possible they are dependent on GAPs to maintain and improve food safety. Even within the processing segment of the supply chain there is a non-traditional area for HACCP application where it is quite difficult to develop CCPs. Many processed foods are handled and consumed raw or undercooked, e.g., fresh fruits, vegetables and their products, as well as raw ground beef and other raw meat and poultry products. The processors of these products must rely mainly on GHPs to maintain and enhance food safety. Long considered as a traditional area for the application of food-safety practices, the distribution and storage segments also offers few opportunities for establishing CCPs, except for facilities that handle refrigerated or frozen foods. Food-safety efforts in these areas also depend heavily on GHPs.

The remaining non-traditional area, the table segment of the supply chain, is much more amenable to the application of HACCP, wherever foods are cooked. This segment can best be considered in two parts: food service operations and the home. Food service operations include many grocery stores, restaurants, hospitals, schools, prisons, nursing homes, etc. In the US, food service operations are largely governed by the Food Code, a compilation of requirements for retail food protection (FDA, 2009). Some of the requirements could be managed as CCPs and others as GHPs. The home part of this segment is rarely mentioned as a place for food-safety management, although many of the same food handling and preparation practices and controls that are used in food service operations could be applied with substantial benefits in the home. No code of practice is in place to guide consumers in the safe handling of foods in the home. A recent publication teaches how HACCP could be applied usefully in the home (Sperber, 2011). Consumers would be well served by the development and use of GCPs, currently the missing PRP element in the supply chain. Even though CCPs and GCPs can be established in food service and home food operations, it is globally acknowledged that poor practices in these venues are the most commonly identified causes of foodborne illnesses. While poor practices are obviously responsible for the illnesses, the difficulty in improving food safety at the table end of the supply chain is driven by the sheer number of its locations. There are millions of food service establishments that serve prepared food to several billion customers each day and several billion homes in which food handling and preparation errors can occur. It is quite a challenge to promote food-safety awareness among the entire global population. Keep in mind that food-safety errors can occur in well-managed centralized food-processing operations in which a mere dozens or hundreds of workers are very well trained for working in a single facility. It would be much more difficult to maintain standards, train millions of food service workers and apply the same culture and practices in several billion homes.

While a great deal of progress has been made in food-safety management practices over the past 40 years, and the practices are continuing to spread across the global food supply chain, there are several important changes that are necessary to produce significant improvements in food-safety outcomes. These are: implementing HACCP correctly and creating, developing and implementing PRPs correctly.

21.4.1 Let's get HACCP right

During its first 25 years, HACCP was developed in the food industry and applied logically to control significant identified hazards at CCPs, and it was supported by GHPs. This development was honed by food processors, in the middle of the supply chain. Solid as it became, this state of affairs began to unravel in the mid-1990s, as attempts were made to apply HACCP to unprocessed or partly processed foods that were distributed and consumed raw or undercooked. The 1992 triumph of NACMCF in developing a HACCP document was undermined, in the US at least, by unsound food-safety regulations that were promulgated by its principal food-safety regulatory agencies – USDA's Food Safety and Inspection Service (FSIS) and FDA's Center for Food Safety and Applied Nutrition (CFSAN). This adverse turn of events occurred in large part because of a lack of understanding of hazard analysis and HACCP (Sperber, 2005a). Applying HACCP to the production of raw products creates a likelihood that the hazard analysis will identify one or more significant hazards, but further efforts by the HACCP team will not be able to identify CCPs that could be developed to control the hazards in raw foods. Pathogenic bacteria in raw milk are easily killed by pasteurization, which is managed as a CCP. However, the same pathogens in raw ground beef or fresh produce cannot be easily killed or removed by any means. When a HACCP plan has no CCP(s), it cannot be called a HACCP plan. At best it is only a hazard analysis, which should be recorded by the HACCP team to document the identification of significant hazards. Two important rules must be applied in the real world of HACCP – when significant hazards are identified and no CCPs are possible to control those hazards, either the process or product must be altered such that a CCP can be established, or the product must not be produced. These very important rules, long applied in conventional food-processing industries, have often been ignored in efforts to apply HACCP to non-traditional food-processing areas.

The difficulties of applying HACCP in non-traditional areas became more complicated when the new regulations reinforced expectations, particularly on the part of consumers, regulators and politicians, that raw meat and poultry products could be pathogen free at the points of consumption. These false expectations were then extended to the production of raw seafood, and fresh fruit and vegetables. Taken together, these developments created confusion and frustration, a situation that will continue until effective educational and regulatory changes are made.

21.4.2 Let's get PRPs right

At a minimum, the foundation of PRPs to support the HACCP system should include three programs:

Good agricultural practices

In the past 20 years, many sets of GAPs have been developed to improve the food-safety profile of food animals and crops. These are not as uniformly developed as GHPs, but have typically been developed as specific guidance for particular crops or animals in individual countries or agricultural industry segments.

- Soil
- Water
- Crop and fodder production
- Crop protection
- Animal production
- Animal health
- Animal welfare
- Harvest and on-farm processing and storage
- Energy and waste management
- Human welfare, health and safety
- Wildlife and landscape.

Good hygienic practices

The most developed PRPs are the GHPs. These provide a broad and effective foundation to enhance food safety in food-processing operations; they need to be used globally.

- Objectives
- Scope, use and definition
- Primary production
- Establishment: design and facilities
- Control of operation
- Establishment: maintenance and sanitation
- Establishment: personal hygiene
- Transportation
- Product information and consumer awareness
- Training.

Good consumer practices

The third necessary PRP, GCPs, does not yet formally exist, but it must be created and implemented if the foundation for food safety is to be completed. Here is a list of prospective GCPs:

- Create a culture of food-safety awareness, consider mini-HACCP program for home use
- Wash hands properly
- Practice good personal hygiene
- Use potable water to rehydrate and prepare foods
- Keep pets off tables and counters
- Prevent contamination from raw to cooked foods
- Wash all cooking utensils properly
- Achieve proper cooking temperatures, especially for raw meat and poultry
- Maintain proper refrigeration temperatures
- Handle and store leftovers properly
- Thaw frozen foods properly
- Do not consume raw milk, raw commercial juices, sprouts or cake batter
- Mind the diet of immunocompromised persons
- Clean and disinfect bottles for infant feeding before filling.

There has been too much reluctance on the part of the food industry, politicians, regulators, consumer advocates, etc. to even suggest that consumers have an important role in food safety. The blame for most foodborne illnesses is usually placed solely upon the food-processing industry. This mindset is a major barrier to extending HACCP beyond traditional areas. Food-safety outcomes cannot be improved unless there is a global recognition that food safety is a shared responsibility, from farm to table. And then, even when HACCP and PRPs are applied as effectively as possible, it must be acknowledged that there will still be foodborne illnesses and outbreaks. Until these points are understood, communicated and agreed, efforts to apply HACCP and improve food-safety outcomes in all parts of the supply chain, traditional and non-traditional alike, will not succeed. Without such advances, the current status of farm-to-table food-safety efforts – analogous to a rather leaky boat – will remain unchanged.

21.5 Necessary and practical food-safety improvements

Applying HACCP properly in non-traditional areas is important to enhance the safety of the food supply for seven to ten billion consumers over the next several decades. Meeting this difficult challenge will require many changes in the current leaky boat system of food-safety management and regulation. Effective food-safety management practices need to be applied globally, effective national food-safety and regulatory agencies need to be put into place, effective international food-safety leadership is required and the broader environment in which food is produced, regulated and consumed must be considered and improved. Increasingly, food-safety programs interface with food-security and food-defense programs. It is perhaps not fully appreciated, but all actions to improve global food safety will need to be taken on a playing field that is constantly shifting in many directions. The expanding human population and its improving economic status require food production to double by 2050. There has been a rapid increase in free trade, facilitating the development of emerging markets. Perhaps least well understood is that food-safety improvements will continue to be driven in large part by food companies, particularly global companies that have introduced their own successful food-safety practices, which protect their businesses, enhance global food trade and substantially strengthen the global food supply chain.

21.5.1 Effective food-safety management in the global supply chain

In the past several decades the international trade in food commodities, ingredients and products has increased dramatically. A food product manufactured in one country can contain ingredients from many countries. This trend seems likely to continue for more decades. As many countries are both importers and exporters of foodstuffs, we are faced with the responsibility of managing food safety in a very large and growing web of commercial activities. This is a daunting task, one that will not be accomplished unless sound management and regulatory activities are developed and maintained globally, from farm to table. To a significant extent, sound management activities are already in place, based upon the application of HACCP and PRPs. As published by Codex, HACCP and one of the PRPs – GHPs – are already in place. GAPs are being adopted in many countries and an international approach has been started (www.globalgap.org). The remaining PRP, GCPs, is not yet developed. This void must be eliminated. GCPs are just as essential as GAPs and GHPs in our efforts to improve safety and better protect public health.

Just as the creation and application of HACCP was a voluntary US food industry effort, the global food industry must now assume the primary responsibility and provide the leadership to ensure effective global food-safety management practices and results from farm to table. The industry cannot succeed on its own; it must collaborate with multiple partners – trade and professional associations, academic institutions and governmental and non-governmental organizations.

21.5.2 Effective national food-safety regulatory capabilities

As the food industry has the primary responsibility for ensuring food safety, governmental regulatory agencies have the primary responsibility for verifying that safe food handling and production practices are followed. Typically, the same regulatory agencies also develop regulations intended to enhance food-safety and public health efforts. Unfortunately, some regulations have unintended, and counterproductive, consequences. In this limited chapter, three examples of the latter in the US will be presented as instructive examples; without awareness, they could be replicated in other countries.

Following an outbreak of *Escherichia coli* O157:H7 illnesses in 1993, FSIS overreacted by publishing a notice declaring this pathogen to be an adulterant in raw ground beef (FSIS, 1994). Since then FSIS has reported test results on about 9000 samples of raw ground beef each year. The meat industry in the US has tested a much larger number of samples. The 17 years of FSIS sampling show little variation and no significant reduction of *E. coli* O157:H7 contamination (Table 21.4). A great deal of resources has been spent on this program with no apparent food-safety or public health benefit. In 2012 FSIS is proposing to declare six additional STEC serotypes to be adulterants in raw ground beef, greatly increasing this counterproductive use of resources. FSIS also overreacted in publishing the *Pathogen Reduction; HACCP Systems* rule for use in the production of raw meat and poultry products (CFR, 1996). While this rule has a number of counterproductive features (Sperber, 2005a), its principal adverse effect has been the distortion of the HACCP concept, by using it in the production of raw products for which conventional CCPs are not available. These points are not merely the opinion or bias of this author. A prominent US food law attorney has claimed that the US food regulatory system is a 'ridiculous system,' overloaded with regulations that are ignored or unenforced, and referred to the FDA's Food Safety Modernization Act as the 'Food Safety Confusion Act' (Tarver, 2012).

Much of the US confusion is caused by the dispersion of food-safety regulatory authority across multiple government agencies. Multiple agencies mean multiple leaders, most of whom are lawyers or medical personnel with little or no food-safety or food industry experience. This confusion and relative ineffectiveness could be minimized by the creation of a single food-safety agency, not only in the US, but in each country. Each food-safety agency should be led by a single food-safety leader, who would be a food-safety expert from the food industry and/or academia. The primary responsibilities of the national food-safety agency

Table 21.4 Incidence of *E. coli* O157:H7 in raw ground beef, 1994 through 2011

Period	Number of samples	Number positive	% Positive
October 1994 through 2001	52 291	171	0.33
2002 through 2011	101 765	292	0.29
All years	154 056	463	0.30

Source: FSIS, 1994.

would be to verify that food-safety requirements were met in the production of food, and to take effective enforcement actions when necessary. In the global scheme, single food-safety agencies in each country would more efficiently collaborate with the single international food-safety organization (below). The following are essential actions to be taken nationally to improve global food safety:

- Organize a single food-safety agency in each country.
- Each country's program to be led by a single, qualified food-safety leader.
- Require the application of HACCP and PRPs in all food handling and processing operations.
- Eliminate pathogen performance standards and adulterant rules as regulatory tools.
- Apply and verify food-safety requirements at all points in the supply chain, farm to table.
- Replace sporadic regulatory plant inspections with verifications of the food-safety management system.

21.5.3 Effective international food-safety leadership

Given the vast web of food commerce and the great number of national food-safety agencies (even if each nation has only one), it is important to establish effective food-safety leadership at the global level. Individual agencies already on the global scene are capable of meeting their intended responsibilities, but none of them are focused on food-safety leadership. In the United Nations, the Food and Agriculture Organization (FAO) is primarily responsible for food security and the WHO primarily for public health. The Codex Alimentarius Commission (CAC) supports both FAO and WHO by developing guidelines, food standards and codes of practice. Outside the UN, the World Organization for Animal Health (OIE) is responsible for animal health and the safety of foods of animal origin. Also interfacing with some food-safety activities are the International Plant Protection Convention (IPPC) and the WTO.

It would be most helpful to create an international organization for food safety, such as the Food Protection Organization (FPO), which would have food safety as its sole focus (Sperber, 2008). The FPO would be an independent UN organization, parallel to FAO and WHO and supported by Codex. What FAO has done for food security and WHO has done for public health, FPO could do for global food safety. Its responsibilities would cover the global food supply chain. It would provide leadership to assist the food industry in developing food-safety management programs. Importantly, FPO would employ or contract boots on the ground to verify compliance with food-safety requirements. It would work closely with the single food-safety agency in each nation; it could not work well with nations whose food-safety agencies were not fully capable.

Some of the principal activities and accountabilities of the FPO would be similar to those of food corporations and national food-safety agencies:

- Ensure farm-to-table coverage, emphasizing points of origination of food commodities and ingredients, and traceability.
- Require the use of HACCP and PRPs insofar as possible throughout the supply chain.
- Facilitate understanding, implementation and verification of effective food-safety measures.
- Create and maintain an environment that will ensure collaboration among governmental and intergovernmental organizations.

An effective system for food-safety verification must be adopted for global use. The Global Food Safety Initiative (GFSI) has developed a system with great promise (www.mygfsi.com). An initiative begun by food-safety experts in food manufacturing, retailing and food service companies, GFSI has developed or recognized requirements that are broader than those contained in the Codex GHPs. It builds food-safety capacity, benchmarks food-safety systems and provides accredited certification of food-safety management systems. For all parties involved in global food safety, the principal benefit of GFSI promises to be its system for third-party certification, so that the certification of a particular operation at any point in the supply chain would be accepted globally: 'if certified, accepted everywhere.' Such a tool is essential for food-safety assurance at any level, but especially at the global level.

21.5.4 Recognition of the environment in which food is produced

Enhancing food safety for seven to ten billion people will never be easy. It is a tall order; the deck seems stacked against us. It is essential to create, change or improve all systems and organizations to become as efficient and effective as possible. There are additional adverse factors that will complicate our ability to produce safe food reliably: overpopulation, wars, climate change, environmental degradation, economic irresponsibility and dysfunctional political processes. In this environment it is possible that better food-safety leadership could be provided by the Food Protection Organization. Global problems require global solutions.

21.6 Barriers to food-safety progress

While many effective food-safety practices and regulations are in effect, a number of barriers are impeding food-safety progress. Sometimes subtle, the barriers are often deeply ingrained.

21.6.1 Who makes the rules?

At least in the US, a great deal of food-safety expertise is ignored because of this barrier. The opportunity and authority to develop and promulgate rules, regulations and laws is typically vested in those who have little or no food-safety management experience. Ironically, the professionals with the most knowledge of and experience with food-safety programs – those in the food industry – are frequently excluded because of their presumed pro-industry bias. It is more likely that those representing consumer advocacy groups are included. Some of these have been most unhelpful when positing bromides such as 'irradiated poop is still poop.' Following that logic, it would not be acceptable to pasteurize raw milk or cook raw meat and poultry. Worse, their position stifles the application of useful technologies such as e-beam ionizing irradiation of raw ground beef and other foods.

21.6.2 Is it science-based?

The FSIS *Pathogen reduction; HACCP systems* rule (CFR, 1996) has been strongly criticized for its abuse of HACCP and for its use of *Salmonella* performance standards. Each criticism was deflected by claims that the decisions were science based. Much of this rule was not science based at all; it was a poor application of statistics (Sperber, 2005a). Beware of the term 'science based.' In the late 1990s an illogical technology to improve produce safety was favorably received. It involved the disinfection of sprouting seeds by using >20 000 ppm hypochlorite. The use of hypochlorite as a food contact equipment sanitizer is restricted to 200 ppm. Why is it permissible to treat a food product with >100 times the amount used to sanitize equipment? In such an environment the terms 'HACCP' and 'science-based' become devalued.

21.6.3 Where is the political courage?

Given the confusion, especially in the US, over the application of pathogen performance standards and adulterant rules, a great deal of uncertainty over the safety of foods that are handled and consumed raw or undercooked has been created. The proven inability of extensive product testing to eliminate pathogens in raw products has magnified the confusion. Politicians and regulators need to examine this situation and accept one of two possible remedies:

- Accept the fact that raw foods cannot be pathogen free, particularly when irradiation is discouraged, and eliminate pathogen performance standards and adulterant rules. They and the public must recognize that a certain minimal risk is inherent in raw foods.
- Alternatively, ban the sale of raw foods, e.g., sprouts and raw ground beef, or require them to be cooked or irradiated at the point of manufacture.

Assuming the continued absence of regulatory courage on these points, consumers must decide for themselves whether to avoid a particular food or to ensure that is properly cooked before consumption.

21.6.4 Can we afford it?

Recommendations for improving global food safety, e.g. by restructuring or creating national and international organizations, are immediately countered by claims that 'We can't afford it!' That is a very weak claim on a planet that wastes trillions of dollars each decade on wars and unsound economic practices. Where is the courage to do what is best? Moreover, restructuring dysfunctional food-safety agencies and programs should greatly improve efficiency while yielding better public health outcomes. We cannot afford not to do it.

21.6.5 Are audits the only tool to verify food safety?

In the midst of prominent food-safety failures, audits and audit scores have come under criticism. Properly developed and administered, audits can become a more important tool to verify food safety. Perhaps another tool – an informal food-safety evaluation by an industry food-safety expert – could be used in conjunction with audits. This author has conducted food-safety evaluations at hundreds of facilities on four continents. There were no formal documents, no scoring, just a list of deficiencies that needed to be corrected. The number of deficiencies ranged from several to >100 per evaluation, from minor to severe, sometimes necessitating the temporary or permanent shutdown of production lines or facilities. This informal procedure immediately corrects identified problems. In contrast, stacks of formal HACCP audits conducted by local company personnel or by third parties often prove to be worthless. Every food company could benefit, while enhancing food safety throughout the supply chain, by using expert corporate food-safety evaluators on a regular basis.

21.7 A plan for global food-safety progress

As discussed throughout this chapter, there are a number of necessary requirements and changes that would facilitate the expansion of HACCP into non-traditional areas and enhance food safety worldwide:

- Creation of a single food-safety agency in each nation.
- Employ an experienced food-safety expert to lead each national agency.
- Create an independent Food Protection Organization in the United Nations.
- Require the use of HACCP and PRPs globally, from farm to table.
- Access food industry expertise for the development of regulatory procedures.
- Develop reliable third-party certification and audit procedures.
- Encourage the use of corporate food-safety evaluators.
- Develop the collective courage to make sound decisions.

Applying HACCP and PRPs throughout the global food supply chain, including non-traditional areas, will not be easy but it can be done. Every country has many experienced food-processing and food-safety experts. We are lacking, however, the combined political will, courage and national and international leadership to improve the existing fragmented global food-safety efforts. Now is the time for all food-safety and related professionals to become part of a serious effort that will improve and sustain food-safety management programs.

21.8 References

CAC (2003) *Recommended code of practice: General principles of food hygiene, and HACCP annex.* www.codexalimentarius.net/web/more_info.jsp?ic_sta=23 [Accessed 22 February 2012].

CFR (1996) *Pathogen reduction; Hazard analysis and critical control point (HACCP) systems; Final Rule*, Code of Federal Regulations, Title 9, part 304, Washington, DC, US Government Printing Office.

CFR (2011) *Current good manufacturing practices in manufacturing, packaging, or holding human food*, Code of Federal Regulations, Title 21, Part 110, www.accessdata.fda.gov/scripts/cdrh/cfdocs/cfcfr/CFRSearch.cfm [Accessed 16 February 2012].

FANTASIA, L. D., SPERBER, W. H. and DEIBEL, R. H. (1969) 'Comparison of two procedures for detection of *Salmonella* in food, feed, and pharmaceutical products.' *Appl Microbiol*, 17, 540–541.

FDA (2009) *Food Code*. www.fda.gov/Food/FoodSafety/RetailFoodProtection/FoodCode/FoodCode2009/ [Accessed 9 February 2012].

FSIS (1994) *Microbiological testing program for Escherichia coli O157:H7 in raw ground beef*, FSIS Notice 50–94, now FSIS Directive 10,010.1. www.fsis.usda.gov/Science/Ecoli_O157_Summary_Tables_1994-2010/index.asp#1994 [Accessed 1 March 2013].

NACMCF (1998) 'Hazard analysis and critical control point principles and application guidelines.' *J Food Protect*, 61, 762–775.

NRC (1985) *An Evaluation of the Role of Microbiological Criteria for Foods and Food Ingredients*, Washington, DC, National Academy Press.

SPERBER, W. H. (2005a) 'HACCP and transparency.' *Food Control*, 16, 505–509.

SPERBER, W. H. (2005b) 'HACCP does not work from farm to table.' *Food Control*, 16, 511–514.

SPERBER, W. H. (2006) 'Rising from the ocean bottom – the evolution of microbiology in the food industry.' *Food Protect Trends*, 26, 818–821.

SPERBER, W. H. (2008) 'Organizing food protection on a global scale.' *Food Technol*, 62, 96.

SPERBER, W. H. (2010) 'Shifting the emphasis from product testing to process testing.' *Food Safety Mag*, 15, 42–46.

SPERBER, W. H. (2011) 'Food safety in the home: a review and case study', in Wallace, C. A., Sperber, W. H. and Mortimore, S. E. (eds), *Food Safety for the 21st Century*, Oxford, Wiley-Blackwell, 303–310.

SPERBER, W. H. and DEIBEL, R. H. (1969) 'Accelerated procedure for *Salmonella* detection in dried foods and feeds involving only broth cultures and serological reactions.' *Appl Microbiol*, 17, 533–539.

SPERBER, W. H. and NAMA (2007) 'Role of microbiological guidelines in the production and commercial use of milled cereal grains: a practical approach for the 21st century.' *J Food Protect*, 70, 1041–1053.

SPERBER, W. H. and STIER, R. F. (2010) 'Happy 50th birthday to HACCP: retrospective and prospective.' *Food Safety Mag*, 15, 42–46.

SPERBER, W. H., STEVENSON, K. E., BERNARD, D. T., DEIBEL, K. E., MOBERG *et al*. (1998) 'The role of prerequisite programs in managing a HACCP system.' *Dairy Food Environ Sanit*, 18, 418–423.

TARVER, T. (2012) 'Shortcomings of food regulatory systems.' *Food Technol*, 66, 48–49.

22

Biotracing in food safety

G. C. Barker, Institute of Food Research, UK

DOI: 10.1533/9780857098740.5.433

Abstract: Biotracing is a method for investigating food chain contamination events that concentrates on identification of probable sources. Biotracing is complementary to existing food safety assessments and is related to similar investigation processes in molecular epidemiology and forensic science. This chapter describes the background to biotracing, identifies important elements and describes an emerging methodology. In addition the chapter includes a review of a detailed biotracing model for *Staphylococcus aureus* in processed milk. The chapter concludes by highlighting options for food chain integration and includes some possibilities for future developments.

Key words: food chain, contamination, sources, biotracing, network model.

22.1 Introduction

Increasingly events that involve microbial contamination, including many well-publicized food safety incidents, are the subject of detailed investigations. Investigations usually center on procedures for gathering and analyzing information but many of the processes also have a specific intention to identify the probable origins of the observed microbial contamination – source tracing. These emerging processes include elements of forensic science, molecular epidemiology and pollution control. Biotracing is a natural extension of this developing inferential process to food chain systems (Barker *et al.*, 2009). Expressing this food chain investigation process as biotracing emphasizes the principal concern about biological contamination but, in addition, it reflects a strong connection with well-established tracing processes that are used when it is necessary to reveal the sources of particular food materials and ingredients. Often tracing, in a food chain context, is facilitated by permanent labels on the food packaging, which can be matched with recorded data at points during the production process. However, in contrast, biotracing operates with biological information that is less permanent, less fixed and not so easy to access; nonetheless it has become clear that, in many cases, biotracing and tracing can work together within a management framework to improve food safety and increase food chain security.

Although biotracing is only loosely defined it can be expressed succinctly as the ability to use downstream information to point to materials, processes or actions within a particular food chain that can be identified as the source of an undesirable agent. In simple terms: the origins of undesirable food contamination are established from detailed, and accessible, food chain observations. To be effective, i.e. to point to contamination sources in an accountable fashion that supports decision-making, biotracing, and the other investigative processes, must be quantitative. Over the last few years a systematic approach to the identification and quantification of microbial contamination sources, based on improved data supplies and novel mathematical modeling, has rapidly developed and has driven the progression of food chain biotracing.

The case of the anthrax letters in 2001, and the subsequent forensic investigation, provided a major impetus for research surrounding source-level inference techniques (Dance, 2008) and spawned a new field of research called microbial forensics (Keim *et al.*, 2008). Microbial forensics highlights the quantity, the control and the specificity of information that is required to support criminal investigations concerning the sources of microbial contamination but, in addition, it indicates that mathematical methods and analyses can be combined to make source-level inferences even in the presence of incomplete evidence and significant uncertainty (Jarman *et al.*, 2008). Microbial forensics is a natural springboard for the development of food chain biotracing.

Developments matching those in microbial forensics have followed in other areas of inference that relate to the tracing of microbial contamination. Molecular epidemiology, i.e. molecular tools applied to outbreak strains of foodborne pathogens, aims to provide evidence that leads to a strong association between public health observations and particular food sources. Matching agent types at distinct spatial and temporal locations is usually the dominant form of evidence. Significant *Escherichia coli* outbreak investigations in the USA in 2006 and in Germany in 2011, employed type matching processes to infer the source of contamination (e.g. Jay *et al.*, 2007; Beutin, 2011). In the United States and elsewhere, the matching process used for outbreak investigations is dominantly based on pulsed field gel electrophoresis and is facilitated by large data networks such as PulseNet. However, based on modern genetic tools increasingly more sophisticated marker types are available for typing bacterial populations and these can lead to improved accuracy and confidence in source tracing; recently a whole genome approach was used to trace a multi-state outbreak of *Salmonella enterica* that occurred in the USA in 2009–2010 (den Bakker *et al.*, 2011). Another example of source-level inference, microbial source tracking (e.g. Hagedorn *et al.*, 2011), has been developed to strengthen discrimination among possible sources of fecal matter in rivers and lakes. In this case sophisticated numerical schemes, including Bayesian classifiers, combine with molecular data about microbial communities to support source identification for polluting material and to help establish responsibility.

These processes show that modern data sources support outbreak investigations or environmental monitors that can identify sources with high levels of resolution and accuracy. Globalization, strengthening controls and security issues ensure that it is highly desirable to extend this methodology to food chain systems.

A food chain process for tracing the source of contamination, biotracing, has many elements in common with microbial forensics and outbreak investigation but also has some important distinctions. Firstly the chain itself provides a strong constraint on agent dynamics – particularly for bacterial agents – so that once a bacterial population enters the system its dynamics, in terms of growth, survival and death, are determined by the physico-chemical conditions imposed by the chain. In this way background knowledge of chain properties contributes strongly to an understanding of potential bacterial population changes and, if uncertainties are sufficiently reduced, this information can be reformulated into a scheme for source identification based on end-point observations of actual population parameters. It is immediately clear that food chain biotracing, to be effective, will include consistent probabilistic reasoning. A second distinct feature of biotracing follows from the food manufacturing process itself; although extrinsic sources of contamination, such as raw materials or food handlers, can be considered in the same way as the common sources that lead to outbreak incidents, there are also intrinsic effects, such as the unobserved failure of thermal processing or other hurdles, which act as additional effective food chain sources. An inventory of sources for a particular food chain is often the starting point for consideration in biotracing.

Highly discriminative agent type information, such as full genotypes that are sometimes used to establish a trace in forensic and epidemiological situations, is often, initially, too expensive to obtain for food chain applications. For food chain biotracing less discriminative types, such as serotypes or resistance patterns, are more realistic as contributors to the source identification process. Again this approach clearly indicates the role of probabilistic reasoning in establishing the biotracing significance of particular type matches. For some agent-chain systems a hierarchical investigation, with increasingly discriminative typing, may be an effective approach for biotracing. Forensic and outbreak investigations are both triggered by evidence of actual incidents and, hence, take place after the actual contamination events (although it is not unusual for an outbreak investigation to coexist with a developing contamination scenario). In contrast food chain biotracing includes opportunities for in-line diagnostic processes and, potentially, pre-emptive interventions that, ideally, contribute to improved food safety. As an early indicator of effective remedial actions, targeted at a particular element in food manufacture, biotracing could provide a commercial incentive in terms of reduced loss of performance, etc.

22.2 Elements of biotracing

Modern food chain systems and the associated sciences have developed in many ways that make the introduction of biotracing timely. Innovative digital technologies aimed at audits, monitoring or control activity (loggers, controllers, thermostats, etc.) automatically record large volumes of data that document normal operations of food chain elements but, in so doing, also indicate a framework for measuring complex out-of-specification behavior. These information supplies routinely engage automated storage and retrieval systems that have structured access and multi-use. In this way they promote the integration of multiple information sources and offer support for new diagnostic opportunities such as biotracing.

Additionally modern food chain surveillance engages novel, rapid, often molecular methods for data generation. In particular, methods based on the polymerase chain reaction have introduced increased sensitivity for detection and enumeration of contamination, and molecular-based typing schemes have introduced strong discrimination and identification of agents. Both these technologies are fast and so give increased opportunities for in-line actions. Improved sensitivity and specificity ensure that simple decision-making processes, such as those based on a threshold, can be replaced by more complex processes with multi-facetted outcomes and action levels. Together the integrated data supplies and molecular technologies are driving a progression away from isolated, uncoordinated, decision-making towards automated decisions that are integrated into broader, operational, food safety management – this includes biotracing.

Bayesian statistical techniques, particularly the systematic application of Bayes' theorem to complex sets of variables, have become an essential tool in the investigation and tracing of sources of evidence (e.g. Taroni *et al.*, 2006). Bayes' theorem inverts a conditional probability – usually transferring a causal description of an uncertain event to an inferential reasoning about potential sources – and it is a natural component of a source identification process. However, the successful application of Bayes' theorem in complex systems requires substantial computation so it is only recently that this technique has become a practical tool for investigation of real-world scenarios (Kjaerulff and Madsen, 2008). Bayesian methods have previously been used to encapsulate the information uncertainties associated with food safety models (e.g. Barker, 2003) and are central to microbial forensics (Jarman *et al.*, 2008).

The application of Bayesian methods for biotracing has been illustrated by two simple Bayesian belief network models (http://bbn.ifr.ac.uk/btmodeller/). One of these models, called SimpleTrace, inverts a series of causal probabilities to reveal posterior beliefs about possible contamination sources, based on end-point evidence for a population size. The forward conditional probabilities define the steps in the population dynamics of the agent, or a marker, within the particular food chain. This model clearly illustrates that reduced information uncertainty is crucial for establishing an inference about active sources. The other model, called SimpleMatch, uses type matching, at two distinct points that are close to the source and close to the end point of a chain, to establish a posterior belief about the true source of an observed agent. The model shows that strong discriminating ability of a match and detailed knowledge of possible sources is crucial for establishing this inference. Many technologies support the matching process for harmful agents in food and the associated inference is likely to become a major consideration for food chain biotracing (e.g. Foley *et al.*, 2009). Together these simple models illustrate two powerful principles that facilitate inference concerning sources and support further developments, and real applications, for food chain biotracing.

Ultimately biotracing provides support for decision-making, e.g. where to initiate remedial actions or whether to search for a hidden source of contamination in a food chain system. In practical applications, and particularly in legal frameworks, output probabilities from biotracing are converted to likelihood ratios to assist with the decision-making process (e.g. Taroni *et al.*, 2006). A likelihood ratio expresses the value of the evidence leading to particular source identification and, to some extent, reduces the dependency of the results on particular probability assignments during modeling. This transformation makes complex information easier to interpret. Additionally Bayesian methodology makes it practical to combine multiple information sources into one consistent inference process so that food chain biotracing can be performed as a structured information gathering process, which does not rely on prescribed data structures or complete data sets. A biotracing screening strategy that successively eliminates high probability sources of observed contamination may be practical for some food chain systems.

22.3 Biotracing in food chain systems

Source-level inference applied to food chain systems, potentially, has a huge impact on chain management. Most explicitly this might involve support for the transfer of liability (Pouliot and Sumner, 2008) but, in practice, biotracing impinges on all aspects of chain operation from analysis of control points to targeted maintenance, process design and specialist communication. Overall biotracing acts as a versatile decision support tool for food chain operations, which can drive improvements in food safety and promote improved efficiency and security.

Food chain biotracing identifies foodborne pathogens and the associated toxins as the dominant agents of concern and includes process failures, external contamination events and consumer misuse, as well as actual food materials, in the set of possible contamination sources. However, crucially, biotracing also harnesses many other associated (statistical) information supplies, concerning coexisting populations, complementary observations or prior knowledge, to generate integrated beliefs. In milk chain biotracing the naturally occurring enzymes provide valuable information about the efficiency of heating processes; in cheese chain biotracing the developing pH adds to the belief concerning process integrity and for the feed chain historical details of persistent bacterial strains associated with individual production locations complement direct sampling results. In many respects the biotracing approach concerns the development of models that successfully account for the relationships between variables so that added value, in terms of source identification, stems from multiple pieces of connected evidence.

Food chain biotracing does not have a uniquely defined methodology but several key elements have emerged. Firstly, for a particular food chain, a list of potential sources (failure modes) is identified. The list should be exclusive but not necessarily exhaustive and may include conveniently grouped sets of independent sources as single elements. In a recent communication concerning an outbreak of *E. coli* originating from vegetables, the UK Health Protection Agency identified potential sources including cross-contamination in storage, inadequate washing of vegetables, insufficient hand washing and cross-contamination from kitchen utensils. In this case inadequate source-level inference hindered effective public health communications (Campbell, 2011). Secondly, a (qualitative) food chain model that identifies causal connections between sources and relevant end-point measures, as well as dependencies for other important measurable quantities, is developed. This modeling step is usually highly interdisciplinary involving food chain experts but, increasingly, could be data driven; Sarkar *et al.* (2008) describe a data-driven latent space model that quantifies occurrences of *Salmonella* at food facilities. Finally, the (domain) model is quantified, invariably using mixed data sources, and is implemented as a structured probability calculation. As an example Jarman *et al.* (2008) used a Bayesian network to integrate disparate analytical measurements as part of a source-level inference for spore contamination. Ultimately a biotracing implementation provides a consistent framework for information integration and evidence propagation, which can be used to address biotracing questions, most visibly: 'For a set of possible sources, which is the most likely cause of actionable evidence?' but, equally valid: 'For a particular source, is the likelihood of initiating a hazard acceptable?' or 'Does the evidence support the existence of unidentified sources?' Barker *et al.* (2009) provide a more complete description of biotracing methodology.

Several prototype biotracing schemes illustrate elements of the methodology; e.g. Cook *et al.* (2006) traced human *Salmonella* infections to feed rather than chickens; Wilson *et al.* (2008) discriminated between animal and environmental sources for *Campylobacter* using multilocus sequence typing information; Lomonaco *et al.* (2009) distinguished between pasteurization failure and cross-contamination as sources of *Listeria monocytogenes* in Gorgonzola cheese and Verhoef *et al.* (2010) differentiated between human and animal origins for *Norovirus* from foodborne outbreaks using genotype markers.

A more complete example of a strongly integrated biotracing scheme concerns the hazards associated with *Staphylococcus aureus* enterotoxin in pasteurized milk (Barker and Goméz-Tomé, 2011). In this hazard domain there is a direct (complex) coupling between the *S. aureus* population and the staphylococcal enterotoxin concentration in milk and there is an additional indirect relationship, mediated by the common thermal process, between the bacterial population and the concentration of heat-sensitive milk enzymes such as alkaline phosphatase (ALP). During milk processing possible staphylococcal hazards are associated with toxin production in raw milk, incomplete pasteurization or post-process contamination and, in a biotracing model, these translate into three effective sources. Rapid and efficient measurements of ALP, following processing, are often used as an accurate indicator of milk quality so that, along with microbiological measures, an ALP test result is an important end-point measure.

The biotracing model, for *Staphylococcus* hazards in processed milk, is represented by the network diagram in Fig. 22.1. Three shaded nodes in a row at the bottom of the diagram terminated by three partially shaded nodes represent elements of the coupled dynamics of *S. aureus*, enterotoxin and ALP as milk passes along the production chain. Unshaded nodes, which are all described in detail by Barker and Goméz-Tomé (2011), represent additional pieces of information (uncertain parameters and control values) that quantify beliefs about the hazard domain. The nodes occur in four groups, or modules, that represent the collection and pooling of raw milk, cooling and storage, thermal processing (pasteurization) and volume partition of final product (filling). The individual modules are largely independent and are quantified by distinct expertise and data supplies. For example the cooling module includes a full predictive model for *S. aureus* population growth and toxin production with uncertain parameters developed from extensive laboratory experiments and the processing module includes a variability distribution, generated from a farm survey, which describes the combinations of time and temperature that are used for pasteurization operations. In total the network model encodes several megabytes of data from many sources.

Two other groups of nodes, all Boolean, are also identified in the network diagram as partly shaded nodes. Three nodes represent variables that are associated with the identified sources and three nodes represent appropriate end-point observations. The source nodes are labeled *Cool Fail?*, $PE < 1$ and *BoolX*. (PE is an abbreviation for 'pasteurization equivalent' so that PE < 1 indicates that the thermal process did not meet the requirements of a standard process such as the short-time high-temperature process that corresponds with 15 seconds at 72°C.) If the rapid cooling that follows milk collection works correctly then *Cool Fail?* is false and there is little opportunity for *S. aureus* population growth and toxin production prior to processing. In this case the source strength for toxin production in raw milk is quantified by a small (prior) probability for the case when *Cool Fail?* is true. Similarly the probability of true values for $PE < 1$ or *BoolX* quantify the source strengths for incomplete pasteurization and post-process contamination. The end-point measures quantify beliefs concerning the coupled populations that are built from combining information about sources with information about the domain; as end points these beliefs are expressed with respect to some carefully chosen critical values. For example, if the variable labeled $c_F > 10^5$ is true there are large concentrations of *S. aureus* bacteria in the filler tank milk and, hence, there is potential for intoxication in the event of consumer negligence. For end points labeled $c_{ToxF} > 0$ and *ALP –ve?* true values correspond with finite amounts of toxin in the filler tank milk and a failed ALP test.

The probabilities associated with the hazard variables and their dependencies are encapsulated within the model during construction but the structured

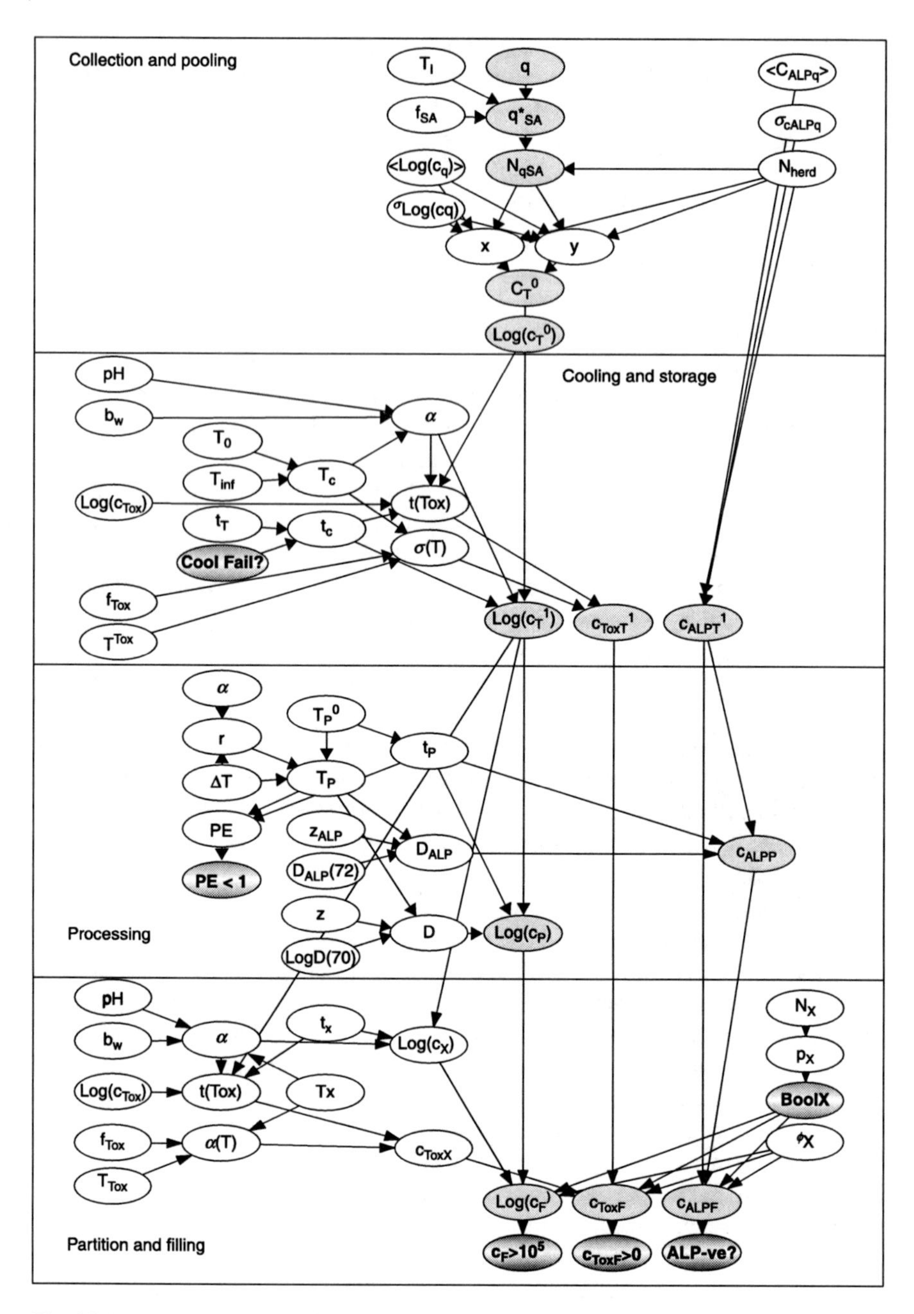

Fig. 22.1 Network diagram for a biotracing model for *Staphylococcus aureus* hazards in processed milk.

implementation ensures continued flexibility. The network representation enables additional evidence and 'what if' type queries to be tested, consistently, against the established beliefs with minimal effort. One role for the end-point observation nodes is immediately clear: the end-point probabilities quantify risks from staphylococcal enterotoxin hazards in processed milk and these values can be monitored in response to added evidence concerning the sources, i.e. how do risks change in response to different failure possibilities? In this form the domain model promotes a very versatile form of risk assessment. However, the biotracing approach goes one step further; rather than propagating information from source nodes to end points (along the causal path) biotracing propagates information from end-point observations towards the source nodes and hence leads to inference about sources. This propagation can only occur because the biotracing model is constructed in a way that allows a consistent application of Bayes' theorem. This is the first indication of biotracing: based on a background knowledge of the milk chain, observed information about the quality of filler tank milk can be used to change beliefs about whether the tank cooling was successful, whether the pasteurization worked as required and whether contaminating material may have entered the chain after the thermal process.

The milk chain biotracing model shows that combining end-point evidence with prior knowledge about the hazard domain leads to modified beliefs about source strengths and provides support for decision-making. Large end-point values for *S. aureus* loads suggest post-process contamination, with a likelihood ratio ~15, while significant amounts of toxin in filler tank milk suggests cooling failures with a likelihood ratio ~30. Propagating multiple end-point observations leads to strong support for the ALP test as a measure of pasteurization effects and as an aid to source discrimination. The milk chain biotracing model is very versatile and, strictly, can be viewed as an intelligent knowledge-based system for evaluation of milk chain operations and *S. aureus* hazards.

Detailed biotracing models are being developed for many food chains and hazard scenarios. These include a scheme to identify the source of *Salmonella* contamination for freshly slaughtered pork (Smid *et al.* 2011, 2012). This model indicates that house flora, on or in the carcass splitter, may be a dominant source of contaminated pork but it also includes solutions from systems of coupled equations and Monte Carlo simulations during the construction process and so highlights the range of suitable approaches.

22.4 Conclusion

Biotracing is an emerging technology, which can support decision-making in relation to food chain management. It combines progressive mathematical aspects, particularly network models and Bayesian reasoning, with new data supplies that have superior accessibility and precision (part of the data deluge). In many ways biotracing is a timely complement to traditional food safety processes, such as control point schemes and risk assessments, because it adds inference about the potential causes of hazards to surveillance information or to expressions of the expected outcomes. However, to become fully operational the statistical inference, which is the core of a biotrace, has to be augmented by strong data management, maintenance of data streams, databases and services including visualization and display. To be effective a biotracing scheme should operate without the need for manual input and should require minimal maintenance or upgrading. The initial design and implementation, within a particular manufacturing process, are crucial. Currently an initial investment in data infrastructure often presents a hurdle to operational biotracing but, as development and control of food manufacture move inexorably to the digital domain, these barriers will steadily reduce.

Development of detailed biotracing models, such as the milk chain model or the pork chain model identified above, includes substantial food chain and microbiological expertise whereas, in other fields, many Bayesian networks are constructed directly from case data using a computer learning process. As the digital environment for food manufacturing expands, the data-driven approach for the development of biotracing will be increasingly practical. Initial exploration has shown that the microbiological hazards associated with cheese-making, where direct causal structures are difficult to establish from first principles, may be suited to data-driven biotracing. A series of bacterial and physical measurements at strategic time points in the cheese manufacturing process may be used to represent a single case and a case database with a few hundred records may be sufficient to capture important correlations (e.g. Delbes *et al.*, 2006). Machine learning methods have previously been used in the development of microbial source tracking systems (Belanche-Munoz and Blanch, 2008).

The milk chain biotracing model described by Barker and Goméz-Tomé (2011) is relatively static because the underlying knowledge base is built from consideration of a population of systems (small dairies) that are analogous to the particular system of interest. In practice a model development that engages local real-time data streams, so that the knowledge base encapsulates previous realizations of the actual system of interest, can ensure the direct applicability of biotracing results and can accommodate slow trends, such as seasonal variations, into the model so that biotracing becomes dynamic. Local data supplies and dynamic biotracing systems are expected to dominate biotracing developments.

Biotracing emphasizes a system-wide picture of complex food chain processes and facilitates the integration of information supplies to support decisions about food safety. In particular food chain biotracing establishes information about the source of a problem simultaneously with information about the size of a problem so that, in an operational scheme, source-level information appears alongside hazard information even when the hazard is at subcritical levels. In this case, directed interventions can be considered. Biotracing itself does not remove an important food safety decision-making process that is central to food chain management but, potentially, it does add to the number of possible decisions and hence improves management flexibility. Development of biotracing models shows that a progression in food chain management that includes source-level inferences could benefit food safety, improve process efficiency and support the security and maintenance of a varied, multi-scale, food supply chain.

22.5 References

BARKER G.C. (2003) Application of Bayesian belief network models to food safety science. In *Bayesian Statistics and Quality Modelling in the Agro-Food Production Chain*, P. van Beek, T. van Boekel, A. van Bruggen and A. Stein. (eds), Kluwer, 117–128.

BARKER G.C. and GOMÉZ-TOMÉ N. (2011) A risk assessment model for enterotoxigenic *Staphylococcus aureus* in pasteurized milk: a potential route to source level inference, *Risk Analysis* 33, 249–269.

BARKER G.C., GOMEZ N. and SMID J. (2009) An introduction to biotracing in food chain systems, *Trends in Food Sci. Tech.* 20, 220–226.

BELANCHE-MUNOZ L. and BLANCH A.R. (2008) Machine learning methods for microbial source tracking, *Environ. Model. Software* 23, 741–750.

BEUTIN L. (2011) Outbreak with aggregative EHEC O104:H4 in Germany: specific characteristics and search for possible sources, www.foodprotection.org/files/annual_meeting/2011-breakout-session-lothar-beutin.pdf.

CAMPBELL D. (2011) Unpublicised *E. coli* outbreak leaves 250 ill and one dead, *Guardian* 30 Sept.

COOK R.L., WILSON R. and HATHAWAY S.C. (2006) Tracking *Salmonella* Typhimurium ST1 from contaminated poultry feed to a cluster of human salmonellosis, *Proc. 11th International Symposium on Veterinary Epidemiology and Economics*.

DANCE, A. (2008) Anthrax case ignites new forensics field, *Nature* 454, 813.

DELBES C., ALOMAR J., CHOUGUL N., MARTIN J.-F. and MONTEL M.-C. (2006) *Staphylococcus aureus* growth and enterotoxin production during the manufacture of uncooked, semihard cheese from cows' raw milk, *J. Food Prot.* 69, 2161–2167.

DEN BAKKER H.C., MORENO SWITT A.I., CUMMINGS C.A., HOELZER K., DEGORICIJA L. *et al.* (2011) A whole genome SNP based approach to trace and identify outbreaks linked to a common *Salmonella enterica* subsp. *enterica* serovar Montevideo pulsed field gel electophoresis type, *Appl. Env. Mico.* 77, 8648–8655.

FOLEY S.L., LYNCH A.M. and NAYAK R. (2009) Molecular typing methodologies for microbial source tracking and epidemiological investigations of Gram-negative bacterial foodborne pathogens, *Infect. Gen. Evol.* 9, 430–440.

HAGEDORN C., BLANCH A.R. and HARWOOD V.J. (eds) (2011) *Microbial Source Tracking: Methods, Applications and Case Studies*, Springer, New York.

JARMAN K.H., KREUZER-MARTIN H.W., WUNSCHEL D.S., VALENTINE N.B., CLIFF J.B. *et al.* (2008) Bayesian-integrated microbial forensics, *Appl. Env. Micro.* 74, 3573–3582.

JAY M.T., COOLEY M., CARYCHAO D., WISCOMB G.W., SWEITZER R.A. *et al.* (2007) *Escherichia coli* O157:H7 in feral swine near spinach fields and cattle, central California coast, *Em. Infect. Dis.* 13, 1908–1911.

KEIM P., PEARSON T. and OKINAKA R. (2008) Microbial forensics: DNA fingerprinting of *Bacillus anthracis* (anthrax), *Anal. Chem.* 80, 4791–4800.

KJAERULFF U.B. and MADSEN A.L. (2008) *Bayesian Networks and Influence Diagrams*, Springer, New York.

LOMONACO S., DECASTELLI L., NUCERA D., GALLINA S., BIANCHI D.M. *et al.* (2009) *Listeria monocytogenes* in Gorgonzola: subtypes, diversity and persistence over time, *Int. J. Food Micro.* 128, 516–520.

POULIOT S. and SUMNER D.A. (2008) Traceability, liability and incentives for food safety and quality, *Am. J. Agric. Econ.* 90, 15–27.

SARKAR P., CHEN L. and DUBRAWSKI A. (2008) Dynamic network model for predicting occurrences of *Salmonella* at food facilities. In *Biosurveillance and Biosecurity*, D. Zeng *et al*. (eds), Springer, 56–63.

SMID J.H., SWART A.N., HAVELAAR A.H. and PIELAAT A. (2011) A practical framework for the construction of a biotracing model: application to *Salmonella* in the pork slaughter chain, *Risk Analysis* 31, 1434–1450.

SMID J.H., HERES L., HAVELAAR A.H. and PIELAAT A. (2012) A biotracing model of *Salmonella* in the pork production chain, *J. Food Prot.* 75, 270–280.

TARONI F., AITKEN C., GARBOLINO P. and BIEDERMANN A. (2006) *Bayesian Networks and Probabilistic Inference in Forensic Science*, J. Wiley, Chichester, UK.

VERHOEF L., VENNEMA H., VAN PELT W., LEES D., BOSHUIZEN H. *et al*. (2010) Use of norovirus genotype profiles to differentiate origins of food borne outbreak, *Emerg. Infect. Dis*. 16, 617–624.

WILSON D.J., GABRIEL E., LEATHERBARROW A.J.H., CHEESBROUGH J., GEE S. *et al*. (2008) Tracing the source of campylobacteriosis, *PloS* 4(9), e1000203.

Part VI

Understanding and modelling pathogen behaviour

23

Advances in single-cell approaches in the study of foodborne pathogens

B. Brehm-Stecher, Iowa State University, USA

DOI: 10.1533/9780857098740.6.447

Abstract: Traditionally, populations of microbes have been characterized rather than individual cells. Methods for examining individual cells have long been available but there were technological barriers. The rapidity and ease of growing microbes to large numbers has limited the motivation to study them individually and terminology reinforced the belief that cell populations comprise assemblages of identical cells. Recognition of the heterogeneity of individual cells, coupled with technology capable of resolving discrete differences has driven single-cell microbiology, enabling the high-resolution study and physical manipulation of individual microbes – even allowing detection and tracking of single molecules within living cells. The list of single-cell techniques or their refinements is growing.

Key words: foodborne pathogens, cellular heterogeneity, single-cell analysis, non-photonic microscopy, lab-on-a-chip, combinatorial analysis, single-cell genomics.

23.1 Introduction

The cell is the basic unit of life, a microscopic victory over entropy, an ordered assemblage of atoms, molecules and structures capable of living autonomously in the world. Microbial cells are ubiquitous in nature, and can be found in almost any habitat, from surface waters and soils to deep subsurface sediments to the roots and leaves of plants and every available animal surface, both interior and exterior (Brehm-Stecher, 2007). Because they are derived from plant or animal sources, the foods we eat can also be thought of as extensions of the natural environment. Depending on the level of processing used, food environments may host populations of microbes having various levels of complexity. At one extreme, foods such as alfalfa and other seed sprouts host rich, not yet fully characterized multitaxon communities, of which only a fraction may be culturable (Bisha and Brehm-Stecher, 2008). At the other end of the spectrum, some foods may be so highly processed that they contain very low microbial loads. These foods may be at risk of post-production contamination with human pathogens, which may then grow unhindered by natural population check valves such as competitive exclusion. Whether our intent as microbiologists is to describe the communities inhabiting and transforming our foods or to surveil our food supply for low numbers of pathogens and to investigate their physiological state and capacity to cause disease, analytical methods capable of single-cell resolution will provide levels of understanding unattainable with the population-scale approaches historically used by food microbiologists. Ultimately, the study of foodborne pathogens at their most elemental level will lead to a greater understanding of their *in situ* behaviors and interactions in foods, enable us to understand the risks associated with their presence, allow us to detect them at very low (if not single-cell) levels and to control or prevent their growth (Rantsiou *et al.*, 2011).

23.2 Single-cell analyses in food microbiology

23.2.1 The importance of single-cell analyses in food microbiology

By definition, bacteria and other microbes are unicellular life forms. In other words, the single cell forms the foundation of our discipline. However, since the inception of modern microbiology, we have relied almost exclusively on bulk-scale methods for growth, manipulation and measurement of microbes (Brehm-Stecher and Johnson, 2004). This has mostly been for practical reasons: these methods are well established, simple, require minimal training or tools and we have ready access to an abundant, self-generating supply of microbes to work with in the lab. Still, echoes of the single-cell basis of our discipline are reflected in the concerns we routinely express as food microbiologists. For example, detection of a single cell within a food sample remains the ultimate goal for the analytical sensitivity of any new detection technique. Although population-scale methods can provide useful and actionable information on the presence and activity of microbes in foods, the results obtained represent averages of all possible values for the parameters being measured. While the composite picture obtained may provide an accurate description of the population, it does little to describe the individuals comprising the population. Reliance on averaged data can blur our view of discrete phenomena occurring in individual cells, the clear observation of which may provide new insight into mechanisms for gene regulation, molecular trafficking or chemical messaging within the cell (Brehm-Stecher and Johnson, 2004; Brehm-Stecher, 2007). The following analogy may help illustrate the effects of such averaging: when the population of the Earth reached an estimated seven billion people recently, a national magazine published a fuzzy-edged computer-simulated composite of the most typical human face, a 28-year-old Han Chinese man (National Geographic, 2011). Based on their personal knowledge of human diversity, readers of the article can easily recognize biases in this aggregate description of the world's human population, almost two-fifths of which lives in Asia. Lost to this and other 30 000-foot view population-scale representations is the depth of variety among individuals – details that can only be seen at ground level and which for foodborne microbes may have tremendous practical significance for pathogen survival in foods and/or their capacity to cause disease.

A key feature of biological systems is heterogeneity, or differentness. Microbial cells exhibit heterogeneity across a number of dimensions: genetic, physiological, biochemical and behavioral. Cellular heterogeneity can serve as an important engine for adaptation and evolution (Brehm-Stecher and Johnson, 2004). As food microbiologists, we know that a single rogue pathogen can wreak havoc in our foods, whether it is a *Salmonella* cell carrying a mutation conferring increased competitiveness (a faster growth rate in low-moisture foods such as peanut butter or resistance to antimicrobials, for example) or a clostridial spore having unpredictable germination characteristics (Fig. 23.1). Population-scale

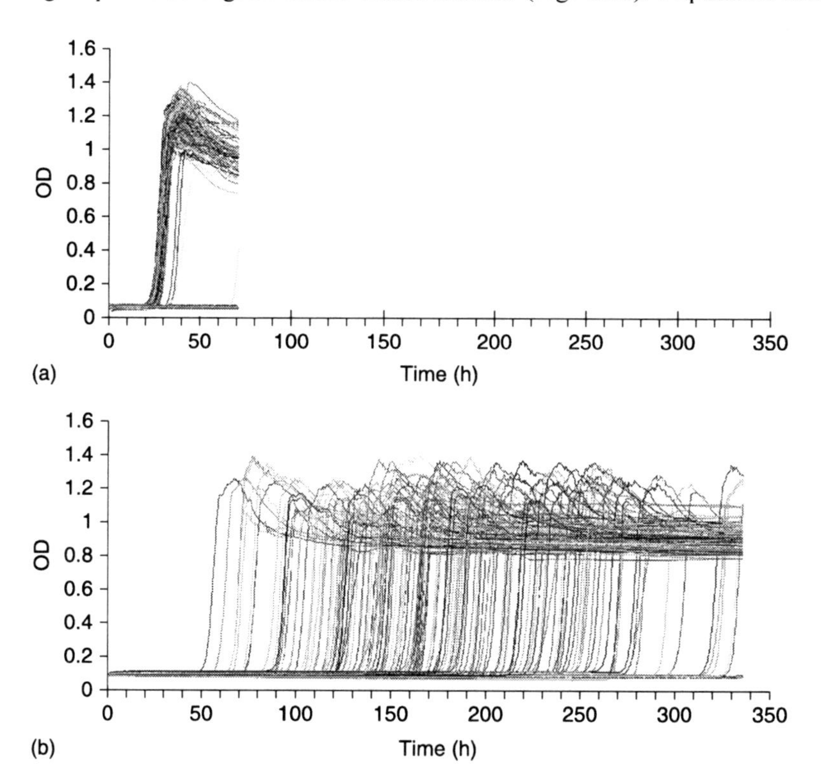

Fig. 23.1 Variability in growth curves initiated from single spores of non-proteolytic *Clostridium botulinum* Eklund 17B as a function of pre-germination heat treatment. Individual spores were obtained through dilution, resulting in one spore per well in a Bioscreen C Microbiological Reader. Panel (a) shows growth curves from unheated spores (outgrowth time 68 h). Panel (b) shows growth curves from spores heated for 1 min at 80°C (a 3D reduction) prior to germination and outgrowth (outgrowth time 338 h). OD = optical density. These results graphically show the impact of prior heat treatment on variability of spore germination and outgrowth, an effect that could have profound implications for food safety. Figure modified from Stringer *et al.* (2011) and reproduced with permission from the publisher.

observations provide a coarse-grained view of microbial life – useful for some purposes, but less descriptive, and possibly misleading for others. Access to data at the single-cell level may help resolve fundamental questions that are rooted in the fine-grained properties and behaviors of individual microbial cells. Examples of fundamental physiological observations made possible by single-cell analysis include the discovery that bacteriophage infection in *Escherichia coli*, *Yersinia pseudotuberculosis* and *Vibrio cholerae* is localized to bacterial cell poles (Edgar *et al*., 2008), that bacterial cell division is functionally asymmetric, with old pole cells showing a diminished growth rate, decreased production of offspring and higher incidence of death compared to new pole cells (Stewart *et al*., 2005) and that a narrow time window of protein induction in supposedly dormant *E. coli* cells can serve as a target for inhibition of antibiotic-resistant persister cells (Gefen *et al*., 2008).

23.2.2 Microbial multicellularity and biofilms

Although we might think of microbial cells existing simply as groups of single cells growing together, living parallel, but individual lives, there are many examples of cooperative, social, multicellular behaviors and adaptations, ranging from coordinated motility to formation of specialized micro- and macrostructures. These include fruiting body formation in *Bacillus subtilis* and *Myxococcus xanthus* (Branda *et al*., 2001; Velicer and Yu, 2003), growth of *Saccharomyces cerevisiae* cells in clumps rather than as individual cells (Koschwanez *et al*., 2011) and formation of supercoiled helical cellular macrofibers (Mendelson, 1999; Mendelson *et al*., 2002). Although they can be costly to individual cells (i.e. by slowing the individual reproduction rate) higher-order communal behaviors like swarming, clumping and fibril formation can benefit the group as a whole, enabling migration into new resource-rich environments (Mendelson, 1999; Velicer and Yu, 2003), effective sequestration and sharing of nutrients that might be otherwise lost to diffusion (Koschwanez *et al*., 2011) and exertion of supracellular mechanical power capable of breaking surface tension boundaries that might otherwise limit growth (Mendelson, 1999). One multicellular behavior of particular interest to food microbiology is the biofilm, a multispecies or multitaxon assemblage of microbes bound together within an extracellular matrix. Biofilms can form on any free, colonizable surface, including food contact surfaces (belts, blades and boards), or in difficult-to-clean areas (drains, fixtures and equipment) that can serve as sources of harborage, dissemination and recontamination. Biofilms present substantial challenges to food safety, as they may consist of or harbor human pathogens, present diffusional barriers to antimicrobials and sanitizers and are difficult to mechanically remove. Microenvironmental differences within the biofilm may also alter the gene expression, physiology, growth and subsequent detectability of organisms of interest, such as pathogens. Methods for single-cell analysis of microorganisms are essential for understanding the role of the individual cell in collaborative or cooperative behaviors, for detection of target cells within heterogeneous microbial aggregates or for determining the biological or biochemical function of an organism within a given environment or community (Valm *et al.*, 2011).

23.3 Advances in single-cell techniques in the study of foodborne pathogens

Although a wide variety of tools and technologies for analysis of microbial cells at the single-cell level have been previously reviewed (Brehm-Stecher and Johnson, 2004; Brehm-Stecher, 2007, 2008), the field is continually expanding, with the introduction of new approaches or improvement of existing methods. A handful of such new approaches or refinements to existing techniques are reviewed below.

23.3.1 Cell viability determination

Of particular interest to food microbiologists is the problem of how to obtain information on the viability of individual microbial cells, preferably in a dynamic fashion and in response to the application of external stimuli (physical or chemical forces, antimicrobials, etc.). Knowing how pathogens respond may provide new information that can be used to develop improved inactivation methods or to glean basic information on how traditional antimicrobial processes work. Currently, the only way to determine a cell's viability is still by culturing it – if it is able to grow, then it was (past tense emphasized) viable. However, the simple binary definition live/dead probably does not accurately describe the complex physiological states possible with microbial cells. Viable-but-not-culturable (VBNC) cells are those that cannot be cultured using traditional recovery regimes, but which exhibit physiological or biochemical evidence of basal life processes. Pathogens in a VBNC state have the potential to be resuscitated at some point, given the right environmental conditions or signals. The threat of their eventual recovery and outgrowth combined with our inability to detect them using traditional culture methods makes VBNC cells a perennial worry among food safety microbiologists. The chapter by Rantsiou and Cocolin provides an additional discussion of the VBNC state and its implications for food safety.

Several black box kits for supposedly determining the live/dead status exist and are available commercially. These typically rely on the principles of dye exclusion, where damage to cellular permeability barriers (especially the cell membrane) enables the entry of dyes such as propidium iodide (PI) into the cell, where it can intercalate into the cell's DNA and fluoresce. Because cell viability is highly correlated with an intact cell membrane, positive staining with PI is interpreted to indicate that the cell is dead. However, if we think of the route from injury to death as a process, at what point does this process reach its end point (actual, irreversible, irrecoverable cell death), and is this route sometimes a two-way street, where moribund cells can eventually recover and divide (Davey, 2011)? Another complicating factor for the use and interpretation of dye exclusion technology is the concept of transient permeability, a process that we know exists, if only from its commonplace use in the lab for rendering *E. coli* artificially competent for foreign DNA. Other means for enhancing the permeability of the cell membrane are well known (Brehm-Stecher and Johnson, 2003) and other cellular barriers, including the Gram-negative outer membrane, also play a role in exclusion of exogenous compounds (Davey, 2011). Regardless of the mechanism for transient permeability, the presence of viable but inherently leaky cells may produce challenges to the effective application of dye exclusion-based techniques for determining the viability of microbial cells.

Real-time methods for investigating physiological processes related to cell viability would be valuable for understanding the dynamics and mechanisms behind microbial inactivation by processes such as acidification or application of bacteriocins (Siegumfeldt and Arneborg, 2011). Cells must maintain internal pH homeostasis in order to live, and while measurement of internal cell pH (pH_i) is not a direct reporter of cell viability, it can provide information on the physical compromise of the cell membrane as well as the microphysiological consequences of such compromise. Fluorescence ratio imaging microscopy (FRIM) provides a valuable, real-time method for measuring pH_i and has been used successfully on a number of food-related organisms, including *Listeria monocytogenes, Campylobacter jejuni* and *Zygosaccharomyces bailii* (Brehm-Stecher and Johnson, 2004; Siegumfeldt and Arneborg, 2011). FRIM has been used to assess microbial responses to various stresses affecting organisms in food environments, including acid, osmotic, disinfectant and bacteriocin-mediated stresses (Siegumfeldt and Arneborg, 2011).

23.3.2 Alternative and non-photonic microscopy

The Greek root *scopos* means 'to watch' or 'to see'. Accordingly, we use a microscope to watch or see very small things, such as individual microbial cells. Although microscopes have traditionally relied on visible light, other regions of the electromagnetic spectrum or non-photonic modalities can also be used to generate images of microbes (Brehm-Stecher, 2007). These include Raman microscopy/microspectroscopy and microbeam methods such as scanning transmission X-ray microscopy (STXM) and high-resolution secondary ion mass spectrometry (NanoSIMS – see the chapter by Fratamico and Gunther for a discussion on the use of mass spectrometry for the proteomics-based detection, identification and characterization of foodborne pathogens) (Behrens, *et al.*, 2012; Brehm-Stecher and Johnson, 2004). These methods can give biochemical, ionic or elemental mapping of cells and cellular substructures, enabling discrete physiological characterization of their biochemical make-up, vitality/viability and activity (e.g. carbon utilization or nitrogen fixation) within an environment (Behrens *et al.*, 2012; Lechene *et al.*, 2006). In theory, if an accurate database of whole-organism spectra can be built, chemical imaging methods could be used for reagentless microscopic identification of specific organisms, such as pathogens, in complex environments such as foods or biofilms (Kalasinsky *et al.*, 2007).

Atomic force microscopy (AFM) and other related cantilever-based scanning probe platforms can be used to form images based on touch. For example, AFM allows us to push, pull, tug or poke individual cells or their structures to gain information on their discrete micro- or nanomechanical properties, including turgor, elasticity or bursting force. AFM and related force spectroscopy techniques also enable measurement of the nanonewton forces governing cell-cell and cell-substrate binding interactions (Brehm-Stecher and Johnson, 2004; Brehm-Stecher, 2007). One advantage of AFM is that it requires relatively little sample preparation and time-course imaging can be carried out on living cells in aqueous media. Figure 23.2 shows an example of how AFM can be used to simultaneously image and physically characterize the action of an antimicrobial against a living bacterial pathogen. In this work, Francius and colleagues incubated *Staphylococcus aureus* with the cell wall lytic agent lysostaphin (of interest as a last-resort agent for use against drug-resistant bacteria such as MRSA) and followed the results as a function of time. Lysostaphin-mediated digestion of the cell wall was accompanied by the formation of increased surface roughness, formation of nanoscale perforations, cell swelling, splitting of the septum and decreases in cell wall stiffness and bacterial spring constant (Francius *et al.*, 2008). The simultaneous collection of images and nanophysical data of living cells during the cell wall digestion process provided high-resolution, artifact-free characterization of lysostaphin's action, potentially yielding novel insights into mechanisms of susceptibility and resistance, which can be used to design or implement new

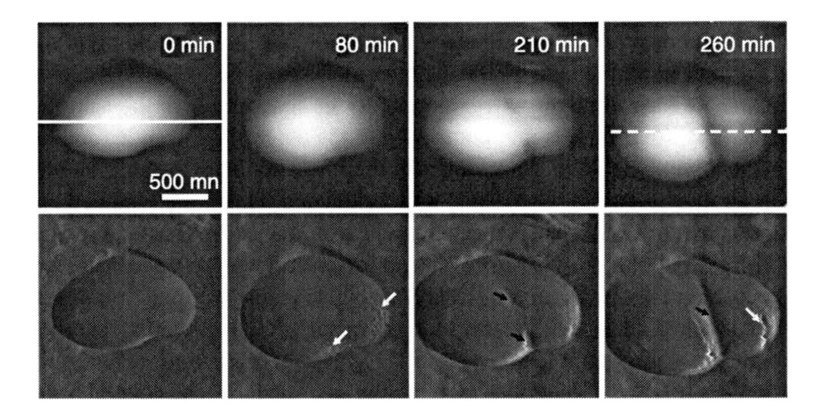

Fig. 23.2 Use of atomic force microscopy (AFM) for nanophysical characterization of the effects of lysostaphin treatment on living *Staphylococcus aureus* cells. Cells were exposed to 16 μg/mL lysostaphin in phosphate-buffered saline and images were obtained in real time over the course of 260 min, using height mode (top) and deflection mode (bottom) imaging. White arrows indicate nanoscale perforations that became larger over time and black arrows show splitting of the septum. Apart from simple imaging, AFM-based characterization enabled determination of other salient cell characteristics related to lysostaphin activity, including changes in cell roughness, bacterial spring constant and cross-sectional height measurements (an indicator of cell swelling). Figure modified from Francius *et al.* (2008) and reproduced with permission from the publisher.

antimicrobial therapies (Francius *et al.*, 2008). In a related development, rooted in the same scanning probe microscopy family of instrumentation that AFM belongs to, is the recent development of nanoscale magnetic resonance imaging (MRI). MRI is a powerful macro-imaging modality used routinely in healthcare for imaging the human body. The same principles at work in macroscale MRI provide the foundation for magnetic resonance force microscopy (MFRM), which uses a microscopic cantilever to mechanically measure attonewton (10^{-18} N) magnetic forces occurring between nuclear spins in a sample and a magnetic scanning probe tip poised in close proximity to the sample (Degen *et al.*, 2009). Degen and colleagues showed that MFRM could be used to image single tobacco mosaic virus particles, mapping the nuclear spin density of protons in these particles at spatial resolutions better than 10 nm – a 100 million-fold improvement over conventional MRI. This advance demonstrates the potential for further applications of MFRM for the characterization of microbial cells. As with NanoSIMS and STXM above, MFRM is an imaging modality capable of providing elementally selective chemical imaging with subcellular resolution, suggesting its use for detailed characterization of discrete regions or domains within microbial cells.

23.3.3 Direct measurement of bacterial cell weight

Although we can see the shape of a microbe under a microscope and we can determine its nanomechanical properties by prodding it with an AFM, there are some measurements that we have not been able to carry out directly on an individual cell, such as weighing it. Indirect estimations of individual microbial cell weight can be made by taking the wet weight of a cell pellet and dividing this by the number of viable cells in the sample, as determined by plating. However, this bulk-scale measurement will undoubtedly be inaccurate, as the sample will contain extracellular liquids that contribute to the determined weight, and not all cells present will be able to grow, and will therefore not be detectable by plating. Even in an ideal scenario where only cell-associated liquids are weighed and all cells are viable, this bulk-phase measurement would only provide an average cell weight for the population. Therefore, this approach still would be unable to answer questions about the distribution of discrete weights across a population or resolve other data rooted in individual cellular characteristics. Resonating cantilevers, related to the core technology used in AFM, have been used successfully to measure very small and light objects, typically in a vacuum. The general principle relies on the fact that a mass applied to a resonating cantilever will modulate the frequency of resonance, in proportion to the mass of that object. Objects as light as 7 zeptograms (7×10^{-21} g) have been successfully measured in this way. However, measurement of objects in liquids is complicated by liquid viscosity, which causes resonance damping, therefore limiting applications of this approach in life science applications, where aqueous solutions are standard. Burg and colleagues developed a unique silicon crystal resonator having a low weight (100 ng) and high quality factor (a measurement of its ability to ring at a pure tone), which was able to overcome the limitations of liquid damping. The resonator contained a liquid channel through which objects such as cells and nanoparticles could flow and be measured during transit. With this approach, these authors were able to directly measure the mass of *Escherichia coli* cells (110 ± 30 fg) and *Bacillus subtilis* cells (150 ± 40 fg) (Burg *et al.*, 2007). Possibilities for this technology include mass-based flow cytometry and non-optical characterization of colloidal particles (Burg *et al.*, 2007).

23.3.4 Lab-on-a-chip systems

In the past decade, improvements in microfluidic and micromechanical design and fabrication have driven the rapid growth of lab-on-a-chip systems – miniaturized analytical devices integrating multiple analytical processing steps. Lab-on-a-chip systems that perform a full suite of analytical steps, taking the sample from its raw state to an analytical end point are often referred to as micro total analysis systems (μTASs). Other systems are simply scaled-down versions of common laboratory tools that are otherwise relatively large and expensive, such as flow cytometers, cell sorters or fermenters. A critical disadvantage of such miniaturized systems is throughput – unless many devices are run in parallel, these instruments simply cannot match the rate of analysis of larger systems capable of accommodating larger volumes of liquids, and therefore, greater numbers of cells. However, these millimeter-scale systems have unique, size-based advantages. These include the use of lower quantities of reagents, and for flow-through systems, the ability to reanalyze the same cells multiple times through reversal of the liquid flow. Other advantages stem from the small dimensions of the device, which are closer to the dimensions of microbial cells and which considerably alter system behavior with regard to diffusion of nutrients, control and alteration of temperature or pH and measurement of end products. Grünberger and colleagues described a disposable picoliter fermentation system for the growth and characterization of populations grown from single microbial cells (Fig. 23.3). Although several reports of microscale bioreactors can be found in the literature, few, if any, incorporate both precise environmental control and microscopic observation of growing cultures for time-course analysis (Grünberger *et al.*, 2012). The performance of large-volume bioreactors depends on many variables, including heterogeneity of the inoculum and environmental heterogeneity within the reactor. When multiple sources of heterogeneity are compounded, the phenotypic analyses typically required in biotechnological applications are complicated. In their work, Grünberger *et al.* (2012) were able to seed their picoliter bioreactors with individual cells, so that each reactor contained an isogenic population (of ~500 cells) after growth. Precise control of environmental conditions also enabled these researchers to instantaneously change the nutrient composition within the reactor. Finally, cell growth under a variety of conditions could be readily monitored using light microscopy and image analysis, processes which were facilitated by the 1-micron vertical dimensions of the growth chamber, which resulted in cells occupying a single focal plane (Grünberger *et al.*, 2012). Experiments with *Corynebacterium*

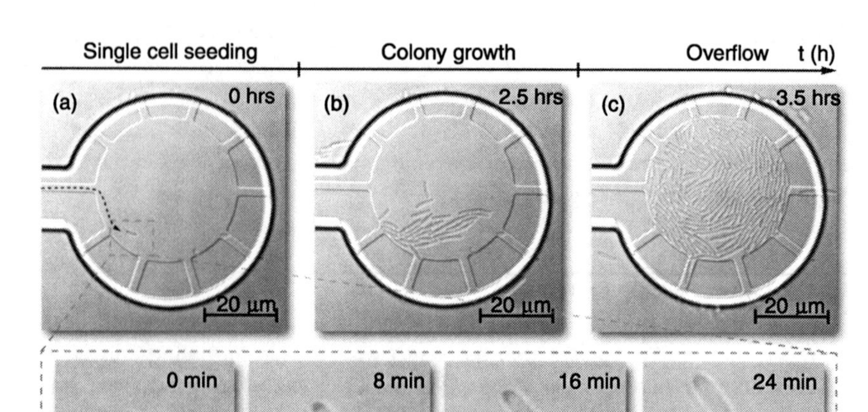

Fig. 23.3 Time-lapse microscopic images showing cultivation of *Escherichia coli* within a picoliter bioreactor (PLBR), beginning with a single-cell inoculum (panel (a)) and ending with a ~500 cell isogenic population after 3.5 h cultivation. Panel (d) is an expanded view of cell division for an individual *E. coli* cell. Instrumentation such as the PLBR shown here may play an important role in tasks such as development of new pathogen-selective media formulations or characterizing the response of individual pathogen cells to environmental stimuli. Figure reproduced from Grünberger *et al.* (2012) with permission from the publisher.

glutamicum in this system yielded several key observations, including an increase in cellular growth rate by a factor of 1.5 compared to that observed in 1-L bulk phase reactors, and emergence of phenotypic differentiation of cells in the picoliter system within one hour, which was not observed in the 1-L system until after an entire week of continuous cultivation (Grünberger *et al.*, 2012). The ability to follow individual cells through multiple divisions, to exert precise control over environmental conditions within the bioreactor and to follow cell growth dynamics via microscopy and image analysis may also be useful outside of traditional biotechnological applications. For example, parallel examination of variant medium formulations combined with image-based quantitation of output could speed the development of new or improved selective media for recovery and detection of foodborne pathogens. This approach could also be useful for the detailed examination of phenotypic changes occurring during outgrowth of differentiated microbial forms, including stress-induced filaments, coccoid cells and spores, when exposed to new environmental conditions.

23.3.5 Combinatorial labeling for microbial community analysis

Fluorescence *in situ* hybridization (FISH) is a method for whole-cell molecular labeling of specific microorganisms and can be used to examine complex assemblages of microorganisms, such as those found in biofilms or attached to environmental surfaces or even in foods themselves, enabling localization of target cells within these samples (Ercolini *et al*., 2003). FISH is a workhorse single-cell analysis tool in environmental microbiology and its use in food microbiology is rapidly expanding (Brehm-Stecher and Johnson, 2004; Bisha and Brehm-Stecher, 2008, 2009; Ercolini *et al*., 2003). When used simply as a detection tool for one or two specific cell types, FISH and the subsequent data analysis are relatively simple and can be carried out using widely available instrumentation. However, if multiple microbial taxa are present in the sample and a global characterization of this complex community is desired, limitations on the number of spectrally resolvable dyes available for multicolor differentiation of unique organisms make this approach untenable without modification. Valm and colleagues developed a combinatorial labeling approach where 15 oligonucleotide probes targeting the major taxa present in the human oral cavity were each uniquely labeled with binary combinations of six commonly available dyes. After hybridization and image acquisition, cells labeled with each binary combination could be differentiated after spectral analysis, with each combination read as a unique color. Since the hybridization reaction itself is robust and supports the simultaneous use of multiple probes, this approach could enable multiplex hybridization and analysis of samples containing many different types of organisms. As an example, the authors cite that 15 fluorophores would give 105 unique binary dye combinations and 455 unique ternary combinations – each combination representing a new channel in the expanded analytical bandwidth made possible using this combinatorial approach (Valm *et al*., 2011). This strategy for multiplexing FISH overcomes critical limitations inherent in traditional single-label FISH and will allow simultaneous differentiation of individual cells from many different taxa, which may be present in the same environmental sample.

23.3.6 Whole-cell reporters

Although most single-cell approaches are focused on application of exogenous tools to detect or characterize cells, some methods also rely on cells themselves as probes. Environmental microbiologists have taken the lead in the use of genetically engineered microbes for use as whole-cell bioreporters. When distributed *in situ*, cells carrying a selectively expressible fluorescent reporter system can be used to reveal information about the prevailing conditions within the microenvironment they inhabit. These include intercellular molecular communication (quorum sensing), water availability, the presence and concentration of specific nutrients, antibiotics, toxic chemicals and heavy metals (Tecon and van der Meer, 2006). Whole-cell bioreporters can perform these functions as either planktonic cells or as components of single- or mixed-species biofilms (Tecon and van der Meer, 2006). Genetic modification of spore-forming bacteria can provide additional insights into key aspects of spore germination and outgrowth, reporting both on expression of germination-specific proteins and on the general outgrowth process. These constructs may be especially useful for probing the effects of heat treatment or other methods for inactivation on spore germination (Hornstra *et al.*, 2009; Ter Beek *et al.*, 2011). Because whole-cell sensors can provide us with a bacterium's-eye view of prevailing conditions in specific food microniches, they may provide new insights into microbial responses to environmental stressors, suggesting mechanisms for how some pathogens can survive or persist in these environments. On a more global scale, similar methods could be used for mapping pathogen distribution and activity on anatomically distinct carcass regions. These data could inform microbial risk assessment studies or aid in the development of more effective methods for carcass decontamination.

23.3.7 Single-cell genomics

Single-cell genetic approaches such as the digital polymerase chain reaction (PCR) enable detailed characterization of DNA within single microbial cells. For example, Tadmore *et al.* (2011) used digital PCR to probe individual environmental bacteria for the viruses they hosted. Taking the DNA-based analysis of individual cells even further, another exciting development from the realm of environmental microbiology is single-cell genomics. The ability to obtain a genome sequence from a single microbial cell enables the comprehensive genetic characterization of non-culturable cells or individual cells from within complex assemblages of microorganisms, such as biofilms. Individual cells may be isolated in any number of ways, including micromanipulation, dilution to extinction and flow cytometry (de Jager and Siezen, 2011; Woyke *et al.*, 2010). In traditional genome sequencing, genomic DNA from 10^6–10^8 cells grown in a pure culture is pooled to provide enough material to carry out the analysis (de Jager and Siezen, 2011). A key challenge for analysis of an individual cell, therefore, is how to proceed when the starting material (a single copy of the genome) is present at only one millionth or one hundred millionth of the material required. This is accomplished presently using multiple displacement amplification (MDA), an amplification method using random primers and a viral DNA polymerase (de Jager and Siezen, 2011; Woyke *et al.*, 2010). Although any method for generating enough DNA for genomic analysis runs the risk of introducing some sort of amplification bias, MDA is less prone to doing so than randomly primed PCR, and it generates larger fragments, which are more suitable for sequencing. Another advantage of single-cell genomics is that the resulting sequence is not a composite of the genome of millions of cells, but rather the unique sequence from an individual cell (Woyke *et al.*, 2010). The ability to detect and pinpoint genomic differences among individual cells may be essential to understanding the behavior or persistence of cells isolated directly from an environment, including complex foods or biofilms, without a requirement for culturing. The emerging field of phenomics studies an organism's phenome – the constellation of physical and biochemical traits that it displays (Houle *et al.*, 2010). Phenomic and genomic data from a single foodborne pathogen may ultimately be linked, providing a more global description of the cell. A further discussion of various omics approaches for the study of foodborne pathogens is provided in the chapter by Fratamico and Gunther.

23.4 Conclusion and future trends

The field of single-cell microbial analysis is continually expanding, with both incremental advances to existing methods and development of entirely new approaches contributing to the growth of this sector. Although we might typically think of natural environments in terms of surface waters, soils and plant or animal surfaces, we also recognize that foods themselves are equally relevant as microbial environments. Methods developed for traditional environmental microbiology can be advantageously applied to problems in food microbiology, based on the theoretical and procedural kinship shared by these disciplines. Examples described above include non-photonic microscopy, combinatorial FISH labeling, digital PCR, single-cell genomics and genetically engineered whole-cell reporters. These and other approaches for single-cell analysis promise to provide fine-scale characterization of foodborne pathogens or spoilage organisms at levels of resolution unattainable with population-based analyses. As a result, we may be able to obtain fundamentally new information on the micro- or nanophysiology of these organisms, how they respond to and persist within food microenvironments and how they interact with other microbes, all of which could have profound practical implications for maintaining the safety of our food supply.

23.5 References

BEHRENS, S., KAPPLER, A. and OBST, M. 2012. 'Linking environmental processes to the *in situ* functioning of microorganisms by high-resolution secondary ion mass spectrometry (NanoSIMS) and scanning transmission X-ray microscopy (STXM)', *Environmental Microbiology*, doi: 10.1111/j.1462-2920.2012.02724.x.

BISHA, B. and BREHM-STECHER, B.F. 2008. 'Flow-through imaging cytometry for characterization of *Salmonella* subpopulations in alfalfa sprouts, a complex food system', *Biotechnology Journal*, vol. 4, no. 6, pp. 880–887, doi: 10.1002/biot.200800360.

BISHA, B. and BREHM-STECHER, B.F. 2009. 'Simple adhesive-tape-based sampling of tomato surfaces combined with rapid fluorescence *in situ* hybridization for *Salmonella* detection', *Applied and Environmental Microbiology*, vol. 75, no. 5, pp. 1450–1455, doi: 10.1128/AEM.01944-08.

BRANDA, S.S., GONZÁLEZ-PASTOR, J.E., BEN-YEHUDA, S., LOSICK, R. and KOLTER, R. 2001. 'Fruiting body formation by *Bacillus subtilis*', *Proceedings of the National Academy of Sciences of the United States of America*, vol. 98, no. 20, pp. 11621–11626, doi: 10.1073/pnas.191384198.

BREHM-STECHER, B.F. 2007. 'New technologies for imaging individual microbial cells' in *Imaging Cellular and Molecular Biological Function*, F. Frischknecht and S. Shorte (eds.), Springer-Verlag, Berlin, doi: 10.1007/978-3-540-71331-9_11.

BREHM-STECHER, B.F. 2008. 'Methods for whole cell detection of microorganisms' in *Structure, Interaction and Reactivity at Microbial Surfaces*, T. Camesano and C. Mello (eds.), American Chemical Society, Washington, DC, doi: 10.1021/bk-2008-0984.ch003.

BREHM-STECHER, B.F. and JOHNSON, E.A. 2003. 'Sensitization of *Staphylococcus aureus* and *Escherichia coli* to antibiotics by the sesquiterpenoids nerolidol, farnesol, bisabolol, and apritone', *Antimicrobial Agents and Chemotherapy*, vol. 47, no. 10, pp. 3357–3360, doi: 10.1128/AAC.47.10.3357-3360.2003.

BREHM-STECHER, B.F. and JOHNSON, E.A. 2004. 'Single-cell microbiology: tools, technologies and applications', *Microbiology and Molecular Biology Reviews*, vol. 68, no. 3, pp. 538–559, doi: 10.1128/MMBR.68.3.538-559.2004.

BURG, T.P., GODIN, M., KNUDSEN, S.M., SHEN, W., CARLSON, G. *et al.* 2007. 'Weighing of biomolecules, single cells and single nanoparticles in fluid', *Nature*, vol. 446, pp. 1066–1069, doi: 10.1038/nature05741.

DAVEY, H.M. 2011. 'Life, death, and in-between: meanings and methods in microbiology', *Applied and Environmental Microbiology*, vol. 77, no. 16, pp. 5571–5576, doi: 10.1128/AEM.00744-11.

DEGEN, C.L., POGGIO, M., MAMIN, H.J., RETTNER, C.T. and RUGAR, D. 2009. 'Nanoscale magnetic resonance imaging', *Proceedings of the National Academy of Sciences of the United States of America*, vol. 106, no. 15, pp. 1313–1317, doi: 10.1073/pnas.0812068106.

DE JAGER, V. and SIEZEN, R.J. 2011. 'Single-cell genomics: unraveling the genomes of unculturable microorganisms', *Microbial Biotechnology*, vol. 4, no. 4, pp. 431–437, doi: 10.1111/j.1751-7915.2011.00271.x.

EDGAR, R., ROKNEY, A., FEENY, M., SZABOLCS, S., KESSEL, M. *et al.* 2008. 'Bacteriophage infection is targeted to cellular poles', *Molecular Microbiology*, vol. 68, no. 5, pp. 1107–1116, doi: 10.1111/j.1365-2958.2008.06205.x.

ERCOLINI, D., HILL, P.J. and DODD, C.E.R. 2003. 'Bacterial community structure and location in Stilton cheese', *Applied and Environmental Microbiology*, vol. 69, no. 6, pp. 3540–3548, doi: 10.1128/AEM.69.6.3540-3548.2003.

FRANCIUS, G., DOMENECH, O., MINGEOT-LECLERCQ, M.P. and DUFRÊNE, Y.F. 2008. 'Direct observation of *Staphylococcus aureus* cell wall digestion by lysostaphin', *Journal of Bacteriology*, vol. 190, no. 24, pp. 7904–7909, doi: 10.1128/JB.01116-08.

GEFEN, O., GABAY, C., MUMCUOGLU, M., ENGEL, G. and BALABAN, N.Q. 2008. 'Single-cell protein induction dynamics reveals a period of vulnerability to antibiotics in persister bacteria', *Proceedings of the National Academy of Sciences of the United States of America*, vol. 105, no. 16, pp. 6145–6149, doi: 10.1073/pnas.0711712105.

GRÜNBERGER, A., PACZIA, N., PROBST, C., SCHENDZIELORZ, G., EGGELING, L. *et al.* 2012. 'A disposable picolitre bioreactor for cultivation and investigation of industrially relevant bacteria on the single cell level', *Lab on a Chip*, doi: 10.1039/c2lc40156h.

HORNSTRA, L.M., TER BEEK, A., SMELT, J.P., KALLEMEIJN, W.W. and BRUL, S. 2009. 'On the origin of heterogeneity in (preservation) resistance of *Bacillus* spores: input for a "systems" analysis approach of bacterial spore outgrowth', *International Journal of Food Microbiology*, vol. 134, pp. 9–15, doi: 10.1016/j.ijfoodmicro.2009.02.011.

HOULE, D., GOVINDARAJU, D.R. and OMHOLT, S. 2010. 'Phenomics: the next challenge', *Nature Reviews Genetics*, vol. 11, no. 12, pp. 855–866, doi: 10.1038/nrg2897.

KALASINSKY, K.S., HADFIELD, T., SHEA, A.A., KALASINSKY, V.F., NELSON, M.P. *et al.* 2007. 'Raman chemical imaging spectroscopy reagentless detection and identification of pathogens: signature development and evaluation', *Analytical Chemistry*, vol. 79, pp. 2658–2673, doi: 10.1021/ac0700575.

KOSCHWANEZ, J.H., FOSTER, K.R. and MURRAY, A.W. 2011. 'Sucrose utilization in budding yeast as a model for the origin of undifferentiated multicellularity', *PLoS Biology*, vol. 9, no. 8, e1001122. doi: 10.1371/journal.pbio.1001122.

LECHENE, C., HILLION, F., MCMAHON, G., BENSON, D., KLEINFELD, A.M., *et al.* 2006. 'High-resolution quantitative imaging of mammalian and bacterial cells using stable isotope mass spectrometry', *Journal of Biology*, vol. 5, 20, doi: 10.1186/jbiol42.

MENDELSON, N.H. 1999. '*Bacillus subtilis* macrofibres, colonies and bioconvection patterns use different strategies to achieve multicellular organization', *Environmental Microbiology*, vol. 1, no. 6, pp. 471–477, doi: 10.1046/j.1462-2920.1999.00066.x.

MENDELSON, N.H., MORALES, D. and THWAITES, J.J. 2002. 'The mechanisms responsible for 2-dimensional pattern formation in bacterial macrofiber populations grown on solid surfaces: fiber joining and the creation of exclusion zones', *BMC Microbiology*, vol. 2, no. 1, doi: 10.1186/1471-2180-2-1.

National Geographic 2011. 'The face of seven billion'. Available at: ngm.nationalgeographic.com/2011/03/age-of-man/face-interactive. Accessed 3 March 2013.

RANTSIOU, K., MATARAGAS, M., JESPERSEN, L. and COCOLIN, L. 2011. 'Understanding the behavior of foodborne pathogens in the food chain: new information for risk assessment analysis', *Trends in Food Science and Technology*, vol. 22, Supp. 1, pp. S21–S29, doi: 10.1016/j.tifs.2011.03.002.

SIEGUMFELDT, H. and ARNEBORG, N. 2011. 'Assessment of survival of food-borne microorganisms in the food chain by fluorescence ratio imaging microscopy', *Trends in Food Science and Technology*, vol. 22, Supp. 1, pp. S3–S10, doi: 10.1016/j.tifs.2011.01.011.

STEWART, E.J., MADDEN, R., PAUL, G. and TADDEI, F. 2005. 'Aging and death in an organism that reproduces by morphologically symmetric division', *PLoS Biology*, vol. 3, no. 2, pp. 295–300, doi: 10.1371/journal.pbio.0030045.

STRINGER, S.C., WEBB, M.D. and PECK, M.W. 2011. 'Lag time variability in individual spores of *Clostridium botulinum*', *Food Microbiology*, vol. 28, pp. 228–235, doi: 10.1016/j.fm.2010.03.003.

TADMORE, A.D., OTTESEN, E.A., LEADBETTER, J.R. and PHILLIPS, R. 2011. 'Probing individual environmental bacteria for viruses using microfluidic digital PCR', *Science*, vol. 333, no. 6038, pp. 58–62, doi: 10.1126/science.1200758.

TECON, R. and VAN DER MEER, J.R. 2006. 'Information from single-cell bacterial biosensors: what is it good for?', *Current Opinion in Biotechnology*, vol. 17, pp. 4–10, doi: 10.1016/j.copbio.2005.11.001.

TER BEEK, A., HORNSTRA, L.M., PANDEY, R., KALLEMEIJN, W.W., SMELT, J.P.P.M. *et al.* 2011. 'Models of the behavior of (thermally stressed) microbial spores in foods: tools to study mechanisms of damage and repair', *Food Microbiology*, vol. 28, pp. 678–684, doi: 10.1016/j.fm.2010.07.003.

VALM, A.M., MARK WELCH, J.L., RIEKEN, C.W., HASEGAWA, Y., SOGIN, M.L. *et al.* 2011. 'Systems-level analysis of microbial community organization through combinatorial labeling and spectral imaging', *Proceedings of the National Academy of Sciences of the United States of America*, vol. 108, no. 10, pp. 4152–4157, doi: 10.1073/pnas.1101134108.

VELICER, G.J. and YU, Y.N. 2003. 'Evolution of novel cooperative swarming in the bacteria *Myxococcus xanthus*', *Nature*, vol. 425, pp. 75–78, doi: 10.1038/nature01908.

WOYKE, T., TIGHE, D., MAVROMATIS, K., CLUM, A., COPELAND, A. *et al.* 2010. 'One bacterial cell, one complete genome', *PLoS One*, vol. 5, no. 4, e10314, doi: 10.1371/journal.pone.0010314.

24

Advances in genomics and proteomics-based methods for the study of foodborne bacterial pathogens

P. Fratamico and N. W. Gunther IV, USDA-ARS, USA

DOI: 10.1533/9780857098740.6.462

Abstract: Omic technologies, including genomics, proteomics and metabolomics, are used to study pathogen behavior at the molecular level and develop improved pathogen detection and typing systems. Omic technologies analyze complete or nearly complete expressions of cell functions. DNA sequencing has resulted in complete genomes of foodborne pathogens. Omic-based technologies explore biological processes in a quantitative and integrative manner. They facilitate identification of genes and proteins that contribute to survival and persistence in food and other environments, that play a role in pathogenesis and that are targets for detection methods and control strategies. Challenges that remain are performing genomic and proteomic studies in food and other complex matrices and interpreting and analyzing the data produced from these investigations to enhance food safety.

Key words: foodborne pathogens, genomics, proteomics, detection, genotyping, PCR, microarrays, subtyping.

24.1 Introduction

The Centers for Disease Control and Prevention estimates that 31 pathogens, including diarrheagenic *Escherichia coli*, *Salmonella enterica* ssp. *enterica* serotypes, *Listeria monocytogenes*, *Campylobacter* and *Toxoplasma*, cause 9.4 million cases of foodborne illness, 55 961 hospitalizations and 1351 deaths per year in the United States (Scallan *et al.*, 2011). Hence, research on molecular characterization of pathogens to better understand their virulence and stress-response mechanisms, as well as on methods for enhanced detection of foodborne pathogens is critical to improving food safety and developing improved surveillance and control strategies. In the past three decades, over 1500 microbial genomes have been sequenced, and sequencing of over 1800 microbial genomes is in progress (www.ncbi.nlm.nih.gov/genomes/lproks.cgi?view=1). This information, together with the application and integration of a variety of cutting-edge omic technologies, will provide the ability to determine and interpret the mechanisms underlying pathogenesis and stress responses of foodborne pathogens and to understand their interactions within bacterial communities, as well as provide the tools for their detection and identification.

Genomic and other omic technologies emerged in the 1990s essentially in parallel to the human genome project and other major sequencing projects (Joyce and Palsson, 2006; Kandpal *et al.*, 2009). Omic technologies include genomics, transcriptomics, proteomics, metabolomics, fluxomics, interactomics, pathogenomics and metagenomics, to name a few. Proteomic-based analyses are described in detail below. Bacterial genomics is the study of an organism's genetic material and the application of the information from genome-scale technologies for comparative and functional genomics, gene expression studies and proteomic analyses. The availability of genome sequences has accelerated the ability to analyze gene transcription (expression) at the level of the whole genome using tools such as DNA microarrays. This omic area is referred to as transcriptomics. The bacterial transcriptome consists of all of the RNA transcripts produced by an organism at any one time, and patterns of gene expression vary considerably under different environmental conditions. Microarrays consist of glass or silicon surfaces spotted with a representative of every gene in the bacterial genome. For gene expression analyses, labeled complementary DNA (cDNA) copies of messenger RNA (mRNA) transcribed by the bacteria are hybridized to the arrays, and hybridization is detected in a quantitative manner so that differences in the expression of genes under two different conditions can be measured (see below).

Metabolomics is the term used to describe comprehensive analyses for identifying and quantifying all of the metabolites of a biological system. The small-molecule metabolites of an organism collectively are known as the metabolome (Cevallos-Cevallos *et al.*, 2009). Metabolomics can be targeted, in which a specific metabolite or group of metabolites is analyzed, or untargeted, in which many groups of metabolites are analyzed to determine relationships or fingerprints. Chen and coworkers (2007) examined metabolic fingerprints to detect spoilage of meat and *E. coli* contamination of spinach. Fluxomics measures the dynamic changes of cellular molecules over time and thus provides a direct measure of time-dependent metabolic fluxes (Sekiyama and Kikuchi, 2007). A combination of metabolomics and genomics/transcriptomics can be used to understand pathogen responses to nutrients and other components in food, as well as under other environmental conditions.

Metagenomics is used to assess bacterial diversity within complex communities and generally involves sequencing the 16S ribosomal RNA (rRNA) gene. A description of metagenomic approaches is provided below. Pathogenomics (Pallen and Wren, 2007) is the omic field that deals with genomic, bacterial evolution and bacterial pathogenesis research, as well as the forces that shape genome evolution, including horizontal gene transfer, gene gain/loss and pathogen/host interactions. Interactomics can be described as the integration of pathogenomics, genomics, transcriptomics, proteomics, metabolomics or other omic data to study the functional linkage of cellular molecules and the consequences of those interactions (Kiemer and Cesareni, 2007).

Bioinformatics is the use of statistical and computational approaches to analyze, manage and store genomic, proteomic and other omic data. Computational biologists develop and utilize computer databases and algorithms to analyze complex biological data and determine relationships within large data sets. They perform bioinformatic techniques such as data mining, sequence alignment, genome assembly, gene annotation, comparative genomics, protein structure alignment and prediction, prediction of gene expression and modeling of evolution based on genome-wide association studies. This chapter discusses molecular methods based on genomic and proteomic approaches for comparative analyses of pathogens under different environmental conditions, as well as for biomarker discovery and pathogen detection, identification and typing.

24.2 Genomic technologies: sequencing, typing and profiling

24.2.1 Genome sequencing and comparative genomics

Sequencing technologies, including whole genome sequencing, are having a great impact on food microbiology and food safety. The omic technologies that have emerged have resulted in tremendous advances in knowledge about foodborne pathogens greatly beyond what could be achievable by classic microbiology techniques. Comparative genomics is the term used to describe techniques employed to study genome structure and function relationships across different species or strains; it provides information for understanding the mechanisms of bacterial pathogenesis and evolution. The information can also be utilized to identify virulence factors and genes involved in fitness and stress responses, as well as to identify biomarkers that can be used as targets for pathogen detection. DNA microarrays have been employed for comparative genomics since they can be used to determine the presence or absence of specific genes by comparing whole genomes of different bacteria, provided that all of the genes are represented on the arrays. However, a limitation of arrays is that the presence of novel genes, mutations, deletions, insertions and gene rearrangements may not be easily discerned. Comparative genomics is used to identify genes that can be used as targets to detect specific pathogens (Yu *et al.*, 2010), to identify regulatory mechanisms associated with heat shock responses in *Staphylococcus aureus* (Chastanet *et al.*, 2003) and to identify virulence markers and the potential mechanisms of evolution of enterohemorrhagic *E. coli* (Ogura *et al.*, 2009). Some examples of available computer software and tools for comparative genome analyses were summarized in a review by Bhagwat and Bhagwat (2008).

Phylogenomics refers to the use of genomics to study evolutionary relationships among organisms (Eisen and Fraser, 2003). Comparative phylogenomic analyses using DNA microarrays and Bayesian-based phylogenies identified only 20.8% of genes that were shared by *Yersinia enterocolitica* strains isolated from humans and animals, and 125 predicted coding sequences were found only in highly pathogenic strains (Howard *et al.*, 2006). The strains grouped into three clusters: non-pathogenic, low pathogenic and highly pathogenic. A similar study investigated *Campylobacter jejuni* strains isolated from animals, humans and the environment, and the strains were placed into two clades: livestock and non-livestock (Champion *et al.*, 2005). Unexpectedly, a large portion of the human isolates grouped with the non-livestock clade, suggesting non-livestock sources for human *C. jejuni* infections.

24.2.2 Virulotyping and single nucleotide polymorphism (SNP) genotyping

Many methods can be used for genotyping, which differentiates between organisms and analyzes the genetic differences among isolates within a species. Several genotyping techniques are described below in Section 24.4.2. Bacterial virulotyping is a form of genotyping that involves determining the presence of a set of virulence genes since there is a direct correlation between the presence of specific virulence genes and the risk and severity of disease (Wassenaar, 2011). Bugarel *et al.* (2010) used an array platform based on the GeneDisc real-time polymerase chain reaction (PCR) system to test for important Shiga toxin-producing *E. coli* (STEC) virulence factors and *nle* (non-locus of enterocyte effacement (LEE) effector) genes found in pathogenicity islands OI-122 and OI-71, as well as genes for identification of the O serogroup and specific H types. They determined that a certain set of virulence genes was found in STEC strains associated with severe human illness, and they suggested that their array platform could be used to identify highly virulent STEC. Multiplex PCR assays were used to assess the presence of virulence factors in *Salmonella* strains isolated from poultry and vegetables and belonging to 38 serotypes (Khoo *et al.*, 2009). Various virulence gene patterns were identified even within the same *Salmonella* serotype. A study using PCR and DNA arrays to examine the virulence gene and antibiotic resistance profiles of 523 *Salmonella* strains (five serovars) isolated from various sources from nine European countries (Huehn *et al.*, 2010) found that strains of the same serovar grouped together, and strains with similar virulotypes were found throughout Europe. However, some variations in strains belonging to specific serovars were due primarily to prophage- or plasmid-encoded genes. A molecular risk assessment for non-O157 STEC was proposed by Coombes *et al.* (2008), which was based on the presence of 16 *nle* genes found within pathogenicity islands, OI-122, OI-57, OI-36 and OI-71. They found that 14 *nle* genes and OI-122, OI-57 and OI-71 were more often associated with strains that caused severe disease and hemolytic uremic syndrome (HUS).

Genotyping based on single nucleotide polymorphisms (SNPs) has been used to measure genetic variation between members of a species. SNPs in 96 loci in 528 *E. coli* O157:H7 strains isolated from clinical cases identified 39 genotypes that differed at 20% of SNP loci, which separated the strains into nine clades (Manning *et al.*, 2008). A strain associated with an outbreak caused by contaminated spinach in 2006 belonged to clade 8, and patients with HUS were also significantly more likely to be infected with clade 8 strains. Thus, genomic content variability among the different clades may explain the differences in clinical manifestations, and a subpopulation of the clade 8 lineage that acquired important factors contributing to more severe disease has emerged. Bono *et al.* (2007) examined SNPs in the LEE *tir* and *eae* genes to identify polymorphisms that correlated with severe illness and isolate source (bovine vs human). Polymorphisms in *tir* rather than *eae* were a greater predictor of *E. coli* O157:H7 strains that can cause human illness, with the *tir* T>A T allele vs the A allele as a marker for virulence. An octamer-based genome-scanning technique was also used to examine the divergence of *E. coli* O157:H7 into different genetic linkages, and as in the study by Bono and coworkers (2007), bovine and human isolates were distinguishable and were found in separate lineages (Kim *et al.*, 1999).

24.2.3 Gene expression profiling

Microarrays can be used to determine the gene expression profile of an organism under a certain set of conditions, and this technology is also referred to as transcriptomics. Gene expression is measured by the level of the corresponding RNA and essentially is determined by isolating bacterial RNA, reverse transcription of the RNA template into cDNA, which is either directly labeled with fluorescent deoxynucleotide triphosphates or is labeled after reverse transcription, followed by hybridizing the labeled cDNA with the probe sequences spotted on the array. Samples representing the two conditions being examined are labeled with different fluorophores, and since the reporters emit light at different wavelengths, the signals generated from the two samples can be quantified separately and compared. The size of the signal generated is proportional to the amount of target bound to the probe. Quantitative real-time PCR is used to confirm gene expression (up- or down-regulation) data from microarray experiments (Morey *et al.*, 2006). DNA microarrays can be used to compare the expression of genes in response to a specific treatment by comparing treated cells and untreated control cells and to assist in determining the function of a gene by comparing gene expression of mutant and wild-type strains.

Transcriptomic studies have been performed with bacteria grown on laboratory media; however, analysis of gene expression of foodborne pathogens in complex matrices such as food are lacking. To address this need, a comparison of gene expression profiles of *E. coli* O157:H7 in a raw ground beef extract (GBE) and tryptic soy broth (TSB) was made using DNA microarrays (Fratamico *et al.*, 2011b). There were 74 up-regulated and 54 down-regulated genes when the pathogen was exposed to GBE compared to TSB. The *asr* (acid shock RNA) gene was up-regulated 6.8-fold, and *E. coli* O157:H7 incubated in GBE for 2 h showed significantly increased survival when exposed to synthetic gastric fluid, pH 1.5, compared to cells exposed to TSB. Allen *et al.* (2008) constructed a microarray spotted with probes for 125 stress-response and virulence genes, and they examined the effect of acid and cold stress and nutrient replenishment with fresh TSB on the expression of these genes in *E. coli* O157:H7. It was concluded from this study that different stress-response networks were induced after exposure to sublethal stress, and this may render the cells more tolerant to subsequent stress. To identify genes that may be controlled by quorum sensing through autoinducer 2 (AI-2), a *luxS* mutant (a gene involved in synthesis of AI-2) and a wild-type strain were studied using transcriptional analysis (DeLisa *et al.*, 2001). There were 242 genes that were differentially expressed in the wild type compared to the *luxS* mutant, thus providing insight into the role of AI-2 in quorum sensing-related gene regulation in *E. coli.*

24.3 Functional genomics

Functional genomics makes use of genomic and transcriptomic data to uncover the function and interaction of genes and gene regulatory networks. Generating mutations, deletions or disruptions in specific genes can provide information on the contribution of the gene product on the organism. Comparing gene expression of the *luxS* mutant to the wild type, DeLisa *et al.* (2001) explored the function of *luxS* and the role of AI-2 in cell-to-cell signaling in *E. coli.* In identifying the function of lmo0036 in *L. monocytogenes*, Chen *et al.* (2011b) found that this gene encoded for a putative carbamoyltransferase by sequence analysis. Transcription of lmo0036 was induced after exposure of *L. monocytogenes* to low pH stress. Furthermore, lmo0036 likely played a role in acid adaptation in *L. monocytogenes*, since absence of the enzyme reduced growth of the pathogen under mild acidic conditions and reduced survival after exposure to gastric fluid (pH 2.5). *E. coli* O157:H7 ATCC 43895OR produces an abundant amount of curli fibers and forms strong biofilms (Uhlich *et al.*, 2009). Two differentially expressed proteins in 43895OR and the wild-type 43895 strain were identified as *csgA* (which encodes for the curli subunit) and *lpp* (which encodes for a lipoprotein). Mutants of *lpp, csgA* and both *lpp* and *csgA* were created, and phenotypic tests showed that both genes contributed to the observed colony morphology, Congo red binding, motility, biofilm formation and Hep-2 cell invasion. Deng *et al.* (2011) began to uncover the regulatory network that may contribute to survival of *E. coli* O157:H7 during exposure to chlorine. Microarray analysis showed that *ycfR* was up-regulated in response to oxidative stress with chlorine exposure. There was greater resistance to chlorine and better survival on spinach leaves of the wild-type Sakai strain compared to the mutant. Thus *ycfR* may be involved in chlorine resistance, and further investigation showed that upstream *ycfQ*, a DNA-binding regulator, may be a repressor of *ycfR*, uncovering a potential regulatory network for O157:H7 survival after chlorine treatment during food processing.

24.3.1 Genetic-based methods for detection, identification and differentiation of foodborne pathogens

Information from genomic and other omic technologies can be utilized to develop methods for detection and identification of foodborne pathogens (Lauri and

Mariani, 2009; O'Flaherty and Klaenhammer, 2011). These methods include PCR, DNA hybridization and DNA microarrays.

PCR

PCR basically involves amplification of specific segments of DNA, and PCR products can be detected in a variety of ways. Over the past 25 years, there have been numerous publications on the use of standard PCR, multiplex PCR assays (amplification of more than one target in the same reaction tube) and more recently real-time PCR assays, which utilize fluorescent dyes to monitor amplification of the product in real time. There are techniques to examine microbial diversity that utilize PCR as the first step of the method, and these include PCR-restriction fragment length polymorphism (PCR-RFLP) and PCR-denaturing gradient gel electrophoresis (PCR-DGGE). These techniques are described in more detail below. As an example, species-specific PCR assays and PCR-RFLP were designed to detect and differentiate different *Campylobacter* species based on the *gyrB* gene (Kawasaki *et al.*, 2008). The *gyrB* gene encodes for the subunit B of DNA gyrase, and because the frequency of base substitutions in *gyrB* generally exceeds that of the 16S rRNA gene, differentiation based on *gyrB* can be more discriminating than that based on 16S rRNA. A real-time multiplex PCR screening assay for detection of strains carrying the stx_1 and stx_2 genes (which encode for Shiga toxins) and the *eae* gene (which encodes for the intimin outer membrane protein), followed by multiplex PCR assays targeting serogroup-specific regions within the O-antigen gene clusters were developed to detect the top six (O26, O45, O103, O111, O121 and O145) non-O157 STEC in ground beef (Fratamico *et al.*, 2011a). Sánchez *et al.* (2012) used one such method to concentrate and recover hepatitis A virus, norovirus, murine norovirus, *E. coli* O157:H7, *L. monocytogenes* and *Salmonella enterica* from fresh vegetables, followed by PCR for simultaneous detection of the different pathogens.

Newer technologies are the lab-on-a-chip systems, which are microfluidics-based devices that perform reactions such as DNA extraction, PCR and microarray hybridization in one single piece of miniaturized equipment with minimal hands-on effort (Auroux, 2008). A completely self-contained miniature integrated biochip device, which performed sample preparation and magnetic bead cell capture followed by PCR and product detection by DNA hybridization, was used to detect *E. coli* inoculated into whole blood samples (Liu *et al.*, 2004b). Digital PCR is a modification of conventional PCR that can be used to accurately quantify DNA, cDNA or RNA because each sample is partitioned into many individual real-time PCR reactions so that there is only one cell (a single copy of the PCR template) per droplet (Pohl and Shih, 2004). The products are hybridized with fluorescent probes to allow detection using software that reads and plots positive results. There are a number of kits that involve nucleic acid amplification techniques and other materials commercially available for detection of specific foodborne pathogens, including *E. coli* O157:H7, *Campylobacter* spp. and *Salmonella* spp.

Besides conventional PCR-based assays, there are other nucleic acid amplification techniques that have been used to detect foodborne pathogens, including nucleic acid-based sequence amplification (NASBA), ligase chain reaction (LCR), rolling circle amplification (RCA) and loop-mediated isothermal amplification (LAMP) (Goodridge *et al.*, 2011). NASBA is an isothermal (performed at a constant temperature without the need for a thermal cycler) nucleic acid amplification method that involves reverse transcription of RNA into cDNA followed by synthesis of the complementary strand, and then production of additional RNA molecules that are used as a substrate for the next cycle. Nadal *et al.* (2007) used a molecular beacon-based NASBA assay for detection of *L. monocytogenes* in meat and salmon products targeting the *hly* gene (which encodes for listeriolysin). LAMP is also an isothermal method; it employs four to six primers and a DNA polymerase with strand-displacing activity. This method was reported to be able to detect STEC targeting the stx_1, stx_2 and *eae* genes in ground beef and human stools with a sensitivity of 1–20 CFU/PCR reaction using pure cultures and 10^3–10^4 CFU/g of ground beef or per 0.5 g of stool after 4 h of enrichment (Wang *et al.*, 2012).

Microarrays

DNA microarray-based platforms are another genetic-based approach for pathogen detection and identification (Rasooly and Herold, 2008). For these applications, labeled target molecules hybridize with the recognition molecules (oligonucleotides, cDNA or PCR products) spotted on the array and the signal generated is detected. A variable region of the 16S rRNA gene was amplified using universal primers, followed by hybridization with probes immobilized on the array specific for different bacterial species (Wang *et al.*, 2007). The assay could discriminate 204 bacterial strains belonging to 13 genera with a sensitivity of 100 bacteria. The low density DNA microarray described by Quiñones *et al.* (2011) targeted the stx_1, stx_2, *eae* and *per* genes (the latter being an O157-specific gene in the O-antigen gene cluster) and used a novel colorimetric method to detect biotinylated PCR products; this is known as the ampliPHOX Detection System and it is based on light-initiated signal amplification through polymerization. The Virochip, a microarray-based assay consisting of thousands of probes derived from over 1500 viral sequences found in GenBank, was designed to detect all known viruses and identify novel viruses (Chen *et al.*, 2011a).

24.4 Molecular serotyping, subtyping and metagenomics

24.4.1 Molecular serotyping

E. coli strains are classified into hundreds of serotypes based on the presence of different O (somatic) and H (flagellar) antigens. Serotyping has long been the basis for *E. coli* diagnostics and for epidemiological studies. However, serotyping using antisera is a laborious process, can only be performed in specialized reference laboratories that possess all of the antisera, and equivocal results are often generated due to cross-reactions, resulting in strains that are mistyped or that are not typeable. Molecular serotyping methods based on targeting unique sequences within the *E. coli* O-antigen gene clusters and within genes that encode for the H flagellar antigens can overcome some of the problems associated with conventional antibody-based serotyping. Many of the *E. coli* O-antigen gene clusters have been sequenced, the genes annotated and serogroup-specific PCR assays designed based on genes, including *wzx* (O-antigen flippase) and *wzy* (O-antigen polymerase) (DebRoy *et al.*, 2005; Fratamico *et al.*, 2003, 2005, 2008, 2010). Multiplex PCR assays targeting serogroup-specific genes, as well as virulence genes such as the Shiga toxin genes or other virulence genes have been developed for detection of pathogenic *E. coli* (Fratamico *et al.* 2005, 2008, 2010). PCR and other genetic-based assays targeting the 53 H antigens in *E. coli* can also be designed since the sequences for most of these are available (Wang *et al.*, 2003). *E. coli* serogroups can also be identified using DNA microarrays (Liu and Fratamico, 2006). Another approach that has been used for molecular serotyping of the six most common *Salmonella* serogroups and Paratyphi A involved use of a combination of PCR and a multiplexed bead-based suspension array using Luminex technology (Fitzgerald *et al.*, 2007).

24.4.2 Subtyping

Information from subtyping, also known as bacterial fingerprinting, is used to define clonal lineages of bacteria, compare isolates from different geographical locations, assess changes in populations over time and perform trace-back and outbreak investigations. There are various methods for subtyping bacteria, and these include phenotypic methods such as serogrouping and serotyping; biotyping based on biochemical utilization profiles, phage typing, antibiotyping and multi-locus enzyme electrophoresis (Gebreyes and Thakur, 2011) and genotypic methods such as pulsed field gel electrophoresis (PFGE), ribotyping, PCR-ribotyping, amplified fragment length polymorphism (AFLP), repetitive element PCR (rep-PCR), arbitrarily primed PCR (AP-PCR) and randomly amplified polymorphic DNA (RAPD) analysis, multilocus sequence typing (MLST) and multilocus variable-number tandem repeat analysis (MLVA) (Cooper, 2011; Fields *et al.*, 2011). The choice of a particular subtyping method depends on why subtyping is being performed and other factors including typeability, reproducibility, discriminatory power and capacity for standardization. The discriminatory power of a subtyping method measures its ability to differentiate strains that are considered unrelated based on epidemiologic data.

PFGE subtyping involves macrorestriction (use of infrequently cutting enzymes to cut the DNA into fragments) of genomic DNA and analysis using a pulsed electric field (alternating in direction), which separates out large genomic fragments. Bacteria are embedded in agarose and digested with an enzyme that produces between eight and 25 fragments (20 kb to >1 Mb). The fragments are then separated, the gel is stained with a fluorescent dye and the resulting banding patterns are visualized and compared using commercially available software. PFGE is the method used in the PulseNet program, a standardized international subtyping network developed by the Centers for Disease Control and Prevention, which facilitates sharing of PFGE results. PulseNet assists in rapid identification of common source outbreaks and has played a valuable role in epidemiological investigations (Swaminathan, *et al.*, 2001). The network consists of state health departments, local health departments and federal agencies, including the CDC, the Food and Drug Administration and the USDA Food Safety and Inspection Service. PulseNet International tracks foodborne outbreaks worldwide.

MLST involves sequencing of a specific set of loci on a genome, which are typically housekeeping genes since these are less likely to undergo random mutations. However, if the housekeeping genes are too conserved for subtyping, less variable genes can also be included to improve the discriminatory power. In MLST, the target genes are amplified, and then both strands of the PCR products are sequenced and analyzed. Nucleotide differences of SNPs among the strains are determined, and the unique sequences are assigned allele numbers, which are then combined into an allelic profile and assigned a specific sequence type (ST). A database with MLST information on various bacterial species is available at www.mlst.net, and other databases also exist. MLST was used to subtype *Campylobacter* strains based on seven housekeeping loci (Sopwith *et al.*, 2010). It was found that there was no epidemiological link between *C. coli* strains isolated from humans and strains isolated from environmental waters. As described above, Manning and coworkers (2008) evaluated SNP loci found in 83 *E. coli* O157 genes using over 500 clinical strains of *E. coli* O157. Phylogenetic analyses separated the *E. coli* into nine clades, with clade 8 isolates associated more often with patients with hemolytic uremic syndrome.

MLVA is a subtyping method based on analysis of variable number tandem repeats (VNTRs), which are short nucleotide sequences organized into clusters repeated a few to a number of times in the genome. Whole genome sequence information has made finding of VNTRs in bacteria possible. The method involves amplification of the VNTR loci by PCR with primers that flank the site, followed by separation and sizing of the products. Because a number of loci can be examined by MLVA, compared to PFGE, using only one or two enzymes, MLVA can be more discriminatory than PFGE. This was demonstrated by Torpdahl and coworkers (2007), who compared the two typing methods by analyzing 1019 *Salmonella* Typhimurium isolates collected over a two-year period of routine surveillance. The isolates separated into 148 and 373 PFGE and MLVA types, respectively. They concluded that MLVA was more discriminatory and superior to PFGE for epidemiological studies.

Subtyping using the rep-PCR technique takes advantage of interspersed repetitive DNA elements in the bacterial genome. Commonly used repetitive sequences include repetitive extragenic palindromic (REP) elements (35–40 bp), enterobacterial repetitive intergenic consensus (ERIC) elements (124–127 bp) and the BOX element (154 bp) composed of the boxA, boxB and boxC sequences. A single PCR primer containing a portion of the repeat is used to amplify regions between the repetitive elements and the PCR products are then separated to generate a fingerprint pattern. Rep-PCR is used as a collective term to describe REP-PCR, ERIC-PCR and BOX-PCR subtyping.

The AFLP technique is based on PCR amplification of restriction fragments from a total digest of genomic DNA generated using one or a combination of two restriction enzymes. After ligating adapters to the fragments, PCR is performed using primers targeting the adapters that have one or two selective bases extending into the restriction fragments. The presence of the selective bases reduces the number of fragments generated since only those fragments in which the nucleotides flanking the restriction sites that match the 3′ primer extensions are amplified. The resulting products are separated by electrophoresis. Two AFLP methods and PFGE were used to subtype a collection of 82 *L. monocytogenes* isolates from environmental and food sources, and results showed that the methods had a very similar discriminatory power (Lomonaco *et al.*, 2011).

AP-PCR and RAPD use short primers to amplify regions in genomic DNA arbitrarily. Strain-specific DNA fingerprints, from sites to which the random primers are partially or perfectly complementary, are generated following PCR and visualized by electrophoresis. PCR-ribotyping is based on amplification of the region between the 16S and 23S rRNA genes. The size of this region varies among strains and the PCR products are visualized following electrophoresis. Ribotyping does not involve PCR amplification, but rather genomic DNA is digested with a restriction enzyme to generate DNA fragments ranging from 1 to 30 kb in size. The DNA is subjected to electrophoresis, and then the fragments are transferred to a membrane, which is then hybridized with probes complementary to ribosomal DNA sequences, generating a ribotype or DNA fingerprint. Ribotyping performed using a commercially available ribotyping system (Qualicon-DuPont Riboprinter) with digestion using the PstI restriction enzyme was used to characterize and determine the genetic relatedness of strains of *C. lari* isolated from human cases, animals and water (Rosef *et al.*, 2008). The optical mapping technique generates genome-wide, high-resolution restriction maps from the complete bacterial genome, generating a genetic fingerprint or bar code for each strain. Optical mapping has been evaluated for distinguishing and differentiating strains of *E. coli* O157:H7 (Kotewicz *et al.*, 2007).

24.4.3 Metagenomics

Only about 0.1% to 5% of bacterial species are culturable. Metagenomics is the study of the diversity within the bacterial communities found in a wide variety of environmental niches, including the analysis of non-culturable microbes. Various methods have been used to study microbial communities and population dynamics, including microarrays, denaturing gradient gel electrophoresis and tag-encoded pyrosequencing (Reynisson *et al.*, 2011). Metagenomic approaches can be used to assess the effects of changes in environmental conditions on microbial community structure or, potentially, the effects of community composition on the growth of pathogens in food or on the shelf life of food products. A common method for metagenomic analysis sequences the 16S rRNA genes found in all bacteria. Universal primers targeting the 16S rRNA molecule can be used for the majority of bacterial species. The species found within microbial communities can be determined by pyrosequencing-based 16S rRNA profiling founded on analysis of the hyper-variable regions within the gene (Humblot and Guyot, 2009; Sakamoto *et al.*, 2011). Bar code-tagging of primers allows analysis of multiple samples in a single sequencing run. Allen *et al.* (2011) used metagenomics to monitor the effect of two antibiotics on swine intestinal microbial communities and phage induction. There was increased prophage induction in swine that were fed the two antibiotics compared to non-medicated swine and the composition of the microbial community was affected. Furthermore, detection of antibiotic-resistance genes in the phage metagenomes suggested that feeding antibiotics to swine may contribute to phage-mediated transfer of antibiotic resistance genes in the swine gut.

24.5 Proteomic-based techniques

Proteomics is the study of the proteins produced by organisms under different environments, conditions and challenges. Proteomic studies have been advanced by and benefited greatly from the maturation and expansion of the complementary field of genomics, particularly the progress of whole genome sequencing. The ever increasing number of organisms with a complete record of their genetic sequences greatly simplifies the task of identifying proteins isolated from any proteomic-based study. However, proteomics should not be thought of as simply an extension of genomics. Genomics focuses on the genetic sequences of organisms and the expression of these gene sequences into messages (mRNA) for translation into proteins. Proteomics studies the resulting expression of the proteins as well as any modifications of the proteins after they are expressed. It has been a well-accepted dogma that the levels of translation of genes into messenger sequences directly corresponds with the levels of expressed proteins found in an organism. However, more recent research has suggested that the relationship between genetic transcription and protein translation is not as simple or direct as had been originally envisioned (Ideker *et al.*, 2001; Griffin *et al.*, 2002). For this reason research seeking to determine the physiological responses of organisms may not simply be able to rely on transcription studies such as microarrays and quantitative reverse transcriptase PCR. Measuring the end product (proteins) of the expression process would help to reduce misleading estimations of an organism's true response. Therefore proteomics should be viewed as an avenue of research complementary to genomics with each of these omic approaches having their own specific advantages and disadvantages within the research discovery process.

Proteomics can be utilized in a host of different types of experiments both quantitative and descriptive. The goal of comparative proteomic experiments is to measure and compare the entire protein complement of an organism when exposed to two different conditions; it allows researchers to determine which proteins are integral to the organism under specific conditions based on the measured increase or decrease in proteins between the two conditions compared. Additionally, proteomics can detect specific organisms within a defined environment. This detection is often facilitated by the observation of known biomarker proteins, which herald the existence of a specific organism within a sample, or an effecter protein such as a toxin. Proteomic techniques can also be used to differentiate and identify a collection of unknown organisms. Specific protein fingerprints can be used to differentiate one organism from another often down to the species level by identifying the components of a mixed population of organisms.

Proteomic experiments can be subdivided into four separate steps: protein extraction, protein/peptide separation (with quantification in some cases), protein identification and data analysis (Pedreschi *et al.*, 2010). Different technologies exist to accomplish each of these steps. Protein extraction methods focus on disrupting cells and solubilizing the liberated proteins. Researchers often use different extraction methods to extract preferentially a sub-fraction of the cells being studied, such as a membrane fraction or a cytosolic fraction. This first step in a proteomic experiment is extremely important given that it is responsible for collecting the proteins that are analyzed in the subsequent steps; however, this step often receives the least focus. The second step, separation, utilizes gel-based and gel-free methods to divide the protein complement into more easily managed portions. The protein identification step relies most heavily on a single technique, mass spectrometry (MS), which can be accomplished by a host of different systems. Finally data analysis, including storing and organizing the data, can be a daunting task given the amount of information produced by even a simple set of proteomic experiments.

Proteomic approaches have been applied to a host of different research areas including the food sciences. Some of this research has focused on investigating the foodborne pathogens that contaminate the food supply and endanger human health. Foodborne pathogens comprise a variety of organisms including bacteria, viruses, parasites, yeasts and molds. Foodborne pathogenic bacteria (prokaryotes) possess varied but relatively small genomes and therefore significantly smaller complements of the proteins making up their proteomes compared to genetically larger and more complex eukaryotic organisms. Proteomic techniques have several desirable advantages when researching food pathogens. Proteins are more durable and have significantly longer half-lives compared to the genetic measures of expression, mRNA. Proteins are the active agents of a food pathogen's phenotype and are directly measured by proteomic techniques. However, there are also disadvantages inherent in proteomics. A small number of extremely plentiful proteins tend to mask the presence of less plentiful proteins. Additionally, the proteins that make up the food matrices in which foodborne pathogens dwell will also make it difficult to observe the pathogen's proteins preferentially. Proteomics provides advantages, as well as challenges for research focusing on foodborne pathogens. Scientists currently researching or contemplating research in this field will hopefully benefit from an overview of the current technologies and techniques available, as well as a review of recent scientific papers that make good use of the proteomic techniques to further the research of foodborne pathogens.

24.6 Mass spectrometry

Mass spectrometry is the central technique for all proteomic-based research. It is the technique by which the proteins being investigated can be identified. Additionally, MS can provide relative or absolute quantification of the amounts of the individual proteins found within a collected sample. Mass spectrometers accomplish these tasks by taking protein or peptide samples, ionizing them, separating out the resulting ions based on their mass-to-charge ratios (m/z) and measuring the relative abundance of each of the ions in the sample to produce a mass spectrum unique to the protein sample (Graham *et al.*, 2007). Many current MS systems are tandem mass spectrometer systems, also known as MS/MS systems, which give superior identification of the individual ions analyzed by the system (Eng *et al.*, 1994; Steen and Mann, 2004). MS/MS systems select a number of the ions originally separated and quantified by m/z ratio and break those ions into several smaller ions through collision with an inert gas, which is known as collision-induced dissociation. The resulting ions are separated based on their m/z ratios, and this gives an additional mass spectrum for each of the precursor ions. The resulting mass spectrum is a unique MS fingerprint for the precursor ion, which can be compared against a collection of theoretical MS fingerprints of known proteins resulting in confident identification of the original ion and protein in the sample being analyzed. Mass spectrometers generally comprise three parts: the ion source, the mass analyzer and the detector (Graham *et al.*, 2011).

The most common techniques used for ion production are electrospray ionization (ESI) and matrix-assisted laser desorption/ionization (MALDI) (Karas and Hillenkamp, 1988; Fenn *et al.*, 1989). These ionization techniques are collectively known as soft ionization since they are able to successfully ionize the protein or peptide of interest without fractionizing the material's structure and are therefore integral to MS-based identification. Electrospray ionization utilizes high-pressure liquid chromatography (HPLC) to pump the material of interest, dissolved within a solvent (acetonitrile/water), through a capillary that has been raised to a high electrical potential. The material to be analyzed is expelled from the charged capillary in a spray of highly charged droplets of solvent directed towards the mass analyzer portion of the mass spectrometer. The solvent in the droplets evaporates, smaller droplets form until finally only multiply charged ions of the materials of interest remain for analysis. These multiply charged gas-phase ions of the material for analysis next move into the mass analyzer portion of the mass spectrometer. Since ESI relies on an initial passage through a capillary, this ionization technique is most commonly utilized with liquid chromatography-based mass spectrometry (LC-MS) systems.

Matrix-assisted laser desorption/ionization produces ions by first co-crystallizing the analyte of interest with a photoactive matrix (i.e. 3,5-dimethoxy-4-hydroxycinnamic acid (sinapinic acid) or α-cyano-4-hydroxycinnamic), which absorbs UV energy. The matrix and analyte are placed on metal plates as dried spots and inserted into an area of the MS system that is under a vacuum; here the spots are then irradiated by a UV laser. The laser bombardment vaporizes the matrix analyte mixture and usually produces single charge ions of the material for analysis. The ions of the analyte in the gaseous form are then directed by an electric field towards the mass analyzer portion of the mass spectrometer.

The mass analyzer portion of a mass spectrometer can comprise a number of different analysis technologies or combinations of several of the individual technologies. Mass analyzers commonly used in MS systems are time-of-flight (TOF), quadrupole (Q) and ion trap (IT) systems. The time of flight is widely used for analyzing mass in MS and determines the mass of ions by observing how fast the ions are able to travel through a vacuum. The ions are accelerated using an electrical field into a long straight tube under vacuum conditions with a detector at the far end of the tube, which is a fixed distance from the point where the ions were accelerated. All of the ions receive the same initial push from the electric field. Since smaller mass ions will be accelerated faster than larger mass ions when given an equal strength push, the smaller mass ions will arrive at the detector in less time than larger mass ions. Therefore, the time it takes each ion to traverse the tube to the detector will be directly proportional to the m/z ratio of the ion. Tandem time-of-flight (TOF-TOF) systems use two separate flight tubes to first measure the m/z ratios of the original ions produced by the mass spectrometer and then to measure the m/z ratios of fragment ions resulting from the selective fragmentation of ions measured by the first TOF system, giving the MS/MS analysis.

Quadrupole mass analyzers utilize four circular poles that produce a rapidly changing electrical field controlled by varying the amplitude of the direct current and radio frequencies applied to the poles. The electrical field only permits ions with certain m/z charge ratios to pass through the quadrupoles to the detector and all other ions collide with the poles. As the quadrupoles scan through a range of different frequencies, ions with different m/z ratios are able to pass through the changing electrical field of the analyzer and reach the detector. Hence, m/z charge ratios can be assigned to the different ions. Like tandem TOF systems, quadrupole systems do not generally utilize a single quadrupole group to analyze the ions produced by the mass spectrometer. More commonly a triple quadrupole system is used (QqQ). The three separate quadrupole analyzers allow for tandem mass spectrometry. As described previously the first quadrupole only allows ions of a certain m/z to pass into the second quadrupole where those ions are further fractionated to create product ions. The product ions are then directed to the third quadrupole, which functions similarly to the first quadrupole allowing only product ions of a certain m/z through to the ion detector. A triple quadrupole analyzer can determine the m/z ratios of the precursor ions and the m/z ratios of the ions produced from the individually fragmented precursor ions. It produces mass spectra fingerprints for the ions, which can be utilized along with DNA and protein databases to identify the precursor ion.

Ion trap mass analyzers function in a manner similar to the quadrupole system but instead of allowing only certain m/z ratio ions to pass through the quadrupole, the ion trap captures and holds ions with specific m/z ratios. Utilizing an electrostatic field the ion trap system captures ions of specific m/z ratios and then expels them in a specific order, again dependent on their *m/z* ratio. As the ions are released from the trapping field, they become free to collide with the ion detector of the mass spectrometer. Similar to the other mass analyzers discussed, the ion trap system is also capable of tandem mass spectrometry (MS/MS). In this process the ion trap captures and holds an ion, further fragmenting this precursor ion into several product ions that are released from the trap into the detector again based on their *m/z* ratios. The product ions form a mass spectra fingerprint for the precursor ion, which can be used to identify the precursor ion by comparing the spectrum against databases of known fragmentation spectra. Additionally, the common mass analyzers can also be combined with one another to form hybrid systems to take advantage of the unique strengths of the different systems. Examples of these include: quadrupole time of flight (Q-TOF), quadrupole ion trap (QIT) and ion trap time of flight (IT-TOF).

After the mass analysis portion of the mass spectrometer comes the ion detector. The function of the ion detector is to determine the numbers of specific ions as they are preferentially passed through the mass analyzer and to transmit this data to an external computer system. From this information the computer system is able to produce mass spectra detailing the *m/z* ratios of the ions detected, as well as their relative abundance.

24.7 Separation techniques

Mass spectrometry techniques are responsible for identifying the proteins and peptides in proteomic-based research. However, the number of proteins requiring analysis in a single proteomic experiment can be problematic. A relatively simple organism such as *Escherichia coli* is expected to be capable of producing upwards of 4000+ proteins. It is therefore often necessary to be able to separate an organism's proteome before the protein components can be measured, compared or identified through MS analysis.

Separation techniques for proteomic research can be divided into two different groupings, known as gel-based and gel-free separation methods. The gel-based techniques utilize gel electrophoresis to separate proteomic samples in one- or two-dimensional formats. The gel-free techniques are all the other techniques that do not utilize gels; the most common type of gel-free technique utilizes liquid chromatography methods for protein separation.

Gel electrophoresis is a well-characterized technique, which has been utilized in protein research for many years, and the separation is either one-dimensional or two-dimensional. One-dimensional separation of denatured proteins utilizes an electrical field to migrate the proteins in the sample through the gel matrix based approximately on their molecular weight (MW). The smaller MW proteins move more rapidly towards the positive end of the gel compared to larger MW proteins. The migration produces a one-dimensional organization of the proteins in the samples stretching from the loading well of the gel down towards the far end of the gel based on the proteins' molecular weights. One-dimensional gels, however, have limited uses in proteomic studies. Given the large number of proteins that are present in a whole cell proteomic sample and the likelihood that multiple proteins in the sample will share relatively similar MWs, one-dimensional separation does not provide sufficient isolation of individual protein bands. One-dimensional gels are more useful after other initial separation techniques have been applied to a sample.

The need for greater separation of proteins led directly to the development of two-dimensional gel electrophoresis (2D-GE) (O'Farrell, 1975). In 2D-GE the proteins in the sample are first separated based on their isoelectric point on a pH gradient gel. A protein migrates through the pH gradient gel until it arrives at the pH value in the gel where the protein's net charge equals zero and it is, therefore, not drawn towards either the negative or positive ends of the gel. After the isoelectric focusing (IEF) of the proteins, the pH gradient gel is placed into the loading well of a sodium dodecyl sulfate (SDS) polyacrylamide gel at 90° to the direction of protein migration in the second dimension gel. The isoelectric focused proteins are then separated by MW, thus distributing the proteins in the sample over two dimensions. The resulting 2D gels can then be treated with a host of different stains (Coomassie blue, silver stain, fluorescent stains) to visualize and quantify the separated proteins as discrete spots of differing sizes (Miller *et al.*, 2006). The different stains vary with regards to their sensitivity, dynamic range and reproducibility. The increased separation and visualization achieved by 2D-GE allows the protein spots to be collected and identified using the previously discussed MS technologies. Two-dimensional gel electrophoresis lends itself well to relative quantification of the proteins in a sample and comparative proteomic experiments. Samples of organisms incubated under different growth conditions, or isogenic mutants and their parent strain, can be separated with multiple gels and then the gel spot patterns and the intensities of the spots can be compared to identify the proteins, which are differentially expressed between the different conditions. Two-dimensional gel electrophoresis can perform whole proteome separations and comparisons with a minimal amount of equipment and therefore at a relatively low cost. Additionally, the technology is reliable, simple and easily maintained. For these reasons 2D-GE remains an extremely popular method for proteomic research particularly in smaller research laboratories. However, 2D-GE does have several problems that limit the technique's effectiveness in proteomic experimentation; for example, there are problems resolving very hydrophobic, basic or large proteins (Westermeier and Schickle, 2009). Additionally, 2D-GE suffers from reproducibility problems in protein spot separation, distribution and quantification from one gel to the next (Molloy *et al.*, 2003; Aittokallio *et al.*, 2005). This variation from gel to gel results in problematically large standard deviation values for multiple proteomic experiments using traditional 2D-GE.

To address this specific shortcoming of 2D-GE, two-dimensional differential in-gel electrophoresis (2D-DIGE) was developed as a separation technique (Unlu *et al.*, 1997). This technique is similar to 2D-GE but differs in that the two samples to be compared are labeled with two different fluorescent markers prior to gel electrophoresis. The two fluorescent markers can be differentiated spectrally and are visualized through excitation at two separate wavelengths. Several protein samples are separated together on the same gel but visualized separately to produce images of the different protein samples for proteomic comparisons. The ability to separate multiple samples on one gel removes the issues of gel-to-gel variation, which made the data from traditional 2D-GE often difficult to interpret. However, this advantage necessitates the additional costs of purchasing fluorescent dyes as well as the systems needed to visualize and record the fluorescently labeled proteins.

The other main group of separation techniques, collectively known as gel-free, use mainly liquid chromatography methods but also include selective protein purification techniques. These techniques involve the preferential purification of sub-cellular compartments (Lopez-Campistrous *et al.*, 2005; Cordwell *et al.*, 2008; Brewis and Brennan, 2010; Desvaux *et al.*, 2010), including a cell's membrane proteome, secretory proteome and cytoplasmic proteome. The reduced sample complexity often results in the enhanced detection and identification of proteins previously found in less abundant quantities in the whole cell proteome.

Liquid chromatography protein/peptide separation can utilize a variety of different column-based systems to separate the protein or peptide samples using one- or two-dimensional procedures. The column-based LC separation systems have a significant advantage in that they can be connected directly in line with an MS system to simplify the subsequent quantification and identification of the samples. Columns used for LC separation include reverse phase, cation exchange, anion exchange, size exclusion, affinity, chromatofocusing and capillary electrophoresis (Lopez-Campistrous *et al.*, 2005; Fournier *et al.*, 2007). Reverse phase columns separate proteins and peptides based on the hydrophobicity of the molecule. Cation and anion exchange columns separate molecules based on their overall charge. Size exclusion columns separate molecules based on their size. Affinity chromatography uses specific interactions such as those between antigen and antibody to separate the molecules in a sample. Chromatofocusing columns separate proteins based on the isoelectric point of the proteins using a pH gradient. Capillary electrophoresis uses an electric field and column to separate proteins and peptides based on their mass-to-charge ratio. Many of these columns can be placed directly in line with an MS system to give a single dimension separation of the proteins and peptides before ionization, which reduces the number of molecules entering the mass spectrometer at individual time points, thus allowing for superior acquisition and analysis of the sample ions. However, in a complex biological sample there can still be a relatively large number of proteins and peptides eluting from a one-dimensional separation and reaching the ionization source at a given time, and the overabundance of sample ions can overwhelm the ability of the mass spectrometer to acquire and analyze the available ions. For this reason two column-based separation methods can be sequentially applied to a protein or peptide sample to accomplish a two-dimensional LC separation. The two-dimensional separation further distributes the proteins and peptides in the sample resulting in smaller numbers of proteins and peptides co-eluting from the column and becoming ionized by the mass spectrometer at a given time. This helps to prevent the mass spectrometer from becoming overwhelmed with regards to its ability to acquire and analyze the co-eluting ions, which could lead to ions being lost. This theoretically will lead to better analysis and identification of proteins in a given sample. There are many different combinations of chromatography separations, which can be used; however, the combination of a strong cation exchange column and a reverse phase column is currently the most commonly used.

The LC-based separation systems lack the immediate quantification capacity that the protein gel electrophoresis techniques can accomplish with visualization dyes. However, when coupled to a mass spectrometer, the proteins and peptides in a sample can still be quantified using two different groups of methods: label-based quantification and label-free quantification (Schulze and Usadel, 2010). The label-based quantification methods use a variety of different tags on different protein or peptide samples, allowing for quantification and comparison. The two main methods for labeling proteins and peptides are metabolic labeling and chemical-based labeling. Metabolic labeling relies on integration of the label into the organism being studied through the growth medium. This type of labeling has been accomplished through the use of ^{15}N or stable isotope-labeled amino acids (SILAC) provided to growing organisms for integration into their proteins. The isotope-incorporated proteins have slightly different masses and are referred to as heavy. They can be compared to their unlabeled counterparts, known as light, in comparative proteomic experiments.

In chemical labeling, isotope-labeled affinity tags are integrated into proteins or peptides. While metabolic labeling takes place before any protein purification, chemical labeling methods are accomplished after protein purification procedures. The isotope-coded affinity tag (ICAT) system utilizes a light tag (hydrogen) and a heavy tag (deuterium) to label specific side chains of protein samples for comparison. There is also an ICAT for affinity purification of labeled proteins and peptides. The general workflow for this system consists of labeling the two samples of interest with either a light or heavy tag, respectively, mixing the two differentially tagged samples, digesting the mixed samples into peptides with trypsin and then isolating the labeled peptides using affinity chromatography. The isolated peptides are then analyzed by MS. A direct comparison of the peak values of matchable peptides from the original two samples gives the relative concentrations of specific proteins. Isobaric tags for relative and absolute quantitation (ITRAQ) are a similar system. This system uses differential isotopic labeling through eight unique tags that can be bound to the free amine group of peptides produced from the tryptic digestion of the proteins contained in samples for analysis (Ross *et al.*, 2004; Aggarwal *et al.*, 2006). In this type of labeling, project samples are first digested with trypsin and then labeled with one of the eight different ITRAQ tags. All of the samples for comparison are mixed together. The mixed sample is then analyzed using an LC-MS/MS system. In the initial MS analysis, the peaks for any specific peptide all appear as a single peak with the same *m/z* value. This occurs because each of the isotopically different ITRAQ tags have a balance group so that the different tags bound to a peptide all shift the peptide *m/z* values by the same amount. After the collision-induced dissociation step and the second MS analysis the ITRAQ tags are removed from the peptides and appear as measurable peaks on the MS/MS spectra. The peak heights of the different tags can then be compared to determine the relative concentration of this peptide and its parent protein for the various conditions being compared.

Label-based quantification methods provide useful evaluations of protein concentrations in proteome samples; however, label techniques can be both technically challenging, as well as expensive for researchers. These concerns have led to the development of label-free quantification techniques for proteomic research. Common label-free quantification methods include spectra counting and comparing ion intensities. Spectra count-based quantification (Liu *et al.*, 2004a) counts the number of MS/MS spectra that can be assigned to a particular protein in one condition and directly compares that value against the number of spectra that can be assigned to the same protein in another condition. The technique works well for abundant proteins, which can be expected to produce a large number of measurable spectra but is less successful with scarce proteins, which may only produce one recorded spectrum. Additionally, the comparative ability of spectra counting works best when successfully identifying protein differences between samples that are greater than 2.5-fold (Old *et al.*, 2005). Ion intensity-based quantification measures the number of ions for a specific peptide in one experimental condition and compares it against the number in another condition (Wang *et al.*, 2006). This is accomplished by comparing the peak heights of a specific peptide, which are a graphic representation of the number of ions that the peptide collected during a certain time. This type of comparison is only valid for the same peptide analyzed from different samples given that there are differences in the ionization potential of different peptides derived from the same protein. The ion intensity method is similar to the spectra count method in that it is most successful at identifying variations between proteins with differences greater than 2.5-fold (Old *et al.*, 2005; Schulze and Usadel, 2010).

24.8 Food-based proteomic investigations

There is a considerable range of technology and techniques available to proteomic researchers including those working in the field of food safety. Proteomics have proven useful to food-safety researchers for detection, identification, comparative proteomics and biomarker discovery (Fig. 24.1). However, the greatest technical challenge to food-safety proteomic research is also the most important element for making this research relevant and impactful: the food environment. Simply put, proteomic food-safety research must take place within relevant food environments. The challenge is that food environments, particularly in terms of protein composition, are extremely complex matrices for experimentation. Food environments are composed of a large number of different proteins, fats and nucleic acids, which are often found in significantly large concentrations. Therefore, proteomic investigations targeting microbes affecting food safety face

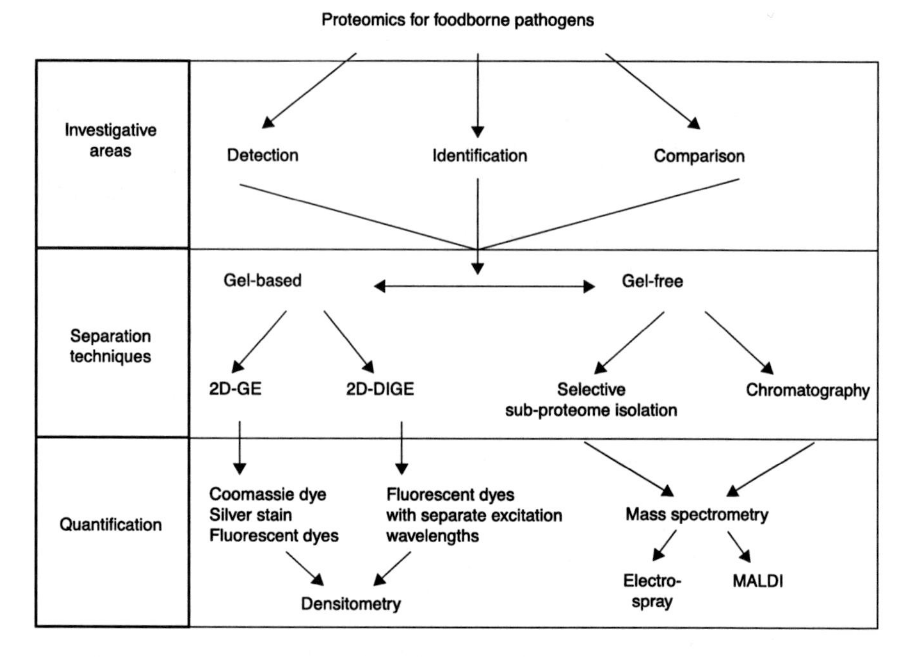

Fig. 24.1 The primary uses for proteomic research in the study of foodborne pathogens and common techniques for the separation and quantification of proteins during these investigations. 2D-GE and 2D-DIGE, two-dimensional gel electrophoresis and differential in gel electrophoresis, respectively; MALDI, matrix-assisted laser desorption ionization.

the challenge of needing to be able to isolate, separate, quantify and identify the microbial proteins, which are often found as a small proportion of the overall protein composition of the food sample. For this reason many researchers perform proteomic investigations on food-safety-relevant microbes that have been incubated on common laboratory media. Since the goal of these experiments, ultimately, is to study the actions of the food-safety-relevant microbes through the changes in their protein expression profiles it is counterintuitive to utilize an experimental environment with only a tenuous relationship to a true food environment. Therefore, when a proteomic investigation is being applied to answer a question about an aspect of food safety, it is important that the environment utilized, if not the exact food environment in question, is at least an attempt to mimic the relevant aspects of the food environment.

24.8.1 Detection

Proteomic techniques have been utilized to detect pathogenic bacteria and virulence factors contaminating food products. Detection is accomplished through the isolation and identification of proteins or peptides specific to the bacteria or bacterial element of interest. Again this can be a challenging process given that the proteins of the contaminating bacteria or virulence factors will make up only a small percentage of the proteins in the food sample that is being analyzed. For this reason most detection assays utilize nucleic acid to detect microbial contaminants in food. Nucleic acids can be amplified using PCR to increase the relative concentration of the portion of nucleic acid being used for detection. Unfortunately, there is no equivalent amplification step available for proteins. However, target proteins in a mixed sample can benefit from enrichment techniques such as affinity chromatography. In affinity chromatography, antibodies specific to the proteins or peptides of interest are used to bind and purify the targets away from the host of other proteins and peptides in the sample, allowing for detection. However, given the amplification power of PCR and the specificity that can be achieved from nucleic acid sequences, it will not at first appear clear why protein-based detection assays are needed. The danger posed by pathogenic bacteria in food is not always found with the bacteria themselves but rather with the toxins they extrude and leave behind. There are examples where the bacteria containing the necessary nucleic acid for detection are no longer present in a food but the toxins from those bacteria still persist in high enough concentrations to pose a risk to human health. In this situation only a protein-based detection assay would be able to detect the threat to food safety. Additionally, inhibitors of PCR amplification can be found in food samples, which remove the primary advantage of nucleic acid detection compared to protein-based detection.

Several research projects have investigated protein-based methods for detecting *Staphylococcus aureus* enterotoxins in foods (Schlosser *et al.*, 2007; Dupuis *et al.*, 2008; Hennekinne *et al.*, 2009). *Staphylococcus aureus* is a common food pathogen responsible for a significant number of food-poisoning cases. Among other virulence factors, the pathogen produces several different enterotoxins, which are heat stable and capable of causing gastrointestinal disease. Research groups have developed methods to detect and in some cases quantify this group of enterotoxins in several relevant food products, including cheese, coco-pearls (a Chinese dessert product) (Dupuis *et al.*, 2008) and milk. The different detection methods follow a similar procedure. The enterotoxins are initially separated from the food matrices by affinity chromatography using antibodies specific to several of the common enterotoxins. The isolated materials are then digested with trypsin and analyzed using mass spectrometry systems – quadrupole time-of-flight (Q-TOF) and matrix-assisted laser desorption/ionization time-of-flight (MALDI-TOF) (Schlosser *et al.*, 2007). The presence and numbers of peptides derived from *S. aureus* enterotoxins measured by mass spectrometry allow for the detection and quantification of enterotoxins in the food. This type of proteomic-based detection works well for detecting many different types of toxins, including botulinum neurotoxin (Kalb *et al.*, 2005). Additionally, research is being conducted on using proteomics to detect whole pathogenic bacteria within food environments. A recent research effort utilized affinity labeling to increase the sensitivity of a protein-based detection assay (Li *et al.*, 2010). Antibodies specific to surface proteins of the food pathogen *Escherichia coli* O157:H7 were first coupled to gold particles. Next the labeled antibodies were mixed into a laboratory sample containing a small concentration of the *E. coli*.

The gold-labeled antibodies bound to the *E. coli* and when the samples were analyzed using mass spectrometry the gold particles produced a large quantity of gold ions, allowing detection by a mass spectrometer of the relatively small concentration of bacteria in the sample.

24.8.2 Identification

In addition to detection, proteomic-based approaches have also been developed to identify and differentiate food microbes. Similar to detection, the value of identification (ID) methods utilizing proteomics compared to ID methods using nucleic acid sequences or biochemical assays is not readily apparent. Nucleic acid sequences generally provide adequate discrimination between bacteria, often to the species level, and target concentrations can be amplified by PCR, while biochemical assays are well characterized and validated. However, proteomic-based identification has some potential advantages such as the speed of identification, as well as the degree of bacterial discrimination possible, with some research labs claiming they are able to separate bacteria to the subspecies or even individual strain level (Kiehntopf *et al.*, 2011; Podlesny *et al.*, 2011). The proteomic-based ID of food related bacteria depends on the use of mass spectrometers to generate fingerprint mass spectra that contain peak(s) with *m/z* ratios unique to a particular bacterial species or strain (Bohme *et al.*, 2010). This sort of research produces libraries of these mass spectra fingerprints, which can then be compared against the spectra of unknown bacteria, hopefully resulting in reliable identification. The successful ID of any microbe using a mass spectrometry-based technique depends heavily on the completeness of the library of unique reference mass spectra (Pinto *et al.*, 2011).

Bacteria can be analyzed by mass spectrometry at the level of whole cells or at the protein level after cell lysis and at the peptide level after trypsin digestion. Proteomic experiments are often referred to as top-down or bottom-up style experiments. Top-down experiments focus on analyzing intact protein samples and are labeled in this manner since the protein is considered the topmost structure in a proteomic analysis with subsequent investigations working down from that point (Loo *et al.*, 1992; Reid and McLuckey, 2002). Alternatively, bottom-up analysis focuses on the level of the peptide, the reduced level of complexity or the bottom of protein structures, and further investigation works upward from that point to identify the whole proteins. Proteomic experiments to identify food-relevant bacteria can be found in both the top-down and bottom-up formats. In the bottom-up form, the bacterial cells for ID are first completely disrupted and the freed proteins are isolated and then digested into peptides. In the top-down method, the bacteria can be directly submitted to MS analysis or chemically disrupted to free internal and membrane proteins producing a protein lysate that is then submitted to MS analysis (Mazzeo *et al.*, 2006; van Veen *et al.*, 2010; Dieckmann and Malorny, 2011). The top-down or bottom-up samples can next be analyzed by liquid chromatography electrospray ionization (LC-ESI) or MALDI mass spectrometry. However, MALDI mass spectrometry has a significant advantage in this type of assay because of the shorter time required for analysis and identification of the target protein. MALDI is able to acquire multiple samples in minutes compared to hours for LC-ESI (Fagerquist *et al.*, 2010). The speed of data acquisition demonstrated by MALDI gives reliable bacteria identification in 24 h. This significant reduction in the time needed for successful ID allows the proteomic ID techniques to compete with biochemical and genetic sequence-based techniques for food bacteria identification. For bacterial ID experiments, regardless of the methods utilized, the bacteria are commonly first isolated and grown as single colonies on laboratory media. However, it is relatively easy to see how the methods in the section on detection research and the current section on ID techniques can and have been applied jointly to accomplish both detection and ID of microbes or their toxins from environmental samples such as food (Ochoa and Harrington, 2005; Pocsfalvi and Schlosser, 2011).

24.8.3 Comparative proteomics

Research utilizing comparative proteomics involves the identification and quantification of at least two separate sets of proteins and then the matching and comparison of individual proteins between the two groups to determine changes in protein expression between them. This allows the researchers to determine what proteins might be more important in one set compared to the other, providing an increased understanding of the physiological role of the proteins being analyzed. The majority of comparative proteomic experiments still utilize two-dimensional gel electrophoresis (2D-GE) for separation and comparison of protein quantities given the relatively low cost of purchasing, using and maintaining 2D-GE systems. However, in more recent years, gel-free systems using liquid chromatography (LC) and advanced mass spectrometers to separate and quantify protein sets have increased in popularity and use. These systems are expensive but provide superior assay automation and increased coverage of the proteomes being investigated.

Comparative proteomic studies of foodborne pathogens can be grouped roughly into three different types of investigations: (i) proteomic comparison of strains with different physiological traits, (ii) identification and comparison of different sub-proteomes and (iii) protein comparison of the same bacteria exposed to the different stresses found in food environments. The proteomic comparison of foodborne pathogens with different physiological traits can include comparing different strains of the same bacterial species or comparing a constructed isogenic mutant with the parent strain from which the mutant was constructed. Experiments comparing the protein production of different strains of the same species are often conducted to compare the more pathogenic strains of the same species against the less pathogenic strains to determine the mechanisms of pathogenesis (Folsom and Frank, 2007; Donaldson *et al.*, 2009; Gunther *et al.*, 2010; Brul *et al.*, 2011). As an example, in one set of experiments a foodborne *Bacillus subtilis* isolate, which was extremely heat resistant, was compared proteomically against a laboratory strain of the same species with normal heat tolerance (Brul *et al.*, 2011). A comparison was able to identify proteins that may play a role in heat resistance in the foodborne organism. Comparative proteomics of isogenic mutants versus parental strains are designed to investigate how the loss of a gene and its protein product might have a broader effect on the expression of other bacterial proteins and pathways (Vidovic *et al.*, 2011; Zhu *et al.*, 2011).

Experiments to identify and compare the protein contents of the sub-proteomes of food pathogens can also by necessity include experiments to identify the protein composition of the entire proteome (Carranza *et al.*, 2009). The investigation and organization of proteins into sub-proteomes aids in assigning a specific function to the organized proteins (Lopez-Campistrous *et al.*, 2005; Cordwell *et al.*, 2008; Alam *et al.*, 2009; Desvaux *et al.*, 2010). Additionally, the sub-fractionation of proteomes before separation and MS analysis has been shown to result in the identification of more proteins than without prior sub-fractionation. This is often the result of fractionating out overly abundant proteins from samples allowing less abundant proteins to be successfully analyzed.

Perhaps the most common use of comparative proteomics with foodborne pathogens is for investigating the effects that different stress conditions have on the pathogenic organisms. These stresses can come in the form of the normal food and food-processing environmental stresses or in the form of stresses brought on by interventions designed to improve food quality and/or reduce pathogenic organisms. Common food environmental stresses include temperature changes commonly referred to as cold and heat shock (Mihoub *et al.*, 2003; Annamalai and Venkitanarayanan, 2005; Tangwatcharin *et al.*, 2006; Chiang *et al.*, 2008; Jones *et al.*, 2008; Cacace *et al.*, 2010; Carranza *et al.*, 2010; Sánchez *et al.*, 2010; Vidovic *et al.*, 2011). Additional stresses common to food environments include osmotic stress (Weber *et al.*, 2006; Riedel and Lehner, 2007), acid stress (Huang *et al.*, 2007) and hydrogen peroxide stress (Kim *et al.*, 2010). Environmental stresses are also encountered in the form of food additives designed to enhance the quality of food (Mbandi *et al.*, 2007; Guilbaud *et al.*, 2008; Calhoun *et al.*, 2010). Finally, comparative proteomic experiments are often used to investigate a bacterium's response to human interventions to the food supply designed to eliminate pathogenic organisms from the food (Jofre *et al.*, 2007; Kim *et al.*, 2009; Martinez-Gomariz *et al.*, 2009; Bieche *et al.*, 2010). This type of research is performed to determine the mode of action by which interventions work and to determine how the bacteria are able to circumvent the interventions so that a new intervention can address the former's shortcomings.

Comparative proteomic studies often illustrate a major shortcoming of current proteomic research into foodborne pathogens, specifically that very little of the proteomic studies are performed under food-relevant conditions. The use of laboratory media to incubate food pathogens will provide little insight into the expression patterns of these bacteria as they exist in their common environment – food and food-processing facilities. A major reason for the lack of relevant food conditions is that food matrices are complex environments where it is difficult to perform proteomic-based experiments. Food matrix proteins are often present in high concentrations and easily drown out the signal of the less plentiful bacterial proteins that are the target of the investigation. Additionally once bacteria are placed into a food environment it is often difficult to separate the bacteria back out of the food. However, successful proteomic experiments using food-relevant environments have been accomplished and rely on the careful selection of an experimentally manageable food environment. Proteomic research within food-relevant conditions often utilizes simple well-characterized liquid foods or the liquid exudates of solid foods. Cow milk is a food-relevant environment, which has been used in proteomic studies of foodborne pathogens (Annamalai and Venkitanarayanan, 2005; Lippolis *et al.*, 2009). Milk is a well-characterized environment containing a small number of highly abundant proteins, which can be preferentially removed from the liquid just prior to a proteomic analysis of the pathogenic bacteria incubated within the milk (Danielsen *et al.*, 2010). This makes the less abundant bacterial proteins available for analysis. Additionally, bacteria incubated in milk can be extracted from the food environment through centrifugation (Lippolis *et al.*, 2009). Meat exudates are also attractive food-relevant conditions for comparative proteomic studies of bacteria (Fadda *et al.*, 2010). Meat exudates are commonly contaminated with pathogenic bacteria and are often responsible for foodborne disease through cross-contamination of other foods, making the exudates a relevant food-safety concern. Additionally, exudates tend to be thin in nature and do not contain high concentrations of meat proteins or other large biomaterials. The bacteria can be separated from the exudate by centrifugation or size exclusion filtration. Certain solid food surfaces have also been successfully utilized for bacterial proteomic studies (Annamalai and Venkitanarayanan, 2005). In this work, bacteria were evenly spread on a smooth food surface, incubated and then removed by gentle scraping designed to minimize the carry-over of food particles.

24.9 Future trends

Knowledge of the complete genome sequences of foodborne pathogens, the subsequent analyses based on the sequence data and the integration of genomic data with other omic technologies is providing valuable information that will impact the ability to ensure food safety. However, in order to identify biomarkers that can be used as targets for detection assays and for intervention strategies, and to better understand the evolution of foodborne pathogens and the mechanisms utilized to survive stress, the genome sequences of many other bacterial species and strains must be determined and analyzed. Systematic studies that integrate omic technologies and targeted gene disruptions will help to identify bacterial genes of unknown function, which may be required for virulence or survival of the pathogens. Gene expression investigations will reveal genes that could potentially be targeted for intervention strategies; however, there are considerable technical challenges in accurately measuring bacterial gene expression in food and other complex matrices, which must be overcome. As various types of omic data on foodborne pathogens become available, there is a critical need for methods and software to visualize, analyze and integrate this information before it can be used to enhance food safety. The development of high-throughput analysis techniques and software will make multi-omic approaches for understanding complete biological systems possible; this field is known as systems (integrated) biology. Omic technologies are becoming commonly used as research tools for pathogen detection and control and to enhance our understanding of the behavior of foodborne pathogens at the molecular level. However, there are still major challenges in using these technologies to enhance food safety, including finding ways to manage the large quantity of complex raw data that is generated so that the data can be effectively analyzed, scrutinized and compared by the scientific community.

The major requirements for future proteomic research of foodborne pathogens are two-fold: (i) better coverage of an entire proteome, so that all proteins in the proteome are analyzed during experimentation, and (ii) proteomic research performed on foodborne pathogens within food-relevant environments and not in laboratory media. Proteomics should study the entire protein complement of an organism and should not be limited to a fraction of the whole. However, with current separation, visualization and analysis technology, only a fraction of the proteome of a foodborne pathogen is available for investigation. Even with the most advanced proteomic technology, a single experiment is able to identify only roughly half the proteins that a particular bacterial species is expected to produce (Becher *et al.*, 2009; Franzel and Wolters, 2011). Gel-based technology, 2D-GE and 2D-DIGE, is limited in the number of proteins that it can separate and quantify in a single experiment and in the future will probably only provide modest increases in proteome coverage. Alternately, gel-free methods, specifically multidimensional protein identification technology (MudPIT) or shotgun proteomic approaches, appear to have room for significant increases in single-experiment proteome coverage. Current research is focused on the LC separation portion of the LC-MS/MS analysis. The goal of the research is to separate the peptides generated from the proteome being analyzed sufficiently to give more efficient ionization of individual peptides, which is negatively impacted by co-elution of multiple peptides at any given time. Two-dimensional LC technology and lengthy one-column separation techniques have succeeded in greatly increasing the coverage of bacterial proteomes during a single experiment (Iwasaki *et al.*, 2010; Tao *et al.*, 2011). Further research into enhanced protein and peptide separation and analysis techniques are needed until all of the proteins actively expressed by an organism can be easily identified and measured during a single experiment.

As was mentioned previously, a major shortcoming of current research into the proteomics of food pathogens is that few experiments are conducted within food-relevant environments. The majority of experiments still use laboratory media for incubation since this gives easier manipulation and re-isolation of the bacteria for proteomic-based analysis. If artificial environments are used, they should at the very least be constructed to mimic relevant real-world environments (Sonck *et al.*, 2009). Moreover, it is not surprising that the majority of research using relevant food environments utilizes liquid food environments that are similar to laboratory media in physical characteristics (Ideker *et al.*, 2001; Annamalai and Venkitanarayanan, 2005; Fadda *et al.*, 2010). Liquid environments are convenient to store, measure and divide. Additionally, the bacteria for analysis can be more easily separated from this type of environment by filtration, centrifugation or affinity chromatography. Future proteomic research into food pathogens needs to prioritize the use of relevant food environments so that the protein expression results have greater relevance to real-world conditions. This will necessitate the development of new fractionation and separation techniques, as well as clearly defining the relevant environmental conditions for a specific food pathogen. Additional research may focus on technology where food pathogens can interact with the molecules of a food matrix while maintaining the food bacteria in a somewhat separated state allowing for easy removal at the appropriate time. An example of this type of technology was a comparative genomic project that utilized a sealed dialysis tube to segregate bacterial cells while allowing affecter molecules from a milk environment to easily pass through to reach the bacteria (Liu and Ream, 2008).

Proteomic research investigating the proteins produced by foodborne pathogens has great potential to discover the means to remove the threat of foodborne diseases. There are significant and varied technologies and methods for all investigations targeting foodborne pathogens. It should be possible for researchers to identify a proteomic technology or technique that will satisfy their particular research needs while not exceeding their laboratory's resources. Additionally, continual progress is being made to improve and expand the available technologies. This progress will hopefully address the needs for increased proteome coverage and encourage the use of relevant food environments to increase the quality of this type of research. To understand how bacteria persist within our food requires an understanding of how the bacteria act, and the actions are illustrated by the proteins that the bacteria express at any given time.

24.10 References

AGGARWAL, K., CHOE, L.H. and LEE, K.H. 2006. Shotgun proteomics using the iTRAQ isobaric tags. *Brief. Funct. Genomic. Proteomic.* **5**:112–120.

AITTOKALLIO, T., SALMI, J., NYMAN, T.A. and NEVALAINEN, O.S. 2005. Geometrical distortions in two-dimensional gels: applicable correction methods. *J. Chromatogr. B Analyt. Technol. Biomed. Life Sci.* **815**:25–37.

ALAM, S.I., BANSOD, S., KUMAR, R.B., SENGUPTA, N. and SINGH, L. 2009. Differential proteomic analysis of *Clostridium perfringens* ATCC13124; identification of dominant, surface and structure associated proteins. *BMC Microbiol.* **9**:162.

ALLEN, H.K., LOOFT, T., BAYLES, D.O., HUMPHREY, S., LEVINE, U.Y. *et al.* 2011. Antibiotics in feed induce prophages in swine fecal microbiomes. *mBio* doi: 10.1128/mBio.00260-11.

ALLEN, K.J., LEPP, D., MCKELLAR, R.C. and GRIFFITHS, M.W. 2008. Examination of stress and virulence gene expression in *Escherichia coli* O157:H7 using targeted microarray analysis. *Foodborne Pathog. Dis.* **5**:437–447.

ANNAMALAI, T. and VENKITANARAYANAN, K. 2005. Expression of major cold shock proteins and genes by *Yersinia enterocolitica* in synthetic medium and foods. *J. Food. Prot.* **68**:2454–2458.

AUROUX, P.-A. 2008. Detection of pathogens by on-chip PCR, pp. 833–853. In: *Principles of Bacterial Detection: Biosensors, Recognition Receptors and Microsystems*, M. ZOUROB, S. ELWARY and A.P.F. TURNER (Eds.). Springer Science+Business Media, Philadelphia.

BECHER, D., HEMPEL, K., SIEVERS, S., ZUHLKE, D., PANE-FARRE, J. *et al.* 2009. A proteomic view of an important human pathogen – towards the quantification of the entire *Staphylococcus aureus* proteome. *PLoS One* 4:e8176.

BHAGWAT, A.A. and BHAGWAT, M. 2008. Methods and tools for comparative genomics of food-borne pathogen. *Foodborne Pathog. Dis.* **5**:487–497.

BIECHE, C., DE LAMBALLERIE, M., FEDERIGHI, M., LE BAIL, A. and TRESSE, O. 2010. Proteins involved in *Campylobacter jejuni* 81–176 recovery after high-pressure treatment. *Ann. NY Acad. Sci.* **1189**:133–138.

BOHME, K., FERNANDEZ-NO, I.C., BARROS-VELAZQUEZ, J., GALLARDO, J.M., CALO-MATA, P. *et al.* 2010. Species differentiation of seafood spoilage and pathogenic gram-negative bacteria by MALDI-TOF mass fingerprinting. *J. Proteome Res.* **9**:3169–3183.

BONO, J.L., KEEN, J.E., CLAWSON, M.L., DURSO, L.M., HEATON, M.P. *et al.* 2007. Association of *Escherichia coli* O157:H7 *tir* polymorphisms with human infection. *BMC Infect. Dis.* 7:98.

BREWIS, I.A. and BRENNAN, P. 2010. Proteomics technologies for the global identification and quantification of proteins. *Adv. Protein Chem. Struct. Biol.* **80**:1–44.

BRUL, S., VAN BEILEN, J., CASPERS, M., O'BRIEN, A., DE KOSTER, C. *et al.* 2011. Challenges and advances in systems biology analysis of *Bacillus* spore physiology; molecular differences between an extreme heat resistant spore forming *Bacillus subtilis* food isolate and a laboratory strain. *Food Microbiol.* **28**:221–227.

BUGAREL, M., BEUTIN, L., MARTIN, A., GILL, A. and FACH, P. 2010. Micro-array for the identification of Shiga toxin-producing *Escherichia coli* (STEC) seropathotypes associated with hemorrhagic colitis and hemolytic uremic syndrome in humans. *Int. J. Food Microbiol.* **142**:318–329.

CACACE, G., MAZZEO, M.F., SORRENTINO, A., SPADA, V., MALORNI, A. *et al.* 2010. Proteomics for the elucidation of cold adaptation mechanisms in *Listeria monocytogenes*. *J. Proteomics* **73**:2021–2030.

CALHOUN, L.N., LIYANAGE, R., LAY JR, J.O. and KWON, Y.M. 2010. Proteomic analysis of *Salmonella enterica* serovar Enteritidis following propionate adaptation. *BMC Microbiol.* **10**:249.

CARRANZA, P., HARTMANN, I., LEHNER, A., STEPHAN, R., GEHRIG, P. *et al.* 2009. Proteomic profiling of *Cronobacter turicensis* 3032, a food-borne opportunistic pathogen. *Proteomics* 9:3564–3579.

CARRANZA, P., GRUNAU, A., SCHNEIDER, T., HARTMANN, I., LEHNER, A. *et al.* 2010. A gel-free quantitative proteomics approach to investigate temperature adaptation of the food-borne pathogen *Cronobacter turicensis* 3032. *Proteomics* 10:3248–3261.

CEVALLOS-CEVALLOS, J.M., REYES-DE-CORCUERA, J.I., ETXEBERRIA, E., DANYLUK, M.D. and RODRICK, G.E. 2009. Metabolomic analysis in food science: a review. *Trends Food Sci. Technol.* 20:557–566.

CHAMPION, O.L., GAUNT, M.W., GUNDOGDU, O., ELMI, A., WITNEY, A.A. *et al.* 2005. Comparative phylogenomics of the food-borne pathogen *Campylobacter jejuni* reveals genetic markers predictive of infection source. *Proc. Natl. Acad. Sci. USA.* 102:16043–16048.

CHASTANET, A., FERT, J. and MSADEK, T. 2003. Comparative genomics reveal novel heat shock regulatory mechanisms in *Staphylococcus aureus* and other Gram-positive bacteria. *Mol. Microbiol.* **47**:1061–1073.

CHEN, E.C., MILLER, S.A., DERISI, J.L. and CHIU, C.Y. 2011a. Using a pan-viral microarray assay (Virochip) to screen clinical samples for viral pathogens. *J. Vis. Exp.* **50**:e2536 doi: 10.3791/2536.

CHEN, H.W., WORTMANN, A. and ZENOBI, R. 2007. Neutral desorption sampling coupled to extractive electrospray ionization mass spectrometry for rapid differentiation of biosamples by metabolomic fingerprinting. *J. Mass Spectrom.* **42**:1123–1135.

CHEN, J., CHENG, C., XIA, Y., ZHAO, H., FANG, C. *et al.* 2011b. Lmo0036, an ornithine and putrescine carbamoyltransferase in *Listeria monocytogenes*, participates in arginine deiminase and agmatine deiminase pathways and mediates acid tolerance. *Microbiology* **157**:3150–3161.

CHIANG, M.L., HO, W.L., YU, R.C. and CHOU, C.C. 2008. Protein expression in *Vibrio parahaemolyticus* 690 subjected to sublethal heat and ethanol shock treatments. *J. Food Prot.* **71**:2289–2294.

COOMBES, B.K., WICKHAM, M.E., MASCARENHAS, M., GRUENHEID, S., FINLAY, B.B. *et al.* 2008. Molecular analysis as an aid to assess the public health risk of non-O157 Shiga toxin-producing *Escherichia coli* strains. *Appl. Environ. Microbiol.* **74**:2153–2160.

COOPER, K.L.F. 2011. Pulsed-field gel electrophoresis and other commonly used molecular methods for subtyping of foodborne pathogens, pp. 157–180. In: *Tracing Pathogens in the Food Chain*, S. Brul, P.M. Fratamico and T.A. McMeekin (Eds.). Woodhead Publishing, Oxford, UK.

CORDWELL, S.J., LEN, A.C., TOUMA, R.G., SCOTT, N.E., FALCONER, L. *et al.* 2008. Identification of membrane-associated proteins from *Campylobacter jejuni* strains using complementary proteomics technologies. *Proteomics* **8**:122–139.

DANIELSEN, M., CODREA, M.C., INGVARTSEN, K. L., FRIGGENS, N.C., BENDIXEN, E. *et al.* 2010. Quantitative milk proteomics – host responses to lipopolysaccharide-mediated inflammation of bovine mammary gland. *Proteomics* **10**:2240–2249.

DEBROY, C., FRATAMICO, P.M., ROBERTS, E., DAVIS, M.A. and LIU, Y. 2005. Development of PCR assays targeting genes in O-antigen gene clusters for detection and identification of *Escherichia coli* O45 and O55 serogroups. *Appl. Environ. Microbiol.* **71**:4919–4924.

DELISA, M.P., WU, C.-F., WANG, L., VALDES, J.J. and BENTLEY, W.E. 2001. DNA microarray-based identification of genes controlled by autoinducer 2-stimulated quorum sensing in *Escherichia coli*. *J. Bacteriol.* **183**:5239–5247.

DENG, K., WANG, S., RUI, X., ZHANG, W. and TORTORELLO, M.L. 2011. Functional analysis of *ycfR* and *ycfQ* in *Escherichia coli* O157:H7 linked to outbreaks of illness associated with fresh produce. *Appl. Environ. Microbiol.* **77**:3952–3959.

DESVAUX, M., DUMAS, E., CHAFSEY, I., CHAMBON, C. and HEBRAUD, M. 2010. Comprehensive appraisal of the extracellular proteins from a monoderm bacterium: theoretical and empirical exoproteomes of *Listeria monocytogenes* EGD-e by secretomics. *J. Proteome Res.* **9**:5076–5092.

DIECKMANN, R. and MALORNY, B. 2011. Rapid screening of epidemiologically important *Salmonella enterica* subsp. *enterica* serovars by whole-cell matrix-assisted laser desorption ionization-time of flight mass spectrometry. *Appl. Environ. Microbiol.* **77**:4136–4146.

DONALDSON, J.R., NANDURI, B., BURGESS, S.C. and LAWRENCE, M.L. 2009. Comparative proteomic analysis of *Listeria monocytogenes* strains F2365 and EGD. *Appl. Environ. Microbiol.* **75**:366–373.

DUPUIS, A., HENNEKINNE, J.A., GARIN, J. and BRUN, V. 2008. Protein standard absolute quantification (PSAQ) for improved investigation of staphylococcal food poisoning outbreaks. *Proteomics* **8**:4633–4636.

EISEN, J.A. and C.M. FRASER. 2003. Phylogenomics: intersection of evolution and genomics. *Science* **300**:706–707.

ENG, J., MCCORMACK, A. and YATES, J. 1994. An approach to correlate tandem mass spectral data of peptides with amino acid sequences in a protein database. *J. Am. Soc. Mass. Spectrom.* **5**:976–989.

FADDA, S., ANGLADE, P., BARAIGE, F., ZAGOREC, M., TALON, R. *et al.* 2010. Adaptive response of *Lactobacillus sakei* 23K during growth in the presence of meat extracts: a proteomic approach. *Int. J. Food Microbiol.* **142**:36–43.

FAGERQUIST, C.K., GARBUS, B.R., MILLER, W.G., WILLIAMS, K.E., YEE, E. *et al.* 2010. Rapid identification of protein biomarkers of *Escherichia coli* O157:H7 by matrix-assisted laser desorption ionization-time-of-flight-time-of-flight mass spectrometry and top-down proteomics. *Anal. Chem.* **82**:2717–2725.

FENN, J.B., MANN, M., MENG, C.K., WONG, S.F. and WHITEHOUSE, C.M. 1989. Electrospray ionization for mass spectrometry of large biomolecules. *Science* **246**:64–71.

FIELDS, P.I., FITZGERALD, C. and MCQUISTON, J.R. 2011. Fast and high-throughput molecular typing methods, pp. 81–92. In: *Rapid Detection, Enumeration, and Characterization of Foodborne Pathogens*, J. HOORFAR (Ed.), ASM Press, Washington, DC.

FITZGERALD, C., COLLINS, M., VAN DUYNE, S., MIKOLEIT, M., BROWN, T. *et al.* 2007. Multiplex, bead-based suspension array for molecular determination of common *Salmonella* serogroups. *J. Clin. Microbiol.* **45**:3323–3334.

FOLSOM, J.P. and FRANK, J.F. 2007. Proteomic analysis of a hypochlorous acid-tolerant *Listeria monocytogenes* cultural variant exhibiting enhanced biofilm production. *J. Food Prot.* **70**:1129–1136.

FOURNIER, M.L., GILMORE, J.M., MARTIN-BROWN, S.A. and WASHBURN, M.P. 2007. Multidimensional separations-based shotgun proteomics. *Chem. Rev.* **107**:3654–3686.

FRANZEL, B. and WOLTERS, D.A. 2011. Advanced MudPIT as a next step toward high proteome coverage. *Proteomics* **11**:3651–3656.

FRATAMICO, P.M., BRIGGS, C.E., NEEDLE, D., CHEN, C.-Y. and DEBROY, C. 2003. Sequence of the *Escherichia coli* O121 O-antigen gene cluster and detection of enterohemorrhagic *E. coli* O121 by PCR amplification of the *wzx* and *wzy* genes. *J. Clin. Microbiol.* **41**:3379–3383.

FRATAMICO, P.M., DEBROY, C., STROBAUGH, T.P. and CHEN, C.-Y. 2005. DNA sequence of the *Escherichia coli* O103 O antigen gene cluster and detection of enterohemorrhagic *E. coli* O103 by PCR amplification of the *wzx* and *wzy* genes. *Can. J. Microbiol.* **51**:515–522.

FRATAMICO, P.M., DEBROY, C. and LIU, Y. 2008. Sequencing of the *Escherichia coli* O22 O antigen gene cluster and detection of *E. coli* serogroups O22 and O91 by multiplex PCR assays targeting virulence genes and the *wzy* gene in the respective O-antigen gene clusters. *Food Anal. Methods* **2**:169–179.

FRATAMICO, P.M., YAN, X., DEBROY, C., BRYNE, B., MONAGHAN, A. *et al.* 2010. *Escherichia coli* serogroup O2 and O28ac O-antigen gene cluster sequences and detection of pathogenic *E. coli* O2 and O28ac by PCR. *Can. J. Microbiol.* **56**:308–316.

FRATAMICO, P.M., BAGI, L.K., CRAY JR, W.C., NARANG, N., YAN, X. *et al.* 2011a. Detection by multiplex real-time polymerase chain reaction assays and isolation of Shiga toxin-producing *Escherichia coli* serogroups O26, O45, O103, O111, O121, and O145 in ground beef. *Foodborne Pathog. Dis.* **8**:601–607.

FRATAMICO, P.M., WANG, S., YAN, X., ZHANG, W. and LI, Y. 2011b. Differential gene expression of *E. coli* O157:H7 in ground beef extract compared to tryptic soy broth. *J. Food Sci.* *76*:M79–87.

GEBREYES, W.A. and THAKUR, S. 2011. Phenotypic subtyping of foodborne pathogens, pp. 141–156. In: *Tracing Pathogens in the Food Chain*, S. Brul, P.M. Fratamico, and T.A. McMeekin (Eds.), Woodhead Publishing, Oxford, UK.

GOODRIDGE, L.D., FRATAMICO, P., CHRISTENSEN, L.S., GRIFFITHS, M., HOORFAR, J. *et al.* 2011. Strengths and shortcomings of advanced detection technologies, pp. 15–45. In: *Rapid Detection, Identification, and Quantification of Foodborne Pathogens*, J. HOORFAR (Ed.), ASM Press, Washington, DC.

GRAHAM, C., MCMULLAN, G. and GRAHAM, R.L. 2011. Proteomics in the microbial sciences. *Bioeng. Bugs* **2**:17–30.

GRAHAM, R.L.J., GRAHAM, C. and MCMULLAN, G. 2007. Microbial proteomics: a mass spectrometry primer for biologists. *Microb Cell Fact* **6**:26.

GRIFFIN, T.J., GYGI, S.P., IDEKER, T., RIST, B., ENG, J. *et al.* 2002. Complementary profiling of gene expression at the transcriptome and proteome levels in *Saccharomyces cerevisiae*. *Mol. Cell. Proteomics* **1**:323–333.

GUILBAUD, M., CHAFSEY, I., PILET, M.F., LEROI, F., PREVOST, H. *et al.* 2008. Response of *Listeria monocytogenes* to liquid smoke. *J. Appl. Microbiol.* **104**:1744–1753.

GUNTHER IV, N.W., PANG, H., NUNEZ, A. and UHLICH, G.A. 2010. Comparative proteomics of *E. coli* O157:H7: two-dimensional gel electrophoresis vs. two-dimensional liquid chromatography separation. *Open Proteomics J.* **3**: 26–34.

HENNEKINNE, J.A., BRUN, V., DE BUYSER, M.L., DUPUIS, A., OSTYN, A. *et al.* 2009. Innovative application of mass spectrometry for the characterization of staphylococcal enterotoxins involved in food poisoning outbreaks. *Appl. Environ. Microbiol.* **75**:882–884.

HOWARD, S.L., GAUNT, M.W., HINDS, J., WITNEY, A.A., STABLER, R. *et al.* 2006. Application of comparative phylogenomics to study the evolution of *Yersinia enterocolitica* and to identify genetic differences relating to pathogenicity. *J. Bacteriol.* **188**: 3645–3653.

HUANG, Y.J., TSAI, T.Y. and PAN, T.M. 2007. Physiological response and protein expression under acid stress of *Escherichia coli* O157:H7 TWC01 isolated from Taiwan. *J. Agric. Food Chem.* **55**:7182–7191.

HUEHN, S., LA RAGIONE, R.M., ANJUM, M., SAUNDERS, M., WOODWARD, M.J. *et al.* 2010. Virulotyping and antimicrobial resistance typing of *Salmonella enterica* serovars relevant to human health in Europe. *Foodborne Pathog. Dis.* **7**:523–535.

HUMBLOT, C. and GUYOT, J.-P. 2009. Pyrosequencing of tagged rRNA gene amplicons for rapid deciphering of microbiomes of fermented foods such as pearl millet slurries. Appl. Environ. Microbiol. **75**:4354–4361.

IDEKER, T., THORSSON, V., RANISH, J.A., CHRISTMAS, R., BUHLER, J. *et al.* 2001. Integrated genomic and proteomic analyses of a systematically perturbed metabolic network. *Science* **292**:929–934.

IWASAKI, M., MIWA, S., IKEGAMI, T., TOMITA, M., TANAKA, N. *et al.* 2010. One-dimensional capillary liquid chromatographic separation coupled with tandem mass spectrometry unveils the *Escherichia coli* proteome on a microarray scale. *Anal. Chem.* **82**:2616–2620.

JOFRE, A., CHAMPOMIER-VERGES, M., ANGLADE, P., BARAIGE, F., MARTIN, B. *et al.* 2007. Protein synthesis in lactic acid and pathogenic bacteria during recovery from a high pressure treatment. *Res. Microbiol.* **158**:512–520.

JONES, T.H., JOHNS, M.W. and GILL, C.O. 2008. Changes in the proteome of *Escherichia coli* during growth at 15 degrees C after incubation at 2, 6 or 8 degrees C for 4 days. *Int. J. Food Microbiol.* **124**:299–302.

JOYCE, A.R. and PALSSON, B.Ø. 2006. The model organism as a system: integrating 'omics' data sets. *Nat. Rev. Mol. Cell. Biol.* 7:198–210.

KALB, S.R., GOODNOUGH, M.C., MALIZIO, C.J., PIRKLE, J.L. and BARR, J.R. 2005. Detection of botulinum neurotoxin A in a spiked milk sample with subtype identification through toxin proteomics. *Anal. Chem.* **77**:6140–6146.

KANDPAL, R.P. SAVIOLA, B. and FELTON, J. 2009. The era of 'omics unlimited'. *Biotechniques* **46**:351–355.

KARAS, M. and HILLENKAMP, F. 1988. Laser desorption ionization of proteins with molecular masses exceeding 10,000 daltons. *Anal. Chem.* **60**:2299–2301.

KAWASAKI, S., FRATAMICO, P.M., WESLEY, I.V. and KAWAMOTO, S. 2008. Species-specific identification of campylobacters using PCR-RFLP and PCR targeting the gyrase B gene. *Appl. Environ. Microbiol.* **74**:2529–2533.

KHOO, C.-H., CHEAH, Y.-K., LEE, L.-H., SIM, J.-H., SALLEH, N.A. *et al.* 2009. Virulotyping of *Salmonella enterica* subsp. *enterica* isolated from indigenous vegetables and poultry meat in Malaysia using multiplex-PCR. *Antonie van Leeuwenhoek* **96**:441–457.

KIEHNTOPF, M., MELCHER, F., HANEL, I., ELADAWY, H. and TOMASO, H. 2011. Differentiation of *Campylobacter* species by surface-enhanced laser desorption/ionization-time-of-flight mass spectrometry. *Foodborne Pathog. Dis.* **8** :875–885.

KIEMER, L. and CESARENI, G. 2007. Comparative interactomics: comparing apples and pears. *Trends Biotechnol.* **25**:448–454.

KIM, J., NIETFELDT, J. and BENSON, A.K. 1999. Octamer-based genome scanning distinguishes a unique subpopulation of *Escherichia coli* O157:H7 strains in cattle. *Proc. Natl. Acad. Sci. USA* **96**:13288–13293.

KIM, K., YANG, E., VU, G.P., GONG, H., SU, J. *et al.* 2010. Mass spectrometry-based quantitative proteomic analysis of *Salmonella enterica* serovar Enteritidis protein expression upon exposure to hydrogen peroxide. *BMC Microbiol.* **10**:166.

KIM, S.R., KIM, H.T., PARK, H.J., KIM, S., CHOI, H.J. *et al.* 2009. Fatty acid profiling and proteomic analysis of *Salmonella enterica* serotype Typhimurium inactivated with supercritical carbon dioxide. *Int. J. Food Microbiol.* **134**:190–195.

KOTEWICZ, M.L., JACKSON, S.A., LECLERC, J.E. and CEBULA, T.A. 2007. Optical maps distinguish individual strains of *Escherichia coli* O157:H7. *Microbiol.* **153**:1720–1733.

LAURI, A. and MARIANI, P.O. 2009. Potentials and limitations of molecular diagnostic methods in food safety. *Genes Nutr.* **4**:1–12.

LI, F., ZHAO, Q., WANG, C., LU, X., LI, X.F. *et al.* 2010. Detection of *Escherichia coli* O157:H7 using gold nanoparticle labeling and inductively coupled plasma mass spectrometry. *Anal. Chem.* **82**:3399–3403.

LIPPOLIS, J.D., BAYLES, D.O. and REINHARDT, T.A. 2009. Proteomic changes in *Escherichia coli* when grown in fresh milk versus laboratory media. *J. Proteome Res.* **8**:149–158.

LIU, H., SADYGOV, R.G. and YATES III, J.R. 2004a. A model for random sampling and estimation of relative protein abundance in shotgun proteomics. *Anal. Chem.* **76**:4193–4201.

LIU, R.H., YANG, J., LENIGK, R., BONANNO, J. and GRODZINSKI, P. 2004b. Self-contained, fully integrated biochip for sample preparation, polymerase chain reaction amplification, and DNA microarray detection. *Anal. Chem.* **76**:1824–1831.

LIU, Y. and FRATAMICO, P. 2006. *Escherichia coli* O antigen typing using DNA microarrays. *Mol. Cell Probes* **20**:239–244.

LIU, Y. and REAM, A. 2008. Gene expression profiling of *Listeria monocytogenes* strain F2365 during growth in ultrahigh-temperature-processed skim milk. *Appl. Environ. Microbiol.* **74**:6859–6866.

LOMONACO, S., NUCERA, D., PARISI, A., NORMANNO, G. and BOTERO, M.T. 2011. Comparison of two AFLP methods and PFGE using strains of *Listeria monocytogenes* isolated from environmental and food samples obtained from Piedmont, Italy. *Int. J. Food Microbiol.* **149**:177–182.

LOO, J.A., QUINN, J.P., RYU, S.I., HENRY, K.D., SENKO, M.W. *et al.* 1992. High-resolution tandem mass spectrometry of large biomolecules. *Proc. Natl. Acad. Sci. USA* **89**:286–289.

LOPEZ-CAMPISTROUS, A., SEMCHUK, P., BURKE, L., PALMER-STONE, T., BROKX, S.J. *et al.* 2005. Localization, annotation, and comparison of the *Escherichia coli* K-12 proteome under two states of growth. *Mol. Cell. Proteomics* **4**:1205–1209.

MANNING, S.D., MOTIWALA, A.S., SPRINGMAN, A.C., QI, W., LACHER, D.W. *et al.* 2008. Variation in virulence among clades of *Escherichia coli* O157:H7 associated with disease outbreaks. *Proc. Natl. Acad. Sci. USA* **105**:4868–4873.

MARTINEZ-GOMARIZ, M., HERNAEZ, M.L., GUTIERREZ, D., XIMENEZ-EMBUN, P. and PRESTAMO, G. 2009. Proteomic analysis by two-dimensional differential gel electrophoresis (2D DIGE) of a high-pressure effect in *Bacillus cereus*. *J. Agric. Food Chem.* **57**:3543–3549.

MAZZEO, M.F., SORRENTINO, A., GAITA, M., CACACE, G., DI STASIO, M. *et al.* 2006. Matrix-assisted laser desorption ionization-time of flight mass spectrometry for the discrimination of food-borne microorganisms. *Appl. Environ. Microbiol.* **72**:1180–1189.

MBANDI, E., PHINNEY, B.S., WHITTEN, D. and SHELEF, L.A. 2007. Protein variations in *Listeria monocytogenes* exposed to sodium lactate, sodium diacetate, and their combination. *J. Food Prot.* **70**:58–64.

MIHOUB, F., MISTOU, M.Y., GUILLOT, A., LEVEAU, J.Y., BOUBETRA, A. *et al.* 2003. Cold adaptation of *Escherichia coli*: microbiological and proteomic approaches. *Int. J. Food Microbiol.* **89**:171–184.

MILLER, I., CRAWFORD, J. and GIANAZZA, E. 2006. Protein stains for proteomic applications: which, when, why? *Proteomics* **6**:5385–5408.

MOLLOY, M.P., BRZEZINSKI, E.E., HANG, J., MCDOWELL, M.T. and VANBOGELEN, R.A. 2003. Overcoming technical variation and biological variation in quantitative proteomics. *Proteomics* **3**:1912–1919.

MOREY, J.S., RYAN, J.C. and VAN DOLAH, F.M. 2006. Microarray validation: factors influencing correlation between oligonucleotide microarrays and real-time PCR. *Biol. Proced.* **8**:175–193.

NADAL, A., COLL, A., COOK, N. and PLA, M. 2007. A molecular beacon-based real time NASBA assay for detection of *Listeria monocytogenes* in food products: role of target mRNA secondary structure on NASBA design. *J. Microbiol. Methods* **68**:623–632.

OCHOA, M.L. and HARRINGTON, P.B. 2005. Immunomagnetic isolation of enterohemorrhagic *Escherichia coli* O157:H7 from ground beef and identification by matrix-assisted laser desorption/ionization time-of-flight mass spectrometry and database searches. *Anal. Chem.* **77**:5258–5267.

O'FARRELL, P.H. 1975. High resolution two-dimensional electrophoresis of proteins. *J. Biol. Chem.* **250**:4007–4021.

O'FLAHERTY, S. and KLAENHAMMER, T.R. 2011. The impact of omic technologies on the study of food microbes. *Ann. Rev. Food Sci. Technol.* **2**:353–371.

OGURA, Y., OOKA, T., IGUCHI, A., TOH, H., ASADULGHANI, M. *et al.* 2009. Comparative genomics reveal the mechanism of parallel evolution of O157 and non-O157 STEC enterohemorrhagic *Escherichia coli*. *Proc. Natl. Acad. Sci. USA* **106**:17939–17944.

OLD, W.M., MEYER-ARENDT, K., AVELINE-WOLF, L., PIERCE, K.G., MENDOZA, A. *et al.* 2005. Comparison of label-free methods for quantifying human proteins by shotgun proteomics. *Mol. Cell. Proteomics* **4**:1487–1502.

PALLEN, M.J. and WREN, B.W. 2007. Bacterial pathogenomics. *Nature* **449**:835–842.

PEDRESCHI, R., HERTOG, M., LILLEY, K.S. and NICOLAI, B. 2010. Proteomics for the food industry: opportunities and challenges. *Crit. Rev. Food Sci. Nutr.* **50**:680–692.

PINTO, A., HALLIDAY, C., ZAHRA, M., VAN HAL, S., OLMA, T. *et al.* 2011. Matrix-assisted laser desorption ionization-time of flight mass spectrometry identification of yeasts is contingent on robust reference spectra. *PLoS One* **6**:e25712.

POCSFALVI, G. and SCHLOSSER, G. 2011. Detection of bacterial protein toxins by solid phase magnetic immunocapture and mass spectrometry. *Methods Mol. Biol.* **739**:3–12.

PODLESNY, M., JAROCKI, P., KOMON, E., GLIBOWSKA, A. and TARGONSKI, Z. 2011. LC-MS/MS analysis of surface layer proteins as a useful method for the identification of lactobacilli from the *Lactobacillus acidophilus* group. *J. Microbiol. Biotechnol.* **21**:421–429.

POHL, G. and SHIH, L.-M. 2004. Principle and applications of digital PCR. *Expert Rev. Mol. Diagn.* **4**:41–47.

QUIÑONES, B., SWIMLEY, M.S., TAYLOR, A.W. and DAWSON, E.D. 2011. Identification of *Escherichia coli* O157 by using a novel colorimetric detection method with DNA microarrays. *Foodborne Pathog. Dis.* **8**:705–711.

RASOOLY, A. and HEROLD, K.E. 2008. Food microbial pathogen detection and analysis using DNA microarray technologies. *Foodborne Pathog. Dis.* **5**:531–550.

REID, G.E. and MCLUCKEY, S.A. 2002. 'Top down' protein characterization via tandem mass spectrometry. *J. Mass Spectrom.* **37**:663–675.

REYNISSON, E., RUDI, K., MARTEINSSSON, V.TH., NAKAYAMA, J., SAKAMOTO, N. *et al.* 2011. Automated and large-scale characterization of microbial communities in food production. In: *Rapid Detection, Identification, and Quantification of Foodborne Pathogens*, J. Hoorfar (Ed.), ASM Press, Washington, DC.

RIEDEL, K. and LEHNER, A. 2007. Identification of proteins involved in osmotic stress response in *Enterobacter sakazakii* by proteomics. *Proteomics* **7**:1217–1231.

ROSEF, O., JOHNSEN, G., STØLEN, A. and KLÆBOE, H. 2008. Similarity of *Campylobacter lari* among human, animal, and water isolates in Norway. *Foodborne Pathog. Dis.* **5**:33–39.

ROSS, P.L., HUANG, Y.N., MARCHESE, J.N., WILLIAMSON, B., PARKER, K. *et al.* 2004. Multiplexed protein quantitation in *Saccharomyces cerevisiae* using amine-reactive isobaric tagging reagents. *Mol Cell Proteomics* **3**:1154–1169.

SAKAMOTO, N., TANAKA, S., SONOMOTO, K. and NAKAYAMA, J. 2011. 16S rRNA pyrosequencing-based investigation of the bacterial community in nukadoko, a pickling bed of fermented rice bran. *Int. J. Food Microbiol.* **144**:352–359.

SÁNCHEZ, B., CABO, M.L., MARGOLLES, A. and HERRERA, J.J. 2010. A proteomic approach to cold acclimation of *Staphylococcus aureus* CECT 976 grown at room and human body temperatures. *Int. J. Food Microbiol.* **144**:160–168.

SÁNCHEZ, G., ELIZAQUÍVEL, P. and AZNAR, R. 2012. A single method for recovery and concentration of enteric viruses and bacteria from fresh-cut vegetables. *Int. J. Food Microbiol.* **152**:9–13.

SCALLAN, E., HOEKSTRA, R.M., ANGULO, F.J., TAUXE, R.V., WIDDOWSON, M.-A., *et al.* 2011. Foodborne illness acquired in the United States – major pathogens. *Emerg. Infect. Dis.* **17**:7–15.

SCHLOSSER, G., KACER, P., KUZMA, M., SZILAGYI, Z., SORRENTINO, A. *et al.* 2007. Coupling immunomagnetic separation on magnetic beads with matrix-assisted laser desorption ionization-time of flight mass spectrometry for detection of staphylococcal enterotoxin B. *Appl. Environ. Microbiol.* **73**:6945–6952.

SCHULZE, W.X. and USADEL, B. 2010. Quantitation in mass-spectrometry-based proteomics. *Annu. Rev. Plant Biol.* **61**:491–516.

SEKIYAMA, Y. and KIKUCHI, J. 2007. Towards dynamic metabolic network measurements by multi-dimensional NMR-based fluxomics. *Phytochemistry* **68**:2320–2329.

SONCK, K.A., KINT, G., SCHOOFS, G., VANDER WAUVEN, C., VANDERLEYDEN, J. *et al.* 2009. The proteome of *Salmonella* Typhimurium grown under *in vivo*-mimicking conditions. *Proteomics* **9**:565–579.

SOPWITH, W., BIRTLES, A., MATTHEWS, M., FOX, A., GEE, S. *et al.* 2010. Investigation of food and environmental exposures relating to the epidemiology of *Campylobacter coli* in humans in Northwest England. *Appl. Environ. Microbiol.* **76**:129–135.

STEEN, H. and MANN, M. 2004. The ABC's (and XYZ's) of peptide sequencing. *Nat. Rev. Mol. Cell. Biol.* **5**:699–711.

SWAMINATHAN, B., BARRETT, T.J., HUNTER, S.B., TAUXE, R.V. and CDC PULSENET TASK FORCE. 2001. PulseNet: the molecular subtyping network for foodborne bacterial disease surveillance, *United States*. *Emerg. Infect. Dis.* **7**:382–389.

TANGWATCHARIN, P., CHANTHACHUM, S., KHOPAIBOOL, P. and GRIFFITHS, M.W. 2006. Morphological and physiological responses of *Campylobacter jejuni* to stress. *J. Food. Prot.* **69**:2747–2753.

TAO, D., ZHANG, L., SHAN, Y., LIANG, Z. and ZHANG, Y. 2011. Recent advances in micro-scale and nano-scale high-performance liquid-phase chromatography for proteome research. *Anal. Bioanal. Chem.* **399**:229–241.

TORPDAHL, M., SØRENSEN, G., LINDSTEDT, B.-A. and NIELSEN, E.M. 2007. Tandem repeat analysis for surveillance of human *Salmonella* Typhimurium infections. *Emerg. Infect. Dis.* **13**:388–395.

UHLICH, G.A., GUNTHER IV, N.W., BAYLES, D.O. and MOSIER, D.A. 2009. The CsgA and Lpp proteins of an *Escherichia coli* O157:H7 strain affect HEp-2 cell invasion, motility, and biofilm formation. *Infect. Immun.* **77**:1543–1552.

UNLU, M., MORGAN, M.E. and MINDEN, J.S. 1997. Difference gel electrophoresis: a single gel method for detecting changes in protein extracts. *Electrophoresis* **18**:2071–2077.

VAN VEEN, S.Q., CLAAS, E.C. and KUIJPER E.J. 2010. High-throughput identification of bacteria and yeast by matrix-assisted laser desorption ionization-time of flight mass spectrometry in conventional medical microbiology laboratories. *J. Clin. Microbiol.* **48**:900–907.

VIDOVIC, S., MANGALAPPALLI-ILLATHU, A.K. and KORBER, D.R. 2011. Prolonged cold stress response of *Escherichia coli* O157 and the role of *rpoS*. *Int. J. Food Microbiol.* **146**:163–169.

WANG, F., JIANG, L. and GE, B. 2012. Loop-mediated isothermal amplification assays for detecting Shiga toxin-producing *Escherichia coli* in ground beef and human stools. *J. Clin. Microbiol.* **50**:91–97.

WANG, G., WU, W.W., ZENG, W., CHOU, C.L. and SHEN, R.F. 2006. Label-free protein quantification using LC-coupled ion trap or FT mass spectrometry: Reproducibility, linearity, and application with complex proteomes. *J. Proteome Res.* **5**:1214–1223.

WANG, L., ROTHEMUND, D., CURD, H. and REEVES, P.R. 2003. Species-wide variation in the *Escherichia coli* flagellin (H-antigen) gene. *J. Bacteriol.* **185**:2936–2943.

WANG, X.W., ZHANG, L., JIN, L.Q., JIN, M., SHEN, Z.Q. *et al.* 2007. Development and application of an oligonucleotide microarray for the detection of food-borne bacterial pathogens. *Appl. Microbiol. Biotechnol.* **76**:225–233.

WASSENAAR, T.M. 2011. Virulotyping of foodborne pathogens, pp. 342–357. In: *Tracing Pathogens in the Food Chain*, S. Brul, P.M. Fratamico and T.A. McMeekin (Eds.), Woodhead Publishing, Oxford, UK.

WEBER, A., KOGL, S.A. and JUNG, K. 2006. Time-dependent proteome alterations under osmotic stress during aerobic and anaerobic growth in *Escherichia coli. J. Bacteriol.* **188**:7165–7175.

WESTERMEIER, R. and SCHICKLE, H. 2009. The current state of the art in high-resolution two-dimensional electrophoresis. *Arch. Physiol. Biochem.* **115**:279–285.

YU, S., CHEN, W., WANG, D., HE, X., ZHU, X. *et al.* 2010. Species-specific PCR detection of the food-borne pathogen *Vibrio parahaemolyticus* using the *irgB* gene identified by comparative genomic analysis. *FEMS Microbiol. Lett.* **307**:65–71.

ZHU, X., LIU, W., LAMETSCH, R., AARESTRUP, F., SHI, C. *et al.* 2011. Phenotypic, proteomic, and genomic characterization of a putative ABC-transporter permease involved in *Listeria monocytogenes* biofilm formation. *Foodborne Pathog. Dis.* **8**:495–501.

25

Next generation of predictive models

C. Pin, A. Metris and J. Baranyi, Institute of Food Research, UK

DOI: 10.1533/9780857098740.6.498

Abstract: In this chapter, we endeavour to predict the trends in predictive microbiology that are going to shape its development as a multidisciplinary field. We integrate the most commonly used, population and single-cell models of bacterial kinetics into a top-down framework. We predict that modellers will need to face various complexities induced by interactions at different levels: between food and microorganism, between cells and species and between molecular elements at the intracellular level. Hence, advances in the area of complex systems (e.g. network science, systems biology and stochastic modelling) will have a significant effect on the future development of predictive microbiology.

Key words: predictive microbiology, food microbiology, dynamic models, stochastic models, genome scale models, metabolic models.

25.1 Introduction

During the last 30 years, predictive microbiology has developed into a discipline in its own right. This has been facilitated by dramatic developments in information technology (IT) and numerical and computational methods, which have revolutionised life sciences primarily via the new tools developed within bioinformatics and systems biology.

One of the most obvious drivers behind these developments was the sudden deluge of data, and the increasingly more effective acquisition, organisation, transfer and processing of data. Once measurements have been systematically recorded in suitable databases, programmes and mathematical models give a much better understanding from the data than has ever been possible before.

Excellent monographs (McMeekin *et al.*, 1993; McKellar and Lu, 2003) have been published about the early periods of predictive microbiology, which was the 1980s and 1990s. During this time, concepts such as primary and secondary models, bias and accuracy, uncertainty and variability, to mention just a few, were

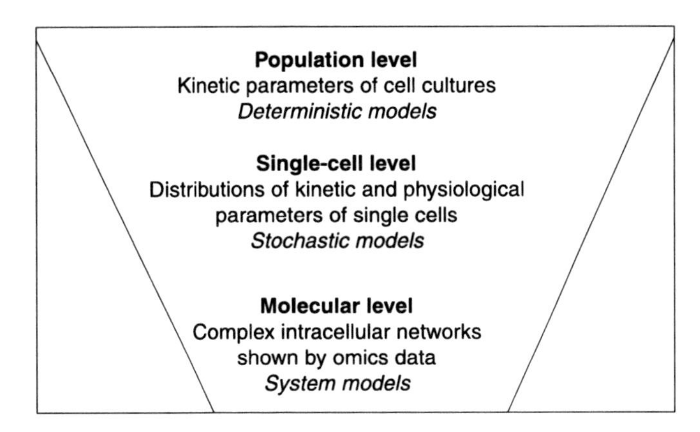

Fig. 25.1 Top-down framework integrating modelling approaches to predictive microbiology.

developed. Today, these terms are part of the standard vocabulary of predictive microbiology used in the relevant literature. More importantly, they set a trend for future developments in parallel with the increasing depth and efficiency of data acquisition. In this chapter, we integrate these modelling trends into a top-down framework (see Fig. 25.1) and extrapolate them to the future.

25.2 Models at the cell-population level

25.2.1 Dynamic models

The first predictive models in food microbiology took a simple approach: certain parameters of microbial responses to a given environment were estimated at the cell-population level by means of primary models, then the effects of the environment on these parameters were modelled by multivariate functions of the most important environmental variables (secondary models). Key to the application of these models was the question of which parameters of the primary model were to be considered in the secondary model and which environmental factors were the most dominant in terms of their effect on the primary model. It has been widely agreed that the maximum specific growth rate (the maximum slope of the curve fitted to the natural logarithm of cell concentration vs. time) is the most important primary parameter, and the temperature is the most crucial explanatory variable in the secondary model (McMeekin *et al.*, 1993).

A generation of dynamic primary models were introduced by Baranyi and Roberts (1994). These models had a more mechanistic approach in the sense that they did not model the state of the system (identified by the actual bacterial concentration) directly; instead the instantaneous change of the system was described as a function of its state at that moment. The main advantage of such an approach, formulated by differential equations, is that its extension to situations where the environment changes with time during the measured response (in a dynamic environment) is relatively straightforward. Its disadvantage is that working with differential equations (e.g. for regression) with commonly used spreadsheets is much more difficult than working with the static models described by algebraic functions. Still, in our opinion, it is likely that in future the predictive tools of food microbiology will be based on dynamic models with variables that quantify:

- the concentration level of a certain foodborne organism
- the level of a metabolic product of interest
- the ability of cells to survive, adapt and proliferate in a given environment (their physiological state).

An important aspect of modelling is the validity region of the model. For example, in terms of temperature, the region of a growth model of a pathogenic organism is approximately 0–45 °C, the analogous maximum region for thermal inactivation models is approximately 45–100 °C. An unsolved problem that the next generation models will need to address is the question of the boundary and crossover between these regions. The question is this: how can we predict the overall microbial response (e.g. probability of survival, cell concentration, total metabolites produced, etc.) as a function of arbitrary temperature profiles, while changing from the growth region to the inactivation region, and then possibly back to the growth region again. The risk assessment of minimally processed foods is a good example that demonstrates the need to predict microbial responses during dynamically changing temperature profiles. Proper quantification of the physiological state of cells (see the last point in the list above) will play a crucial role in developing appropriate methods for such dynamic modelling.

An example of such a scenario was reported by Le Marc *et al.* (2008). Figure 25.2 shows the variation in the concentration of *Clostridium perfringens* in bulk turkey after the temperature increased to 80 °C, then decreased slowly back

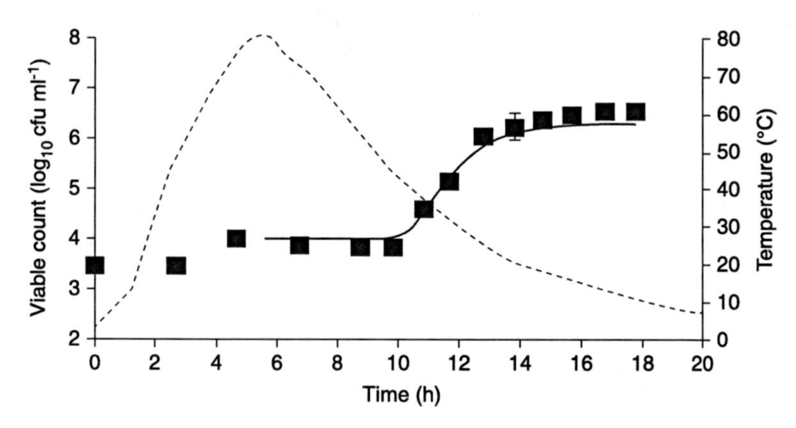

Fig. 25.2 Growth of *Clostridium perfringens* (symbols) during the cooling of bulk turkey from 80 °C to 10 °C (dashed line).

below 10 °C. The dynamic temperature profile affected the physiological state of the cells, which determined the duration of the lag time during cooling. Note that existing software tools handling such situations (e.g. the Perfringens Predictor on www.combase.cc) restrict the dynamics of the temperature profiles; for example, users cannot get predictions if there are multiple crossovers between growth and inactivation regions. The problem becomes simpler if the lag time is not considered and only pure inactivation or growth processes are modelled. An example of this was given by Pin *et al.* (2010), who combined logistic regression models for the probability of growth with dynamic models for the rates of growth and inactivation of *Salmonella* Typhimurium, when the environment varied from growth to no growth conditions.

25.2.2 The effect of microbial competition

Competition (or coexistence) models focus on a more complex but also more realistic scenario. The most frequent types of interactions between bacterial populations in a co-culture can be put into two very general categories.

The first category consists of interactions caused by modification of the medium where the populations coexist. The medium can be modified by the production of substances inhibiting growth, such as organic acids or bacteriocins. End metabolites can also modify the physical-chemical characteristics of the medium such as the pH. In addition, essential nutrients could be completely consumed by some species. This type of microbial competition has been modelled using systems of differential equations that account for the modification of the medium by one or more species. Such modification can have an effect on the species itself and on the other species present (Breidt and Fleming, 1998; Janssen *et al.*, 2006).

The second category consists of interactions involving the Jameson effect, which is the simultaneous cessation of growth of all bacterial species when the dominant bacterial population reaches its stationary phase (Jameson, 1962). The cessation of the growth of all populations could result from the competition for a common limiting nutrient resource. The Jameson effect is frequently observed in food products (Ross *et al.*, 2000), but our understanding regarding its mechanistic reasons is rather limited. The phenomenon was empirically modelled by Gimenez and Dalgaard (2004) and Le Marc *et al.* (2009) using Lotka–Volterra type models, extended with some new empirical parameters accounting for the strength of the interactions between the species in the mixed culture. The model was able to predict the cessation of growth of a species well before it would have reached its maximum population density (Dens *et al.*, 1999).

25.2.3 Shift from deterministic to stochastic models

Deterministic predictive models are valid for homogeneous cell populations, for which bacterial responses are measurable and reproducible as long as the (averaged out) responses are produced by at least hundreds of cells. This means that the variability of individual cells diminishes due to the population effect.

When using deterministic models, which discard variability among single cells, the generated predictions are valid only for the responses of homogeneous, relatively large, cell populations. The inevitable, stochastic error of the predictions has nothing to do with the stochastic diversity of single cells but it can be attributed to: (i) the variability of the environmental conditions, (ii) the inherent biological variability of the bacterial responses and (iii) model error, where the mathematical equations used are not accurate.

Food processing aims to ensure food safety while preserving quality and frequently results in very low contamination levels of foods with few bacterial cells. The outcome of such contamination is greatly affected by the variability among single cells. The demand for predictive tools based on single-cell kinetics resulted in the development of stochastic models. A discussion of such models, where the probability is defined for single cells, follows.

25.3 Models at the single-cell level

25.3.1 Relationship between single cells and populations

A recent advance in predictive microbiology is stochastic modelling of the individual cells within a bacterial population. The integration of experimentation with mathematical models describing single-cell behaviour has revealed novel features of bacterial behaviour that could not have been discovered by analysing the cell population.

Lag phase has been a frequently investigated topic at the single-cell level. This is because the apparent lag time of a population is affected by single-cell variability when the initial concentration of bacteria is very low (Augustin *et al.*, 2000; Pin and Baranyi, 2006). Theoretical work comparing population and single-cell parameters (Baranyi and Pin, 2001; Baranyi, 1998; Kutalik *et al.*, 2005a) proved that if cells of a homogeneous culture can be in two states only – the lag phase or the exponential phase – then the lag time observed at the population level (defined via the geometry of the population growth curve, namely as the intersection between the inoculum level and the extension of the exponential phase) is shorter than the average of the single-cell lag times within that population. This result was derived by comparing stochastic models for single cells with both stochastic and deterministic models for the population. This model was a two-state stochastic birth process, based on the assumption that, after the first division, the cells are in the exponential phase, i.e. their generation times (the time for the cell cycle from birth to division) follow the same probability distribution regardless of whether the cells are in the first, second or in further successive generations. We call this the two-state model.

It is important to clarify that there is no known signal that would indicate the end of the lag phase for a single cell. What can be measured for a single cell is the time to the first division, which includes both its physiological lag period and the first generation time. If the traditional, geometric lag definition is used (i.e. the single-cell lag time ends at the intercept with the time axis of the extension of the exponential phase of the logarithm of the single-cell-generated subpopulation vs.

time growth curve), then the estimated lag time is affected by both the first generation time and the stochastic kinetics of the successors of the original cell. This is why Baranyi *et al.* (2009) distinguished between physiological and geometrical lag for a single cell; the latter is affected by the shape of the growth curve generated by the successors of the cell rather than being a property of the original cell. For example, both negative and positive geometrical lags can be obtained for situations when in fact there was no lag but simply the time to the first division was random, following the same distribution as in the exponential phase. However, if the physiological lag time does exist and is much longer than the subsequent generation times in the exponential phase, then the geometrical lag can be used as a surrogate for the physiological lag.

The distinction between geometric and physiological lag is very important in situations when, after the first division, the cells are not yet in the exponential phase. This was proven by the image analysis technique developed by Elfwing *et al.* (2004), which allowed the observation of the growth and division of large numbers of individual bacterial cells. Using this technique, Pin and Baranyi (2006) found that cells needed to divide several times to reach the exponential phase. Other studies have also demonstrated that the time intervals between successive divisions decrease gradually after the first division (Elfwing *et al.*, 2004; Pin and Baranyi, 2006, 2008; Metris *et al.*, 2005; Kutalik *et al.*, 2005b; Shachrai *et al.*, 2010). The cell cycles of the first, second, third and sometimes even the fourth generations are longer than those in the exponential phase, which is why, in real situations, the population lag time is in fact longer than the average physiological lag time of the initial single cells.

Figure 25.3 compares the lag time of a cell population, dividing according to the two-state birth process model, with that of a growing population where the

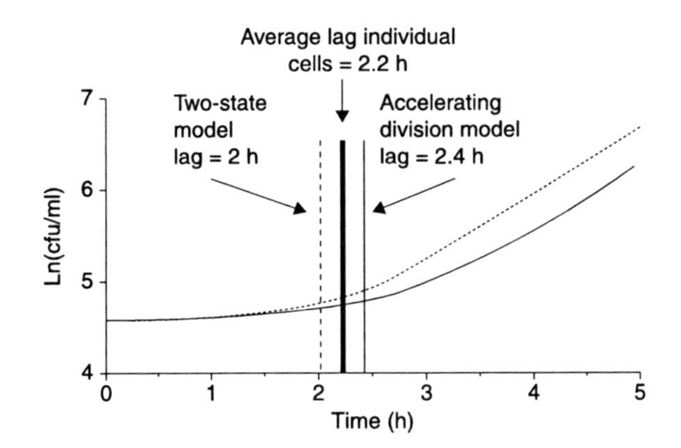

Fig. 25.3 Simulation results: the average lag time of single cells within a homogeneous population (2.2 h) was longer than the population lag time (2.0 h; dashed line) if the cells entered the exponential growth phase immediately after the lag. However, the average lag time of single cells within a population (2.2 h) was shorter than the population lag time (2.4 h; continuous line) if the cells accelerate their division cycles progressively.

distribution of successive division times is not constant: after the relatively long first division time, cell division progressively accelerates over successive cycles, just as experimentally measured (Pin and Baranyi, 2008). We call the latter the accelerating division model. The figure shows that the average single-cell lag time (2.2 h) is longer than the population lag time (2.0 h) defined by the two-state model, and shorter than the population lag time defined by the accelerating division model (2.4 h).

These findings indicate that a necessary step forward in predictive microbiology is to account for the stochastic dynamism of bacterial populations explicitly and a refined definition should be used for the lag time, to make bacterial growth models more sophisticated. Several studies have suggested that the response of a population and its ability to adapt to sudden changes in environmental conditions efficiently may be linked to its phenotypic or non-genetic heterogeneity (Veening *et al.*, 2008; Sumner and Avery, 2002). Population models do not consider this heterogeneity but describe average bacterial behaviour. However, microbes exhibit surprisingly diverse behaviour and even when one is interested only in their average dynamics, simple population models may result in incorrect predictions (McAdams and Arkin, 1998).

25.3.2 The effect of history on single cells

As has been seen, the relationship between the population lag time and the lag time of the single cells within that population is not straightforward. Therefore, it is difficult to infer the properties of single cells by parametric methods, once observations have been made at the population level only. This is why the variability of single-cell responses to stress in a contemporary environment (Francois *et al.*, 2005; George *et al.*, 2008; Li *et al.*, 2006; Metris *et al.*, 2003, 2006), over their history (Aguirre *et al.*, 2011; D'Arrigo *et al.*, 2006; Metris *et al.*, 2008; Niven *et al.*, 2008; Smelt *et al.*, 2002, 2008) or both combined (Guillier and Augustin, 2006; Guillier *et al.*, 2005; Stringer *et al.*, 2009; Webb *et al.*, 2007) has been intensively studied for various pathogens. Pin and Baranyi (2008) showed that the first division times were longer for cells inoculated from the late stationary phase than for cells inoculated from the early stationary phase, while the following division cycles seemed to be unaffected by the previous history. Adaptation to the current growth environment followed the accelerating division model whether the cells were inoculated from the early or late stationary phase.

Note that these results needed single-cell measurements. From population measurements, only complex Bayesian inference methods were able to derive properties of single cells (Malakar and Barker, 2008).

In the same way as for the population level, primary and secondary models can be defined also for single-cell lag times (Francois *et al.*, 2005). The primary model uses the probability distribution of the lag time while the secondary model uses the relationship between the parameters of that distribution and the previous and/or contemporary environmental conditions. As a general rule, the more stringent the stress either in the history or in the contemporary environment, the longer and more spread the distribution of the lag times of a single cell. For instance, the variance and average of single-cell lag times were greater in heated cells, with this effect more pronounced at higher temperatures, even for treatments with identical bactericide effects (Elfwing *et al.*, 2004; Kutalik *et al.*, 2005b; D'Arrigo *et al.*, 2006). It has been reported that as the stress becomes more stringent the probability distributions of the lag time tend to be skewed to the right (like the lognormal or gamma distributions), although there is no consensus on which distribution function best fits the data (McKellar and Hawke, 2006).

Only a few secondary models have been proposed so far and most assume a constant coefficient of variation for a specific stress (George *et al.*, 2008; Guillier and Augustin, 2006; Metris *et al.*, 2006, 2008; Niven *et al.*, 2008; also see Table 25.1). This assumption reduces the number of distribution-identifying parameters and increases the robustness of the model when used for regression or prediction. However, the coefficient of variation does vary greatly (from ca. 0.02 to more than 1) from one type of stress to another (Aguirre *et al.*, 2011; Francois *et al.*, 2005). These results were typically obtained from indirect measurements, such as optical density methods, where the distribution of the lag time was identified with that of the detection time, i.e. the time needed for the subpopulation generated by an initial single cell to reach a detectably high cell concentration. Confidence in the findings for the individual lag times is inevitably small. Moreover, the distribution in question is very sensitive to the population growth rate (Metris *et al.*, 2003), to the measurement method (e.g. how the first division is defined by image analysis) and to the model used to fit these distributions (for example, whether the selected distribution has a shift parameter or not; see Aguirre *et al.*, 2011). Hence currently there is no consolidated agreement on the most suitable distribution.

Single-cell lag times have been modelled as a function of the probability of growth in two studies. Metris *et al.* (2008) found that the F-value concept can be extended to sublethal injuries, too, but instead of the log-kill, the recovery time after the heat treatment should be used to quantify its effectiveness. The authors also found a simple linear relationship between the mean and the standard deviation of the lag time distribution and the decrease in viable cell number due to the treatment.

However, a systematic relationship between the distribution of the single-cell lag times and the probability of growth is not a general rule. It has been reported for other stresses combined and applied successively that the relationship between the distribution of single-cell lag times and the inactivation parameters is not straightforward (Guillier and Augustin, 2006; Dupont and Augustin, 2011). Some other approaches, including the concept of work to be done and relative lag times when modelling single-cell lag times, are summarised in Table 25.1.

For spores, the process is more complicated as they undergo several transformations to become vegetative cells before division, which include germination, emergence and elongation (Rolfe *et al.*, 2012). For *Clostridium botulinum*, no correlation was found between germination stages (Stringer *et al.*, 2005) and hence modelling requires further understanding of the physiology of

Table 25.1 Published information on the relationship between the parameters of the distributions for the single-cell lag times and the previous and/or current environmental conditions

Bacteria	Current environmental stress	Previous stress (history)	Primary model (distribution)	Secondary model	Coefficient of variation	Reference
Listeria	Temperature, pH, water activity	NA	Weibull	Non-zeros mean lag time fitted to an hyperbolic surface. Variance assumed proportional to the mean.	Variable; 0.41 ± 0.27	Standaert *et al.*, 2007
Listeria	Various	Various	Extreme value type II	Coefficient of variation independent of stress and environmental conditions.	0.63 ± 0.12	Guillier and Augustin, 2006
Listeria	Acetic acid	NA	Gamma	Constant shape parameter, logarithm of scale parameter increases linearly with undissociated acid concentration.	0.5 (assumed)	Metris *et al.*, 2008
Listeria	Lactic acid, acetic acid and sorbic acid	Stationary phase vs. exponential phase	Gamma	Constant shape parameter, average of work to be done increases linearly with undissociated acid concentration.	0.5 (assumed)	George *et al.*, 2008
Bacillus spores	NA	Sublethal heat treatment	Weibull	Logarithm of delay in germination linear with temperature and length of heat treatment; rate of germination linear with temperature and length of heat treatment.	NA	Smelt *et al.*, 2008
Listeria	NA	Sublethal to lethal heat treatment	Shifted gamma (shift is repair time, distribution is adjustment time)	Repair time follows the F-value extended to sublethal heat treatment; constant shape parameter and mean of the adjustment increase with severity of sublethal heat treatment becoming constant for lethal treatments.	0.68 ± 0.08	Metris *et al.*, 2008
Enterococcus, Listeria, Salmonella, Pseudomonas	NA	Electron beam irradiation	NA	Linear relationship between the mean/standard deviation of individual lag time and inactivation.	Variable; 0.018–0.59	Aguirre *et al.*, 2011

the cell, which may be found at the molecular level. For *Bacillus subtilis*, a model was proposed by Smelt *et al.* (2008) for the germination times after heat treatment (see Table 25.1).

25.3.3 Implementation of the models

To be able to apply single-cells models in practice, as with population models, we need to understand how well the distributions of the single cells measured in culture media describe distributions in foods. To the best of our knowledge, no systematic studies have been undertaken so far although a good correlation between culture medium and food has been shown in particular cases (Aguirre *et al.*, 2011; D'Arrigo *et al.*, 2006).

To implement these models in a user-friendly way, as with population models, the data will need to be collected into databases so that a large pool of data is available for parameter estimation and model validation. To model the lag phase or a cell's physiological state, the previous environment needs to be recorded as well as the contemporary environment. Existing modelling platforms such as ComBase require significant development to accommodate the required information. Single-cell models are not implemented in user-friendly software yet but particular sets of data can already be integrated in risk assessments (Augustin, 2011; Barker *et al.*, 2005). Although single-cell models are more complex and have a better predictive performance than population models in many cases they are not sufficient. Certain bacterial responses to food environments can be modelled only by understanding and imbedding, in the model, the molecular mechanisms underlying the response. Thus, further steps towards molecular levels are required.

25.4 Molecular level

25.4.1 Modelling gene expression

Many studies on bacterial responses to the environment focus on individual phenotypes rather than developing a more comprehensive cell-wide picture. Holistic approaches investigate systems and processes as a whole without focussing on their individual elements in isolation (Dennis *et al.*, 2004; Kell, 2004). Nowadays, these approaches are being applied to understand bacterial responses to environmental stimuli. New modelling approaches are required, which can handle the increased complexity. Many investigations aiming to exploit the large datasets available from high-throughput parallel assay technologies recognise the suitability of networks for modelling complex systems. Very few modelling approaches meet the important goal of integrating all the layers that can be derived from a genome-scale response. By integrating high-throughput and computational data in bacterial networks, biologists will be able to reconcile heterogeneous data types, find inconsistencies and generate systematic hypotheses.

Network science is widely applied in many different fields (Barabási, 2002). The study of networks has a long tradition in graph theory, discrete mathematics, sociology, communications, bibliometrics, scientometrics, webometrics, cybermetrics, physics and, recently, biology (Oltvai and Barabási, 2002). It has already been used to model the regulatory network of *Escherichia coli* (Balázsi *et al.*, 2005), based on the publicly available network of *E. coli* MG1655, which originally contained 418 operons and 519 interactions. Protein-protein interactions in *E. coli* have also been identified and compiled in a large network (Arifuzzaman *et al.*, 2006; Su *et al.*, 2008). More generally, the construction of metabolic networks in bacteria has been shown to be a valuable tool for elucidating the components and pathways of biological processes (Jeong *et al.*, 2000; Covert *et al.*, 2004; Guimera and Nunes Amaral, 2005; Ouzounis and Karp, 2000).

A bipartite partially directed genome-scale network (Fig. 25.4) for *E. coli* integrating information on signalling, metabolic pathways, transcriptional control and specific cellular activities has been applied to model the molecular responses occurring during the lag phase, in relation to the duration of the immediately preceding stationary phase (Pin *et al.*, 2009). In this work, the network of genes differentially transcribed in early stationary phase cells had a larger number of nodes and connections than the network of starved cells for long periods during the stationary phase. This contradicted the anticipated greater transcriptional activity in old cells as a mechanism for counteracting the more severe senescence during the stationary phase. In addition, network properties, structure and

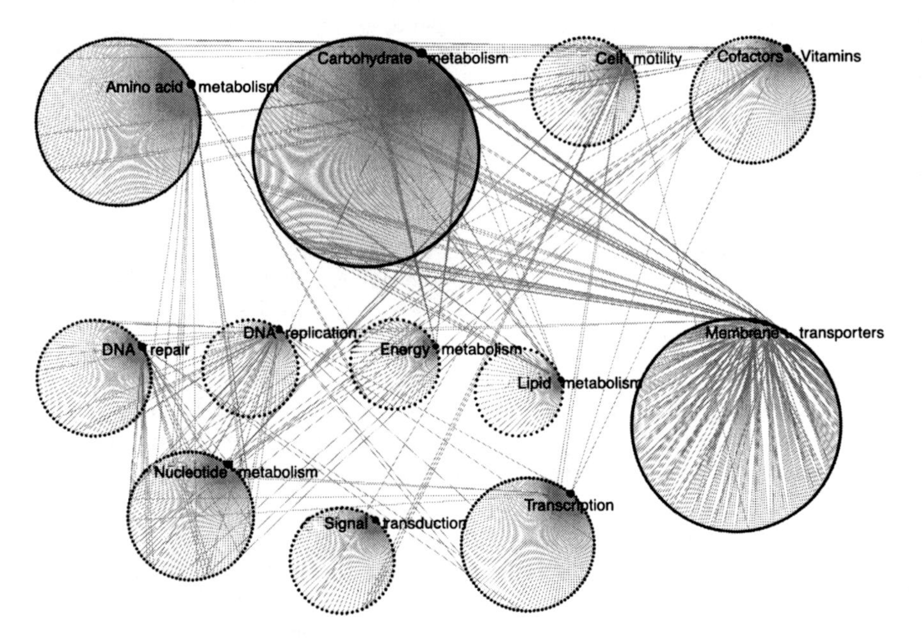

Fig. 25.4 Partial representation of the bipartite genome-scale network of *Escherichia coli* with nodes grouped according to cellular functional categories.

composition did not change during the lag phase (Pin *et al.*, 2009). This means that the transcription pattern holds fairly constant during the lag period for both cultures, unlike other intracellular activities. This result has been corroborated by a recent study on gene transcription during the lag phase of *Salmonella* Typhimurium, in which lag phase profiles showed similarities and clustered together while they were distinct from stationary and exponential phase profiles (Rolfe *et al.*, 2012). This study demonstrated that senescence is not wholly reversed during the lag phase; rather cells start the division cycle by adopting a different metabolic strategy depending on their initial condition. The lack of up-regulation of aerobic respiration during the lag phase of cells kept for 17 days in the stationary phase showed poor adaptation to the new aerobic conditions of a fresh medium and this less efficient energy production strategy resulted in longer lag times than those measured in young cells respiring aerobically (Pin *et al.*, 2009). Molecular mechanisms underlying the effect of history on bacterial responses to current conditions have been reported in a study in which the transcriptional profile of *Salmonella* Typhimurium was analysed during, as well as immediately and 30 minutes after, heat shocks (Pin *et al.*, 2012). This study demonstrated that the transcriptional response of *S.* Typhimurium to heat shock was hysteretic, i.e. it is a process exhibiting memory. Such systems switch between two distinct stable steady states (bistability), and switching from one state to the other happens when a stimulus exceeds a threshold. Once switched, the system remains at that steady state until the stimulus decreases to a level below the original switching level. In between these two switching stimulus levels, the state of the system depends on the previous history (Markevich *et al.*, 2004). Most of the genes induced by the heat shock in *Salmonella* Typhimurium remained induced 30 minutes after resetting to the original temperature (Pin *et al.*, 2012).

25.4.2 Modelling metabolic networks

A reconstruction of the metabolic network of whole cells targets the full understanding of the molecular mechanisms of a particular organism in a given environmental niche. Metabolic networks can provide quantitative predictions at the molecular level to feed single-cell and population models. The metabolic network at the genomic scale of some foodborne pathogens has recently been reconstructed (Baumler *et al.*, 2011; Metris *et al.*, 2011; Thiele *et al.*, 2011) and with techniques such as constraint-based analysis, the growth rate of *E coli* can be predicted as a function of the rates of uptake of chemical entities (Feist *et al.*, 2007). Indeed, in a continuous reactor, it has been shown that it is sufficient to assume that the fluxes of the metabolic reactions in the steady state optimise biomass production with some given constraints on the uptakes of chemicals. In mathematical terms, the stoichiometric matrix containing all stoichiometric coefficients for each chemical involved in the metabolic network and the vector of the fluxes define an underdetermined system of equations. The system is underdetermined because the number of fluxes exceeds the number of metabolites; however, as each flux is bounded, the solution space is convex. Whilst the growth rate can be predicted accurately, as the system is underdetermined, there is more than one solution for each flux (Kauffman *et al.*, 2003). Measurements with ^{13}C have showed that the actual fluxes are contained in the solution space (Palsson, 2006). In other experimental conditions, such as in a batch culture (Fischer and Sauer, 2003; Schuetz *et al.*, 2007), for the Gram-positive bacterium *Lactobacillus plantarum* (Teusink *et al.*, 2006) or under osmotic stress (Metris *et al.*, 2012), it has been proposed that the biomass is not optimised. Hence alternative objective functions must be sought or additional levels of data integrated into the models. Hierarchical analysis, or the integration of additional rules like regulatory rules, which can also be used to determine potential antimicrobials (Lee *et al.*, 2009; Metris *et al.*, 2011), has been attempted in different ways (Brul and Westerhoff, 2007). A different way of analysing the metabolic network has been proposed in a study of elementary flux modes (Schuster *et al.*, 2000); instead of applying constraints to the systems, the biochemical pathways are decomposed into sets of reactions that follow a set of mathematical rules. This has not been used quantitatively to predict the growth rate so far but it may be a tool for finding objective functions (Schuster *et al.*, 2011) and it can be used in combination with flux balance analysis (Teusink *et al.*, 2009).

Changes to the environment or the growth phase are currently rarely studied in a quantitative fashion. This is because dynamic simulations of the kinetics of metabolites requires a knowledge of many parameters including the topology of the regulatory enzymes (Klipp, 2007) and/or the objective function has to be modified to accommodate the changes.

In food safety, the problem is complicated further by the fact that there are many species potentially competing for the substrate while modifying the environment as a result of their growth and metabolic activity. This creates local modifications to the environment. In ecology, this phenomenon is well known and the systems are referred to as complex adaptive systems (Levin, 1998). Whilst optimisation may be valid for organisms in an optimum close system, the interactions with others and the environment require different modelling approaches such as game theory (Ruppin *et al.*, 2010). So the unification of the scales has to take place not only at the cell level, but at the level of the food environment as well. An additional evolutionary level might exist within the food chain; between farm and fork there are potentially enough generations for genetic transfer and selection of agent phenotypes, such as virulence or the acquisition of antibiotic resistance, which might influence risk estimations (Hunter *et al.*, 2008).

25.4.3 Towards multi-scale models

Modelling molecular processes requires substantial changes to the conventional hypothesis-dependent and reductionist thinking that has been commonly applied in predictive microbiology for foods. Advances in the use of large datasets, networks and *in silico* models are required for reconstructing the biological system behind bacterial responses, handling the complexity and gaining an integrative view of the responses.

The review presented here is not comprehensive and different approaches are being developed as research progresses in these fields. Most of the modelling techniques at the molecular level introduced above concentrate on specific examples. They do, however, represent attractive long-term development possibilities and we foresee that the advances in systems biology and modelling complexity will drive the development of predictive microbiology.

25.5 Conclusions

More than ever, it is clear that predictive microbiology is an increasingly multidisciplinary field. Inspiration for further developments can be sought from biotechnology, ecology, statistical physics and elsewhere. One of the challenges modellers are facing is to discover the general laws behind the complexity of bacteria at the molecular level and their interactions with the food environment.

The fundamental questions to be answered in predictive microbiology are similar to ones in other areas of biology. We need to reveal the molecular influence that defines the bacterial responses and population kinetics throughout the food chain. The mathematical models of the future need to find a trade-off between robust parsimony and detailed versatility at different scales.

25.6 References

AGUIRRE, J. S., RODRIGUEZ, M. R. and GARCIA DE FERNANDO, G. D. 2011. Effects of electron beam irradiation on the variability in survivor number and duration of lag phase of four food-borne organisms. *Int J Food Microbiol*, 149, 236–46.

ARIFUZZAMAN, M., MAEDA, M., ITOH, A., NISHIKATA, K., TAKITA, C. *et al.* 2006. Large-scale identification of protein-protein interaction of *Escherichia coli* K-12. *Genome Res*, 16, 686–91.

AUGUSTIN, J. C. 2011. Challenges in risk assessment and predictive microbiology of foodborne spore-forming bacteria. *Food Microbiol*, 28, 209–13.

AUGUSTIN, J. C., BROUILLAUD-DELATTRE, A., ROSSO, L. and CARLIER, V. 2000. Significance of inoculum size in the lag time of *Listeria monocytogenes*. *Appl Environ Microbiol*, 66, 1706–10.

BALÁZSI, G., BARABÁSI, A. L. and OLTVAI, Z. N. 2005. Topological units of environmental signal processing in the transcriptional regulatory network of *Escherichia coli*. *Proc Natl Acad Sci USA*, 102, 7841–6.

BARABÁSI, A. L. 2002. *Linked*, New York, Penguin Group (USA) Inc.

BARANYI, J. 1998. Comparison of stochastic and deterministic concepts of bacterial lag. *J Theor Biol*, 192, 403–8.

BARANYI, J. and PIN, C. 2001. A parallel study on bacterial growth and inactivation. *J Theor Biol*, 210, 327–36.

BARANYI, J. and ROBERTS, T. A. 1994. A dynamic approach to predicting bacterial growth in food. *Int J Food Microbiol*, 23, 277–94.

BARANYI, J., GEORGE, S. M. and KUTALIK, Z. 2009. Parameter estimation for the distribution of single cell lag times. *J Theor Biol*, 259, 24–30.

BARKER, G. C., MALAKAR, P. K., DEL TORRE, M., STECCHINI, M. L. and PECK, M. W. 2005. Probabilistic representation of the exposure of consumers to *Clostridium botulinum* neurotoxin in a minimally processed potato product. *Int J Food Microbiol*, 100, 345–57.

BAUMLER, D. J., PEPLINSKI, R. G., REED, J. L., GLASNER, J. D. and PERNA, N. T. 2011. The evolution of metabolic networks of *E. coli*. *BMC Syst Biol*, 5, 182.

BREIDT, F. and FLEMING, H. P. 1998. Modeling of the competitive growth of *Listeria monocytogenes* and *Lactococcus lactis* in vegetable broth. *Appl Environ Microbiol*, 64, 3159–65.

BRUL, S. and WESTERHOFF, H. V. 2007. Systems biology and food microbiology. In: Brul S., van Gerwen, S., Zwietering M. (eds.), *Modelling Microorganisms in Food*, Woodhead Publishing Limited and CRC Press LLC.

COVERT, M. W., KNIGHT, E. M., REED, J. L., HERRGARD, M. J. and PALSSON, B. O. 2004. Integrating high-throughput and computational data elucidates bacterial networks. *Nature*, 429, 92–6.

D'ARRIGO, M., GARCIA DE FERNANDO, G. D., VELASCO DE DIEGO, R., ORDONEZ, J. A. *et al.* 2006. Indirect measurement of the lag time distribution of single cells of *Listeria innocua* in food. *Appl Environ Microbiol*, 72, 2533–8.

DENNIS, P. P., EHRENBERG, M. and BREMER, H. 2004. Control of rRNA synthesis in *Escherichia coli*: a systems biology approach. *Microbiol Mol Biol Rev*, 68, 639–68.

DENS, E. J., VEREECKEN, K. M. and VAN IMPE, J. F. 1999. A prototype model structure for mixed microbial populations in homogeneous food products. *J Theor Biol*, 201, 159–70.

DUPONT, C. and AUGUSTIN, J. C. 2011. Relationship between the culturability of stressed *Listeria monocytogenes* cells in non-selective and selective culture media and the cellular esterase activity measured by solid phase cytometry. *J Microbiol Methods*, 87, 295–301.

ELFWING, A., LEMARC, Y., BARANYI, J. and BALLAGI, A. 2004. Observing growth and division of large numbers of individual bacteria by image analysis. *Appl Environ Microbiol*, 70, 675–8.

FEIST, A. M., HENRY, C. S., REED, J. L., KRUMMENACKER, M., JOYCE, A. R. *et al.* 2007. A genome-scale metabolic reconstruction for *Escherichia coli* K-12 MG1655 that accounts for 1260 ORFs and thermodynamic information. *Mol Syst Biol*, 3, 121.

FISCHER, E. and SAUER, U. 2003. A novel metabolic cycle catalyzes glucose oxidation and anaplerosis in hungry *Escherichia coli*. *J Biol Chem*, 278, 46446–51.

FRANCOIS, K., DEVLIEGHERE, F., SMET, K., STANDAERT, A. R., GEERAERD, A. H. *et al.* 2005. Modelling the individual cell lag phase: effect of temperature and pH on the individual cell lag distribution of *Listeria monocytogenes*. *Int J Food Microbiol*, 100, 41–53.

GEORGE, S. M., METRIS, A. and STRINGER, S. C. 2008. Physiological state of single cells of *Listeria innocua* in organic acids. *Int J Food Microbiol*, 124, 204–10.

GIMENEZ, B. and DALGAARD, P. 2004. Modelling and predicting the simultaneous growth of *Listeria monocytogenes* and spoilage micro-organisms in cold-smoked salmon. *J Appl Microbiol*, 96, 96–109.

GUILLIER, L. and AUGUSTIN, J. C. 2006. Modelling the individual cell lag time distributions of *Listeria monocytogenes* as a function of the physiological state and the growth conditions. *Int J Food Microbiol*, 111, 241–51.

GUILLIER, L., PARDON, P. and AUGUSTIN, J. C. 2005. Influence of stress on individual lag time distributions of *Listeria monocytogenes*. *Appl Environ Microbiol*, 71, 2940–8.

GUIMERA, R. and NUNES AMARAL, L. A. 2005. Functional cartography of complex metabolic networks. *Nature*, 433, 895–900.

HUNTER, P. R., WILKINSON, D. C., CATLING, L. A. and BARKER, G. C. 2008. Meta-analysis of experimental data concerning antimicrobial resistance gene transfer rates during conjugation. *Appl Environ Microbiol*, 74, 6085–90.

JAMESON, J. E. 1962. A discussion of the dynamics of *Salmonella* enrichment. *J Hyg (Lond)*, 60, 193–207.

JANSSEN, M., GEERAERD, A. H., LOGIST, F., DE VISSCHER, Y., VEREECKEN, K. M. *et al.* 2006. Modelling *Yersinia enterocolitica* inactivation in coculture experiments with *Lactobacillus sakei* as based on pH and lactic acid profiles. *Int J Food Microbiol*, 111, 59–72.

JEONG, H., TOMBOR, B., ALBERT, R., OLTVAI, Z. N. and BARABASI, A. L. 2000. The large-scale organization of metabolic networks. *Nature*, 407, 651–4.
KAUFFMAN, K. J., PRAKASH, P. and EDWARDS, J. S. 2003. Advances in flux balance analysis. *Cur Opin Biotech*, 14, 491–6.
KELL, D. B. 2004. Metabolomics and systems biology: making sense of the soup. *Curr Opin Microbiol*, 7, 296–307.
KLIPP, E. 2007. Modelling dynamic processes in yeast. *Yeast*, 24, 943–59.
KUTALIK, Z., RAZAZ, M. and BARANYI, J. 2005a. Connection between stochastic and deterministic modelling of microbial growth. *J Theor Biol*, 232, 285–99.
KUTALIK, Z., RAZAZ, M., ELFWING, A., BALLAGI, A. and BARANYI, J. 2005b. Stochastic modelling of individual cell growth using flow chamber microscopy images. *Int J Food Microbiol*, 105, 177–90.
LE MARC, Y., PLOWMAN, J., ALDUS, C. F., MUÑOZ-CUEVAS, M., BARANYI, J. *et al.* 2008. Modelling the growth of *Clostridium perfringens* during the cooling of bulk meat. *Int J Food Microbiol*, 128, 41–50
LE MARC, Y., VALIK, L. and MEDVEDOVA, A. 2009. Modelling the effect of the starter culture on the growth of *Staphylococcus aureus* in milk. *Int J Food Microbiol*, 129, 306–11.
LEE, D. S., BURD, H., LIU, J., ALMAAS, E., WIEST, O. *et al.* 2009. Comparative genome-scale metabolic reconstruction and flux balance analysis of multiple *Staphylococcus aureus* genomes identify novel antimicrobial drug targets. *J Bacteriol*, 191, 4015–24.
LEVIN, S. A. 1998. Ecosystems and the biosphere as complex adaptive systems. *Ecosystems*, 1, 431–6.
LI, Y., ODUMERU, J. A., GRIFFITHS, M. and MCKELLAR, R. C. 2006. Effect of environmental stresses on the mean and distribution of individual cell lag times of *Escherichia coli* O157:H7. *Int J Food Microbiol*, 110, 278–85.
MALAKAR, P. K. and BARKER, G. C. 2008. Estimating single-cell lag times via a Bayesian scheme. *Appl Environ Microbiol*, 74, 7098–9.
MARKEVICH, N. I., HOEK, J. B. and KHOLODENKO, B. N. 2004. Signaling switches and bistability arising from multisite phosphorylation in protein kinase cascades. *J Cell Biol*, 164, 353–9.
MCADAMS, H. H. and ARKIN, A. 1998. Simulation of prokaryotic genetic circuits. *Annu Rev Biophys Biomol Struct*, 27, 199–224.
MCKELLAR, R. C. and HAWKE, A. 2006. Assessment of distributions for fitting lag times of individual cells in bacterial populations. *Int J Food Microbiol*, 106, 169–75.
MCKELLAR, R. C. and LU, X. 2003. *Modelling Microbial Responses in Foods*, Boca Raton, FL, CRC Press.
MCMEEKIN, T. A., OLLEY, J. N., ROSS, T. and RATKOWSKY, D. A. 1993. *Predictive Microbiology: Theory and Application*, Exeter, England, Research Studies Press Ltd.
METRIS, A., GEORGE, S. M., PECK, M. W. and BARANYI, J. 2003. Distribution of turbidity detection times produced by single cell-generated bacterial populations. *J Microbiol Methods*, 55, 821–7.
METRIS, A., LE MARC, Y., ELFWING, A., BALLAGI, A. and BARANYI, J. 2005. Modelling the variability of lag times and the first generation times of single cells of *E. coli*. *Int J Food Microbiol*, 100, 13–19.
METRIS, A., GEORGE, S. M. and BARANYI, J. 2006. Use of optical density detection times to assess the effect of acetic acid on single-cell kinetics. *Appl Environ Microbiol*, 72, 6674–9.
METRIS, A., GEORGE, S. M., MACKEY, B. M. and BARANYI, J. 2008. Modeling the variability of single-cell lag times for *Listeria innocua* populations after sublethal and lethal heat treatments. *Appl Environ Microbiol*, 74, 6949–55.
METRIS, A., REUTER, M., GASKIN, D. J., BARANYI, J. and VAN VLIET, A. H. 2011. *In vivo* and *in silico* determination of essential genes of *Campylobacter jejuni*. *BMC Genomics*, 12, 535.
METRIS, A., GEORGE, S. and BARANYI, J. 2012. Modelling osmotic stress by flux balance analysis at the genomic scale. *Int J Food Microbiol*, 152, 123–8.

NIVEN, G. W., MORTON, J. S., FUKS, T. and MACKEY, B. M. 2008. Influence of environmental stress on distributions of times to first division in *Escherichia coli* populations, as determined by digital-image analysis of individual cells. *Appl Environ Microbiol*, 74, 3757–63.

OLTVAI, Z. N. and BARABÁSI, A. L. 2002. Systems biology. Life's complexity pyramid. *Science*, 298, 763–4.

OUZOUNIS, C. A. and KARP, P. D. 2000. Global properties of the metabolic map of *Escherichia coli*. *Genome Res*, 10, 568–76.

PALSSON, B. O. (ed.) 2006. *Systems Biology*, Cambridge University Press.

PIN, C. and BARANYI, J. 2006. Kinetics of single cells: observation and modeling of a stochastic process. *Appl Environ Microbiol*, 72, 2163–9.

PIN, C. and BARANYI, J. 2008. Single-cell and population lag times as a function of cell age. *Appl Environ Microbiol*, 74, 2534–6.

PIN, C., ROLFE, M. D., MUÑOZ-CUEVAS, M., HINTON, J. C. D., PECK, M. W. *et al.* 2009. Network analysis of the transcriptional pattern of young and old cells of *Escherichia coli* during lag phase. *BMC Sys Biol*, 3, 108.

PIN, C., AVENDANO-PEREZ, G., COSCIANI-CUNICO, E., GOMEZ, N., GOUNADAKIC, A. *et al.* 2010. Modelling *Salmonella* concentration throughout the pork supply chain by considering growth and survival in fluctuating conditions of temperature, pH and a_w. *Int J Food Microbiol*, 145, S96–102.

PIN, C., HANSEN, T., MUÑOZ-CUEVAS, M., DE JONGE, R., ROSENKRANTZ, J. T. *et al.* 2012. The transcriptional heat shock response of *Salmonella* Typhimurium shows hysteresis and heated cells show increased resistance to heat and acid stress. *PLoS ONE*, 7(12): e51196.

ROLFE, M. D., RICE, C. J., LUCCHINI, S., PIN, C., THOMPSON, A. *et al.* 2012. Lag phase is a distinct growth phase that prepares bacteria for exponential growth and involves transient metal accumulation. *J Bacteriol*, 194, 686–701.

ROSS, T., DALGAARD, P. and TIENUNGOON, S. 2000. Predictive modelling of the growth and survival of *Listeria* in fishery products. *Int J Food Microbiol*, 62, 231–45.

RUPPIN, E., PAPIN, J. A., DE FIGUEIREDO, L. F. and SCHUSTER, S. 2010. Metabolic reconstruction, constraint-based analysis and game theory to probe genome-scale metabolic networks. *Curr Opin Biotechnol*, 21, 502–10.

SCHUETZ, R., KUEPFER, L. and SAUER, U. 2007. Systematic evaluation of objective functions for predicting intracellular fluxes in *Escherichia coli*. *Mol Syst Biol*, 3, 119.

SCHUSTER, S., FELL, D. A. and DANDEKAR, T. 2000. A general definition of metabolic pathways useful for systematic organization and analysis of complex metabolic networks. *Nat Biotechnol*, 18, 326–32.

SCHUSTER, S., DE FIGUEIREDO, L. F., SCHROETER, A. and KALETA, C. 2011. Combining metabolic pathway analysis with evolutionary game theory: explaining the occurrence of low-yield pathways by an analytic optimization approach. *Biosystems*, 105, 147–53.

SHACHRAI, I., ZASLAVER, A., ALON, U. and DEKEL, E. 2010. Cost of unneeded proteins in *E. coli* is reduced after several generations in exponential growth. *Mol Cell*, 38, 758–67.

SMELT, J. P., OTTEN, G. D. and BOS, A. P. 2002. Modelling the effect of sublethal injury on the distribution of the lag times of individual cells of *Lactobacillus plantarum*. *Int J Food Microbiol*, 73, 207–12.

SMELT, J. P., BOS, A. P., KORT, R. and BRUL, S. 2008. Modelling the effect of sub(lethal) heat treatment of *Bacillus subtilis* spores on germination rate and outgrowth to exponentially growing vegetative cells. *Int J Food Microbiol*, 128, 34–40.

STANDAERT, A. R., FRANCOIS, K., DEVLIEGHERE, F., DEBEVERE, J., VAN IMPE, J. F. *et al.* 2007. Modeling individual cell lag time distributions for *Listeria monocytogenes*. *Risk Anal*, 27, 241–54.

STRINGER, S. C., WEBB, M. D., GEORGE, S. M., PIN, C. and PECK, M. W. 2005. Heterogeneity of times required for germination and outgrowth from single spores of nonproteolytic *Clostridium botulinum*. *Appl Environ Microbiol*, 71, 4998–5003.

STRINGER, S. C., WEBB, M. D. and PECK, M. W. 2009. Contrasting effects of heat treatment and incubation temperature on germination and outgrowth of individual spores of nonproteolytic *Clostridium botulinum* bacteria. *Appl Environ Microbiol*, 75, 2712–19.

SU, C., PEREGRIN-ALVAREZ, J. M., BUTLAND, G., PHANSE, S., FONG, V. *et al.* 2008. Bacteriome. org – an integrated protein interaction database for *E. coli*. *Nucleic Acids Res*, 36, D632–6.

SUMNER, E. R. and AVERY, S. V. 2002. Phenotypic heterogeneity: differential stress resistance among individual cells of the yeast *Saccharomyces cerevisiae*. *Microbiol*, 148, 345–51.

TEUSINK, B., WIERSMA, A., MOLENAAR, D., FRANCKE, C., DE VOS, W. M. *et al.* 2006. Analysis of growth of *Lactobacillus plantarum* WCFS1 on a complex medium using a genome-scale metabolic model. *J Biol Chem*, 281, 40041–8.

TEUSINK, B., WIERSMA, A., JACOBS, L., NOTEBAART, R. A. and SMID, E. J. 2009. Understanding the adaptive growth strategy of *Lactobacillus plantarum* by *in silico* optimisation. *PLoS Comput Biol*, 5, e1000410.

THIELE, I., HYDUKE, D. R., STEEB, B., FANKAM, G., ALLEN, D. K *et al.* 2011. A community effort towards a knowledge-base and mathematical model of the human pathogen *Salmonella* Typhimurium LT2. *BMC Syst Biol*, 5, 8.

VEENING, J. W., STEWART, E. J., BERNGRUBER, T. W., TADDEI, F., KUIPERS, O. P. *et al.* 2008. Bet-hedging and epigenetic inheritance in bacterial cell development. *Proc Natl Acad Sci USA*, 105, 4393–8.

WEBB, M. D., PIN, C., PECK, M. W. and STRINGER, S. C. 2007. Historical and contemporary NaCl concentrations affect the duration and distribution of lag times from individual spores of nonproteolytic clostridium botulinum. *Appl Environ Microbiol*, 73, 2118–27.

Index

CPSIA information can be obtained at www.ICGtesting.com
Printed in the USA
BVOW02*0400030214

343683BV00006B/184/P

9 780857 094384